The Tectonics and Metallogenesis of Asia

Tianfeng Wan

The Tectonics and Metallogenesis of Asia

Geological Publishing House

Springer

Tianfeng Wan
China University of Geosciences
Beijing, China

ISBN 978-981-15-3031-9 ISBN 978-981-15-3032-6 (eBook)
https://doi.org/10.1007/978-981-15-3032-6

Jointly published with Geological Publishing House
The print edition is not for sale in China (Mainland). Customers from China (Mainland) please order the print book from: Geological Publishing House.
ISBN of the China (Mainland) edition: 978-7-116-12013-6

This Springer imprint is published by the registered company Springer Nature Singapore Pte Ltd.
The registered company address is: 152 Beach Road, #21-01/04 Gateway East, Singapore 189721, Singapore

Preface

There are lots of special characteristics in the Asian continental lithosphere, and however, it was not paid attention to in early time when the Plate Tectonic Theory was proposed. The author considers that the Asian continental plate is composed of 28 giant blocks and more than hundreds of small blocks, which have undergone multi-periodic subductions, collisions and convergences to form 38 collision zones and widespread intraplate deformations by more than 14 tectonic events. As a result, those tectonic events could destroy or preserve the mineral deposits, control the topographic evolutions, change the ecological environments, form several tectonic detachments in the lithosphere and show two special types of lithospheric tectonics in Asia.

Undoubtedly, it is possible and reasonable for us to understand the formation and evolution of continental tectonic units and their tectono-metallogenesis by the Plate Tectonic Theory that focuses on the horizontal migration. It is a great pity, some researchers have ever thought that the Plate Tectonic Theory would not be suitable for the continental tectonics.

Based on the collected data of 242 giant or supergiant ore deposits, fields and provinces in Asia, the author synthetically analyzes the formation and distribution of endogenic and exogenic deposits and researches the tectono-metallogenesis in the Asian continent. The importance of intraplate extension metallogenesis in the Asian continental lithosphere has been recognized. Finally, from the view of tectono-metallogenesis, it will be useful for the future exploration of mineral resources in the unknown areas of the known tectono-metallogenic belts, and some reasonable common views have been obtained for readers' reference.

It is so difficult to study the tectono-metallogenesis in the Asia continent because the Asian continental lithosphere is the largest and extremely special continental lithosphere in the world. In addition, the predecessors have never made a profound study on the tectono-metallogenesis in the Asia continent. Therefore, we must open our minds to avoid blind work, only in this way can we get the truths.

In recent years, China is known as "the Workshop of the World." As to the poor and backward old China in the past 100 years, it seemed to be very proud of China to be called as "the Workshop of the World." In fact, it is unfair to all Chinese people and not worthy bragging about. On the one hand, a large number of mineral and energy resources in China have been consumed to serve global development, and on the other hand, it has resulted in such serious environmental pollution in China. This resource-consumed development mode will not last for a long time in the future. Therefore, it is necessary to scientifically exploit the global mineral, energy and other resources, reasonably integrate the global mining enterprises, effectively control the environment pollution, prevent the giant natural disasters from threatening our human beings and finally make the global people enjoy a good life. For geologists, we must strengthen the basic research on the mineral resources, energy and ecological environment to make our contributions to all mankind. This monograph of *The Tectonics and Metallogenesis of Asia* has just taken a very small step. I hope that the more young geologists will make their great contributions to society.

In the monograph, there may be some mistakes. Comments and suggestions are most welcome!

Acknowledgements Many leading scholars or authors have provided useful information for this monograph, including Academicians Tingdong Li, Xuchang Xiao, Rongfu Pei, Zhiqin Xu, Jishun Ren, Jingwen Mao, Wenjiao Xiao; Profs. Xiufu Qiao, Jingyi Li, Zuoheng Zhang, Fengxiang Lu, Hong Zhu, Shaofeng Liu, Changhou Zhang, Yu Wang, Yalin Li, Junlai Liu, Xinqi Yu, Deliang Liu and Xiaohong Ge. They have also made some beneficial comments on this monograph, introduced lots of data and discussed some important professional issues with me.

Professor Xiangnan Liu with his graduate students Da Liu and Xiaopen Xia, as well as Dida Jinghua Working Group helped me draw the Asian tectonic unit map and all figures in the text. Mrs. Xiuhua Cao helped me do many experiments and routine researches. Their hardworking helped me complete the monograph, and here, I would like to express my sincere thanks to them.

Professor Barber A. J. and Prof. Hall R. (Royal Holloway, University of London, UK), Dr. Pospelov I. I. (Geological Institute, Russian Academy of Sciences, Commission for the Geological Map of the World), Dr. Tay Thye Sun (Far East Gemological Laboratory, Singapore), senior geologist Yu Jie (Shell Co.) and Prof. Chang J. H. (South Korea) sent their new monographs or maps to me. Their support is very useful for this monograph. This research has been supported by the China Geological Survey and the State Key Laboratory of Geological Processes and Mineral Resources (China University of Geosciences).

The editors at Geological Publishing House, Mr. Zhiru Wei and Kaiming Li, have made their efforts for this monograph. Lisa Fan, the editor of Springer Beijing Representative Office, has given a lot of help in co-publishing, and here, I would like to express my sincere thanks to them.

Here, I must give my earnest thanks to my wife Guangxi Zhao. If she had not provided me a cozy and comfortable life, I could not do my further research, yet complete this monograph.

At last, at the time of completing this monograph and reviewing my research life, I must thank my teacher, Prof. Peiren Zhuang, specially. In 1960, he taught me the structural geology and tectonics and pointed out that I should pay attention to details and microstructures first and then focus on large-scale tectonics; first, study the Mesozoic and Cenozoic structures and then the more ancient structures. Thus before more than fifty years, I have learned the research methods for the rock deformations of geometry, kinematics and dynamics from a small area to a great region. Under his guidance, as well as valuable help from many old professors, experts, colleagues and students, I am able to get on with a rather smooth research way, in addition, during the period of China's reform and opening up, I can complete this monograph.

Beijing, China Tianfeng Wan

Contents

1 Introduction

The Asian continent is the main part of the greatest continental plate—Eurasian plate, on the Earth. The Asian continent covers the tremendous area and has a very long geological history, complex intraplate deformations, a lot of mineral resources and the changeable ecologic environment. Researching the Asian tectonics, which plays a dominant role in the resources and environments, is an interesting subject. With the rapid development of society and economy, it is becoming more and more important, interesting and urgent to re-recognize and understand the relationship between the Asia tectonic evolution and giant ore deposit (including giant or supergiant ore fields and provinces) from the perspective of lithospheric plate tectonics.

Since the late period of 1960s, the Plate Tectonic Theory (Wilson 1970; Le Pichon et al. 1973; Press and Siever 1974), based on the research of oceanic floor, has been well developed. Since that period, the abundant practical data have been accumulated in the research of continental tectonics. Obviously, how to understand the tectonics of the Asian continent has the great theoretical significance and application value (Jin and Yao 2004; Guo et al. 2004; Zhang et al. 2006; Wan et al. 2008). Thus, based on the research of China continental tectonics (Wan 2011), the author will make further investigation and discussion on the Asian continental lithospheric tectonics and the relative formation background of main mineral resources.

For the tectonic research of China continental lithosphere, the monograph is mainly originated from the original data of 1:200,000 regional geological surveys and petroleum geological surveys for the most of Chinese areas. For the Xizang (Tibet) and Xinjiang areas, it is from the original data of 1:1,000,000 regional geological surveys and some important results of 1:250,000 regional geological surveys (Wan 2011). In recent years, the author has collected many involved data of Asia tectonics and the reference materials of 242 giant ore deposits, fields and provinces, to discuss the relationship between tectonics and mineralization. As a result, the author points out their dominant tectonic-controlling factors of giant deposits, metallogenesis and then tries to discuss and propose the further exploration suggestions.

In terms of the research of Asia tectonics and its map complication, C.Y. Li and the Geology Institute, Chinese Academy of Geological Sciences firstly published the *Asia Tectonic Map* (1:8,000,000) in 1982. They collected and synthesized the data of geological evolution and then to compile the tectonic map based on the *Asia Geological Map* (Chinese Academy of Geological Sciences 1980), which firstly used the Plate Tectonic Theory to explain the Asia tectonics (Li et al. 1982; Li 2004). It is the important reference data for recent research. However, limited to the research level at that time, there are some unclear viewpoints in the tectonic unit's division principle and research method. Their research on continental collision zones has still been at the initial period; thus, the deficiencies are unavoidable. Petrov et al. (2008) published the *Atlas of Geological Map of Central Asia and Adjacent Areas* (1:2,500,000), in which they showed the Tectonic Zoning of Central Asia and Adjacent Areas (1:20,000,000). But they negated the "Kazakhstan plate," which was considered to exist for a long time, and pointed out that the Central Asia is composed of a series of Paleozoic collision zones and mixed up with many small old blocks. Their re-division of the Central Asian tectonic units may be rather reasonable. In this monograph, the author refers to their viewpoints about the division of tectonic units in the Central Asia.

As to the tectonic data of West Asia and Middle East areas, the author not only refers the research data of Li et al. (1982) and Chinese Academy of Geological Sciences (1980), but also adopts Pubellier (2008) result and many data published on journals and Web sites.

The tectonic data on China continent, East Siberia, Korean Peninsula, Japan Islands and Indochina Peninsula are mainly based on the author's monograph (Wan 2011), the papers of Karsakov et al. (2008), Parfenov et al. (1995, 2009) and my field investigations for IGCP 224 and IGCP 321. The geological data of Indochina Peninsula are mainly

T. Wan, *The Tectonics and Metallogenesis of Asia*,
https://doi.org/10.1007/978-981-15-3032-6_1

cited from Lan et al. (2003), Cheng et al. (2010), and Ridd et al. (2011). The geological data of Malay Peninsula mainly refer to the monograph of Hutchison and Tan (2009). The tectonic data of Philippines–Sunda Islands are mainly collected from the monograph of Hall et al. (2011), the map compiled by Pubellier (2008) and the thesis of Smyth et al. (2007). The geological ages in the ocean floor are referenced to the map of Pubellier (2008).

The basic geological data in the Asian continent, such as crystalline basements, intrusions, volcanics, ophiolite suites and sedimentary basins, are mainly derived from the data, maps or monographs of the Geology Institute, Chinese Academy of Geological Sciences (1980, 2004), Ma (2002), China Geological Survey (2004), Li et al. (2008) and Ren (2013).

As to some newly Asia tectonic maps and opinions (such as Miao and Zhou 2010; Shan et al. 2011), there are many great differences from most of researchers, so their results are for reference only.

To sum up, the author had synthesized and contrasted near 1000 Chinese and English monographs or papers on Asia tectonics and metallogenesis, then proposed some new ideas. It should be said that there are still some understandings that are not mature enough. Criticism and correction are warmly welcomed!

Thus, the author has drawn an *Asia Tectonic Unit Map* (1:10,000,000, and showed the small one as Fig. 1.1 in the text), which is compiled by the ArcGIS 10.0 software, applying the Lambert Azimuthal Equal Area method, with the central meridian of 90° E and central latitude of 40° N, respectively. On the geographic basic map, it does not show national boundaries, but mainly show important essential geographic elements, such as beach lines, large rivers, big lakes, giant mountain ranges and the main capital cities. In the very long geological history, the division of tectonic units must be changed as time goes on. So it is not very well to compile a comprehensive tectonic map for whole geological history. However, in the practice and many cases, it needs a comprehensive tectonic map to show the main tectonic characteristics. This map does not show the tectonic evolution processes and details and only demonstrates the main tectonic units formed in geological history. The detailed figures of every tectonic unit at different geological time will be shown in Part 3 and discussed in detail.

In the *Asia Tectonic Unit Map*, the tectonic units are divided into two types: **Plate (or Block)**, with weaker tectonic activity; **Collision Zone**, with strong tectonic activity. The geological ages are marked for different tectonic units.

In the geological history, after the lithosphere plate forming, the tectonic activity in those areas will trend weaker. In the early foundation period of Plate Tectonic Theory, some researchers believed that plate was of "rigid." However, in the last forty years, many continental geologists have recognized that after the strong displacement of continental lithosphere plates, rock deformations also could occur with magmatism and dynamic metamorphism near the faults (Wan 2011), detachments near some tectonic boundaries or surfaces, such as the regional faults, some soft layers in sedimentary system, the crystalline basements, the low-velocity and high-conductivity layers in middle crust, near Moho discontinuity, or the bottom of lithosphere (Zhang 1984).

The boundary of lithosphere plate is surely penetrated through the bottom of lithosphere, which is the collision zone between continental crusts, or the subduction zone between oceanic crust and continental crust. The main detachment of oceanic lithosphere plate is usually located at the bottom of plate (Wan 2011). Many paleo-blocks in the Asian continent are belonged to "micro-plate." However, when data are rare and it is impossible to determine whether those faults penetrate the entire lithosphere at the boundary of the block or not, it will not know that the bottom of the small block is a detachment. So the author proposes that the stable area should be called "block," which is a common term. Despite the stable area is subsided into a sedimentary basin in the later stage, it should be called "block" as for its insufficient basement data.

The author considers that "terrain" is not a suitable tectonic term, because its definition is not clear. In 1988, the author and his colleagues asked Howell (who first defined the term of terrain) what the correct definition of the "terrain" is. He did not give the answer. Howell's term of "terrain" is neither a block detached from the bottom of lithosphere, nor a block detached from a thrust fault plane in the lithosphere. What is a terrain and where does it detach from on earth?

In fact, it is enough to use the definition of micro-plate or nappe (for thrust), so the "terrain" is an unnecessary term. Later, in the early period of 1990, IGS and UNESCO rejected Howell's project (on the Pacific Rim Terrains International Cooperation Project). Thus, the "terrain" is hardly used, except for those researchers who do not know the fact.

The plate should be divided mainly based on the period and region of plate formation, i.e., the period of forming uniform crystalline basement. Its boundary should penetrate through the whole lithosphere, and it must be a lithospheric fault zone, i.e., a collision zone or a subduction zone. Beneath the lithosphere is the asthenosphere, which is a major detachment layer for a lithosphere plate. In the figures of this monograph, the similar lighter colors on common international geological maps are used to show the different formation periods of plates. The plates formed in different geological periods and areas usually have different regional geochemical characteristics, where different mineral resources could be concentrated and formed. Using the boundaries

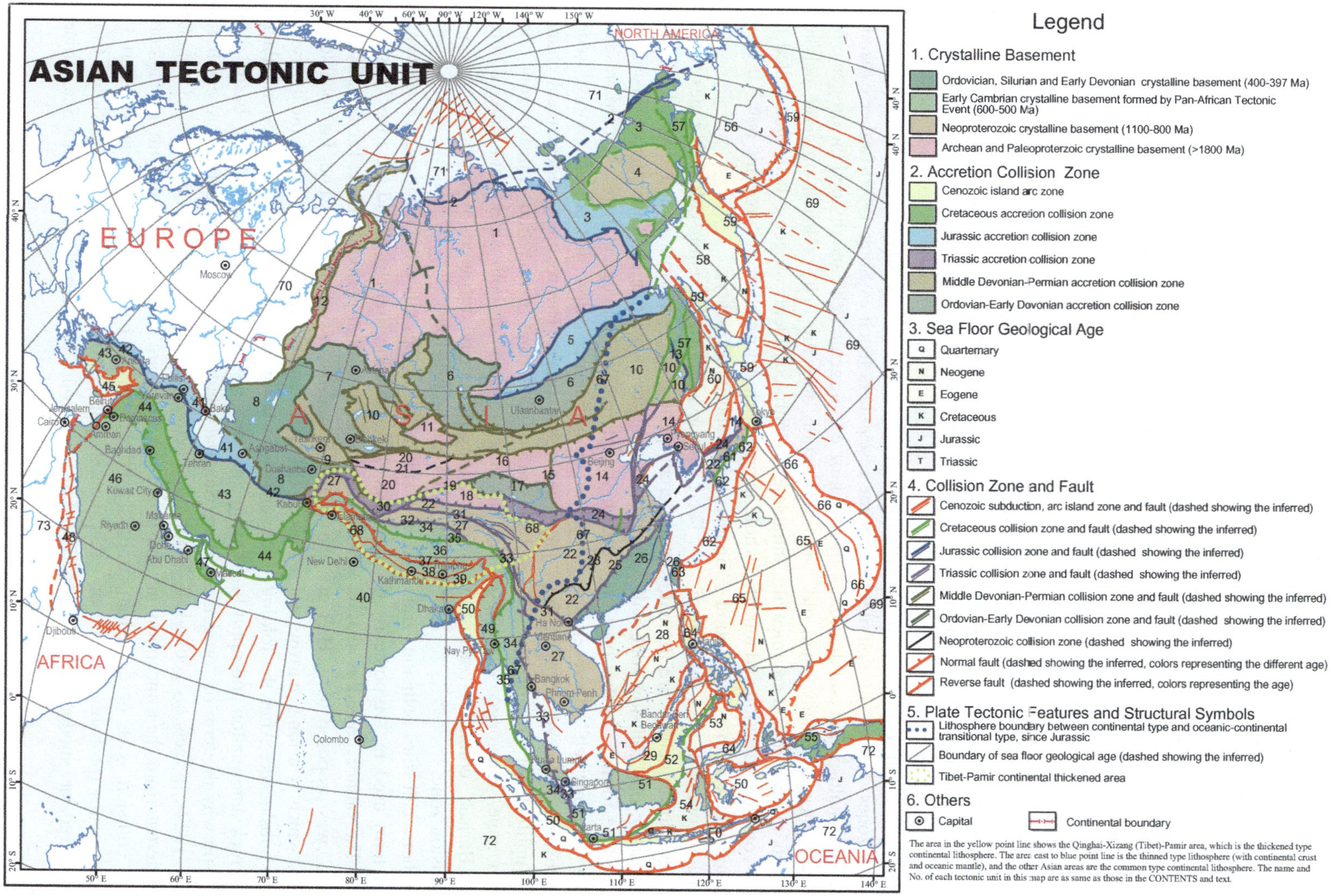

Fig. 1.1 Asian tectonic unit map

of different plates to divide the tectonic units is rather helpful for compiling the metallogenic map, which could summarily reflect the different characteristics of metallogenesis in different blocks. In this monograph, 28 plates, blocks or sedimentary basins are divided in total.

It is reasonable that plates or blocks are independent and far apart in geological history. When compiling a series of paleo-tectonic maps, the paleo-magnetic data should be used to reconstruct position of paleo-continental plates or blocks. However, due to the complete lack of data about the paleo-oceans, and the wide coverage of them, the most part of map will be left blank, like the map compiled by Yin and Nie (1996). In terms of available data, the paleo-continental plate reconstruction maps are only suitable to compile the small-scaled figures (as it is shown in Part 3 of this monograph). At last, the paleo-oceans are only shown in the collision or subduction zone.

The suture zone between two continental plates is the continental-continental collision zone. Some collision zones are rather narrow, with the width less than 20 km, and the inner structure is a quite simple lithosphere fault zone, which can be just shown as a thick line on maps scaled at 1:10,000,000. Some collision zones formed in complex evolution history must undergo multi-period subduction, collision or strike-slip, in which there are many small old continental blocks and fragments of oceanic and continental crust or mantle, as a result, the complex, accretion and mixed zone was formed. So the author names them as "Accretion–Collision Zone." In the collision zone, if a large number of granitic magma intrusions with a lower density are formed, under the effect of gravity isostasy, this collision zone will be uplifted and mountains will be formed. In this case, it seems suitable to be called the "orogenic belt". However, if a large number of granitic intrusions had never formed, the collision zone would not be uplifted, for example the Jinshajiang–Red River collision zone, Bangongco–Nujiang–Mandalay–Phuket–Northern Barisan collision zone and Shaoxing–Shiwandashan collision zone.

Many geological researchers call the collision zone "orogenic belt," which is a term used in "Hypothesis of Geosyncline and Platform." The author considers that is unsuitable. The "Geosyncline" was proposed by J. Hall in 1859 and J.D. Dana in 1873, which is a long, narrow and shallow sea with a long-term subsidence. After the folding inversion, it will form strong rock deformation and uplift to the mountains, this is the reason why the term of "orogenic belt" to be called by Stille in (1924). In facts, its geological implication is completely different from the conception in Plate Tectonic Theory. The collision zone or accretion–collision zone means that two plates (probably oceanic or continental plates) will converge or subduct along the long-distance horizontal displacement; in the zone, the paleo-ocean or marginal sea exists, at last continental and continental collision occurs, forming a tectonic mélange. In the collision zone or accretion–collision zone, there are many big or small tectonic blocks at different periods, including continents, shallow seas, islands, deep ocean floor sediments, magmatic and metamorphic rocks, ophiolite suits, strong tectonic magmatism, high and ultra-high metamorphism. Sometimes the collision zone or accretion zone could be uplifted to form the mountains. So the term "orogenic belt" is not used in this monograph.

Some continental geologists prefer the term of "intra-continental orogenic belt" (Hsu 1988; Hsu et al. 1988; Ge 1989; Zhao 1995; Neves and Mariano 2004; Shao et al. 2007; Shu et al. 2008; Ge and Ma 2014). When two continental plates collide with each other, the collision zone must be preserved in the jointed continent after collision. All the collision zones or so-called orogenic belts in geological history are all preserved in the continent, so it seems reasonable to persist on "intra-continental orogenic belt." In fact, it is a copy from "Hypothesis of Geosyncline and Platform" and is not worth popularizing.

Some researchers (such as Song 1999) further expanded the meaning of "orogenic belt" and divided the orogenic belt into subduction orogenic belt, collision orogenic belt and intraplate (or intra-continent) orogenic belt, then called the whole intense rock deformation as "orogenic belt." It seems inappropriate and unnecessary.

The collision zones or accretion–collision zones formed at different periods are determined and divided by their geological times or isotopic ages of the main collision period. On the international geological maps, they are presented with the similar deeper colors that are commonly used to show geological times. The regional geochemical characteristics in collision zone or accretion–collision zone show mixture characteristics of both sides, and thus, different combinations of mineral resources with characteristics of both sides can be generated.

The subduction zones since the Cenozoic are shown as the units of "subduction (trench)–island arc zone." It should be noted that in the initial period of Plate Tectonic Theory, the "trench–island-basin system" was rather fashionable (Xu 1980). J.H. Xu recognized that the most typical case for "trench–island-basin system" is in the Western Pacific Ocean area and to say the typical "back-arc basin" is in Japan Sea and South China Sea. After detailed research, it is found that neither the Japan Sea basin nor South China Sea basin are the "back-arc basins;" both of them are the continental border extension basins (see more details in Part 3 of this monograph; Tectonic Group, Institute of South China Sea, Tectonic Group, Institute of South China Sea, Chinese

Academy of Sciences 1988; Yoon 2001; Tamaki et al. 1992; Jolivet and Tamaki 1994). So the tectonic and structural geologists only discuss "trench–island arc system," not to use the term of "back-arc basin." It is a pity, that till now many sedimentary and paleogeographic researchers still persist that "back-arc basin" must exist behind the arc.

In this monograph, the other two boundaries—rift and transform fault, will be listed alone. In the oceanic basin, their divisions are usually done according to the geological age of rocks.

To sum up, in this monograph, 38 collision zones, subduction zones and rift belts with strong tectonic activity are divided. Figure 1.1 mainly demonstrates the division of plates and collision zones during their formation periods. However, multi-period tectonic events after the crystalline basement formed cannot be reflected on this figure, and it is very difficult to guide the exploration of mineral deposits only by using this figure. It should research the relationship between the detailed tectonic evolution after the formation of plates, collision zones and metallogenesis (shown in Parts 3 and 4).

In this monograph, the tectonic units, with similar formation age of crystalline basement or close relationship of tectonic evolution, are defined as "Tectonic Domain." Part 2 will explain summarily the division principle of tectonic domains and tectonic units, as well as their main characteristics. In this monograph, six tectonic domains and 68 tectonic units are divided. In addition, there are five tectonic units beyond the Asian continent. In the monograph, the Arabic numeral codes of tectonic units in the text and figures are consistent, which are put in square brackets in text. The isotopic ages or geological ages of collision zone and crystalline basement are put in the parentheses behind the name of each tectonic unit.

In recent years, Sengör et al. (1993) and Xiao et al. (2009) also put forward their proposal concerning the division of Asia tectonic unit. Some of their viewpoints are similar to those of this monograph, such as the division of Siberian plate. However, there are many differences: (1) they proposed the Central Asia Orogenic Belt (CAOB) or Altaids, which actually is equivalent to four collision zones: the Altay–Middle Mongolia–Hailar Early Paleozoic accretion–collision zone [6], the Karaganda–Kyrgyzstan (Qirghiz) Early Paleozoic accretion–collision zone [7], the Western Tianshan Late Paleozoic accretion–collision zone [9] and the Balkhash–Tianshan–Hingganling Late Paleozoic accretion–collision zone [10] in this monograph. Although they were located in similar tectonic units in the early period, their forming periods were different, and they were broken obviously in recent. The author still persists that they should be separated into four collision zones. (2) They named the Sino–Korean plate, Tarim block to Turan–Karakum plate and Middle Caspian Sea area as "Intermediate Unit." Its main problem is about the basement of Turan–Karakum plate (probably formed in Early Paleozoic), for example, in the eastern part, the crystalline basement of Sino–Korean plate was formed at the end of Paleoproterozoic, so it is not suitable to be called a tectonic unit. (3) They listed the Indian plate and Arabia plate separately, and named the northern parts of the two plates and southern parts of the "Intermediate Unit" as "Tethysides." This division confuses the Yangtze and Indosinian plates formed in the middle period of Neoproterozoic (850 Ma), many blocks and collision zones formed in the latest period of Neoproterozoic ($\sim$509 Ma, Pan-African Tectonic Event) near Gondwana. Of course, they mainly discussed the "Central Asia Orogenic Belt," other parts might not be paid close attention in their research.

Liu et al. (2012) named the "tectonic domain" (defined in this monograph) as "plate," the second-class tectonic unit as "micro-continental block," "basin" or "orogenic belt." Their "plate" is the ancient block and collision zone, not a complete plate. Their "micro-continental block" or "basin" is as same as the block in this monograph. Their "orogenic belt" is as same as the "collision zone" or "accretion–collision zone" in this monograph. Fortunately, the recognitions are getting closer each other. Based on the formation periods of crystalline basement, the plates or blocks in the Asian continent were all developed in the following four periods: Late period of Paleoproterozoic (1800 or 1600 Ma), Middle period of Neoproterozoic ($\sim$850 Ma), Latest Neoproterozoic–Early Cambrian (570–509 Ma) and Latest Early Paleozoic–Early Devonian (419–410 Ma) (Table 1.1; Fig. 1.1). They are the key and foundation of the Asian continental lithosphere plate.

In the late period of Paleoproterozoic, the Siberia [1], Balkhash–Tianshan–Hingganling [10], Sino–Korea [14], Alxa–Dunhuang [16], Qaidam [18], Junggar [11], Tarim [20] plates or blocks and some small paleo-blocks in Indian plate were formed. Outside Asia, many plates else formed uniform crystalline basement at the period of 1800 Ma, which was the main formation period of Columbia supercontinent (Rogers and Santosh 2002, 2004). In the Middle period of Neoproterozoic ($\sim$850 Ma), the Kolyma–Omolon [4], Yangtze [22], Indochina [27], South China Sea [28], Palawan–Sarawak–Zengmuansha [29], Eastern Hindukush–Karakoram–Northern Qiangtang [27], Oman [47] and Okhotsk [58] plates or blocks were formed. About 1000 Ma ago, the Rodinia supercontinent was converged at the Western Hemisphere; and during the period of 850 Ma, the Rodinia supercontinent begun to break up in the Western Hemisphere. However, in the Eastern Hemisphere the above crystalline basements were formed, some collisions occurred, such as the ones between the North and the South Yangtze plates, the North and the South Tarim blocks. In the latest period of Neoproterozoic–Early Cambrian (570–

Table 1.1 Formation period of crystalline basement of plate or block in Asia

Plate or block	Formation period of crystalline basement			
	Before 410 Ma	Before 509 Ma	Before 850 Ma	Before 1600 Ma
Siberia (1600 Ma) [1], Sino–Korea (1800 Ma) [14], Dunhuang–Alxa [16], Qaidam [18], Song–Neng [10], Junggar [11], Tarim [20]. Outside: Baltica [69], North America [70]				
Kolyma–Omolon [4], Yangtze [22], Indosinian [27], South China Sea [28], Palawan–Zengmuansha [29], Eastern Hindukush–Northern Qiangtang-Indosinian [32], Oman [47], Okhotsk [58]				
Southern Qiangtang-Sibumasu [34], Gangdise [36], Himalayan [38], Indian [40], Turkey–Iran–Afghan [43], Arabian [46], Western Burma [49], Sunda [51], Celebes Sea [53], East Argo [54], Small blocks in Altay–Middle Mongolia–Hailar [6] and Karaganda–Kyrgyzstan [7]. Outside: Australia [71], Africa [72]				
Turan–Karakum [8] Cathaysian [26]				

509 Ma), it is the period when the Gondwana supercontinent was formed, and the great tectonic event was called as "Pan-African Tectonic Event" (Kennedy 1964). The green schist system in that supercontinent was formed, exposed on the surface in recent times. In Asian continent, Southern Qiangtang–Sibumasu [34], Gangdise [36], Himalaya [38], India [40], Turkey–Iran–Afghanistan [43], Arabia [46], Western Burma [49], Sunda [51], Sulawesi (or Celebes) Sea [53], Eastern Argo [54] plates or blocks were formed, as well as more than several tens small crystalline blocks in Altay–Middle Mongolia–Hailar [6] and Karaganda–Kyrgyzstan [7] collision zones (Fig. 2.7). However, those small and metamorphic crystalline blocks are hardly to be determined by paleo-magnetism data, till now it cannot be reconstructed in detail. In that period, the Australian plate [72] and African plate [73] were formed and jointed into Gondwana supercontinent.

In the latest period of Early Paleozoic–Early Devonian (419–410 Ma), only the Turan–Karakum block [8] and Cathaysian plate [26] were formed. It is not far away in time when the crystalline basement of the above two blocks were formed and the erosion did not stretch to the deep, so rocks of low-grade green schist facies are outcropped on these blocks at present. The most areas of Turan–Karakum block are covered by the desert, and research has not well conducted. The block is one part of so-called Kazakhstan plate in the past. Petrov et al. (2008) divided the Kazakhstan plate independently, mostly belonging to collision zones. Some researchers (such as Faure et al. 2009; Xiang and Shu 2010) considered that in this period, orogenic events occurred only. But in those blocks the metamorphism and rock deformation are widely distributed and shown as a narrow belt, so it should be expressed as "block with metamorphism crystalline basement."

In terms of the collision zones or accretion–collision zones in Asian continent, they should be divided according to their formation period. Since Neoproterozoic, all the collision zones or accretion–collision zones have undergone eight periods: Middle period of Neoproterozoic ($\sim$850 Ma), the late period of Early Paleozoic ($\sim$419 Ma), Late Paleozoic ($\sim$252 Ma), Middle and Late Triassic ($\sim$201 Ma), Jurassic-early period of Early Cretaceous ($\sim$135 Ma), Late Cretaceous-Paleocene ($\sim$56 Ma), Oligocene ($\sim$23 Ma), Neogene–Early Pleistocene (23–0.78 Ma), as well as the

Table 1.2 Formation period of accretion–collision zones in Asia

Accretion collision zones	Collision period							
	23–0 Ma	56–23 Ma	Before 56 Ma	Before 135 Ma	Before 201 Ma	Before 252 Ma	Before 419 Ma	>850 Ma
Middle Tarim [21], South Anhui–Northeast Jiangxi–Xuefeng–East Yunnan (Jiangnan) [23]								■
Altay–Middle Mongolia–Hailar [6], Karaganda [7], Qilian [17], Altun [19]							■	
Balkhash–Tianshan–Hingganling [10], Ural [12], West Tianshan [9], Helanshan–Liupanshan [15]						■		
Qinling–Dabie–Jiaonan–Hida [24], Shaoxing–Shiwandashan [25], Western Hindukus–Pamir–Kunlun [30], Jinshajiang–Red River [31], Shuanghu [32], Changning–Menglian–Chiangrai–Central Malaya [33]					■			
Southern Margin of East Siberian Sea [2], Verkhojansk–Chersky [3], Transbaikalia [5], Wandashan [13]				■				
Bangongco–Nujiang–Mandalay–Phuket [35], Kavkaz–Alburz [41], Anatolia–Dehran [42], Oman [47], E. Kalimantan–SuluIslands [52], Sikhote–Alin–Koryak [57], Japan Median Tectonic Line [61]			■					
Yarlung Zangbo–Myitkyina [37], Aleutian–Kamchatka–Kurile–NE Japan [59], South Honshu–South Shikoku–Ryukyu [62], Izu–Bonin–Mariana [66]		■						
Himalayan [39], Zagros–Kabul [44], Toros [45], Red Sea Rift [48], Arakan–Sunda [50], Northern New Guinea [55], East Taiwan [63], Philippines–Moluccas [64]	■							

Middle Pleistocene subduction zone (Table 1.2; Fig. 1.1). More formation and evolution characteristics will be discussed in Parts 2 and 3.

To sum up, the six tectonic domains and their numbers of block (or plate) and collision zone in Asia are shown in Fig. 1.2.

In this monograph, the isotopic ages of tectonic-strata unit are revised and corrected according to "International Stratigraphic Chart" compiled by International Commission on Stratigraphy (Cohen et al. 2013).

Part 2 dilates the main characteristics of every tectonic domain and unit.

Part 3 expounds the tectonic evolution of Asian continental lithosphere, summarizes 14 tectonic events and discusses the Asian continental blocks how to converge gradually by plate migration, compression, subduction and collision, how to form the strong intraplate deformations, related magmatism, metamorphism and how to change the

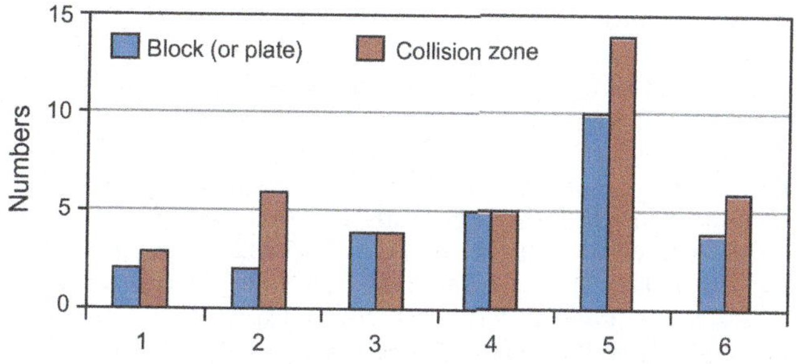

Fig. 1.2 Numbers of block (or plate) and collision zone in every tectonic domain of Asia. 1. Siberian tectonic domain (including 2 blocks and 3 collision zones); 2. Central Asia–Mongolia tectonic domain (including 2 blocks and 6 collision zones); 3. Sino–Korean tectonic domain (including 4 blocks and 4 collision zones); 4. Yangtze tectonic domain (including 5 blocks and 5 collision zones); 5. Gondwana tectonic domain (including 10 blocks, 14 collision zones and 1 boundary of lithosphere type); 6. Western Pacific tectonic domain (including 4 blocks, 6 collision zones and 1 boundary of lithosphere type)

topography and environment. At last, some important theory problems are discussed: the growth process of the Asian continent, the dynamic mechanism of great-scale intraplate deformation, three lithosphere types of Asian continent and their forming reasons, the formation mechanism of basin and mountain evolution. Based on recent data, the dynamic mechanism of global lithospheric plate tectonic formation and migration since Mesozoic is discussed.

As to Part 4, the main tectono-metallogenesis of 242 giant and supergiant ore deposits, fields or provinces, the typical mineral deposits in the different tectonic domains, the main tectono-metallogenesis in different tectonic periods are analyzed. At last, some problems of important tectono-metallogenesis are discussed: the influence of fault activity on endogenic metal mineralization, the relationship between rock deformation and the preservation space of deposits, the influences of later tectonics and moderate uplift and settlement on preservation condition of ore deposits. The author recognizes that the intraplate extension plays a dominant role in the tectono-metallogenesis of the Asian continent. At last, some exploration proposals are given based on the tectono-metallogenesis research.

For reader's convenience, at the end of this monograph, the giant ore deposits or fields in tectonic units of the Asian continent are listed in Appendix.

References

Cheng Y Q, Liu J L, Feng Q L et al. (2010) Geology and Ore Deposits Related to Granite in Indochina Peninsula, Southeast Asia. Beijing: Geological Publishing House, 1–192 (in Chinese).

China Geological Survey (2004) Geological Map of People's Republic of China (1:2 500 000, attached Instructions). Beijing: Sinomap Press (in Chinese).

Chinese Academy of Geological Sciences (1980) Asia Geological Map. Beijing: China Geography Publishing House (in Chinese).

Cohen K M, Finney S, Gibbard P L (2013) International Chronostratigraphic Chart, International Commission on Stratigraphy. http://www.stratigraphy.org/ICSchart/ChronostratigraphicChart2013.01.pdf.

Faure M, Shu L S, Wang B et al. (2009) Intracontinental subduction: A possible mechanism for the Early Paleozoic Orogen of SE China. Terra Nova: 1–10. DOI: 10.1111/j. 1365-3121-2009.00888.x

Ge X H (1989) The formation history of North China orogenic belt. Geological Review, 35(3): 254–261 (in Chinese).

Ge X H, Ma W P (2014) A Course for Regional Tectonics of China. Beijing: Geological Publishing House, 1–466 (in Chinese).

Geology Institute, Chinese Academy of Geological Sciences (2004) China Geological Map (1:4 000 000, Second edition). Beijing: Geological Publishing House (in Chinese).

Guo A L, Zhang G W, Cheng S Y (2004) Beyond plate tectonics: Discussion on the new opportunity of continental geology research. Progress of Nature Science, (7): 10–14 (in Chinese with English abstract).

Hall R, Cottam M A, Wilson M E J (eds.) (2011) The SE Asia gateway: History and tectonics of Australia-Asia collision. Geological Society of London, Special Publication, 355: 1–381.

Hsu K J (1988) Relict back-arc basins: Principles of recognition and possible new examples from China. In: Kleinpell K L, Paola C (eds), New perspective in Basin Analysis. New York: Springer Verlag, 245–263.

Hsu K J, Sun S, Li J L et al. (1988) Mesozoic overthrust tectonics in South China. Geology, 16: 418–821.

Hutchison C S, Tan D N K (eds.) (2009) Geology of Peninsular Malaysia. The University of Malaya and the Geological Society of Malaysia. Kuala Lumpur: Murphy, 1–479.

Jin Z M, Yao Y P (2004) Beyond plate tectonics: What should be done for Chinese structural geology? Earth Science, (6): 644–650 (in Chinese with English abstract).

Jolivet L, Tamaki (1994) Japan Sea, opening history and mechanism: A synthesis. J. Geophys. Res., 99: 22237–22259.

Karsakov L P, Zhao C J, Malyshev Y et al. (2008) Tectonics, Deep Structure, Metallogeny of the Central Asia–Pacific Belts Junction Area (Explanatory Notes to the Tectonic Map Scale of 1: 1 500 000). Beijing: Geological Publishing House, 1–213.

Kennedy W Q (1964) The structural differentiation of Africa in the Pan-African (500 m.y.) tectonic episode. Res. Inst. Ar. Geol., University of Leeds, 8th Ann. Rep., 48–49.

Lan C Y, Chung S L, Long T V et al. (2003) Geochemical and Sr-Nd isotopic constraints from the Kontum massif, Central Vietnam on the crustal evolution of the Indochina block. Precambrian Research, 122: 7–27.

Le Pichon S, Francheteau J, Bonin J (1973) Plate Tectonics. New York: Elsevier Publishing Company, 1–300.

Li C Y (2004) The Selections of Plate Tectonic. Beijing: Geological Publishing House, 1–279 (in Chinese).

Li C Y, Wang Q, Liu X A et al. (1982) Asia Tectonic Map (1:8 000 000, attached instructions, 45 p.). Beijing: Geography Publishing House (in Chinese).

Li T D, Uzhkenov B S, Mazorov A K et al. (2008) Central Asia and Adjacent Geological Map (1:2 500 000). Beijing: Geological Publishing House (in Chinese with English abstract).

Liu X, Li T D, Geng S F et al. (2012) Geotectonic division of China and some related problems. Geological Bulletin of China, 31(7): 1024–1034 (in Chinese with English abstract).

Ma L F (Editor in chief) (2002) Atlas of China Geology (electronic version 2004). Beijing: Geological Publishing House (in Chinese).

Miao P S, Zhou X Q (2010) The Map of Global Tectonic System (1:25 000 000, attached explanation, p. 1–41). Beijing: Earthquake Publishing House (in Chinese with English abstract).

Neves S P, Mariano G (2004) Heat-producing elements-enriched continental mantle lithosphere and Proterozoic Intracontinental orogens: Insights from Brasiliano/Pan-African Belts. Gondwana Research, 7(2): 427–436.

Parfenov L M, Prokopiev A V, Gaiduk V V (1995) Cretaceous frontal thrusts of the Verkhoyansk fold belt, eastern Siberia. Tectonics, 14 (2): 342–358.

Parfenov L M, Badarch G, Berzin N A et al. (2009) Ogasawara Mand Yan H, Summary of Northeast Asia geodynamics and tectonics, Stephan Mueller Spec. Publ. Ser., 4: 11–33.

Petrov O, Leonov Y, Li T D et al. (Editors in chief) (2008) Tectonic Zoning of Central Asia and Adjacent Areas (1:20 000 000). In: Atlas of Geological Maps of Central Asia and Adjacent Areas (1:2 500 000). VSEGEI Cartographic Factory.

Press F, Siever R (1974) Earth. W H Freeman and Company, 1–613.

Pubellier M (2008) Structural Map of Eastern Eurasia (1:12 500 000). Paris: CGMW.

Ren J S (Editor in chief) (2013) International Asia Geological Map (1:5 000 000). Beijing: Geological Publishing House (12p.) (in Chinese with English abstract).

Ridd M F, Barber A J, Crow M J (2011) The Geology of Thailand. London: The Geological Society, 1–626.

Rogers J J W, Santosh M (2002) Configuration of Columbia: A Mesoproterozoic supercontinent. Gondwana Res., 5: 3–22.
Rogers J J W, Santosh M (2004) Continents and Supercontinents. New York: Oxford Press, 1–289.
Sengör A M C, Nal'in B A, Burtman U S (1993) Evolution of the Altaid tectonic collage and Paleozoic crustal growth in Eurasia. Nature, 364: 209–304.
Shan Y N, Kang YS, Yue L Q et al. (2011) Eastern Asia regional tectonic evolution and division of tectonic domain. Journal of Geomechanics, 17 (3): 211–222 (in Chinese with English abstract).
Shao J, He G, Zhang L (2007) Deep crustal structures of the Yanshan intracontinental orogeny: A comparison with pericontinental and intercontinental orogenics. Geological Society, London, Special Publications, 280: 189–200.
Shu L S, Faure M, Wang B et al. (2008) Late Paleozoic-Early Mesozoic geological features of South China: Response to the Indosinian collision events in Southeast Asia. Tectonics, 340: 151–165.
Smyth H R, Hamilton P J, Hall R et al. (2007) The deep crust beneath island arcs: Inherited zircons reveal a Gondwana continental fragment beneath East Java, Indonesia. Earth and Planetary Science Letters, 258: 269–282 (www.sciencedirect.com).
Song H L (1999) The basic characteristics of Yanshan type intraplate orogenic belt and its dynamic. Earth Science Frontier, 6 (4): 309–316(in Chinese with English abstract).
Stille W H (1924) Grundfragen der Vergleichenden Tektonik. Berlin: Borntraeger, 1–443.
Tamaki K, Suyehiro K, Allan J et al. (1992) Tectonic synthesis and implications of Japan Sea ODP drilling. Proc. Ocean Drill. Program Sci. Results, 127–128: 1333–1348.
Tectonic Group, Institute of South China Sea, Chinese Academy of Sciences (Liu Z S et al.) (1988) Tectonic and Continental Extension in South China Sea. Beijing: Science Press.
Wan T F (2011) The Tectonics of China: Data, Maps and Evolution. Beijing, Dordrecht Heidelberg, London and New York: Higher Education Press and Springer, 1–501.
Wan T F, Wang Y M, Liu Q L (2008) The tectonic detachment and magma origin depth in Yanshanian and Sichuanian periods for Eastern China. Earth Science Frontier, 15(3): 1–35 (in Chinese with English abstract).
Wilson J T (ed.) (1970) Continents Drift: Readings from Scientific American. San Francisco: W. H. Freeman and Company.
Xiang L, Shu L S (2010) Pre-Devonian tectonic evolution in the eastern part of South China: Evidences from detrital zircons. Scientia Sinica (Earth Sciences), (10): 1377-1388 (in Chinese with English abstract).
Xiao W J, Kröner A, Windley B F (2009) Geodynamic evolution of Central Asia in the Paleozoic and Mesozoic. International Journal of Earth Sciences, 98: 1185–1188.
Xu J H (1980) The thin plate tectonics of collision orogenic belt. Science of China (B), (11): 1081–1089 (in Chinese with English abstract).
Yin A, Nie S Y (1996) A Phanerozoic plain plastic reconstruction of China and its neighboring regions. In: Yin A, Harrison M (eds.), The Tectonic Evolution of Asia. Cambridge University Press, 442–485.
Yoon S (2001) Tectonic history of the Japan Sea region and its implications for the formation of the Japan Sea. Journal of Himalayan Geology, 22 (1): 153–184.
Zhang G W, Quo A L, Yao A P (2006) On the China continental geology and continental tectonic basement. Natural Science Progress, (10): 12–17(in Chinese with English abstract).
Zhang W Y (1984) The Introduction of Fault Block Tectonics. Beijing: Petroleum Industry Press, 1–385 (in Chinese).
Zhao Z F (1995) On the Yanshanian Movement. Geological Review, 19 (8): 339–346 (in Chinese).

2 Tectonic Domains and Tectonic Units in Asian Continent

2.1 Siberian Tectonic Domain

The Siberian tectonic domain (Fig. 2.1) is located in the northern part of Asian continent. The center is Siberian plate [1], around which there are several blocks and collision zones, such as Southern margin of East Siberian Sea Jurassic collision zone (200–135 Ma) [2], Verkhojansk–Chersky Jurassic accretion–collision zone (200–135 Ma) [3], Kolyma–Omolon plate ($\sim$850 Ma) [4] and Transbaikalia (or Mongolia–Okhotsk) Jurassic accretion–collision zone ($\sim$170 Ma) [5].

In Jurassic and Cretaceous, the Siberian plate [1] collided hard against the North American plate [71] in the northeast, and against the Kolyma–Omolon plate [4] in the east, and they finally were converged together. In addition, the Siberian plate [1] crashed with the Central Asia–Mongolia tectonic domain in the southeast, forming the Transbaikalia (or Mongolia–Okhotsk) Jurassic accretion–collision zone ($\sim$170 Ma) [5] (Fig. 2.1).

2.1.1 Siberian Plate (1600 Ma) [1]

This plate includes the East Siberian and West Siberian plates, in which the Archean and Paleoproterozoic metamorphic systems are widely spread. The East Siberian plate (Fig. 2.2) is composed of the Kheta–Olenek (I), Olenek (II), Tyung (III), Vilyui (IV), Angara (V) and Aldan–Stanovik (VI) continental nuclei. Aldan–Stanovik continental nucleus is constituted of belts of granitic granulite–greenstone at the age of 3.3–2.8 Ga and zones of metamorphic rocks at the age of 3.5–2.7 Ga. Besides, Baikal uplift is outcropped to the south of it. In the above blocks, Pre-Cambrian mafic intrusions also exist. The eclogite inclusions formed in the process of kimberlite eruption, and mantle differentiation provides the important clues for Archean evolution (Pearson et al. 1994). Because their basaltic composition may be the relict of magmatic ocean of 4.0 Ga and the subduction products of Archean oceanic crust or the later crystallized high-pressure mantle melt. The Re-Os isotopic data, acquired from the eclogite with diamond inside the Siberia Udachnaya kimberlite pipe, show that they were formed in Mesoarchean (2.9 $\pm$ 0.4 Ga). These age data are much younger than those of eclogite formed in the process of differentiation and retained in early period of the Earth. However, those data are rather suitable for the age (2.85–3.2 Ga) of crust formation and craton consolidation of Aldan and Anabar continental nuclei. Because the 150 km depth is the lowest limit for preserving diamond, the above data and the data of Udachnaya dunite with Re-Os model age of 3.1 Ga and diamond inside show that the Mesoarchean Siberian lithosphere is 150 km thick at least.

In the Siberian plate, 15% of kimberlite contains diamonds. According to the seismic exploration near the crust and mantle, it is discovered that all known kimberlite pipes with the diamonds were produced on Archean blocks with high density (i.e., super-mafic rock), and that the kimberlite pipes were formed and took place at last in Late Paleozoic (about 360 Ma) or Cretaceous (127–90 Ma), and controlled by intraplate rift magmatism. In most cases, the isotopic age of ultramafic rocks such as kimberlite enriched in diamonds and eclogite inclusions is Archean and Paleoproterozoic (from 3.5–3.2 to 2.0 Ga). Generally speaking, the high-density blocks in lower crust or mantle are the residue of Pre-Cambrian high-pressure eclogite (with diamonds) and dunite that had been partially melted; obviously, they are the main source of diamond-bearing kimberlite. From the view of ore field space distribution, they are controlled by central radiation faults in silica-alumina continental nuclei, and it is believed that they are the most ancient tectonic remain fragments of continental crust. It seems that the "Continental Nuclei Model" of Pre-Cambrian geodynamics is the most suitable theory to explain the space distribution of kimberlite fields in Siberia (Moralev and Glukhovsky 2000; Fig. 2.2).

In the Paleoproterozoic at 2.50–1.60 Ga, the whole Siberian plate formed the uniform crystalline basement, i.e., Siberian plate was also called "Angara plate" by some researchers, a large number of magmatism, metamorphism

T. Wan, *The Tectonics and Metallogenesis of Asia*,
https://doi.org/10.1007/978-981-15-3032-6_2

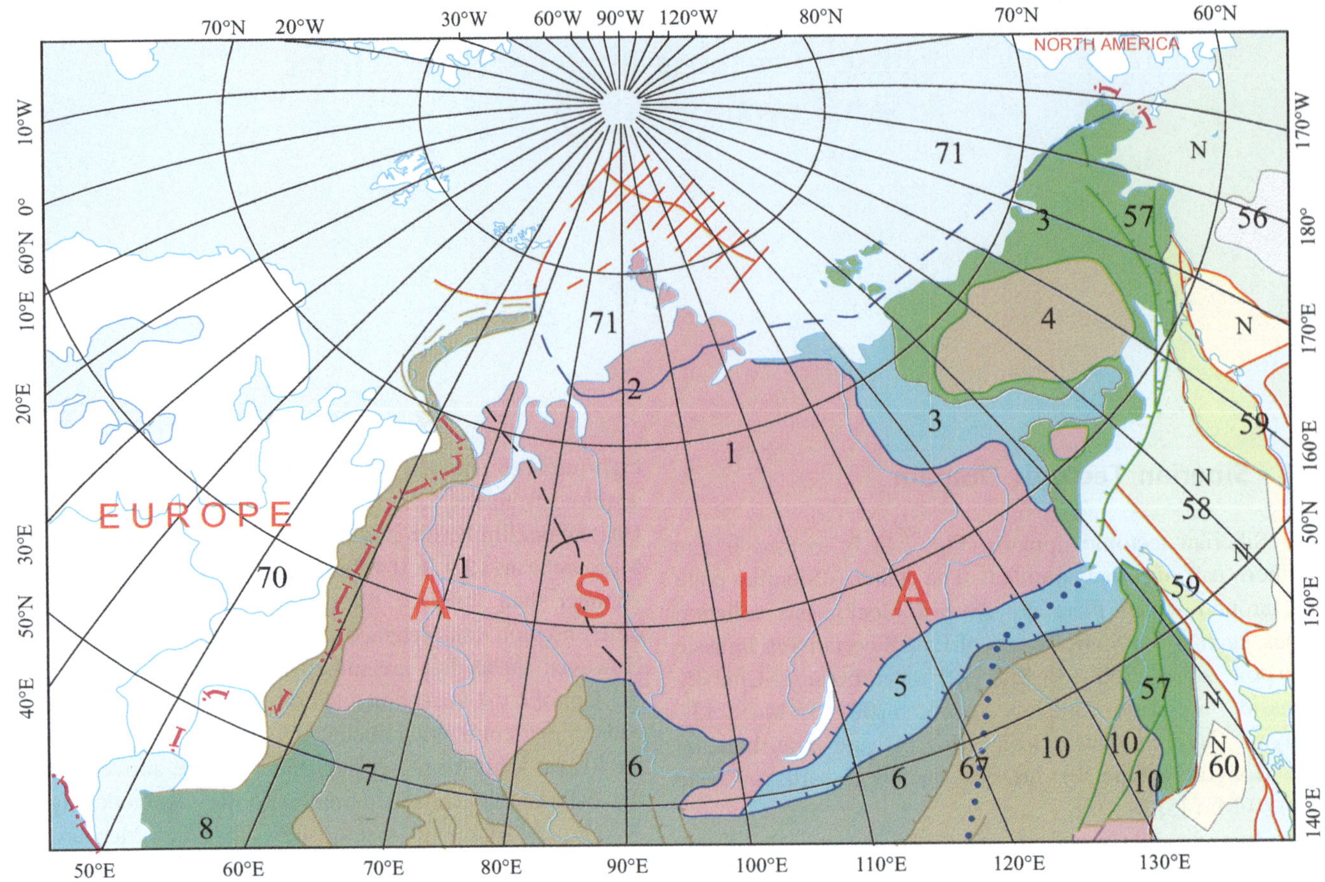

Fig. 2.1 Siberian tectonic domain [1–5]. The black numbers in the figure show the tectonic units, as same as those in the CONTENTS and Fig. 1.1

and rock deformation (a very strong tectono-thermal event) reformed the ancient Archean crust (Glukhovaky 2009). Generally, the evolution characteristics of Siberian plate in Archean and Paleoproterozoic are very different from those of Sino–Korean plate. However, the periods to form the uniform crystalline basement are rather similar. In the Neoproterozoic (1000–541 Ma, Vendian Period), the rifts were developed between continental nucleus in East Siberian plate (Fig. 2.2). At the southern and western margins of East Siberian plate, large-scale extension and fault-depression were developed, with the generation of oceanic crust, ophiolite suite and arc accretion wedge. It was a period when obviously different tectonic characteristics took shape between East and West Siberian plates (Khain et al. 2003).

After Neoproterozoic, a large amount of basalt magmas erupted and the fault-depression developed in West Siberian plate, and later, they were covered by Paleozoic and Mesozoic–Cenozoic basic volcanic rocks and sedimentary rocks. The thickness of sedimentary rocks reaches 14 km. Due to the great quantity high-density basalts by the gravity balances, the West Siberia became a subsidence area in the long periods (shallow yellow area in Fig. 2.3). In Late Permian (252 Ma), on the western area of that rather stable block, the great-scale continental basalt eruption occurred to form the large volcanic province (LVP) from Noril'sk, Taimyr and Tunguska of North Siberia to western area on eastern side of Urals (Fig. 2.3; Saunders and Reichow 2009). The ancient basalts are still preserved in horizontal attitude. The basalt's volume was estimated as 2×10^6 to 5×10^6 km^3 (Dobretsov et al. 2008). As to the mechanism of this large volcanic province, usually it is believed that the volcanic province may be caused by the heat from super-critical fluid of mantle plume. The results of petrology and geochemistry show that temperature of hot mantle might reach 1600 °C, causing the interaction between magma and lithosphere. The head of the mantle plume directly affected the relatively stable lithosphere. The ascending of a small amount of materials brought about the underplating at bottom of lithosphere and extension faults at some areas and boundaries, thus to produce the magma intrusion or eruption (Dobretsov et al. 2008). However, some researchers (such as Xiao et al. 2009a) considered the West Siberia to belong to "Central Asia Late Paleozoic Orogenic Belt," which recognition may be improper.

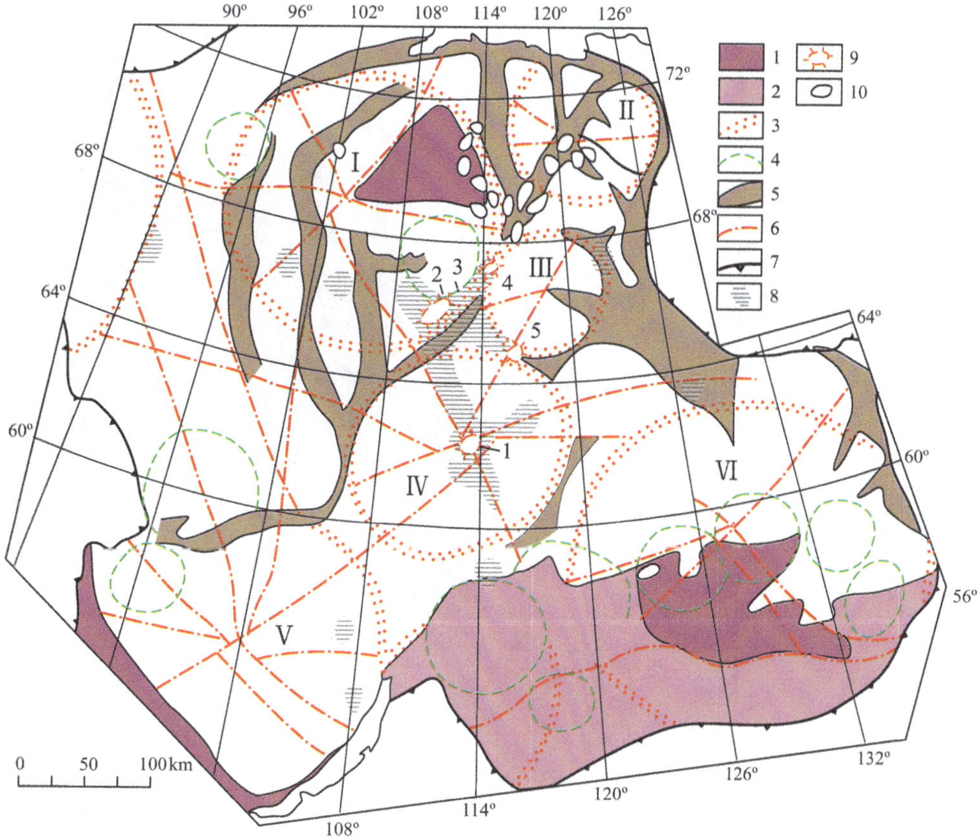

Fig. 2.2 Siberian craton kimberlite field and their main dome structure of basement. **Legend**: (1) Early Pre-Cambrian metamorphic complex, with basement dome; (2) Early Pre-Cambrian complex, reformed by Proterozoic or Phanerozoic tectonic-thermal events; (3) boundaries of contemporaneous continental nucleus; (4) boundaries of main Archean and Paleoproterozoic granitic gneiss dome; (5) Neoproterozoic (Vendian Period) rift; (6) main faults; (7) boundaries of Siberian plate; (8) zones of high-density rocks, formed at the boundary of crust and mantle ($v_p \geq 8.4$ km/s); (9) kimberlite pipe with diamonds; (10) Other kimberlite pipes. **Continental nuclei**: (I) Kheta–Olenek; (II) Olenek; (III) Tyung; (IV) Vilyui; (V) Angara; (VI) Aldan–Stanovik. **Diamond-bearing kimberlite distribution areas**: (1) Malobatuoba; (2) Alakit; (3) Daldyn; (4) Mun; (5) Nakyn (Modified from Moralev and Glukhovsky 2000)

According to the principle that the distribution of dyke swarm in large volcanic province points out central position of magma uplift, Czamanske et al. (1998) and Ernst and Buchan (2001) considered the central of West Siberian large volcanic province is located on the northern side of Siberian plate (Fig. 2.3) and as same as the central magmatism of Mackenzie dyke swarm in Canada (Hou 2012; Fig. 3.2). Their recognition can explain why the large volcanic province has never broken the West Siberian plate. However, the above analyses did not use the paleo-magnetism data to reconstruct. Moreover, the seismic research has not been completed yet. So the distribution of intrusive rocks and the detailed structures of crust and lithosphere are not very clear. Thus, it is better not to make the final conclusion. Depending on the paleo-magnetism research, before the Early Paleozoic the central area of Siberian plate was located in the south of 31° S and in the end of Paleozoic it migrated to near 40° N (Figs. 3.6–3.8 and 3.10–3.12). In that migration process, the thickness of marine sedimentation was not very huge. In the end of Triassic, the central referential point reached 64° N. From Paleozoic to Triassic, lasting about 310 million years, the average latitude changing velocity was about 4 cm/year. However, in the later periods the Siberian plate was only moved or rotated in a rather small range at the high latitude of the Northern Hemisphere (Figs. 3.16, 3.20, 3.24, 3.27, 3.32 and 3.34; Khramov et al. 1981; Wan 2011b).

At the Baikal–Tuva region on the southern border of the Siberian plate, influenced by the stronger Cenozoic tectonics, earthquakes and global weather changed. They controlled the sedimentation, neotectonics and the formation of basins (Gladkochub and Donskaya 2009). In the Neogene, the direction of regional shortening and compression was NNE, as same as the main trending of Lake Baikal; near E–W extension took place, and the main boundary faults showed

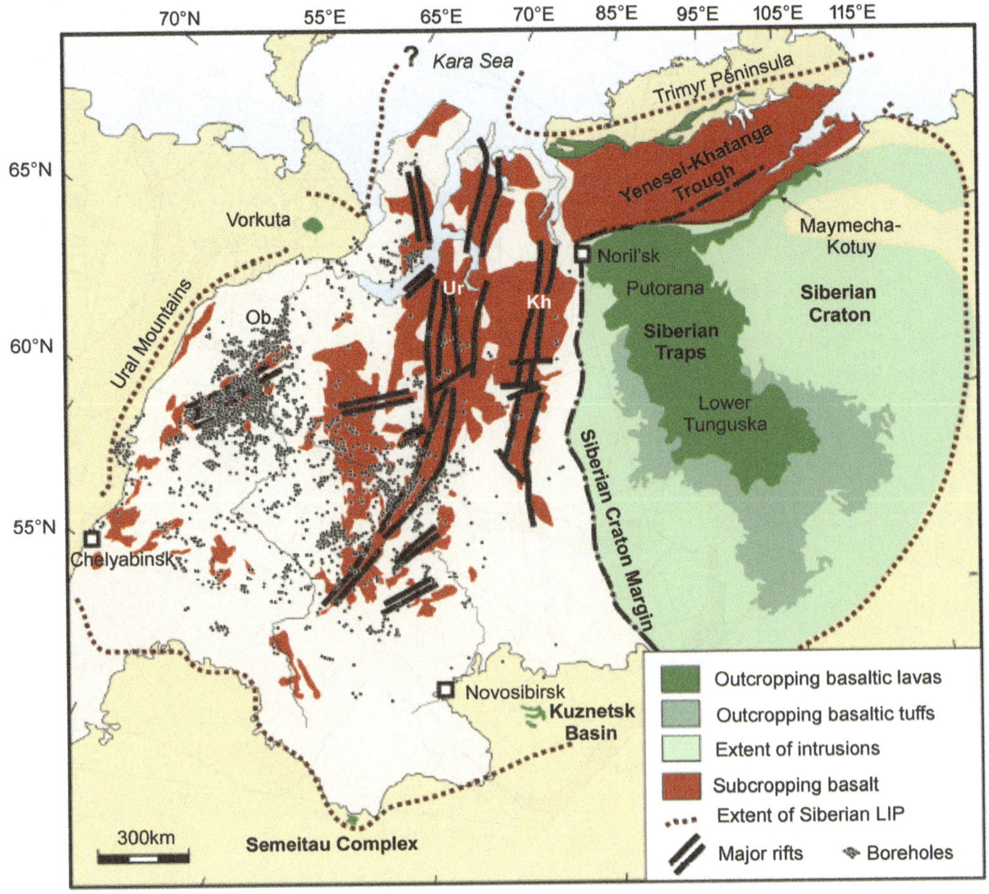

Fig. 2.3 Continental basalt distribution in Siberian great volcanic province (Modified from data of Saunders and Reichow 2009)

the characteristics of normal faults, forming the Baikal rift, which is the main rift period of Baikal. However, since Middle Pleistocene, the orientation of regional maximum principal compression changed to near E–W trending. The extension-shear active faults and earthquakes are concentrated at northern and southern sides of Lake Baikal with near E–W trending, i.e., along the two Siberian railways. On the eastern and western sides of Lake Baikal, the faults activities are rather weak (Bulgatov 1990, personal communication). Thus, it can well explain why the earthquakes are stronger along the two Siberian railways in E–W trending.

Since the Siberian plate got into shape at the period of 1600 Ma, it has undergone long stable and migration, from the Southern Hemisphere to high latitude of the Northern Hemisphere. In the Late Paleozoic, on the west side of that plate, the Ural collision zone was formed, i.e., collided with Baltic plate, and the Siberian plate turned into a part of the Pangea supercontinent. After the Triassic, the Pangea supercontinent was broken, and the Siberian plate was still located at the northern part of Eurasian continent. In Jurassic, under the influences of southwestward compression and collision of North American plate, the plate rotation and the intraplate deformations on the boundary of the Siberian plate were formed.

2.1.2 Southern Margin of East Siberian Sea Jurassic Collision Zone (200–135 Ma) [2]

It is a collision zone among the North American plate [71], Siberian plate [1], Verkhojansk–Chersky Jurassic accretion–collision zone [3] and Kolyma–Omolon plate [4]. It is an important boundary between North American plate and Siberian tectonic domain. The data related to this collision zone are rather rare. However, from New Siberian Islands to Bering Strait, the fitful outcrops of ophiolite suits and strong rock deformations show that the existence of the Jurassic collision zone is irrefutable. Some Japanese and American researchers considered that the Verkhojansk–Chersky Jurassic accretion–collision zone [3] and Kolyma–Omolon plate [4] belong to North American plate. Probably, they did not know the existence of East Siberian Sea Jurassic collision zone, or they did not know the uniform crystalline basement of Kolyma–Omolon and Okhotsk plates were all formed at about 800 Ma, which characteristic is not shown in crystalline basement of North American plate.

The uniform crystalline basement of North American plate was formed in the late period of Paleoproterozoic (1600 Ma). The North American plate was the main part of Columbia supercontinent, and later was broken up (Rogers and Santosh 2002, 2004; Figs. 3.1 and 3.2). In the late period of Mesoproterozoic (about 1000 Ma), the North American plate participated in the conversion of the Rodinia supercontinent. Thereafter, the North American plate, with the Canada shield as its core, has kept stable for a long time and the sediment with horizontal attitude covered a great area (Condie 1994, 1997). The obvious intraplate deformations in Paleozoic and later were only limited at the boundary of the plate, with a width of 200–300 km. The tectonic activity inner plate is still very weak (Bally and Albert 1989). It is the reason why the plate was called as "rigidity" by the American researchers in the early period of Plate Tectonic Theory. On the western boundary of North American plate, some micro-plates exist, which may come from the Gondwana.

In the late period of Early Paleozoic, the North American plate and Baltic plate conversed to form the Appalachian–Caledonian collision zone (Brenchley and Rowson 2006; Gee et al. 2008; Fig. 3.8). In the late period of Late Paleozoic and the Hercynian, the Ural [12] and Balkhash–Tianshan–Hingganling [10] collision zones were formed, and thus, the Siberian plate, North American plate and Baltic plate converged to be one part of the Pangea supercontinent (Fig. 3.12). However, since Triassic, with the extension of Atlantic Ocean (Condie 2001), the North American, Eurasian and African plates had been separated radioactively (Fig. 3.15), and the North American plate migrated northwestward. When the western part of North American plate moved to west by south, it crashed with the northeast part of Siberian plate [1] and Kolyma–Omolon plate [4], forming the Southern margin of East Siberian Sea Jurassic collision zone [2] and Verkhojansk–Chersky Jurassic accretion–collision zone [3], and the Siberian plate rotated anticlockwise, migrated a little bit southward (Wan 2011b), and collided with the Altay–Middle Mongolia–Hailar Early Paleozoic (541–419 Ma) zone [6], forming the Transbaikalia (or Mongolia–Okhotsk) Jurassic accretion–collision zone ($\sim$170 Ma) [5] (Figs. 3.21 and 3.40).

Besides the features of compression, collision, the migration of Southern margin of East Siberian Sea Jurassic collision zone shows the feature of sinistral strike-slip as well, and quite obvious in north part of N–S trending Ural accretion–collision zone occurred (Fig. 2.1). After Jurassic, the Siberian tectonic domain was rather stable.

2.1.3 Verkhojansk–Chersky Jurassic Accretion–Collision Zone (200–135 Ma) [3]

The Verkhojansk–Chersky Jurassic accretion–collision zone is a collision zone among the Siberian plate [1], Kolyma–Omolon plate [4] and Okhotsk plate [58] (Fig. 2.1). It had a great influence on the intraplate deformation of East Siberian plate and East Asian continent. Parfenov et al. (2009) considered that the collision zone is the boundary zone of Siberian Craton. According to the data of sedimentology and paleogeography (in rare of the volcanism), it infers that there are abundant oceanic and arc sedimentary formations, such as Neoproterozoic–Paleozoic shallow marine carbonates and fine-grain clastic sediments, Devonian and Carboniferous rift-type continental sedimentary system, Triassic–Jurassic alkali basaltic flow, sill and dyke, which consist of sedimentary complex in the Verkhojansk basin, with the thickness of sedimentary system up to 20 km.

The western part of this collision zone was formed in Jurassic, and the eastern part was the Cretaceous tectonic zone. The Verkhojansk–Chersky Jurassic accretion–collision zone is mainly composed of the Middle and Late Jurassic ophiolite suit with greenschist facies, mud-gravel strata, thrust debris and nappe. In the end of Late Jurassic, the thrusts and strike-slip faults connected with collision were formed. In the southern Verkhojansk granite zone, its isotopic age is 157–93 Ma (Parfenov et al. 1995b; Parfenov et al. 2009; Oxman 2003).

This collision zone was formed by the southwestward migration and compression and may be caused by the long-distance effect of Northern Atlantic Ocean radiative extension, with the northern part of North American plate migrating and compressing toward WSW and the Tethys Ocean extending toward NE. Thus, it caused the obvious anticlockwise rotation (Khramov et al. 1981), resulting in the Siberian plate having an anticlockwise rotation of 36.2°, and the North and South China, Korean Peninsula having an anticlockwise rotation of 30°–20° (Wan 2011b; Fig. 3.20). Since Jurassic, the East Asia, including Siberian plate, the north orientation of paleo-magnetism has been kept almost the same.

In the southern part of this collision zone, Aniuy area, undergone an Archean–Paleoproterozoic small block (Parfenov et al. 1995a; Parfenov et al. 2009). It may be a fragment of Siberian plate at first. That micro-block became one part of the Verkhojansk–Chersky Jurassic collision zone. This collision zone is the boundary between Omolon and Pre-Kolyma blocks. The extension of Aniuy Paleo-ocean in the southern boundary of the Triassic–Jurassic Pacific also formed Verkhoyansk–Kolyma fold zone (Parfenov et al.

1995b). At last, the movement of Northeastern Russia became fixed and unchangeable in the Early Cretaceous, which also belongs to the collision zone between the Aniuy Paleo-ocean and Western Pacific trench–arc zone. Thus, strong deformation complex in the Mesozoic Verkhoyansk system occurred to form thrusts and overthrusts at the corner of Kolyma block (Parfenov et al. 1995b, 2009).

In 2000–2003, by means of the seismic reflection and refraction data (Gebhardt et al. 2006), the first complete El'gygytgyn crater, with a diameter of 18 km, was discovered. The crater has five layers: On the surface, it is covered by water, beneath it is lake sediments. In the bowl-type impact crater, the upper part is the downfallen breccia, with the P-wave velocity of 3 km/s; the lower is the breccia basement (the P-wave velocity is more than 3 km/s); and at the bottom it shows a typical anticline, which has the similar scale to the crater of other continents or the Moon. This crater not only has the intact geomorphic characteristics, but also is a typical crater formed in the crystalline rock.

2.1.4 Kolyma–Omolon Plate (~850 Ma) [4]

The Kolyma–Omolon plate (Fig. 1.1) and Okhotsk plate [58] may be the same plate originally. Both of them have some Archean metamorphic crystalline blocks, forming uniform crystalline basements in Neoproterozoic (~850 Ma) (Parfenov et al. 1995b, 2009), which characteristics are very similar to those of the Yangtze plate (Yang et al. 1994; Liu et al. 1995; Tang and Zhou 1997), but very different to the evolution characteristics of North American plate (Engebretson et al. 1985). Thereafter, the Kolyma–Omolon plate developed many Early Paleozoic sediments. In the Silurian–Devonian continental accretion period, the oceanic, arc compositions were developed, forming the Kolyma Bay and Kolyma volcanic zone, which was provided with the complex of basaltic arc and marginal sea. This plate migrated and collided in the WSW trending in Jurassic, forming the Verkhojansk–Chersky Jurassic accretion–collision zone [3]. According to the result of $^{40}Ar/^{39}Ar$ dating, the ages of subduction and collision of the Kolyma–Omolon plate toward southwest are 160–140 Ma and 143–138 Ma, respectively. The collision age for the northern boundary of Kolyma–Omolon plate is 130–123 Ma. It means that plate was completely combined together with the Verkhojansk–Chersky accretion–collision zone till the Cretaceous. 106–92 Ma ago, magma intrusions were triggered by extension of nearly E–W direction (Parfenov et al. 1995b, 2009).

In addition, on the east side of the plate, the Sikhote–Alin–Koriak accretion–collision zone [57] was formed (Layer et al. 2001). It was a subduction–island arc zone in Cretaceous, caused by northward migration and subduction of the Pacific plate. In Paleogene, it evolved into accretion–collision zone and became fixed and unchangeable, and it is proved to be the northeastern area of Eurasian continent.

2.1.5 Transbaikalia (or Mongolia–Okhotsk) Jurassic Accretion–Collision Zone (~170 Ma) [5]

The Transbaikalia accretion–collision zone (Fig. 2.1) located on the southeastern boundary of East Siberian plate. Some researchers sometimes call it the Mongolia–Okhotsk collision zone or Jablonov–Stanov collision zone (or Trans-Hingganling in China). The center of that collision zone is near Chita of Russia and extends toward southwest to the North and East Mongolia. Between the Siberian plate [1] and Altay–Middle Mongolia–Hailar Early Paleozoic accretion–collision zone (541–419 Ma) [6], it formed the Transbaikalia accretion–collision zone in Jurassic (Figs. 2.4 and 2.5; Zolin et al. 2001). After Jurassic, the Siberian plate [1] then jointed with the Altay–Middle Mongolia–Hailar Early Paleozoic accretion–collision zone, Balkhash–Tianshan–Hingganling Late Paleozoic accretion–collision zone and Sino–Korean plate completely. This accretion–collision zone only has a little bit of Archean–Paleoproterozoic crystalline basement rocks, and some parts are covered by Neoproterozoic sedimentary strata, mainly the Paleozoic and Jurassic strata, where strong deformations occurred, the thrusts and folding were developed, and a lot of Late Paleozoic granitic magmatism was active, as well as super-shallow bodies and the volcanic rocks were erupted. In many places on both sides for this collision zone, ophiolite suits exist.

According to the research on paleo-magnetism, since Early Permian to Early Jurassic, the paleo-ocean, which is one part of Tethys Ocean between the Mongolia and Siberia, had been existing. However, the ocean shrank gradually; the Siberian plate moved a little in a SE direction, and the blocks of Altay–Middle Mongolia–Hailar Early Paleozoic accretion–collision zone migrated northward to the areas of 50° N–60° N (Figs. 3.6–3.8 and 3.10–3.12). In Late Jurassic (140 Ma), the Siberian plate crashed with the Altay–Middle Mongolia–Hailar Early Paleozoic accretion–collision zone and formed the Transbaikalia collision zone (Zolin et al. 2001; Figs. 2.4 and 2.5). Parfenov et al. (2009) considered that the collision went on from the Middle Jurassic (175 Ma) to the Cretaceous (96 Ma). Till that time, the Siberian tectonic domain and Central Asia–Mongolia tectonic domain have combined together completely.

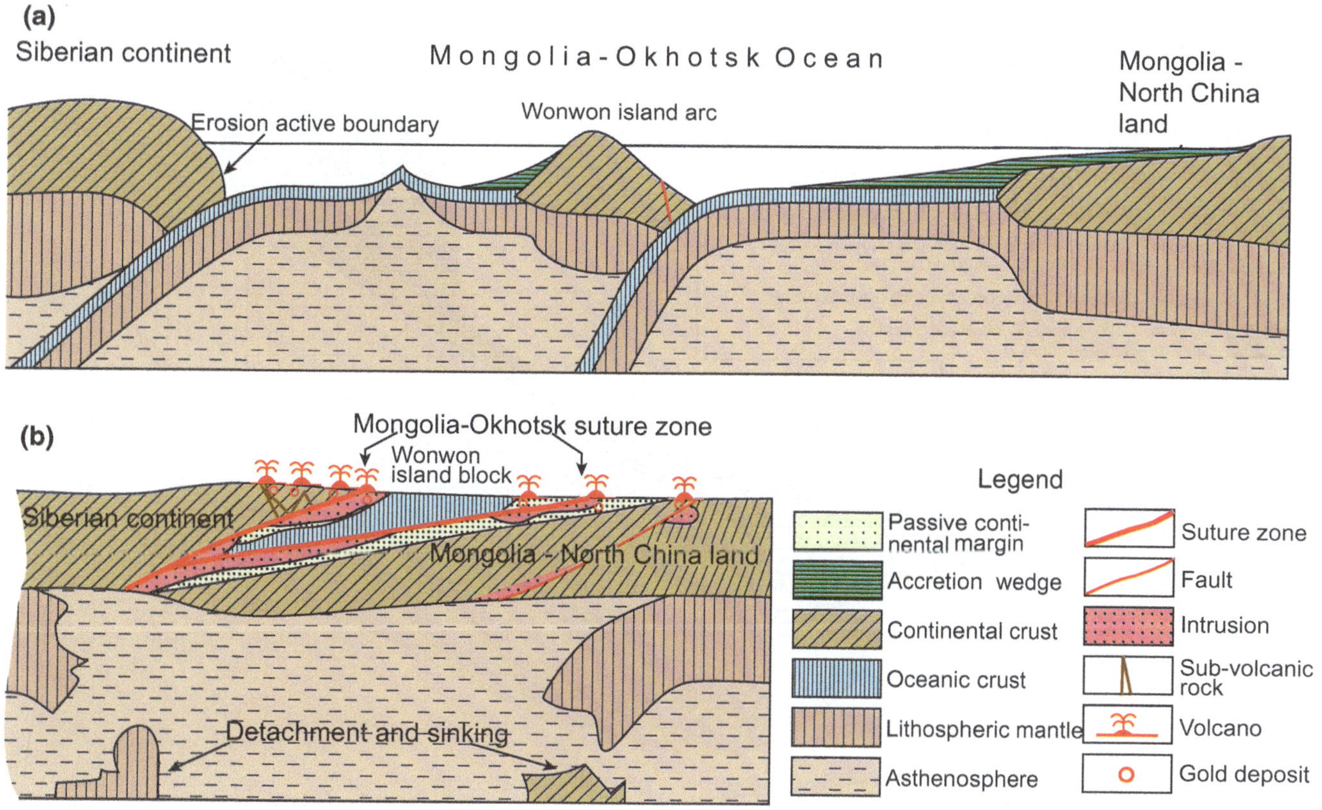

Fig. 2.4 Evolution model of Transbaikalia (or Mongolia–Okhotsk) Jurassic accretion–collision zone. **a** Model before collision, in Late Permian–Early Jurassic; **b** model after collision, in Middle–Late Jurassic (Modified from Zolin et al. 2001)

2.2 Central Asia–Mongolia Tectonic Domain

The Central Asia–Mongolia tectonic domain is mainly composed of Paleozoic accretion–collision zones (Fig. 2.6) and includes many crystalline blocks at different sizes (Fig. 2.7) and collision zones formed in different periods. It consists of the following tectonic units: The Altay–Middle Mongolia–Hailar Early Paleozoic accretion–collision zone [6], Karaganda–Kyrgyzstan (Qirghiz) Early Paleozoic accretion–collision zone [7], Turan–Karakum plate [8], Western Tianshan Late Paleozoic accretion–collision zone [9], Balkhash–Tianshan–Hingganling Late Paleozoic accretion–collision zone [10], Junggar block [11], Wandashan Jurassic collision zone [13] and Ural Late Paleozoic accretion–collision zone [12].

Xiao et al. (2015) considered that the Central Asia–Mongolia tectonic domain and Transbaikalia (or Mongolia–Okhotsk) Jurassic accretion–collision zone at the south margin of Siberian tectonic domain should be called "Southern Central Asia Orogenic Belt," including many collision zones formed in different periods between Siberian plate and Sino–Korean plate. It may be not very suitable because their forming periods are too long and the areas are too large.

In the Central Asia–Mongolia tectonic domain, there are many small Early Cambrian crystal blocks, such as Bureyat, Jiamus, Song–Nen, Totoshan–Xilinhot, Hailar, Middle Mongolia and Altay and many small Pre-Cambrian crystal blocks, such as Northern Bayannur, Hongshishan, Yagan, Turpan, Xingxingxia and Kuruktag blocks (Fig. 2.7). Originally, they may be formed near the Siberian plate. They moved along with the Siberian plate from the Southern Hemisphere to the middle and high latitudes of the Northern Hemisphere and collided successively in the Early Paleozoic or Late Paleozoic. They can be considered as the Paleozoic accretionary block group on the edge of the Siberian plate (Wan 2011b). In recent years, some researchers have got some younger isotopic ages from those blocks, so it needs conducting further discussion.

It is these crystalline blocks of varying sizes that play an important role in the mineralization of this tectonic domain. Due to the relatively high strength of the older metamorphic crystalline blocks and the significantly lower strength of the Paleozoic sedimentary and volcanic rock series, the stress distribution of this tectonic domain during the accretionary–

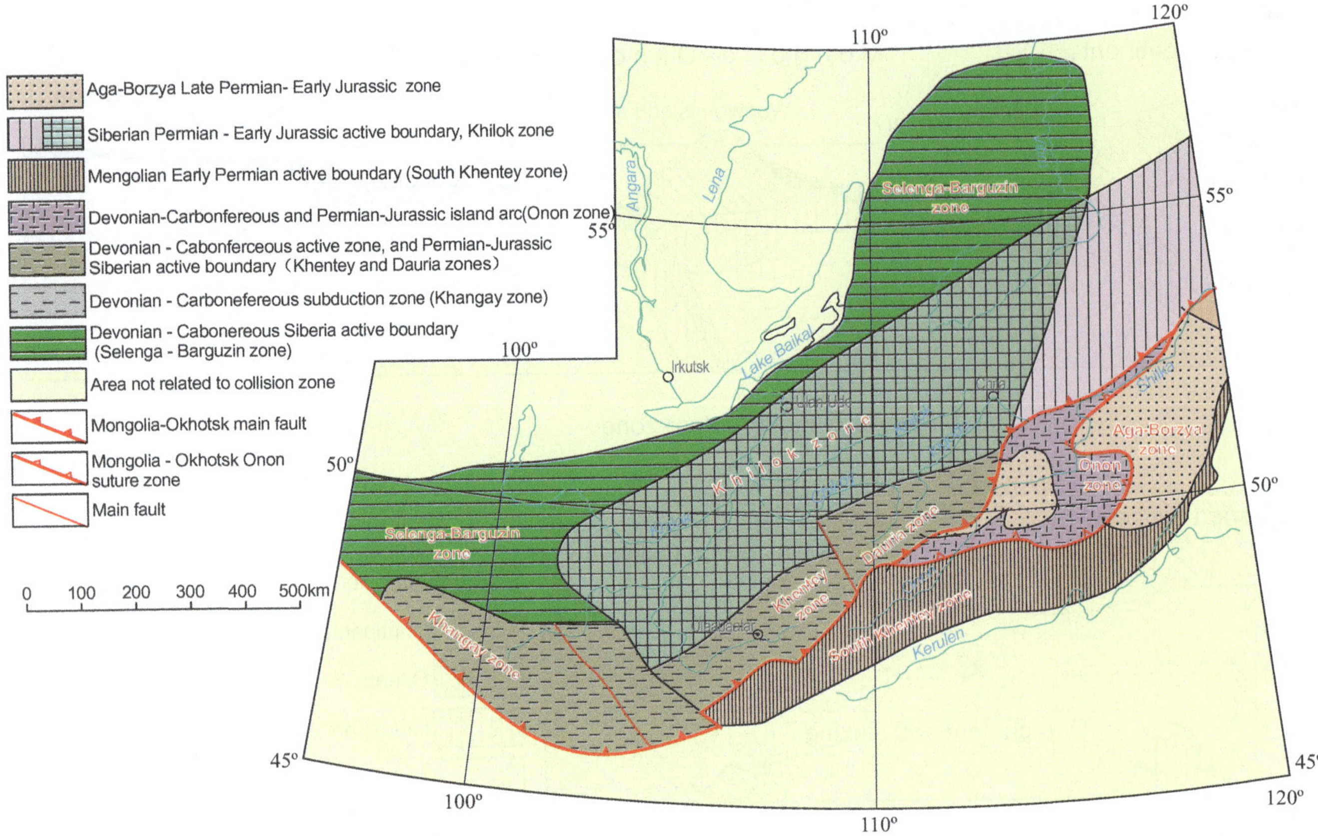

Fig. 2.5 Tectonic sketch of western part and its adjacent area in Transbaikalia collision zone. The center of figure shows the Devonian–Early Jurassic active continental margin, partially locating on the Devonian–Carboniferous sedimentary strata (Modified from the geoscience section data of Siberia–Mongolia of Zolin et al. 2001)

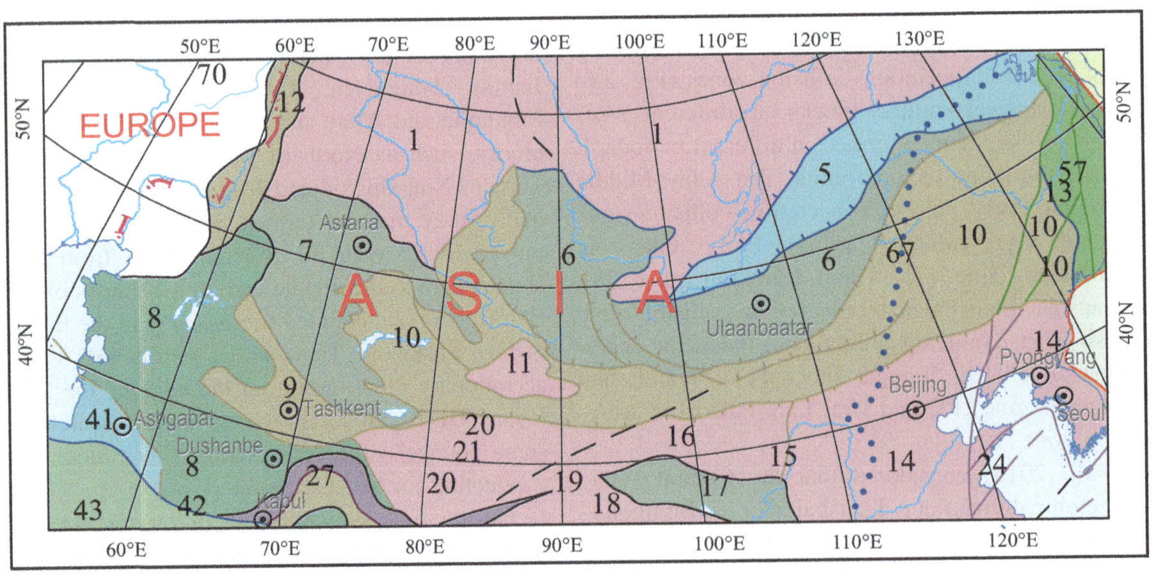

Fig. 2.6 Sketch of Central Asia–Mongolia tectonic domain [6–13]. The black numbers in the figure show the tectonic units, as same as those in the CONTENTS and Fig. 1.1

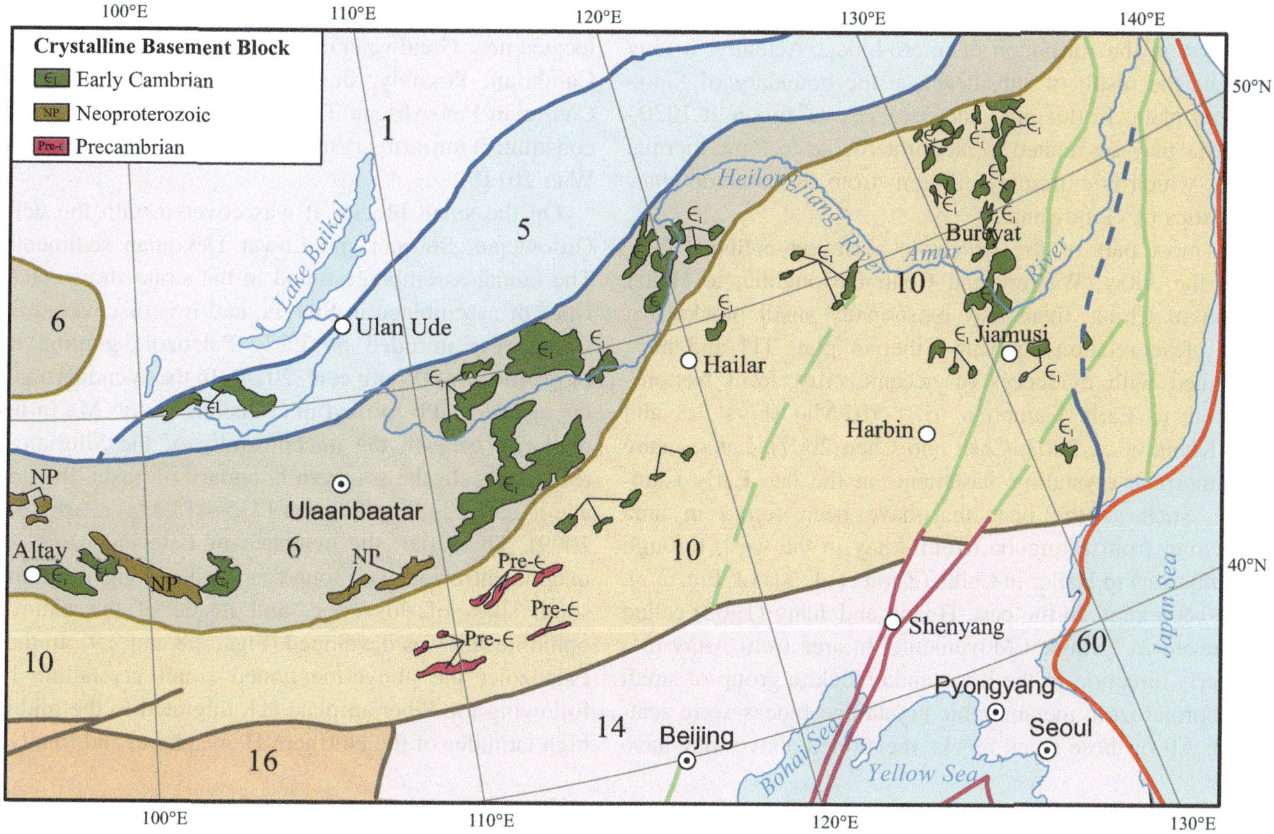

Fig. 2.7 Distribution of Pre-Cambrian small crystalline blocks in Central Asia–Mongolia tectonic domain

collisions is very uneven and local faults, especially transverse splitting faults, are easily produced in areas with different lithology (such as the edges of crystalline blocks). The fractures are very favorable for the large-scale penetration or accumulation of magmatic or ore-bearing fluids. It is very likely that this is an important reason for the formation of many super-large endogenous metal deposits in this structural domain. As to the tectonic evolution after Triassic, the middle and eastern parts of this domain and East Asian continent underwent six intraplate deformations with different characteristics and strengths, which will be discussed in Part 3 of this monograph. As to the endogenous metal deposits, they will be expounded in Part 4 of this monograph.

2.2.1 Altay–Middle Mongolia–Hailar Early Paleozoic Accretion–Collision Zone (541–419 Ma) [6]

This is an Early Paleozoic arc accretion–collision zone, distributing at the borders among Kazakhstan, China, Russia and Mongolia. It extends through the Central Mongolia, toward northeast to Hailar region of Northeastern Inner Mongolia, China (Fig. 2.6). Parfenov et al. (2009) divided this zone into: Yenisey–Transbaikalia tectonic zone and Altay–Wendulhan tectonic zone. They are all composed of the sedimentary–metamorphic strata, formed in the Vendian period of Neoproterozoic to Ordovician.

According to the recent research on detrital zircon ages (Rojas-Agramonte et al. 2011), the peak ages are mainly in Neoproterozoic (1020–700 Ma) and Paleozoic (600–350 Ma) and a small amount of data in the Archean–Mesoproterozoic (2570–1240 Ma). The Mesoproterozoic–Neoproterozoic strata were reformed by Paleozoic dolomitization and also suffered from the Neoproterozoic rifting and the crust evolution event in Grenville period.

The zircons in Middle Mongolia all have the peak age in Neoproterozoic (1020–700 Ma), which is completely different from that of the Siberian plate [1]. This feature effectively indicates that the potential source area of basement rocks is located in this area. The characteristic of detrital zircon is somewhat related to the Neoproterozoic (~850 Ma) rift and collision events in Tarim block or the subsidence in Sino-Korean plate boundary. Depending on the age dating results of detrital zircons and inclusions, Rojas-Agramonte et al. (2011) considered that the Tarim block [20] may be the main source area of sediments in this area and was split up later and participated in the Paleozoic collision in the formation of micro-block. Actually, it may also be the result of subsidence at the boundary of Sino–Korean plate. In this area, the peak age of zircon at 1020–700 Ma may be related to Neoproterozoic tectonic-thermal event, which is extremely different from the tectonic characteristics of Gondwana.

In most part of the Paleozoic accretion–collision belt, from the Altay, Western and Central Mongolia, to Hailar region of China, there may exist many small blocks that have close relationship with Siberian plate [1] and have remained with evidences of oceanic crust from Neoproterozoic to Early Cambrian (670–510 Ma) (Figs. 2.8 and 2.9; Khain et al. 2003; Chen and Chen 2007). Later, many metamorphic crystalline basements in the late Early Cambrian, such as the ones that have been found in area stretching from Mongolia (from Altay in the west, through Ulaanbaatar) to Hailar in China (Zhou et al. 2011a; Fig. 2.7), have been kept. In the past, Huang and Jiang (1962) called this event as "Xingkai Movement." In area from Altay in a westerly direction to the Karaganda block, a group of small Mesoproterozoic metamorphic crystalline blocks were scattered. All of these small blocks mentioned above may have been located near the Siberian plate. They were originally located near Gondwana in the Southern Hemisphere in Early Cambrian. Possibly, due to the Neoproterozoic to Early Cambrian Pan-African Tectonic Events, these small blocks constitute a uniform crystalline basement (Zhou et al. 2011b; Wan 2011b).

On the small blocks, it was covered with the deformed Ordovician, Silurian and Lower Devonian sedimentations. The faunal assemblage found in the strata shows characteristics of assemblage in Siberia, and it is discovered that the strata were intruded by Early Paleozoic granitic magma (450–400 Ma) (Wang et al. 2014). In the Wenduermiao area, the age of U–Pb formed in collisions is 466 Ma in the granodiorite beneath the unconformity of the Silurian clastic rock series. In the southern boundary of basin, the collision age is dated as Late Silurian (435–415 Ma) (Parfenov et al. 2009). Thereafter, the evidence of Late Paleozoic granitic magma intrusion was found as well. Along the north and south sides of this zone, and inside of the main faults, ophiolite suit was developed (Figs. 2.8 and 2.9). In the Early Paleozoic, the above-mentioned small crystalline blocks, following the Siberian plate [1], migrated to the middle and high latitudes of the Northern Hemisphere, and finally in the

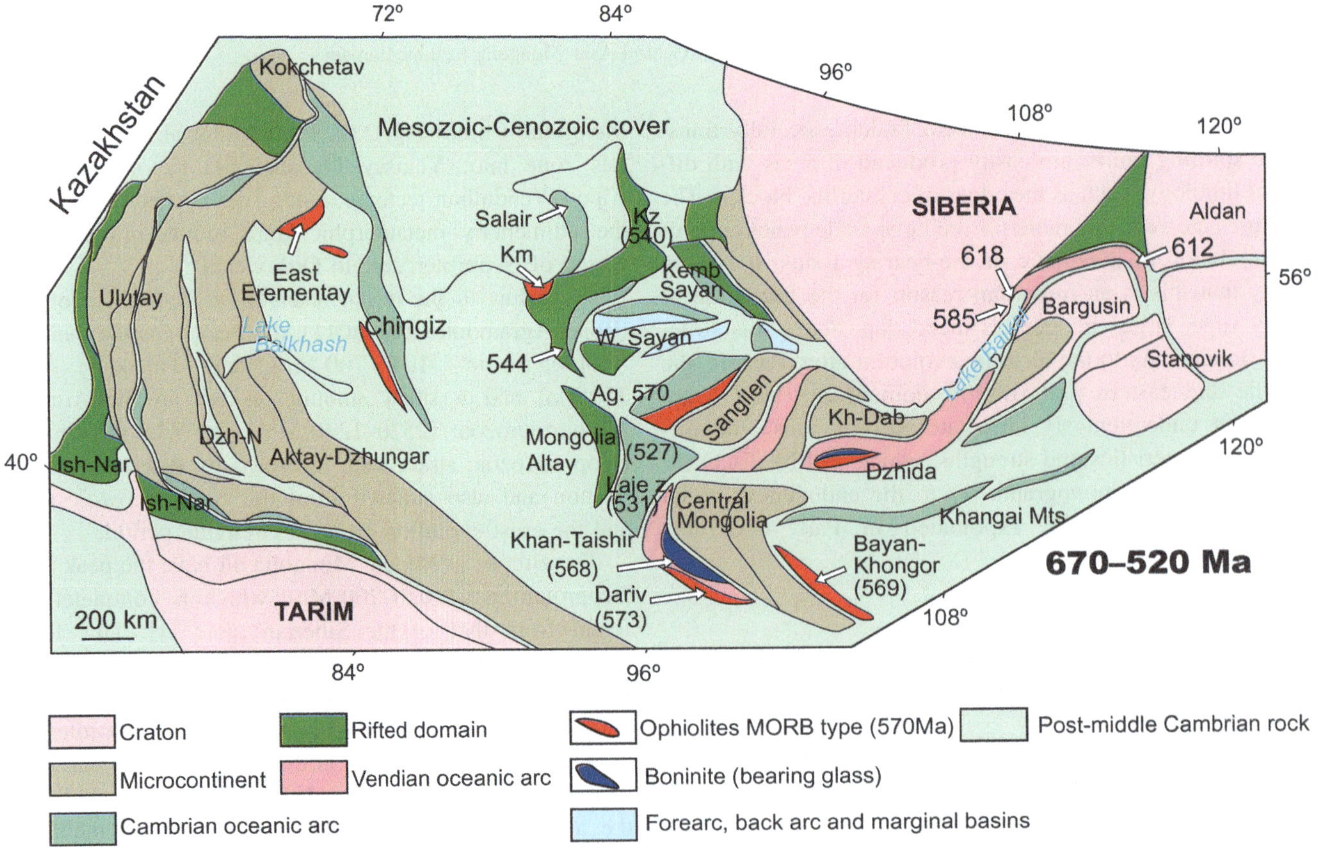

Fig. 2.8 Neoproterozoic–Early Cambrian (670–520 Ma) oceanic crustal tectonics in Western Mongolia-Altay (Modified from Khain et al. 2003)

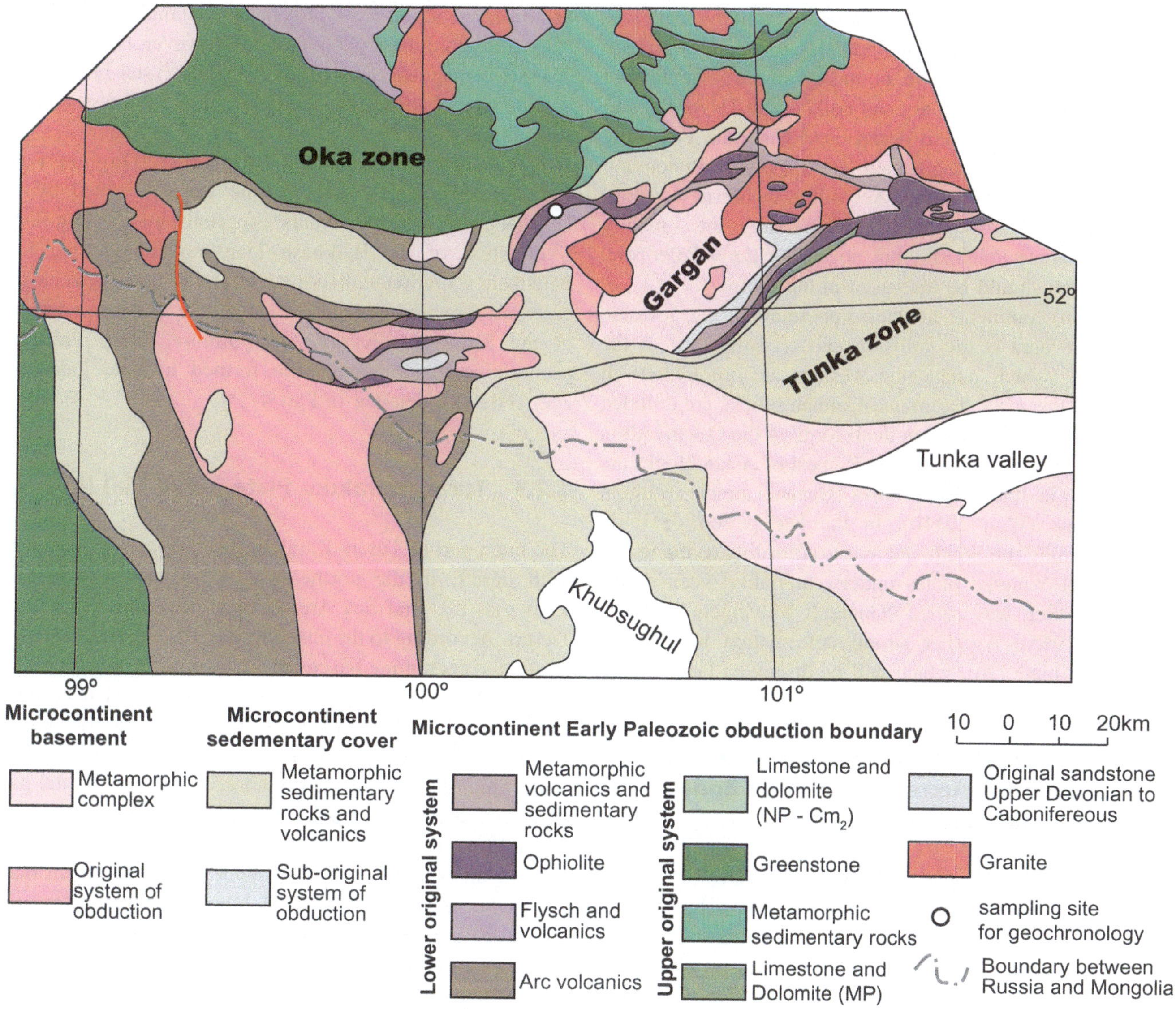

Fig. 2.9 Tectono-lithos sketch of Early Paleozoic collision zone in Southeastern Sayan (Modified from Khain et al. 2003)

Early Devonian (~393 Ma), the Altay–Middle Mongolia–Hailar Early Paleozoic accretion–collision zone [6] was formed.

Sengör et al. (1993) considered that the Altay collision zone produced the accretion arc and complex system in Asian continent in Late Devonian (359 Ma), forming the new-born continental crust, covering a relatively large area of 215 × 10^6 km^2 and with an average increase velocity of 0.3 km^3/year. This recognition has been supported by Sr–Nd–O isotopic data.

The Altay–Middle Mongolia–Hailar Early Paleozoic accretion–collision zone [6] has a limited distribution in China, only distributing in Southern Altay of the Northern Xinjiang and Hailar area of Northeast China. In those areas, strong tectonics are found. The most areas of Altay–Middle Mongolia–Hailar Early Paleozoic accretion–collision zone had been suffered from the transformation of Late Paleozoic tectonic movements, i.e., the collision effect of Balkhash–Tianshan–Hingganling Late Paleozoic accretion–collision zone, and the further squeezing effect in N–S direction (in terms of modern magnetic orientation). Deposits mostly were formed in Late Paleozoic and even in Triassic (details are given in Part 4). The Late Paleozoic–Triassic tectonism may be slightly weaker than that of the Early Paleozoic, but it is more favorable for the ore-bearing fluid to preserve (details are given in Part 4).

Safonova (2014) summarized the tectonic evolution of South Russia–Kazakhstan–Altay area. She believes that there exist three main tectonic regions in this area: (1) Altay–Mongolia block; (2) subduction–accretion block zone, included Rudny Altay, Gorny Altay and Kalba–Narym collision zone; (3) Kural, Charysh–Terekta, NE Irtysh and

Char shear suite zone. She also considered that there are five main tectonic periods: (1) subduction and accretion in Late Neoproterozoic; (2) passive boundary in Ordovician and Silurian; (3) active boundary and collision in Devonian and Carboniferous; (4) ancient ocean closed in Late Paleozoic and formed collision magmatism; and (5) post-collision and deformation since Mesozoic. According to her new recognition, the collisions of Kazakhstan–Altay area mainly occurred in Late Paleozoic, but not in the Early Paleozoic. This problem should be discussed in the future.

In Asian continent, another important strong tectonic deformation area is the collision and consolidation of the Xiyu plate (which means the western area of China). It means that the Alxa, Tarim and Qaidam blocks are collided with the Xiyu plate, between them forming the Qilian–Altun collision zone (Fig. 3.9). However, most of Asian blocks are all spread in the Paleo-Tethys Ocean, almost without deformations (Wan 2011b). In the late period of Early Paleozoic, only the South and West (according to the recent magnetism) Yangtze plates underwent rather week deformations, which was called "Guangxi Movement" (Huang and Yin 1965). Another strong deformation area is the Cathaysian plate [26], which will be discussed later.

2.2.2 Karaganda–Kyrgyzstan (Qirghiz) Early Paleozoic Accretion–Collision Zone (541–419 Ma) [7]

The Karaganda–Kyrgyzstan (Qirghiz) Early Paleozoic accretion–collision zone (Fig. 2.6) is located on the southern edge of West Siberian plate, whose sedimentary characteristics are as same as those of the Altay–Middle Mongolia–Hailar Early Paleozoic accretion–collision zone [6]. It is probable that they originally all belonged to the same collision zone in the Early Paleozoic, but were later destroyed, modified and distorted by the Balkhash–Tianshan–Hingganling Late Paleozoic collision zone, making it look like an independent collision zone (Li et al. 2008). For a long time in the past, many researchers thought that it is a stable block, called "Kazakhstan plate." However, most researchers (such as Sengör et al. 1993; Li et al. 2008; Petrov et al. 2008; Xiao et al. 2009a; Safanova 2014) denied the recognition.

The small Mesoproterozoic and Early Cambrian (~513 Ma) blocks with metamorphic crystalline basements exposed in Erguna, Middle Mongolia to Karaganda, presumably were products of the deformation and metamorphism at the boundary of Siberian plate or Gondwana. Due to that in Cambrian, the Siberian plate was located in middle latitude area of the Southern Hemisphere near Gondwana and the tectonic event showed similar characteristics of Pan-African Tectonic Event and it is possible that they had the same crystalline basement resulted from Pan-African Tectonic Event and were broken and entangled in the Early Paleozoic accretion–collision zone. However, due to the fact that paleo-magnetic data of these small crystal blocks have not yet been obtained, accurate paleo-structural restoration is still difficult.

Most areas of Karaganda–Kyrgyzstan (Qirghiz) Early Paleozoic accretion–collision zone were reformed by the Late Paleozoic tectonic events, especially by crashing in N–S direction of the Balkhash–Tianshan–Hingganling Late Paleozoic accretion–collision zone and remote squeezing in eastward direction of Ural collision zone. Most ore deposits in the Karaganda–Kyrgyzstan (Qirghiz) Early Paleozoic accretion–collision zone were formed in Late Paleozoic, even Triassic (detailed in Part 4).

2.2.3 Turan–Karakum Plate (~420 Ma) [8]

The main part of Turan–Karakum plate (Fig. 2.6) is covered with desert; in the northeast is KyzylKum Desert (also known as the Aral Sea Area) and in the south is Karakum Desert. According to the characteristics of exposed boundary strata, the crystalline basement was formed in late period of Early Paleozoic (~400 Ma). The basement is uncomfortably covered with Late Paleozoic (mainly Carboniferous and Permian) and Mesozoic–Cenozoic marine sedimentary systems which have weak deformations (Li 2008). In the past, this plate and the area around Kazakhstan are zoned as the Kazakhstan plate. Now the collision zones are removed, leaving this smaller and more stable Turan–Karakum plate. It can be seen as the residual block after the destruction caused by adjacent plates and accretion–collision zone. Due to the covering desert, there is still no data for studying the deep structure of this plate.

Brookfield (2000), Garzanti and Gaetani (2002) and Luo et al. (2005) all considered the crystalline basement of Turan–Karakum plate was spliced together by multiple micro-blocks in late period of Early Paleozoic. The plate was suffered from rather strong reformation in the Late Triassic. The real basin evolution period started from Jurassic, and Mesozoic–Cenozoic marine facies hydrocarbon rock series were formed; its oil-generating potential is much greater than that of Tarim basin. However, Petrov et al. (2008) and Cheng et al. (2010) still believed that the Turan–Karakum plate and Tarim block originally belonged to the same block, and they all had the Pre-Cambrian crystalline basement which later was broken into pieces. As to the tectonic location, the two seems to be somewhat similar. However, according to the data listed in the text, their crystalline basements are formed in different periods. Although the Mesozoic and Cenozoic sedimentary basins and tectonic locations are similar, it seems that they cannot be treated as a single block.

2.2.4 Western Tianshan Late Paleozoic Accretion–Collision Zone (385–260 Ma) [9]

The Western Tianshan (Toshkent–Karatau) Late Paleozoic collision zone (Fig. 2.6) could be interpreted as the western extension of Balkhash–Tianshan–Hingganling Late Paleozoic accretion–collision zone (385–260 Ma) [10] and located on the west side of Fergana dextral strike-slip fault (Li et al. 2008; Brookfield 2000). Balkhash–Tianshan–Hingganling Late Paleozoic accretion–collision zone and the Western Tianshan Late Paleozoic accretion–collision zone were originally supposed to be the same collision zone, which was later cut off by the Fergana dextral strike-slip fault. The Western Tianshan Late Paleozoic accretion–collision zone is located on northeast side of Turan–Karakum plate [8]. The direction of the zone is generally NW. In the northern part of this collision zone, some Paleoproterozoic crystalline rock series and Mesoproterozoic–Neoproterozoic sedimentary strata are outcropped in the Karatau Mountains, and Late Paleozoic rock series developed in other areas. The Late Paleozoic and previous rock series all experienced strong rock tectonic deformations. However, till now ophiolite suits have not been discovered on two sides of the collision zone.

2.2.5 Balkhash–Tianshan–Hingganling Late Paleozoic Accretion–Collision Zone (385–260 Ma) [10]

The Balkhash–Tianshan–Hingganling Late Paleozoic accretion–collision zone (Fig. 2.6) is a great arc zone, locating in the eastern part of Central Asian continent. In that zone, there are a series of ancient blocks. Some of them are the small crystalline blocks; from west to east, there are the Balkhash–Yili, Junggar, Turpan–Xingxingxia, Kuruktag, Hongqi Mount, Yagan, Northern Bayannur, Tuotuoshang–Xilinhot, Song–Nen and Jiamusi–Bureyat blocks (Fig. 2.6) (Zhou et al. 2011a; Wan 2011b). Most of them are covered uncomfortably by sedimentary strata. Among them, the Junggar [11] and Song–Nen blocks are the largest.

Xiao et al. (1995) discovered high-grade metamorphic rock series and got the Rb–Sr isochronal age (1.7–1.9 Ga), which belongs to the Paleoproterozoic. In recent years, through large-scale geological mapping and structural analysis, it is discovered that there are geological bodies of different rocks, metamorphic grades, deformation styles and formation periods in the Xilinhot block. Ge et al. (2011) proposed to make that block disassemble, and called the surface crust rock as "Xilinhot Group," and determined the isotopic age of metamorphic rock and mud clastic rocks, which metamorphic core ages are concentrated at 1005–1026 Ma. That group was intruded by gabbro (SHRIMP U–Pb age of 739.6 Ma) and Ordovician gneiss–granite, and the formation period of "Xilinhot Group" is inferred to be Mesoproterozoic, which block maybe a small one at first, belonging to the boundary part of Sino–Korean plate. In that zone, the crystalline basement formation period of Jiamusi–Bureyat block (Fig. 2.6) is Early Cambrian (~513 Ma). The tectonic event was called "Xingkai Movement" by Huang and Yin (1965). Recently, there are many similar test results (Wan 2011b). Wilde et al. (2003) got the isotopic ages of crystalline basement block in the Jiamusi–Bureyat and other blocks in Hailar–Middle Mongolia [6], and Balkhash–Tianshan–Hingganling [10] collision zones are most of in the Early Cambrian (~513 Ma). In the granitic body of Jiamusi block, they got the SHRIMP zircon isotopic ages of 525–515 Ma. Those small crystalline blocks could be similar cause, and they were all formed in the Pan-African Event near Gondwana in early period, and followed the Siberian plate migrating to the Northern Hemisphere and crushed into many small blocks. In those small blocks, some are the fragments of the boundary area of Sino–Korean plate, and later, they were all mixed into the Hailar–Middle Mongolia [6] and Balkhash–Tianshan–Hingganling [10] collision zones.

Parfenov et al. (1995b, 2009) expanded the Jiamusi–Bureyat block to the vast majority of Northeast China, including the most areas of Eastern Balkhash–Tianshan–Hingganling collision zone. Some researchers of Jilin University also have kept these similar views in recent years. However, this assumption has many problems and should be discussed more detail. According to the information we have obtained, it is obvious that the crystalline blocks are smaller and surrounded by the products of Late Paleozoic collision belts. It would be inappropriate to deny the existence of wider range of Late Paleozoic collisions.

In the Balkhash–Tianshan–Hingganling accretion–collision zone, the biggest block is the Song–Nen block (the Daqing oil field is located on the upper part of it). According to recent dating ages of detrital zircon from the deep metamorphic rocks and the sedimentary cover, the crystalline basement may be formed in Paleoproterozoic and later experienced multi-period tectonic-thermal events (Pei et al. 2006; Zhang et al. 2008). The accurate formation age of crystalline basement for other small blocks is lacked.

All the above small blocks with the crystalline basement of high strength were formed in Early Cambrian or Proterozoic, and the other shallow metamorphic sedimentary rocks were formed in Late Paleozoic, which rock strength are weaker. On those boundaries between metamorphic and sedimentary rocks, there are obvious differences in material properties, and thus, it is easy to form faults or fractures and possible to form the hidden ore deposits. Therefore, in those areas it will be the important potential prospect. It must be considered and paid close attention specially in the prospecting of Inner Mongolia desert and grassland areas.

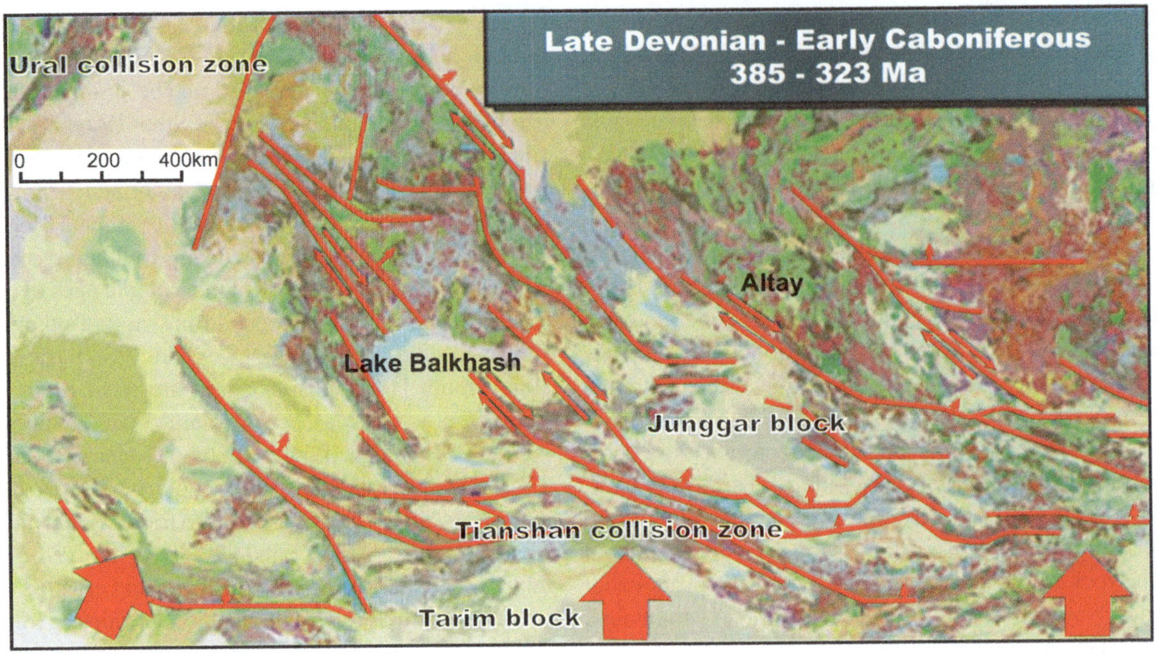

Fig. 2.10 Regional stress field and fault activity in main collision period for the Balkhash–Tianshan–Altay areas. The big red arrows show the regional compression and shortening orientation; the small red arrows show the reverse fault, and a pair of small red arrows show the fault with dextral strike-slip. The northwest part of the geological body of Balkhash Lake is the Balkhash Orocline (Complied from Buslov et al. 2004; Wang et al. 2008; Li et al. 2002a; Xiao et al. 2010)

The Balkhash–Tianshan–Hingganling Late Paleozoic accretion–collision zone is a giant arc tectonic belt, with the western part (Balkhash–Tianshan) stretching in a WNW direction, and the middle part (near the border of China and Mongolia) nearly in an E–W direction and the eastern part (Hingganling) in a NE direction. The accretion–collision zone is exposed rather completely in China, showing an arc tectonic zone (Figs. 2.6 and 3.13). In the Tianshan–Altay areas, there are many Late Paleozoic A-type granitic intrusions. Their ε_{Nd} normal value shows that they all came from mantle (Han et al. 1998); the strongest magmatism was concentrated in Late Devonian–Early Carboniferous (385–323 Ma) and Late Carboniferous–Early Permian (323–260 Ma) (Figs. 2.10 and 2.11). Those are two of the strongest tectonic-thermal events. The accretion–collision zone was formed by the accretion and collision around the Siberian plate in Late Paleozoic. Thus, in the Late Devonian–Early Carboniferous (385–323 Ma, the formation period of collision zone), its maximum principal compression stress orientations, in western section is mainly northeast direction, in the middle section is near N–S direction and in the eastern section is NW direction.

In the Balkhash–Tianshan–Altay areas, the width of rock deformation zone, formed in Late Paleozoic collision, could be about several hundred kilometers. One of the earlier tectonic event occurred in the Late Devonian–Early Carboniferous (385–323 Ma), resulting in a series of NW regional faults, such as Charysh–Terckta and Talas–Ferganan faults with dextral strike-slips. As to the near E–W regional faults, most researchers considered them to be the characteristics of reverse faults (Xiao et al. 1992, 2010; Allen et al. 1992; Che et al. 1994; Li et al. 2002a; Buslov et al. 2004; Charvet et al. 2007; Wang et al. 2008; Fig. 2.10).

In recent years, Lin et al. (2011) got the zircon U–Pb age of 385–360 Ma in the garnet–sericite schist of the northern margin of Middle Tianshan collision zone. According to the above data, it is possible that the tectonic movement of Late Devonian–Early Carboniferous (385–323 Ma) may be caused by the northward (based on modern magnetic orientation) migration and collision of the Altay and Tianshan blocks (Wan et al. 2015). However, the tectonics in the Late Carboniferous–Early Permian (323–260 Ma) formed many NW trending faults (such as the Chara, Irtysh and Krail Kuznetsk–Teletsk–Bashkauss and the faults of northeast area of Altay), with sinistral strike-slip obviously (Fig. 2.11; Buslov et al. 2004). In other hand, Li et al. (2002a), Wang et al. (2008) and Han et al. (2011) discovered that the E–W trending faults, in the Late Carboniferous–Early Permian (323–260 Ma), were all changed to dextral strike-slip (Figs. 2.11 and 2.12). According to the above data, it just shows that the regional compression and shortening orientation (namely the orientation of maximum principal compression stress) in the Late Carboniferous–Early Permian (323–260 Ma) periods should be located at the two sectors of intersection angle between those two faults, i.e., WNW–ESE orientation. It is obviously not to explain that action force was caused by N–S trending

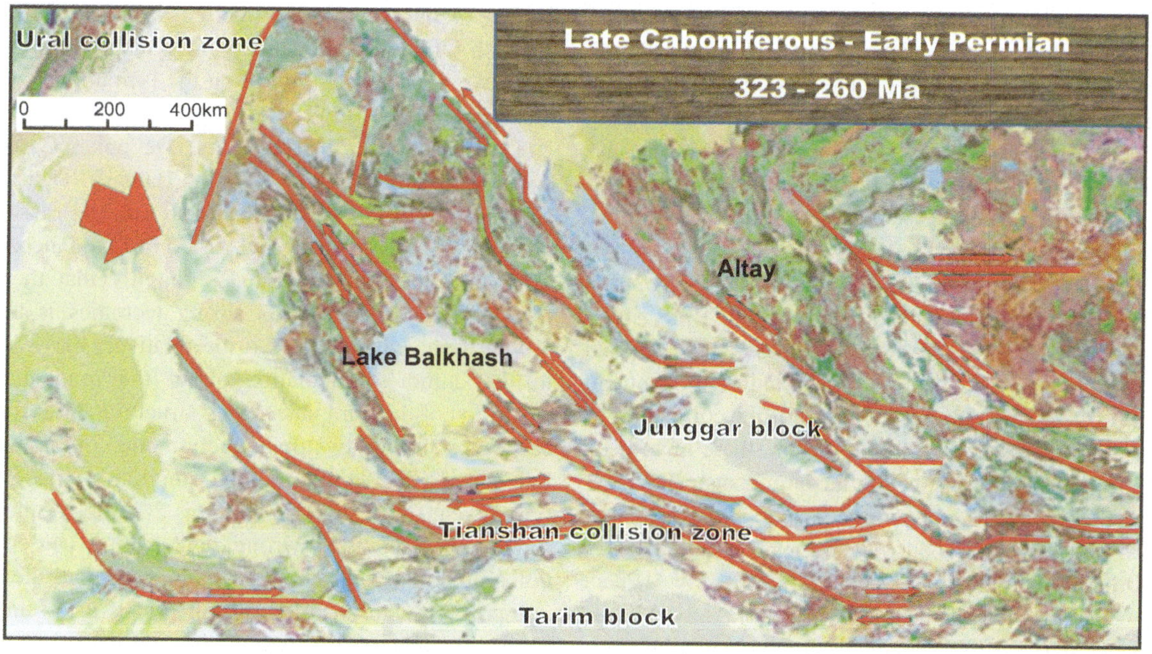

Fig. 2.11 Regional stress field and fault activity after the collision period for the Balkhash–Tianshan–Altay areas. The big red arrows show the regional compression and shortening orientation, and a pair of small red arrows show the fault with dextral or sinistral strike-slip. The southeast corner of this figure is the Turpan–Hami–Kuruktag area, for details, to see the satellite image in Fig. 2.12 (Complied from Buslov et al. 2004, Wang et al. 2008; Li et al. 2002a; Han et al. 2011)

Fig. 2.12 Satellite image of Turpan–Kuruktag area. It suggests that the near E–W faults show features of dextral strike-slip (From Wang Xiaoniu)

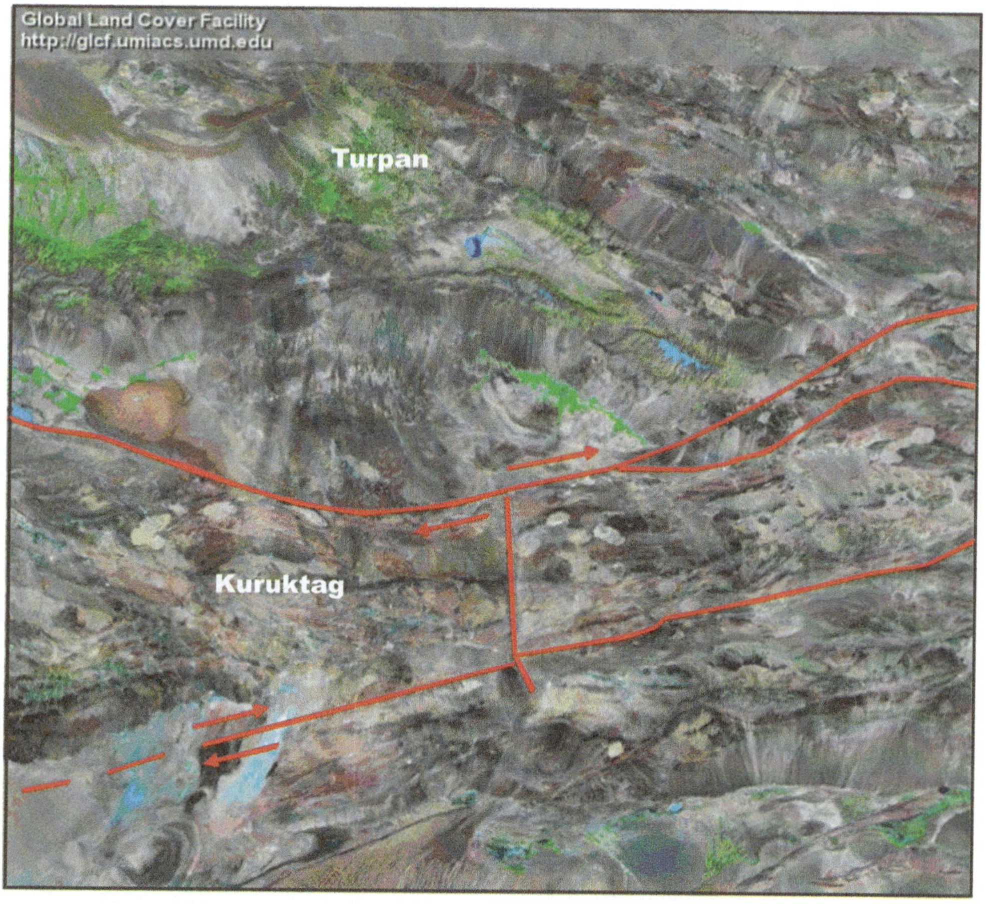

collision, but it could be the result of near eastward compression by the long-distance effect for the Ural Late Paleozoic accretion–collision zone [12]. The tectonic stress in Late Carboniferous transmitted to the Balkhash–Tianshan–Altay areas, caused the WNW-ESE trending compression and shortening and resulted in the NW trending faults changing to sinistral strike-slip and the E–W trending faults changing to the dextral strike-slip (Wan et al. 2015).

In recent years, Liu et al. (2011) researched the rhyolite in Shawan area of North Tianshan and got the LA-ICPMS zircon U–Pb isotopic age of 310 ± 2 Ma and found that the original magma of that volcanic rocks was formed in the environment of intraplate extension, but not the result of the collision.

In the tectonic research of Central Asia, there have been some doubts for a long time: Under the circumstances of near N–S direction collision and convergence, why could the meniscus-shaped tectonic belts, such as the so-called Kazakhstan Orocline and Balkash Orocline by Xiao W. J. near Kazakhstan and Balkash in the western part of Altay–Middle Mongolia–Hailar Early Paleozoic accretion–collision zone (541–419 Ma) and Balkhash–Tianshan–Hingganling Late Paleozoic accretion–collision zone (385–260 Ma), be extremely inconsistent with the direction of the overall tectonic line? That is to say, why could the tectonic line change near NW to N–S to NE and finally to E–W under the circumstances of N–S collision and convergence? According to the research of Xiao et al. (2003, 2009a, b), Xiao and Santosh (2014), the Kazakhstan and Balkash Oroclines (Fig. 2.8) begun to form at the Carboniferous and mainly became stable in Middle Permian to Early Triassic. It is likely that the compression of the Carboniferous–Permian WNW–ESE direction bended the strata into near-NNE–NE-directional folds and the Triassic shortening of near N–S direction further bended them into meniscus shape. In fact, the NW trending faults changed to sinistral strike-slip, the near E–W trending faults changed to dextral strike-slip, as well as the formation of "Kazakhstan and Balkash Orocline," it needs to use the long-distance effect of ESE trending compression derived from the Ural Late Paleozoic collision zone to explain (Wan et al. 2015). This long-distance effect toward to the east trending is often neglected. It is commonly and wrongly considered that the Balkash–Tianshan collision zone had two-period near N–S trending shortening in "orogeny stage" or multiple productions of orogeny. In the previous research, the Late Carboniferous–Early Permian near E–W trending compression was easy to ignore. However, it is just this tectonic action that has exerted an important influence on the great metallogenesis in the Balkash–Tianshan–Altay areas.

After all, the long-distance effect of the Ural Late Paleozoic collision is not very strong enough to form a new fold rock system extensively. It does not fundamentally change the entire area's foliation, but changes the sliding direction of the pre-existing fault along the fault surface and foliation surface. A number of new oroclines such as Kazakhstan and Balkash Oroclines are derived from the sliding movements and have important influence on the formation of endogenic metal deposits.

Due to the fact that the Ural Devonian–Carboniferous collision zone is far away from the Balkash–Tianshan–Altay areas with a distance of about 1000 km, the eastward compression is delayed to the late period of Late Paleozoic (323–260 Ma). The author estimates roughly that the eastward transmission speed of this strong tectonics is about 2.5–3.0 cm/year, which seems to be relatively slow. According to the author's calculation about the gradual westward migration of the intraplate strong structural deformation zone caused by the westward subduction and compression of the Paleogene Pacific plate, the migration rate is about 65 cm/year (Wan 2011b), which is 20 times of the eastward migration velocity of strong deformation caused by Ural collision. Thus, it can be recognized that the eastward compression caused by Ural collision is a less intense tectonization that occurs within the continental crust. It just made the rocks be crushed and deformed moderately and changed the nature of the fault, then provided very favorable structural conditions for the formation of endogenic metal deposits in Balkash–Tianshan–Altay.

Here, there is another problem about the tectonics of Southwestern Tianshan in China. In recent ten years, many Late Carboniferous–Permian isotopic ages from the ophiolites, eclogites, glaucophane and other metamorphic rocks have been obtained (Zhang et al. 2005; Han et al. 2011). Based on these data, some researchers considered that the main collision period of Southwestern Tianshan occurred in the Late Carboniferous–Permian, even in Triassic. Zhang et al. (2009) believed that the above characteristics should be related to collision and the two metamorphic thermal events at 326–308 and 263–243 Ma. The author considers that two metamorphic thermal events all occurred after the Tianshan main collision period and were derived from eastward squeezing and embedding of blocks between Southern and Northern Tianshan. The strikes of Southwestern Tianshan tectonic zone and main faults are all ENE or WNW trending. In the Late Carboniferous–Permian, they were suffered from the long-distance effect of eastward compression of Ural collision zone. As a result, a series of ENE or WNW trending faults showed the characteristics of compression-shear and the metamorphism-magmatism systems were caused by similar to collision-compression, and in addition, a series of faults also showed the characteristics of strike-slip. So it is the intraplate deformations after main collision of the Balkash–Tianshan areas. The main collision of Balkash–Tianshan zone occurred in Devonian–Early Carboniferous (385–323 Ma), with the main characteristics of N–S convergent and shortening. The recognition that the collision of Southwestern Tianshan mainly occurred in Late Carboniferous–Permian

(Zhang et al. 2009) may be debatable. Recently, Ju and Hou (2014) also believed that the Late Devonian–Late Carboniferous in the tectonic evolution stage of Southwestern Tianshan was the collision stage, the Early Permian was the post-collision magmatic activity and rift development stage and the Late Permian–Triassic was an intraplate orogenic stage. Their opinions are similar to the author's. After the near N–S collision in Late Devonian–Early Carboniferous and near E–W compression and shear in Late Carboniferous–Early Permian, the tectonic pattern of Balkash–Tianshan–Altay areas basically became stable and unchangeable. In Triassic and Jurassic, that area experienced rather weak N–S compression and further intraplate tectonic deformations (E–W reverse fault activity), magma intrusion and large-scale endogenic mineralization (Zuo et al. 2011; Wang et al. 2006). In Cretaceous–Paleogene, the crust was rather stable in those areas, and there was not rock deformations discovered. In Miocene, the areas underwent the E–W compression with rather weak deformation.

The great uplift and orogeny of the Tianshan area, caused by the subduction of Tarim and Junggar beneath the Tianshan area, had started since Neogene, which is considered as the result of the long-distance effect of northward migration and collision of Indian plate (Wan 2011b). The near E–W active faults in the Tianshan area are all normal faults (Bai 1993). Sun et al. (2009) recognized that the great uplift of Tianshan area mainly occurred in the past 6.5 Ma, according to the comprehensive researches of tectono-magmatism and sedimentation. It is worth noting that the Mesozoic–Cenozoic tectonic characteristics are extremely different from those of Paleozoic.

2.2.6 Junggar Block (~1400 Ma) [11]

The Junggar block (Fig. 2.6) is a bigger block located in the Balkash–Tianshan–Hingganling collision zone. To the northeast, northwest and south of the block, there are the ophiolite suites. In the interior of Junggar block, a deep well was drilled to discover the Carboniferous rocks at the bottom. No scholar has yet obtained direct evidence about its crystalline basement rocks. However, the age data of detrital zircon gotten in the borehole and at the border of block have been obtained as 2.5–0.7 Ga.

On the basement characteristics, the forming period and evolution process of Junggar block, there is still controversy. Some researchers considered that there are the oceanic crust (Jiang 1984; Carroll et al. 1990; Hu and Wei 2003), continental crust or double-layered basements (Zhao et al. 2008) and the island arc system (Wang et al. 2002; Li et al. 2012). Hanet al. (1999) and Gao (2013) proposed that the basin basement is mainly young crust according to the evidences of geochemical characteristics and the young model age. These different opinions are all supported by some evidences.

However, based on the distribution of detrital zircon ages on the east border of the basin, Li et al. (2007) considered that there is a Pre-Ordovician metamorphic basement and pointed out that it is dominated by the Mesoproterozoic–Neoproterozoic basement. Su et al. (2019) confirmed this point by her discovery of the Ordovician and Late Cambrian magmatic zircons in volcanic rocks of the Batamayineishan Formation in the Sancan-1 well on the Sangequan uplift. A large number of Proterozoic zircons have been found successively in volcanic rocks in different regions of West Junggar (Zhu et al. 2007; Fan et al. 2012), suggesting that there are also Proterozoic basements in West Junggar.

Yang et al. (2014) conducted the U–Pb dating, microelements and the Lu-Hf zircon isotopic analysis of volcanic clastic rocks of Batamayineishan Formation in Db-1 well and Y-1 well in the middle and northern parts of Junggar land uplift. The zircon internal structure, Th/U ratio and rare-earth element distribution pattern show that: (1) In the Db-1 well and Y-1 well, the minimum age of zircon sample is, respectively, 303 and 306 Ma, which can represent the forming age of Batamayineishan Formation. (2) The zircon U–Pb age shows that there are the age records of 1447–1410 and 885–559 Ma (Mesoproterozoic–Neoproterozoic), which provides the foundations for Pre-Cambrian crystalline basement in the middle and northern parts of Junggar basin; there are age records (536–420, 401–360, 359–303 Ma) for early and middle periods of Paleozoic as well; the age records indicate that the Pre-Permian basement in the middle and northern parts of Junggar basin experienced a multi-stage continental crust evolution process. (3) The microelements analysis of zircon show that the Pre-Permian basement was the active crust composed of granite, syenite, basalt and diabase in the middle and northern part of the basin and intruded by the granite and intermediate-basic rocks. (4) The zircon Hf isotopic analysis show that the zircon ages have the positive ε_{Hf} (*t*) value (+4.6 to +19.0), so the source rocks of zircon mainly were originated from melting of asthenosphere mantle or deficit lithosphere. During the magma intruding upward, the magma was contaminated with ancient basement material.

Many researchers seem to believe that the Junggar area does not have crystalline basement, but a collage of the oceanic crust and island arc. These inferences and assumptions are not appropriate. The author agrees with Yang et al. (2014) that the possibility of the Pre-Cambrian crystalline basement in Junggar block is rather large. That is, the Junggar block is similar to that of Sino–Korean plate or

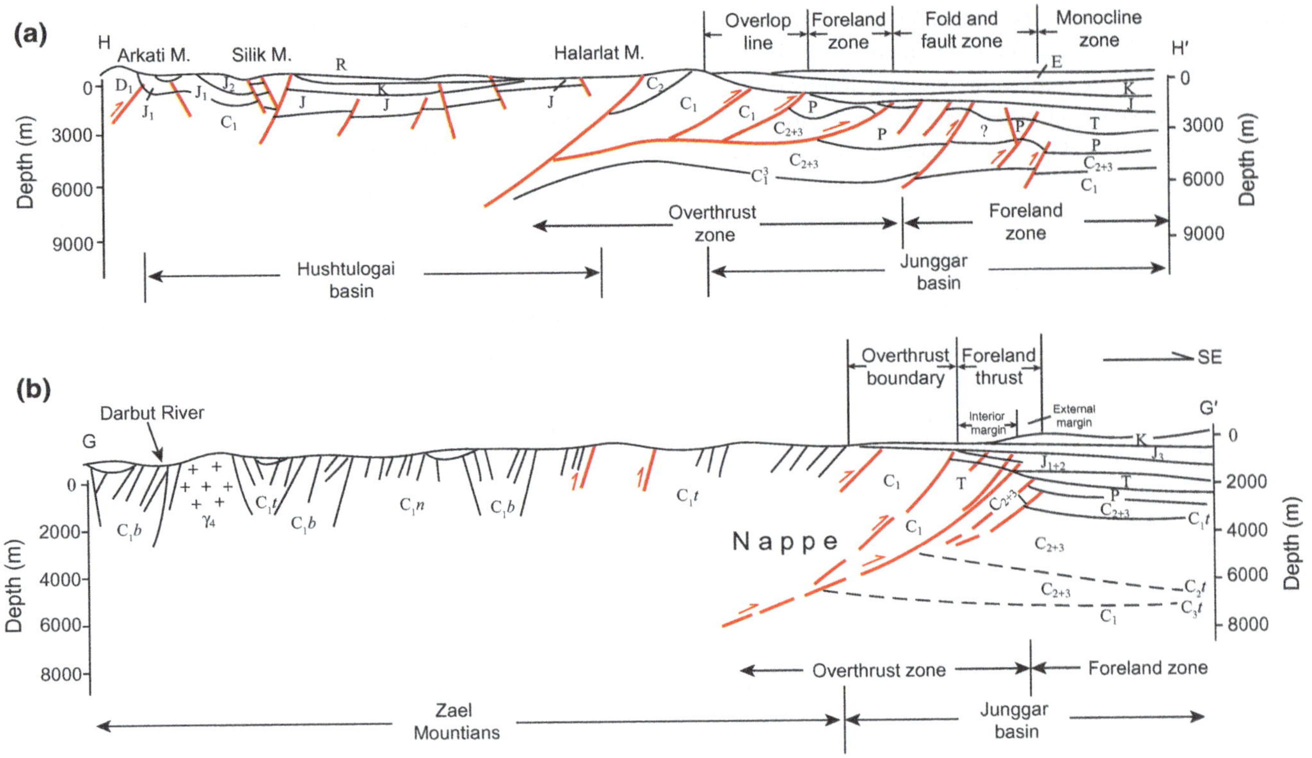

Fig. 2.13 Structural section of the southern border of Junggar (After Deng et al. 2011)

Tarim block. However, the Junggar block is obviously influenced by the collision in the Early Paleozoic and early period of Late Paleozoic.

On the southern border of Junggar block, there is a near E–W overthrust nappe fault zone which cuts off the Triassic or Early–Middle Jurassic strata. It is assumed to form in the late period of Middle Jurassic (Fig. 2.13; Deng et al. 2011), from N–S shortening. So the reverse fault in Karamay oil field was also formed in the late period of Middle Jurassic (to see Part 4).

In the past, the tectonic research of Tianshan and Junggar areas only focuses on the Late Paleozoic tectonics, rather than the tectonics of Triassic and Jurassic. Due to the Asian continental crust anticlockwise rotation in Jurassic, the Junggar block had an anticlockwise rotation about 30° and migrated relatively southward (Li et al. 1991). However, the tectonic action in Mesozoic was not stronger than that of Paleozoic. On the whole, the tectonics of Junggar block after Jurassic became weaker, so the mountains formed in Paleozoic were gradually eroded as the planation. In Cretaceous and Paleogene, the Junggar, Tarim and Qaidam areas were formed as a uniform sedimentary basin (Wang 1985; Yin 1988), and the rock deformations were rather weak. In recent years, the weak folds with N–S axis formed in late age of Miocene and Pliocene were discovered in South and West Junggar blocks (Zheng Y. D. personal communication; Wang et al. 2006; Fig. 4.20), which were caused by long-distance effect of westward compression and subduction of Pacific plate (Wan 2011b).

2.2.7 Ural Late Paleozoic Accretion–Collision Zone (400–260 Ma) [12]

The Ural Mountains caused by Ural Late Paleozoic accretion–collision zone (Fig. 2.6) are the boundary of Asian continent and European continent, starting from the Kara Sea in the Arctic Ocean to the north, reaching Kazakhstan grasslands in the south, and stretching more than 2000 km between East Europe and West Siberian plains. The modern Ural Mountains are not very high, with an average elevation of about 500–1200 m. The Ural Late Paleozoic accretion–collision zone is between the Baltic plate and Siberian plate (Dobretsov 2003; Bea et al. 2002). The Baltic plate (1800 Ma, Fig. 1.1), also known as East Europe plate, is a continental plate with the Archean–Paleoproterozoic crystalline blocks of Scandinavia in North Europe as its core. In the Early Paleozoic, eastern and central parts of Europe and most of North Europe all belonged to this plate (Figs. 3.6–3.8), with a thin sedimentary cover. At the end of Early Paleozoic, the Baltic plate collided with North American plate on the west, forming the Caledonian collision zone

(Fig. 3.8), thus forming a giant Laurussia supercontinent (Gee et al. 2008). As to the Late Paleozoic, on the southern side border of the Laurussia supercontinent, the Hercynian collision zone was formed (Ferrara et al. 1978; Kröner and Greiling 1984), and on the eastern side, the Ural accretion–collision zone [12] was formed, which was connected with the Siberian plate. Thus, they were all united together to form the main part of Pangea supercontinent (Fig. 3.12). At the end of Triassic, the Pangea supercontinent began to break into the North American, South American, African and Eurasian plates, and began to form the Paleo-Atlantic Ocean (Figs. 3.15, 3.19). Since Paleogene, the collision occurred between Baltic and African plates, forming the Alps–Carpathian collision zone, and then forming the recent Europe continent (Cavazza et al. 2004; Fig. 3.27).

The collision orientation of Ural Late Paleozoic accretion–collision zone was mainly near E–W direction (in terms of modern magnetic orientation). This collision zone, Hercynian (or Variscan) collision zone (Ferrara et al. 1978; Kröner and Greiling 1984), and Western Tianshan and Balkhash–Tianshan–Hingganling Late Paleozoic accretion–collision zone [9, 10] were almost formed in the same time (Figs. 3.12–3.14).

According to the data of paleo-tectonics, from Mesoproterozoic to late period of Permian (1600–252 Ma) there was a paleo-ocean in the Ural area. In that geological period, there were the syn-tectonic pulsation cycles lasting about 30 Ma, which is characterized by the rearrangement of island arc, occurrence of regional collisions and formation of featured eclogite and glaucophane schist (Dobretsov 2003).

Bea et al. (2002) got the zircon Pb–Pb evaporation age of 360–330 Ma in the Chelyabinsk intrusion, which could roughly represent the collision age. The main tectono-magmatism transition during the formation of Ural accretion series occurred in the formation process of the Alexandrinka system. Re-Os age data from 13 ore deposits are all Late Devonian–Early Permian (347–288 Ma) (Tessalina et al. 2000). Their dating results are highly reliable. Those age dating results show that the formation process of Ural collision zone (with E–W trending shortening), Balkhash–Tianshan–Hingganling collision zone (with N–S trending shortening) and the eastward slippage were formed almost in same time. The forming ages of the Hercynian (Variscan) collision zone are: 340–341 Ma (Kröner and Greiling 1984) according to ^{207}Pb–^{206}Pb isotopic data in South Germany, 340 Ma (Ferrara et al. 1978) according to the Rb–Sr data in the northeast area of Sardinia. These data explain that the Hercynian (Variscan) collision zone was formed in Late Devonian, a little bit earlier than that of the Ural and Balkhash–Tianshan–Hingganling collision zone. To sum up, in the European–Asian continental blocks, the near N–S conversion occurred mainly in Late Devonian–Early Carboniferous, while the eastward compression of Ural collision zone mainly happened in Late Carboniferous–Early Permian. The above three Late Paleozoic collisions are the key tectonic events for the formation of Pangea supercontinent (first proposed by Wegener 1912) (Figs. 3.12 and 3.13).

However, why did the Pangea supercontinent form in Late Paleozoic? The dynamic mechanism of its formation still lacks sufficient evidence so far, and there is no reasonable explanation. Steinberger and Torsvik (2012), Torsvik et al. (2014) and Domeier and Torsvik (2014) used the uplift mantle plume beneath the African continent to explain formation and disintegration of the supercontinent. It seems that there are still many problems and some contradictions that cannot be overcome. It may be rather reasonable that the uplift of mantle plume forces the lithosphere to crack. But, why can the mantle plume uplift cause the convergence of supercontinent? This is difficult to explain.

2.2.8 Wandashan Jurassic Collision Zone (170–135 Ma) [13]

The Wandashan Jurassic collision zone (Fig. 2.6), at first belonged to the southern extension of Verkhojansk–Chersky Jurassic accretion–collision zone (200–135 Ma) [3] and in the Cretaceous, was cut off by the Sikhote–Alin–Koryak Cretaceous–Paleogene accretion–collision zone (130–23 Ma) (Karsakov et al. 2008). It became the only one Jurassic collision that was remained in Northeast China and caused by near E–W compression and collision. In the Wandashan area, the Raohe ophiolite suites are developed, with basic pillow-like lava and cumulate complex as the main body, which represent the submerged oceanic crust or islands during the Triassic (228 Ma) (Tian 2007). Besides, there are Carboniferous–Permian limestone, Triassic lamellar silexite, Middle Jurassic siliceous shale and Late Jurassic–Early Cretaceous terrestrial sandstone and shale, etc. They represent the surface sediments and later related sedimentary rocks of paleo-oceanic plate (He 2009). Most of researchers recognized that there was an Early Jurassic (188–173 Ma) oceanic sedimentary series in Raohe area of the Wandashan, and the Wandashan collision zone was formed in Middle–Late Jurassic–Early period of Early Cretaceous. This understanding may be rather reasonable (Shao et al. 1991, 1995; Zhao et al. 1996; Tian 2007). In that collision zone, the isotopic age of Raohe granite is 130 Ma, which was formed after collision. The Wandashan collision zone used to be called the Nadahada collision zone (Mizutani et al. 1986; Kojima 1989). This name is improper, which could not represent the whole area of that collision zone. Shao et al. (1991, 1995) and Mizutani et al. (1986) inferred that block in this collision zone was located at the Southern Hemisphere in the Early Paleozoic and was transferred to this place. That recognition has attracted the attention and shock of many scholars. Now it is known that many small blocks of Bureyat of Russia, Jiamusi of China and the borders between China and Mongolia, all migrated to the Northern Hemisphere from the Southern Hemisphere following Siberian plate. The great migration of the blocks in Wandashan area is no exception.

2.3 Sino–Korean Tectonic Domain

The Sino–Korean tectonic domain (Fig. 2.14) is dominated by the Sino–Korean plate [14], including the Helanshan–Liupanshan Late Paleozoic collision zone [15], Alxa–Dunhuang block (1800 Ma) [16] Qilian Early Paleozoic accretion–collision zone (541–400 Ma) [17], Qaidam block (1800 Ma) [18], Altun Early Paleozoic sinistral strike-slip collision zone (541–400 Ma) [19], Tarim block (1800 Ma) [20] and the Central Tarim Neoproterozoic collision zone (~850 Ma) [21].

The Sino–Korean tectonic domain starts from Tarim block [20] in the west and ends at the Hida Peninsula of Japan. The north boundary is the south border of Balkhash–Tianshan–Hingganling Late Paleozoic accretion–collision zone [10], and the south boundary is the north edge of Qinling–Dabie–Jiaonan–Hida Marginal Triassic accretion–collision zone [24], West Yangtze plate [22] and the Western Hindukush–Kunlun–Pamir Late Paleozoic–Triassic accretion–collision zone [30] (Fig. 2.14).

All the blocks in Sino–Korean tectonic domain are with the crystalline basement (>1800 Ma). They could be an ancient great plate at that time (Wang 1979; Li et al. 1982; Yang et al. 2014). In recent years, most of researchers have recognized that the formation time of the unified crystalline basement for the Sino–Korean tectonic domain is at about 1.8 Ga (Cheng 1994; Bai et al. 1996; Wang and Mo 1995; All China Commission on Stratigraphy 2000, 2002; Wang et al. 2014; Qiao et al. 2014). Only the Tarim block was separated from Sino–Korean plate since Early Neoproterozoic and split into the North and South Tarim blocks; they were reformed and collided at Mesoproterozoic, forming the Central Tarim collision zone [21] and constituting the Tarim block (Wu et al. 2006).

The paleontological combination characteristics of Tarim [20] and Qaidam [18] blocks are obviously similar to that of Yangtze plate in the Neoproterozoic to Early Paleozoic. The biota of Alxa–Dunhuang block [16] in Middle Cambrian to Silurian was also similar to that of Yangtze plate (Peng 2003, personal communication; Lu 1976; Mu 1983). The Tarim, Qaidam and Alxa–Dunhuang blocks merged in the Early Paleozoic to form the Qilian Early Paleozoic collision zone [17] and the Altun Early Paleozoic sinistral strike-slip fault and collision zone [19]. They merged together, forming the Xiyu plate (Gao et al. 1983). The whole Sino–Korean tectonic domain was moved to north and collided with the Tianshan–Hingganling collision zone [10]; thus about half of Asian continental blocks was merged into the Pangea supercontinent (Figs. 3.12 and 3.14). In Triassic, affected by the northward migration and collision of Yangtze tectonic domain, the Proterozoic–Triassic strata were all developed with the E–W intraplate deformations with broader folding (according to modern magnetic orientation). Later, in the tectonic evolution since Jurassic, the tectonic domain had experienced five intraplate deformations with different active features and intensities of activity, which will be elaborated in Part 3.

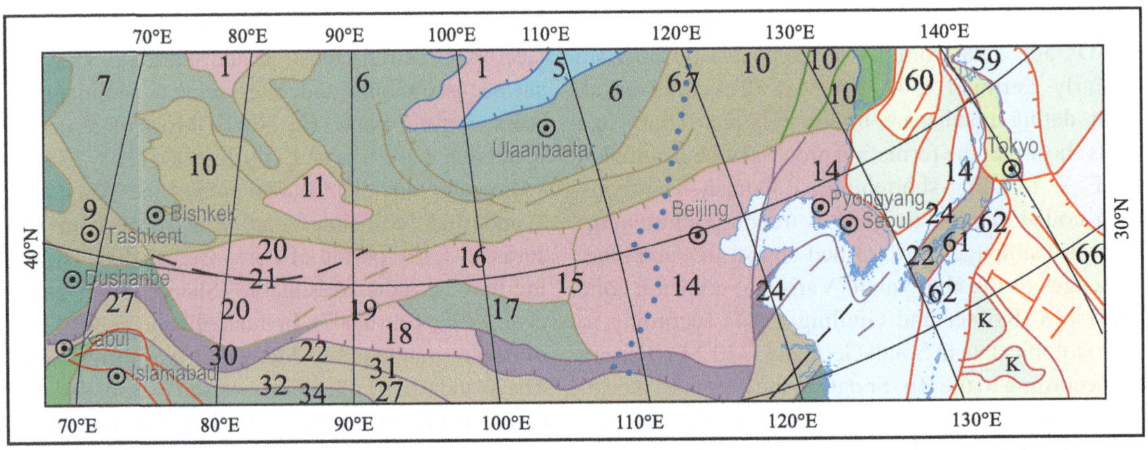

Fig. 2.14 Sino–Korean tectonic domain [14–21]. The black numbers in the figure show the tectonic units, as same as those in the CONTENTS and Fig. 1.1. The Sino–Korean tectonic domain is mainly shown as the roseate color, while the Qilian Early Paleozoic accretion–collision zone [17] is shown as green color

2.3.1 Sino–Korean Plate (1800 Ma) [14]

The Sino–Korean plate (Figs. 1.1 and 2.14) in the end of Proterozoic (1800 Ma) formed the uniform crystalline basement. According to recent data, the peak period for the formation of the high-pressure granulite was 1950–1900 Ma, while the period of 1900–1800 Ma was the process of retrogressive metamorphism of the middle pressure granulite and amphibolite with the uplift after the collision (Zhang et al. 2009). And accompanied with the great-scale mantle uplift, the lower crust uplifted near the surface and resulted in the stronger migmatization to form the six paleo-continental nuclei (gneiss domes; Dixon 1987) and their surrounding ductile shear zone (Bai et al. 1996; Wan 2011b). Almost at the same time, many researchers proposed the similar division schemes. Among the most significant ones are the division schemes proposed by Widle et al. (2002) and Zhao (1998, 2001, 2002, 2004, 2005,2007, 2014; Faure et al. 2007). At first, Zhao G. C. et al. divided the Sino–Korean (i.e., North China) continental plate into two continental blocks, which has a great influence both at home and abroad. Depending on the outcrops at the middle part of North China, in the Hengshan–Wutaishan–Taihangshan, their crystalline basement rocks are really developed very well, and the collision is most obviously. Later, they divided the paleo-continental nucleus into four or five blocks, gradually closer to Bai's division scheme. Regarding the division of continental nucleus of Sino–Korean plate, many division schemes have been proposed in recent years, for example, Wang and Mo (1995) and Wu et al. (1988) divided it into five blocks, Deng et al. (1999) divided it into ten and Zhai M. G. divided it into six. Although their understanding is not exactly the same, the basic outlines are still relatively close.

According to the results of the deep magnetic interface isobaths deduced from regional geology and aeromagnetic extension data (Guan et al. 1987), Bai et al. (1996) divided the part of the Sino–Korean block in China (i.e., North China) into six continental nucleuses: Dongsheng, Chifeng, Liaoji, Linfen, Jining and Bohai separately (Fig. 2.15). The center of each continental nucleus is corresponding to the uplift area of the deep magnetic interface (the depth of the interface is 10–18 km), and the depression area of the deep magnetic interface (the depth of the interface is 16–24 km) becomes the boundary of the continental nucleus. The Archean granulite–gneiss region is located in the middle of the land core (the uplift zone of the deep magnetic interface).

As to the formation of continental nuclei, it is generally assumed that they are induced by the large-scale meteorite impact, bringing about upwelling of hot mantle in large scale and upraising of the lower crust entirely to near the surface, and accompanied with stronger migmatization to form ancient continental nuclei (i.e., gneiss domes) (Bai et al. 1996; Wan 2011a, b). This formation mechanism may be similar to that of the Siberian plate.

The granulite and amphibolite in the central part of the ancient continental nuclei were formed in the Fuping Period (3.0–2.8 Ga), and the surrounding greenstone zones were usually formed in the Wutai Period (2.7–2.5 Ga) or Hutuo Period (2.5–1.8 Ga). In the past, the Archean systems were recognized. However, in those systems, many researchers found the different isotopic ages (Guan et al. 2002; Liu et al. 2002; Santosh et al. 2006; Wang et al. 2003; Wilde et al. 2002, 2004; Xia et al. 2006; Zhao 2001, 2002, 2004, 2005, 2007, 2014). Thus, most of researchers believe that the uniform crystalline basement of Sino–Korean plate should be formed in Paleoproterozoic.

The characteristics of collision among the paleo-continental nucleuses are very different from that of the later collision and subduction between plates. The Archean greenstone belts are distributed around the continental nucleuses. The gneissosity or foliation of ductile shear zone at the margin of each large continental nuclei and their secondary nuclei all have the characteristics of ring-like or arc-shaped rotation, which indicates that the continental nucleuses may be rotated in collision process and finally converged together.

Bai et al. (1996) used the aeromagnetic data to reflect the deep crystalline basement and combined a large number of field geological data to judge the distribution of continental nucleus, in the case where the outcrop area of the crystalline basement of the Sino–Korean plate is small. This is currently a relatively feasible and relatively realistic method, which is more reliable than the method of delineating ancient cores using only geological outcrop observations and rock chemistry data. Of course, other scholars have made great achievements in the research of rock geochemistry and isotopic chronology.

In North China, the retrogressive metamorphism of intermediate pressure granulite and hornblende facies is located in high-pressure granulite dykes outcrops, with ages ranging from 1820 to 1800 and 1790 to 1760 Ma. The *P-T-t* paths are the types of isothermal depressurization and cooling depressurization, with some autometamorphism structure. Those dykes do not have the characteristics of continental collision metamorphism. After the uplift of Sino–Korean crystalline basement, the rift event occurred. The rift of Xiongr–Angou–Southern Shanxi and Yanshan rift may belong to the same period. The earliest eruption of the Xionger Group occurred at 1791 ± 20 Ma, the age of volcanic rock series basement in the Xiyanghe Group, Southern Shanxi and in the Xionger Group, Western Henan ranges from 1776 to 1800 Ma (Zhao et al. 2004; He et al. 2009). The LA-ICP-MS U-Pb zircon age of volcanic rock in

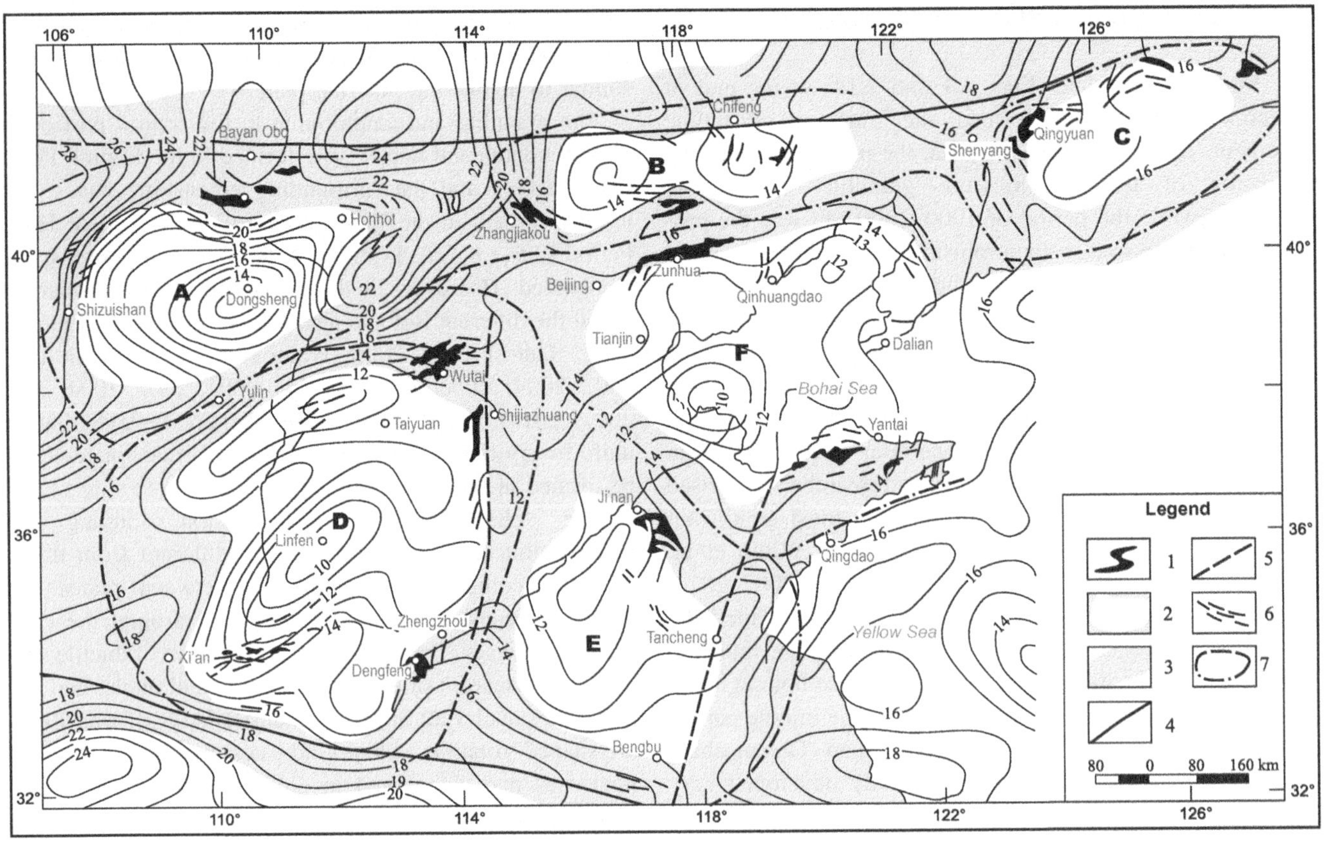

Fig. 2.15 Distribution of Archean continental nuclei in Sino–Korean plate in North China and average depth contours of the deep magnetic interface. Data of the deep magnetic interface from the aeromagnetic survey, with the filtering magnetic anomaly, are from Guan et al. (1987). **Legend**: (1) Archean greenstone belt; (2) uplift areas of the deep magnetic interface, i.e., Archean continental nuclei; (3) depression zones of the deep magnetic interface, i.e., collision and amalgamation zones; (4) border of Sino–Korean plate; (5) inferred fault zone; (6) Archean rift; (7) inferred border of Archean continental nuclei. **Continental nuclei**: (A) Dongsheng; (B) Chifeng; (C) Liaoji; (D) Linfen; (E) Jining; (F) Bohai (After Bai et al. 1996)

Xiaoliangling Formation of Lvliangshan, Shanxi is 1779 ± 20 Ma (Xu et al. 2007). In recent years, Qiao and Wang (2014) got zircon SHRIMP age of 1778 ± 20 Ma at the bottom of volcanic rock of Xiaoliangling Formation.

As a result, it is proper and reasonable that many Chinese researchers (Cheng 1994; Bai et al. 1996; Wang et al. 1995, 1997; All China Commission on Stratigraphy 2000, 2002; Zhai 2004, 2007, 2010; Wang et al. 2014; Qiao and Wang 2014) all define the base of Mesoproterozoic as 1800 Ma. Thus, the suggestion that Cohen et al. (2013) and the International Commission on Stratigraphy define the base of Paleoproterozoic as 1600 Ma is not suitable for tectonic evolution of Sino–Korean plate.

The ages of 1800–1600 Ma are the periods for the formation of sedimentary cover of Sino–Korean plate, and also, the rifting period of the craton (Qiao and Wang 2014). Zhai (2004, 2007, 2010) considered that Sino–Korean plate was suffering from the splitting period at 1800–1600 Ma, rather than the collision or orogenic periods. In the period of 1600–1400 Ma, it turned into the depositional period for the wide epeiric sea basin (Qiao et al. 2007, 2014).

Zhao (2007) recognized the beginning period of North China plate with the plate tectonic characteristics was 2560 Ma marked by the Wutai granite intrusion, and at that time, the eastern and western blocks of North China converged. However, most of researchers recognized that the beginning period with the plate tectonic character, i.e., the period of formation, the basement may be the end of Paleoproterozoic (~1800 Ma) (Zhai 2004, 2007, 2010).

As to the distribution area for Sino–Korean plate, it had been recognized to include the Tarim [20], Qaidam [18], Alxa–Dunhuang [16], the North China and Liaoning, eastward to Northern Yellow Sea, the entire Korean Peninsula and Hida Peninsula of Japan. From Mesoproterozoic to Cambrian, the Tarim [20], Qaidam [18] and Alxa–Dunhuang [16] blocks had separated gradually from the main part of Sino–Korean plate. The entire Korean Peninsula and Hida Peninsula of Japan had always been a part of Sino–Korean plate (Wan 2011b). The north border of Sino–Korean plate is the southern margin of Tianshan–Hingganling Late Paleozoic accretion–collision zone [10]; the south border of Sino–Korean plate is northern margin of Qinling–Dabie–Jiaonan–Hida Marginal Triassic accretion–collision zone [24] and Eastern Kunlun block on the northwestern boundary of Yangtze plate; the west border is the west side of Tarim basin; the east is cut through by Tanakura Tectonic Line (the common view of site survey for PIGC 321). On the east side, it is the northeastern island arc of Japan, which is the southern part of Aleutian–Kamchatka–Kurile–Northeast Japan Cenozoic subduction and island arc zone. Some Japanese researchers thought that this zone belonged to North American plate. This view lacks of evidence and may not be correct. The Hida block could be a small residues block, at first belonged to Sino–Korean plate, later compressed by the Yangtze plate [22] and Philippine Sea plate [65] and at last influenced by the extension of Japan Sea.

On the northern border of Sino–Korean plate, a series of Mesoproterozoic–Neoproterozoic metamorphic rocks are distributed in the area from Ondor to Bainaimiao in Inner Mongolia (Wu et al. 1998). The original rocks of Bainaimiao Group are intermediate-basic and acid volcanic complex, with bimodal volcanic features. The Sm–Nd isotopic isochron age is 1107 ± 28 Ma. In the northern part, the original Urausu amphibolite series are basic volcanic rocks with some clastic rocks, whose Sm–Nd isotopic isochron age is 607 ± 46 Ma. East to Ondor, in the Deyanqimiao, migmatitic hornblende system is developed, whose original rock is also basic volcanic rocks with some clastic rocks, and whose Sm–Nd isotopic isochron age is 638 ± 14 Ma. They are all formed by the boundary extension of Sino–Korean plate and become the volcanic–metamorphic systems. Peng et al. (2010) discovered Neoproterozoic acid volcanic rocks in Langshan, Inner Mongolia, with a zircon U–Pb age of 805–849 Ma. To sum up, the magmatism of the Mesoproterozoic–Neoproterozoic in East Asia should belong to the Yangtze plate or Gondwana (such as Li et al. 2003). The above facts prove that the magmatism from Mesoproterozoic–Neoproterozoic boundary extensions really exists in the northern boundary of Sino–Korean plate.

The controversy of the eastern extension part of south border of the Sino–Korean plate has existed for a long time. Many researchers (Huang 1945, 1960; Ren et al. 2000) thought that the south border for the Sino–Korean plate started from Zhucheng–Rongcheng of Shandong, went straightly toward to east and directly extended to around the Imjin River, Korea. In recent years, according to many field works and researches, the scholars both home and abroad have recognized any collisions not to exist (Zhai 2007) and found that the crystalline basements for northern and southern parts of Korean Peninsula are all the same, and the characteristics of sedimentation and paleontology in Korean Peninsula are as same as those in North China (Wan and Zeng 2002; Zhai 2007). In the last ten years, at the southwest sea-beach of South Korea, the high-pressure metamorphic block with the eclogite was discovered (Oh et al. 2005, 2006), its foliation is NNE trending, which may be formed near the collision zone and then migrated by the Eastern Yellow Sea dextral strike-slip fault zone and remained at the sea-beach of Eastern Yellow Sea (Wan and Hao 2010).

In the southern margin of Sino–Korean plate, there is Neoproterozoic Baishugou Formation of the lower part of Luanchuan Group, which is composed by the black shale, sandstone, phyllite, schist and slate. According to the data of Geological Bureau of Henan (1989), the Rb–Sr isotopic age is about 902 Ma for the above rocks. That formation is mainly distributed in the Luanchuan, Lushi Counties of Henan and Luonan County of Shaanxi, which belongs to the sedimentation of continental margin extension basins, and in the later, suffered from shallow metamorphism.

Kim and Cho (2003), Kwon et al. (2003) and Chang (2015) got the age data of 1900 Ma in the granite bodies at the northeast part of Yeunnan block. Sagong et al. (2003) used the Sm–Nd and U–Pb methods to date the metamorphic ages ranging from 1989 to 1835 Ma in the Kyonggi and Yeunnan blocks. Those data can well compare with the age data in North China crystalline basement. Thus, they all recognized that the Korean Peninsula and North China areas belong to same plate, i.e., the Sino–Korean plate. Through the arguments over the past twenty years or so, it seems that the recognition of Sino–Korean plate or "quasi-platform" (Huang 1945) is correct (Zhai et al. 2007). Some domestic scholars, in order to avoid commenting on the structural attribution of the Korean Peninsula, only talk about "North China plate (block)," which seems unnecessary.

However, Lee et al. (2003) reported the age data of 742 Ma in metamorphic granite at the Kyonggi block of southern part of Korean Peninsula. The age data of this tectonic-thermal event are close to the collision period of Yangtze plate when the south and north parts spliced together (Jinning Period, 860–750 Ma). So Li et al. (2003) considered that the Kyonggi block in Neoproterozoic had the characteristics close to those of Australian plate. However, from the Paleozoic to Mesozoic, the sedimentary and paleontological characteristics are very close to those of North China. In the Mesoproterozoic–Neoproterozoic, the Sino–Korean plate had undergone a series of marginal extension and led to the tectono-magmatism at 1100–700 Ma, forming many sedimentary systems of low-grade metamorphic marine facies. With this in mind, it is not surprising that the Neoproterozoic magma intrusion occurred in the southern part of Korean Peninsula, which maybe also an expression of the marginal extension of Sino–Korean plate. With only some Neoproterozoic age data, it cannot be said that the southern part of Korean Peninsula belongs to Australia block of Gondwana or Yangtze plate. In recent years, Wang et al. (2014) also got the isotopic age data of 2154–1530 Ma and 920–730 Ma in the granitic rocks of southern part of Korean Peninsula, which are fully compatible with the magmatic periods on the north border of Sino–Korean plate, which could support the recognition that southern part of Korean Peninsula still belongs to Sino–Korean plate.

After the formation of basement of the Sino–Korean plate in the end of Paleoproterozoic, the tectonic environment of the plate was rather stable in Mesoproterozoic–Paleozoic. In the plate tectonic evolution periods, locating in the Paleo-Tethys Ocean, the Sino–Korean plate almost had not collided with other plates. A Mesoproterozoic–Paleozoic carbonite and clastic rock series was deposited on the crystalline basement. During the period, influenced by tectonic activities of surrounding blocks, some magmatism from the Mesoproterozoic–Neoproterozoic marginal rifting in the Sino–Korean plate happened, and some large-scale uplifts and subsidence occurred, resulting in some sedimentary interruption and the parallel unconformity of strata, such as the sedimentary discontinuity between the Xiamaling Formation (formed at 1368 Ma) in Mesoproterozoic Xishan Period and Luotuoling Formation (formed at 930 Ma) (Qiao et al. 2007; Gao et al. 2009b); the sedimentary interruption between the Neoproterozoic Jingeryu Formation and Early Cambrian Mantou Formation; the parallel unconformity between the Early Ordovician Majiagou Formation and Middle Carboniferous Benxi Formation etc. (Wan 2011a).

In the Tarim block in the western part of Sino–Korean plate, there were actually two blocks (the Northern and the Southern Tarim blocks) in the early period of Neoproterozoic; in the Central Tarim, there was an anomalous belt with high positive gravity and magnetic anomalies; in the depth, there existed a metamorphic and deformed magmatic belt. The $^{40}Ar/^{39}Ar$ ages of granodiorite are between 932 and 892 Ma in the borehole of Tacan 1 (Li et al. 2003), and the $^{40}Ar/^{39}Ar$ plateau ages are between 923.3 and 891 Ma in the Central Tarim. At the western part of the Central Tarim basin, the peak of forming age is 800 Ma in the Aksu glaucophane schist, which contains many zircons with the peak age of 1940 Ma. At the Eastern Korla, the LA-ICP-MS zircon age of gneiss is 1042 ± 21 Ma (Zhu et al. 2011). Their results provide the important evidences for the existence of the collision zone in Central Tarim during the middle period of Neoproterozoic.

In several marginal depressions of Sino–Korean plate, the sedimentary action of the Early Paleozoic was relatively continuous, with short of depositional hiatus. For example, in the Taizi River area of Liaoning, North Qinling, Imjin River and Wochuan zone in Korean Peninsula, usually the parallel unconformity, are discovered between the Lower Ordovician and Middle Carboniferous, including the Upper or Middle Ordovician and even some Silurian, and the upper strata could be Devonian System (Wan and Zeng 2002).

On the southern border of Ordos in Sino–Korean plate (the coal field of Pucheng–Tongchuan areas, Shaanxi), Tan (1992) even discovered the folds of Lower Paleozoic rock series under the influence of regional compression of the surrounding Early Paleozoic blocks, resulting in the unconformity between Upper Ordovician and Permian Systems.

At this time, the maximum principal compression orientation was NNE or near N–S (Xu et al. 2006). The above data show that partial uplift or horizontal compression also could occur in the rather stable Paleozoic tectonic environment of the Sino–Korean plate.

The center reference point of the Sino–Korean plate moved from the latitude of 20.2° S to the north of the latitude 14.2° S at the ancient latitudes of the Paleozoic. The block migrated significantly northward with a latitude migration velocity of approximately 1.15 cm/year and a clockwise rotation of 4.3° (Figs. 3.6–3.8 and 3.10–3.12) (Huang et al. 1999; Ma et al. 1999; Wan 2011b). In the most period of Paleozoic, the Sino–Korean plate was located in the Paleo-Tethys Ocean and was basically in the free state. There is few sign of collision with other plates except for regional folds in the southern margin of Ordos due to the compression of the neighboring Late Paleozoic blocks (Xu et al. 2006).

The Middle Permian is the critical moment for the northward compression and collision of the eastern part of Sino–Korean plate, moving to the north trending and colliding with the Balkhash–Tianshan–Hingganling Late Paleozoic accretion–collision zone [10]. Thus, the Sino–Korean tectonic domain became a part of Pangea supercontinent and made the Sino–Korean plate form the wider basin deposits of clastic rock series: the coarse detritus in the north and fine detritus in the south.

The middle and late period of Triassic is the collision period for Yangtze plate [22] and Sino–Korean plate, forming the Qinling–Dabie–Jiaonan–Hida Marginal Triassic accretion–collision zone [24]. In the interior of the Sino–Korean plate, a gentle and wide fold cuts through the sedimentary strata of Mesoproterozoic to Triassic, forming an E–W axis as well (Wan 2011a, b). However, the southwestern margin of Ordos area were strongly influenced by the SW trending compression (Xu et al. 2006), becoming important target horizons for the development of deep gas reservoirs, coal-forming gas fields and shale gas fields. The PZM position of the Sino–Korean plate shifted from NW329.6° in the Early Triassic to NE30° in the Late Triassic, and the latitudinal reference point latitude shifted from latitude 18.2° to north latitude 27° (Fang et al. 1988); the northern paleo-magnetic orientations of Sino–Korean plate shifted from NW329.6° in Early Triassic to NE30° in Late Triassic, and the latitude of center reference point of that shifted from 18.2° N to 27° N (Fang et al. 1988; Ma and Yang 1993).

As to the background data, the regional characteristics and data basis of Mesozoic–Cenozoic tectonic evolution will be discussed in Part 3.

In the Yanshanian tectonic event (Jurassic–Early Cretaceous) controlled by the anticlockwise rotation of East Asian continental crust, the East Sino–Korean plate mainly formed the axial NNE-oriented folds and overthrusts with moderate strength together, with strong magmatic intrusion and eruption activities. The Yanshanian tectonic event made the East China crust slightly thicker and is formed a huge endogenous metal ore accumulation zone. For example, in the northern margin of the plate, the polymetallic metallogenic province is formed and dominated by molybdenum, gold, etc., in the south, the Xiaoqinling gold metallogenic province and in the east, Qinling polymetallic metallogenic province (Wan et al. 2008; Wan and Zhao 2012).

In the Sichuanian tectonic event (Middle Cretaceous–Paleocene), the whole plates on the Earth surface migrated to the North trending, the East Sino–Korean plate formed a series low or intermediate angle folds with WNW trending axis, as well as related thrusts, and then, the NNE trending thrusts changed into normal faults. Controlled by the Eastern Taihangshan, Eastern Changzhou-Tianjin, Tancheng–Lujiang faults in North China were formed, and also, some NNE trending metamorphic core complex were formed to make up many NNE trending fault-depression basins, such as the Bohai Bay, Song–Liao plain and Northern Jiangsu–Southern Yellow Sea (Wan and Hao 2010). Along the above regional faults, magma intrusion and eruption occurred, forming many metallogenic provinces, such as Jiaodong gold province and Eastern Taihangshan iron and polymetallic metallogenic province (Wan et al. 2008; Wan and Zhao 2012).

In Eocene–Oligocene, due to the long-distance effect of subduction and compression of Pacific plate in the WNW direction, the near E–W compression resulted in the N–S extension. Thus, on the northern and southern boundaries of Sino–Korean plate, near E–W normal faults, fault-block mountains were formed, such as Yinshan–Yanshan and Qinling–Dabie and metamorphic core complex. Between the mountains, the North China plain is located, where the warm–moisture air mass from the Pacific is driven straightly in, creating a warm and humid climate zone. At this stage, the basin became an excellent site for the mass reproduction and accumulation of organic matter. Under the control of the S–N extension in eastern part of Asia and Sino–Korean plate, the East China basins produced a series of normal faults and hydrocarbon-rich fault basins on the basis of the pre-existing faults, forming great oil and gas fields, such as the Shenli, Jidong (North Hebei Province), Liaohe, Dagang oil fields and so on.

In Miocene–Early Pleistocene, influenced by the long-distance effect of northward subduction and compression of the Australian plate, the East Sino–Korean plate experienced the near N–S trending shortening effect and caused above-mentioned NNE trending faults (Eastern Taihangshan and Tancheng–Lujiang) to extend, penetrating into the bottom of lithosphere. Many near N–S trending alkali basalt erupted and intruded into fractured dyke (Wang and Wan 2008; Wan 2011b). In the Eastern Ordos, it only shows the weaker NNE directional shortening (Xu et al. 2006).

Since Middle Pleistocene, the Sino–Korean plate had been affected mainly by the WSW directional compression, causing the once split NNE directional faults to close and to develop large hydrocarbon reservoirs in some sedimentary basins, such as Penglai 19-3 and Zhongyuan oil and gas field. The pre-existing NEE directional fractures were shear-strengthened and became a good channel for the fluid migration or spillage of oil and gas resources, coalbed methane, underground thermal fluids and gas bursts.

In recent years, the "destroy of North China Craton" has become a very hot topic of discussion (Zhu et al. 2008; Wu et al. 2008; Zhang 2009; Zhu et al. 2012). In fact, if the craton in North China actually broke down, it could mainly be occurred in the early Mesoproterozoic, not in the Mesozoic–Cenozoic. In Mesozoic–Cenozoic, only some stronger intraplate deformation as cracks, reverse faults and folds occurred, but the integrity of Sino–Korean plate was not destroyed, which is so different from that of the Gondwana. The statement that the North China Craton was "destroyed" in the Mesozoic–Cenozoic era sounds surprising, but not appropriate. This is a question worthy deliberating.

Besides the above-mentioned tectonic evolution features, the Sino–Korean plate has regional geochemical characteristics that it is relative rich in the follow elements: Fe (ΣFeO 12.0–12.28%), MgO (5.84–7.75%), Mo (0.64×10^6–0.58×10^6), with rather higher Zr/Hf (54.4–44), which are significantly higher than those of Yangtze plate; in the basement, CaO is 0.22%, and Al_2O_3 0.2–2.0%, which are lower than those of Yangtze plate. The P-wave velocity in crust is slightly less than that of the Yangtze plate (5.0–7.8 km/s). On the diagram of ε_{Nd}–t (Ga), the ε_{Nd} values (which age of 3.0–2.0 Ga, samples from basement of South Sino–Korean plate) are almost constant (+2.9 to +2.2), but in the basement of Northern Yangtze plate, the ε_{Nd} values gradually increase along the DM line (+2 to +7) (Zhang et al. 1998; Zhang et al. 2001).

To sum up, since the crystalline basement was shaped at the end of Paleoproterozoic, the Sino–Korean plate had experienced the very complex, multiple-phase intraplate deformations. It is just those intraplate deformations to control a large number ore deposits and greatly influence the ecological environment.

2.3.2 Helanshan–Liupanshan Late Paleozoic Collision Zone [15]

The Helanshan–Liupanshan Late Paleozoic collision zone (Fig. 2.14) had extended since Early Cambrian, separating the Sino–Korean plate [14] from the Alxa–Dunhuang block [16]. The paleontological data such as Trilobites show that west to this zone, the biota of the Alxa–Dunhuang block in Early–Middle Cambrian is basically the same as that of the North China in Sino–Korean plate. However, the biota characteristics in the Late Cambrian and Ordovician–Silurian Systems are very similar to those of Yangtze plate (according to the existing fossil name list, Peng 2003, personal communication; Lu 1976; Mu 1983). In the Middle Permian, the Xiyu plate (including [16] to [20]) and Sino–Korean plate [14] collided northward, and at the same time, the long-distance eastward compression effect of Ural collision zone compression to the eastward resulted in the formation of the Helanshan–Liupanshan Late Paleozoic collision zone.

However, due to the poor outcrops and the strong retrofitting effect of late deformations, it should be said that the evidences so far are not sufficient. There have been controversies regarding the origin of that zone. Some researchers thought that zone is just an aulacogen in the continent, but it is contradictory to the evidence of paleontological data. In the Late Cambrian–Silurian Systems, the biota of Alxa–Dunhuang block all has the characteristics of Yangtze plate. It means that at that time, the Alxa–Dunhuang block and Sino–Korean plate were not connected. As a result, the Helanshan–Liupanshan areas were supposed to locate in the ocean. However, till now there are insufficient evidences for the Paleozoic collision.

According to the variety data, the recognition of the existence of Helanshan–Liupanshan Late Paleozoic collision zone may be rather reasonable. According to the seismic exploration data (such as the Majiagou section) (Wang 1995), it is known that underneath the surface the Jurassic System developed very strong ramp thrusts. After Cretaceous, there were very weak intraplate deformations. At the end of Paleogene, due to the westward compression and subduction of the Pacific plate, high-angle reverse faults were formed on both sides of Liupanshan, making it uplift into a mountain (Bureau of Geology and Mineral Resources of Ningxia 1990). That is to say, the modern Liupanshan–Helanshan Mountains was formed in the end of Paleogene.

In recent years, on the western side of Liupanshan–Helanshan collision zone in the Alxa metamorphism crystalline basement, a large number of Middle Permian granites with weak deformations have been discovered. The zircon U–Pb ages of the diorite gneiss, quartz-mica diorite gneiss with the garnet, quartz-mica diorite, biotite plagioclase gneiss and gneissic granite, sampled from the eastern part of Alxa are 270 ± 1.6 Ma, 276 ± 1.8 Ma, 269 ± 2.4 Ma, 276 ± 2.4 Ma and 287 ± 2.5 Ma, respectively. The zircon LA-ICP-MS and U–Pb ages collected in the granitic dioritic gneiss, dioritic gneiss, coarse granitic dioritic gneiss and dioritic gneiss in the Western Alxa metamorphic basement are 284 ± 3 Ma, 289 ± 3 Ma, 276 ± 2 Ma and 279 ± 2 Ma (Gen and Zhou 2012). Although the rock types and chemical compositions of the Early–Middle Permian granites were different, they all formed in the relatively short time range of 289–269 Ma and belonged to the same tectono-magmatic thermal event. The formation age of the above Permian granites is close to the $^{39}Ar/^{40}Ar$ plateau age of 288–277 Ma in argillite in the basement metamorphic rocks.

In the Alxa metamorphism basement, the discovery of a large number of Middle Permian granitic intrusions shows that metamorphic basement may be influenced by the long-distance effect of eastward compression of Ural collision. It means that the tectono-magmatism associated with the collision from west to east in the Late Paleozoic Liupanshan–Helanshan collision zone occurred. It is also an important circumstantial evidences for the existence of Liupanshan–Helanshan collision zone.

2.3.3 Alxa–Dunhuang Block (1800 Ma) [16]

The Alxa–Dunhuang block (Fig. 2.14) belonged to Sino–Korean plate [14] before the Middle Cambrian, and its original crystalline basement (>1800 Ma) was the same as that of Sino–Korean plate (Cheng 1994). In the rocks of Dibusg Group hornblende phase, the whole rock Rb–Sr isochron method age is 3219 Ma. In granodioritic gneiss of the Bayan'ula Mountains the zircon U–Pb age is 2082 ± 22 Ma, and in the amphibolite of Dunhuang, Gansu the zircon U–Pb age is 1900 Ma (Bureau of Geology and Mineral Resources of Gansu 1989). Shen et al. (2005) researched the protolith of amphibolite in the Dibusg Group and confirmed that it was formed in Neoarchean, got the $^{39}Ar/^{40}Ar$ plateau age and isochron ages of 1918 Ma and 1919 Ma, respectively, suggesting that those rocks had been suffered from the amphibolite metamorphic superposition in

Paleoproterozoic. In the Bayan'ula Formation, the amphibolite was formed at 2271–2264 Ma. In the Bolustanmiao gneiss complex, the amphibolite was intruded by granitic gneiss aging at 1818 and 1839 Ma. According to the comparison of geochemical features of amphibolite found both in the complex and Bayan'ula Formation, it is inferred to form in the Early Paleoproterozoic. In the Qingeletu area of Alxa, the strata outcropped are mainly gneiss intercalated with some amphibolite, and however, the precise isotopic age data have never been got in the past. Zhou et al. (2007) got U–Pb isotopic age of 1826 ± 13 Ma from single zircon, which means that it is a Paleoproterozoic rock. This is the more reliable direct dating data obtained so far in this area. In the Bayan'ula Mountains, Alxa, Li et al. (2002b) got the zircon U–Pb age of 2082 ± 22 Ma, which is the believable data, which suggests that the rock system was formed in the Paleoproterozoic. Li (1994) obtained the isotopic age in Dunhuang Group metamorphic rocks and got the Sm–Nd isochron ages of 2935, 2946 and 3487 Ma, which all belong to Archean. In the Southern Beishan, according to the detail geological survey, they got the Sm–Nd isochronism isotopic ages of 2949, 2956, 3237 and 2203, 2059 Ma in the metamorphic rocks, which all belong to Archean–Paleoproterozoic. The Sm–Nd isochronism ages of 1622 and 1624 Ma and Baicalia stromatolites were got, and it suggests that they belong to the bottom of Mesoproterozoic (Bureau of Geology and Mineral Resources of Gansu 1989). All of the above information shows that in the Archean–Early Proterozoic, the tectonic events and rock features of this block were similar to those in the Sino–Korean plate. Therefore, it may be reasonable to state that the characteristics of the Alxa–Dunhuang block and the crystalline basement of the Sino–Korean plate are consistent. In recent years, in the Habudaha gneiss and deformed granite, Gen et al. (2002, 2007) got the ages of 1077 ± 11 Ma, 928 ± 7 Ma and 845 Ma using single zircon evaporation dating method. However, the above data may be influenced by the results of the fluid action in late periods. It cannot show that the Alxa–Dunhuang block and Sino–Korean plate have the different basement characteristics. In last ten years, in the extension zone at the margin of the Sino–Korean plate, many Neoproterozoic tectonic-thermal events have been discovered. It cannot suggest that only the Yangtze plate has the Neoproterozoic tectonic-thermal events. It seems that the above information is not sufficient to explain that the Alxa–Dunhuang block has the characteristics of the Yangtze plate in the Neoproterozoic, but it can only suggest that its characteristics are similar to those of the Sino–Korean plate.

Zhang et al. (2012) used the clastic zircon age data of Eastern Alxa and North China to research the relationship between above two blocks. They discovered that Alxa block and northern part and Langshan area of Sino–Korean plate all have the age data of about 2500 Ma, 1800 Ma and 850–950 Ma. But the Alxa area lacks of the age data of 1350–1400 Ma. Thus, it could be considered that the Alxa area in the Proterozoic belonged to the Sino–Korean plate (Fig. 2.16). To sum up, the tectono-magmatism of the Alxa block in the whole Proterozoic period showed almost no difference from the Sino–Korean plate, and they were never separately, not same as some researcher's recognition (e.g., Ge 1989; Ge and Ma 2014). Judging from the characteristics of the paleo-biological combination, Pen (2003, personal communication), using the collected fossils data provided by Duan and Ge (1992) for analysis, concluded that the Trilobites of the Early and Middle Cambrian in the Alxa area had the similar characteristics to those of the Sino–Korean plate. Since Late Cambrian, the paleontological and paleogeographic characteristics of Alxa area had been rather similar to those of Yangtze plate and different from those of Sino–Korean plate. His understanding is also consistent with the opinions of Lu (1976) and Mu (1983). Thus, the author infers that the breaking up between Alxa area and Sino–Korean plate occurred in Late Cambrian. In the late period of Early Paleozoic–Early Devonian, the Alxa–Dunhuang block [16], Qaidam [18] and Tarim block [20] collided together and formed the Qilian collision zone [17] and Altun Early Paleozoic sinistral strike-slip collision zone [19], further formed the Xiyu plate (it means "western area" in Chinese) [16–20] (Gao and Wu 1983; Wang and Chen 1987), whose paleontological characteristics of Early Paleozoic are similar to those of Yangtze plate, but not exactly the same. However, the existence period for Xiyu plate did not last very long. In the Late Paleozoic (about the Late Carboniferous–Early Permian), the Xiyu plate was pieced to gather with Sino–Korean plate to form the Helanshan–Liupanshan Late Paleozoic collision zone [15] and in Late Devonian–Early Carboniferous, moved to the north and formed the Tianshan–Hingganling Late Paleozoic accretion–collision zone [10] and the intraplate deformation in them (Figs. 3.13 and 3.14). Thus, Alxa–Dunhuang [16], Qaidam [18] and Tarim block [20] and Sino–Korean plate [14] all were merged into the Pangea supercontinent. Later, in the Indosinian Period (251–201 Ma), the above blocks were all influenced by N–S shortening, resulting in a series of tectono-magmatism. Shi (1987) discovered a lot of granites in the Alxa area. He thought that magnetism occurred in Permian Period. But he got the isotopic ages of 237.8 Ma in the Yablai Mount, 201 and 216.5 Ma in granites in the Jilantai area. However, the above data all exhibited the stronger Triassic tectono-magmatism (Figs. 3.16 and 3.17), and it explained

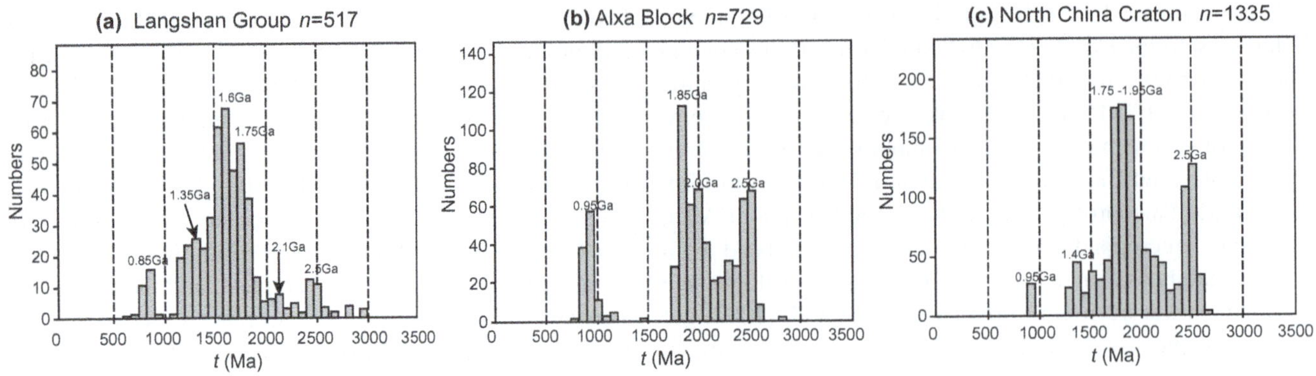

Fig. 2.16 Clastic zircon ages in the Langshan, Alxa areas and Sino–Korean plate (After Zhang et al. 2012)

Fig. 2.17 Pillow basalt in Ordovician ophiolite suites near the north of Qilian Mountains. Near the arc-like surface of every small stratum, the blowholes are bigger; in the inner of the stratum, the blowholes are smaller obviously. Those pillow basalt strata are overturned (Photo taken by author in 1998)

that the determined ages got by Shi (1987) are collected, but he had the mistake on deciding the geological periods. In Jurassic, influenced by WNW trending maximum principal compression stress, the block had a counterclockwise rotation and near N–S extension, forming many WNW trending fault-depression basins, such as the Jiuquan–Wuwei, Tengli, Jilantai and Dunhuang basins (Fig. 3.23). Since Cretaceous, the rock deformations of that area have become weaker.

According to the data of seismic reflection section, near the Moho discontinuity, in the Xiyu plate there is a transitional zone; but in the Sino–Korean plate, there is a sudden change zone (Wan 2011b). It means that there are obvious differences in deep crust structure for the above two plates, i.e., there is an evidence for the existence of the Xiyu plate.

2.3.4 Qilian Early Paleozoic Accretion–Collision Zone (541–400 Ma) [17]

The Qilian Early Paleozoic accretion–collision zone (Figs. 2.14 and 3.9) is the splicing zone between the Alxa–Dunhuang block [16] and Qaidam block [18]. After the collision, those tectonic units became a part of Xiyu plate. That collision zone is a belt with strong rock deformation, metamorphism and magmatism, and the collision period could last 400–380 million years. In the zone, there are a series of Ordovician–Silurian ophiolite suites (Xu et al. 1994; Feng 1995; Xia et al. 1995; Zhang et al. 1997, 1999; Liu et al. 1998; Zhang et al. 1998). In that accretion zone, there are the Paleoproterozoic Middle Qilian and Hualong blocks, which were all reformed by Early Paleozoic stronger tectonic magmatism. Later, they suffered the reformation of Triassic tectonics, and the whole strata were almost vertical. The Northern and Southern Qilian thrusts are the boundary of that collision zone, and there are series of ophiolite suites and eclogites (Fig. 2.17) in the zone. The thrusts were originally lithosphere fault, and later, they were all cut off by the detachments near Moho discontinuity (Gao et al. 2011). Under that collision zone, the lithosphere thickness may reach about 120 km.

Zhang et al. (2000) got the zircon SHRIMP U–Pb metamorphic ages of 477–489 Ma (Early Ordovician). In the Qilian and Altun area, Zhang et al. (2009) got many Neoproterozoic magmatic-metamorphic age data. It means that the preliminary tectono-thermal event occurred in Neoproterozoic. The Qilian Early Paleozoic accretion–collision zone is now surrounded by the Xiyu plate, giving the impression that it is just a small Early Paleozoic intra-continental ocean basin before the collision (Ge et al. 2000, 2014). However, according to the reconstruction from paleo-geomagnetic data, in the early period of Early Paleozoic, in the Qilian area an ocean existed, which was a part of Paleo-Tethys Ocean (Fig. 3.8).

2.3.5 Qaidam Block (1800 Ma) [18]

The Qaidam block (Fig. 2.14) was originally a part of Sino–Korean plate, which crystalline basement was formed in Late Paleoproterozoic (Bai et al. 1996). On the north border, a great Paleoproterozoic magma complex zone, Dakendaban Group, was found, and was identified as a product of the formation of the crystalline basement, in which isotopic ages of many sheets with intruded granitic gneiss are 1020 $\pm$ 41 Ma and 803 $\pm$ 8 Ma, suggesting partial converge in Neoproterozoic (Lu et al. 2011). In Late Neoproterozoic (800 Ma), the Qaidam block was separated from the Sino–Korean plate, and Neoproterozoic Nantuo glacial strata (Quanji Group, with the isotopic age of basic volcanic rock at 738 $\pm$ 28 Ma) (Lu 2001) were developed, similar to Yangtze plate.

In Paleozoic, the paleontological composition of Qaidam block was very near to that of Yangtze plate (Ge et al. 1990, 2014), so it means that they existed in the same paleontological geographic region, with similar latitude (Figs. 3.6–3.8 and 3.10–3.12).

In the late period of Early Paleozoic–Early Devonian, it was pieced together with the Alxa-Dunhuang block [16] and Tarim block [20] to form the Xiyu plate [16–20] (Gao et al. 1983; Wang and Chen 1987) (Fig. 3.9). The collision zone is the Qilian Early Paleozoic collision zone [17] and Altun sinistral strike-slip collision zone [19]. And in the late period of Late Paleozoic (Late Carboniferous–Middle Permian), they collided northeastward into the Pangea supercontinent (Figs. 3.11 and 3.12).

Through systematical research and measurement of the isotopic ages for the granite in Northern Qaidam, Wang et al. (2014) considered that a series of granite intrusions could reflect different origins: The granites formed in 475–460 Ma were products of the subduction of oceanic crust; the granites formed in 450–440 Ma (Late Ordovician) were products in the starting period of continental collision; 410–395 Ma (Early Devonian) granites were formed in the main collision period; and the granites of 385–370 Ma belonged to post-collision product. It proves further that the deformation of Northern Qaidam and the main collision of Qilian occurred in Early Devonian.

Although the Qaidam block suffered from repeated near N–S compression, the intraplate deformations are not obvious and till now the crust thickness is about 38 km. Compared with Kunlun and Qilian on south and north sides, the Moho discontinuity of Qaidam is a little bit uplift (Ge et al. 1990, 2014). But no one thinks that the Qaidam basin was caused by mantle uplift and extension. It is believed that on both sides of collision mountain zones are rich in granitic rocks, crust thickness is huge and the density is lower totally. However, deep in the Qaidam basin, the rocks is relatively dense and its surface sinks relatively, while the crust thickness is relatively thin, resulting in a relatively higher Moho discontinuity (Wan 2011b).

2.3.6 Altun Early Paleozoic Sinistral Strike-Slip Collision Zone (541–400 Ma) [19]

The Tarim [20], Qaidam [18] and Alxa–Dunhuang [16] were pieced together in the late period of Early Paleozoic to Early Devonian (Fig. 3.9), with the peak of collision in the late period of Early Paleozoic. The Altun Early Paleozoic sinistral strike-slip collision zone [19] and Qilian collision zone [17] were formed (Fig. 2.14), thus forming a uniform Xiyu plate (Fig. 3.9). At the Western Altun fault zone, there is the granulite in the khondalite series. Zhang et al. (1999) measured U–Pb and Pb–Pb isotopic age of metamorphic zircons in the khondalite series with a granulite facies in the west section of the Altun Early Paleozoic sinistral strike-slip collision zone and obtained the age of 447–462 Ma, representing the metamorphic age of the khondalite series, i.e., the initial active period of this strike-slip collision zone. They also estimated the formation temperature and pressure of metamorphic mineral and got the temperature of metamorphic peak of about 700–850 °C and the pressure of 0.8–1.2 GPa.

The U–Pb ages were measured by Zhang et al. (2011) of zircons from the deep metamorphic rocks in Central Altun block and Southern Altun subduction and collision complex, recording three tectono-thermal events of different time: early–middle periods of Neoproterozoic (1000–850 Ma), late period of Neoproterozoic (about 760 Ma) and Early Paleozoic (450–500 Ma). Thus, this finding demonstrates that the Altun fault zone was tectonically active in the collision period between Northern and Southern Tarim blocks and in the Early Paleozoic. Ma et al. (2011) got the isotopic ages of 461 ± 2 Ma to 471 ± 2 Ma in the mafic-ultramafic rock bodies in Qingshuiquan, Southern Altun zone and further confirmed the activity of Altun fault zone in Early Paleozoic.

The Altun sinistral strike-slip collision zone shows features of compressive shear. It obviously cuts off the Qilian collision zone [17] to form late period of Late Paleozoic. The geological and tectonic units on both sides could be compared rather well. Its total fault displacement is about 400 km, which was the result of multiple activities of different extents in the late period of Early Paleozoic, Triassic (260–220 Ma) (Li et al. 2001), Cretaceous (112–83 Ma), Neocene (sinistral displacement velocity is 16–20 mm/year) and Quaternary (sinistral displacement velocity is 6.4 mm/year) (China Earthquake Administration 1992; Ge et al. 1990, 1998, 2000, 2014: Xu et al. 1999; Yang et al. 2001; Chen et al. 2001). As to the estimation of strike-slip distance and velocity in each time, there is still much controversy and it needs to be further studied in detail.

2.3.7 Tarim Block (1800 Ma) [20]

The crystalline basement of Tarim block (Fig. 2.14) was formed in Late Paleoproterozoic, as same as the Sino–Korean plate [14]. It may have been a part of ancient Sino–Korean plate (Bai et al. 1996). Xin et al. (2011) conducted the SHRIMP zircon U–Pb dating research in Aktashitak area on the southeast boundary of Tarim block and got the crystalline age of gneiss diorite, gneiss quartz diorite: 2135 ± 110 Ma, 2051.9 ± 9.9 Ma and 2050 ± 16 Ma, respectively, and the crystalline age of quartz syenite is 1873.4 ± 9.6 Ma. Besides, in the Eastern Tarim, the isotopic age of volcanic rock in Dunhuang Group is 2140.5 ± 9.5 Ma. The other Neoarchean metamorphic rocks all had two metamorphic ages, 2.27–2.38 Ga and 1.9–2.05 Ga. During Late Paleoproterozoic (1.60–2.50 Ga), strong deep melting of crust-derived rocks formed igneous carbonates, quartz diorites and mixed lithology of potassic rocks. Pre-existing rocks were subjected to amphibolite facies metamorphism and strong ductile shear deformation. The uniform crystalline basement was formed in the end of Paleoproterozoic.

Recently, it is demonstrated again that the formation period of TTG rocks in the Kuluktag (Eastern Tarim) is between 2.6–2.8 Ga, and the metamorphic ages of Xindig Group and Altun Group are concentrated in 1.8–2.0 Ga (Zhang et al. 2012). In the depth of Tadong 2 Well of Tarim basin, the U–Pb zircon dating age is 1908.2 ± 8.6 Ma for hornblende granite (Wu et al. 2012). According to the rock samples in the well, Yang et al. (2014) further confirmed that in the Kuluktag of Eastern Tarim basin, Western Kunlun (Tiklik and Puluogou in Yutian), and in the MB1, XH1 and TD2 drilling well inside the Tarim basin, the composition and age of the Paleoproterozoic and Archean crystalline basement have some differences. They pointed out that the basin basement could be divided into: Yutian–Milan Archean granulite–migmatite area, Bachu–Yecheng–Hetian Paleoproterozoic para-gneiss area, Northern Tarim basin Mesoproterozoic–Neoproterozoic schist area and Eastern Tarim Paleoproterozoic granite area.

All these have undoubtedly proved that the Tarim basin has a Paleoproterozoic crystalline basement. Like the eastern part of the Sino–Korean plate, it has a similar ancient crystalline basement. The establishment of geochronological framework indicates that the Paleoproterozoic and its previous geological evolution of the Tarim block have more affinity with Sino–Korean plate. It seems that the original Sino–Korean plate, with a unified crystalline base, should include vast areas in the Paleoproterozoic from the Hida Peninsula in Japan, the Korean Peninsula, North China, Alxa–Dunhuang, Qaidam and Tarim block.

In Mesoproterozoic, the Tarim block had sedimentary rocks, sedimentary facies and sedimentary structures that were very similar to those in the eastern part of Sino–Korean plate (Gao and Wu 1983). However, from Neoproterozoic, the Tarim block had been separated from Sino–Korean plate to form two crystalline blocks in Northern and Southern Tarim (Fig. 2.14). In the middle period of Neoproterozoic (Ar–Ar ages of 825–837 Ma in the basalt), Northern and Southern Tarim blocks converged again and formed the Central Tarim collision zone [21] (Fig. 3.5; Wu et al. 2006). It means that in the splitting period of the global Rodinia supercontinent, some of the China continental blocks showed the phenomenon of reaggregation (Lu 2001). After Neoproterozoic, the Tarim blocks were migrated to near the equator, resulting in the similar paleontological and sedimentary characteristics of Yangtze plate [22] (Li et al. 1998; Zhang et al. 2000; Jin et al. 2010; Wan 2011b) (Figs. 3.6–3.8 and 3.10–3.12). In the late period of Early Paleozoic–Early Devonian period, the Tarim and Alxa–Duhuang, Qaidam blocks converged together into the Xiyu plate [16–20] (Fig. 3.9). At the middle period of Late Paleozoic, the

Xiyu plate migrated to north to form the western part of the Tianshan Late Paleozoic collision zone[10] and to be a part of Pangea supercontinent, while the Sino–Korean plate migrated and collided northward in the Middle Permian (Fig. 3.14).

In the schist of the Tugermin anticline core in Kuche depression of North Tarim basin, He et al. (2011) got the zircon U–Pb age of 775 ± 5.8 Ma to 787 ± 6.8 Ma (Neoproterozoic). It means that the Neoproterozoic tectono-thermal event not only happened in the Central Tarim but also in the North Tarim.

The Tarim block later suffered the weaker intraplate deformation in Indosinian tectonic period. However, from Jurassic to Paleogene, the tectonics was very weak, and only the areas near the faults developed weak deformations. The Junggar, Tarim and Qaidam blocks were made up a uniform sedimentary basin with the alluvial and lacustrine facies (Li et al. 2000). In Neogene, influenced by the long-distance effect of Indian plate northward subduction and compression (Fig. 3.33), the Tarim block subducted beneath the Tianshan and Kunlun Mountains separately and formed a series of near E–W thrust–folding system along the northern margin. Beneath the thrust plane the "triangle zone," the very good oil and gas reservoirs were reserved in the Northern Tarim basin. In the Northern Tarim oil field, the reservoir layers could be the Cambrian–Ordovician, Carboniferous–Permian, Jurassic and Paleogene Systems (Li et al. 2000; Zhao et al. 2004; Liu 2006; He et al. 2006; Liao et al. 2010; Zhao et al. 2011; Wang et al. 2012).

However, in the Central Tarim area, a large number of geese-type, steeply inclined small faults and joints are dominant in the carbonate rock series, forming fissure-type reservoirs (Xu 2011). In the above oil and gas field, the oil-bearing structures were shaped mainly in Neogene (Wan 2011a, b).

According to the analysis and data of zircon U–Pb ages in the basin, the basement of Tarim block is inferred to experience the tectono-thermal events in 2950–3100, 2100–2400, 1900–2000, 1300–1600, 900–950, 700–800, 540–560, 400–500 and 270–290 Ma. It means that the Tarim micro-plate experienced the tectono-thermal events in the Archean, at the end of Paleoproterozoic, in the Early, Middle and Late Mesoproterozoic, and in the Early and Late Paleozoic (Wu et al. 2012). To sum up, the formation and evolution characteristics of the Tarim oil- and gas-bearing basin should be completely different from those of Turan–Karakum plate on the west side, but they are similar in the characteristics and morphology of the depressions in recent times.

According to the latest seismic reflection profiles, the Moho discontinuities underneath the Tianshan and Tarim areas are almost the flat detachment surfaces, which depth is about between 60 and 53 km. It means that we cannot see the phenomenon of the lithosphere fault and the "root" phenomenon. That is, the ancient lithosphere fault has been reconstructed by the slip surface near the Moho; the existing faults are crustal faults; the thrust faults at the northern margin of Tarim are the intra-crustal faults (Gao et al. 2011).

2.3.8 Central Tarim Neoproterozoic Collision Zone (~850 Ma) [21]

In the middle of Tarim (Figs. 2.14 and 3.5), the normal high-anomalous zone of gravity and magnetism has been discovered for a long time. The drilling confirms that it is a hidden basic magmatic rock zone, where the Neoproterozoic metamorphic and deformed basic volcanic–intrusive rocks are found. It means to have the paleo-ocean crust, with Ar–Ar isotopic ages of 825–837 Ma, providing the important evidence for the formation of Central Tarim Neoproterozoic collision zone, as well as the Northern and Southern Tarim blocks being pieced together (Wu et al. 2006).

After the formation of Central Tarim Neoproterozoic collision zone, since Paleozoic there have been the uplift structure (hidden uplift of Central Tarim) for a long time, which become propitious for the accumulation of the reservoirs of oil and gas (Xu 2011). Judging from the aeromagnetic anomaly data, the Central Tarim is a multi-stage tectonic zone. In the Mangar depression, near the Central Tarim collision zone, an extension environment existed from Late Neoproterozoic to Late Ordovician (He 2011).

2.4 Yangtze Tectonic Domain

The Yangtze tectonic domain (Fig. 2.18) includes: Yangtze–Southwest Japan plate (~850 Ma) [22], Southern Anhui–Northeastern Jiangxi–Xuefeng Mountains–Eastern Yunnan Neoproterozoic collision zone (i.e., Jiangnan collision zone, ~850 Ma) [23], Qinling–Dabie–Jiaonan–Hida Marginal Triassic accretion–collision zone (250–200 Ma) [24], Shaoxing–Shiwandashan Triassic collision zone (250–237 Ma) [25], Cathaysian plate (~400 Ma) [26], Eastern Hindukush–Northern Qiangtang–Indosinian plate (~850 Ma) [27], South China Sea Cenozoic fault-depression basin [28], Palawan–Sarawak–Zengmuansha block [29], Western Hindukush–Pamir–Kunlun Late Paleozoic–Triassic accretion–collision zone (360–200 Ma) [30] and Jinshajiang–Red River Triassic collision zone (252–201 Ma) [31].

The united crystalline basement for the above blocks was formed in two periods: Middle period of Neoproterozoic (~850 Ma) for the Yangtze–Southwest Japan plate [22] and Eastern Hindukush–Northern Qiangtang–Indosinian plate

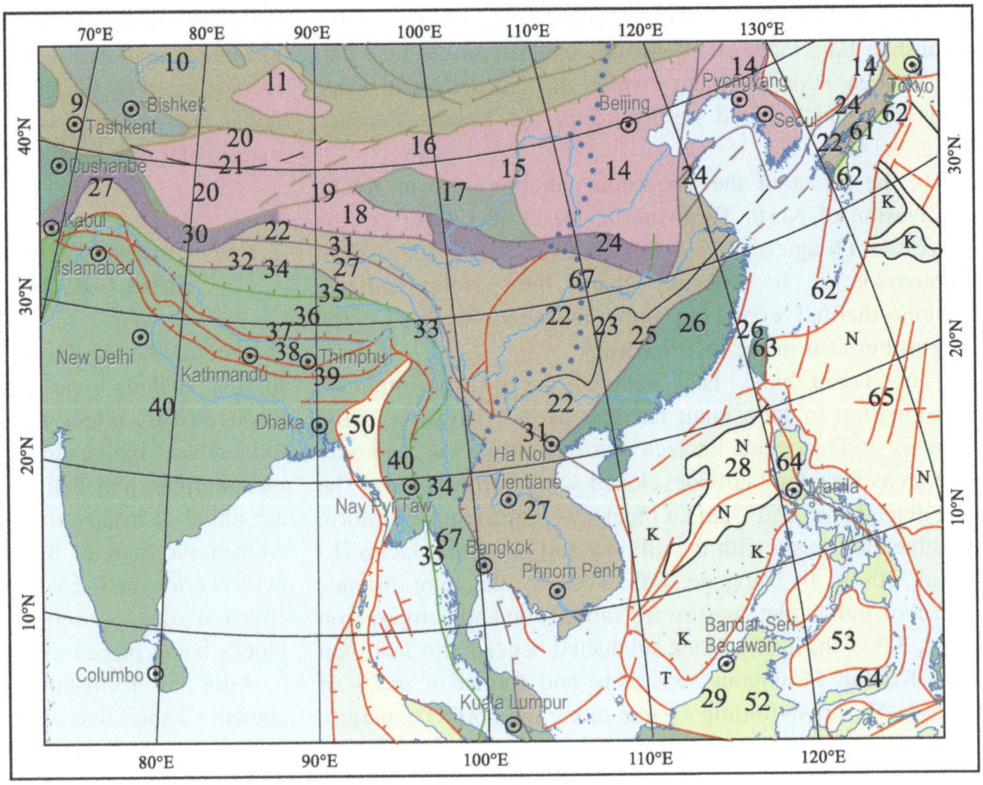

Fig. 2.18 Yangtze tectonic domain [22–31]. The numbers in the figure show the tectonic units, as same as those in the CONTENTS and Fig. 1.1. Brown color shows the Yangtze tectonic domain; green color shows the Cathaysian plate [26]; purple color shows the Qinling–Dabie–Jiaonan–Hida Marginal Triassic accretion–collision zone [24] and the Western Hindukush–Pamir–Kunlun Late Paleozoic–Triassic accretion–collision zone (360–200 Ma) [30]

[27]; and late period of Early Paleozoic ($\sim$400 Ma) for the Cathaysian plate [26], South China Sea Cenozoic fault-depression basin [28], Palawan–Sarawak–Zengmuansha block [29]. At the end of Early Paleozoic, partial collision occurred at the northwest boundary of Yangtze plate, the Eastern Kunlun area. The South China Sea Cenozoic fault-depression basin [28] and Palawan–Sarawak–Zengmuansha block [29] with the similar crystalline basement to that of the Cathaysian plate were subsided to be sea area and oceanic basin until Cenozoic and then separated a little bit with the Cathaysian plate.

As to the collision zones, the Southern Anhui–Northeastern Jiangxi–Xuefeng Mountains–Eastern Yunnan collision zone (i.e., Jiangnan collision zone) [23] was formed in Neoproterozoic ($\sim$850 Ma), with the North and South Yangtze plates converging into a great Yangtze plate. Besides, in the Yangtze tectonic domain there are many other Triassic collision zones, such as the Qinling–Dabie–Jiaonan–Hida Marginal Triassic accretion–collision zone (250–200 Ma) [24], Shaoxing–Shiwandashan Triassic collision zone (250–237 Ma) [25], Western Hindukush–Pamir–Kunlun Late Paleozoic–Triassic accretion–collision zone (360–200 Ma) [30] and Jinshajiang–Red River Triassic collision zone (252–201 Ma) [31] (Figs. 2.18 and 3.16). In addition, between the Yangtze tectonic domain and Gondwana tectonic domain there also were two collision zones almost at the same time as Triassic: Shuanghu Triassic collision zone (252–201 Ma) [32] and Changning–Menglian–Chiangrai–Central Malaya Triassic collision zone (252–201 Ma) [33]. This made the eastern and northern continental blocks of Southern Qiangtang–Sibumasu plate [34] all be incorporated into Pangea supercontinent, i.e., two-thirds of Asian continental blocks became a part of the Pangea supercontinent.

In the paleo-crystalline blocks of Dabie–Qinling collision zone, many gneiss domes have been discovered (Cai 1965, 1978; Xu et al. 2015 personal communication). Their formation was before Paleoproterozoic when the paleo-continental nuclei was formed and had no relationship with the collision.

Some researchers called the Qilian Early Paleozoic collision zone, Qinling–Dabie Triassic collision zone and Eastern Kunlun Triassic collision zone as the "Central Orogenic Belt," which was first proposed by Yin et al. (1998) and then used by many institutes of China Geological Survey. Actually, the above three collision zones have obvious different characteristics. The NW trending Qilian collision zone was only formed at the end of Early Paleozoic–Early Devonian and cut off by Western Qinling at Southern Gansu. The Northern Qinling and Eastern Kunlun in the Early Paleozoic just showed some partial ocean–continent subduction. A continent–continent collision between them occurred in Triassic, and they are not directly connected. The E–W trending Western Qinling could extend to the eastward of Dulan in Eastern Qinghai Province. However, the Eastern Kunlun collision zone is distributed over the southern boundary of Qaidam basin, southern border of Qinghai Province. It is not scientific significance and does not fit the facts to call the three collision zones as the "Central Orogenic Belt," which were formed in different periods though in adjacent positions.

The great accretion of Asian continent was related with the northward extension of East Tethys Ocean in the Eastern Hemisphere in Triassic, which was the main period for the Asian continental accretion. However, in the Western Hemisphere the Pangea supercontinent began its extension and break-up, forming the Pan-Atlantic Ocean (Wan 2011b).

When large-scale collisions occurred in the Asian continent during Triassic, the Yangtze tectonic domain also experienced rather strong and widespread intraplate deformations (Fig. 3.17). Later, the tectonic units in the Yangtze tectonic domain all suffered from various intraplate deformation at five tectonic periods, similar to the Sino–Korean tectonic domain (explained in details in Part 3, Figs. 3.23, 3.25, 3.31, 3.33 and 3.35).

2.4.1 Yangtze–Southwest Japan Plate (~850 Ma) [22]

The Yangtze–Southwest Japan plate (abbreviated as Yangtze plate; Fig. 2.18), from west to east, includes the Hohxil–Bayanhar–Garzi–Aba areas to the Yangtze River drainage basin, Southern Yellow Sea and Southwest Japan areas, is a residual plate suffered from very strong intraplate deformations. The areas to the south of Hida marginal belt in the Southwest Honshu of Japan, all belong to Yangtze plate (Yoshikura et al. 1990; Osozawa 1994, 1998). The Yangtze and Sino–Korean plates are so different in the formation period of crystalline basement, tectonic evolution, paleontology, paleogeography and regional geochemistry that they can be distinguished rather easily (Wan and Zeng 2002).

In the Yangtze plate, there are some small paleo-continental nuclei formed in the Archean–Paleoproterozoic, such as on the east border of Yunnan, Diancangshan–Ailaoshan (Liu et al. 2013), Fanjinshan of Eastern Guizhou and Huangling of Western Hubei, where multiple tectono-thermal events occurred at 2500, 1800 and 1000 Ma. In recent years, a lot of measurement data have shown that the formation of the united crystalline basement and amalgamation of the South and North Yangtze plates took place in Neoproterozoic (about 850 Ma), which is also the period to form the Jiangnan collision zone (Southern Anhui–Northeastern Jiangxi–Xuefeng Mountains–Eastern Yunnan collision zone [23]). In the early time, it would be determined that the collision occurred in 1000 Ma. Recently by using the zircon SHRIMP method, researchers have got a lot of precise age data showing the collision period of about 850 Ma (Gao et al. 2009a; Sun et al. 2012). This is just the break-up period for Rodinia supercontinent instead of conversion (Fig. 3.3). It means that the convergence of the Yangtze plate and the evolution of the Rodinia supercontinent are not synchronized (Lu 2001).

Some blocks, inside the Qinling–Dabie accretion–collision zone, have similar characteristics to the Yangtze plate, such as the formation age of crystalline basement, lead isotopic age and geochemistry (Zhang 1995). It means that these small blocks may be a part of Yangtze plate in the Neoproterozoic. Also, the Cathaysian plate [26] ever had weakly connected with the Yangtze plate in the Neoproterozoic (Shui et al. 1986). However, according to the differences in aspects of Paleozoic paleo-magnetism, metamorphism and tectonic line, the author (Wan 2011b) considered that the above two plates were separated in Early Paleozoic, although they were located in short distance and in the same climate belt (details in Part 3, Figs. 3.6–3.8 and 3.10–3.12)

The central reference point of the Yangtze plate was located at 11°S in the Cambrian. The long axis of the plate was nearly N–S trending, and later migrated northward and gradually rotated to nearly E–W trending (Figs. 3.7, 3.8, 3.10 and 3.11). In the Late Permian, the plate reached near the equator (Fig. 3.12) and later reached 18° N in the Middle Triassic. The plate moved to 27° N during the Late Triassic when collided with the Sino–Korean plate, thus forming its location and shape similar to the recent (Fig. 3.16). Later those plates had some minor displacement and rotation and then reached recent positions (Figs. 3.27, 3.31 and 3.33; Wan 2011b; Wan and Zhu 2011).

In the northwest border of North Yangtze plate, Eastern Kunlun area, Meng (2015, personal communication) considered that in the Early Paleozoic the Yangtze plate may be subducted beneath the Qaidam block, and the related granite intrusion isotopic ages were between 445–382 Ma (the personal communication from Qinghai Geological Survey 2003). However, it needs further research to determine whether the above granite originated from an island arc or interpolate magmatism.

In the South Yangtze plate (Central Hunan, Guangxi and Eastern Yunnan) in the late period of Early Paleozoic, the intraplate deformation occurred with a nearly E–W trending fold axis (by recent magnetic orientation), and this tectonic event was called "Guangxi Movement" (first proposed on the 6th Annual Meeting of China Geological Society by Ding 1929). That tectonic event was showed by the angular unconformity between Devonian system and those underneath systems. However, in the most areas of North Yangtze plate those folds and unconformity are never found at all, where between the Early and Late Paleozoic systems there was only conformity or a little bit parallel unconformity

Table 2.1 Comparison of the tectonic characteristics among the Cathaysian, North and South Yangtze plates in Late Paleozoic

Characteristics of plate	North Yangtze plate	South Yangtze plate	Cathaysian plate
Biogeographical province	South China	South China	South China
Formation period of united crystalline basement	~850 Ma	~850 Ma	~400 Ma
Mean paleo-magnetic declination	129.1°–83.8° in Early Paleozoic	129.1°–83.8° in Early Paleozoic	205.4°–201.3° in Cambrian
Paleo-latitude for central reference	11.7° S–2.8° N	11.7° S–2.8° N	10.8° S–11° S
Strata contact between Upper and Lower Paleozoic	Disconformity	Unconformity	Unconformity
Fold axis trending in Lower Paleozoic (based on recent magnetism)	Few folding	Near E–W trending	Near N–S trending
Crust shortening orientation in Lower Paleozoic (based on recent magnetism)	None	Near N–S	Near E–W
Regional metamorphism	None	None	Greenschist
Magmatism	Almost none	Weak	Strong
Typical metallogenic types (in a descending order of amount)	Fe, Cu, V, Hg, Au, REE	Sn, Cu, Pb, Zn, Sb, W	W, Ag, Pb, Zn, Cu, U, Sn, Au, F, REE

Notes The data of folds and paleomagnetism are shown in Appendixes 3 and 6 in Wan (2011b)

(detailed to see Table 2.1; Fig. 2.19). Only in the northern border of North Yangtze plate, some partial deformations and angular unconformity were found.

In the Triassic, the Yangtze–Southwest Japan plate moved northward and collided with Sino–Korean plate [14], leading to the formation of the Qinling–Dabie–Jiaonan–Hida Marginal Triassic accretion–collision zone (250–200 Ma) [24], and the incorporation of the Yangtze–Southwest Japan plate into the Pangea supercontinent. At the same time, the Western Hindukush–Pamir–Kunlun Late Paleozoic–Triassic accretion–collision zone (360–200 Ma) [30], Jinshajiang–Red River Triassic collision zone (252–201 Ma) [31], Shuanghu Triassic collision zone (252–201 Ma) [32] and Changning–Menglian–Chiangrai–Central Malaya Triassic collision zone (252–201 Ma) [33] were all formed. Thus, in the eastern parts of Sino–Korean plate and the inner parts of Yangtze plates, widespread intraplate deformations and magmatism occurred during the Triassic. From the Mesoproterozoic to Triassic, sedimentary strata were all affected by this tectonic event, which made the fold axes mainly E–W trending (according to recent magnetic orientation). Due to the inhomogeneous strength of strata, nearly E–W trending arc fold systems were formed, of them the large-scale ones such as Huaiyang and Guangxi arcuate structures (Fig. 3.17).

In Late Permian–Triassic, extension occurred between the main part of the Yangtze plate and its western part, Eastern Kunlun–Hohxil Shan–Bayanhar–Garze block, i.e., near Longmenshan area. In Late Permian (260 Ma), in Eastern Kunlun area, the granodiorite intruded, which was formed in the extension environment (Luo et al. 2015). In Triassic, the Eastern Kunlun–Hohxil Shan–Bayanhar–Garze block, i.e., West Yangtze plate, experienced the subsidence, formation of widespread bathyal areas and huge flysch deposition with continental slope. Underneath the deposition, marine sediments might exist (Zhao et al. 2014). And at the end of Triassic, a rather strong collision occurred on the north side of the Eastern Kunlun area. It has been considered that the Eastern Kunlun was a collision zone developed in the Paleozoic and Triassic, and the Qaidam block was subducted underneath the Eastern Kunlun area in that period. However, deep seismic data show that it is a high-angle normal fault toward the north (Zhao et al. 2014; Fig. 2.20). In Fig. 2.20, the blue layer of middle crust (v_p = 6.2–6.3 km/s) may be a ductile deformation zone, formed by oceanic crust metamorphism in the Late Paleozoic. Karplus et al. (2011) considered that it is a "channel flow," not the fluid flow, but the solid-state ductile deformation, in which the P-wave can pass through and its velocity can be estimated. Zhao et al. (2014) inferred that it is the result of metamorphism of ancient oceanic sedimentation (Fig. 2.20), which may be relatively reasonable. This understanding is similar to that underneath the continental crust there existed the oceanic crust or mantle, such as in Alps (Cavazza et al. 2004) or in East Asia. It indicates that the structure of continental lithosphere is more complex than that of the oceanic lithosphere.

If there had been the convergence, some part of the Eastern Kunlun–Hohxil Shan–Bayanhar–Garze block could be subducted underneath the Qaidam block. According to

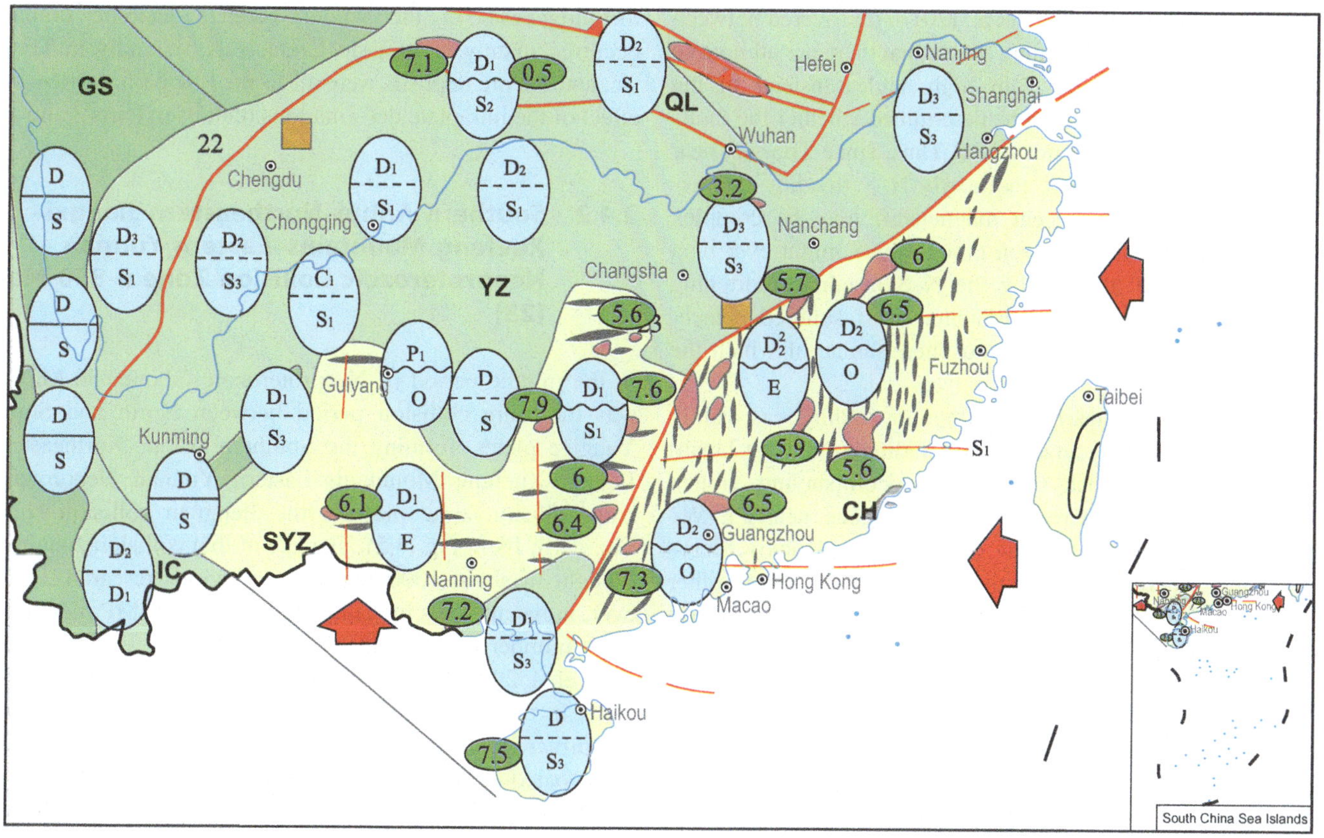

Fig. 2.19 Tectonic sketch of late period of Early Paleozoic in Yangtze and Cathaysian plates. The blue oval shows the stratum contact for top of Late Paleozoic strata, which code is as same as the international stratum code. Angular unconformity is shown as a wavy line, disconformity as broken line, conformity as continuous line. The shallow green area shows disconformity, the deep green area shows conformable strata, the shallow yellow area shows angular unconformity, and the black short line shows anticline fold axis (data from Wan 2011b). The red dot dash line is maximum principal compression stress (σ_1) trending, the red block area shows the Early Paleozoic granitic intrusion, and the big red arrow shows the migration orientation of plate. The deep green oval and number show the velocity of intraplate deformation inferred by petrochemical data; "-" is extension velocity, others are shortening velocity (unit: cm/year, data from Wan 2011b). The thick red line shows plate boundary (including oceanic crust). Late period of Early Paleozoic fault belts and their serial numbers: No. 22 is Qinling–Longmenshan fault zone; No. 23 is the probably existed Shaoxing–Yunkaidashan oceanic crust belts. Tectonic units: YZ. Yangtze plate (GS. Garze–Songpan block, SYZ. South Yangtze block); CH. Cathaysian plate; QL. Qinling–Dabie collision zone; IC. Indosinian plate. Location and boundary of each block without doing tectonic reconstruction

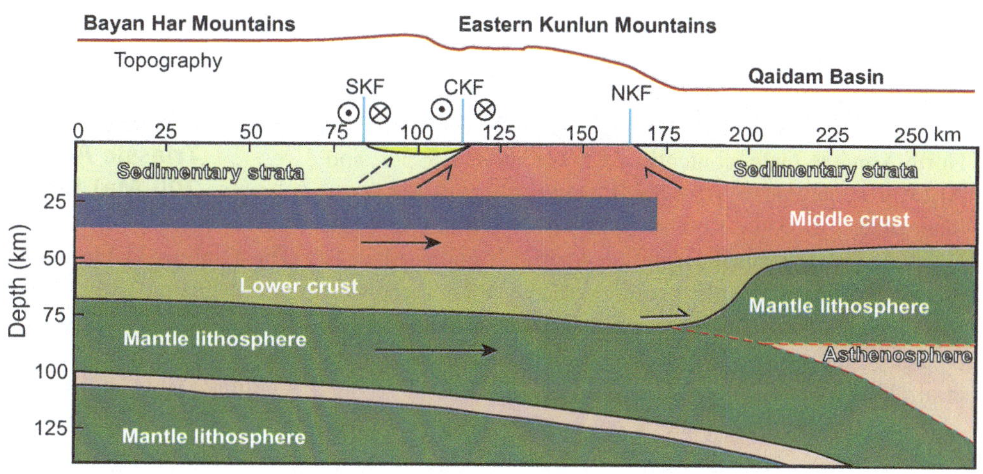

Fig. 2.20 Deep structural model in Eastern Kunlun convergence belt. The blue layer (v_p = 6.2–6.3 km/s) is inferred to be the ductile zone, which may be resulted from the metamorphism of the Late Paleozoic ancient ocean crust (After Zhao et al. 2014)

the deep seismic data, Zhao et al. (2014) discovered between both sides of crusts of Kunlun block, that there are all normal crustal strike-slip faults with high angles instead of the lithosphere faults. It cannot be determined whether the faults are plate boundaries or not so far. Thus, since Triassic, near the Eastern Kunlun area, the nearly N–S trending convergence (according to recent magnetism) may be complex intraplate deformations. Even though there might ever be a lithosphere fault affected by the N–S trending shortening since Triassic, the inner continental crust suffered detachments, reforming the paleo-lithosphere faults, of which the evidences are not to find.

In Jurassic, the nearly NNE trending folding was mainly developed at East Yangtze plate (including Southwest Japan) and Garze–Aba block (Fig. 3.23), accompanying violent magma intrusion. In Cretaceous and Neogene, nearly E–W–WNW trending folds were developed in Hohxil Shan–Bayanhar where the magmatism was strong while in other areas weak (Figs. 3.25, 3.31 and 3.33). In Paleogene–Early Miocene, at the Eastern Sichuan basin and Western Yunnan areas there developed many NNE trending folds and reverse faults, and however in the Eastern Kunlun, Hohxil and Northern Qiangtang, there formed a series of E–W trending normal faults with some strike-slip structures. These are all the results of long-distance effect of westward migration, compression and subduction of the Pacific plate, but not probably caused by the northward migration and subduction of the Indian plate.

In the East Yangtze plate, i.e., east to the Wuling–Eastern Yunnan, the Mesozoic–Cenozoic stronger intraplate deformations and magmatism were accompanied with series of endogenic metallogenesis. However in the North and South Yangtze plates, there formed different types of ore deposits, which may be caused by the different geochemical characteristics of their paleo-continental nuclei. The relatively rich elements in North Yangtze plate are: Li (26.3×10^{-6}), Rb (27×10^{-6}–30×10^{-6}), Sc (34×10^{-6}–46.7×10^{-6}), Cu (80×10^{-6}–126×10^{-6}), with a high ratio of Nb/Ta (16–25), a low ratio of Zr/Hf (40). Compared with Sino–Korean plate, the North Yangtze plate has low ΣFeO (9.14%), MgO (5.19–6.84%) and Mo (0.3×10^{-6}–0.54×10^{-6}). In the North Yangtze plate, contents of CaO is about 2.5–5%, and Al_2O_3 is about 3–5%. Also, the P-wave velocities are faster than that of Sino–Korean plate (6.0–8.0 km/s). The ε_{Nd}–t (Ga) values on the diagram for the Yangtze plate is along the deficit mantle evolution line (DM), increasing from 2 to 7, while the values of Sino–Korean plate do not change in the geological history (Zhang et al. 1998; Zhang et al. 2001).

The North Yangtze plate is characterized by the formation of giant ore deposits with Fe, Cu, Au and Hg, while the South Yangtze plate is characterized by the formation of giant ore deposits with Sn, Cu, Pb, Zn and Sb. The metallogenesis in the whole Yangtze tectonic domain occurred mainly in Jurassic–Cretaceous, secondly in Triassic, and the ore deposits were all formed by partial extension effect of the intraplate deformations (details in Parts 3 and 4).

2.4.2 Southern Anhui–Northeastern Jiangxi–Xuefeng Mountains–Eastern Yunnan Neoproterozoic Collision Zone (~850 Ma) [23]

In the middle period of Neoproterozoic (about 850 Ma), it was the main collision period between North and South Yangtze plates, forming the Southern Anhui–Northeastern Jiangxi–Xuefeng Mountains–Eastern Yunnan Neoproterozoic collision zone (namely the Jiangnan collision zone, Fig. 2.18; JN in Fig. 3.5). The age of that collision zone was thought in about 1000 Ma in the previous research. However, in recent years by using the zircon SHRIMP method, a large number of age data of about 850 Ma were obtained (Gao et al. 2008, 2009a, b, 2011; Sun et al. 2012). Along this zone, in the Southern Anhui and Northeastern Jiangxi, the Neoproterozoic ophiolite suites were discovered, showing that a paleo-ocean once existed (Shui et al. 1986; Zhou et al. 1989; Chen 1991; Yang et al. 1994; Liu et al. 1995; Tang and Zhou 1997).

However, some researchers (such as Li et al. 1996, 1998, 2003, 2007) considered that the Southern Anhui–Northeastern Jiangxi–Xuefeng Mountains–Eastern Yunnan Neoproterozoic collision zone [23] (Fig. 2.19; JN in Fig. 3.5) and the Shaoxing–Shiwandashan Triassic (250–237 Ma) collision zone [25] (SS in Fig. 3.5) were the same one. It seems improper to confuse these two collision zones in different periods and positions. Although these two collision zones are rather near, South Yangtze plate [22] exists between them, which includes the Northwestern Zhejiang–Northeastern Jiangxi–Central Hunan, and most of Guangxi and Eastern Yunnan areas.

2.4.3 Qinling–Dabie–Jiaonan–Hida Marginal Triassic Accretion–Collision Zone (250–200 Ma) [24]

The Qinling–Dabie–Jiaonan–Hida Marginal Triassic accretion–collision zone (Figs. 2.18 and 2.21, or shortly the Qinling–Dabie collision zone) is a collision zone between the Yangtze plate [22] and Sino–Korean plate [14]. The Qinling–Dabie collision zone extends eastward to Dabieshan, where it is cut off by the Tancheng–Lujiang sinistral strike-slip fault zone and continues to Jiaonan (Sulu). Its east side is again cut off by Eastern Yellow Sea dextral strike-slip fault zone and extends across the southward Cheju Island to

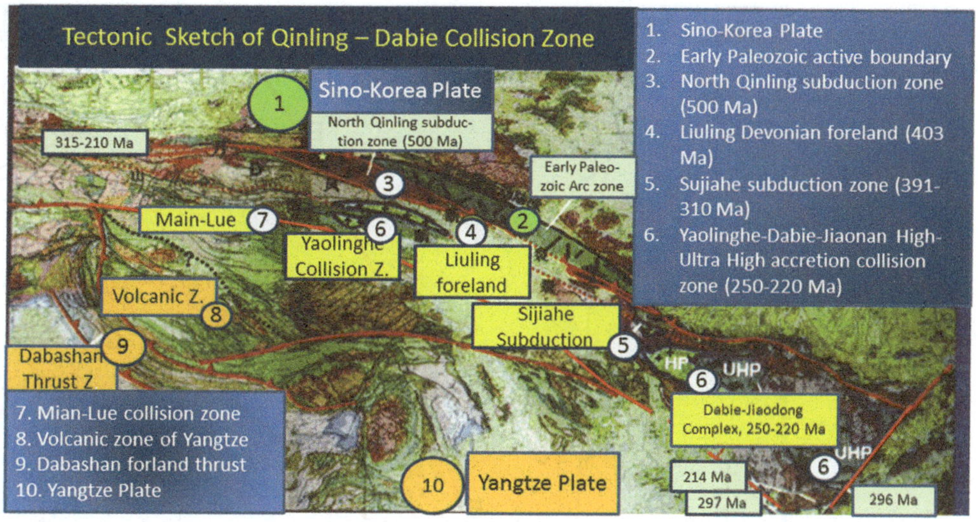

Fig. 2.21 Tectonic sketch of Qinling–Dabie collision zone (After Yang et al. 2009)

Hida southern marginal zone. Tsujimori et al. (2000) and Kunugiza et al. (2001) discovered glaucophane in a rock sample with partial eclogite mineralogy in the Hida marginal belt. Three groups of metamorphic ages (350–300, 270–210 and 210–180 Ma) were determined from single grains of zircon, monazite and uranite using SHRIMP and U–Th–Pb methods and from amphibole, mica and whole rock using Rb–Sr and K–Ar methods. They considered that the 270–210 Ma tectono-thermal event could be correlated with the similar events in the Dabie collision zone. The Hida southern marginal zone continued to extend eastward and at last was cut off by the Tanakura Tectonic Line. The east side of the Tanakura Tectonic Line belongs to the Aleutian–Kamchatka–Kurile–Sakhalin–Northeast Japan Cenozoic subduction and island arc zone (~40 Ma) [59].

In the Qinling–Dabie collision zone, there existed the evidence for oceanic crust in the Neoproterozoic and Paleozoic (Liu et al. 1993; Suo et al. 1993; Li et al. 1993; Zhang et al. 1996, 2001; Yang et al. 2000, 2002). In the Neoproterozoic (~850 Ma), some blocks in the Qinling–Dabie–Jiaonan collision zone may be connected with the northern area of Yangtze plate, for they had the same trending of tectonic line, structure style and similar isotopic characteristics. In this collision zone, there are many small ancient crystalline blocks, such as the Middle Qinling, Wudang, Dabie and Jiaonan, in which there developed some gneiss domes (Cai 1965; Ma and Cai 1965; Xu, Z. Q. personal communication).

Based on the tectonic evolution research, in the Early Paleozoic, in the northern part of the Qinling–Dabie–Jiaonan collision zone there existed the Northern Qinling (Shanxian–Danfeng) subduction complex zone (500–403 Ma; ⑥ in Fig. 2.21, in that zone the isotopic age of granite is 500 Ma; Wang et al. 2009). In the Middle Devonian–Early Carboniferous (393–323 Ma), there formed Sujiahe subduction complex zone (⑤ in Fig. 2.21). However, there is no evidence showing that the collision between Sino–Korean and Yangtze plates had already occurred during these two periods (Yang et al. 2009). In that zone, much evidence for Paleozoic oceanic crust was found, and the ages of ophiolite suite separated between 408 and 264 Ma. It means that the oceanic crusts never disappear after the above two subductions (Zhang et al. 1996).

On the southern boundary of the Qinling–Dabie–Jiaonan accretion–collision zone, i.e., the Mianxian–Lueyang–Yaolinghe–Dabie–Jiaonan high to ultra-high-pressure accretion–collision zone (⑦ in Fig. 2.21), also on the northern boundary of Yangtze plate (Zhang et al. 1996; Dong et al. 1999; Dong and Zhao 2002), the initial collision age should be between 264 and 240 Ma, about 250 Ma, and the completed time of collision was between 220–210 Ma, i.e., formed mainly in Triassic (Li et al. 1996, 1997, 2001). Recently, Wang (2015, personal communication) considered that the middle part of the Mianxian–Lueyang–Yaolinghe–Dabie–Jiaonan collision zone is through the middle of Wudang block, which is a problem for consideration. South to this boundary, the volcanic zone of north boundary for Yangtze plate (⑧ in Fig. 2.21) and Dabashan foreland thrust zone (⑨ 9 in Fig. 2.21) should belong to the Yangtze plate (Yang et al. 2009).

In Triassic, the paleo-magnetic poles of Sino–Korean and Yangtze plates could not coincide completely, which means after that period the continental crusts still had some rotation and migration. Nevertheless, according to the research about near 30 years, most researchers considered that the Qinling–Dabie–Jiaonan Hida Marginal Triassic zone is a multiple

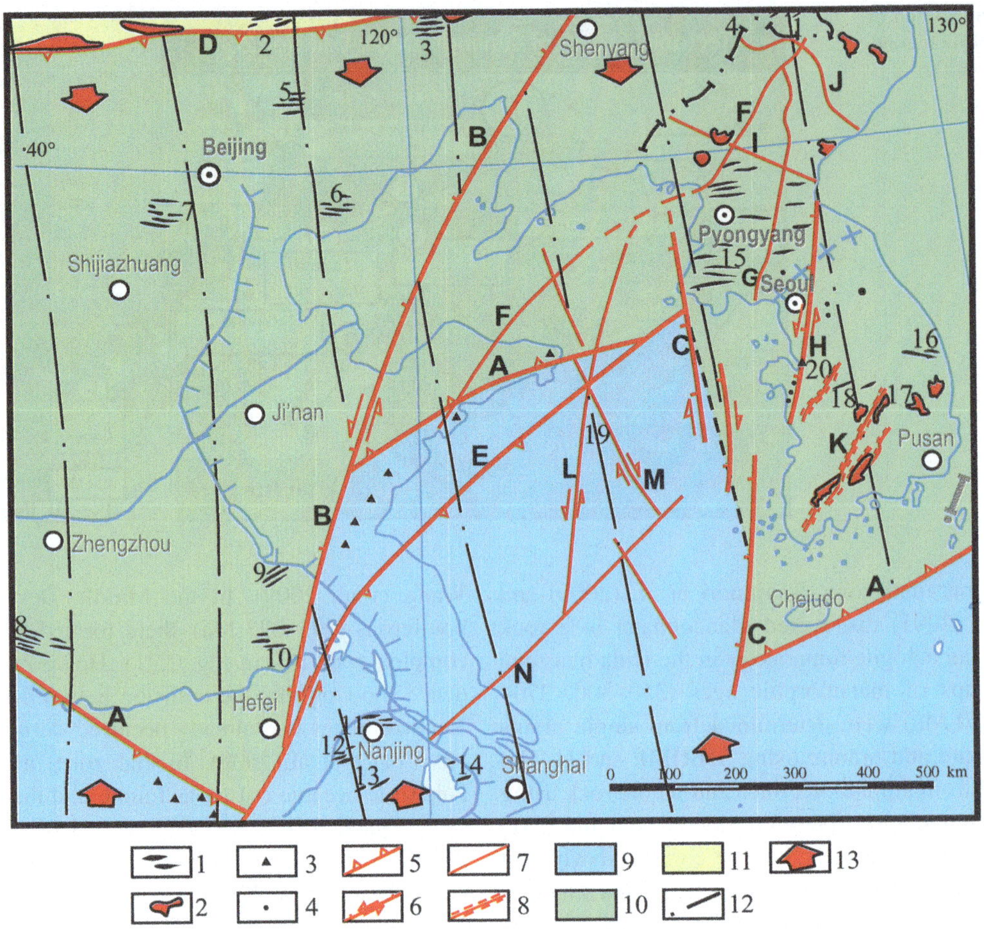

Fig. 2.22 Qinling–Dabie–Jiaonan Triassic collision zone shown in Southern Yellow Sea. **Legend**: (1) folding; (2) granite body; (3) ultra-high-pressure rock and eclogite; (4) small granite body; (5) collision zone or overthrust; (6) normal strike-slip fault; (7) regional fault; (8) ductile shear zone; (9) Yangtze plate; (10) Sino–Korean plate; (11) Hingganling–Tianshan collision zone; (12) maximum principal compression stress trace; (13) crust shortening. The numbers in this Fig. (1–20) are the observation points collecting the paleo-stress data (the original data shown in Wan and Hao 2010). **Fault's names**: (A) Dabie–Jiaonan–Cheju collision zone; (B) Tancheng–Lujiang sinistral strike-slip fault zone; (C) east border of Yellow Sea dextral strike-slip fault zone; (D) north margin of Sino–Korean plate overthrust; (E) Kaesong–Zhangjin fault; (F) Chengcengang–Yantai–Jiaozhouwan fault; (G) Kaesong–Zhangjin fault; (H) Suwon–Wonsan fault; (I) south border of Rangrim block fault; (J) north margin of Rangrim block fault; (K) Honam ductile shear zone; (L) western branch of Southern Yellow Sea central conjugate fault (NNE trending); (M) eastern branch of Southern Yellow Sea central conjugate fault (NNW trending); (N) Jingjiang–Wunansha fault (After Wan and Hao 2010)

accretion–collision zone between the Sino–Korean and Yangtze plates (Figs. 2.18 and 2.22).

Some researchers (such as Huang 1960; Li 1998; Ren et al. 2000) once considered that the above zone extended from Jiaonan of Shandong directly to the middle of Korea Peninsula and recognized that the South Korean Peninsula is a part of the Yangtze plate. However, until now at Imjin River, middle Korean Peninsula, no evidence has been found for any collision event (Zhai 2007). In addition, the north and south sides of the Imjin River area share the same formation age of united crystalline basements (~1600 Ma) and similar Proterozoic, Paleozoic and Mesozoic paleontology, paleogeographic and tectono-magmatic characteristics. But these characteristics are different from those of the Yangtze plate.

Thus, the assumption of Huang (1960), Li (1998) and Ren et al. (2000) is not credible (Wan and Zeng 2002; Zhai et al. 2007). In recent years, at Southern Korean Peninsula, some Neoproterozoic isotopic ages (1000–850 Ma) in granites have been obtained (Li et al. 2003), supporting many researchers' assumption that Southern Korean Peninsula belongs to the Yangtze plate. However, on the northern border of Sino–Korean plate, many Neoproterozoic tectonic-thermal events have been discovered (Wu et al. 1998; Gen et al. 2002, 2007), which were resulted by the magmatism related with plate boundary extension instead of crystalline basement formation.

At the late period of Triassic, the Dabie and Jiaonan collision zone was cut off by the Tancheng–Lujiang sinistral strike-slip fault zone, migrating northward for over 300 km and migrated southward for over 300 km by the dextral strike-slip fault zone of the eastern border of Yellow Sea, thus making the collision zone extend near the Cheju Island and then to the Hida Marginal zone (Tsujimori et al. 2000; Kunugiza et al. 2001; Fig. 2.18). In recent years, at the western seacoast of the Southern Korean Peninsula, NNE trending eclogite residue schist has been found. This might be high-pressure metamorphic schist in the collision zone dislocated to this place by the nearly N–S trending dextral strike-slip fault zone of the eastern border of Yellow Sea (Oh et al. 2005, 2006; Zhai et al. 2007; Wan and Hao 2010; Chang 2015). So it cannot be said that all the Southern Korean Peninsula belongs to the collision zone or the Yangtze plate.

Although the collision between the Sino–Korean and Yangtze plates was completed in the Triassic, it is not to say that the collision zone was stable without deformation at all. In the Jurassic, the East China continental crust rotated (Figs. 3.21 and 3.23), and the above zone suffered extension of the WNW trending faults or joints influenced by the WNW trending maximum principal compression stress, forming WNW trending strike-slip faults and secondary tectonic deformations, as well as a lot of ore deposits. In the Cretaceous, the intraplate deformations were still rather strong with the extension of the NNE trending faults, forming many WNW trending overthrusts, accompanied with large-scale granitic intrusions (Fig. 3.25). The period of Jurassic and Cretaceous is an important intraplate metallogenic period in East Asia.

Based on the deep seismic exploration data in Western Qinling and its two sides, Gao et al. (2011) discovered that the Moho discontinuity of the recent Western Qinling and its nearby areas is almost rather flat with a depth of 30–40 km. It shows that the detachments occurred to form the ramp thrusts in the crust, i.e., duplex structure, but the lithosphere and the mountain root are not discovered in recent.

2.4.4 Shaoxing–Shiwandashan Triassic Collision Zone (250–237 Ma) [25]

The Shaoxing–Shiwandashan Triassic collision zone is a collision zone between the Yangtze [22] and Cathaysian [26] plates (Fig. 2.18; SS in Figs. 3.5 and 3.17), which were formed in the end of Middle Triassic (237 Ma). Because along that zone, the outcrops are rare, and the area is covered with many Cretaceous fault-depression sedimentary basins, thus till now no ophiolite suite has been found. As a result, some researchers thought that there is not a collision zone in there. Also, the characteristics of paleontological geography for the two sides of that collision zone are almost same, and thus, Shu et al. (2008) considered that the Yangtze and Cathaysian plates are the same one, and called it "South China Plate."

However, based on the differences in the formation periods of crystalline basements, in the strata contact relationships between the Upper and Lower Paleozoic, in the fold axes trending in the Early Paleozoic, in the paleo-magnetism, and in the characteristics of metamorphism, it shows that the two sides of that zone should belong to different plates and there should be a collision zone in the Shaoxing–Shiwandashan areas (Table 2.1; Wan 2011b). After the collision in the Middle Triassic, the characteristics of sedimentations, rock deformations and magmatism between the Yangtze [22] and Cathaysian [26] plates are rather similar.

According to the data of rock deformation and petrology, many researchers recognized that in Neoproterozoic (1000–800 Ma), collision occurred along this zone (Shui et al. 1986; Zhou et al. 1989; Yang et al. 1994; Zhang et al. 2003) and discovered near that zone granites with high ε_{Nd} and low T_{DM} values, which indicates mantle influence and provides an evidence for the collision in the Neoproterozoic (Gilder et al. 1996; Hong et al. 2002).

Thus, the above recognition could be correct. Zhang et al. (2003) considered that the Yangtze and Cathaysian plates completed the collision and splicing in Neoproterozoic (1000–800 Ma). However, for the different structural characteristics of the two blocks on two sides of the collision zone during Early Paleozoic, the researchers recognized that they were not connected together but separated. They are located not too far away, and thus, it can explain why their characteristics of paleontology geography are rather similar (Table 2.1). In recent years, along that zone some researchers have discovered siliceous layer formed in the bathyal water during the Late Paleozoic (Li 2013, personal communication). Wang et al. (2015) discovered the Shaoxing–Shiwandashan fault with sinistral strike-slip events in Early Devonian, which means that tectonic fault events occurred between Yangtze and Cathaysian plates in Late Paleozoic–Early Devonian (398.4–401.1 Ma).

It seems reasonable that it was until the Middle Triassic the Yangtze and Cathaysian plates completed their collision and convergence and started to possess similar tectonic characteristics (Wan 2011b).

The data of deep geophysical exploration show the existence of the Shaoxing–Shiwandashan collision zone, with a main fault plane dipping to NW from Shaoxing to Zhejiang–Jiangxi belt, and a shallow main fault plane

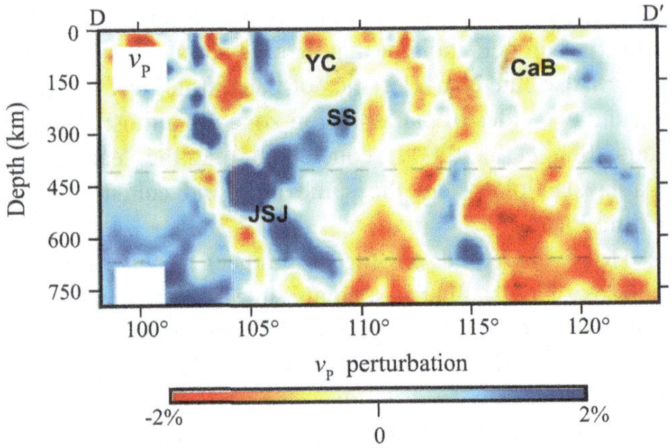

Fig. 2.23 Seismic tomography at 25.8° N in South China. YC. Yangtze plate [22]; CaB. Cathaysian plate [26]; SS. Shaoxing–Shiwandashan collision zone [25]; JSJ. Jinshajiang collision zone [31]. The red area in left-upper corner of the figure is the low-velocity perturbation area. It means that the hot mantle underneath the Emeishan large volcanic province must be originated from the depth of more than 200 km, belonging a "mantle diapir," but not a mantle plume (After Zheng et al. 2012)

dipping to the east near the Jiangxi–Hunan border. However, according to recent seismic tomography data (Fig. 2.23; Zheng et al. 2012), the Shaoxing–Shiwandashan collision zone [25] is a lithosphere fault dipping to the west with an intermediate angle and is penetrated downward to about 450 km, where it was cut off by the east dipping Jinshajiang collision zone [31]. The Jinshajiang collision zone is the boundary between the Yangtze and Gondwana tectonic domains, which could penetrate deep into more than 700 km, i.e., to the middle mantle.

The Jiangnan Neoproterozoic collision zone [23] and Shaoxing–Shiwandashan Triassic collision zone [25] seemed as one orogenic belt (Li et al. 1996, 1998, 2003, 2007). Indeed, the above two collision zones are not far away in Zhejiang and Jiangxi areas and collided in Neoproterozoic; but they are far apart in Hunan and Guangxi areas, with the South Yangtze plate (Table 2.1) between them, of which the tectonic characteristics are extremely different from those of the North Yangtze and Cathaysian plates.

It notes that the research of Shaoxing–Shiwandashan Triassic collision zone does have huge imperfections, due to the failure of discovery of ophiolite suite or high-pressure metamorphic rocks until now, for the zone is covered by many Cretaceous sedimentary fault-depression basins. However, based on the many differences in tectonic characteristics between the two sides of the collision zone (Table 2.1), it is hard to take Su et al. (2009)'s proposal to negate the existence of the Shaoxing–Shiwandashan Triassic collision zone.

2.4.5 Cathaysian Plate (About 400 Ma) [26]

The Cathaysian plate (Fig. 2.18) was at first proposed by Grabau (1940), who called the metamorphic rock systems along the Fujian southeastern coast "Cathaysian Paleo-continent." Huang (1945) continued to use this term and thought they were all Pre-Cambrian systems. However, later based on the isotopic age data, it was found that besides Pre-Cambrian metamorphic rock systems, there were also Triassic and Jurassic ones (Bureau of Geology and Mineral Resources of Fujian Province 1985).

In the regional geological survey, many scattered Archean–Proterozoic paleo-continental blocks were discovered, making the united crystalline basement at the end of Early Paleozoic (about 400 Ma) to form an independent plate —"Cathaysian plate" (Wan 2011b). In recent years, the Cathaysian plate has been thought to include the continental shelf of East Sea, almost whole South China Sea and most of Taiwan Island (westward to the East Taiwan sinistral strike-slip fault, i.e., East Taiwan longitudinal valley) and include the nearby oceanic zones of the Diaoyu Island uplift (Fig. 2.24; Wan 2011b), which east side is the Ryukyu Trench.

Some researchers like to call the above tectonic unit "orogenic belt." The author considers that the formations of a united crystalline basement and an "orogenic belt" are indeed a little bit similar, namely both discussing the continental conversion. In the Pre-Cambrian geological research, when a lot of paleo-continental blocks exist usually and the splicing areas are rather wide, it will be called to form the "united crystalline basement." And the orogenic belt or collision zone should be distributed in a narrow belt. As to the Cathaysian area, we'd better call it "plate." However, the united crystalline basement was formed relatively late, in the end of

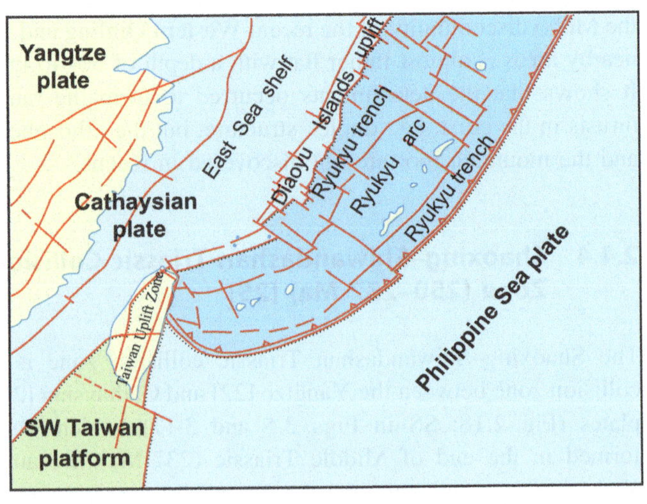

Fig. 2.24 Tectonic division on east side of the Cathaysian plate (After Ying 2012, personal communication; redrawn)

Early Paleozoic, and the erosion depth was rather shallow, so the metamorphic rock systems caused by amalgamation in this block only show the characteristics of low greenstone facies.

The northwest side of Cathaysian plate is the Shaoxing–Shiwandashan collision zone [25] and Yangtze–Southwest Japan plate [22]; east side is the South Honshu–South Shikoku–Ryukyu subduction and island arc zone [62], Okinawa Trough, East Taiwan sinistral strike-slip fault (Neogene–Quaternary) [63] and Philippines–Moluccas Cenozoic subduction and island arc zone [64]; the southern plate is connected with the South China Sea Cenozoic fault-depression basin [28] and the southwest side is adjacent to the Indosinian plate [27]. The east border of Cathaysian plate is the east border of Diaoyu Island uplift and East Taiwan longitudinal valley (Fig. 2.24). The Okinawa Trough is a boundary area, the east of which is the Ryukyu island arc zone [62], now controlled by Japan.

The paleo-magnetism data of Cathaysian plate are not enough (Wan 2011b). Most of researchers recognized that the Cathaysian and Yangtze plates collided in Late Proterozoic. However, in Paleozoic the above two plates were obviously separated (Figs. 3.6–3.8), although both plates were located near the equator with tropical climate and had similar paleontological geographic characteristics. According to available data, in Late Paleozoic the Cathaysian plate was located about 10° S, south to the Yangtze plate (Figs. 3.10–3.12). In Middle Triassic, the Cathaysian plate arrived to the similar latitude to the Yangtze plate, then collision occurred and the Shaoxing–Shiwandashan collision zone was formed [25]. Thus, the Cathaysian plate was merged into the Pangea supercontinent.

From the comparison of a series of Early Paleozoic geological characteristics (Table 2.1), it is clear to consider that the tectonic characteristics during the late period of Early Paleozoic between Yangtze and Cathaysian plates were fundamentally different, so they should be independent plates. The tectonic difference of the North and South Yangtze plates were caused by the different boundary action forces, i.e., they all were mainly affected by the southward tectonic compression (in recent magnetic orientation), to north, the compression action forces decreased, so in the North Yangtze block there were almost no east–west trending folds for the late period of Early Paleozoic and, only showing the disconformity between the Upper and Lower Paleozoic Systems. Since Triassic, collision and amalgamation occurred between plates in South China. Thus, the Sino–Korea, Yangtze and Cathaysian plates began to have homogeneous intraplate tectono-magmatism characteristics, and they all became parts of the Pangea supercontinent.

As to the most areas with intraplate deformations of Sino–Korean and Yangtze plates, the Indosinian period (Triassic) folding was the first most widespread one after the formation of sedimentary cover strata. In the middle and southern Sino–Korean plate, the Indosinian folds involved strata from the Mesoproterozoic to Triassic. In the Northeast Yangtze plate, the Indosinian folds involved strata from Jinjing or Nanhua Periods of Neoproterozoic to Upper Triassic. Among the above strata, the Fentou Formation shale of Silurian, the Longtan coal system of Permian and the gypsum-halite formation of Middle Triassic were easy to form detachments, and led to different folding and structural styles between the upper and lower strata. It means that the Indosinian tectonic event for the Sino–Korean and Yangtze plates both occurred at the end of Late Triassic ($\sim$200 Ma). However, in the Cathaysian and Indosinian plates the Indosinian Tectonic Event took place a little bit earlier, at the end of Middle Triassic ($\sim$237 Ma). In other words, Indosinian Tectonic Event happened earlier in the Southern Asian continent and later in the northern part.

In the southern part of Yangtze plate (such as Guangxi area), the folding with E–W trending were formed in the Early Paleozoic and Triassic Tectonic Events, thus resulted in the Triassic Tectonic Event not too obvious, only showing a little bit unconformity. So some researchers negated the existence of tectonic deformations in the Indosinian Period in Guangxi (Guo 1998). But in the Cathaysian plate between the Early and Late Paleozoic, Middle and Late Triassic Systems, the unconformity are shown distinctly, the folds of Early Paleozoic and older systems are orientated to N–S trending, and from the Late Paleozoic to Middle Triassic Systems are orientated to E–W trending. There are more than thousands folds shown very clearly on the regional geological maps.

In the Cathaysian plate (including East Sea and Northern South China Sea), the characteristics of Mesozoic–Cenozoic intraplate deformations are similar to those of the East Asia continent and China continent. The folding of Indosinian Tectonic Events involved the Paleozoic and Lower and Middle Triassic Systems. After Jurassic, the Cathaysian plate underwent the formation of lithosphere with continental crust and oceanic mantle, strong tectono-magmatism and metallogenesis. In the Cathaysian plate, there are large tungsten hydrothermal ore deposits, while in the South Yangtze plate, are large tin hydrothermal ore deposits. In the Indosinian plate, Malay Peninsula and North Sumatra, large tin deposits with similar characteristics to the South Yangtze plate have been discovered. The metallogenic difference between the Cathaysian and South Yangtze plates might be due to different element enrichment types in the original particles at the plate formation periods. It seems that the

South Yangtze, Indosinian, Eastern Malay Peninsula and North Sumatra blocks might belong to same plate, back to Neoproterozoic (about 800 Ma).

2.4.6 Eastern Hindukush–Northern Qiangtang–Indosinian Plate (~850 Ma) [27]

The northern and eastern boundaries of Eastern Hindukush–Northern Qiangtang–Indosinian plate (Fig. 2.18) are connected with the Jinshajiang–Red River Triassic collision zone [31], its southern boundary connected with the Shuanghu Triassic collision zone [32] and its western boundary connected with the Changning–Menglian–Chiangrai–Central Malaya Triassic collision zone [33] (Li 1997, 2006; Li et al. 2010). Several years ago, it was recognized that the Karakoram, Kunlun and Northern Qiangtang belonged to Gondwana in the Paleozoic (Metcalfe, 1991). According to recent paleontological data, the above recognition may not be proper.

The crystalline basement formation period of Eastern Hindukush–Northern Qiangtang plate might be similar to that of the Yangtze plate formed in Neoproterozoic ($\sim$850 Ma). Due to coverage of rather thick Late Paleozoic–Mesozoic sedimentary systems on the surface, till now there has not been found any outcrop of crystalline basement and its accurate isotopic age. However, this plate has similar Late Paleozoic paleontological geographic characteristics of the Yangtze plate [22] including the growth of warm water fauna mixed with a few cold water fauna molecules (Wan 2011b). In the Mesozoic–Cenozoic, there formed a series of fault-depression basins in this plate. The Indosinian block has a crystalline basement formation period of Neoproterozoic ($\sim$850 Ma) evidenced by isotopic age data.

The Eastern Hindukush–Northern Qiangtang plate extends southeastward to the Changdu area and Eastern Xizang (Tibet), which is the block between Jinshajiang, Shuanghu and Changning–Menglian collisions and continues to extend southward to the Lanping–Simao of Yunnan and further southward to the Indochina block. The Eastern Hindukush–Northern Qiangtang plate and Indosinian plate should be a complete plate in earlier period with roughly E–W trending. Due to the westward compression of the Pacific plate and northward compression of the Indian plate in Paleogene, the above plates changed to a fold-line distribution. Unfortunately, many researchers lost sight of the strong long-distance effect of westward compression of the Pacific plate. In Paleogene, the Indian plate subducted northward, but the deformations for the Asian continent were not so strong.

When the Eastern Hindukush– Karakoram–Northern Qiangtang plate extends southward across the Lanping–Simao of Yunnan, it arrives the Indosinian plate (Fig. 2.25), which includes most areas of the Indochina Peninsula. The NE side of the Indosinian plate connects with the Yangtze plate [22] with the Jinshajiang–Red River Triassic (252–201 Ma) collision zone [31] as the boundary, and its southwest border connects with the Southern Qiangtang–Sibumasu plate [34] with Shuanghu [32] and Changning–Menglian–Chiangrai–Central Malaya Triassic collision zone [33] as the boundary. Its east side is joined to the South China Sea Cenozoic fault-depression basin [28], and the south border is connected to the Sunda plate [51].

The lithosphere thickness of the main part of the Indosinian plate is similar to that of the East China, which is about 80 km (Cai et al. 2002), and it is inferred to be lithosphere with the continental crust and oceanic mantle.

Pubellier (2008) considered the Sunda [51] and Indosinian plates as the same ones. However, recent data showed that the crystalline basement of the Sunda plate was formed at about 500 Ma (Hall et al. 2011), which was possibly related to the Pan-African Tectonic Event, but the Indosinian plate was formed at about 850 Ma (Lan et al. 2003). Thus in this monograph, the point of view of Pubellier is not employed (2008).

There are also metal deposits formed in the Indosinian plate, which are similar to the Yangtze plate. In Paleozoic, Early and Middle Triassic, the Indosinian and Yangtze plates had similar paleontological geographic characteristics for they were both located in the Tethys Ocean (Figs. 3.6–3.8, 3.10–3.12 and 3.15). However, the Southern Qiangtang–Sibumasu plate [34] had the paleontological geographic characteristics of Gondwana (Figs. 3.6–3.8). During the Indosinian Tectonic Event at the end of Middle Triassic ($\sim$237 Ma), the Indosinian plate, Yangtze and Southern Qiangtang–Sibumasu plates were merged into Eurasian continent (Wan 2011b). Since Jurassic, the intraplate deformation characteristics and tectono-magmatism of the Indosinian plate were rather similar to those of the East Asia (detailed in Part 3).

The Song Ma Suture exists in the Northeast Indosinian plate, and its northern side is the Dien Bien Phu fault (Fig. 2.25). These two faults (suture) were the boundary accretion parts of Indosinian plate in Early Paleozoic, in which the Ordovician–Silurian ophiolite suites were formed. It means that the Tuojiang block (north to the Song Da Suture) and area west to the Dien Bien Phu fault were amalgamated to the Indosinian plate in Early Paleozoic (the author got the recognitions from participation in the field excursion of IGCP Project 321).

In the West Indosinian plate, there are a series of Permian ophiolite suites (Fig. 2.25), from north to south distributing over the Jinghong area in China, Nan and Sra Kaco areas in Thailand. In the late period of last century, some researchers thought those were the west border of Indosinian plate. But west to the above boundary, i.e., in

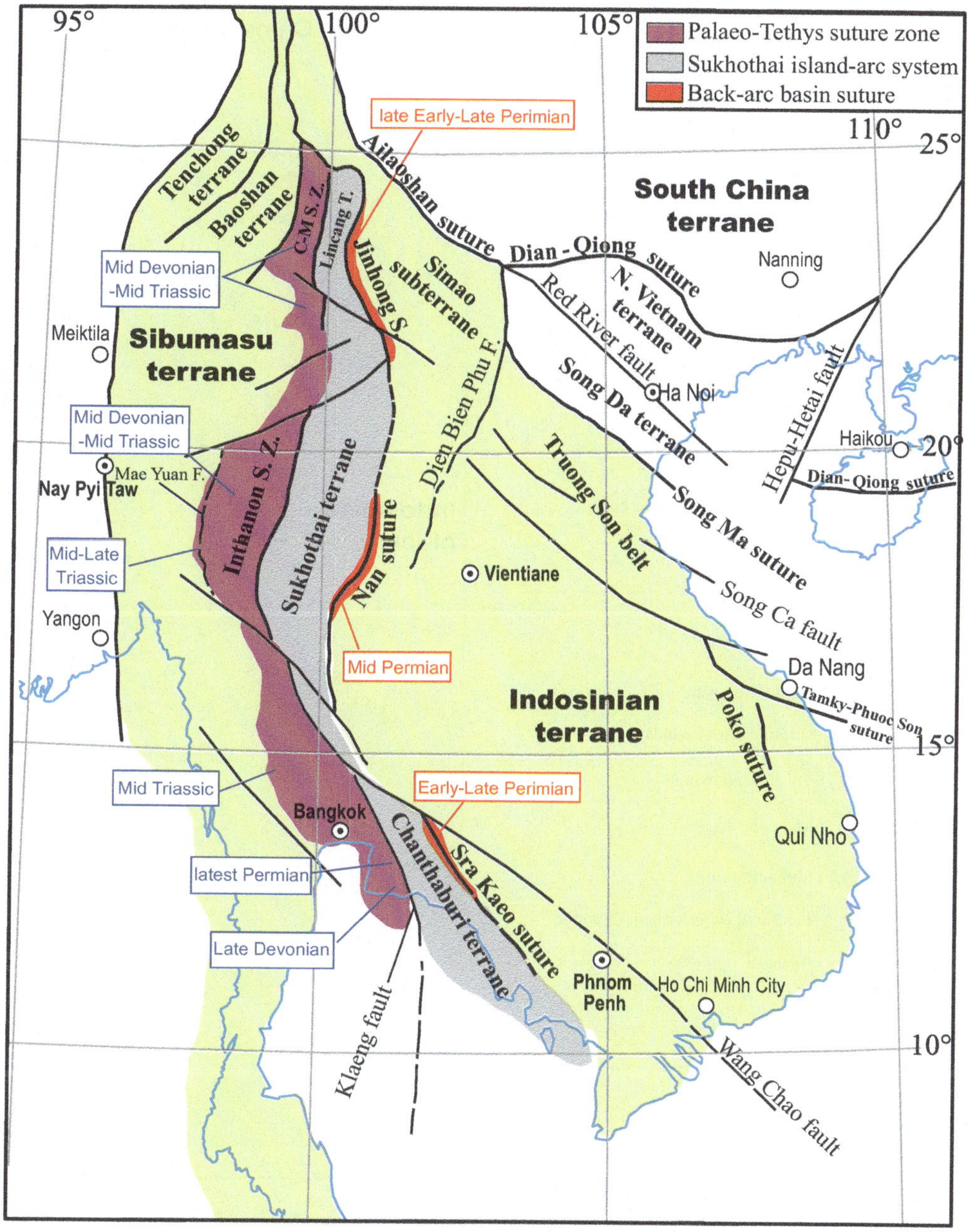

Fig. 2.25 Tectonic sketch of Indochina Peninsula (Modified from Sone and Metcalfe 2008)

the Khun Tan block in Middle Thailand (Fig. 2.25), there are the sylvite deposits formed in the hot dry climate, which is extremely different to the characteristics of Gondwana. So the author and many scholars all considered that the collision zone between the Gondwana and Yangtze–Indosinian plates is the Changning– Menglian–Chiangrai–Central Malaya Triassic collision zone [33] (Figs. 2.18, 2.25, 3.16 and 4.83; Wan 2011b). The Jinghong, Nan and Sra Kaco ophiolite suites are the Permian collision zone between the Indosinian and Khun Tan–Linchang blocks (Fig. 2.25), which also could be believed as the result of accretion of Indosinian plate in Permian.

2.4.7 South China Sea Cenozoic Fault-Depression Basin [28]

To the north, the South China Sea Cenozoic fault-depression basin (Figs. 2.18 and 2.26) is connected to the Cathaysian plate [26], and till now between them no lithosphere fault has been found, but submarine landslides are obviously developed at the continental slope. To the west, it is jointed with the Indosinian plate [27], with north–south trending faults cutting off between them. The faults are probably lithosphere faults (still lacking deep exploration data) for they are connected with the Red River lithosphere fault. In

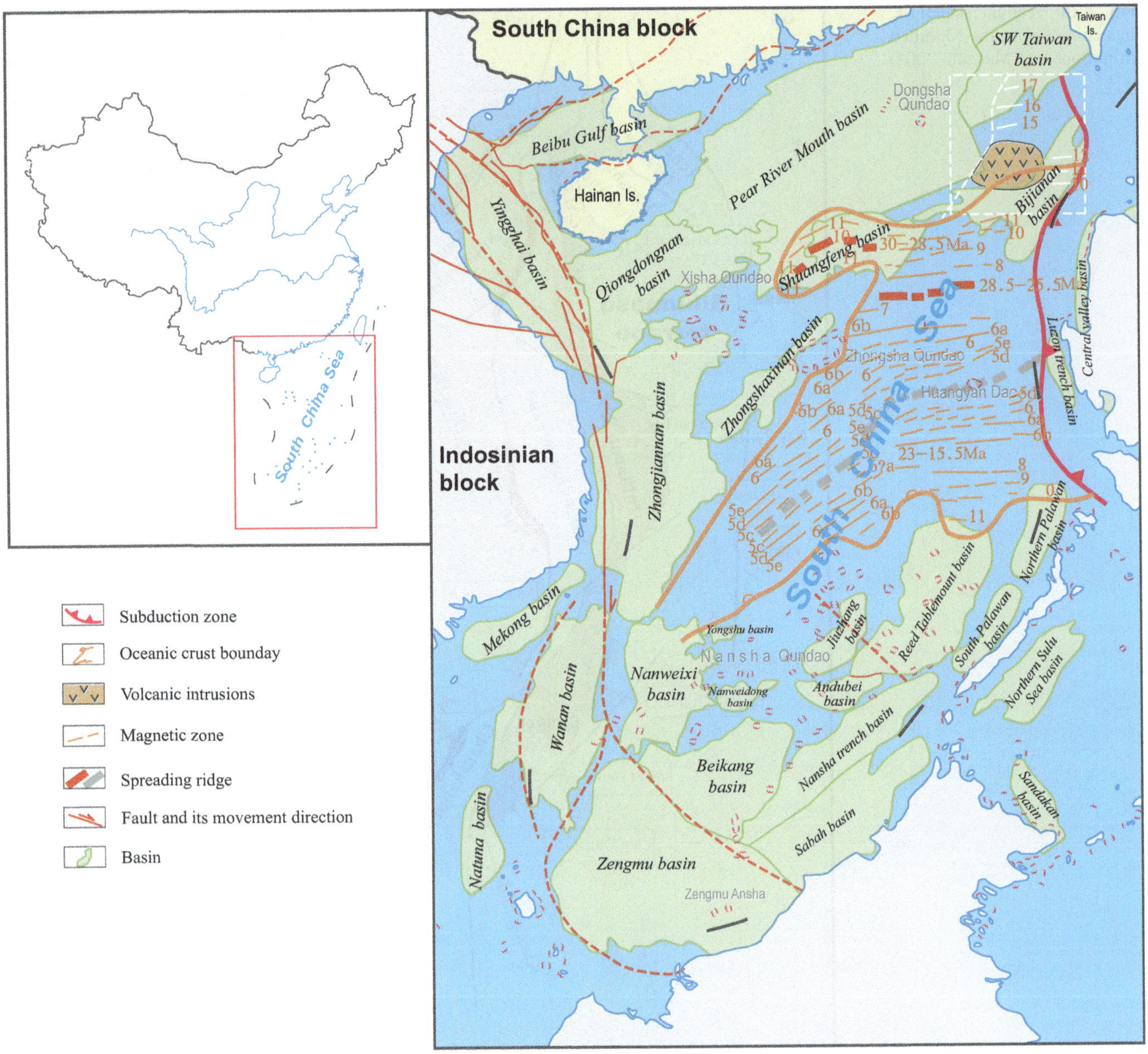

Fig. 2.26 Recent extension of South China Sea basin (Modified from Sun et al. 2009)

the south part of this basin, it is connected with the Palawan–Sarawak–Zengmuansha block [29], and in the east border, it is jointed with the Philippines–Moluccas Cenozoic subduction and island arc zone (showed the deep sea trench) [64].

It is obviously that the boundary between deep and shallow water sea, the U-shaped boundary between deep and shallow blue color, which is nearly to the border of China and surrounding countries, that was formed in the history, due to the deep water sea had the big wave, in ancient time the smaller boats of surrounding countries were not easy to go to those areas. So the U-shaped boundary of South China Sea was formed in the history for a long time (Fig. 2.26). The eastern boundary of South China Sea block is a subduction zone to the eastward and formed a deep trench—West Luson Trench (with more than 3000 m depth), the Western Philippines–Moluccas Cenozoic subduction and island arc zone.

In Paleogene, the Philippine Sea plate [65] subducted and compressed westward, thus making the eastern part of South China Sea Cenozoic fault-depression basin suffer nearly N–S trending extension to form the E–W trending oceanic rift basin. In the late period of Paleogene (33–23 Ma, between magnetic anomalies 11 and 7), the oceanic basin extended with a velocity of 5 cm/year (Briais et al. 1993). However in Neogene, due to the SW trending compression of western Philippine Sea plate [65] (Hall et al. 1995, 2011, Fig. 2.26), the NW–SE trending extension and rift occurred in the southwest part of South China Sea, thus forming the NE trending wedge-shaped fault-depression oceanic basin. In recent years, according to many geophysical and oceanic surveys, it has been recognized that the above conclusion is rather correct (Fig. 2.26; Sun et al. 2009; Wan 2011b). Why is the west side shallow sea of South China Sea (i.e., the Vietnam continental shelf, Fig. 2.26) so narrow? Because there is a large N–S trending fault cutting off the crust, which fault is a continuing part of the Red River lithosphere fault; when that fault turned to the N–S trending, its east side could be deeply subsided to form the very narrow continental shelf. It is the natural result of tectonic action.

As to the formation of South China Sea, the author's recognition was very different to that of Tapponnier et al. (1986, 1990) that had rather great influence in the academics. They thought that the South China Sea Cenozoic fault-depression basin was formed by northward compression of the Indian plate [40], leading to the southeastward escape of the Indosinian plate [27] and South China area. They ignored the influences of Philippine Sea plate [65] westward or southwestward subduction and compression completely and also did not notice the relatively weak influence of northeastward subduction and compression of the Australian plate [72]. However, the NE trending wedge-shaped fault and the NW–SE trending extension of the southwestern part of South China Sea are hard to be explained by the southeastward escape of the Indosinian plate [27]. It suggests that it should open our minds to take the northward compression of the Indian plate as the sole driving force to explain the whole Asian tectonics.

Sun et al. (2009) proposed that the extension center of South China Sea basin had migrated gradually from north to south since Cenozoic (Figs. 2.26 and 2.27). The extension center at 60–50 Ma was located at Sanshui basin, Guangdong Province; that of 50–40 Ma was located at the mouth of Zhujinag basin; that of 39 Ma was located at the oceanic crust of SW Taiwan basin; that of 32 Ma was located at Northeastern South China Sea basin; that of 30–28.5 Ma was located at Northern Xisha Trough basin; that of 28.5–25.5 Ma was located at the Eastern Zhongsha extension ridge in the central basin of South China Sea; that of 23–15.5 Ma was located at southwest secondary basin of South China Sea (Sun et al. 2009).

There are many explanations for the extension center of South China Sea basin gradually migrating southward in the Cenozoic. Hypothesis of Tapponnier et al. (1986, 1990)—the Indian plate compressing northward and the South China Sea being squeezed out, sounds farfetched. Some researchers (such as Sun et al. 2009; Hall et al. 2011) proposed that the Australian plate's gradually northward subduction and compression led to the gradual southward migration of the extension center of South China Sea basin. The trouble is that, if the back-arc extension for the Sunda Trench–arc island zone was used to explain the migration, the extension zone should migrate gradually from south to north, instead of migrating from north to south. What is more, the trending of South China Sea extension zones is not coordinated with the boundary and migration trending of the Australian plate. Therefore, the above explanation seems to be debatable.

In the late period of Cretaceous–Paleocene, the migration orientation of Indian plate was mainly northeast trending, while at 80–70 Ma mainly NE50°–40°, and at 70–45 Ma mainly NE 30°–20°, and then changed to basically north (Lee and Lawver 1995). Thus, the orientation of maximum principal compression stress shows the ENE trending in the Late Cretaceous–Paleocene for East China continent. So it can be inferred that the formation of ENE trending Sanshui Guangdong–Beibu Bay and the mouth of Zhujiang basin may be influenced by the above stress field. At that time, the Red River fault zone showed the sinistral strike-slip characteristics, and the ENE trending faults were extension-shear ones.

The formations of the oceanic crust of Southwest Taiwan basin in the Eocene (about 39 Ma), the Northeastern South China Sea basin at 32 Ma and Northern Xisha Trough basin at 30–28.5 Ma might be influenced by the long-distance effect of ENE trending compression of the Indian plate (Fig. 2.27). In Eocene, beginning from 43 Ma, the Caroline, Samoa and Easter, Foundation Islands all migrated toward WNW direction with a velocity of 7.7 cm/year (Engebretson et al. 1985; Koppers et al. 2001, 2003; Northrup et al. 1995). This migration influenced areas in the South China Sea and Luzon Islands suffered from nearly E–W trending compression, resulting in the formation of some E–W trending eastern basins of South China Sea, west to the Luzon Islands at the end of Oligocene (25–20 Ma) (Fig. 2.27).

In Neogene (after 23 Ma), the Philippine Sea plate [65] underwent nearly N–S trending extension, and the Philippine islands arc [64] subducted with its compression center moving southward gradually (Fig. 2.28). According to the distribution of ocean floor magnetic orientations and the paleo-magnetic data in adjacent blocks, Hall et al. (1995, 2011) inferred that the Philippine Sea plate experienced gradual accretion and some clockwise rotation at 50–5 Ma. When the Western Philippine Sea plate compressed northwestward in Neocene, it underwent NNE–SSW trending extension. Thus, the SW side of Philippine Sea plate suffered from SW trending compression, and the Philippine (Luzon) islands arc and trench experienced dextral strike-slip oblique compression (Figs. 2.27 and 2.28), leading to the gradual southward migration with the strongest tectonic compression. As a result, the Neogene South China Sea basin changed to nearly NE orientation. As for the area between mouth of Red River and west to the Hainan Island, i.e., in the Beibu Bay, due to the Red River fault zone's changing to be dextral strike-slip in the Neogene, the Red River–

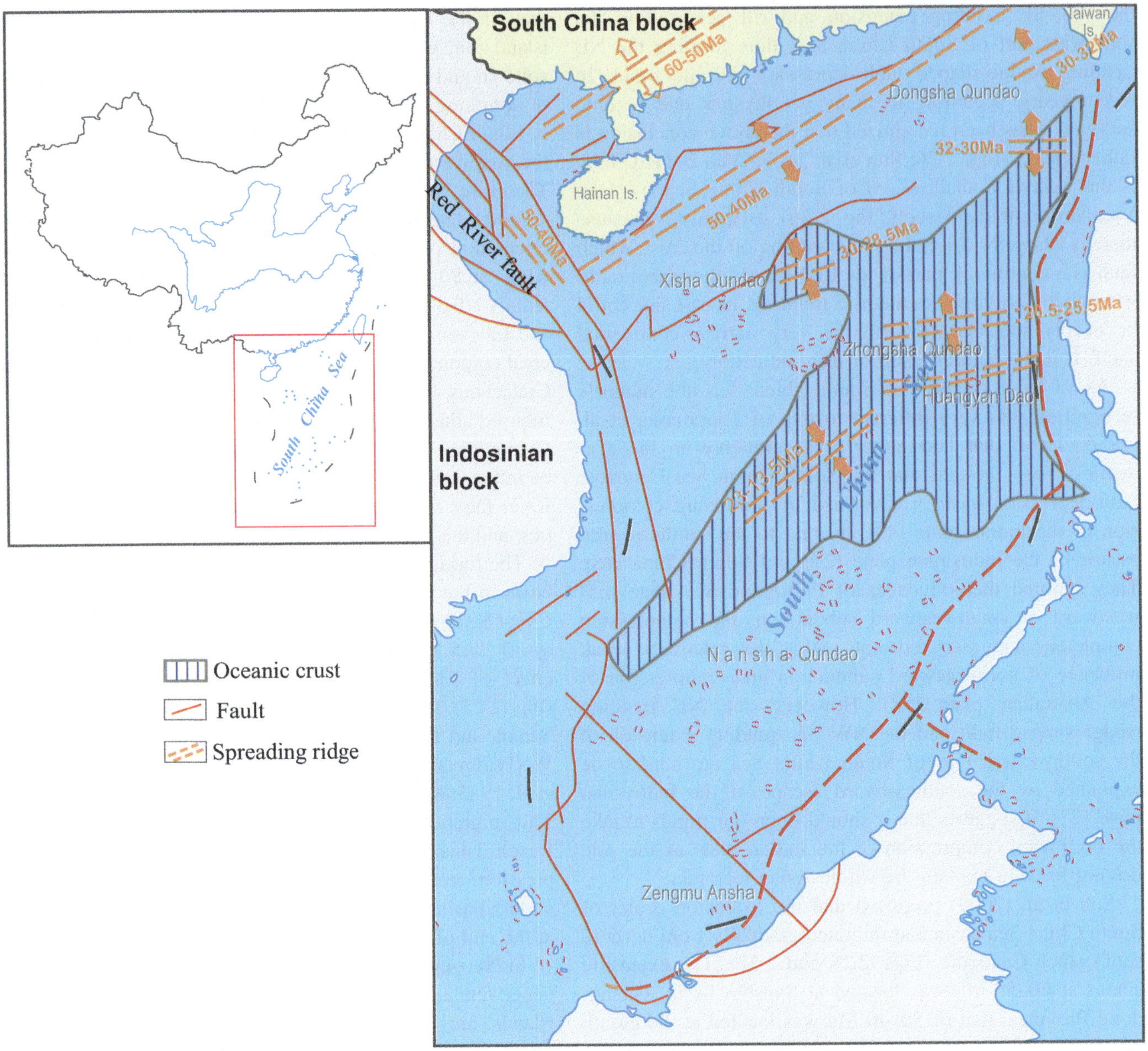

Fig. 2.27 Southward migration of extension center of South China Sea in Cenozoic (Modified from Sun et al. 2009)

Yinggehai fault-depression basin was derived. According to the regional tectonic research, the Red River fault zone was sinistral strike-slip in Paleogene, but changed to be dextral strike-slip in Neogene.

When taking a closer view of the ocean floor extension areas of the South China Sea, Sulu Sea, Sulawesi Sea and Banda Sea (purple areas in Fig. 2.29) since Neogene, they all show the wider east parts and narrower west parts. It might indicate that the formation of these ocean basins was affected by stronger compression from the east side and weaker compression from the western side. Thus, the author considers that the WSW trending compressions of the Philippine Sea plate in the Philippine island arc zone might be the main dynamics to form the above small oceanic basins with wider eastern side and narrower western side. The above deduction may be relatively reasonable and can explain the mechanism of southward migration of the South China Sea basin extension center and the change in the oceanic basin extension orientation.

Thus, under the control of the E–W trending compression, the middle of South China Sea basin had extended to become oceanic crust since Neogene, with its north and south boundaries, i.e., the ocean and continent transition belts, showing the nearly N–S extension models. The north

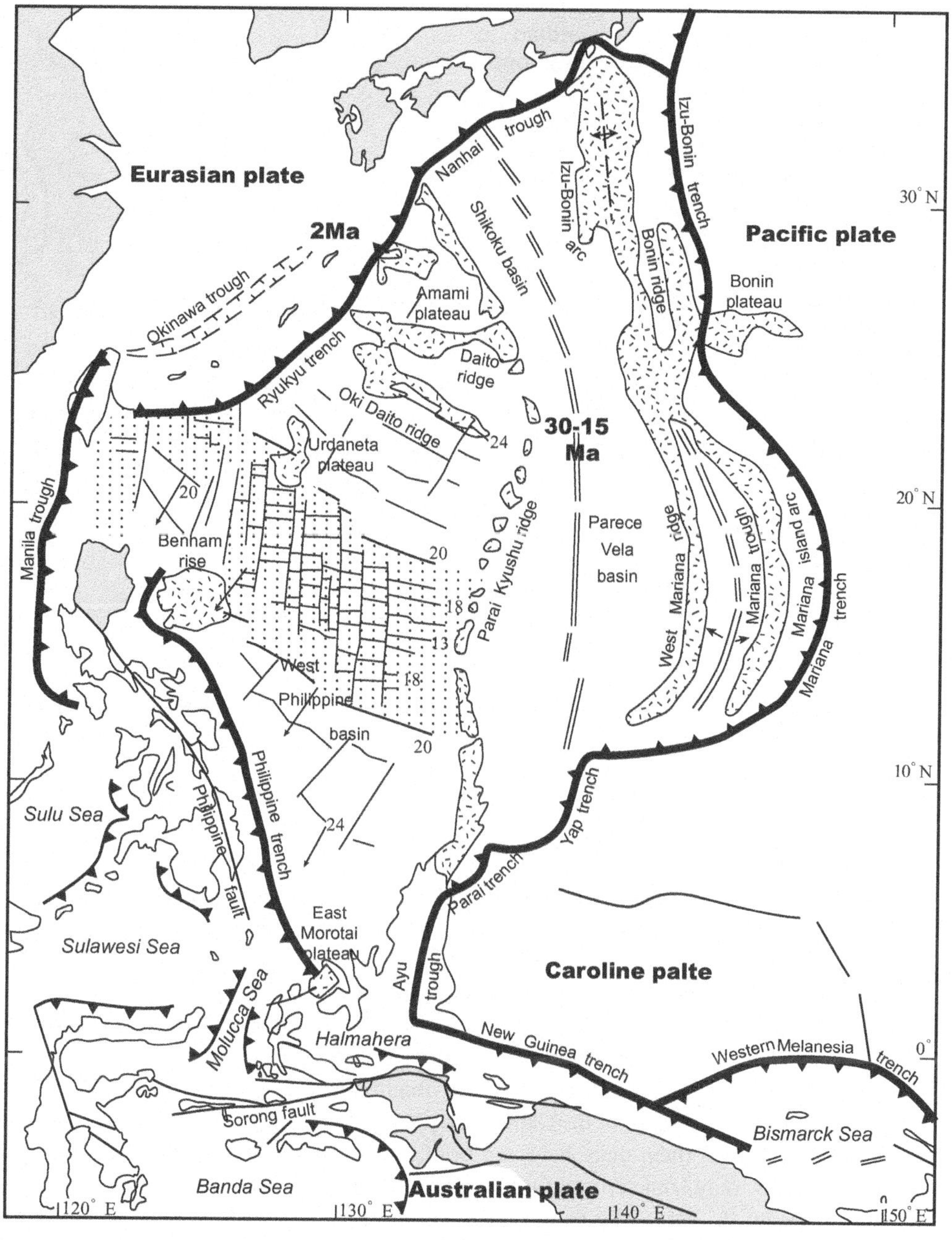

Fig. 2.28 Tectonic outlines of Philippine Sea plate. The arrow shows the SSW trending compression by the Philippine island arc in Neogene (After Hall et al. 2011)

limb of the South China Sea basin is separated with the middle oceanic crust with a listric fault zone as their boundary and developed fault-volcanic zone and fault blocks dipping to the ocean, as well as high-velocity layer in the lower crust; the south limb is characterized by dip fault blocks, with the listric faults (Gao et al. 2015).

The recognition that the South China Sea is a back-arc basin, proposed by Hsu (1981, 1988), has been negated by many facts. In Oligocene (32–20 Ma), the South China Sea underwent nearly N–S trending extension, and the E–W distribution extension zone was vertical to that of the Western Philippine Trench nearly N–S. According to Hsu (1981, 1988), the trending of trench–arc and back-arc basin should be almost parallel. In addition, the basalts on the oceanic floor of South China Sea are not the oceanic tholeiite basalts, but continental alkali basalts and transition-type basalts (Tayler et al. 1980, 1983; Briais et al. 1993). The South Sea Institute of China Academy of Science (Liu et al. 1988) considered that the South China Sea is not a back-arc basin, but a continental boundary extension basin. Hsu (1981, 1988) thought that the South China Sea and Japan Sea are the most typical back-arc basins on the Earth. If

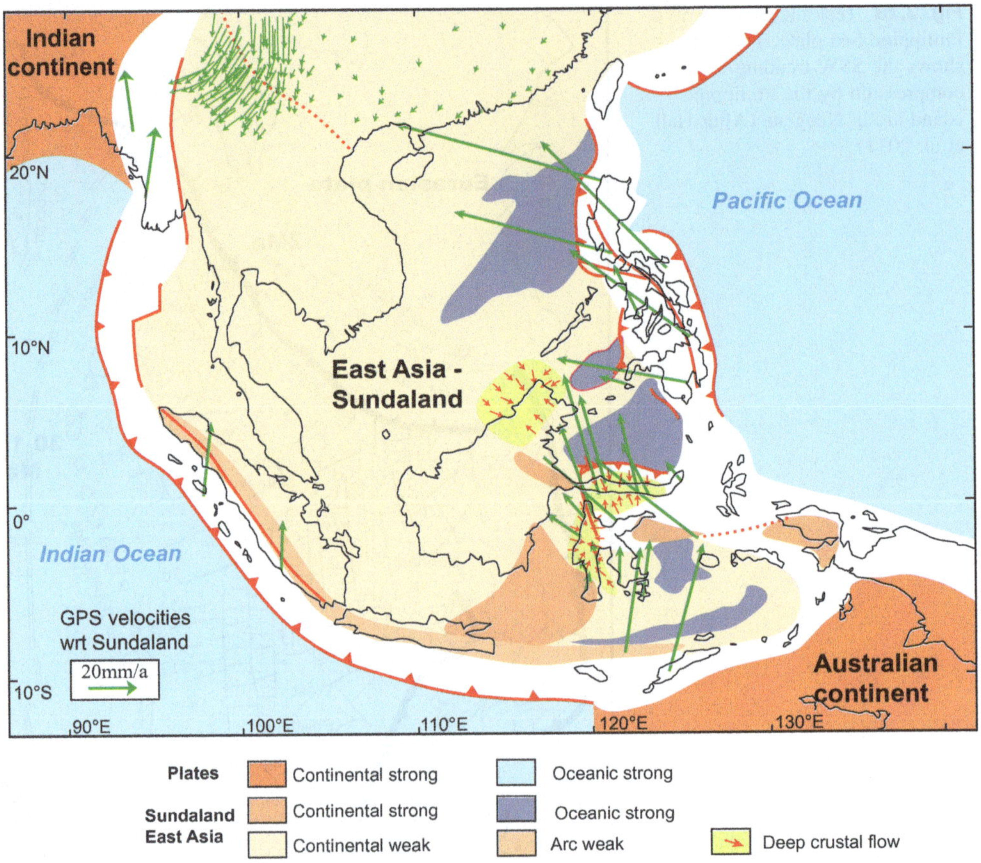

Fig. 2.29 Neogene and recent deformation and oceanic extension in Southeast Asia. The purple color areas show the partial oceanic basin formed by continental boundary extension in Neogene. They all show wider east part and narrower west part, which means the dynamics of forming the oceanic basins came from westward moving of Philippine Sea plate. The green arrows show the recent migration orientation of upper crust, measured by GPS. The small red arrows show the flow orientation in the deep crust measured by the rock heterogeneity (After Hall et al. 2011)

these two seas are not back-arc basins, it will be hard to consider that "the back-arc basins are the inevitable derivations of the plate subduction."

Does the South China Sea fault-depression basin belong to the Cathaysian plate or the Indosinian plate (Chen et al. 2010)? Depending on the inseparable connection with the Cathaysian plate, the South China Sea basin might have similar characteristics to those of the Cathaysian plate, and in the boundary between them there have not been found any lithosphere faults, only some crustal fault due to lower crust rheology (Wu et al. 2014). As to the west side of South China Sea, there is a nearly N–S trending fault, which is the southern extension of Jinshajiang–Red River lithosphere fault zone (Figs. 2.26 and 2.27), cutting off the Indosinian plate and South China Sea basin. So the author infers that the South China Sea basin should have originally belonged to the Cathaysian plate. Later in Cenozoic, influenced by multiple plate compressions, many extension fault systems occurred in the lithosphere of South China Sea, causing upwelling and eruption of the basalts with relatively high density. Thus, the lithosphere density of the South China Sea area increased, leading its divergence and depression into sea basins and oceanic basins.

2.4.8 Palawan–Sarawak–Zengmuansha Block [29]

The southeast side of the Palawan–Sarawak–Zengmuansha block is the Eastern Kalimantan–Sulu Sea Cretaceous accretion–collision zone [52], and the north side of this block is connected with the South China Sea fault-depression basin [28] (Fig. 2.18). The Palawan–Sarawak–Zengmuansha block may be a part of Cathaysian plate [26] previously, but the characteristic data of deep crystalline basement have not been obtained. According to the surface geological characteristics, this block should belong to the Cenozoic intraplate deformation area. Due to the Cenozoic fault-depression of South China Sea, this block was separated from the Cathaysian plate and later influenced by the northward subduction and compression of Australian plate, then rather strong deformations occurred in this area. So that part of this area uplifted to mountains and in the southern boundary faults developed ophiolite suite (Hall et al. 2011).

2.4.9 Western Hindukush–Pamir–Kunlun Late Paleozoic–Triassic Accretion–Collision Zone (360–200 Ma) [30]

This collision accretion zone is among the Turan–Karakum plate [8], Tarim block [20], Qaidam block [18] and Shuanghu Triassic collision zone [32] (Fig. 2.18), and the collision occurred in Late Paleozoic and Triassic (about 300–200 Ma) (Jin et al. 1999; Fan et al. 2009; Qi 2013).

This zone extends northward to the famous NW trending Fergana dextral strike-slip fault. Between the Turan–Karakum plate [8] and Tarim block [20], the Fergana fault zone is SW dipping on the surface, but in the depth, the fault plane is NE dipping. The above collision zone underwent relatively weak intraplate deformations after Triassic. Since Neogene, influenced by the nearly N–S trending shortening effect derived from the northward migration and collision of the Indian plate, it was uplifted to high mountains. This zone has bad transportation conditions with a small population, so that the tectonic research is relatively weak.

2.4.10 Jinshajiang–Red River Triassic Collision Zone (252–201 Ma) [31]

The Jinshajiang–Red River Triassic collision zone (Fig. 2.18) is a collision zone between the Yangtze–Southwest Japan plate (including the Hohxil–Bayanhar–Garzi–Aba areas) [22] and Northern Qiangtang–Indosinian plate [27]. The main fault plane at Northern Qiangtang is dipping to the south, near the Jinsha River dipping to the east and near the Red River dipping northeastward. According to the data of seismic tomography, the Jinshajiang–Red River collision zone is inclined eastward with a medium angle to a depth of 600 km, cutting off the Shaoxing–Shiwandashan collision zone[25] at a depth of 450 km (Fig. 2.20; Zheng et al. 2012). Along this collision zone, any large-scale intermediate-acid magmatic intrusion has not been found, and thus, no high mountains were formed. In recent years, Qi et al. (2010) researched the Ailaoshan mylonitic granites on the east side of Red River collision zone and got the zircon LA-ICP-MS U–Pb ages between 250–247 Ma, showing that they were formed at the early periods of Triassic.

Depending on the deep structure, the Jinshajiang collision zone has the more important division significance, which means it is thc boundary between the Gondwana and Eurasia continental blocks. Whereas near the Lancangjiang, till now only the crustal faults have been found without any lithosphere fault. However, according to the paleontological paleogeography data on the surface, the boundary between the Gondwana and Eurasian continental blocks in Late Paleozoic should be located at the Shuanghu [32] (Li et al. 1997, 2006) and Changning–Menglian–Chiangrai–Central Malaya Triassic collision zones (252–201 Ma) [33] (Liu et al. 1993), i.e., near the Lancangjiang. It seems that the boundaries of the above two blocks are different in the crust surface and depth.

In Eocene and Oligocene, the southern part of Jinshajiang–Red River zone was influenced by the westward compression and migration of the Pacific [69], Philippine Sea [65] and Yangtze [22] plates, and the fault zone was a sinistral strike-slip one with displacement of about 400–500 km (Zhong 1998). However since Neogene, due to the northward collision and compression of Indian plate [40], that fault zone obviously showed the dextral strike-slip characteristics. Tapponnier et al. (1986, 1990) thought that since Paleogene, the Red River fault zone had been the result of southeastward escape of the Indosinian plate which was derived by the northward collision and compression of Indian plate [40]. It is obvious that their opinion is not consistent with facts. In Paleogene, the Indosinian plate migrated southeastward and the Jinshajiang–Red River zone showed sinistral strike-slip characteristics, it seems passable at this time to say that the Indosinian plate was escaping southeastward. But in Neogene, when the Jinshajiang–Red River zone changed to be dextral strike-slip, the Indosinian plate was migrating northwestward relatively, which means that the Indosinian plate was not escaping, but changing to squeeze into the Asian continent. For a long time, Tapponnier et al. (1986, 1990) have always been using the northward migration and compression of Indian plate to explain all the changes of the Asian continental tectonics since Triassic, which is obviously not appropriate. Because they mainly used remote sensing imagery data, it is rather suitable to explain the active structures or Neogene tectonics, but not suitable for the Paleogene and Mesozoic ones.

The Red River fault zone extends southeastward to the South China Sea and turns to nearly N–S trending faults at the southern part of Hainan Island, which are the west border faults of the South China Sea fault-depression basin [28] (Wan 2011b; Fig. 2.24). This boundary is almost the boundary between the shallow sea (continental shelf) and bathyal zone of South China Sea, which is also very near to the boundary between China and Vietnam in history.

2.5 Gondwana Tectonic Domain

The Gondwana tectonic domain was formed in Early Cambrian, including the most plates or blocks in the Southern Hemisphere, such as Africa, South America, Antarctica, Australia, Indian subcontinent, most areas of Southeast Asia, areas south to Southern Qiangtang of China, Afghanistan, Iran, Turkey and Middle East areas, i.e., the blocks of the original Gondwana. The term of Gondwana was first proposed by E. Suess (1831–1914, an Austrian geologist) in *The Face of the Earth* published in 1885.

In Asia, the Gondwana tectonic domain includes the majority of blocks and collision zones [32–55] in the southern part of Asia (Fig. 2.30): the Shuanghu Triassic collision zone (252–201 Ma) [32], Changning–Menglian–Chiangrai–Central Malaya Triassic collision zone (252–201 Ma) [33], Southern Qiangtang–Sibumasu plate (~510 Ma) [34], Bangongco–Nujiang–Mandalay–Phuket–Northern Barisan Cretaceous collision zone (100–66 Ma) [35], Gangdise plate (~510 Ma) [36], Yarlung Zangbo–Myitkyina Paleogene collision zone [37], Himalayan plate (~510 Ma) [38], Southern Himalayan main boundary thrust (since Neocene) [39], Indian plate(~510 Ma) [40], Kavkaz–Alborz Late Paleozoic–Late Jurassic accretion–collision zone [41], Anatolia–Tehran Middle Cretaceous–Paleocene collision zone (100–56 Ma) [42], Turkey–Iran–Afghanistan plate (~510 Ma) [43], Zagros–Kabul accretion–collision zone (since Cretaceous) [44], Toros accretion–collision zone (since Neocene) [45], Arabian plate (~510 Ma) [46], Oman Cretaceous accretion–collision zone [47], Red Sea rift zone since Neocene [48], Western Burma (Pegu Mountains–Rangoon) plate (~510 Ma) [49], Arakan–Sunda Cenozoic subduction and island arc zone [50], Sunda plate (~500 Ma) [51], Eastern Kalimantan–Southern Sulu Sea Cretaceous accretion–collision zone [52], Sulawesi Sea block (500 Ma) [53], Eastern Argo block (500 Ma) [54] and Northern New Guinea island arc zone (since Neocene) [55].

In the Gondwana tectonic domain, from Archean to Proterozoic there developed many small paleo-continental blocks, and however, almost all the blocks shared united crystalline basements with greenschist system in Late Neoproterozoic–Early Cambrian (600–509 Ma, except the Oman block), which was called "Pan-African Tectonic Event" (Kennedy 1964). They were all formed in the southern part of Southern Hemisphere, later gradually migrating northward and splitting, and underwent the subductions, collisions and amalgamations near the equator (Figs. 3.6–3.8 and 3.10–3.12). The collision occurred mainly since Mesozoic (Figs. 3.15, 3.17, 3.18, 3.21, 3.23 and 3.24; Klootwijk and Radhakrichnamurty 1981; Schettino and Scotese 2005).

The term "Gondwana" came from the Carboniferous to Jurassic Systems located at the Gondwana area of the Central India. Suess (1885) recognized that the Indian and African continents used to be a united continent for they had the same geological evolution and paleo-flora. In Carboniferous, large-scale glaciation occurred in Gondwana which had been confirmed by the tillite found in Africa, South America, Australia and India, as well as small cold water fauna. The paleo-magnetic data showed that the above continents were all near the ancient South Pole (Antarctica), and the massive ice sheet was distributed within the 60° S paleo-latitude. In Permian, the typical flora were the Glossopteris of gymnosperms, which were distributed over the South America, Middle Africa, South Africa, Antarctica, Australia and India, and however, in the northern continents, such as North America, Eurasia and Greenland, those flora fossils are not found. The Gondwana started to break up in Mesozoic and gradually migrated to the recent position during Cenozoic.

2.5.1 Shuanghu Triassic Collision Zone (252–201 Ma) [32]

The Shuanghu Triassic zone is a collision zone between the Eastern Hindukush–Northern Qiangtang–Indosinian plate (~850 Ma) [27] and Southern Qiangtang–Sibumasu plate (~510 Ma) [34] (Fig. 2.30). The Shuanghu collision zone is the boundary between Gondwana cold water fauna and Eurasian warm water fauna in Late Paleozoic, and along this line a series of Triassic (230–210 Ma) ophiolite suites and high-pressure metamorphic zone were developed (Li 1997, 2006). In recent regional geological surveys, it has been discovered that the Shuanghu Triassic collision zone extended southeastward (Li et al. 2010) and in the late period of Cretaceous was cut off by the sinistral strike-slip fault of the Bangongco–Nujiang collision zone [35]. Thus, the Shuanghu Triassic collision zone [32] and the Changning–

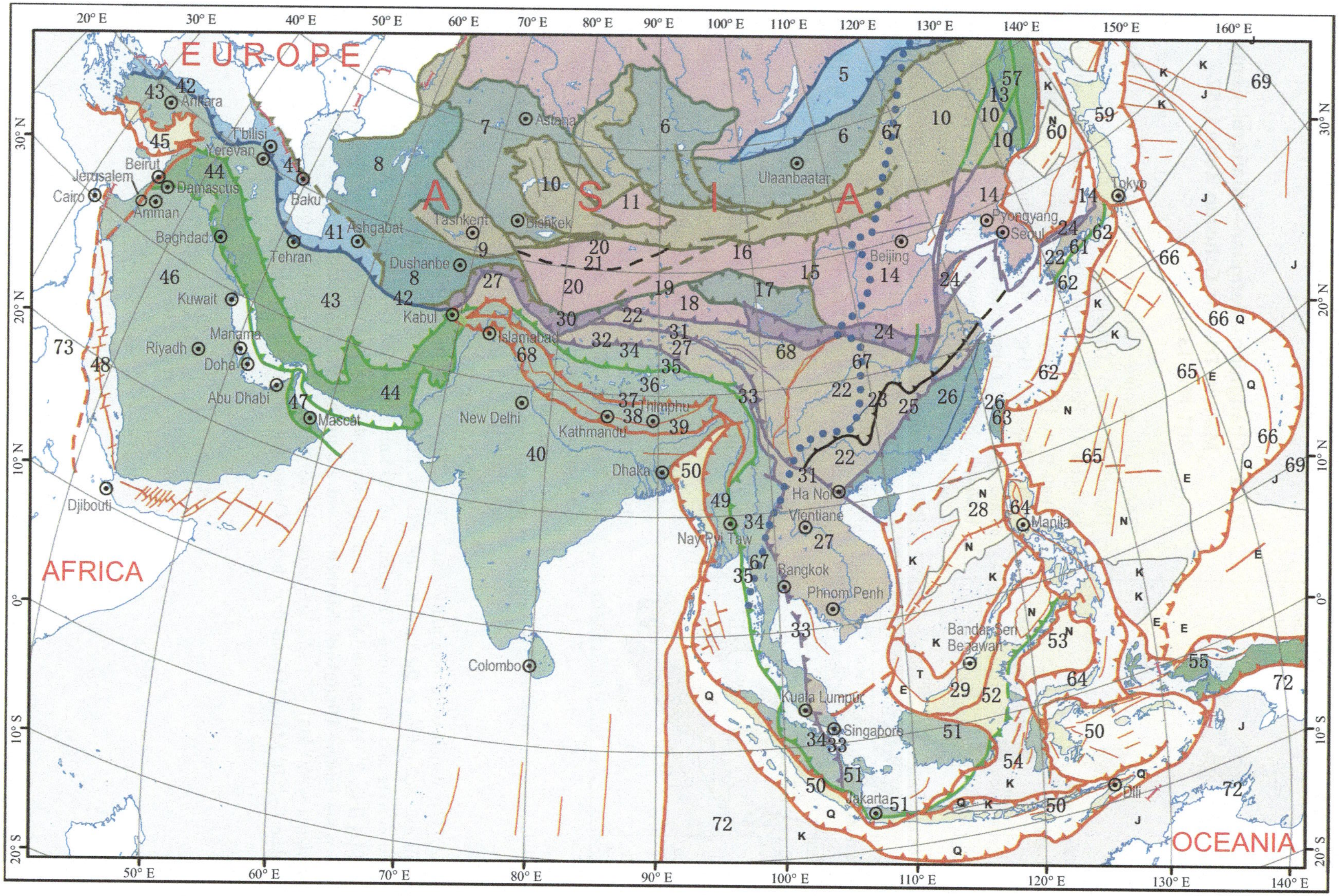

Fig. 2.30 Gondwana tectonic domain [32–55]. The numbers in the figure show the tectonic units, as same as those in the CONTENTS and Fig. 1.1

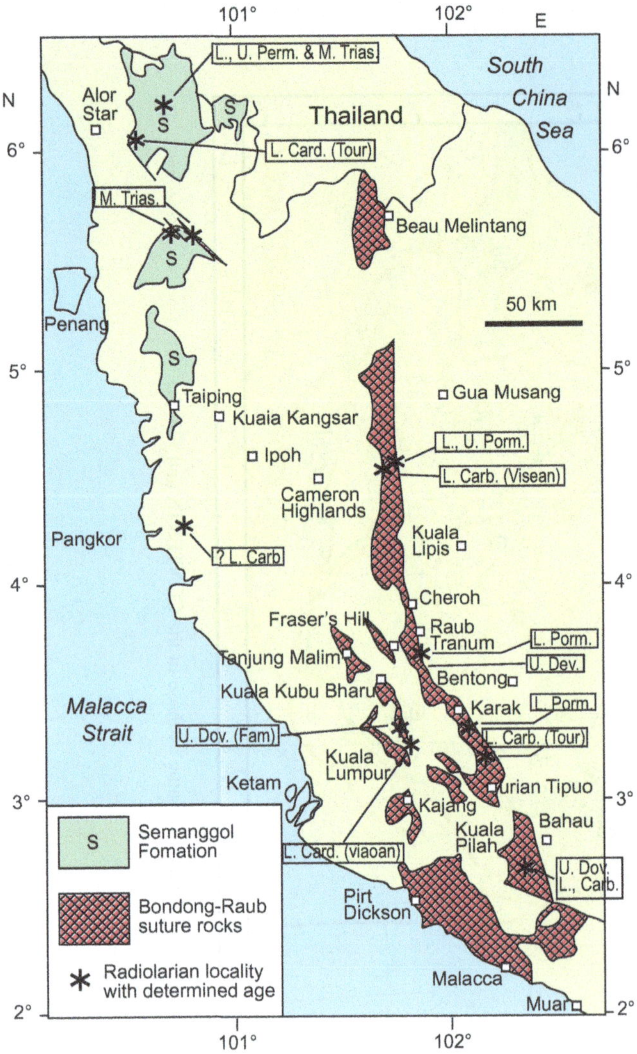

Fig. 2.31 Bondong–Raub collision zone (Modified from Hutchison and Tan 2009)

Menglian collision zone [33], which were originally connected together, were cut off. According to the fact that the Shuanghu Triassic collision zone and Changning–Menglian Triassic collision zone [33] (Fig. 2.31) are both located in a very narrow area between the Eastern Hindukush–Northern Qiangtang–Indosinian plate (~850 Ma) [27] and Southern Qiangtang–Sibumasu plate (~510 Ma) [34] and that they have similar tectonic characteristics and formation time, it could be inferred that they should be connected together in Triassic and were later cut off by the Bangongco–Nujiang collision zone [35] in Cretaceous.

2.5.2 Changning–Menglian–Chiangrai–Central Malaya Triassic Collision Zone (252–201 Ma) [33]

The Changning–Menglian–Chiangrai–Central Malaya (Bondong–Raub) Triassic collision zone is the boundary between the Eastern Hindukush–Northern Qiangtang–Indosinian plate [27] and Southern Qiangtang–Sibumasu plate (~510 Ma) [34] (Fig. 2.30). According to the existence of Late Paleozoic fauna characteristics, it is inferred to the boundary between Gondwana cold water fauna and Eurasia warm water fauna, and along the boundary, a series of Paleozoic ophiolite suites (473–439 Ma, Deng et al. 2014a) and the Triassic (230–210 Ma) high-pressure metamorphic zone are intermittently developed (Liu et al. 1991, 1993); the rock is mixed with the Late Paleozoic Radiolarian siliceous rocks in the Central Malay Peninsula. However, at the westward of this zone, it shows the widespread Late Paleozoic Radiolarian siliceous rocks, cold water fauna and the shale with pebble. They all had the typical Gondwana characteristics. The Semanggol Formation is the main rock system to form the Bondong–Raub suture zone. In the east side region, the warm water fauna are developed, with the Asian continental characteristics (Zhong 1998; Hutchison and Tan 2009; Metcalfe 2011; Deng et al. 2014b). The Changning–Menglian collision zone extends southward across the Chiangrai–Inthanon–Gulf of Siam of Thailand to the Bondong–Raub collision zone of Central Malay Peninsula [33] (Figs. 2.30 and 2.31).

The Triassic Bondong–Raub suture zone (Hutchison and Tan 2009) is a strong rock deformation mélange zone (Figs. 2.31 and 2.32), with a width from several kilometers to 40 km. The strata of western block of this zone show typical Late Paleozoic Gondwana cold water fauna complex and are mainly the shale with moraine pebble and Radiolarian siliceous rocks, which belong to the Southern Qiangtang–Sibumasu plate (~510 Ma) [34]. However on

Fig. 2.32 Mélange of Triassic Bondong–Raub suture zone. The Mélange is composed of many lens-shaped structural sheets, including the shales with moraine pebbles, Radiolarian siliceous rocks, carbonates, shales and sandstones (Photograph taken by Author in 1991)

the east side of that zone, it shows very clear Late Paleozoic warm water fauna and sedimentary characteristics, forming Late Paleozoic carbonate systems with almost vertical attitudes. These carbonate systems are similar to the Qixia and Maokou carbonate systems in the Yangtze plate and belong to the Indosinian plate [27]. However, the Late Triassic coal system (very similar to the Anyuan coal system in the Yangtze plate) overlapped on them as an unconformity (by the recognition for IGCP 224 field excursion in 1991; Hutchison and Tan 2009). The above collision zone can extend southward directly to the Northern Sumatra Island (Metcalfe 1991, 1995).

2.5.3 Southern Qiangtang–Sibumasu Plate (~510 Ma) [34]

The Southern Qiangtang–Sibumasu plate (Fig. 2.30) is a long arc stable block among the Shuanghu Triassic collision zone (252–201 Ma) [32], the Changning–Menglian–Chiangrai–Central Malaya Triassic collision zone (252–201 Ma) [33] and the Bangongco–Nujiang–Mandalay–Phuket–Northern Barisan Cretaceous collision zone (100–66 Ma) [35]. This plate extends southeastward from Southern Qiangtang, and is cut off by the Bangongco–Nujiang sinistral strike-slip fault zone. It continues to extend southward across Baoshan–Gengma areas of Yunnan, Shan State of Burma, mountain areas west to Chiang Mai-Tak of West Thailand, Western Malay Peninsula all the way to the Northern Sumatra. It is a twisted long plate that should originally be nearly east–west trending. However, its eastern part has been transformed to be nearly N–S trending now. The Sibumasu plate was first named by Metcalfe (1991, 1995), who used the first two letters of each of the four area names (Sino, Burma, Malaysia and Sumatra) to call a new plate name, later used by many geologists.

The isotopic age of crystalline basement of this plate is about 510 Ma. It was formed by the Pan-African Tectonic Event (Bureau of Geology and Mineral Resources of Xizang 1993; Wang et al. 2001). It is obvious that it is a part of Gondwana and began to break up from the Gondawa since Middle Cambrian. In Early Paleozoic (502–455 Ma), in Baoshan and Tengchong blocks there existed thermal magmatic events related with subduction, and in the 421–401 Ma period, the Simao block might also undergo thermal magmatic events related to the Paleo-Tethys Ocean subduction (Deng et al. 2014a).

In Late Paleozoic, this plate still had typical characteristics of Gondwana cold water fauna and sedimentary shale with moraine pebble. In Triassic, it was merged into Eurasian continent (Metcalfe 2011; Ridd et al. 2011). Since Cretaceous, suffered from the influences of westward subduction and collision of the Pacific plate and the northward migration of Indian plate, the Southern Qiangtang–Sibumasu plate was flattered, extended, rotated and grinded gradually to form the recent shape with nearly N–S trending (Fig. 2.30).

Some researchers took both the Southern Qiangtang block and Shuanghu collision zone as "collision accretion mélange terrains" and did not recognize the existence of Southern Qiangtang block, and until now this viewpoint has not been accepted by most researchers. Although in that area, isotopic age data of the metamorphic system (500 Ma) have not been obtained, depending on the stable shallow sea sedimentary environment in this area from the Ordovician to Permian, rather stable crystalline basement should have existed at that time. It seems reasonable to take Northern Qiantang block as a relatively stable block based on existing data.

2.5.4 Bangongco–Nujiang–Mandalay–Phuket–Northern Barisan Cretaceous Collision Zone (100–66 Ma) [35]

The Bangongco–Nujiang–Mandalay–Phuket–Northern Barisan Cretaceous collision zone (Fig. 2.30) may be formed in Late Jurassic at first with the upper reaches of Nujiang formed in the Middle Cretaceous–Early Paleogene (Guo et al. 1991). This collision zone extends eastward from the Bangongco, crossed Gaize–Dingqing, Kangsha–Shizika of Eastern Xizang (Tibet) and heads southward through the Mandalay Burma, Phuket Thailand to the Barisan Indonesia. The zone was formed by the collision among the Southern Qiangtang–Sibumasu plate [34], Gangdise plate ($\sim$510 Ma) [36] and Western Burma (Pegu Mountains–Rangoon) plate ($\sim$510 Ma) [49].

The overall deformation characteristics of collision zone show that the strata changed into steeply dipping overthrusts with strong folding and fan-shaped profiles, and the faults show ramp characteristics. Most of the faults are dipping to north or east orientation at intermediate or high angles, which indicate that the SW wall of collision zone could be the subduction block and the NE wall be the obduction block. The faults in that collision zone can penetrate the Jurassic, Cretaceous and some Paleogene Systems. Along the zone, there developed many ophiolite suits, of which the Bangongco ophiolite suits were formed in Early Cretaceous (Guo et al. 1991), the Gaize–Dingqing ophiolite suits formed in the Jurassic (Bureau of Geology and Mineral Resources of Xizang 1993), the Kangsha–Shizika ophiolite suits formed in Early Cretaceous (Wang et al. 1996) and on the east side of Gaoligongshan, Western Yunnan, the cold emplacement of the super-mafic rocks, was in the Lower Jurassic strata (Zhong 1998). The extension velocity of oceanic basin was 1.2 cm/year at that time, almost equal to that of recent Atlantic Ocean. The sealed-off time of that oceanic basin was mainly at the end of Early Cretaceous. The collision time of Gongdise–Tengchong block and Southern Qiangtang–Taniantawong block (at that time, which was already a part of Eurasian continent), according to the secure data, was in Late Cretaceous–Early Paleogene.

The formation time of collision granites near the Bangongco was from 100 to 55 Ma (Guo et al. 1991) and that in the Kangsha–Shizika zone was in 75–86.4 Ma (Wang et al. 1996). Along the whole Bangongco–Nujiang collision zone, till now the tectonic research has not been deep enough, and however, tectonic deformation research in the SW side of Taniantawong got some significant results (Wang et al. 1996). They discovered that the mylonites in Kangsha–Shizika fault zone, besides compression and collision, showed relatively strong sinistral strike-slip in 75–86.4 Ma, relatively weak dextral strike-slip after 30 Ma. The Bangongco–Nujiang collision zone also showed characteristics of sinistral strike-slip at 75–86.4 Ma, which fit with the ENE trending characteristics of maximum compression principal stress in East China at the late period of Cretaceous. Under the influence of this tectonic stress field, the nearly E–W trending part of the Bangongco–Nujiang collision zone with the sinistral strike-slip is quite normal. Just in the strong sinistral strike-slip, the Bangongco–Nujiang main fault cut off the collision zone of Shuanghu Triassic (250–210 Ma) [32] and Changning–Menglian–Central Malaya [33]. It is obvious that the maximum principal compression stress orientation was connected with NE40°–50° trending migration of Indian plate [40] in the late period of Cretaceous (70–80 Ma) (Wan 2011b). In the middle and late periods of Cretaceous, the dynamic source of the widespread intraplate deformations in China continent and the sinistral strike-slip of the Bangongco–Nujiang main fault was obviously related to the tectonic events derived from the NE trending migration of the Indian plate.

According to the regional magmatism and the division of paleo-plate, the Cretaceous collision zone was from the Bangongco–Nujiang–Mandalay–Phuket, to the Indonisian Northern Barisan. It is inferred that this zone can across the Java Sea, extending northeastward to the east side of Kalimantan (Borneo) and Sulu Archipelago, i.e., the Eastern

Kalimantan–Southern Sulu Sea Cretaceous accretion–collision zone [52]. This was the Cretaceous collision zone formed by the northward migration of Australian plate [72] colliding with the Eurasian continent. Just because outcrops are rare in the Java Sea, we could only temporarily infer that the above two collision zones might be united.

2.5.5 Gangdise Plate (~510 Ma) [36]

The Gangdise plate is also called the Lhasa block (Fig. 2.30), which crystalline basement was formed at about 510 Ma, and it is also the result of Pan-African Tectonic Event (Zhang 2009; Xu et al. 2010). It is obvious that it originally belonged to the Gondwana and began to break up with Gondwana gradually since Middle Cambrian. According to the paleo-magnetic data, this plate was located stably at the area of about 30°–20° S in the Paleozoic–Triassic period. However in the late period of Paleozoic, there developed cold water fauna in the cold climate zone. From Late Jurassic to Late Cretaceous, this plate migrated from 11.8° S to 11.8° N, with an average velocity of latitude of 3.3 cm/year. After the Cretaceous, this plate continued to move northward with an average velocity of latitude of 2.6 cm/year, reaching the recent position (the central referential point is in 29.7° N; Wan 2011b).

Zhang et al. (2008) researched systematically the ages for the basement metamorphic system of the Gangdise plate and obtained the zircon U–Pb ages of 496, 367 and 56 Ma in the orthometamorphic system and the fragment zircon U–Pb ages of 1555, 1141, 981, 576, 341, 110–80, 55–50 and 35–25 Ma in para-metamorphic system. These data show that this block suffered the Pan-African Tectonic Event (576–496 Ma) of Gondwana. However, the data of 1141 and 981 Ma show that this block had characteristics of both West Australian plate and the Indian plate.

Zhu et al. (2010) researched the characteristics of fragment zircon U–Pb ages in different sediment rocks in and before Permian for the Qiangtang, Gangdise, Tethys Himalaya, High Himalayan and West Australian blocks and found that the North and South Gangdise blocks and West Australian block had fragment zircon U–Pb age data of about 1200 Ma; while in the Qiangtang and Himalayan blocks, fragment zircon U–Pb age data of about 1000 Ma were found, similar to those of the Indian plate[40]. Thus, they inferred that the Gangdise block had similar characteristics with West Australia block in Paleozoic, while the Qiangtang and Himalayan blocks were similar to the Indian plate. Their recognition was different from that of Zhang et al. (2008), which needs further discussion.

Yang et al. (2009) discovered eclogite in the Songduo of Eastern Gangdise block and considered that there was a collision zone which divided the Gangdise block into the North and South blocks. In the middle part of this collision zone, the isotopic age of Songduo eclogite is 261.7 Ma, and the isotopic age of granite (rich alluvial) nearby is 263 Ma, i.e., responding to Late Permian. However, it needs further researching whether this collision zone could extend westward through the whole Gangdise block. But Wang et al. (2008) recognized that the Songduo eclogite may be indication of an arc-type volcanism caused by the Qiangtang plate subducted beneath the Gangdise block. Their viewpoints are completely different and need further discussion.

2.5.6 Yarlung Zangbo–Myitkyina Paleogene Collision Zone [37]

The Yarlung Zangbo–Myitkyina Paleogene collision zone is a Late Paleogene collision zone between Gangdise plate [36] and Himalayan plate [38] (Fig. 2.30). In Paleogene, this zone was also an important boundary between Gondwana continental blocks (including the Indian and Himalayan blocks) and Eurasian continent. On this strong deformation zone, the dip angles of fault surfaces and strata are all rather vertical, mainly north dipping, but also south dipping, and there are outcrops of many tectonic sheets (including the Triassic bathyal facies deposits, Jurassic, Cretaceous–Late Eocene oceanic, shelf facies and mélanges) and ophiolite suites. There is a large Yarlung Zangbo ophiolite suites outcrop near Xigaze, forming an east–west trending belt zone which is 170 km long and 2–20 km wide (Tapponnier et al. 1981; Allègre et al. 1984; Wang et al. 1999, 2002).

In recent geological research, the youngest Late Eocene Radiolaria were discovered in the Yarlung Zangbo ophiolite suite fragments that were in the siliceous rocks, indicating oceanic deposition. In Late Eocene, the oceanic crust finally disappeared. Thus, it can be inferred that the collision between Indian plate and Eurasian continent (the immediate contact in the Gangdise block) at the Yarlung Zangbo zone actually began in the Oligocene (about 34 Ma, Aitchison and Davis 2001; Aitchison et al. 2007; Wang et al. 2002). However, recent popular opinions held that the beginning time of collision was in Paleocene (about 60–50 Ma) or Eocene (40 Ma) (Tapponnier et al. 1981; Allègre et al. 1984; Besse et al. 1984; Mo et al. 2009; Xu et al. 2011). On the beginning time of collision, it is a very complex problem and should be discussed more detail. The beginning time of collision should be after the remnant oceanic crust disappearing, before that time it was the period of convergence

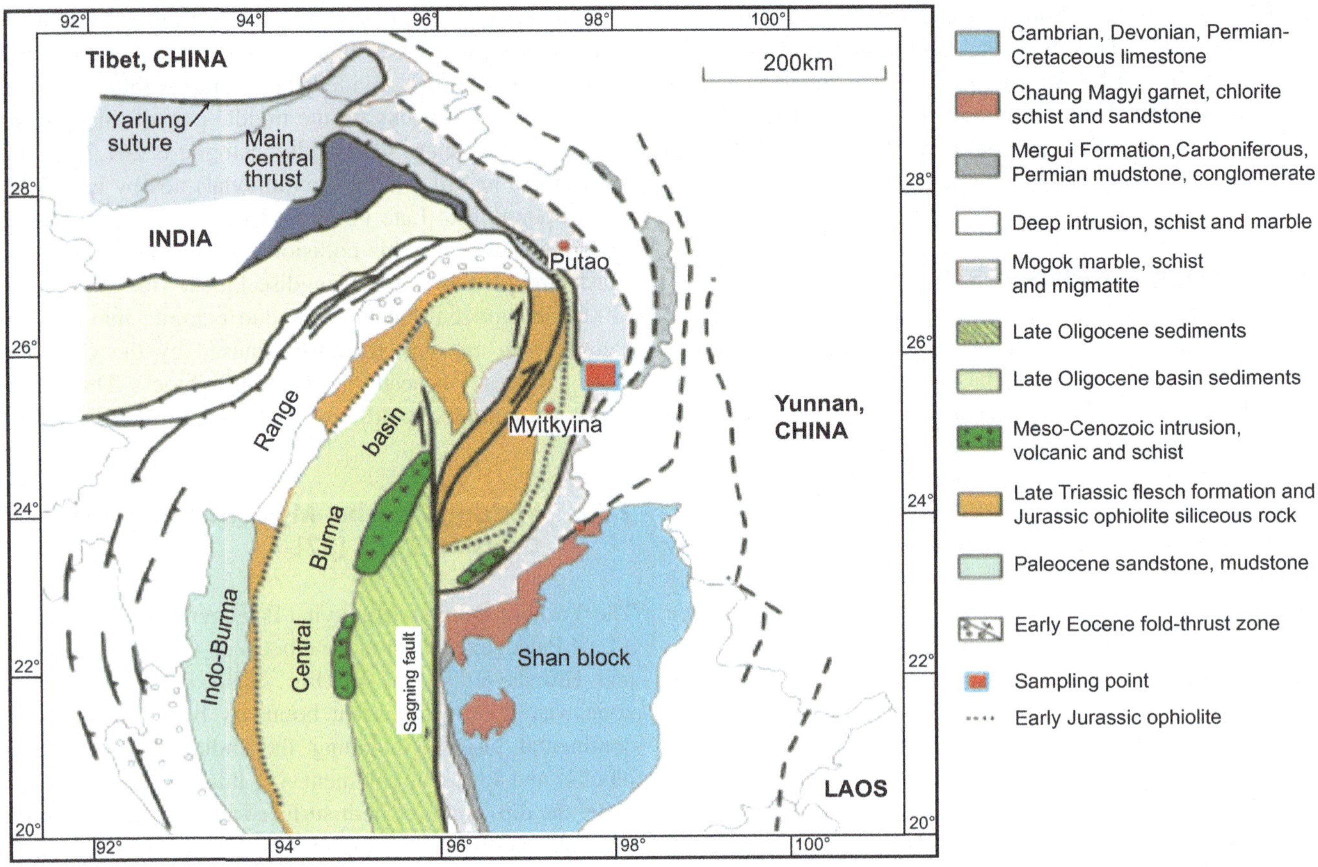

Fig. 2.33 Sketch of geological structure for Burma and its adjacent areas (After Yang et al. 2012)

and subduction (Wan 2011b). According to the deep geophysical exploration data (Fig. 2.35), the beginning period of Indian continental plate subducting beneath the Eurasian continent may be rather later.

Near Myitkyina, Burma, there developed two Jurassic ophiolite zones. The east zone (east to Myitkyina, Aka Sagaing fault; Fig. 2.33) is the collision zone between Sibumasu [34] and Western Burma [49] plates. At the east margins of Arakan Yoma, there formed a lot of SSZ-type ophiolite mélanges and ultramafic intrusions, in which the results of zircon U–Pb ages are: the andesite–basalt is 166 ± 3 Ma, the light-colored gabbro is 177 ± 1 Ma, the olivine pyroxenite is 171 ± 2 Ma, and the plagio-granite is 176 ± 1 Ma. The lithophile elements from the lava and other basic rocks are obviously rich, while the Nb, Th, Ta, Zr, Ti elements are insufficient, being originated from the depleted mantle. The above lava are typical SSZ-type lava (i.e., formed by the rollback of subduction plate). It is different from the western zone of Myitkyina and Yarlung Zangbo ophiolite suite, which were formed in the mid-ocean ridge by the plate extension (MOR type). The western zone ophiolite suites are located at the subduction zone between the Central Burma basin and Indo–Burma Range (Fig. 2.33; Yang et al. 2012). In this monograph, the Arakan–Sunda Cenozoic subduction and island arc zone [50], i.e., the Indian–Burma subduction zone, will be discussed later.

2.5.7 Himalayan Block (~510 Ma) [38]

The crystalline basement of Himalayan block (Fig. 2.30) was formed at about 510 Ma and used to belong to Gondwana which was the production of Pan-African Tectonic Event. Based on fragment zircon U–Pb ages, it has been made clear that tectono-thermal events occurred at 2500, 1650, 1000, 500, 65, 30 and 5 Ma (Zhang et al. 2008; Xu et al. 2011), which were rather similar to the Pre-Cambrian evolution of Indian plate. Since Middle Cambrian, this block began to break up from the Gondwana gradually. Before Triassic, it had been always located near 30° S in the Tethys Ocean (Figs. 3.6–3.8 and 3.10–3.12). It arrived near the equator at the end of Cretaceous and underwent the

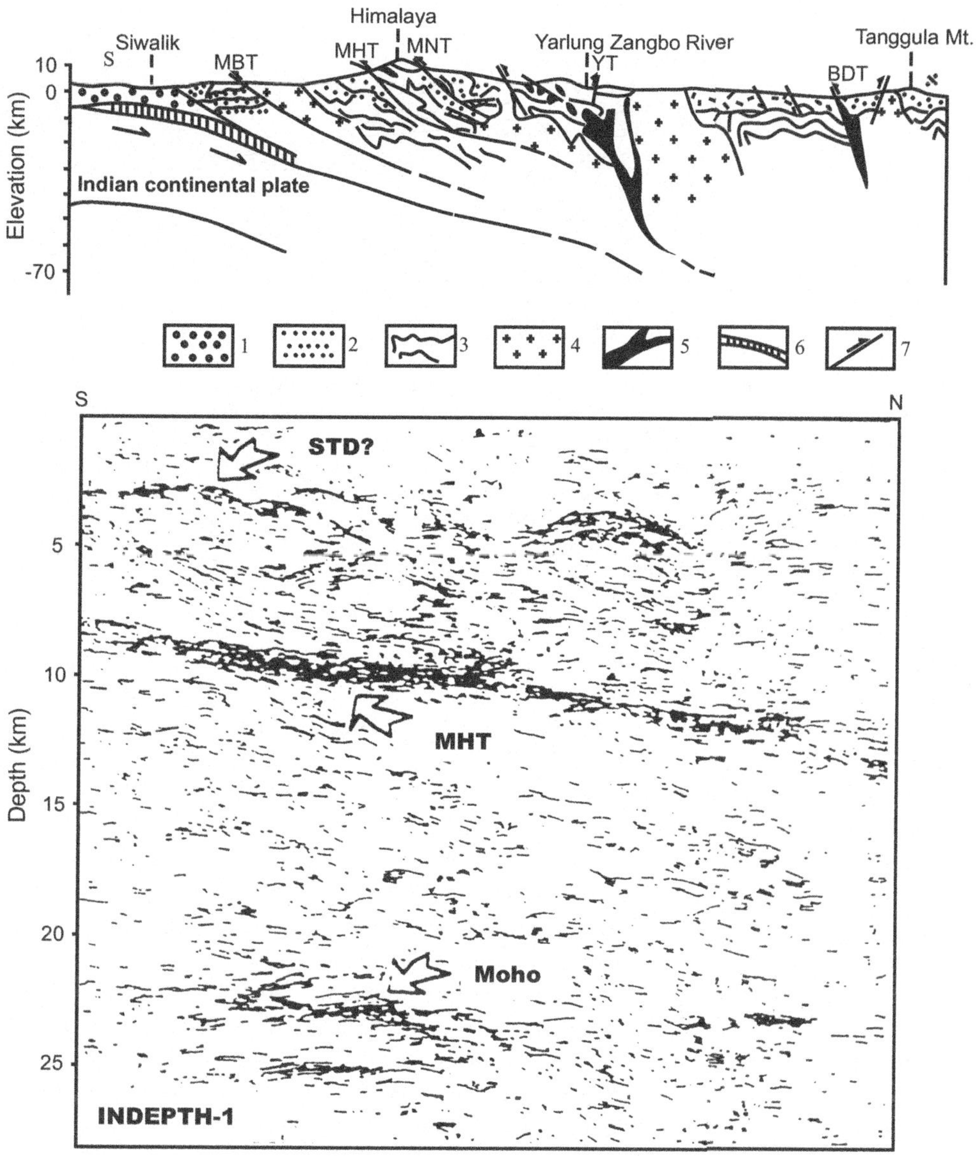

Fig. 2.34 Geological and seismic sections in the Himalayan thrust zone. **Upper section**: (1) Cenozoic molasses formation; (2) sedimentary cover; (3) crystalline basement; (4) granite; (5) ultramafic rocks; (6) subduction zone; (7) overthrust zone. MBT. Main boundary thrust; MHT. Main Himalayan thrust; MNT. Main north thrust; YT. Yarlung Zangbo fault zone; BDT. Bangongco–Dingqing thrust. Lower section: STD?. Top layer of possible partial melting; MHT. Main Himalayan thrust; Moho. Moho discontinuity (After Zhao et al. 1997; redrawn by Wan 2011b)

tectono-magmatism (Fig. 3.24). At the end of Paleogene, there formed the Yarlung Zangbo–Myitkyina collision zone [37] on its north side, integrating it with the Eurasian continent, and its south side was the Southern Himalayan main boundary thrust [39]. In the late period of Paleogene (about 30 Ma), it was the continent–continent collision period between Indian plate and Eurasia continent, and the Himalayan block reached to the position of 28° N (Wan 2011b).

The Himalayan block is now in fact a series imbricate overthrust system (Figs. 2.34 and 2.35), in which there are the main central thrust zone and Kangmar thrust zone. Its south margin is the main boundary overthrust zone, and its north margin is the Yarlung Zangbo–Myitkyina collision zone (overthrust). The fault planes of the above faults are all north dipping, which was caused by the northward subduction and collision of the Indian plate. The fault zones in the block were mainly formed in Miocene, between 16.8 and 23.5 Ma (U/Pb method, 23.5 Ma, by Tapponnier et al. 1990; Ar/Ar method, 17 Ma, by Copeland and Harrison 1990; Ar/Ar, AFT, 18.5 Ma, by Copeland et al. 1987; AFT, 16.8 Ma, by Corrigan and Crowiey 1992; Bureau of Geology and Mineral Resources of Xizang 1993), i.e., they occurred in the intraplate deformation period after the collision. However, north to the Himalayan Mountains, some overthrusts show the normal faults slip, which means that the north walls show characteristics of slipping down relatively (Searle 1996, 2007). These characteristics changes are very similar to the activities of Alps thrust zones.

As to the total displacement of main Himalayan thrust, many researchers used the shear strain method and estimated values of about 80–115 km (Sinha-Roy 1982). According to the paleo-magnetic data, in the last 22 Ma, the central

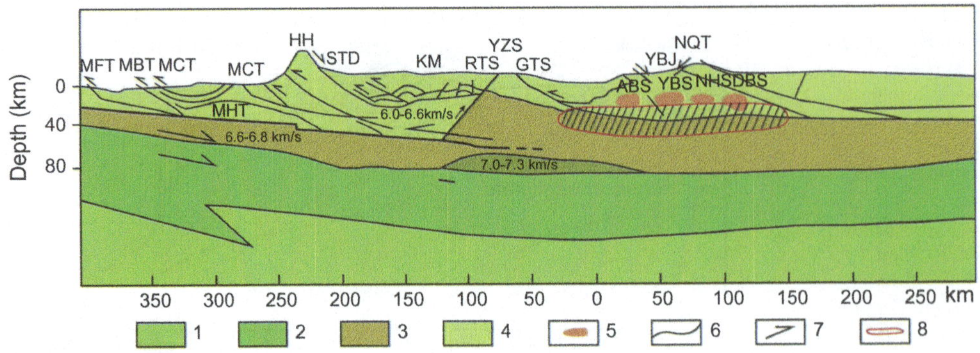

Fig. 2.35 Deep tectonic interpretation in Southern Xizang (Tibet). **Legend**: (1) asthenosphere; (2) Xizang lithosphere mantle (its main part may belong to Indian oceanic lithosphere mantle, about 40 km thick, and its left part may belong to the Indian continental lithosphere mantle, about 60–70 km thick); (3) lower crust (used to be the crust of Indian plate); (4) crust of Xizang area; (5) the bright point of seismic reflect; (6) geological border; (7) migration orientation of fault; (8) partial melted layer. **Alphabetic code**: MFT. Main front thrust; MBT. Main boundary thrust; MCT. Main central thrust; HH. High Himalayan crystalline block; STD. South Tibet detachment; KM. Kangmar dome; YZS. Yarlung Zangbo–Myitkyina suture (collision zone); GTS. Gangdise thrust system; RTS. Ringzhou thrust System; YBJ. Yangbajing depression; NQT. Nyainqentanglha Mts; MHT. Deep main central thrust; ABS, YBS, NHS, DBS. All the "bright points" of seismic wave, i.e., possible partial melt body (After Zhao Wenjin, 2015, personal communication modified from Zhao et al. 2004b)

reference point of Indian plate migrated from 11° N to 21° N, and thus the total convergence and shortening of Himalayan overthrust was about 1000 km (Klootwijk and Radhakrichnamurty 1981; Lee and Lawver 1995). Since Neogene, the north–south trending shortening rate of this block has been about 66% (Li 2010, personal communication).

In Fig. 2.35, it notes that the lithosphere mantle (dark green color) of the Indian plate is beneath the Moho discontinuity of Southern Qingzang (Qinghai–Tibet) Plateau, with a thickness only 40–50 km in most areas. The author insists that this is an oceanic-type Indian lithosphere mantle subducting to the Southern Xizang. However, on the left side of this figure, the thickness of Indian lithosphere mantle is about 70–80 km, indicating that it is the Indian continental lithosphere, with the total thickness more than 100 km (Mishra and Kumar 2014). According to this figure, the Indian continental lithosphere plate subducted to the southern side of Himalayan Mountains only in rather recent periods.

According to the above data shown in Fig. 2.35, it can be calculated that the Indian continental plate was subducted to a depth of only 350 km beneath the Xizang area. The recent northward migration velocity of Indian plate is 5 cm/year. Thus, the collision time between Indian plate and Eurasia continent should be 7 Ma ago. According to the data of Nábelek et al. (2009), it is inferred that Indian continental plate was subducted to a depth of 450 km beneath the Xizang area, so the collision time should be 9 Ma ago (Late Miocene). Before this collision period when the oceanic-type Indian plate was subducted beneath the Xizang area, it could only be called the "subduction period." However, it needs further accurate determination as to the thickness of Indian oceanic and continental lithospheres.

Due to the subduction of Indian plate, the crust thickness of Xizang (Tibet)–Pamir areas was thickened obviously. Its thickness usually reached more than 60 km, with a maximum of 70 km; and the thickness of whole lithosphere could reach 120–180 km. It is a continental lithosphere [68] (Fig. 1.1, in the area limited by the yellow point line) with significant thickness increment. In the Himalayan block and its surrounding areas, due to the rather thick crust and not so enough rock strength, the intraplate deformations were very violent and thus derived the quite strong magmatism. According to the INDEPTH seismic data, there exist magma reservoirs which are partial melting (e.g., the four seismic "brights" in Fig. 2.35) in some areas, providing good conditions for developing endogenic ore deposits and high-temperature geothermal fields (such as Yangbajin and Yangyixiang) (Zhao et al. 2004a, b).

The sections I, II, III in Fig. 2.36 are the seismic tomography sections (Replumaz et al. 2004). According to the results, in the deep part of Himalayan area, after the Indian plate subducting forward and downward obliquely for 200–300 km, it started to subduct with very dip angle downward to the middle mantle, to the depth of about 670 km, where the boundary between the subduction plate and Eurasian continent mantle is not very clear.

Xu et al. (2015) discovered in the N–S trending shortening and collision process of the Qinghai–Tibetan Plateau, besides the thickening of its lithosphere, the E–W trending strike-slip events occurred, i.e., there formed three-dimensional deformation in the collision. They found that the eastward strike-slip occurred mainly in 28–26 Ma, and the westward strike-slip occurred mainly in 25–22 Ma. This contributed to more profound new understandings of the deformations of Qinghai–Tibetan Plateau.

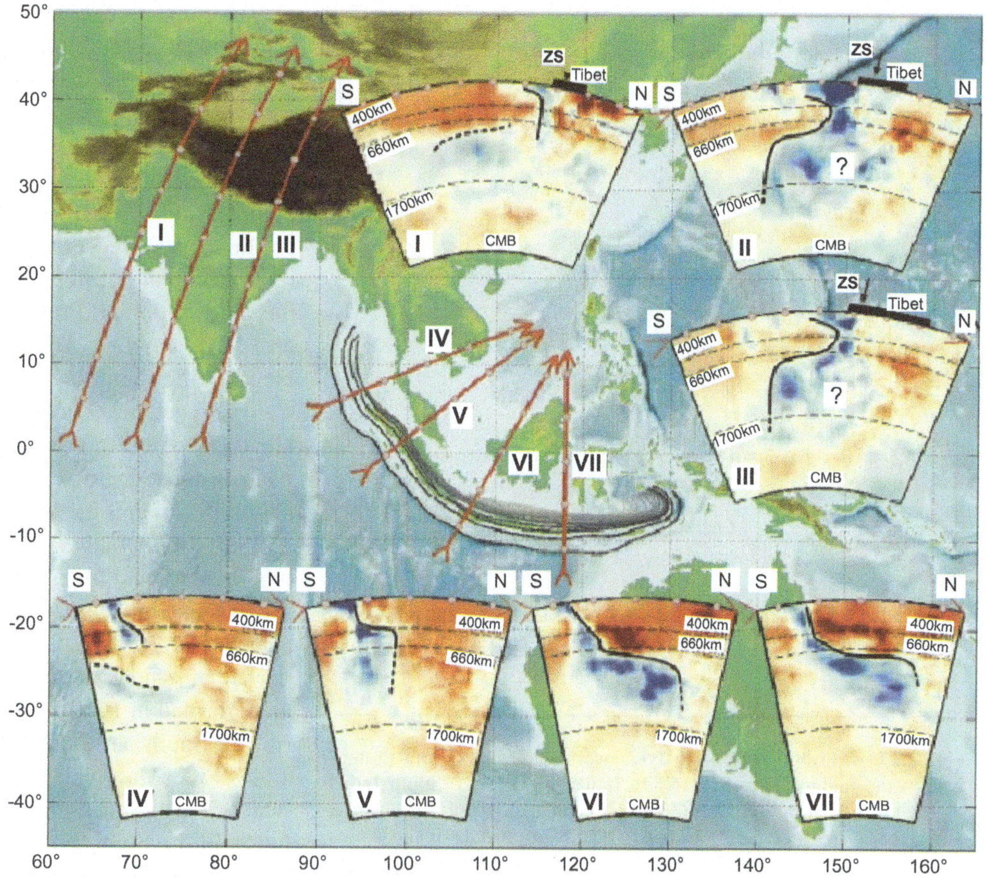

Fig. 2.36 Seismic tomography sections at Himalaya (I–III) and Sunda (IV–VII) Cenozoic subduction-arc zones (After Replumaz et al. 2004)

2.5.8 Southern Himalayan Main Boundary Thrust (Since Neogene) [39]

The surface of Southern Himalayan main boundary thrust is dipping to the north with low or intermediate angles, which mean that the footwall of thrust is subducted northward and downward (Figs. 2.34 and 2.35), and a series of intermediate- or low-grade dynamic metamorphic zone are shown on the Earth surface. This thrust has been a border between the Indian plate [40] and Eurasian continental plate since Neogene (Figs. 2.30 and 2.35), after which the Indian plate was merged into the Eurasian continental plate. So far, this main thrust has been migrating and converging to the north with a velocity of about 5 cm/year (Lee and Lawver 1995).

2.5.9 Indian Plate (~510 Ma) [40]

According to the data of fragment zircon age in the Indian plate (Fig. 2.30), there might be tectonic-thermal events in 2500, 1800, 1650, 1000 and 600–500 Ma, at last forming the united crystalline basement under the influences of Pan-African Event (~510 Ma) (Zhang et al. 2008). The Indian plate can be divided into the North and South blocks, between which there is the Central Tectonic Zone (CTZ, in Fig. 2.37).

The South Indian block consists of large Dharwar and Singhbhum cratons and Eastern Ghats basement granulites deformed by the Late Pan-African Tectonic Event (~500 Ma). The Dharwar craton is composed of Neoarchean Dharwar greenstone belts (2.7 Ga) and TTG gneiss intrusive rocks (~2.55 Ga). In the Dharwar greenstone belt, there developed giant gold deposits and iron ore, forming banded iron formations (BIF). In the Dharwar craton, there also formed some Paleoarchean older metamorphic group and older metamorphic tonalitic gneiss (Jayananda et al. 2000, 2008; Ravikant 2010).

The North Indian block consists of Archean basement gneiss and Paleoproterozoic volcanic–sedimentary strata with hornblende–granulite facies. Between the North and South Indian blocks, there is the Paleoproterozoic suture zone, also known as the Central Tectonic Zone (CTZ, Fig. 2.37; Radhakrishna and Naqvi 1986), in which the shallow volcanic rock system and TTG granitic intrusions in many granitic greenstone belts have similar geochemical characteristics to the subduction island arc system. It indicates that this area might have started to develop similar tectonic characteristics to the modern plate tectonics since

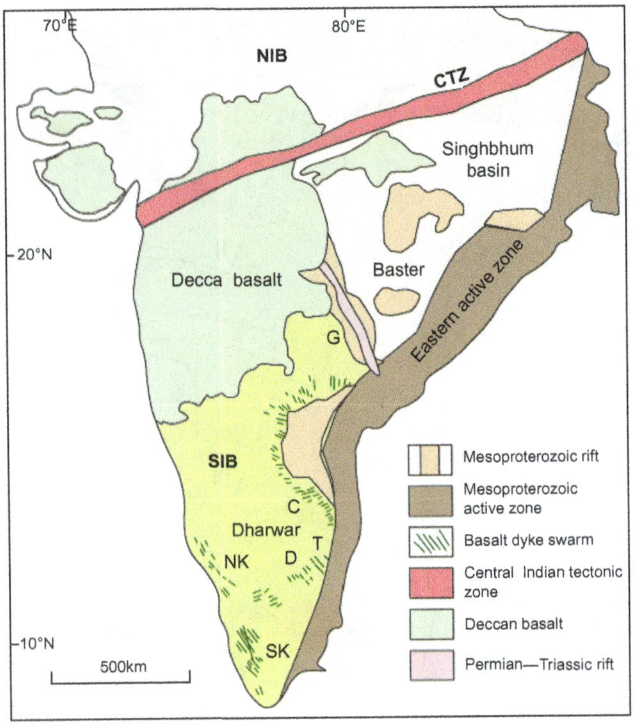

Fig. 2.37 Distribution of basic dyke swarms and rifts during Proterozoic in the Indian plate. NIB. North Indian block; SIB. South Indian block; CTZ. Central Tectonic Zone; C. Cuddapah basin; D. Dharmapuri dyke swarms; G. Godavari basin; NK. North Kerala dyke swarms; SK. South Kerala dyke swarms; T. Tiruvannamalai dyke swarms (Modified from Radhakrishna and Naqvi 1986)

Archean. Some researchers (such as Condie and Richard 2009; Rogers and Santosh 2002, 2009) considered that the Indian plate and Sino–Korean plate had similar formation process in the Archean–Paleoproterozoic, according to similar characteristics in terms of magma, metamorphism and zircon geochronology in this period. However, the Indian, African, Australian and Antarctic plates all underwent the Pan-African Tectonic Event in about 510 Ma, thus forming the united Gondwana continent; but the most of Asian continental blocks, such as Siberian, Sino–Korean, Yangtze, Cathaysian, Tarim and Indosinian blocks, never suffered the Neoproterozoic (about 1000 Ma) and Pan-African Tectonic Event (about 510 Ma), which means that these blocks do not belong to Gondwana. Some researchers classified all the continental blocks of China into Gondwana, which needs further discussion.

Based on the aeromagnetic data (Rajaram and Anand 2014), in the crystalline basement at the south end of Indian subcontinent, a series of nearly east–west trending magnetic anomalies have been discovered, which are deep source related and highly intense. The magnetic anomaly was caused by the high-pressure and ultra-high-temperature ductile shear and subduction effect in the deep metamorphic rocks, with granulite in the shallow part and eclogites buried in the depth.

In the Indian continental plate, there developed a lot of rift basins and dyke swarms in Late Paleoproterozoic–Mesoproterozoic (Fig. 2.37; Radhakrishna and Naqvi 1986). They were obviously derived from the northwestward compression of East Indian Proterozoic Tectonic Zone (according to the recent magnetic orientation), which may be related with the rifting of Columbia supercontinent, in which the most important Proterozoic tectonic zones and rift zones are: NNE trending Aravalli–Delhi Tectonic Zone, south to New Delhi; Satpura Tectonic Zone, also known as the Central Tectonic Zone (CTZ, Fig. 2.37); and the Eastern Ghat Tectonic Zone on the east margin of Indian subcontinent. In those tectonic zones, the thickness of crust reaches about 45 km and the thickness of lithosphere is about 120–130 km. In the faults, there exist the middle crust rocks with high density and high conductivity, which indicates that the lower curst has characteristics of extensional rifts (Mishra and Kumar 2014).

The intrusions in the Singhbhum basin of Northeast India were formed in 1660–1638 Ma. In the Cuddapah continental rift of Southeast India, there developed basic basalts in 1841–1583 Ma, with characteristics of plate margin rift, which is a little bit similar to the Xionger Group rift evolution in the South Sino–Korean plate. The isotopic ages of dyke swarms in South India were in 1870–1170 Ma.

According to many researchers (such as Radhakrichnamurty 1981; Lee and Lawver 1995; Klootwijk et al. 1992; Schettino and Scotese 2005), in Paleozoic, the central reference point of Indian plate basically remained at lower and middle latitude (21° S–45° S) of the Southern Hemisphere. The blocks rotated many times, but the latitude change was small (Figs. 3.6–3.8 and 3.10–3.12). At the end of Jurassic–Early Cretaceous, the Indian plate reached the most southern position, 45° S (Figs. 3.20 and 3.22), thereafter the plate migrated very fast northward, with the highest migration velocities of 17–18 cm/year. In Paleocene–Early Eocene, the velocities reduced to 9–10 cm/year, and since Late Oligocene had reduced to 5–6 cm/year (Klootwijk and Radhakrichnamurty 1981; Lee and Lawver 1995; Acton 1999; Besse and Courtillot 2002; Schettino and Scotese 2005).

The above results are confirmed by reliable data of Deep Sea Drilling and isotopic ages (Fig. 2.41), which is one of the significant evidences of the once large-scale migration of several thousand kilometers. The recent Northern Indian plate has already been subducted underneath the Qinghai–Tibetan Plateau. In Cretaceous, it was the oceanic part of Indian plate that underwent subduction. And it was until the Late Paleogene that subduction changed to collision between the Indian continental plate and Eurasian continental plate (Wan 2011b; Fig. 2.38).

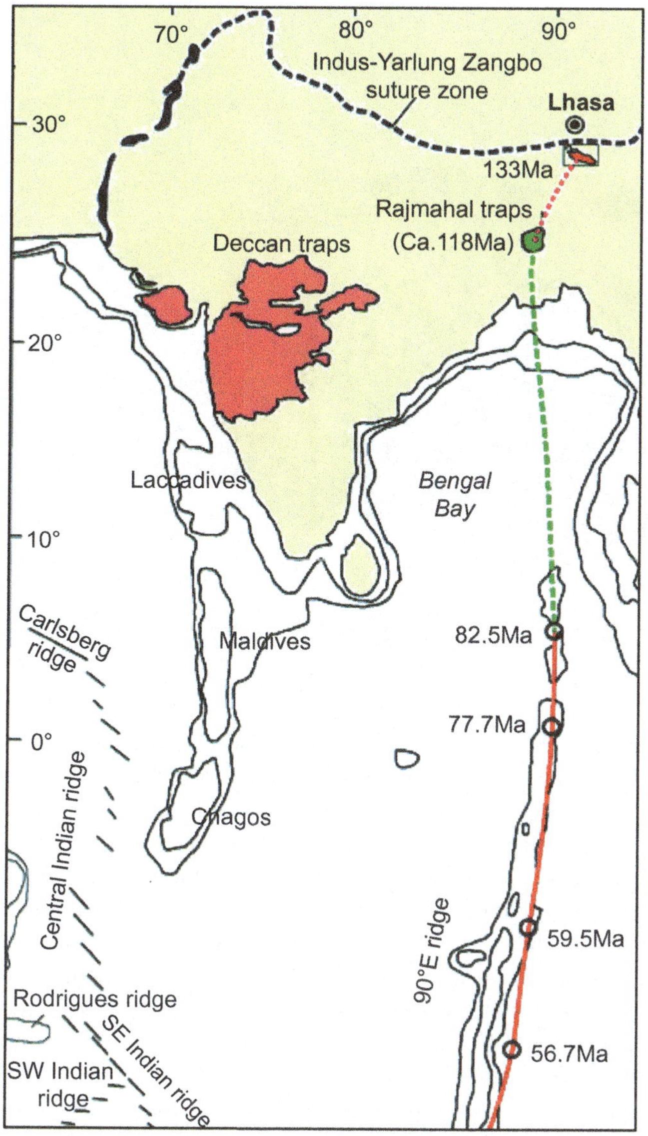

Fig. 2.38 Isotopic ages of hot spots for Deccan large volcanic province in the Indian plate (including its southern ocean) and 90° E ridge. The red line is the hot spot migration trace along the 90° E ridge, and the green line is the estimated hot spot migration trace (Modified from the ODP data)

On the Indian subcontinent, there developed the Deccan basalt large igneous province (shallow green area in Fig. 2.37; red area in Fig. 2.38), which was mainly formed at 65 Ma (Late Cretaceous). Due to the long-term northward migration since Cretaceous, the Indian plate suffered rather strong north–south trending compression and shortening, so there developed very clear NE and NW trending conjugate shear fractures, which are clearly shown on the satellite images.

As to the dynamic mechanism of the Indian plate's fast migration, till now no explanation has been generally accepted, especially no explanation of why the Indian plate moved northward particularly fast in Cretaceous. The author infers that it may be related to the formation of Antarctica mantle plume in Middle Jurassic and the radiated extension of plates. It is also possible that it was caused by the oblique impact of meteorite in Middle Jurassic; the rather small Indian plate was just along the oblique impact orientation, thus leading to its faster migration than other plates, such as South American and African plates, but so far, there are not enough evidences (Wan 2011b). Lee and Lawver (1995) used to explain the Indian plate's northward migration rate change in Cenozoic as "soft collision" first, and "hard collision" later. The author recognizes that the so-called soft collision is actually the subduction of Indian Oceanic plate. After Late Paleogene, when the whole Indian Oceanic plate was subducted underneath the Eurasia continental plate, the actual continent–continent collision period started, i.e., the actual collision period was just rather recent (Wang et al. 2002; Aitchison and Davis 2001; Aitchison et al. 2007; Wan 2011b).

2.5.10 Kavkaz–Alborz Late Paleozoic–Late Jurassic Accretion–Collision Zone [41]

The Kavkaz–Alborz Late Paleozoic–Late Jurassic accretion–collision zone is located between the Turan–Karakum plate [8] and Anatolia–Tehran Middle Cretaceous–Paleocene collision zone [42] (Fig. 2.30). The tectono-magmatism was relatively weak in Late Paleozoic. After the plate collision, the continental strata began to be deposited since Carboniferous. In Late Triassic–Jurassic this area underwent extension, forming rather huge marine sediments. In the Late Jurassic (Kimmeridgian epoch, 155.7–150.8 Ma), the Iran plate [43] and on its north side Turan–Karakum plate [8] collided, causing strong rock deformations and magmatism and the formation of the Kavkaz Mountains north to the Alborz Mountains (China Geological Academy 1980).

2.5.11 Anatolia–Tehran Middle Cretaceous–Paleocene Collision Zone (100–56 Ma) [42]

The Anatolia–Tehran Middle Cretaceous–Paleocene collision zone (100–56 Ma) [42] was formed in the Middle Cretaceous–Paleocene (Fig. 2.39, the gray area in the south of Alborz Mountains, Fig. 2.40). In the late period of Early Paleozoic, regional angular unconformity existed between Ordovician and Silurian, or Silurian and Devonian. However in general, the Early Paleozoic tectonics was not very strong, and strata in most areas are all with conformity. In the Late Paleozoic–Jurassic, on the southern side of this collision zone, the Turkey–Iran–Afghanistan plate [43] subducted

Fig. 2.39 Typical recumbent fold of Dragon Mount, North Garmsar, Semnan, Iran (From Alireza Amrikazemi, www.Hmdfor.me)

northward in the Anatolia–Tehran zone, which was causing rather strong tectonic deformations, magmatism and regional metamorphism.

The Anatolia–Tehran collision zone formed the southern part of Alborz Mountains. In the Cretaceous–Paleocene, there deposited huge continental slope (flysch) facies sediments and volcanic systems and also formed a series of Cretaceous ophiolites with many middle- and small-scale chromite deposits, indicating the existence of the paleo-ocean and plate collision. Since Oligocene, the nearly N–S trending pulsating shortening and convergence continued and angular unconformities between the Oligocene and Miocene, the Miocene and Pliocene were formed (China Geological Academy 1980), which means that this area suffered pulsation tectonic deformation events in Paleogene and Neogene. Under the influence of Neogene tectonics, on the southern margin of this zone, there formed a typical recumbent fold in Gragon Mount, in the Garmsar area of Semnan Province, west to Tehran. It is a geological historical heritage in Iran.

In recent years, zircon U–Pb ages have been obtained (83.1 ± 2.2 and 74.6 ± 4.4 Ma; Karaoğlan et al. 2013) of the rhyolites in the oceanic volcanic arc widely spread at the southeastern part of Anatolia. Kaygusuz et al. (2013), using the LA-ICP-MS U–Pb method, also obtained isotopic age of 78.07 Ma in the Turnagöl granodiorite, in SE Anatolia, Northeast Turkey. However, at the Pontides area on the southern beach of Black Sea, Northeast Turkey, Eyuboglu et al. (2013) got the zircon U–Pb age of 48.71 ± 0.74 Ma in adakite related to the subduction and collision, and zircon U–Pb age of non-adakite is 44.68 ± 0.84 Ma.

In Northeast Iran, i.e., the eastward extension part of Tehran collision zone, the Kopet Dagh area is a NE trending Cenozoic folding zone developed on the Late Paleozoic metamorphic crystalline basement of Turan–Karakum plate [8]. In this zone, there are the Mesozoic and Paleogene carbonate with a thickness of about 10 km. Similar to the Zagros folding zone, Kopet Dagh zone also developed Cenozoic NW–SE trending folds, but almost without any magmatic rock outcrops. The folding in this zone can well indicate the migration and compression from southwest to northeast in the area near Iran plate in Cenozoic (Nezafati 2006).

2.5.12 Turkey–Iran–Afghanistan Plate (~510 Ma) [43]

The Turkey–Iran–Afghanistan plate (~510 Ma, Figs. 2.30 and 2.40; Mansour 2013) formed its united crystalline basement when the Pan-African Tectonic Event (about 500–600 Ma) happened. The older tectono-thermal event was at 1100 Ma, almost the same time with the convergence of Rodinia continent. This was different from the Indian plate, which underwent the tectono-thermal event at 1000 Ma, not at 1100 Ma. It may indicate that the Turkey–Iran–Afghanistan and Indian plates did not undergo tectono-thermal events concurrently, or that they were located in different

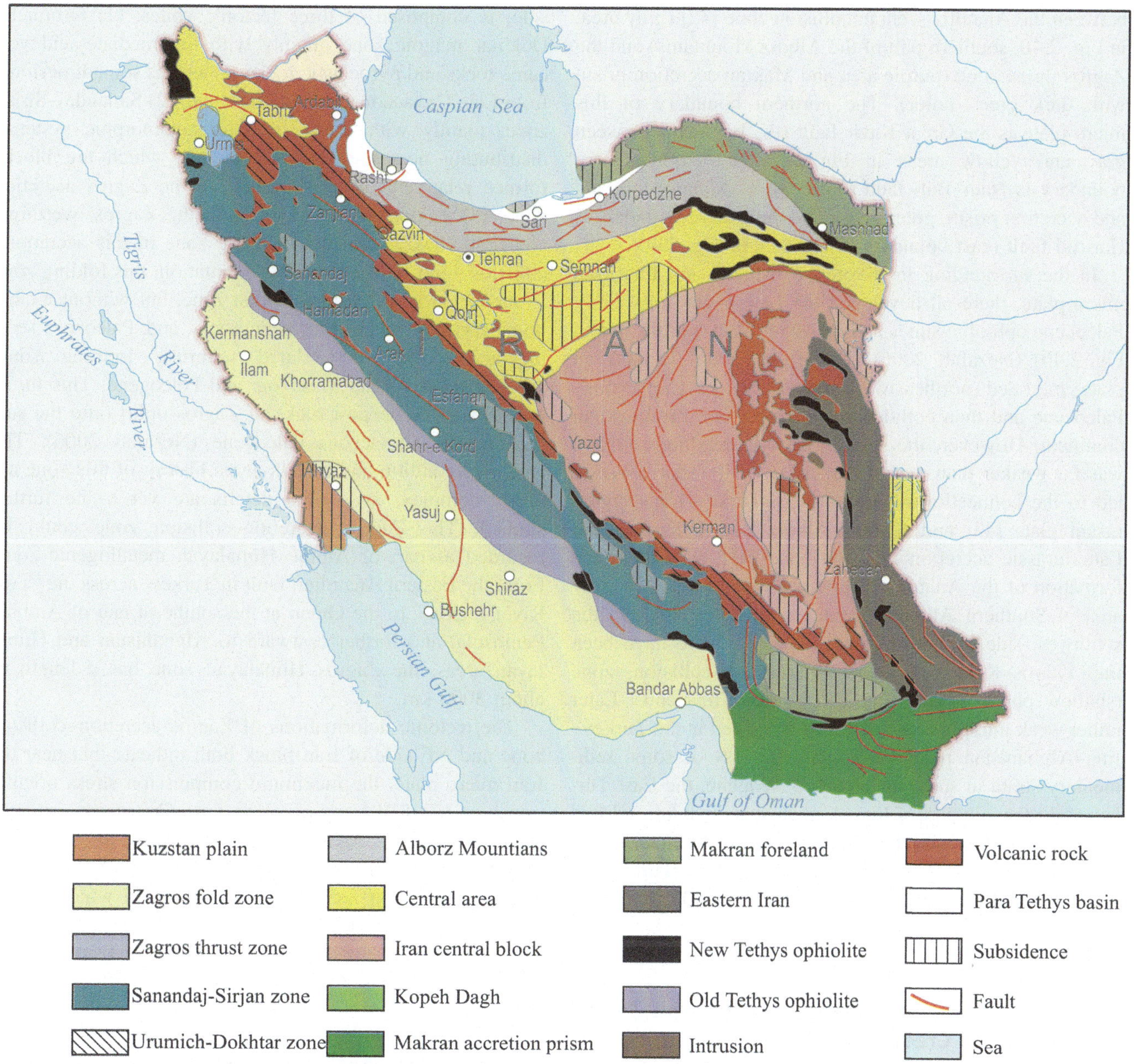

Fig. 2.40 Sketch of Iran tectonics (Modified from Nezafati 2006)

plates at that time. It was in the Pan-African Tectono-thermal Event (500–600 Ma) that the Turkey–Iran–Afghanistan plate was cohered into Gondwana.

In Paleozoic, this plate showed characteristics of forming sea facies sedimentary cover and was in a rather stable tectonic state (Figs. 3.6–3.8 and 3.10–3.12). In the late period of Paleozoic, there were ocean to the north of Turkey area, and since Permian, the Turkey block began to break up from the Gondwana and stayed in the Tethys Ocean (Fig. 3.12). In the Eastern Mediterranean Sea areas subduction occurred in Triassic, which can be confirmed by the Triassic eclogites found in the Northwest Turkey. Paleogeographic reconstruction of the Late Paleozoic showed that the recent Eastern Mediterranean Sea was then a wide sea area (Arallokay and Ma 1997). In Mesozoic, the main part of Turkey–Iran–Afghanistan plate continued to keep characteristics of shallow sea sediments. In Late Jurassic (Kimmeridgian epoch, 155.7–150.8 Ma), first collision happened between the Turkey–Iran–Afghanistan plate and Turan–Karakum plate [8] (Mansour 2013).

The Middle–East Iran micro-plate, i.e., the deep yellow and red parts in Fig. 2.40, including the recent Kuzstan plain and Iran Central area, has become a part of Eurasian continent plate since Paleocene. This micro-plate is located between the Anatolia–Tehran collision zone [42] (gray area, in Fig. 2.40, southern part of the Alborz Mountains) and the Zagros thrust zone (purple area and Makran accretion prism with dark green color). The northern boundary of this micro-plate is the Great Kavir fault (the boundary between gray and yellow areas in Fig. 2.40), the southwestern boundary is Nain–Baft fault (northeast of Makran foreland and accretion prism, green area), and the eastern boundary is Harirud fault (east boundary of red area in Fig. 2.40).

In the surrounding fault zones of Middle and East Iran micro-plate, there all developed the Late Cretaceous–Early Paleocene ophiolite suites and mélange (black-purple area in Fig. 2.40) (Nezafati 2006). The Turkey–Iran–Afghanistan plate migrated northeastward rather fast in Cretaceous–Paleocene and then collided and merged with the Eurasian continent. However, the strength of NE trending collision was far weaker than that of the Indian plate. That collision led to the connection of the northern Turkey–Iran–Afghanistan plate [43] and Kavkaz–Alborz Late Paleozoic and Late Jurassic accretion–collision zone [41], as well as the formation of the Anatolia–Tehran collision zone [42] (gray area of Southern Alborz Mountains, in Fig. 2.40). On the southwest side of Iran–Afghanistan plate, there have been the Zagros–Kabul thrust and accretion–collision zone (shallow purple area in Fig. 2.40) since Cretaceous. Later rather weak intraplate deformations happened in the Turkey–Iran–Afghanistan plate, leading to the loss of some sedimentary strata in some areas. Since Neogene, the East Turkey and the main part of Iran–Afghanistan crust has uplifted about 2000–3000 m and formed the volcanic rock zone and plateau. However, at some areas of the West Turkey the crust suffered subsidence of about 2000–3000 m (China Geological Academy 1980).

2.5.13 Zagros–Kabul Accretion–Collision Zone (Since Cretaceous) [44]

The Zagros–Kabul accretion–collision zone (Figs. 2.30 and 2.40), formed in Cretaceous–Miocene, is between the Turkey–Iran–Afghanistan plate [43] and Arabian plate [46]. The Zagros–Kabul zone is located at the southern part of Turkey–Iran–Afghanistan plate. Since Paleozoic, this zone had been in rather stable tectonic environment for a long time, forming a series of huge sea facies strata and developing extremely abundant source rocks for oil and gas, mainly in the Paleogene, Cretaceous and Jurassic Systems. Since Cretaceous, rather strong collision and tectonic deformations occurred (Mansour 2013).

The Zagros accretion–collision zone (including the light purple and dark green areas, Fig. 2.40), from NE side to SW side, is composed of three tectonic zones: (1) Orumieh–Dokhtar magma zone, mainly with intermediate-acid volcanic rocks and pyroclastic rocks, as well as some limestone, from Late Cretaceous to modern period; (2) Sanandaj–Sirjan zone, mainly with Early Jurassic metamorphic systems, distributing near the intrusion bodies, which are blocks formed relatively in early periods in the Zagros accretion zone; (3) Zaglos folding zone, with the Zagros overthrust and ophiolites, is the main folding zone in this accretion–collision zone. Toward the NE orientation, this folding zone is connected with the main thrust zone, but without a clear boundary. The older Mesozoic rocks and Paleozoic sedimentary covers are related upthrust, forming some schist-blocks of Late Mesozoic and Paleogene. This thrust zone made the deepest rocks of Zagros uplift onto the surface in the Cretaceous–Paleogene (Nezafati 2006). The tectonic evolution and metallogenic history of this zone are rather complex, and some details are yet to be further studied. The Zagros accretion–collision zone could be regarded as part of Alpine–Himalayan metallogenic zone. From the Eastern Anatolian fault in Turkey, across the "Two River region," to the Oman at the southeast end of Arabian Peninsula, and further eastward to Afghanistan and Himalayan areas, the Zagros–Himalayan zone has a length of about 3000 km.

The tectonic deformations of Zagros accretion–collision zone and NE side of Iran block both indicate that near the Iran micro-plate, the maximum compression stress orientation was NE–SW trending since Late Cretaceous, and the block migration orientation was toward NE generally (Nezafati 2006). The total crust shortening of the Zagros accretion–collision zone is about 70 km, in which 20 km maybe belong to the shortening of Arabian plate. That shortening is far smaller than that of Himalayan collision zone.

2.5.14 Toros Accretion–Collision Zone (Since Neogene) [45]

At the Southern Turkey–Cyprus, there formed the Toros accretion–collision zone since Neogene (Fig. 2.30). This zone used to be the western part of Zagros–Kabul accretion–collision zone, but cut by the Acapa–Dead Sea dextral strike-slip fault (northern part of Red Sea rift), and it was separated from the Zagros–Kabul accretion–collision zone and developed rather unique tectonic characteristics. Under the influence of rather fast northward migration and collision of African plate [73], the Toros collision zone underwent strong tectonic deformations, leading to many outcrops of ophiolites in the Southern Turkey and Cyprus Island, as well as deposits for chromite and Platinum family elements (China Geological Academy 1980; McElduff and Stumpfl 1990; Laurent et al. 1991).

2.5.15 Arabian Plate (~510 Ma) [46]

The formation of united crystalline basement of Arabian plate (Fig. 2.30) also took place in the Pan-African Tectonic Event (570–535 Ma). Before that event, there were several tectonic events at 960, 785, 650–600 Ma, showing that the Arabian plate used to belong to the Gondwana (Al-Shanti 2009). Since Permian, this plate began to break up with the Gondwana (Fig. 3.12). In Triassic, the Arabian plate underwent great migration to the north (about 3500 km) with a little bit dextral rotation, and together with the African plate reached the position that was south to the Pre-Jurassic original Atlantic Ocean. Since Cretaceous, the Arabian plate moved northeastward and gradually merged with the Turkey–Iran–Afghanistan plate [43]. Since Paleogene, there formed Zagros–Kabul accretion–collision zone [44], and thus, the Arabian plate was merged into Eurasian continent. Since Neogene, fault-depression occurred between the Zagros and Arabian blocks, leading to the formation of Persian Gulf. At the same time, dextral strike-slip with extension happened in the Red Sea–Dead Sea fault, leading to the formation of oceanic crust and the Red Sea rift zone [48] (Fig. 2.30).

The oil and gas reserves in Arabian plate (mainly including the Arabian Peninsula, the Gulf countries, Southeast Turkey and Southwest Iran, which is south to Taurus/Zagros mountains) account for 66.4 and 33.9% of the global oil and gas reserves, respectively, in which over 98% of the reserves are distributed on the northeastern margin of Arabian continent between Iraq and Oman. The main reason for enrichment of oil and gas in this area relies on the widespread existence of many oil-gas strata systems from Late Paleozoic to Mesozoic–Cenozoic and the coexistence of carbonate and sandstone. During the slow northeastward migration, this plate was in stable sedimentary environment for a long period, contributing to the formation of unparalleled widespread NE trending continental shelf (nearly 2000 km wide and 3000 km long) and a series of large anticline trap structures with low fold amplitude (Beydoun et al. 2000).

2.5.16 Oman Cretaceous Accretion–Collision Zone [47]

On the southeastern end of Arabian Peninsula, i.e., the eastern part of Sultanate of Oman, there formed the NE trending Oman Cretaceous (since 145 Ma) accretion–collision zone (Fig. 2.30). This collision zone was the residual block when the Arabian plate [46] first collided northeastward with the Turkey–Iran–Afghanistan plate [43]. This collision zone was cut off in Neogene by the Zagros–Kabul accretion–collision zone [44] and recent Gulf of Oman fault–depression and making it remain on the boundary of Arabian plate (Clarke 2006). In this block, the isotopic age of the oldest crystalline rock system was about 800 Ma, almost the same time when the united crystalline basement of the Yangtze plate was formed. In this residual block, there developed the Late Neoproterozoic, Early Paleozoic, Permian–Triassic and Jurassic–Cretaceous sedimentary systems and the Cretaceous ophiolite outcrops. Thus, it can be inferred that the collision should occur at the end of Cretaceous. Till now this area has kept the mountain terrain (Clarke 2006).

According to the paleo-magnetic research (Clarke 2006), the Oman area was located between 10°– 20° S in Cambrian, later it migrated southward to 50° S, forming glacier deposits during the end of Carboniferous (~300 Ma), until it began to have characteristics of Gondwana. Later, this block migrated northward gradually and moved back to the north to the equator in Jurassic–Cretaceous. The development of Oman collision zone had a great influence on the formation of the oil-gas reservoir structures such as folds and salt domes in the southeastern strata of Arabian Peninsula (Clarke 2006).

2.5.17 Red Sea Rift Zone (Since Neogene) [48]

After Neogene (23 Ma), the Red Sea rift zone was formed on the southwestern margin of Asian continent (Fig. 2.30), and is the northern extension of Eastern African rift zone (Delvaux and Barth 2009). It is distributed in the center of the Red Sea, with NNW trending, and is divided into two branches when extending northward: The east branch is NNE trending and extends along the bay of Aqaba to the Dead Sea, across the east border of Lebanon to West Syria, cutting off the Zagros and Toros collision zones, with obvious characteristics of dextral strike-slip and extension (Garfunkel et al. 2014); while the west branch is NNW trending and there formed the sinistral strike-slip Bay of Suez fault into the Mediterranean Sea. The Red Sea rift zone is now the boundary between African plate [73] and Arabian plate [46].

In the 1970 s, the geophysical and geological section researches on the transcurrent faults vertical to the Red Sea (Garson and Miroslav 1976) discovered that in the transcurrent faults, there existed the tholeiite dyke parallel to the linear anomaly and sinistral strike-slip shear zone along the strike of Red Sea, which was similar to the characteristics discovered in the Arabian Peninsula. Deep data across the

Red Sea show that when the Gulf of Aden extended in Late Cretaceous–Paleocene, the Red Sea underwent a NE trending sinistral strike-slip of about 75–80 km. Until the Late Paleocene, in the Red Sea appeared oceanic floor extension, of which the directions were affected by the Pre-Cambrian ENE trending faults on the continent. They extended to the sea area, became the transcurrent structures of the Red Sea and in this positions, there deposited metallic minerals. Recent deduction about the late extension period of the Red Sea was from Late Miocene to Pliocene. In the last three million years, the Sinai block in Egypt has migrated for 25 km sinistrally along the Bay of Suez fault.

West to the Red Sea rift zone [48] is the African plate [73] (Fig. 1.1), which used to be part of Gondwana in the Southern Hemisphere. There formed the united crystalline basement in the Pan-African Tectonic Event ($\sim$510 Ma) and in the Permian the African plate separated from the Gondwana (Fig. 3.12), later in Triassic this plate migrated to the north and was cohered to the Arabian plate [46]. Since Neogene, the Red Sea rift zone separated the African plate with the Arabian plate, which means that only the area west to the Bay of Suez belongs to the African plate (Figs. 1.1 and 2.30).

In the Late epoch of Paleogene, the residual oceanic basins between African and European plates existed in 65–44 Ma, and the collision between these two plates was in Oligocene (35–33 Ma), forming the Alps great-scale overthrusts, nappe structures and ultra-high-pressure metamorphism (Cavazza et al. 2004; Martin et al. 2004). The Alps collision zone extended to the east, across the Carpathians, directly to the Toros accretion–collision zone [45] at the north side of Mediterranean Sea.

Delvaux and Barth (2009) used 347 fault plane solutions of earthquakes data, recognized that area suffered nearly N–S trending horizontal tectonic stress mainly. So the Red Sea rift zone develops to continue in recent.

2.5.18 Western Burma (Pegu Mountains–Rangoon) Plate (~510 Ma) [49]

The Western Burma (Pegu Mountains–Rangoon) plate (Fig. 2.30) used to be a part of Gondwana and formed the united crystalline basement in the Pan-African Tectonic Event (about 510 Ma). On the east side of this plate is the Bangongco–Nujiang–Mandalay–Phuket–Northern Barisan Cretaceous collision zone (100–66 Ma) [35], and on the west side of this plate is the Arakan–Sunda Cenozoic subduction and island arc zone [50] (Fig. 2.30). Since Middle Cambrian, this plate began to break up with the Gondwana and migrated northward in the Tethys Ocean (Figs. 3.7, 3.8, 3.10–3.12, 3.15 and 3.20). In Oligocene, it merged with the Eurasian continent (Fig. 3.27) and formed the Yarlung Zangbo–Myitkyina Paleogene collision zone [37]. The Western Burma plate might belong to the same plate with the Himalayan plate [38] originally and was cut off by Namjabarwa in Eastern Himalaya. Under the long-distance effect of the northward collision and compression of Indian plate and westward compression of Pacific plate (Wan 2011b), the Western Burma plate underwent clockwise rotation of nearly 90°, extending from nearly E–W trending to nearly N–S trending. Later in Neogene, there happened the fault-depression and relatively weak intraplate deformations.

In the Western Burma plate, there developed Pre-Ordovician metamorphic systems, with schists and migmatites, forming the crystalline basement of this plate. These metamorphic systems are rather similar to the Gaoligong Group in Yunnan, China (Lin et al. 2012), of which the 14 groups of zircon U–Pb isotopic ages were between 454.4–546.7 Ma. The average age was 489 $\pm$ 16 Ma, which means that these metamorphic systems were resulted from Pan-African Tectonic Event in Gondwana.

In Devonian–Triassic, there developed the sedimentary cover, which was mainly clastic rock system. In Jurassic–Middle Cretaceous, there developed volcanic and pyroclastic systems mainly with calc alkaline series (Chi et al. 1996). In Late Cretaceous, there formed the asphalt limestone. In Cenozoic, the formation of the Namjabarwa, Eastern Himalaya, caused the landform higher in the north and lower in the south as well as huge sediments almost covering the whole block in that period. At the bottom of Paleogene System, there developed the molasses accumulation. In Paleogene, there mainly developed marine facies turbidite sedimentary with oil and the continental facies coal-bearing strata. Between Eocene and Pleistocene, there existed obvious angular unconformity (Chen et al. 2010; Liu 2012).

The southwestern sea area of the Western Burma plate, Andaman Sea, is a NNE trending recent extension zone and now is a bathyal area. It was affected by the plate migration rate differences. The Indian plate migrated northward rather fast, while the Australian plate migrated rather slowly. This tectonic zone was developed under the control of the NNE trending subduction of the Indian plate and Sagaing fault (Fig. 2.33), and thus, there developed a series of nearly NW–SE trending extension-shear faults and formed fault-depression basins in the Andaman Sea (Fig. 2.30; He et al. 2011).

2.5.19 Arakan–Sunda Cenozoic Subduction and Island Arc Zone [50]

The Arakan–Sunda Cenozoic subduction and island arc zone (Figs. 2.30, 2.36 and 2.41) was mainly caused by the northeastward subduction of West Australian plate in

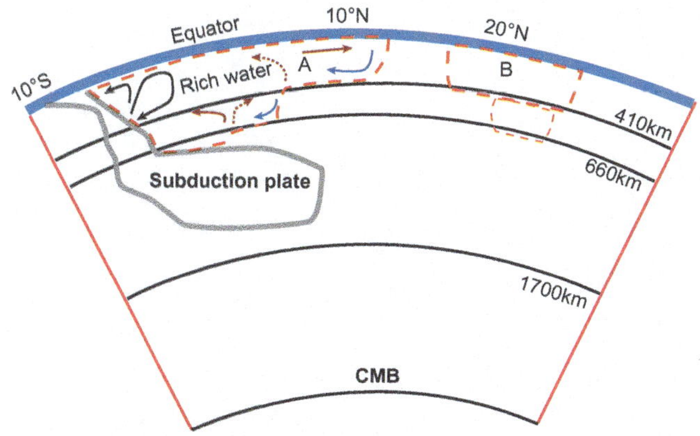

Fig. 2.41 Interpretation of seismic tomography for VI section of Sunda Cenozoic subduction and island arc zone (Modified from Replumaz et al. 2004)

Neogene. The eastern Arakan–Sunda Cenozoic subduction and island arc zone, i.e., the Banda basin and its surrounding islands, besides being affected by northward subduction of Australian plate, was influenced by the migrations to the west by north of Philippine Sea plate [65] and Pacific plate [69], and thus, there formed vertex structure at the eastern end of this zone.

Under the joint influence of recent Australian plate [72] and Pacific plate [69], the Southern Arakan–Sunda Cenozoic subduction and island arc zone, i.e., Sumatra Island area, is migrating mainly northeastward with a velocity of 2–3 cm/year according to GPS survey, while the eastern end of this zone undergoes northwestward migration with a velocity of 4–5 cm/year (Hall et al. 2011).

Based on the seismic tomography data (Replumaz et al. 2004), I, II and III sections in the Fig. 2.36 are for the Himalayan areas, which have been discussed before. IV–VII sections in Figs. 2.36 and 2.41 are the results of seismic tomography for the Sunda Cenozoic subduction and island arc zone. They show that the Australian plate was driving toward the NNE orientation for a depth of 410 km, and the subduction angle changed somewhat. Down below the depth of 660 km, this plate's rock density tends to be gradually consistent with that of the middle and lower mantle. These characteristics are similar to the deep parts of the world's most subductions or collision zones, in which most subductions only reached the middle mantle. According to the accurate wide angle seismic sections data, to the south of the middle Java there existed the fore-arc basin with a width of more than 50 km and a depth of 4 km (Kopp 2011). In this zone, there formed a low angle subduction zone, which subducted northward for 250 km with a depth only 40 km. This subduction and island arc zone is a recent strong volcanic and earthquake zone and is also a main section resulting tsunamis (seismic tidal wave).

South to the Arakan–Sunda Cenozoic subduction and island arc zone is the Australian plate (formed in about 500 Ma, [72] in Fig. 1.1). The Yalgon block of West Australian plate completed its cratonization at 2.7–2.6 Ga, and there happened rather strong tectono-thermal events at about 1200 and 800 Ma, which is different from the Indian plate. At the period of 600–500 Ma, the Australian plate and the other plates (Indian, Africa, South America and Antarctica) underwent the Pan-African Tectono-thermal Event, formed united crystalline basement and the Gondwana (Kennedy 1964).

From the whole Paleozoic to Middle Jurassic period, the Australian plate was a part of Gondwana. In Early Paleozoic, the horizontal migration of the Australian plate was very small with a little bit rotation, located near the equator (Figs. 3.6–3.8). Since the late period of Early Carboniferous, this plate began to migrate southward fast, thereafter the central reference point of this plate was always located between 50° S–60° S (Figs. 3.10–3.12). It began to extend and separate with the Gondwana in Late Jurassic–Early Cretaceous (150–100 Ma) and then migrated northward gradually, thus separating from the Indian and Antarctic plates (Figs. 2.42 and 3.20; Van der Voo 1993; Metcalfe 2011; Hall et al. 2011).

Based on the data of seismic tomography and seismic reflection profile data of multi-channel pre-stack migration, the Australian plate is now subducting northward with very low angle for more than 6.6 km deep. The thickness of oceanic crust of the subducting Australian plate is between 10 km and 20 km, and the plate could be subducted northward for more than 200 km stably, which was the record for the last ten million years. On the top of the subduction plate, traces of sea mounts and bottom of the ocean floor could be found. Beneath the oceanic crust is the oceanic mantle with water (Kopp 2011).

According to the sketch of paleo-continents reconstruction during the Late Jurassic (165 Ma) and Middle Eocene (45 Ma) for the Southeast Asia–Sunda areas (Fig. 2.42; Hall et al. 2011), it can be seen that this period was when the Indian plate migrated northward fast, and it was until the Late Cretaceous (80 Ma) that the Australian plate began to be separated from the Antarctic continent and migrated northward relatively slowly. From the Eocene, the Australian plate subducted partially underneath the Indonesian areas and gradually closed the channel between the Indian and Pacific Oceans (Figs. 2.42 and 2.43; Van der Voo 1993). In the Middle Eocene (45 Ma), the Australian plate began to be subducted with the Southeast Asian continent, thus forming the Arakan–Sunda Cenozoic subduction and island arc zone [50].

In Neogene, the paleo-continent reconstruction for the Australian plate and Southeast Asia could be referred to the

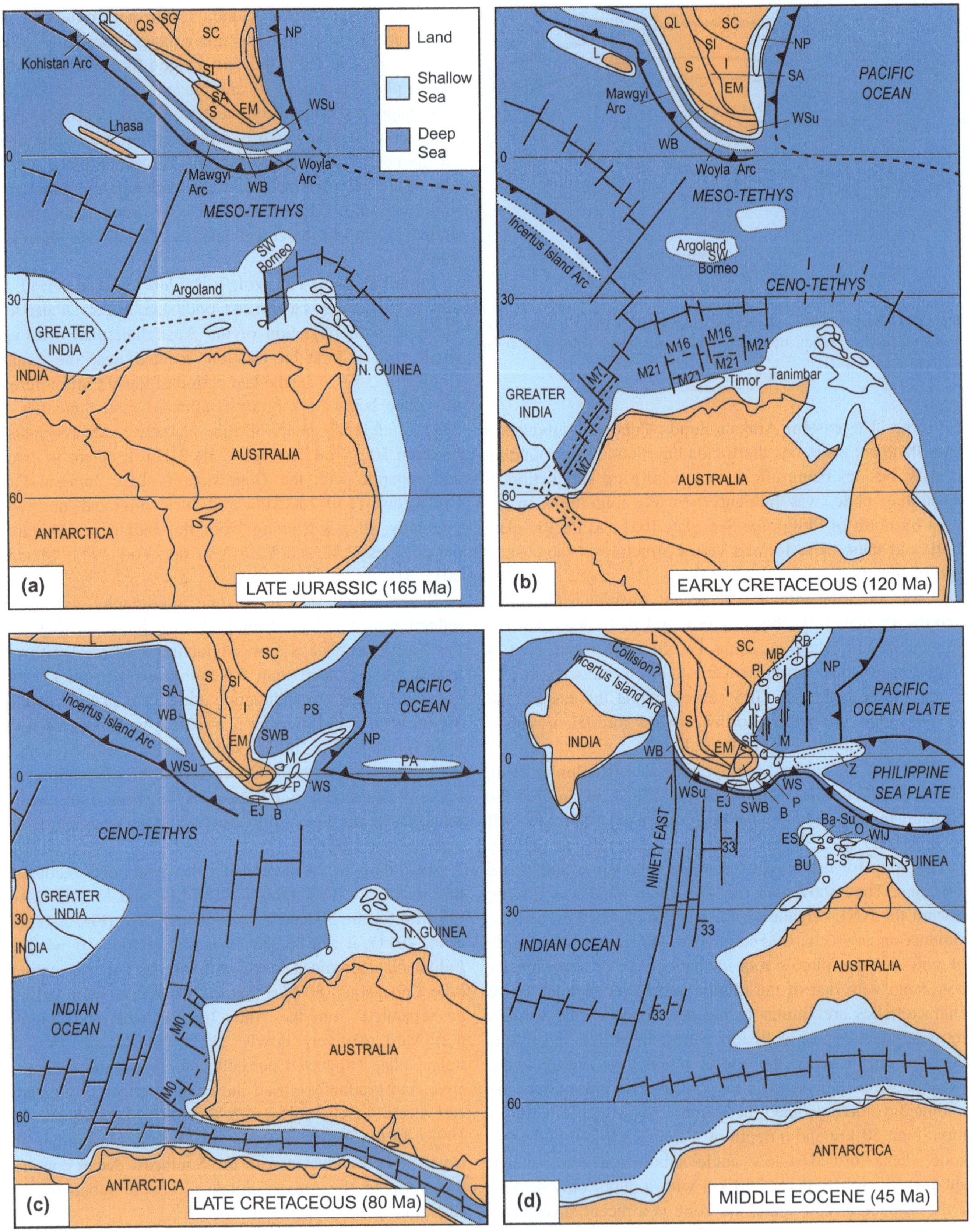

Fig. 2.42 Sketch of paleo-continent reconstruction in Jurassic–Paleogene for Southeast Asia–Sunda areas. SG. Songpan–Garze block; SC. South China; QS. Qamdo–Simao block; SI. Simao block; QL. Qiangtang block; S. Sibumasu plate; SA. Sukhothai block; I. Indosinian plate; EM. Eastern Malay block; WSu. Western Sumatra block; L. Lhasa block; WB. Western Burma block; SWB. Southwest Borneo block; NP. Northern Palawan and nearby small blocks; M. Mangole, Indonesia; WS. Western Sulawesi block; P. Pagununga; B. Balabac; PA. Eastern Philippine island arc; PS. Primitive South China Sea; Z. Zambales ophiolite; ES. Eastern Sulawesi; O. Obi–Bakan; Ba-Su. Bangei–Sula; Bu. Butong; WU. Western Irian–Java; M. No. of magnetic anomaly belt in the Indian Ocean (After Hall et al. 2011)

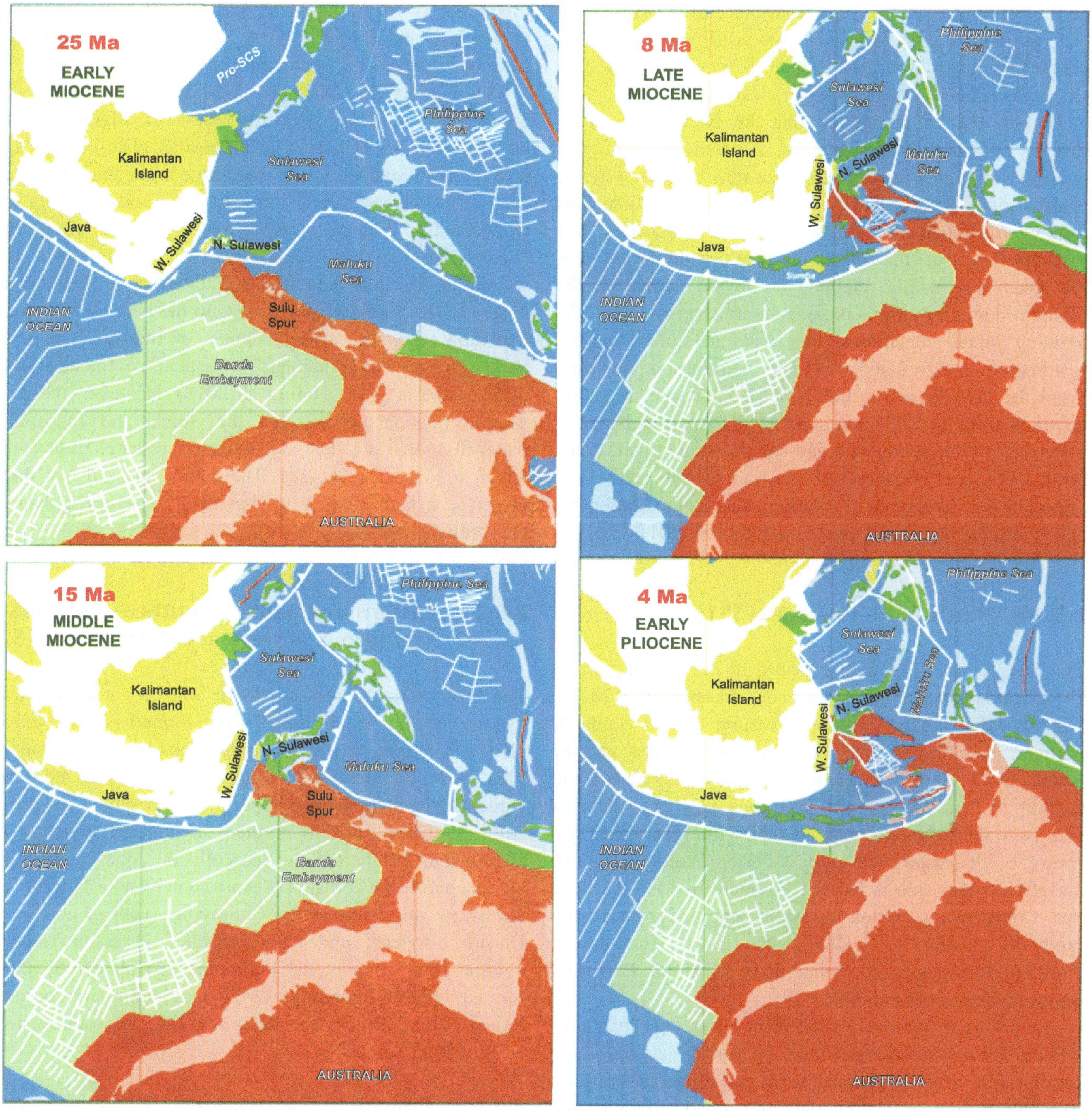

Fig. 2.43 Sketch of paleo-continent reconstruction in Early Miocene (25 Ma), Middle Miocene (15 Ma), Late Miocene (8 Ma) and Early Pliocene (4 Ma) for Southeast Asia–Sunda areas (After Hall et al. 2011)

results of Hall et al. (2011) (Fig. 2.43). In that period, the velocity and force of the northward migration of Australian plate were both relatively small. Since Neogene, the northward subduction velocity of Australian plate was only 2 cm/year. So it was in the last 8 million years, i.e., until Late Miocene, that the northward migration of Australian plate has played significant role in the tectonics of Southeast Asia. However, in the tectonic research of Sunda–Southeast Asia, some scholars only considered the influences of Australian plate, which is not enough obviously. It is necessary to pay attention to the westward migration and subduction of the Pacific plate. From Fig. 2.43, in Neogene (25–4 Ma) the northward migration velocity and distance of the Australian plate are not so great. However, it is just this plate convergence that makes the water channel between the Indian Ocean and Pacific Ocean going to closing gradually.

Hall et al. (2011) inferred that the Neogene collision strongly affected the shape of Australian continental boundary, of which the characteristics may be caused by the Jurassic rift and the rollback of the plate subduction, thus forming Australian continent surrounded by the ocean.

Since Cenozoic, the Australian plate (including its oceanic areas) has mainly migrated northward. However, the northern boundary of this plate is an arc shape, so the NW trending Sunda Trench make this island arc suffer from NE trending compression and strike-slip, i.e., transcompression. From the Java Trench to the Timor Trench, the migration direction turns to ENE trending, so this arc island mainly suffered from the subduction and compression toward to the north. This arc boundary of Australian plate was very important for the tectonic evolution and shape of Indonesia and Philippine areas in Cenozoic (Hall et al. 2011).

2.5.20 Sunda Plate (~500 Ma) [51]

The Sunda plate (Fig. 2.30), located at the middle part of Great Sunda Islands, includes the southeastern end of Malay Peninsula (including Singapore), eastern end of Sumatra, northern boundary of Java Island, southwestern part of Borne, and the Java Sea. The western side of this plate is the Central Malaya Triassic collision zone [33], the southern side is Arakan–Sunda Cenozoic subduction and island arc zone [50], the eastern boundary is Eastern Argo block [54], the northeastern side is Eastern Kalimantan–Southern Sulu Sea Cretaceous accretion–collision zone [52], while the northern boundary is jointed with Eastern Hindukush–Northern Qiangtang–Indosinian plate [27] and Palawan–Sarawak–Zengmuansha block [29].

According to the data of fragment zircon age in Southwestern Borneo and Eastern Java, the Sunda plate has many isotopic ages from the Archean to the end of Neoproterozoic, so it can be inferred that this plate was a part of Gondwana, and the united crystalline basement was formed in the Pan-African Tectonic Event (Hall et al. 2011). The age of crystalline basement for Sunda plate is quite different from that of Indosinian plate [27] ($\sim$800 Ma).

Metcalfe (2011) recognized that it was since Jurassic that the Sunda plate began to be separated with Australian plate (Fig. 2.42; Hall et al. 2011). In most areas of the Sunda plate, the Late Cretaceous–Paleocene strata are missing, which may be the result of uplifting and erosion of this block due to the northward subduction of Australian plate (Clements et al. 2011). Since Eocene, this area began to subside and deposit, which may be related with the further subduction of Australian plate, making the uplift zone move to Sunda island arc zone on its southern side. The Sunda plate merged into Eurasian continent after Eocene.

Pubellier (2008) considered that both the Indosinian and Sunda plates used to belong to Gondwana, which is not in line with the actual data. The Indosinian plate has the age data of about 800 Ma for its crystalline basement (Lan et al. 2003), but the crystalline basement of Sunda plate is the Gondwana type and might have suffered from the Pan-African Tectonic Event ($\sim$500 Ma). According to the seismic tomography data, the crust thickness of Sunda plate is about 20–25 km, which belongs to the ocean–continental transition type. The crust thickness of Indosinian plate is about 35 km which belongs to typical continental crust (Teng and Lin 2004; Hall et al. 2011). However, the connection relationship between the Sunda and Indosinian plates till now is not clear, due to lack of research of this area which is located in the sea. The author guessed that there might be a fault (Fig. 2.30), but there is no more data to prove this assumption.

2.5.21 Eastern Kalimantan–Southern Sulu Sea Cretaceous Accretion–Collision Zone [52]

The Eastern Kalimantan–Southern Sulu Sea accretion–collision zone [52] (in Fig. 2.30) was formed in Cretaceous, along which there existed many outcrops of ophiolite suites. This collision zone may be connected with the Bangongco–Nujiang–Mandalay–Phuke–Northern Barisan Cretaceous collision zone [35] at the northern Java Island. The Eastern Kalimantan–Southern Sulu Sea collision zone was formed by the continental collision of Australian plate [72] during its northward migration in the Cretaceous (Hall et al. 2011). During Neogene, under the influence of southwestward oblique subduction of the Philippine Sea plate [65], there formed NE–SW trending wedge fault-depression oceanic basins (Fig. 2.29) at the Sulu Sea and the southwestern South China Sea [28].

2.5.22 Sulawesi Sea Block (500 Ma) [53]

The northwestern side of Sulawesi (or Celebes) Sea block ([53], Fig. 2.30) is the Eastern Kalimantan–Southern Sulu Sea accretion–collision zone [52] and Eastern Argo block [54], while the southeastern side is the Philippines–Moluccas Cenozoic subduction and island arc zone [64]. The Sulawesi Sea block used to belong to the Australian plate and began to separate with the Australian plate since Jurassic (Metcalfe, 2011). In Paleogene, they were completely separated. Since Neogene, being influenced by the southwestward migration and subduction of Philippine Sea plate and the relatively weaker northward subduction of Australian plate, this block suffered rather strong rock deformations (Hall et al. 2011) and merged into Eurasian continent. Based on the recent GPS data, the recent plate migration orientations are mainly north or NW trending (Fig. 2.29).

It should be mentioned that the shapes of oceanic floor strong deformation areas are all large in the east and narrow in the west (Fig. 2.29) for the South China, Sulu, Sulawesi and Benda Seas. However according to the isotopic ages of oceanic floor basalts, those oceanic floors extended and become fixed in Paleogene–Neogene. The Sunda–Southeast Asia areas between the Philippine Sea plate and Indian Oceanic plate obviously suffered from the NE–SW trending maximum principal compression stress in the Neogene and the extension orientation was NW–SE trending. Depending on the shapes of oceanic floor strong deformation areas, it infers that the WSW compression of the Philippine Sea plate was stronger, while the influence of Australian plate was weaker at that time. However, according to the recent GPS data, the influence of Australian plate is relatively stronger (Fig. 2.29).

2.5.23 Eastern Argo Block (500 Ma) [54]

The Eastern Argo block (Fig. 2.30) is distributed near the Molucca Passage, its northwestern side is connected with the Sunda plate [51] and Eastern Kalimantan–Southern Sulu Sea Cretaceous accretion–collision zone [52] and its northeastern side is connected with the Sulawesi Sea block [53] and Philippines–Moluccas Cenozoic subduction and island arc zone [64], while its southeastern and southern side are both connected with the Arakan–Sunda Cenozoic subduction and island arc zone [50]. This block used to belong to the Australian plate [72] began to separate from the Australian plate since Jurassic (Metcalfe 2011; Figs. 2.42 and 2.43) and merged into Eurasian continent in Cenozoic.

According to the data of seismic sections, underneath the unconformities of Cenozoic strata, in the depth of Eastern Argo block there may be exist sedimentary systems from Cambrian to Triassic, with a thickness of about 8.5 km, and those sedimentary systems should belong to the Australian plate (Granath et al. 2011). In Cretaceous, this block was subducted and collided with the Sunda plate [51] and Palawan–Sarawak–Zengmuansha block [29], forming the Eastern Kalimantan–Southern Sulu Sea accretion–collision zone [52]. Since Neogene under the influences of northward compression and subduction of Australian plate [72] and its nearly E–W trending extension, there have formed the nearly N–S trending Makassar Strait, which was composed of three NNE trending en echelon extension fault zone with some dextral strike-slip. It means that the northward migration and its velocity of the west wall of fault zone are greater than those of the east wall. There is no problem that the intraplate fault-depression of Eastern Argo block was derived by the compression of Australian plate. However, north to the Eastern Argo block, such as the fault-depression oceanic basins of South China, Sulu, Sulawesi and Benda Seas (Fig. 2.29), do not seem to be results of the subduction and compression of the Australian plate.

2.5.24 Northern New Guinea Island Arc Zone (Since Neogene) [55]

The New Guinea block (Figs. 2.30 and 2.44) had belonged to the Australian plate [72] (Fig. 2.43) until the early epoch of Pliocene (4 Ma). The Northern New Guinea island arc zone is located at the northern part of New Guinea block. Since Neogene, under the influences of westward migration of the Philippine Sea plate [65] and the Pacific plate [69] and northward migration of the Australian plate [72], the Northern New Guinea island arc zone was formed and sinistral strike-slip and compression occurred between this arc zone and the New Guinea block, making the Northern New Guinea block and Halmahera island possess tectonic characteristics of island arc zone (Metcalfe 2011), and there also developed some ophiolite suites. In the western island arc zone, influenced by the westward migration of the Pacific and Philippine Sea plates, the WNW toward horizontal migration velocity reached 7–8 cm/year at the western end of this arc zone by GPS data (Hall et al. 2011).

2.6 Western Pacific Tectonic Domain

The Western Pacific tectonic domain (Fig. 2.44) includes the following tectonic units: Bering Sea basin (Jurassic–Paleogene) [56], Sikhote–Alin–Koryak Cretaceous–Paleogene accretion–collision zone (130–23 Ma) [57], Okhotsk plate ($\sim$850 Ma) [58], Aleutian–Kamchatka–Kurile–Northeast Japan Cenozoic subduction and island arc zone ($\sim$40 Ma) [59], Japan Sea Neogene fault-depression basin (23 Ma–) [60], Japan Median Tectonic Line (Cretaceous sinistral strike-slip zone) [61], Neocene South Honshu–South Shikoku Ryukyu subduction and island arc zone [62], East Taiwan sinistral strike-slip fault (Neogene–Quaternary) [63], Philippines–Moluccas Cenozoic subduction and island arc zone [64], Philippine Sea plate (since Paleogene) [65], Izu–Bonin–Mariana (IBM) Cenozoic subduction and island arc zone [66], the lithosphere-type transformation line of Okhotsk–Western Dahingganling–Middle Shanxi–Wuling Mountains–Tak of Thailand (since Jurassic) [67] and the western part of the Pacific plate. They are all influenced and developed by the formation and migration of the Pacific plate. Therefore, it is necessary to discuss the formation and migration process of the Pacific plate here.

Fig. 2.44 Sketch of Western Pacific tectonic domain [56–67]. The black numbers show the tectonic units, as same as those in the CONTENTS and Fig. 1.1

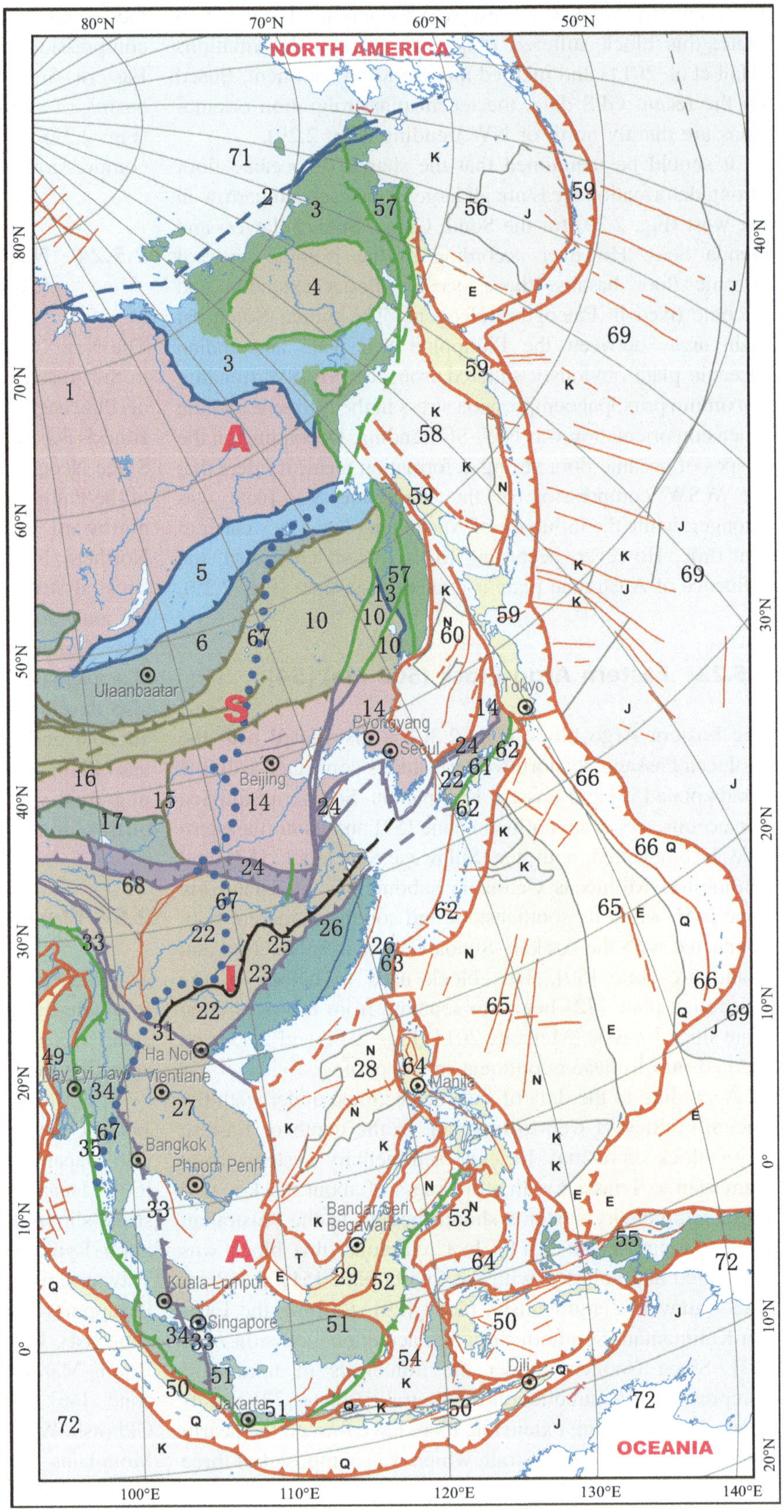

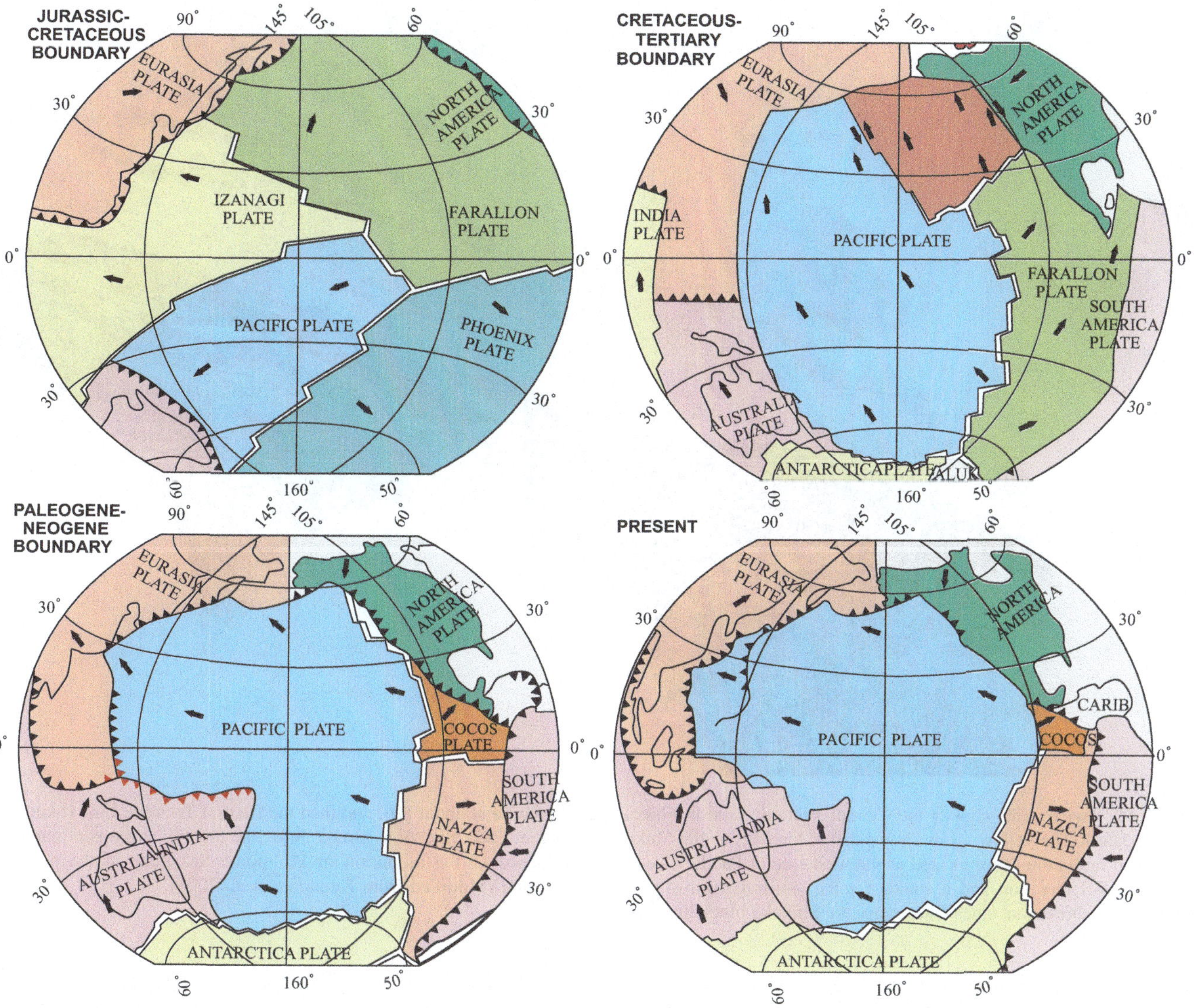

Fig. 2.45 Paleo-tectonic reconstruction since Jurassic in Paleo-Pacific areas (After Moore 1989; from Wan 2011b)

Before Cretaceous, there was no the Pacific Ocean in the eastern part of Asian continent, nor should it be called the Paleo-Pacific Ocean. During and before Triassic, the ocean on the east side of Asian continent should belong to the Tethys Ocean. In Jurassic, the Izanagi plate was formed in the east of the Asian continent. Most of Izanagi plate had subducted underneath the Japan Islands gradually since Jurassic (Maruyama et al. 1997).

According to the results of magnetic anomaly research of the third generation (Moore 1989), the Pacific plate was formed initially in the Southern Hemisphere, subducted to underneath the Australian plate during Late Jurassic (upper left of Fig. 2.45). In Late Jurassic, roughly in the current Pacific areas, each plate showed a radial movement pattern. That is, the Pacific plate (Pacific) in the Southern Hemisphere was a small new oceanic plate that was swept southwest and dipped into the Australia plate (upper left of Fig. 2.45). In that period the Izanagi plate subducted toward NW, underneath to the Asian continent, the Farallon plate toward NE, underneath to the North American plate, the Phoenix plate toward SE, underneath to the South American plate.

In Late Cretaceous, being influenced by the plate radiate extension whose center was located near Antarctic, the Pacific plate and all the other continental and oceanic plates were migrated to the north and expanded their areas. The Pacific plate then extended northward and migrated to the Northern Hemisphere (Moore 1989) (upper right of Fig. 2.45). Between the Pacific Ocean and Asian continent, there developed a great sinistral strike-slip, with a length of

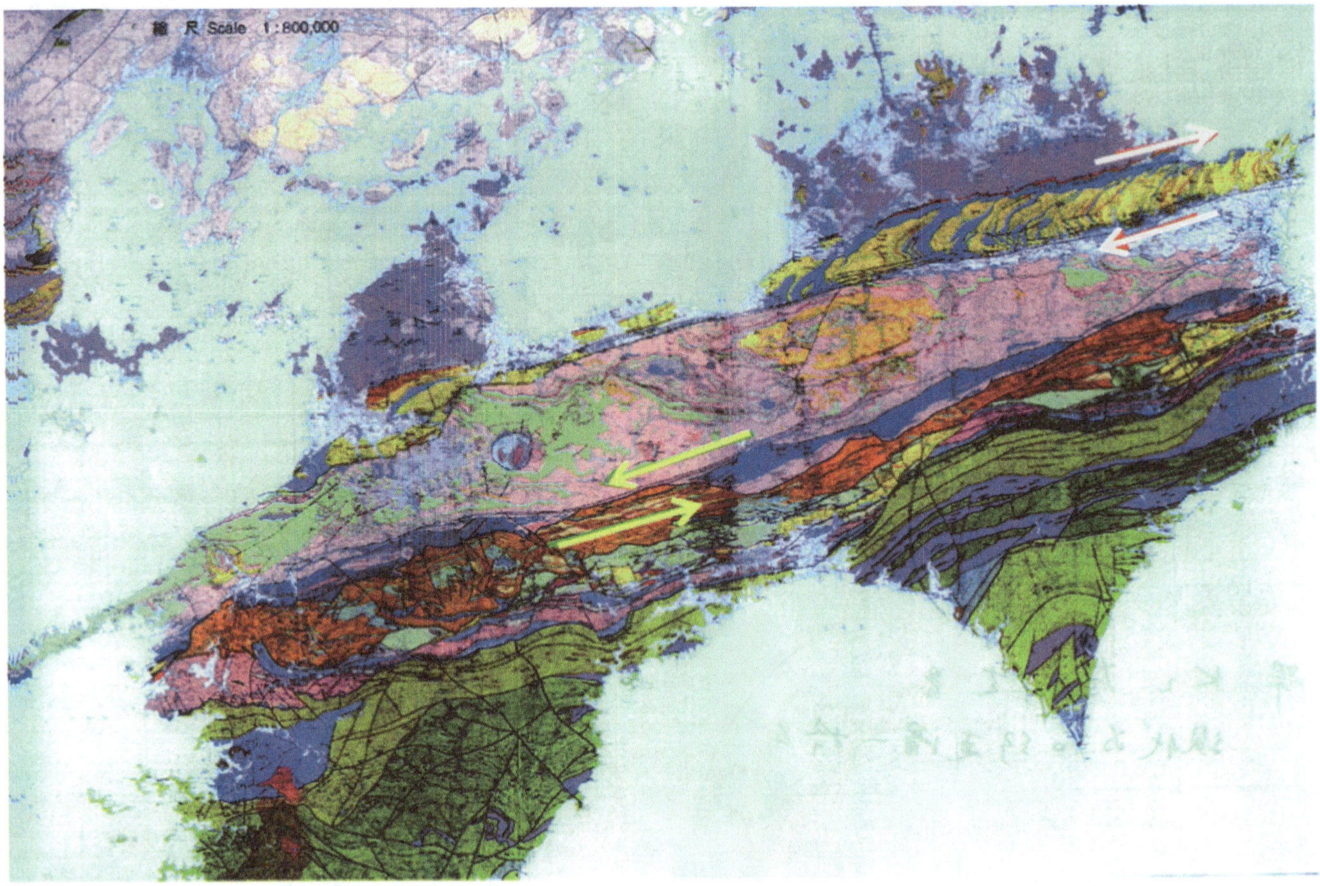

Fig. 2.46 Twice strike-slips of the Central Tectonic Line in Shikoku, Japan Islands. The yellow arrow shows that in Cretaceous the Shikoku block and fault zone were located at the west side of Pacific Ocean in the near N–S direction and it stands for the sinistral strike-slip. The white arrow shows that since Quaternary the Shikoku block has rotated clockwise at about 80°, and then the Central Tectonic Line (fault zone) of the ENE direction turned into the dextral strike-slip under the influence of the migration of Philippine Sea plate in the northwest direction (Modified from Kunugiza et al. 2001)

about 1000 km (Yoshikura et al. 1990), and its typical result was the Central Tectonic Line [61] in the Shikoku, Japan Islands (Fig. 2.46, the yellow arrow showing the sinistral strike-slip) (Kunugiza et al. 2001). As a result, the Western Pacific tectonic domain described herein existed only in and after Cretaceous.

In Middle–Late Eocene (43–35 Ma), influenced by the North American–Caribbean meteor impact event (Glass 1982; Yin 1996; Wan et al. 1997; Wan 2011b), the Pacific plate suddenly changed its migration orientation to the WNW. The stretching direction of Emperor–Hawaii seamount chain abruptly turns from NNW to WNW (Raymond et al. 2000) (lower left of Fig. 2.45 and the migrating trace of 43–0 Ma in Fig. 2.47). This orientation of plate migration caused the stronger compression and deformations in 30 Ma at the Japan Islands and Taiwan Island. In Japan, it is called "Takachiho Movement" and in the Taiwan Island it is called "Puli Orogeny."

After that period, the Pacific plate has basically maintained its characteristics of migration and compression in the WNW direction (Fig. 2.47) and extension in the NNE direction, and then, the recent Pacific plate has been formed gradually (lower right of Figs. 2.45 and 2.47).

Synthesizing the research data by many scholars, the author gets the following results concerning the migration orientations and velocities since Late Jurassic:

At the period of 140–125 Ma, the Shatsky uplift and Typhoon Island subducted into the Australian plate, with a velocity of 10–15 cm/year; at 125–110 Ma, the Hess uplift and Japan Islands migrated toward NW direction with minimum velocity in the 125–95 Ma, and in 110–100 Ma, the hot spot migration velocity was also very slow. During Middle Cretaceous–Eocene (100–43 Ma) Wentworth, Musicians and Imperial Islands migrated to the NNW direction, i.e., the Pacific plate underwent great extension and migrated from the Southern Hemisphere to the Northern Hemisphere. During the period of 95–81 Ma, the migration velocity reached 19.8 (−0.8/+1.2) cm/year, while at 43 Ma the velocity was only 3.8 cm/year. Since Middle Eocene (43 Ma), the Hawaii, Samoa, Easter and Foundation Islands

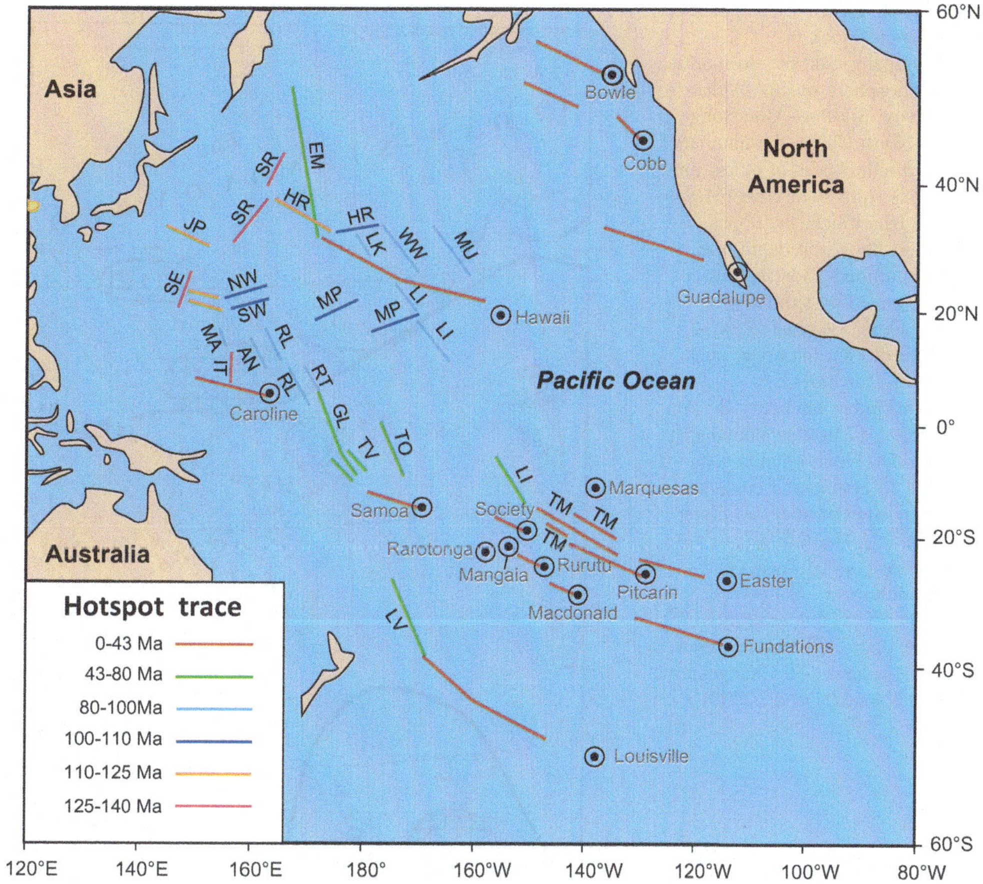

Fig. 2.47 Hot spot migration trace and the change of orientations in the Pacific plate since Jurassic (Synthetically redrawn from Bartolini and Larson 2001; Engebretson et al. 1985; Koppers et al. 2001, 2003; Northrup et al. 1995; Tarduno and Cottrell 1997)

have all migrated to WNW direction. The migration velocities were 7.7 cm/year at 40–20 Ma, 6.0 cm/year at 20–10.5 Ma, 10.6 cm/year since the period of 10 Ma. The above recent results are rather similar to the earlier magnetic anomaly data (Moore 1989; Norton 1995).

The north boundary of Pacific plate is now Aleutian Trench; the west boundary is Kanchak–Kuril Trench, Japan Trench and Izu–Bunin–Mariana (IBM) Trench; its south boundary is the trench of Melanesia Islands (Fig. 2.44); and the east boundary is the western border of North American plate and the Eastern Pacific Ocean Ridge. Since Quaternary, it has maintained an overall westward velocity of about 10 cm/year, which is the main dynamic source to form the current Western Pacific subduction zone, volcano activities, island arc and the earthquakes at the boundary of plate or intraplate. Now the thickness of oceanic crust in Western Pacific is only 6–8 km (Rodnikov et al. 1985).

In Middle Cretaceous–Eocene (100–43 Ma), being influenced by great northward migration of the Pacific plate, the Midian Tectonic Line of Japan [61] (with sinistral strike-slip) (shown as yellow arrow in Fig. 2.46), Sikhote–Alin–Koryak Cretaceous–Paleogene accretion–collision zone [57] and the two fault-depression oceanic basins of Bering Sea basin [56] and Okhotsk plate [58] (Fig. 2.44) were formed.

Now it can be found that the Aleutian–Kamchatka–Kurile–Sakhalin–Northeast Japan Cenozoic subduction and island arc zone [59], South Honshu–South Shikoku–Ryukyu Neogene subduction and island arc zone [62], Philippines–Moluccas Cenozoic subduction and island arc zone [64], Izu–Bonin–Mariana Cenozoic subduction and island arc zone [66], etc. were all formed in Late Oligocene–Neogene. The formation and fault-depression for Philippine Sea plate were the result of extensions in Paleogene and Neogene (Hall et al. 1995), and the oceanic basin of Japan Sea has been formed by the fault-depression since Neogene (Tamaki et al. 1992; Jolivet and Tamaki 1992, 1994; Yoon, 2001).

The west boundary of Sikhote–Alin area is also a transition line of East Asian lithosphere type, i.e., the boundary between the continental type and continental crust–oceanic mantle type (Okhotsk–Western Dahingganling–Middle Shanxi–Wuling Mountains–Tak of Thailand [67]).

Karsakov et al. (2008) listed five relatively hot areas and four relatively cold areas based on the magnetotelluric data (Fig. 2.48). The above phenomena sometimes may be related to the temperature in the lithosphere, or may be related to

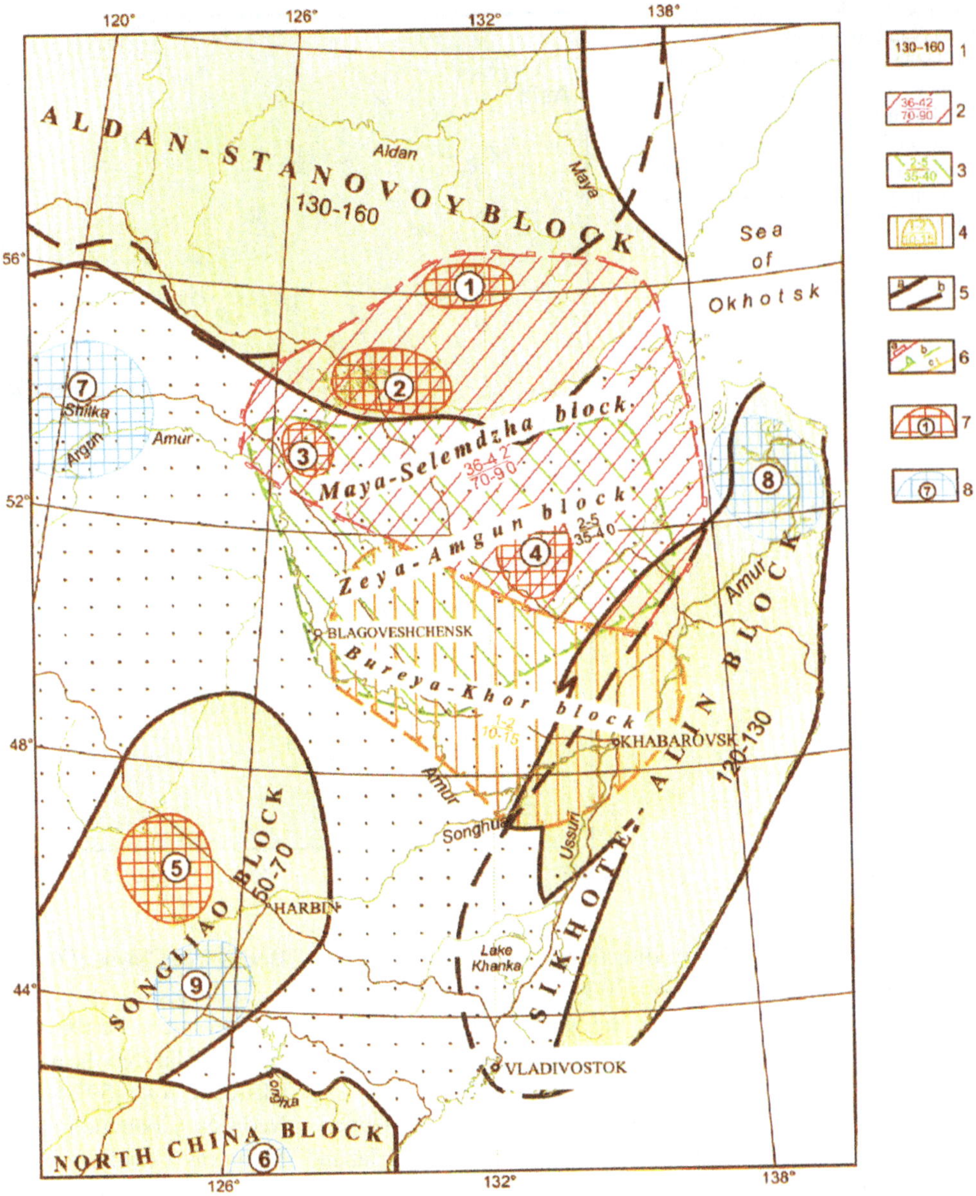

Fig. 2.48 Distribution sketch of lithosphere block in Heilongjiang–Sikhote–Alin areas. (1) Lithosphere blocks (Aldan–Stanovoy, Sikhote–Alin, Song–Liao, North China, the numbers show the thickness of block, unit: km), (2) equivalent crust (Maya–Selemdzha block, the upper numbers show top depth, the lower numbers show bottom depth, unit: km, similarly hereinafter), (3) crust (Zeya–Amgun block); (4) upper crust (Bureya–Kor block); (5) lithosphere boundary (a. Top boundary, b. Bottom boundary); (6) hidden block boundary (a. similar to continental crust, b. crust, c. upper crust); (7) hotter section (① Toko, ② Verkhnezeya, ③ Zeya, ④ Verkhnebureya, ⑤ Lower Nenjiang); (8) colder section (⑥ Yalong Jiang, ⑦ Verkhneamur, ⑧ Nizhneamur, ⑨ Changchun) (After Karsakov et al. 2008)

the containing water or conductivity. Obviously, the problem has multi-solution.

2.6.1 Bering Sea Basin (Jurassic–Paleogene) [56]

The Bering Sea basin (Fig. 2.44) is located in the north of Aleutian-Kamchatka-Kurile-Sakhalin-Northeast Japan Cenozoic subduction and island arc zone [59]. After the northward subduction of the Kula plate during Cretaceous (upper right of Fig. 2.45), the Pacific plate [69] continued to subduct and squeeze northward. The Bering Sea basin was a residual oceanic basin of the Kula plate. Nowadays, at the bottom of ocean basin there are mainly the Jurassic rocks, and a small part of the Cretaceous and Paleogene oceanic crust. The magnetic strips on the oceanic ridge appear as the N–S direction, which could be the Mesozoic residues (117–132 Ma). Using those magnetic anomalous strips, the migration orientation of the Kula–Farallon Pacific plate in Late Mesozoic–Paleogene could be reconstructed. In that period, the Kula plate subducted northward to the Aleutian subduction and island arc zone. At about 70 Ma (Late Cretaceous) that subduction zone was located on the south side of modern Aleutian Trench. Since Paleogene, the Bering Sea basin has been a rather stable sea basin on the eastern margin of Eurasian continent (Cooper et al. 1976).

The continental margin of Bering Sea is adjacent to the East Siberia and Alaska Peninsula. In the lower part (with depths at 1500–2000 m) on the basement slope, the Late Jurassic shallow water sandstone system was deposited with the thickness about 7–10 km, which was covered by the Paleogene–Miocene shallow water mudstone unconformably. The fauna fossil data show that in the sea basin, there had been a depth of several kilometers in Late Paleogene, with the characteristics of an oceanic plate (Marlow et al. 1982).

2.6.2 Sikhote–Alin–Koryak Cretaceous–Paleogene Accretion–Collision Zone (130–23 Ma) [57]

The Sikhote–Alin–Koryak Cretaceous–Paleogene accretion–collision zone (Fig. 2.44) is located at the east end of Eurasian continent. That zone is the collision zone at the eastern and southern margins of Kolyma–Omolon plate [4]. Now the oldest granite isotopic age is about 300–290 Ma (Early Permian) in this area (Wang et al. 2014). That collision accretion zone is a subduction–island arc zone derived from the northward migration and subduction of the Pacific plate [69] during Cretaceous, which has the characteristics of the Andes-type active continental boundary. It formed the island arc during Middle Paleogene ($\sim$40 Ma), thereafter it became the collision accretion zone (Parfenov et al. 2009). In Russia, some researchers call that zone “the Pacific Tectonic Zone” (Karsakov et al. 2008) (Fig. 2.49). The Koryak collision accretion was formed rather earlier, in Late Cretaceous (Parfenov et al. 2009). The crust thickness of Sikhote–Alin is about 30 km (Rodnikov et al. 1985).

In addition, the southern section of Koryak is the split zone between the Kolyma–Omolon plate [4] and Okhotsk plate [58], and however, till now researchers never found any outcrops (Fig. 2.44). The lithosphere thickness of Southern Koryak is usually 60–70 km, which is an ocean–continental transitional-type lithosphere. The crust thickness of Sikhote–Alin areas is about 35–40 km, and the lithosphere thickness is usually 120–130 km, belonging to continental-type lithosphere (Karsakov et al. 2008).

2.6.3 Okhotsk Plate (~850 Ma) [58]

The Okhotsk plate and Kolyma–Omolon plate [4] used to belong to the same plate (Figs. 2.44 and 2.49). They collided to the Eurasian plate in Jurassic and Cretaceous. Since Late Cretaceous, being suffered from the near N–S compression and shortening, the Sikhote–Alin–Koryak fault zone [57] turned into a normal fault and a strong basic magmatism developed in the deep (Tessensohn and Roland 1998). The deep density of Okhotsk crust increased, the fault-depression became a sea area in the Okhotsk area, but no oceanic basin has been found. Later, influenced by the Aleutian–Kamchatka–Kurile–Sakhalin–Northeast Japan Cenozoic subduction and island arc zone [59], the Okhotsk plate suffered from mild intraplate deformations.

The thickness of Okhotsk lithosphere plate is generally greater than 50–60 km, and only four sections are as thick as 100 km (Fig. 2.49), which may also belong to the ocean–continental transitional-type lithosphere. Although the above thickness data might not be very accurate (according to the gravity, magnetism and magnetotelluric data; Karsakov et al. 2008), it can be inferred that the Okhotsk plate is a fault-depression continental block. The average depth of sea water in the Okhotsk Sea is only 821 m, and the maximum depth of sea water is 3916 m. There is no oceanic characteristic completely, and it is a continental fault-depression basin. The Okhotsk plate [58] and Kolyma–Omolon plate [4] may be the same plate in early time (Figs. 2.44 and 2.49). They were cohered into the Eurasian continent together during Jurassic–Cretaceous.

Some Japan scholars consider the Okhotsk plate and Kolyma–Omolon plate as a part of North American plate, but the evidences are insufficient. The Russian researchers do not agree with that point of view (Petrov et al. 2008; Pospelov 2008).

Parfenov et al. (2009) recognized that the crystalline basement is composed of Archean (oldest U–Pb age is 3.7 Ga) and Proterozoic gneiss and schist, and above those systems it is covered with the Neoproterozoic clastic and carbonate rocks, Early Cambrian limestone, marble and sandstone, Early Ordovician conglomerate, sandstone and limestone. Above them, the Middle Devonian limestone, sandstone, shale and conglomerate, and Late Devonian rhyolite, fusion tuff, andesite and tuff cover unconformably, upward are the Carboniferous–Late Jurassic clastic rock systems of non-marine and local marine facies. The Okhotsk plate is similar to the stratigraphic development of the Korima–Omoron plate on the north side, which is a fragment of the Siberian plate. In Late Devonian or Early Carboniferous, the Okhotsk area was in the rift state. Wang et al. (2014) got the ophiolite isotopic age of 260–250 Ma on the north border of Okhotsk plate. It means that area had the oceanic crust evidences during Late Permian. The Okhotsk plate did not overgrow to the edge of the East Asian continent until Late Jurassic. In recent years, some scholars have considered that the Okhotsk plate was floating from a

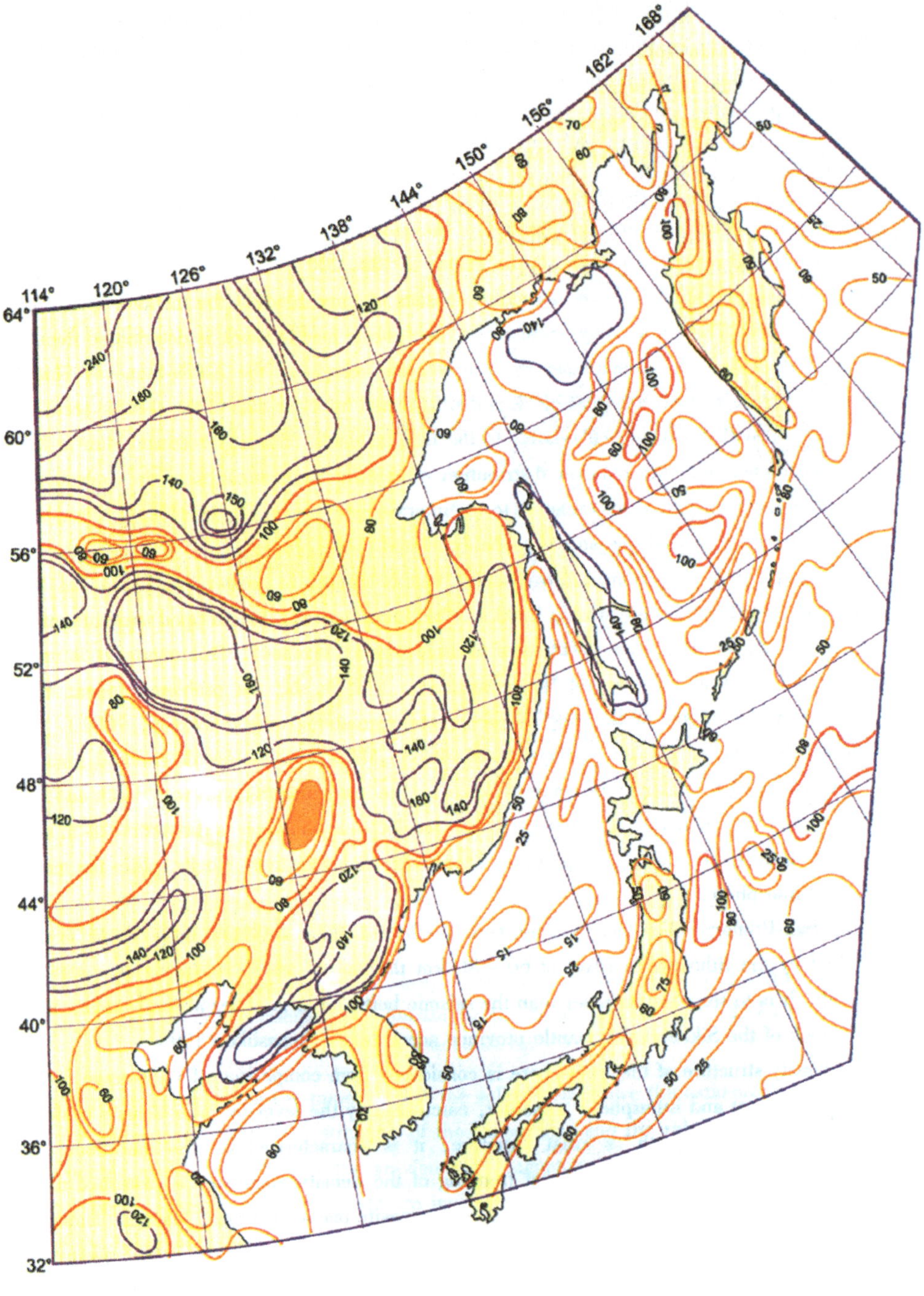

Fig. 2.49 Contours of lithosphere thickness (unit: km) nearby the Okhotsk Sea and Japan Sea (After Karsakov et al. 2008)

distance in Cretaceous and collided with the Siberian land. However, there is still not enough evidence to prove the Okhotsk plate has been moved from a distance.

2.6.4 Aleutian–Kamchatka–Kurile–Sakhalin–Northeast Japan Cenozoic Subduction and Island Arc Zone (~40 Ma) [59]

The Aleutian–Kamchatka–Kurile–Sakhalin–Northeast Japan (east to the Tanakura Tectonic Line, which is the boundary between the Northeastern and Southwestern Honshu, Japan) island arc zone is located on the west side of Pacific plate. The Tanakura Tectonic Line is the boundary between the Asian plate and the Western Pacific island arc zone (Fig. 2.44), and the subduction zone is about 200 km east of Western Pacific island arc zone.

The Pacific plate [69] migrated to NNW before the Cretaceous–Paleocene, and in the Late Cretaceous–Early Paleocene (90–50 Ma), the southern section of the zone showed the sinistral strike-slip feature (Fig. 2.46, shown as yellow arrow). Since Middle and Late Eocene (40 Ma), the Pacific

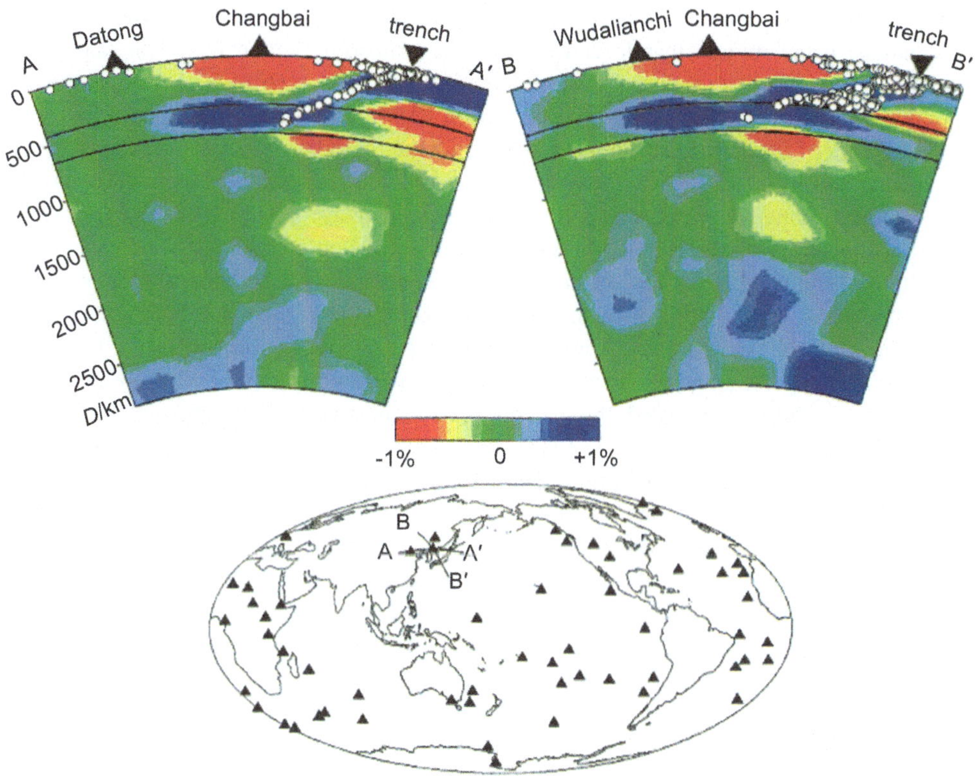

Fig. 2.50 Seismic tomography section for Western Pacific plate subduction (After Zhao and Liu 2010)

plate turned to WNW migration, and the subduction and island arc zone was the manifestation of the Pacific plate's westward migration and subduction to the Eurasia plate. Near the Central Tectonic Line (i.e., Shimandogawa) at Shikoku, Japan Islands, the tectonic mélange and oceanic fragments are observed clearly, and the strongest deformation period is about 40–30 Ma (Fig. 2.46). In Japan, it was formerly called "Takachiho Movement." The crust thickness of Honshu, Japan is about 30 km, which is obviously thicker than that of the Japan Sea and the Pacific Ocean (Rodnikov et al. 1985).

The westward subduction velocity of Pacific plate is about 10 cm/year, and the subduction is still continuing. The Sendai East M_s 9.1 earthquake (2011-3-11) was the most recent activity for that subduction–island arc zone. The Kamchatka island arc was formed in Late Cretaceous to Oligocene and covered by the Late Paleocene–Miocene sedimentary systems (Parfenov et al. 2009).

Using the seismic tomography data, Zhao and Liu (2010) compiled the section of Western Pacific plate subduction zone (Fig. 2.50), from the Western Pacific Trench across Changbai Mountains to Datong of Shanxi or to Wudalianchi. The cold subduction zone (high density, blue in Fig. 2.50) penetrated to the middle mantle about 660 km underneath Changbai Mountains, and it displaced horizontally along the middle mantle, extending forward over 1000 km until the image not clear. That is, the temperature and the density of the cold oceanic lithosphere inserting into the middle mantle increased to the level close to those of the deep mantle. Thus, from this image it can be found that the influence of the Pacific plate subduction zone on the shallow tectonic deformation of the East Asian continental lithosphere is not significant. Many researchers considered that the subducting plate could bring about hot mantle uplift which leads to the thinning of the East Asian lithosphere (Zhao and Liu 2010; Zhu et al. 2012). This statement seems to be plausible in the Changbai Mountains area in Northeast China (Fig. 2.50). However, in North China there has not been any phenomenon of hot mantle uplift, or the indication of mantle plume. Recently, the ages of the mantle inclusions that have been measured are all Archean–Paleozoic, and there is no evidence of any Mesozoic or Cenozoic inclusions and strong activity.

Jiang (2008) researched the 3D velocity structure form the Moho discontinuity to the depth of 700 km in the Kamchatka area, using 768 far earthquakes arrival time data got from 16 seismic stations, and the seismic tomography data made by Zhao and Liu (2010). The imaging results clearly show two major velocity anomalous characteristics: First, the Pacific plate with high-velocity subducts into 660 km deep underneath the Southern Kamchatka area; from south to north, the subduction depth gradually becomes shallower, and near the joint area of Kamchatka and Aleutian, that phenomenon almost disappears. Second, materials of the low-velocity asthenosphere with high temperature exist in the Northern Kamchatka and underneath the subduction zone. He also discovered two high-velocity anomalies in the middle mantle and beneath the transition zone and recognized that they are the fragment of Komandorsky plate, which subducted in the period of 10 Ma and the fragment of Pacific plate dropped down during the period of 2 Ma.

Although many scholars have conducted many researches on the subduction of the Pacific plate in Japan, the detailed structure of the plate is not very clear, such as the change of plate thickness, the variation of seismic velocity, the detailed subduction state and the existence problem of olivine sub-stable wedge. Using the 3D beam trace positive modeling, Jiang (2008) calculated the Pacific plate's average thickness of 85 km according to 333 far seismic data. Using about 3283 seismic data (the source depths are more than 40 km), the velocity anomaly distribution in the plate is tested in sections. The results show that the velocity anomaly decreases with the increase of depth, which is related to the temperature variation in the mantle. According to the above research, he got the oceanic crust subduction depth of 110 km beneath the Northeast Japan and Hokkaido, the oceanic crust's thickness between 7.5 and 5 km, and the velocity anomalies of 1 and –3%. This shows that when the oceanic crust is subducted to a depth of 110 km, it will be dehydrated and degraded due to temperature and pressure until it merges with the plate. By analyzing the location relationship of earthquake focus and oceanic crust, he considered that the earthquakes near the upper part of the plate are triggered by the oceanic crust dehydration and brittleness of the oceanic crust. It is found that most of the deep earthquakes occurred inside the sub-stable wedge.

The crust thickness of Kamchatka–Kurile–Sakhalin–Northeast Japan is about 30 km. The lithosphere thickness of those is often 50–60 km, but some partial areas up to 75–100 km (Karsakov et al. 2008; Fig. 2.49). It belongs to the ocean–continental transition-type lithosphere.

In the Kurile–Sakhalin area, there is a Cenozoic active tectonic zone on the west side of Okhotsk plate [58], and the lithosphere thickness is increased obviously to 140 km. However, the above data were determined by gravity, magmatism and magnetotelluric sounding, but not very accurate (Karsakov et al. 2008; Fig. 2.49). Here, it belongs to continental-type lithosphere. In those areas, it developed a near S–N fault, with sinistral strike-slip feature; its southern part could extend to the Tanakura Tectonic Line in Japan. It controlled the opening of oceanic basin of Japan Sea and promoted the Southwestern Honshu (Japan) block to rotate clockwise about 90° (Tamaki et al. 1992; Yoon 2001). The above fault zone also controlled the formation of oil-gas-bearing depression basins in the Western Sakhalin.

Recently, the "rollback" phenomenon near the subduction zone has been paid attention by many scholars. That is a term cited from political economics to describe the geological phenomena. Its original intention is "gradual decrease," "gradual weakening," "flip" or "price reduction." Geologically, the term is used to describe the partial convolute folding of the upper wall of a subduction zone when a plate or block subducts. This phenomenon is limited and only occurs in the range of tens to hundreds of kilometers on the upper subduction zone. In recent years, some researchers expect to use it to explain the extension in the East China continent (thousands of kilometers away from the Japanese island arc), which is very debatable.

2.6.5 Japan Sea Neogene Fault-Depression Basin (23 Ma–) [60]

In the initial period of the Plate Tectonic Theory, the Japan Sea fault-depression basin (Fig. 2.44) was regarded as the "arc back basin" (Hsu 1988) behind the Cenozoic subduction–island arc belt in Northeast Japan. However, the Deep Sea Drilling Project (DSDP) and geophysical data showed that Japan Sea basin was formed by fault-depression with near E–W extension which is vertical to the subduction zone; the fault-depression occurred at the period of 16–1 Ma (Neogene–Early Pleistocene), later about 10 Ma than the formation of subduction and island arc. It means the subduction and Japan Sea fault-depression basin were completely not formed at the same time (Tamaki et al. 1992; Jolivet 1994; Yoon 2001). Chen et al. (2015) using the volcanic rock sample in ODP 794 drilling hole at the East Japan Sea got the newer basalt age of 13–17 Ma and the older age of 17–23 Ma; they were all formed in Neogene. It is obvious that the fault-depression basin was formed in Neogene, rather than at the same time as island arc and subduction zone.

For the past 20 years, the Japanese researchers (Tamaki et al. 1992; Jolivet 1994; Yoon 2001) have considered that the Japan Sea fault-depression basin was the product of sinistral strike-slip faulting of Kurile–Sakhalin (Tanakura Tectonic Line). At the same time, the Philippine Sea plate strongly subducted northward and led to the clockwise rotation about 90° of the Southwestern Honshu block (Japan, eastern end of Yangtze plate). That is, the southwestern part of block turned from near N–S trending to near E–W trending (Fig. 2.51). The above recognition more fits the geological fact. Thus, it seems that the assumption of the "arc back basin" proposed by Hsu (1988) was inconformable with the geological data. Therefore, in recent years the

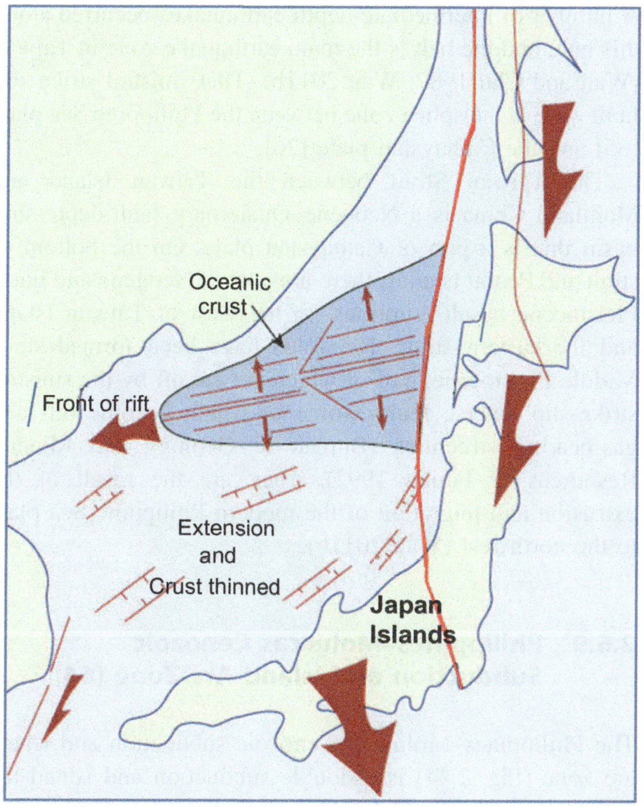

Fig. 2.51 Formation mechanism of Japan Sea. The red arrows show the blocks migration trending. Due to the Philippine Sea plate was stronger subducting to the northward, the southwest Honshu (Japan) block changed orientation from near N–S trending to the E–W trending. However, this figure does not show the Philippine Sea plate subduction orientation, but only the extension trending for the Japan Sea basin (After Tamaki et al. 1992; Yoon 2001)

international tectonic geologists have only used the term "trench and island arc system," instead of the "trench, island arc and basin system." However, some sedimentary scholars still believe that behind the trenches and island arcs, there must be a so-called back-arc basin.

The crust thickness of Japan Sea is only 12–15 km, which lithosphere thickness is about 25 km (depending on the gravity, magnetism and magnetotelluric sounding data), and on the marginal side of Japan Sea, the lithosphere thickness can reach 50 km (Fig. 2.48). In its central part, it is exposed with the oceanic crust. Restricted by the Paleoproterozoic metamorphic blocks, it is a Neogene extension and depression at the continental margin, instead of the back-arc basin of the Kamchatka–Kurile–Sakhalin–Northeast Japan Islands (Rodnikov et al. 1985; Karsakov et al. 2008).

Using the explosion seismic body wave to research the upper mantle structure of Western Pacific areas, Rodnikov et al. (1985) discovered that the low-velocity layer of mantle is at depth of 100 km, with the thickness of 30–40 km and velocity of 8.4–8.6 km/s. The temperature in the low-velocity layer of mantle is about 1200 °C. Beneath the Japan Sea, in the same depth the seismic velocity is only 7.7 km/s. It means that the "ductile" asthenosphere observed under the Japan Sea may not exist at all in the Northwestern Pacific Ocean.

2.6.6 Japan Median Tectonic Line (Cretaceous Sinistral Strike-Slip Zone) [61]

The Japan Median Tectonic Line is a sinistral strike-slip fault zone located in the middle of Shikoku, Japan Islands (Fig. 2.44 and the yellow arrows in Fig. 2.46), and is a boundary fault zone between the Southwest Japan (eastern extension of Yangtze plate [22]) and South Honshu–South Shikoku–Ryukyu Neogene subduction and island arc zone [62]. Osozawa (1994) considered it to be the Japan Middle Cretaceous accretion complex zone. It was the "high-pressure low-temperature metamorphic zone" first proposed by Miyashiro Akiho (1961). At that time, many researchers considered it to be a derivative of the oceanic plate subduction. However, in recent years, the Japanese researchers have changed their original understanding, and reconsidered that the "high-pressure low-temperature metamorphic zone" is actually the Cretaceous sinistral strike-slip fault zone. The so-called low-pressure high-temperature metamorphic zone on its north side is the Jurassic tectono-magmatic zone. The two metamorphic zones were not formed in the same period, nor were they direct originated from the oceanic plate subduction (Yoshikura et al. 1990; Osozawa 1994, 1998).

Before Neogene, the Japan Median Tectonic Line was in the near N–S direction. In Middle and Late Cretaceous, the Pacific plate extended rapidly and migrated northward, while the Asian continental plate was rather stable with slow northward migration, leading to the formation of the Japan Median Tectonic Line (Cretaceous sinistral strike-slip zone) (Osozawa 1998). In Neogene following the extension of Japan Sea and the northward subduction of Philippine Sea plate, the Southwest Japan area rotated clockwise by almost 90°, and the Median Tectonic Line turned from S–N to E–W direction. Continually influenced by the NW subduction of the Philippine Sea plate [65], the Japan Median Tectonic Line shows features of a dextral strike-slip fault zone (shown as white arrows in Fig. 2.46).

2.6.7 South Honshu–South Shikoku–Ryukyu Neogene Subduction and Island Arc Zone [62]

The South Honshu–South Shikoku–Ryukyu Neogene subduction and island arc zone (Fig. 2.44) is the subduction–island arc zone between the Philippine Sea plate [65] and

Eurasian continent plate. Since Permian, there were 3–4 oceanic ridges subducting and colliding with this area. In the period of 250 Ma, the oceanic ridge between Akiyoshi and Farallon plates underwent dextral strike-slip with the Japan block. In the period of 115 Ma the Izanagi plate migrated northward and formed the sinistral strike-slip fault with the length of more than 1000 km between the arc zone and the Japanese block. In the period of 90 Ma the Izanagi plate subducted and squeezed toward the northwest. In the periods of 65–30 Ma, 65–30 Ma, the northern part of the Japanese block was affected by the northward subduction of the Kula plate, while the southern part was affected by the westward subduction of the Pacific plate and the Northern New Guinea block (formerly known as the Takachiho Movement, represented by the unconformity between Paleogene and Neogene at the South Kyushu, the main tectonic event occurred in about 45–30 Ma). Since Neogene, the Northern Japan block has been affected by the westward subduction of Pacific plate with the velocity about 10 cm/year, while the southern part has been controlled by the Philippine Sea plate subduction to the NNW direction (Osozawa 1994). Till now, the Philippine Sea plate continually subducts northwestward with a velocity of 6 cm/year. This subduction–island arc zone is a modern strong earthquake and volcanic active belt. The island arc is also the main occurrence of Kuroko deposit.

On the west side of Ryukyu Neogene island arc zone, there is the Ryukyu Trough which is also a fault-depression basin. On the west of the Ryukyu Trough, it is the Diaoyu Island uplift. The Diaoyu Island uplift is obviously integrated with the Taiwan Island, and both islands belong to a same tectonic unit, i.e., the east boundary area of the Cathaysian plate (Fig. 2.24).

2.6.8 East Taiwan Neogene–Quaternary Sinistral Strike-Slip Fault Zone [63]

The main part of Taiwan Island belongs to the Cathaysian plate [26], and its eastern margin is developed with the East Taiwan sinistral strike-slip fault zone (Fig. 2.44).

In Paleogene, the zone used to be a subduction zone that sloped westward, making the eastern part of Taiwan characterized by island arc. That tectonic event was originally called "Puli Orogeny" (Cheng 1994; Stephan et al. 1986; Bureau of Geology and Mineral Resources of Fujian 1992), which is corresponded to the Takachiho Movement in Japan. The climax of the island arc tectonic-thermal event is at 30 Ma. Later, the latitude of fault surface gradually became steeper and almost became erect. Since Neogene, the extension direction of Philippine Sea plate has turned from east to NW, making the fault a sinistral strike-slip fault zone. A number of intermediate-depth earthquakes occurred along this belt, and the belt is the main earthquake zone in Taiwan (Wan and Chu 1987; Wan 2011b). That sinistral strike-slip fault zone is the splice zone between the Philippine Sea plate [65] and the Cathaysian plate [26].

The Taiwan Strait between the Taiwan Island and Mainland China is a Neogene–Quaternary fault-depression basin that is a part of Cathaysian plate. On the bottom of strait and Penhu Islands, there are a lot of Neogene and Early Pleistocene basalt eruptions. In the west of Taiwan Island and the eastern strait, 105 folds have been formed since Middle Pleistocene, half of which are cut off by the sinistral strike-slip reverse faults, some of which become oil- and gas-bearing structures (Bureau of Geology and Mineral Resources of Fujian 1992). They are the result of the extrusion and migration of the modern Philippine Sea plate to the northwest (Wan 2011b).

2.6.9 Philippines–Moluccas Cenozoic Subduction and Island Arc Zone [64]

The Philippines–Moluccas Cenozoic subduction and island arc zone (Fig. 2.44) is a double subduction and island arc zone among the Southern Philippine Sea plate [65], South China Sea fault-depression basin [28], Palawan–Sarawak–Zengmuansha block [29] and Sulawesi Sea block [53]. On the east side of the zone, the Philippine Sea plate subducted westward, while on the west side, the South China Sea fault-depression basin, Palawan–Sarawak–Zengmuansha block and Sulawesi Sea block subducted eastward. As a result, the Philippines–Moluccas Cenozoic Double subduction and island arc zone was formed, which has been a strong deformation zone since Paleogene and also a strong modern volcanic and earthquake zone. According to the data of GPS, that zone is moving toward the NW 290° with the velocity of 6–7 cm/year (Hall and Blundell 1995).

2.6.10 Philippine Sea Plate (Since Eocene) [65]

The Philippine Sea plate (Figs. 2.28 and 2.44) is an extension-type oceanic plate. In Paleogene, due to the migration and subduction of the Pacific plate [69] toward the WNW direction, the Izu–Bonin–Mariana (IBM) Cenozoic subduction and island arc zone [66] was formed, and the subduction zone was very steep (70°–80°). In the period of 30–15 Ma, in the west side of Izu–Bonin–Mariana (IBM) island arc zone, it formed the oceanic crust extension zone in N–S direction. The extension may be related to the oceanic mantle hot fluid uplift, forming the original Philippine Sea basin. It is a typical arc back basin (Fig. 2.28; Hall et al. 1995).

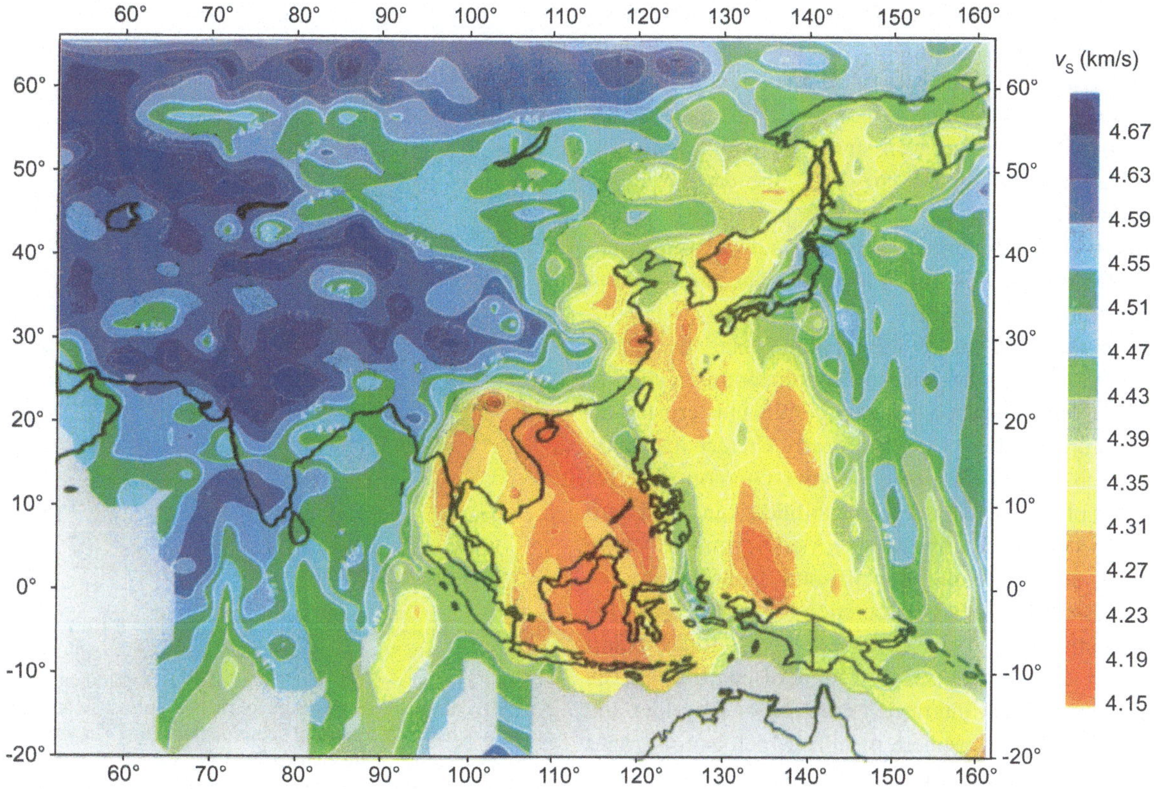

Fig. 2.52 Giant low-velocity belt (depth of 130 km) in the upper mantle in the East Asian–Philippine Sea plate (After Zhu et al. 2005)

It now appears that the "arc back basin" can only be formed when the plate subduction zone is steep.

Since Neogene, the Philippine Sea plate has continually migrated toward NW and has been obliquely subducted to the South Honshu–South Shikoku–Ryukyu Neogene subduction and island arc zone [62]. At the same time, the NW transcurrent extension faults were developed with extension in NNE–SSW direction in the ocean basin of the Philippine Sea, making the Philippine Sea plate form the rhombus boundary (Fig. 2.28) and subduct in the direction of northwest at the Northwestern Philippine Sea plate, but subduct in the direction of SSW on the southwest side of Philippine Sea plate (Hall et al. 1995, 2011).

In the surrounding areas of Philippine Sea plate and South China Sea, the transition velocities of S-wave are between 4.35–4.15 km/s in the depth of 130 km below the lithosphere. It is a giant low-velocity zone (Fig. 2.52) (Zhu et al. 2005). Perhaps this may explain why the Philippine Sea plate and the areas around the South China Sea still maintain the active oceanic plate subduction.

2.6.11 Izu–Bonin–Mariana (IBM) Cenozoic Subduction and Island Arc Zone [66]

In Cretaceous, as the Pacific plate expanded and moved northward, the IBM zone was once a sinistral strike-slip fault. Since Paleogene, this IBM zone has been a high-angle subduction and island arc zone (Figs. 2.28 and 2.44; Hall et al. 1995).

2.6.12 Lithosphere-Type Transformation Line of Okhotsk–Western Dahingganling–Middle Shanxi–Wuling Mountains–Tak of Thailand [67]

In the Asian continental lithosphere, there are widespread intraplate deformations and magmatisms, which is a very peculiar phenomenon on the Earth. Usually, the intraplate deformations of continental plates are only limited in the range of 200–300 km at the edge of the plate. However, in the East Asian continent, especially east to the Okhotsk–Western Dahingganling–Middle Shanxi–Wuling Mountains–Tak of Thailand areas (east to the blue point line; Figs. 1.1 and 2.44) developed intraplate deformation over thousands of kilometers, as well strong tectono-magmatism and metallogenesis (Wan 2011b). This line can be called "East China Magmatic Line" for the petrologists (Chen 2004). Most of the endogenic deposits and old mines are located in the east of this line. This particular phenomenon needs to be discussed in detail.

Based on the analysis of the existing data, the author believes that this singular tectonic phenomenon is related to the Jurassic plate migration and the continental crust rotation. In the Jurassic–Early epoch of Early Cretaceous (200–135 Ma), North American plate, Kolyma–Omolon plate [4] and Okhotsk plate [58] migrated and collided southwestward with the Siberian plate, forming the Verkhojansk–Chersky Jurassic accretion–collision zone [3] and Transbaikalia (or Mongolia–Okhotsk) Jurassic accretion–collision zone [5] and making the East Asian continental crust rotate counterclockwise for 36°–20° above Moho discontinuity. The Asian continental crust migrated to the top of the older and more stable oceanic lithosphere mantle (which isotopic ages of mantle inclusions are all Archean–Paleozoic, with only slight or without perturbations) (Wan 2011b; Wan and Lu (2014), turning the lithosphere of East Asian continent (in the east of the Okhotsk–Western Dahingganling–Middle Shanxi–Wuling Mountains–Tak of Thailand line and the Western Pacific Trench) into a very special continental crust–oceanic mantle-type lithosphere, also known as the ocean–continent transition-type lithosphere. The thickness of that lithosphere is only 70–80 km (including continental crust about 35 km and oceanic mantle about 40–50 km) (Wan and Lu 2014) (Figs. 2.53 and 3.37).

Based on the analysis of the existing data, the author believes that this singular tectonic phenomenon is related to the Jurassic plate migration and the continental crust rotation. In the Jurassic–Early epoch of Early Cretaceous (200–135 Ma), North American plate, Kolyma–Omolon plate [4] and Okhotsk plate [58] migrated and collided southwestward with the Siberian plate, forming the Verkhojansk–Chersky Jurassic accretion–collision zone [3] and Transbaikalia (or Mongolia–Okhotsk) Jurassic accretion–collision zone [5] and making the East Asian continental crust rotate counterclockwise for 36°–20° above Moho discontinuity. The Asian continental crust migrated to the top of the older and more stable oceanic lithosphere mantle (which isotopic ages of mantle inclusions are all Archean–Paleozoic, with only slight or without perturbations) (Wan 2011b; Wan and Lu 2014), turning the lithosphere of East Asian continent (in the east of the Okhotsk–Western Dahingganling–Middle Shanxi–Wuling Mountains–Tak of Thailand line, and the Western Pacific Trench) into a very special continental

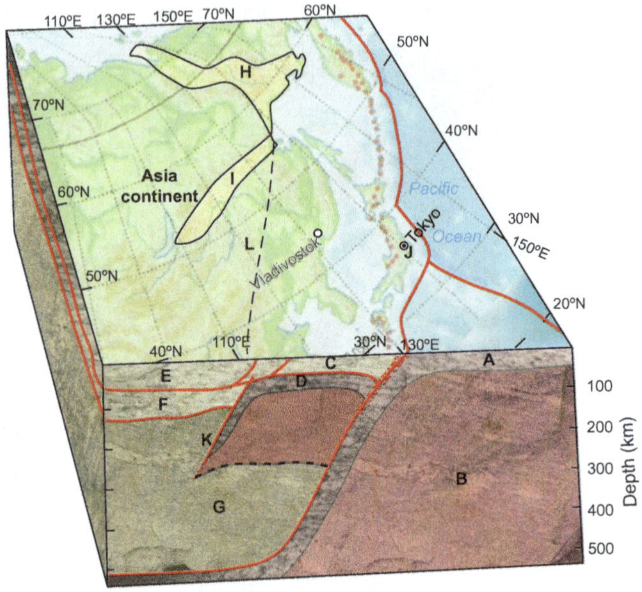

Fig. 2.53 Lithosphere tectonics model for East Asia. **A**. Oceanic lithosphere (40–50 km thick); **B**. upper and middle mantle, beneath oceanic lithosphere; **C**. moderately thinned continental crust in East Asia; **D**. paleo-oceanic lithosphere mantle in East Asia (including some oceanic crust); **E**. normal continental crust in Asia; **F**. normal continental lithosphere mantle in Asia; **G**. upper and middle mantle, beneath continental lithosphere, with a little bit disturbance; **H**. Verkhojansk–Chersky Jurassic (200–135 Ma) accretion–collision zone; **I**. Transbaikalia (or Mongolia–Okhotsk) Jurassic (about 140 Ma) accretion–collision zone; **J**. Japan island arc; **K**. inferred paleo-subduction zone, between paleo-continent and paleo-ocean in mantle; **L**. the surface boundary line between the thin lithosphere (with continental crust and oceanic lithosphere mantle) and the common continental lithosphere in East Asia (After Wan and Lu 2014)

crust–oceanic mantle-type lithosphere, also known as the ocean–continent transition-type lithosphere. The thickness of that lithosphere is only 70–80 km (including continental crust about 35 km and oceanic mantle about 40–50 km) (Wan and Lu 2014) (Figs. 2.53 and 3.37).

However, in the Asian continent to the west of Okhotsk–Western Dahingganling–Middle Shanxi–Wuling Mountains–Tak of Thailand line, it still maintains the characteristics of normal continental lithosphere (the continental crust is 35–40 km thick, the continental lithosphere mantle is 70–100 km thick and the total thickness is 100–150 km). The area to the east of the Western Pacific Trench (Aleutian–Kamchatka–Kurile–Sakhalin–Japan–Ryukyu–East Philippines) is of course a typical oceanic lithosphere (the thickness of oceanic crust is less than 10 km and the thickness of oceanic mantle is about 40–50 km) (Karsakov et al. 2008).

That is to say, during Jurassic, through the rotation of the surface crust of the lithosphere, part of the continental crust of East Asia moved and covered the paleo-oceanic mantle, forming a special continental crust–oceanic mantle-type lithosphere in East Asia. The above recognition is very different from that of many researchers (such as Cai et al.

2002; Zhu et al. 2002; Wu et al. 2008; Zhu et al. 2008; Zhu et al. 2011, 2012). Most of them emphasized the active mantle, which caused the bottom of the lithosphere underplating or being disassembled, resulting in the decrease in the lithosphere thickness. According to their viewpoints, the deep mantle in East Asia should be hot and more active since Jurassic, so the mantle source in the crust should have the Mesozoic–Cenozoic isotopic ages. However, the current data cannot support their assumptions. The recent research of mantle rocks suggested that the mantle magma xenoliths in the crust were formed in Archean or Paleoproterozoic and only experienced a slight disturbance (Lu et al. 2006, 2010).

In East Asia, due to the fact that the temperature of asthenosphere is basically uniform, about 1280 °C, the geothermal gradient inside the thinner lithosphere is obviously increased and higher, making the intraplate deformation easy to enhance. Under the influence of plate migration in the surrounding area, the Moho discontinuity, detachment in middle crust and regional faults in the continental crust–oceanic mantle-type lithosphere in East Asia are easy to induce local decompression and warming. As a result, the magma source area is easy to form, and the magma activity is obviously intense, resulting in the formation of a large number of endogenous metal deposits closely related to magmatism during Jurassic–Cretaceous (Wan and Lu 2014).

According to the research results of Zhu et al. (2002), it seems that the west side of lithosphere-type transitional line is still the crust in the depth of 40 km, so the v_s is low, less than 4 km/s (yellow and red areas in Fig. 2.54). However, the east side of that line is the oceanic mantle, which v_s is higher than 4 km/s (green areas in Fig. 2.54). The above information may indicate that in East Asia, the boundary line of lithospheric type reflected by S-wave seismic tomography data is the lithosphere-type transformation line of Okhotsk–Western Dahingganling–Middle Shanxi–Wuling Mountains–Tak of Thailand in this monograph (Fig. 2.53). Here,

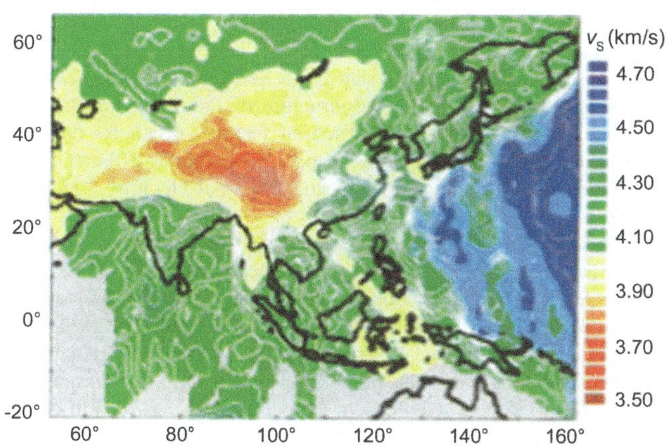

Fig. 2.54 S-wave seismic tomography of Asia and Western Pacific Ocean at the depth of 40 km (After Zhu et al. 2002)

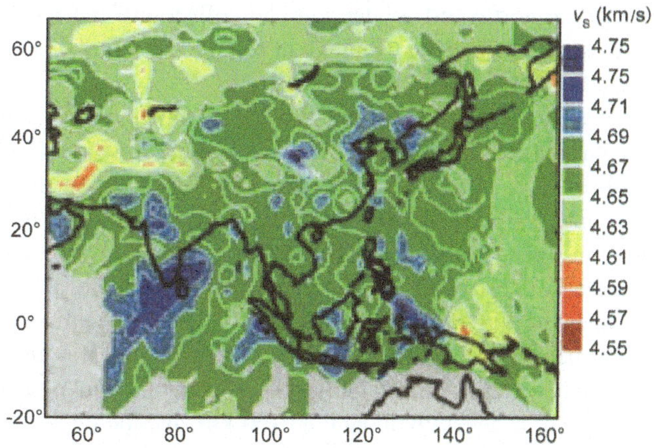

Fig. 2.55 S-wave seismic tomography of Asia and Western Pacific Ocean at the depth of 400 km (After Zhu et al. 2002)

the results of seismic tomography are very consistent with the geological analysis.

According to the seismic tomography data in the depth of 130 km (Fig. 2.52), the interface is in the upper mantle below the lithosphere. In the west of the lithosphere transformation line, the v_s is more than 4.43 km/s; however in the east of that line, the v_s less than 4.43 km/s. It means that in the upper mantle at the depth of 130 km, beneath the East Asia and its adjacent ocean, it may have stronger thermal activity or be rich in the ultra-critical fluid which may result from the stronger subduction of oceanic plate in this area; however, in the inner area of Asian continent, the thermal activity is weak and rare in the ultra-critical fluid. As to the depth of 400 km, the v_s seismic tomography data shows almost no difference between East and West Asian continents (Fig. 2.55), indicating that the physical properties have become uniform in the depth of the mantle.

Just for the special lithospheric tectonics of the East Asian continent, and the influence of the movement of the surrounding plates, a series of special and strong intraplate deformations tectono-magmatism and giant metallogenesis have taken place in the upper part of the lithosphere in East Asia since Jurassic.

2.7 Thickened Continental Lithosphere Region in Qinghai–Xizang (Tibet)–Pamir [68]

The Qinghai–Xizang (Tibet)–Pamir thickened continental lithosphere region [68] does not belong to the Western Pacific tectonic domain. The main extent of this region is in the Gondwana tectonic domain, and the northern and eastern parts of the thickened continental lithosphere have involved the Sino–Korean and Yangtze tectonic domains. It is due to the sharp thickening of the continental lithosphere caused by the strong northward subduction and collision of the Indian plate since Cretaceous (limited in the yellow point line in Figs. 1.1, 2.30, 3.36 and 3.39).

The East Asian lithospheric-type transformation line (continental and continental crust-type lithospheric boundaries [67]) has been discussed in the previous section. That is to say, there is a thin (70–80 km thick), continental crust-type lithosphere in the transition zone between the eastern continent of Asia and the ocean. Most of the Asian continental lithosphere, with a thickness of between 100 and 160 km, is a normal continental lithosphere. However, in the Qinghai–Tibet–Pamir areas, there is the thickened continental lithosphere. The total thickness of lithosphere is generally more than 160 km, and the thickest is up to about 200 km. The thickness of crust can also reach about 60–70 km. The huge and low-density rock systems caused uplift of the area, forming the greatest and highest plateau on the Earth (with average elevation above 4 km; Himalayan Mountains, about 7–8 km).

References

Acton G D (1999) Apparent polar wander of India since the Cretaceous with implication for regional tectonics and true polar wander. In: Radhakrishna T (ed.), The Indian Subcontinent and Gondwana: A Paleomagnetic and Rock Magnetic Perspective. Geol. Soc. India Mem., 44: 129–175.

Aitchison J C, Davis A M (2001) When did the India–Asia collision really happen? In: International Symposium and Field Workshop on the Assembly and Breakup of Rodinia, Gondwana and Growth of Asia. Osaka City University, Japan. Gondwana Research, 4: 560–561.

Aitchison J C, Jason R A, Davis A M (2007) When and where did India and Asia collide? J. G. R., 112: B05423.

All China Commission on Stratigraphy (2000) Direction of China Strata and its Instructions. Beijing: Geological Publishing House. 1–59 (in Chinese).

All China Commission on Stratigraphy (2002) Direction of China Strata and its Instructions. Beijing: Geological Publishing House. 1–72 (in Chinese).

Allègre C J, Courtillot V, Tapponnier P et al. (1984) Structure and evolution of the Himalaya–Tibet orogenic belt. Nature, 307: 17–22.

Allen M, Windley B, Zhang C (1992) Paleozoic collisional tectonics and magmatism of the Chinese Tianshan, Central Asia. Tectonophysics, 220: 89–115.

Al-Shanti M S (2009) Geology of the Arabian Shield of Saudi Arabia. Scientific Publishing Center, King Abdulaziz University, 1–190.

Arallokay, Ma J H (1997) Early Mesozoic subduction in the eastern Mediterranean Sea: The evidences of Triassic ophiolite in Northwestern Turkey. Oceanic Geology Trends, (11):13–15 (in Chinese).

Bai J, Huang X G, Wang H C et al. (1996) The Precambrian Crustal Evolution of China (2nd.) Beijing: Geological Publishing House. 1–223 (in Chinese with English abstract).

Bai M X (1993) The neo-tectonics characteristics in the Tianshan, China. In: Collection of Workshop for Commemoration of Centenary Birthday for Prof. Yuan F L (December 21–23, 1993) Beijing: Earthquake Press, 179–184 (in Chinese).

Bally A, and Albert W (eds.) (1989) The Geology of North America: An Overview. Boulder, Colo: Geological Society of America, 1–629.

Bartolini A, Larson R L (2001) The Pacific Microplate and Pangea Supercontinent in the Early to Middle Jurassic. Geology, 29(8), 735–738.

Bea F, Fershtater G B, Montero P (2002) Granitoids of the Uralides: Implications for the evolution of the Orogen. In: Mountain Building in the Uralides: Pangea to the Present Geophysical Monograph. 132. The American Geophysical Union, 211–232.

Besse J, Courtillot V (2002) Apparent and true polar wander and the geometry of the geomagnetic field over the last 200 Myr. J. Geophys. Res., 108 (B10): 2469.

Besse J, Courtillot V, Pozzi J P et al. (1984) Paleomagnetic estimates of crustal shortening in the Himalayan thrusts and Yarlung Zangbo suture. Nature, 311 (5987): 621–626.

Beydoun Z R, Xu J L, Yang W D et al. (2000) Why is the oil-gas so rich and high production in Arabian Plate? Oil and Gas Exploration in Abroad, (1): 4–13, 65 (in Chinese).

Brenchley P J, Rowson P E (2006) The Geology of England and Wales (2nd Edition). The Geological Society of London.

Briais A, Patriat P, Tapponnier P (1993) Updated interpretation of magnetic anomalies and seafloor spreading stages in the South China Sea: Implications for the Tertiary tectonics of Southeast Asia. Journal of Geophysical Research, 98 (B4): 6299–6328.

Brookfield M E (2000) Geological development and Phanerozoic crustal accretion in the western segment of the south Tien Shan (Kyrgyzstan, Uzbekistan and Tajikistan). Tectonophysics, 328: 1–14.

Bureau of Geology and Mineral Resources of Henan (1989) Regional Geology of Henan. Beijing: Geological Publishing House, 1– 772 (in Chinese with English abstract).

Bureau of Geology and Mineral Resources of Fujian (1992) Regional Geology of Taiwan. Beijing: Geological Publishing House. 1–244 (in Chinese with English abstract).

Bureau of Geology and Mineral Resources of Fujian Province (1985) Regional Geology of Fujian Province. Beijing: Geological Publishing House, 1–671 (in Chinese with English abstract).

Bureau of Geology and Mineral Resources of Gansu (1989) Regional Geology of Gansu. Beijing: Geological Publishing House, 1– 691 (in Chinese with English abstract).

Bureau of Geology and Mineral Resources of Ningxia (1990) Regional Geology of Ningxia. Beijing: Geological Publishing House. 1–522 (in Chinese with English abstract).

Bureau of Geology and Mineral Resources of Xizang (1993) Regional Geology of Xizang. Beijing: Geological Publishing House. 1–707 (in Chinese with English abstract).

Buslov M M, Watanabe T, Fujiwara Y et al. (2004) Late Paleozoic faults of the Altai region, Central Asia: Tectonic pattern and model of formation. Journal of Asian Earth Sciences, 23: 655–671.

Cai X L (1965) The geological structure of Pre-Sinian in Dabieshan area: Also discussing the structural features of gneiss dome. In: Beijing College of Geology. Papers on Scientific Research. No.4 (Special of Structure Geology and Regional Geology), 15–26 (in Chinese).

Cai X L (1978) Preliminary research on early structure behaviors during crustal development. In: Teaching Group of Exploring Geology, Chengdu College of Geology. Papers on Structure Geology, 69–84, Geological Society of China (in Chinese).

Cai X L, Zhu J S, Cao J M et al. (2002) Structure and dynamics of lithosphere and asthenosphere in the gigantic East Asian–West Pacific rift system. Geology in China, 29 (3): 234–245 (in Chinese with English Abstract).

Carroll A R, Liang Y H, Graham S A et al. (1990) Junggar basin, Northwest China: Trapped Late Paleozoic ocean. Tectonophysics, 18:1–14.

Cavazza W, Roure F M, Spakman W et al. (2004) The Transmed Atlas: The Mediterranean Region, from Crust to Mantle. Berlin, Heidelberg: Springer, 1–141.

Chang J H (2015) Yellow Sea Transform fault (YSTF) and the development of Korean Peninsula. ТИХО ОКЕАНСКАЯ ГЕОЛОГИЯ, 34(2): 3–7.

Charvet J, Shu L S, Laurent-Charvet S (2007) Paleozoic structural and geodynamic evolution of eastern Tianshan (NW China): Welding of the Tarim and Junggar Plates. Episodes, 30: 163–186.

Che Z C, Liu H F, Liu L (1994) The Formation and Evolution of Middle Tianshan Orogeny. Beijing: Geological Publishing House. 1–135 (in Chinese with English abstract).

Chen B W, Chen T Y (2007) The main characteristics and metallogene in Trans-Asia super-tectonic zone. Acta Petrologica Sinica, 23(5): 865–876 (in Chinese with English abstract).

Chen H (1991) Preliminary research of collision orogenic belt of Early Neoproterozoic in Northwest Zhejiang. Geological Review, 37 (3): 203–213 (in Chinese).

Chen P R (2004) Geodynamic setting of Mesozoic magmatism and its relationship to uranium metallogenesis in Southeastern China. Uranium Geology, 20(5): 266–270.

Chen S S, Liu J Q, Guo Z F et al. (2015) The Geochemical characteristics of volcanic rocks from site 794, Yamoto Basin: Implications for deep mantle processes of the Japan Sea. Acta Geologica Sinica, 89 (4): 1189–1212.

Chen Y Q, Liu J L, Feng Q L et al. (2010) Southeast Asia Indochina Peninsula Geology and Ore Deposits related to Granites. Beijing: Geological Publishing House, 1–192 (in Chinese).

Chen Z L, Zhang Y Q, Chen X H et al. (2001) The sedimentary respond of middle section of Altun Fault in Late Cenozoic strike-slip. Science in China, D31 (supplement): 90–96.

Cheng S D, Liu T, Wang S W (2010) Resume of tectonic unit division for five countries in Central Asia. Geology of Xinjiang, 28(1): 16–21 (in Chinese).

Cheng Y Q (chief editor) (1994) The Introduction of China Regional Geology. Beijing: Geological Publishing House. 1–517 (in Chinese with English abstract).

Chi J S, Lu F X, Zhao L et al. (1996) The Kimberlite in North China Platform and the characteristics of Paleozoic lithosphere mantle. Beijing:Science Press, 1–29 (in Chinese).

China Geological Academy (1980) Asian Geological Map. Beijing: China Geography Publishing House (in Chinese).

Clarke M H (2006) Oman's Geological Heritage (Second edition). Oman: Petroleum Development Oman, 1–247.

Cohen K M, Finney S, Gibbard P L (2013) International Chronostratigraphic Chart, International Commission on Stratigraphy. http://www.stratigraphy.org/ICSchart/ChronostratigraphicChart2013.01.pdf.

Condie K C (ed.) (1994) Archean Crustal Evolution. Amsterdam: Elsevier Scientific Publishers.

Condie K C (1997) Plate Tectonics and Crustal Evolution (4th Edition). Oxford, UK: Butterwoth–Heinemann, 1–282.

Condie K C (2001) Mantle Plumes and Their Record in Earth History. Cambridge University Press, 1–306.

Condie K C, Richard C A (2009) Zircon age periodicity and growth of continental crust. EOS, Transactions American Geophysical Union, 90 (41): 364, https://doi.org/10.1029/2009eo410003.

Cooper A K, Scholl D W, Marlow M S (1976) Plate tectonic model for the evolution of the eastern Bering Sea Basin. Bulletin of Geological Society of America, 87(8): 1119–1126.

Copeland P, Harrison T M (1990) Episodic rapid uplift in the Himalaya revealed by $^{40}Ar/^{39}Ar$ analysis of detrital K-feldspar and muscovite, Bengal Fan. Geology (Boulder), 18 (4): 354–357.

Copeland P, Harrison T M, Parrish R R et al. (1987) Constraints on the age of normal faulting, north face of Mt. Everest: Implications for Oligo-Miocene uplift. Eos Transactions, 68 (44): 1444. A G U. Washington, DC, USA.

Corrigan J D, Crowley K D (1992) Unroofing of the Himalayas: A view from apatite fission-track analysis of Bengal Fan sediments. Geophysical Research Letters, 19 (23): 2345–2348.

Czamanske G, Gurevitch E, Fedorenko V et al. (1998) Demise of the Siberian plume: Paleogeographic and paleotectonic reconstruction from the prevolcanic and volcanic record, north-central Siberia. International Geology Review, 40: 95–11.

Delvaux D, Barth A (2009) Second and third order African stress pattern from formal inversion of focal mechanisms data. Implications for rifting dynamics Royal Museum for Central Africa, Geology - Mineralogy, Tervuren, Belgium; Geophysical Institute, University of Karlsruhe, Germany, (03): 11:7480.

Deng J F, Wu Z X, Zhao G C et al. (1999) Precambrian granites, continental crust evolution and the formation of Craton. Acta Petrologica Sinica, 15(2): 190–198 (in Chinese).

Deng J, Wang Q F, Li G J et al. (2014a) Tethys tectonic evolution and its bearing on the distribution for important mineral deposits in the Sanjiang region, SW China. Gondwana Research, 26: 419–437.

Deng J, Wang Q F, Li G J et al. (2014b) Cenozoic tectono-magmatic and metallogenic processes in the Sanjiang region, southwest China. Earth-Science Reviews, 138: 268–299

Deng Y, Lv H T, Yu B L et al. (2011) The Seismological data disposed, explained and their effect for complex structure in the southern Junggar Basin. Sino Petrological Exploration (Z1): 31–36, +8 (in Chinese with English abstract).

Ding W J (1929) Orogenic movements of China. Journal of Geological Society of China, 8(2): 151–170.

Dixon J M (1987) Mantled gneiss domes, that first used by Professor Pentti Eskola in 1949. Springer, 10.1007/3-540-31080-0_63.

Dobretsov N L (2003) Evolution of structures of the Urals, Kazakhstan, Tien Shan, and Altai-Sayan region within the Ural-Mongolian fold belt (Paleo-Asian Ocean). Russian Geology and Geophysics, 44(1–2): 5–27. UDC 551.242.51.

Dobretsov N L, Kirdyashkin A A, Kirdyashkin A G et al. (2008) Modelling of thermochemical plumes and implications for the origin of the Siberian traps. Lithos, 100: 66–92.

Domeier M, Torsvik T H (2014) Plate tectonics in the Late Paleozoic. Geoscience Frontiers, 5(3):303–350.

Dong Y P, Zhang G W, Lai S C et al. (1999) Determining the ophiolite complex in Huashan, Suizhou, and its tectonic significance. Science in China, D 29 (3): 222–231 (in Chinese with English abstract).

Dong Y P, Zhao X (2002) Pre-Cambrian magmatism-tectonic event and crust growth in South Qinling. Journal of Northwest University, (2): 172–176 (in Chinese with English abstract).

Duan J Y, Ge X H (1992) Discussion on the Tarim-Yangtze plate and its paleogeography framework. Journal of Changchun Geological Institute, 22 (3): 260–268 (in Chinese).

Engebretson D C, Cox A, Gordon R G (1985) Relative motions between oceanic and continental plates in the Pacific basin. The Geological Society of America (Special Paper 206), 1–59.

Ernst R E, Buchan W, (2001) Evidence from magnetic fabric for the flow pattern of magma in the Mackenzie giant radiating dyke swarm. Nature, 356: 511–513

Eyuboglu Y, Santosh M, Dudas F O et al. (2013) The nature of transition from adakitic to non-adakitic magmatism in a slab window setting: A synthesis from the eastern Pontides, NE Turkey. Geoscience Frontiers, 4(4): 353–375.

Fan L K, Cai Y P, Liang H C et al. (2009) Eastern Kunlun geological structure and geodynamic evolution. Geological Survey and Research, 33(3): 181–186 (in Chinese with English abstract).

Fan T T, Zhou X H, Liu Y Q (2012) The basalt zircon U-Pb age and its geological sense in Northwestern boundary of Junggar Basin and Buksal. Acta of Northwest University (Nature Science), 42(6): 989–994 (in Chinese with English abstract).
Fang D J, Guo Y P, Wang Z L et al. (1988) The tectonic significance of Triassic, Jurassic Paleomagnetism results in Ningwu basin, Shanxi. Science Bulletin, 33(2): 133–135 (in Chinese).
Faure M, Trap Pierre, Lin W et al. (2007) Polyorogenic evolution of the Paleoproterozoic Trans-North China Belt: New insights from the Lüliangshan–Hengshan–Wutaishan and Fuping massifs. Episodes, 30(2): 95–107.
Feng Y M (1995) Geochemistry research in North Qilian ophiolite. Acta Petrologica Sinica, 11(Ophiolite Special Issue): 125–146 (in Chinese).
Ferrara G, Ricci C A, Rita F et al. (1978) Isotopic age and tectono-metamorphic history of metamorphic basement of North-Eastern Sardinia. Contribution of Mineral and Petrology, 68: 99–106.
Gao J W, Wu S G, Peng X C et al. (2015) The tectonic characters of conjugation passive continental boundary ocean-continent transition zone. Geotectonica et Metallogenia, 39(4): 555–570 (in Chinese with English abstract).
Gao L Z, Zhang C H, Yin C Y et al. (2008) The SHRIMP zircon age for Meso-Neoproterozoic strata frame in North China ancient continent. Earth Science, 29 (3): 366–376 (in Chinese with English abstract).
Gao L Z, Zhang C H, Liu P J et al. (2009a) New recognition of Meso-Neoproterozoic strata frames of North China and South Yangtze areas. Journal of Earth, 30 (4): 433–446 (in Chinese with English abstract).
Gao L Z, Zhang C H, Liu P J et al. (2009b) Reclassification of the Mesoproterozoic-Neoproterozoic chronostratigraphy of North China by SHRIMP zircon ages. Acta Geologica Sinica, 83(6): 1074–1084.
Gao L Z, Ding X Z, Pan W H et al. (2011) The revise of China Meso-Neoproterozoic strata – Restrained by the zircon U-Pb age. Journal of Stratigraphy, 35 (1):1–7 (in Chinese with English abstract).
Gao R, Wang H Y, Zhang Z J et al. (2011) Crust and upper mantle, discovery continent deep structure and resource environment effects: Deep exploration technique experiment and integration (Sino Probe-02). Journal of Earth, 32 (S1): 34–48 (in Chinese with English abstract).
Gao S L, Ma Q Y (2013) LA-ICPMS zircon U-Pb age and geochemistry characteristics of Autaukerxi granite in Kalameli, Junggar. Xinjiang Geology, 31(1):1–5 (in Chinese with English abstract).
Gao Z J, Wu S Z (1983) The tectonic evolution of Xinjiang Tarim ancient continent. Science Bulletin, 28 (23): 1448–1450 (in Chinese).
Garfunkel Z, Ben-Avraham Z, Kagan E (eds.) (2014) Dead Sea Transform Fault System: Reviews Series, Modern Approaches in Solid Earth Sciences, Vol. 61–359. Dordrecht: Springer, DOI. 1007/978-94- 017-8872- 4-1.
Garson M S, Miroslav K R S (1976) Geophysical and geological evidence of the relationship of Red Sea transverse tectonics to ancient fractures. GSA Bulletin, 87(2): 169–181.
Garzanti E, Gaetani M (2002) Unroofing history of Late Paleozoic magmatic arc within the "Turan Plate" (Tuarkyr, Turkmenistan). Sedimentary Geology, 151: 67–87.
Ge M C, Zhou W X, Yu Y et al. (2011) The dismantle of Xilingule complex and determined the age of surface crust rock system in Inner Mongolia. Earth Science Frontier, 18(5):182–190 (in Chinese with English abstract).
Ge X H (1989) The formation history of North China orogeny. Geological Review, 35(3): 254–261 (in Chinese).
Ge X H, Duan J Y, Li C et al. (1990) The formation and evolution of Qaidam Basin. Qinhai Bureau of Petrology, Changchun Geological College, 1–151 (in Chinese with English abstract).
Ge X H, Zhang M S, Liu Y J et al. (1998) The science problem and research thinking for Altun fault. Geoscience, 12 (3): 295–301 (in Chinese with English abstract).
Ge X H, Liu J L (2000) Be dismembered "Xiyu Craton". Acta Petrologica Sinica, 16 (1): 59–66 (in Chinese).
Ge X H, Liu J L, Ren S M et al. (2014) Eastern China Mesozoic-Cenozoic tectonic formation and evolution. Geology in China, (1): 19–28 (in Chinese with English abstract).
Ge X H, Ma W P (2014) China Regional Tectonics Course. Beijing: Geological Publishing House, 1–466 (in Chinese).
Gebhardt A C, Niessen F, Kopsch C (2006) Central ring structure identified in one of the world's best-preserved impact craters. Geology, 34: 145–148.
Gee D G, Fossen H, Henriksen N et al. (2008) From the Early Paleozoic platforms of Baltica and Laurentia to the Caledonide orogen of Scandinavia and Greenland. Episodes, 31(1): 44–51.
Gen Y S, Wang X S, Sheng Q H et al. (2002) The discovery and its geological significance for deformed granite in Neoproterozoic of Alxa area. Journal of Petrology and Mineralogy, 21 (4): 413–420 (in Chinese with English abstract).
Gen Y S, Wang X S, Sheng Q H et al. (2007) The preliminary research for the formation of Precambrian metamorphic system. China Geology, 34 (2): 251–261 (in Chinese with English abstract).
Gen Y S, Zhou X W (2012) Early Permian Magmatism-thermal event in Alxa metamorphic basement: The evidence come from isotopic age. Acta Petrologica Sinica, (9): 3–21(in Chinese).
Gilder S A, Gill J B, Coe R S et al. (1996) Isotopic and paleomagnetic constraints on the Mesozoic tectonic evolution of South China. J. Geophys. Res. 101 (B7): 16,137–16,155.
Gladkochub D P, Donskaya T (2009) Overview of geology and tectonic evolution of the Baikal–Tuva. Prog. Mol. Subcell Biol., 47: 3–26.
Glass B P (1982) Tektites. In: Introduction to Planetary Geology. Cambridge University Press, 145–172.
Glukhovaky M Z (2009) Paleoproterozoic thermo-tectogenesis: A rotation-plume model of the formation of the Aldan Shield. Geotektonika, (3): 51–78.
Grabau A W (1940) The Rhythm of the Ages. Beijing: H. Wetch, 1–561.
Granath G, Strengbom J, Rydin H (2011) Direct physiological effects of nitrogen on Sphagnum: A greenhouse experiment. Functional Ecology, 26(2). https://doi.org/10.1111/j.1365-2435.2011.01948.x
Guan H, Sun M, Wilde S A et al. (2002) SHRIMP U-Pb zircon geochronology of the Fuping Complex: Implications for formation and assembly of the North China craton. Precambrian Res., 113: 1–18.
Guan Z N, An Y L, Wu C J (1987) Magnetic surface inversion and North China deep geological structure. In: Wang M J, Cheng J Y (eds.), The Research Special Issue for Regional Geophysics in Eastern China, No. 6. Beijing: Geological Publishing House, 80–101 (in Chinese).
Guo F X (1998) The tectonic characters of Meso-Cenozoic in South China and the folding process of South China orogenic belt. Acta Geologica Sinica, 72 (1): 22–33 (in Chinese).
Guo T Y, Liang D Y, Zhang Y Z et al. (1991) Geology of Ali. Wuhan: China University of Geosciences Press. 1–464 (In Chinese with English abstract).
Hall R, Blundell D J (1995) Reconstructing Cenozoic SE Asia. Geological Society Special Publications, 106: 153–184.
Hall R, Ali J R, Anderson C D (1995) Cenozoic motion of the Philippine Sea Plate: Paleomagnetic evidence from Eastern Indonesia. Tectonics, 14(5): 1117–1132.

Hall R, Cottam M A, Wilson M E J (eds.) (2011) The SE Asian gateway: History and tectonics of Australia–Asia collision. Geological Society of London, Special Publication, 355: 1–381.

Han B F, He G Q, Wang S G et al. (1998) Post-collision magmatism with mantle source and continental crust longitudinal direction grow. Geological Review, 44(4): 396–406 (in Chinese).

Han B F, He G Q, Wang S G (1999) Post-collision magmatism with mantle source, underplating and the characteristics of Junggar Basin basement. Science in China (D), 29 (1): 16–21 (in Chinese).

Han B F, He G Q, Wang X C et al. (2011) Late Carboniferous collision between the Tarim and Kazakhstan-Yili terranes in the western segment of South Tian Shan Orogen, Central Asia, and implications for the Northern Xinjiang, Western China. Earth-Science Reviews, 109: 74–93.

He B Z, Jiao C L, Cai Z H et al. (2011) The new explain for Aero-magnetic anomaly zone in Tarim Basin. China Geology, 38 (4): 961–969 (in Chinese).

He D F, Fan C, Lei G L et al. (2011) The chronology and tectonic sense for Tugermin anticline core schist. China Geology, 38(4): 809–819 (in Chinese).

He G Y, Zhao Q, Li S X et al. (2006) Mesozoic prototype analysis of Tarim Kuche basin. Geological Science, 41 (1):44–53 (in Chinese with English abstract).

He H (2009) Geology-Geochemistry Compose and their Evolution in Wandashan Orogeny. Master thesis in the Jilin University, 1–77 (in Chinese with English abstract).

He W G, Mei L F, Zhu G H et al. (2011) The research on tectonics and evolution characters of Andaman Sea Basin. Oil and Gas in Fault Block, (2): 46–50.

He Y H, Zhao G C, Sun M et al. (2009) SHRIMP and LA-ICP-MS zircon geochronology of the Xiong'er volcanic rocks: Implications for the Paleo-Mesoproterozoic evolution of the southern margin of the North China Craton. Precambrian Research, 168: 213–222.

Hong D W, Xie X L, Zhang J S (2002) Discussion of the geological significance for High ε_{Nd} granite zone in Hangzhou-Zhuguangshan-Huashan. Journal of Geology, 21 (6): 348–354 (in Chinese with English Abstract).

Hou G T (2012) The Basic Swan in North China. Beijing: Science Press, 1–177 (in Chinese).

Hsu K J (1981) Thin-skinned plate-tectonic model for collision-type orogenesis. Scientia Sinica, Vol., XXIV, (1): 100–110.

Hsu K J (1988) Relict back-arc basins: principles of recognition and possible new examples from China, New perspective in Basin Analysis. In: Kleinpell K L, Paola C (eds.), New York: Springer Verlag, 245–263.

Hu A Q, Wei G J (2003) On the discussion of basement forming period of Junggar Basin: According to the result of isotopic age. Xinjiang Geology, 21(4): 398–406 (in Chinese).

Huang B C, Yang Z Y, Otofuji Y et al. (1999) Early Paleozoic paleomagnetic poles from the western part of the North China Block and their implications. Tectonophysics, 308: 377–402.

Huang T K (Jiqing) (1945) On the Major Structural Forms of China. Geological Memoirs, A (20): 1–165. (in English with Chinese summary of 11 pages)

Huang J Q (1960) Basic characteristics of China geological structure: Preliminary summarize. Acta Geologica Sinica, 40(1):1–37 (in Chinese).

Huang J Q, Jiang C F (1962) Preliminary investigation of the evolution of the Earth's crust from the point of the view of polycyclic tectonic movement. Scientia Sinica, 11(10): 1377–1442 (in Chinese).

Huang J Q, Yin Z X (1965) Some commons for the nomination of China crust movement. Geological Review, 23(supplement): 2–4 (in Chinese).

Hutchison C S, Tan D N K (eds.) (2009) Geology of Peninsular Malaysia. Kuala Lumpur: Murphy. The University of Malaya and the Geological Society of Malaysia, 1–479.

Jayananda M, Moyen J F, Martin H et al. (2000) Late Archaean (2550–2500 Ma) juvenile magmatism in the Eastern Dharwar craton: Constraints from geochronology, Nd-Sr isotopes and whole rock geochemistry. Precambrian Research, 99: 225–254.

Jayananda M, Kano T, Peucat J-J et al. (2008) 3.35 Ga komatiite volcanism in the western Dharwar craton: Constraints from Nd isotopes and whole rock geochemistry. Precambrian Research, Elsevier. doi: 10.1016 / J. precamres. 2007.07.010, v.162: 160–179.

Jiang C F (1984) The connection of plate tectonics and geosyncline-platform theory is the recent development orientation. China Geology, (11): 28-30 (in Chinese).

Jiang G M (2008) The detail structure research of subduction plate in Western Pacific. The Ph D thesis of China University of Geosciences, 1–125.

Jin X C, Wang J, Ren L D (1999) Some problems of geological structure for Western Kunlun. In: Ma Z J et al. (eds.), Research Progress of Structure-lithosphere Dynamics. Beijing: Seismology Press, 105–113 (in Chinese).

Jin Z J, Zheng H R, Cai L G et al. (2010) The structure-sedimentation condition of China Pre-Mesozoic marine source rock development. Journal of Sedimentary, (5): 875–883 (in Chinese).

Jolivet L (1994) Japan Sea, opening history and mechanism: A synthesis. J. Geophys. Res., 99: 22237–22259.

Jolivet L, Tamaki K (1992) Neocene kinematics in Japan Sea region and the volcanic activity of the northeast Japan arc. Proc. Ocean Drill, Program Sci. Results, 127–128: 1311–1331.

Ju W, Hou G T (2014) Late Permian to Triassic intraplate orogeny of the southern Tianshan and adjacent regions NW China. Geoscience Frontiers, 5(1): 83–93.

Karaoğlan F, Parlak O, Klötzli U et al. (2013) Age and duration of intra-oceanic arc volcanism built on a suprasubduction zone type oceanic crust in southern Neotethys, SE Anatolia. Geoscience Frontiers, 4(4): 399–408.

Karplus M S, Zhao W J, Klemperer S L et al. (2011) Injection of Tibetan crust beneath the south Qaidam Basin: Evidence from INDEPTH IV wide-angle seismic data. J. Geophys. Res., 116, B07301. 10.102/2010 JB007911.

Karsakov L P, Zhao C J, Malyshev Y et al. (2008) Tectonics, Deep Structure, Metallogeny of the Central Asian–Pacific Belts Junction Area (Explanatory Notes to the Tectonic Map Scale of 1:1 500 000). Beijing: Geological Publishing House, 1–213.

Kaygusuz A, Sipahi F, Ilbeyli N et al. (2013) Petrogenesis of the Cretaceous Turnagöl intrusion in the eastern Pontides: Implications for magma genesis in the arc setting. Geoscience Frontiers, 4(4): 423–438.

Kennedy W Q (1964) The structural differentiation of Africa in the Pan-African (500 m.y.) tectonic episode. Res. Inst. Ar. Geol., University of Leeds, 8th Ann. Rep., 48–49.

Khain E V, Bibikova E V, Salnikova E B et al. (2003) The Palaeo-Asian Ocean in the Neoproterozoic and early Palaeozoic: New geochronologic data and palaeotectonic reconstructions. Precambrian Research, 122: 329–358.

Khramov A N, Petrova G N, Peckersky D M (1981) Paleomagnetism of Soviet Union. In: McElhinny M W, Valencio D A (eds.), Paleoconstruction of the Continents. Geodynamic Series. Boulder, Colorado: Geological Society of America, 177–194.

Kim J, Cho M (2003) Low-pressure metamorphism and leucogranite magmatism, northeastern Yeongnam Massif, Korea: Implication for Paleoproterozoic crustal evolution. Precambrian Research, 122: 235–251.

Klootwijk C T, Gee J S, Peirce J W et al. (1992) An early India-Asia contact: Paleomagnetic constraints from Ninety east ridge, ODP leg 121. Geology, 20: 395–398.

Klootwijk C T, Radhakrichnamurty (1981) Phanerozoic paleomagnetism of the Indian Plate and India–Asia collision. In: McElhinny M W, Valencio D A (eds.), Paleoconstruction of the Continents. Geodynamics Series. Boulder, Colorado: Geological Society of America, 93–105.

Kojima S (1989) Mesozoic terrane accretion in Northeast China, Sikhote-Alin and Japan regions. Palaeogeography, Palaeoclimatology, Palaeoecology, 69: 213–232.

Kopp H (2011) The Java convergent margin: Structure seismogenesis and subduction process. In: Hall R et al. (ed.), The SE Asian Gateway: History and Tectonics of the Australia–Asia Collision. London: The Geological Society of London Special Publication, 355: 111–137.

Koppers A P, Morgan J P, Morgan J W et al. (2001) Testing the fixed hotspot hypothesis using ^{40}Ar/ ^{39}Ar age progressions along seamount trails. Earth and Planetary Science Letters, 185: 237–252.

Koppers A P, Staudigel H, Duncan R A (2003) High-resolution ^{40}Ar/ ^{39}Ar dating of the oldest oceanic basement basalts in the western Pacific basin. Geochemistry Geophysics Geosystems, 4 (11): 8914.

Kröner A, Greiling R (1984) Precambrian tectonics illustrated. Stuttgart, 1–419.

Kunugiza K, Tsujimori T, Kano T (2001) Evolution of the Hida and Hida marginal belt. In: ISRGA Field Workshop (FW-A), Geotraverse across the Major Geologic Unit of SW Japan, 75–131.

Kwon Y W, Oh C W, Kim H S (2003) Granulite-facies metamorphism in the Punggi area, northeastern Yeongnam Massif, Korea and its tectonic implications for east Asia. Precambrian Research, 122: 253–273

Lan C Y, Chung S L, Long T V et al. (2003) Geochemical and Sr-Nd isotopic constraints from the Kontum massif, central Vietnam on the crustal evolution of the Indochina block. Precambrian Research, 122: 7–27.

Laurent R, Dion C, Thibault Y (1991) Structural and petrological features of peridotite intrusions from the Troodos ophiolite, Cyprus. Petrology and Structural Geology, 5:175–194.

Layer P W, Newberry R, Fujita K (2001) Tectonic setting of the plutonic belts of Yakutia, northeast Russia, based on ^{40}Ar/^{39}Ar geochronology and trace element geochemistry. Geology, 29: 167.

Lee S R, Cho M Cheong C S et al. (2003) Age, geochemistry, and tectonic significance of Neoproterozoic alkaline granitoids in the northwestern margin of Gyeonggg massif, South Korea. Precambrian Research, 122: 297–310.

Lee T Y, Lawver L A (1995) Cenozoic plate reconstruction of Southeast Asia. Tectonophysics, 251 (1–4): 85–138.

Li C (1997) ^{40}Ar/^{39}Ar dating for Glaucophane in the blueschist and its geological significance. Science Bulletin, 42 (4):488 (in Chinese).

Li C, Zhao Q G, Chen W et al. (2006) Ar-Ar dating for eclogite in Middle Qiangtang, Qingzang Plateau. Acta Petrologica Sinica, 22 (12): 2843–2849 (in Chinese with English abstract).

Li C Y, Wang Q, Liu X A et al. (1982) Asian Tectonic Map (1:8 000 000, attached explanation). Beijing: Geography Publishing House (in Chinese).

Li D, He D F, Fan C et al. (2012) Geochemistry characteristics and its tectonic significance for Carboniferous basalt in Klameili, Junggar Basin. Acta Petrologica Sinica, 28 (3): 981–992 (in Chinese).

Li H B, Yang J S, Xu Z Q et al. (2001)The geology and chronology evidences for Altun fault zone in Triassic strike-slip. Science Bulletin, 16 (16):1333–1338 (in Chinese).

Li J C, Jin Z J (1998) The sedimentary-erosion process and related to oil-gas in Tarim Basin. Acta Sedimentary, 16 (1): 81–86 (in Chinese with English abstract).

Li J Y, Wang K Z, Li W Q et al. (2002a) The tectonic and mineralization exploration since Late Paleozoic in Eastern Tianshan. Xinjiang Geology, 20 (4): 295–301 (in Chinese).

Li Q Y, Wang K Z, Li W Q et al. (2002b) Eastern Tianshan tectonic and miniral deposits exploration since Late Palaeozoic. Xinjiang Geology, 20 (4): 295–301 (in Chinese).

Li S G, Zhang Z M, Zhang Q D et al. (1993) Zircon U-Pb age for Qindao eclogite and Jiaonan Group gneiss. Science Bulletin, 38 (19): 1773–1777 (in Chinese).

Li S G, Jagoutz E (1996) Super high pressure metamorphic chronology in Dabieshan-Sulu block (I): Sm-Nd isotopic system. Science in China, D 26 (3): 249–257 (in Chinese).

Li S G, Li H M, Chen Y Z et al. (1997) Super high pressure metamorphic chronology in Dabieshan-Sulu block(II), Zircon U-Pb isotopic system. Science in China, D 27(3): 200–206 (in Chinese).

Li S G, Huang F, Li H (2001) Lithosphere detachment after the collision of Dabie-Sulu orogenic belt. Science Bulletin, 46 (17): 1487–1491(in Chinese).

Li T D, Chen B W, Dai W S et al. (2010) Tectonic Map of Qingzang Plateau and its adjacent (1: 3 500 000). In: Atlas of Qingzang Plateau Geology. Guangzhou: Guangdong Science and Technology (in Chinese).

Li T D, Uzhkenov B S, Mazorov A K et al. (2008) Central Asia and Adjacent Geological Map (1:2 500 000). Beijing: Geological Publishing House (in Chinese with English abstract).

Li W F, Wang C S, Gao Z Z et al. (2000) Mesozoic sedimentary evolution in Kuche depression, Tarim Basin. Acta Sedimentary, 18 (4): 514–538 (in Chinese with English abstract).

Li Y A, Li Q, Zhang H (1991) Jurassic-Cretaceous Paleomagnetism research in Junggar Block. In: Collection of Seventh Conference of China Geophysical Society, 171. http://cpfd.cnki.com.cn/Area/CPFDCONFArticle List- ZGDW199110001.htm (in Chinese).

Li Y J, Sun L D, Hu S L et al. (2003) ^{40}Ar/^{39}Ar age of granodiorite and diorite in Tacan1. Acta Petrologica Sinica, 19 (3): 530–536 (in Chinese).

Li Y P, Li J Y, Sun G H et al. (2007) Discussion on the basement of Junggar basin: The evidences come from detrital zircon in Devonian Calameli formation sandstone. Acta Petrologica Sinica, 23: 1577–1590 (in Chinese).

Li Z S (1994) New recognition for metamorphic system in Dunhuang Block. China Regional Geology, (2):131–134 (in Chinese).

Li Z X (1998) Tectonic history of the major East Asia lithospheric blocks since the mid-Proterozoic: A synthesis. In: Flower M F J et al. (eds.), Mantle Dynamics and Plate interactions in East Asia, Geodynamics Series. Washington D C: AGU. Vol 27, 221–243.

Li Z X, Zhang L, Powell C M (1996) South China in Rodinia: Part of the missing link between Australia-East Antarctic and Laurentia? Geology, 23: 407–410.

Li Z X, Cho M, Li X H (2003) Precambrian tectonics of East Asia and relevance to Supercontinent evolution. Precambrian Research, 122: 1–6.

Li Z X, Li X H (2007) Formation of the 1300-km-wide intracontinental orogeny and post-orogenic magmatic province in Mesozoic South China: A flat-slab subduction model. Geology, 35: 179–182.

Liao L, Cheng X G, Wang B Q et al. (2010) The Mesozoic sedimentary paleo-environment recovery, in southwest ward of Tarim Basin. Acta Geologica Sinica, 84(8): 1195–1207 (in Chinese).

Lin S L, Cong F, Gao Y J et al. (2012) LA-ICP-MS zircon U-Pb ages in Tengchong block, SE side of Gaoligong Mountains Group gneiss and its geological significance. Science Bulletin, 31(2–3):258–263 (in Chinese).

Lin W, Li Q L, Su W et al. (2011) SIMS U-Pb rutile age of low-temperature eclogites from southwestern Chinese Tianshan, NW China. Xinjiang Geology, 10, doi:1016/j.lithos.2010.11.007 (in Chinese).

Liu B G, Zheng G C, Chen S S et al. (1995) The zircon U-Pb isotopic age and its implication in Precambrian volcanic rock of western Zhejiang. Science Bulletin, 40(21): 2015–2016 (in Chinese).

Liu B P, Feng Q L, Fang N Q (1991) Tectonic evolution of the Paleo-Tethys in Changning–Menglian and Lancangjiang belts, Western Yunnan. In: Proceedings of 1st International Symposium on Gondwana Dispersion and Asian Accretion (IGCP Project 321), Kunming, China. Beijing: Geological Publishing House, 189–192.

Liu B P, Feng Q L, Fang N Q et al. (1993) The tectonic evolution of Paleo-Tethys multiple islands ocean in Changning-Menlian and Lanchangjiang Zones, Western Yunnan. Earth Science, 18 (5): 529–539 (in Chinese with English abstract).

Liu F, Yang J S, Li T F et al. (2011) Late Carboniferous volcanic geochemistry characteristics and its geological significance in Shawan area, North Tianshan, Xinjiang. China Geology, 38(4): 859–889 (in Chinese with English abstract).

Liu F L, Wang F, Liu C H (2013) Multiple metamorphic events revealed by zircons from the Diancang Shan–Alao Shan metamorphic complex, southeastern Tibetian Plateau. Gondwana Research, 24: 429–450.

Liu G H, Zhang S G, You Z D et al. (1993) Main Metamorphic Group and its Evolution in Qinling Orogenic Belt. Beijing: Geological Publishing House, 1–190 (in Chinese with English abstract).

Liu J S (2012) Exploration history and geological characteristics in the Monyuwa copper ore field, in Burma. http://blog.sina.com.cn/yuelugj. 2012-11-2 (in Chinese).

Liu L, Che Z C, Wang Y et al. (1998) Sm-Nd isochronism age evidence in Early Paleozoic ophiolite at Manya area, Altun. Science Bulletin, 43(8): 880–883 (in Chinese).

Liu L F (2006) Oil-gas reservoir system research in Tabei uplift. PhD thesis of China University of Geosciences (Beijing), 1–138 (in Chinese with English abstract).

Liu S W, Pan P M, Li J H et al. (2002) Geological and isotopic geochemical constrains on the evolution of the Fuping Complex, North China Craton. Precambrian Research, 117: 41–56.

Lu F X, (2010) Multi geological events in the paleo-lithosphere mantle for North China Craton: The evidences come from the peridotite xenolith in the kimberlite. Petrologica Acta Sinica, 26 (11): 3177–3188 (in Chinese).

Lu F X, Zheng J P, Shao J A et al. (2006) The asthenosphere uplift and lithosphere thinner in Late Mesozoic–Cenozoic for Eastern North China. Earth Science Frontier, 13(2): 86–92 (in Chinese).

Lu H J, Wang R C et al. (2011) Geochemistry and genesis on saturated hydrocarbon of crude oils in Gas Hure Oilfield in Qaidam. Global Geology, (1). Supported by the National Natural Science Foundation of China (No. 40973041) (in Chinese).

Lu S N (2001) From Rodinia to Gondwana: Some problems thinking for Neoproterozoic supercontinent research. Earth Science Frontier, 8 (4): 441–448 (in Chinese).

Lu Y H (1976) The Biological strata and ancient animal geography in the China. In: Collected papers of Nanjing Paleontology Institute, Academy of China, No. 7 (in Chinese).

Luo J H, Zhou X Y, Qiu B et al. (2005) The Oil-gas geological characteristics and regional geological evolution, in Tarim-Karakum areas. Geological Review, (4): 409–415 (in Chinese with English abstract).

Luo M F, Mo X X, Yu X H et al. (2015) LA-ICP-MS zircon U-Pb dating for Late Permian granodiorite in Wulong Gou, Eastern Kunlun and its significance. Earth Science Frontier, 22(5): 182–195 (in Chinese with English abstract).

Ma X Y, Cai X L (1965) Structural deformation in Early Archean, Eastern China. In: The China Tectonics. Beijing: Science Press, 141–150 (in Chinese).

Ma X H, Yang Z Y (1993) The collision of three plate in China and reconstruction of Paleo-Eurasia Continent. Acta Geophysica Sinica, 36(4): 476–488 (in Chinese).

Ma Z J, Zhu W, Wang X et al. (1999) Crustal motion of Chinese Mainland monitoring by GPS. Science in China, Series D, 43(4): 394 (in Chinese).

Ma Z P, Li X M, Xu X Y et al. (2011) La-ICP-MS zircon U-Pb isotopic age and its significance for Qingshuiquan mafic-ultramafic intrusion in the South Altun. Geology in China, 38(4): 1071–1078 (in Chinese with English abstract).

Mansour Ghorbani (2013) The Economic Geology of Iran, Mineral Deposits and Natural Resources. Springer: 1–450.

Marlow M S, Cooper A K Scholl D W et al. (1982) Ancient plate boundaries in the Bering Sea region. Geological Society, London, Special Publications, 10: 201–211. https://doi.org/10.1144/gsl.sp.1982.010.01.13.

Martin S, Godard G, Rebay G (2004) The subducted Tethys in the Aosta Valley (Italian Western Alps). In: The 32nd International Geological Congress, Pre-Congress Field Trip Guide Book– B02, 1–18.

Maruyama S, Isozaki Y, Kimura G et al. (1997) Paleogeographic map of the Japan Islands: Plate tectonic synthesis from 750 Ma to the present. The Island Arc, 6: 121–142.

McElduff B, Stumpfl E F(1990) Platinum-group minerals from the Troodos ophiolite, Cyprus. Mineralogy and Petrology, 42(1–4): 211–232.

Metcalfe I (1991) Gondwana dispersion amalgamation and accretion of Southeast Asian terrenes: Progress, problems and prospects. In: Proceedings of 1st International Symposium on Gondwana Dispersion and Asian Accretion (IGCP Project 321), Kunming, China. Beijing: Geological Publishing House, 199–204.

Metcalfe I (1995) Gondwana dispersion and Asian accretion. In: Proceedings of the IGCP Symposium on Geology of SE Asia, Hanoi (Vietnam), XI / 1995. Journal of Geology (B), (5–6): 223–266.

Metcalfe I (2011) Paleozoic-Mesozoic History of SE Asia. In: Hall R et al. (eds.), The SE Asian Gateway: History and Tectonics of the Australia-Asia Collision. London: The Geological Society of London Special Publication, 355: 7–36.

Mishra D C, Kumar M R (2014) Proterozoic orogenic belt and rifting of Indian cratons: Geophysical constraints. Geoscience Frontiers, 5(1): 25–41.

Mizutani S, Kojima S, Shao J A et al. (1986) Mesozoic radiolarians from the Nadanhada area, Northeast China. Proc. Japan Acad. 62 (B): 337–340.

Mo X X, Zhao Z D, Yu X H et al. (2009) Cenozoic Collision–Post Collision Volcanic Rocks in Qingzang Plateau. Beijing: Geological Publishing House, 1–396 (in Chinese with English abstract).

Moore G W (1989) Mesozoic and Cenozoic paleogeographic development of the Pacific region. Abstract 28th International Geological Congress, Washington D C, USA. 2, 455–456.

Moralev V M, Glukhovsky M Z (2000) Diamond-bearing kimberlite fields of the Siberian Craton and the Early Precambrian geodynamics. Ore Geology Reviews, 17 (3): 141–153.

Mu E Z (1983) The type of ecology strata and bios-geographic area in Ordovician, for China. China Paleontology geography division. Beijing: Science Press, 16–31(in Chinese).

Nábelek J, Hetnyi G, Vergne J et al. (2009) Underplating in the Himalaya-Tibet collision zone, Revealed by the Hi-CLIMB Experiment. Science, 325: 1371–1374.

Nezafati N (2006) Au-Sn-W-Cu-Mineralization in the Astaneh-Sarband Area, West Central Iran (including a comparison of the ores with ancient bronze artifacts from Western Asia) der

Geowissenschaftlichen Fakultät, der Eberhard- Karls-Universität Tübingen, 1–116.

Northrup C J, Royden H, Burchfiel B C (1995) Motion of the Pacific Plate relative to Eurasia and its potential relation to Cenozoic extension along the eastern margin of Eurasia. Geology, 23: 719–722.

Norton I O (1995) Plate motions in the North Pacific: The 43 Ma nonevent. Tectonics, 14 (5): 1061–1234.

Oh C W (2006) A new concept on tectonic correlation between Korea, China and Japan: Histories from the Late Proterozoic to Cretaceous. Gondwana Research, 9: 47–61.

Oh C W, Kim S W, Choi S G et al. (2005) First finding of eclogite facies metamorphic event in South Korea and its correlation with the Dabie-Sulu collision belt in China. The Journal of Geology, 113: 226–232.

Osozawa S (1994) Plate reconstruction based upon age data of Japanese accretionary complexes. Geology, 22: 1135–1138.

Osozawa S (1998) Major transform duplexing along the eastern margin of Cretaceous Eurasia. In: Flower M F J et al. (eds.), Mantle Dynamics and Plate Interactions in East Asia, Geodynamics Series, Washington D C: AGU, 27: 245–257.

Oxman V S (2003) Tectonic evolution of the Mesozoic Verkhoyansk–Kolyma belt (NE Asia). Tectonophysics, 365 (1–4): 45–76.

Parfenov L M, Bulgatov A N, Gordienko I V (1995a) Terranes and the history of Mesozoic orogenic belts in East Yakutia. doi:10.1080/00206819509465425.

Parfenov L M, Badarch G, Berzin N A et al. (2009) Summary of Northeast Asia geodynamics and tectonics, Stephan Mueller Spec. Publ. Ser., 4, 11–33.

Parfenov L M, Prokopiev A V, Gaiduk V V (1995b) Cretaceous frontal thrusts of the Verkhoyansk fold belt, eastern Siberia. Tectonics, 14 (2): 342–358.

Pearson D G, Snyder G A, Shirey S B et al. (1994) Archaean Re-Os age for Siberian eclogites and constraints on Archaean tectonics. Nature, 374: 711–713.

Pei F P, Xu W L, Yang D B et al. (2006) Zircon U-Pb age in basement metamorphism of Song-Liao Basin and its significance. Science Bulletin, 51 (24): 2881–2887 (in Chinese).

Peng R M, Zhai Y S, Wang J P et al. (2010) Discovery of Neoproterozoic acid Volcanic rock in Inner Mongolia and its significance. Science Bulletin, 55(26): 2611–2620 (in Chinese).

Petrov O, Leonov Y, Li T D et al. (Editors in chief) (2008) Tectonic Zoning of Central Asia and Adjacent Areas (1:20000000). In: Atlas of Geological Maps of Central Asia and Adjacent Areas (1:2500000),VSEGEI Cartographic Factory.

Pospelov I (2008) Tectonic evolution of northeast Russia in Paleozoic. International Geological Congress, ASI-01, Geodynamic evolution of Asia - Part 2.

Pubellier M (2008) Structural Map of Eastern Eurasia (1:12500000). Paris: CGMW.

Qi S J (2013) Metallogenic characteristic in the Xindukushi–Western Kunlun region. Xinjiang Geology, 31(4): 313–317(in Chinese).

Qi X X, Zhu L H, Li H Q et al. (2010) LA-ICP-MS U-Pb dating for mylonitic granite in Ailaoshan-Jinsha River zone, eastern boundary of Qingzang Plateau. Acta Geologica Sinica, 84(3): 1–12 (in Chinese).

Qiao X F, Gao L Z, Zhang C H (2007) The new thinking of Mesoproterozoic-Neoproterozoic age-strata and tectonic environment. Geology Bulletin, 26 (5): 503–509 (in Chinese).

Qiao X F, Wang Y B (2014) The base age of Mesoproterozoic and discussion of basin character in North China. Acta Geologica Sinica, 88(9): 1623–1637 (in Chinese).

Rajaram M, Anand S P (2014) Aeromagnetic signatures of Precambrian shield and suture zones of Peninsular India. Geoscience Frontiers, 5 (1): 3–15.

Radhakrishna B P, Naqvi S M (1986) Precambrian continental crust of India and its evolution. Jour. Geol., 94:145–166.

Ravikant V (2010) Palaeoproterozoic (~1.9 Ga) extension and rifting along the eastern margin of the Eastern Dharwar Craton, SE India: New Sm-Nd isochron age constraints from anorogenic mafic magmatism in the Neoarchean Nellore greenstone belt. Journal of Asian Earth Sciences, 37(1): 67.

Raymond C A, Stock J M, Cande S C (2000) Fast Paleogene motion of the Pacific hotspots from revised global plate circuit constraints. In: The History and Dynamics of Global Plate Motions. Geophysical Monographys, 121: 359–375.

Ren J S, Wang Z X, Chen T Y et al. (2000) From Globe to Look China Tectonics: Tectonic Map and Explanation of China and Its Adjacent Areas. Beijing: Geological Publishing House, 1–50 (in Chinese).

Replumaz A, Hrafnkell Karason, Rob Dvander Hilst et al. (2004) 4-D evolution of SE Asia's mantle from geological reconstructions and seismic tomography. Earth and Planetary Science Letters, 221: 103–115.

Ridd M F, Barber A J, Crow M J (2011) The Geology of Thailand. London: The Geological Society, 1–626.

Rodnikov A G, Kato T et al. (1985) Sikhote-Alin-Japan Sea-Honshu-Pacific geoscience transect. Marine Geophysical Researches, 7: 379–387.

Rogers J J W, Santosh M (2002) Configuration of Columbia: A Mesoproterozoic supercontinent. Gondwana Res., 5: 5–22.

Rogers J J W, Santosh M (2004) Continents and Supercontinents. New York: Oxford Press, 1–289.

Rojas-Agramonte, Kröner A, Demoux A et al. (2011) Detrital and xenocrystic zircon ages from Neoproterozoic to Palaeozoic arc terranes of Mongolia: Significance for the origin of crustal fragments in the Central Asian Orogenic. Gondwana Research, 19 (3): 751–763.

Safanova I (2014) The Russian-Kazakh Altai orogeny: An overview and main debatable issues. Geoscience Frontiers, 5(4): 537–552.

Sagong H, Cheng C S, Kwon S T (2003) Paleoproterozoic orogeny in South Korea: Evidence from Sm-Nd and Pb step-leaching garnet ages of Precambrian basement rocks. Precambrian Research, 122: 275–295.

Santosh M, Sajeev K, Li J H (2006) Extreme crustal metamorphism during Columbia Supercontinent assembly: Evidence from North China Craton. Gondwana Research, 10: 256–266.

Saunders A, Reichow M (2009) The Siberian Traps and the End-Permian mass extinction: A critical review. Chinese Science Bulletin, 54 (1): 20–37.

Schettino A, Scotese C R (2005) Apparent polar wander paths for the major continents (200 Ma to the present day): A paleomagnetic reference frame for global plate tectonic reconstructions. Geophys. J. Int. 163: 727–759.

Searle M P (1996) Cooling history, erosion, exhumation, and kinematics of the Himalaya-Karakoram-Tibet orogenic belt. In: Yin A, Harrison T M (eds.), The Tectonic Evolution of Asia. Cambridge University Press, 110–137

Searle M P (2007) Geological Map of the Mount Everest-Makalu Region, Nepal-South Tibet Himalaya (Scale 1: 100000). Allen Hochreiter: University of Oxford.

Sengör A M C, Nal'in B A, Burtman U S (1993) Evolution of the Altaid tectonic collage and Paleozoic crustal growth in Eurasia. Nature, 364, 209–304.

Shao J A, Wang C Y, Tang K D (1991) Tectonic exploration in the Wusuli area. Geological Review, 38 (1): 33–39.

Shao J A, Tang K D (1995) The NE China Terrain and Evolution of NE Asia Continental Boundary. Beijing: Seismology Press, 1–185 (in Chinese).

Shen Q H, Gen Y S, Wang X S et al. (2005) The Precambrian amphibolite petrology, geochemistry formation environment and chronology in Alxa area. Journal of Petrology and Mineralogy, 24 (1): 21–31(in Chinese with English abstract).

Shi M L (1987) Some recognitions of Alxa area structural problem. China Regional Geology (3): 268–273 (in Chinese).

Shu L S, Faure M, Wang B et al. (2008) Late Paleozoic-Early Mesozoic geological features of South China: Response to the Indosinian collision events in Southeast Asia. Tectonics, 340: 151–165.

Shui T, Xu B T, Liang R H et al. (1986) The suture zone of Shaoxing-Jiangshan ancient land. Science Bulletin, 31 (6): 444–448 (in Chinese).

Sinha-Roy S (1982) Himalayan main central thrust and its implications for Himalayan inverted metamorphism. Tectonophysics, 84 (2–4): 197–224.

Sone M, Metcalfe I (2008) Parallel Tethyan sutures in mainland SE Asia: New insights for Palao-Tethys closure. Compte. Rendus. Geoscience, 340, 166–179.

Steinberger B, Torsvik T H (2012) A geodynamic model of plumes from the margins of Large Low Shear Velocity Provinces. Geochem. Geophys. Geosyst.13 (1):1–17.

Stephan J F, Blanchet R, Rangin C et al. (1986) Geodynamic evolution of the Taiwan-Luzon-Mindoro belt since the Late Eocene. Tectonophysics, 125(1–3): 245–268.

Su Y P, Zheng J P et al. (2019) Derivation of A1-type granites by partial melting of newly underplated rocks related with the Tarim mantle plume. Geological Magazine, 156: 409–429.

Sun H Q, Huang J Z, Guo L Q et al. (2012) Division of Lengjiaxi Group and isotopic age restrain in Hunan. South China Geology and Mineral Resources, (1): 22–28 (in Chinese).

Sun J M, Li Y, Zhang Z Q et al. (2009) Magnetostratigraphic data on Neogene growth folding in the foreland basin of the southern Tianshan Mountains. Geology, 37(11): 1051–1054.

Suo S T, San L K, Han Y Q et al. (1993) Petrology and tectonics in the Dabieshan Pre-Cambrian metamorphic block. Wuhan: China University of Geosciences Press, 1–259 (in Chinese with English abstract).

Tamaki, K., Suyehiro, K., Allan, J. et al. (1992) Tectonic synthesis and implications of Japan Sea ODP drilling. Proc. Ocean Drill. Program Sci. Results, 127–128: 1333–1348.

Tan Y J (1992) The Structure Deformations of Southern Border of Ordos Basin and Its Evolution. PhD thesis of China Mining Industry University, Beijing Institute, 1–145 (in Chinese with English abstract).

Tang H F, Zhou X M (1997) The geochemical constraint for the origin of two basalt type in eastern Jiangnan Paleo continent. Science in China (D), 27(4): 306–311(in Chinese).

Tapponnier P, Mercier J L, Proust F et al. (1981) The Tibetian side of the Indian-Eurasian collision. Nature, 294 (5840): 405–410.

Tapponnier P, Peltzer G, Armijo R (1986) On the mechanics of the collision between India and Asia. In: Coward M P, Ries A C (eds.), Collision Tectonics. Geological Society Special Publications, 19. London: The Geological Society of London, 115–157.

Tapponnier P, Lacassin R, Leloup P H et al. (1990) The Ailao Shan / Red River metamorphic belt: Tertiary left-lateral shear between Indochina and South China. Nature, 343 (6257): 431–437.

Tarduno J A, Cottrell R D (1997) Paleomagnetic evidence for motion of the Hawaiian hotspot during formation of the Emperor seamounts, Earth Planet. Sci. Lett., 153 (3–4): 171–180.

Tectonic Group, Institute of South China Sea, Academy of China (Liu Z S et al.) (1988) South China Sea Geological Structure and Continental Boundary Extension. Beijing: Science Press, 1–283 (in Chinese).

Teng L S, Lin A T (2004) Cenozoic tectonics of the China continental margin: in slights from Taiwan. In: Malpas J et al. (eds.), Aspects of the Tectonic Evolution of China, 313–332. London: The Geological Society. Special Publication 226.

Tessalina S, Guerrot C, Gannoun A et al. (2000) Isotopic Indicators of Subduction Process in South Urals. Journal of Conference Abstracts, Cambridge Publications, 5(2): 993.

Tessensohn F, Roland N W (1998) A Preface: Third International Conference on Arctic Margins. Polarforschung, 68: 1–9.

Tian D J (2007) The Geology-Geochemistry Composition and its Evolution in Wandashan Orogene. The Master thesis of Jilin University, 1–77 (in Chinese).

Torsvik T H, Van der Voo R, Doubrovine P V et al. (2014) Deep mantle structure as a reference frame for movements in and on the Earth. PNAS, Cross Mark. Edited by Suppe J, National Taiwan University, Taipei, Taiwan, and approved May 8, 2014.

Tsujimori T, Ishiwatari A, Banno S (2000) Ecologic glaucophane schist from the Yunotani Valley in Omi town, the Range metamorphic belt, the inner zone of southwest Japan. J. Geol. Soc. Japan, 106 (5): 353–362.

Van der Voo R (1993) Paleomagnetism of the Atlantic, Tethys and Iapetus Oceans. Cambridge University Press, 1–273.

Wan T F (2011a) On the collision time. Earth Science Frontier, 18(3): 48–56 (in Chinese with English abstract).

Wan T F (2011b) The Tectonics of China: Data, Maps and Evolution. Beijing, Dordrecht, Heidelberg, London and New York: Higher Education Press and Springer, 1–501.

Wan T F, Chu M J (1987) The listric active fault in Fujian and Taiwan. Earth Science, 12(1): 21–29 (in Chinese).

Wan T F, Zeng H L (2002) The distinctive characteristics of the Sino-Korean and the Yangtze plates. Journal of Asian Earth Sciences, 20 (8): 881–888.

Wan T F, Hao T Y (2010) Mesozoic-Cenozoic tectonics of the Yellow Sea and oil-gas exploration. Acta Geologica Sinica, 84 (1): 77–90.

Wan T F, Lu H F (2014) The continental crust and oceanic mantle type lithosphere and its mechanics in Eastern China. Geotectonica et Metallogenia, 38 (3): 495–511 (in Chinese).

Wan T F, Zhu H (2011) Chinese continental blocks in global paleo-continental reconstructions during the Paleozoic and Mesozoic. Acta Geologica Sinica, 85 (3): 581–597.

Wan T F, Zhao Q L (2012) The origin of tectonic-magmatism in Eastern China. Science in China (Earth Science), 42 (2):155–163 (in Chinese).

Wan T F, Yin Y H, Zhang C H (1997) On the extraterrestrial impact and plate tectonic dynamics: A possible interpretation. Proceedings of 30th International Geological Congress, VSP 26: 87–95.

Wan T F, Wang Y M, Liu J L (2008) Tectonic detachment and magma origin depth in Eastern China during Yanshanian and Sichuanian Periods. Earth Science Frontier, 15(3): 1–35 (in Chinese with English abstract).

Wan T F, Zhao Q L, Wang Q Q (2015) Paleozoic Tectono-Metallogeny in Tianshan-Altay Region, Central Asia. Acta Geologica Sinica (English Edition), 89 (4): 1801–1814.

Wang C S, Liu Z F, He Z W (1999) Recognization and discussion on the paleo-ophiolite at Yarlung Zangbo River for Early Cretaceous in southern Xizang. Acta Geologica Sinica, 73(1): 7–14 (in Chinese with English abstract).

Wang C S, Li X H, Hu X M et al. (2002) Latest marine horizon north of Qomolangma (Mt. Everest): Implications for closure of Tethys seaway and collision tectonics. Terra Nova, 14: 114–120.

Wang F Z, Yang M Z, Zheng J P (2002) The geochemistry evidences for arc volcanic terrain composed basement. Journal of Petrology and Mineralogy, 21(1): 1–10 (in Chinese).
Wang G H, Zhou X, Pubuziren et al. (1996) The structural deformation and its evolution in Taniantawong Mountains, Xizang. Beijing: Geological Publishing House, 1–80 (in Chinese with English abstract).
Wang G Z, Wang C S (2001) The break up and age dating for Qiangtang basement metamorphic system. Science in China (D), 31 (supplement):77–82 (in Chinese).
Wang H Z (1979) The main period of Asian tectonic evolution. Science in China, (12): 55–65 (in Chinese with English abstract).
Wang H Z (1985) The Atlas of China Paleogeography. Beijing: Geography Publishing House (in Chinese).
Wang H Z, Mo X X (1995) An outline of the tectonic evolution of China. Episodes, 18 (1–2):6–16.
Wang H Z, Li X, Mei S L et al. (1997) Pangean cycles, earth rhythms and possible earth expansion. In: Wang H Z, Jahn Borming, Mei S H (eds.), Origin and History of the Earth. Proc. 30th Intern. Geol. Congr., 1: 111–128. VSP, Utrecht, The Netherland.
Wang T, Wang X X, Tian W et al. (2009) North Qinling Paleozoic granite complex, magma evolution in time and space and the significance to orogeny. Science in China (D), (7): 119–141 (in Chinese).
Wang T, Liang Y, Ma Z W et al. (2012) Mesozoic tectonic evolution to control the sedimentary in Yuqi areas of Tarim Basin. Popular Science and Technology, 14 (11): 48–50 (in Chinese).
Wang T, Tong Y, Wu C L et al. (2014) Temporal and spatial evolution of granodiorites from the main orogenic belt in Asia and their implication for tectonism and metallogenesis. China Geological Survey, 1(2): 58–64 (in Chinese with English abstract).
Wang T H (1995) Shanxi and Shaanxi tectonic evolution and oil-gas assembled, North China. Journal of Geology and Mining Industry, 10(3): 283–398 (in Chinese).
Wang Y, Li J Y, Sun G H (2008) Post-collision eastward extrusion and tectonic exhumation along the eastern Tianshan orogen, Central Asia: Constraints from dextral strike-slip motion and $^{40}Ar/^{39}Ar$ geochronological evidence. Journal of Geology, 116: 599–618.
Wang Y J, Fan W M, Zhang Y et al. (2003) Structural evolution and $^{40}Ar/^{39}Ar$ dating of the Zanhuang metamorphic domain in the North China Craton: Constrains on Paleoproterozoic tectonothermal overprinting. Precambrian Research, 122: 159–182.
Wang Y M, Wan T F (2008) Cenozoic lithosphere tectonic detachment, magmatism and earthquake in Eastern China. Geosciences, 22(2): 207–229 (in Chinese with English abstract).
Wang Y S, Chen J N (1987) The Metamorphism and Metamorphic Areas in Qinghai and Its Adjacent Area. Geological Specialty No. 6. Beijing: Geological Publishing House (in Chinese).
Wang Z J, Hunag Z G, Yao J X et al. (2014) China stratigraphy explanation and its main progress. Acta Geologica Sinica, 35(3): 271–276 (in Chinese).
Wang Z L, Mao J W, Zhang Z H et al. (2006) The geological characteristics of porphyry copper-molybdenum deposit, time-space distribution and its evolution of metallogene dynamic. Acta Geologica Sinica, (7): 21–33 (in Chinese).
Wang Z X, Li C L, Wang D X et al. (2015) Discovery of the Early Devonian Sinistral shear in the Jiangshan- Shaoxing fault zone and its tectonic significance. Acta Geologica Sinica, 89 (4): 1412–1413.
Wilde S A, Zhao G C, Sun M (2002) Development of the North China Craton during the Late Archaean and its final amalgamation at 1.8 Ga: Some speculations on its position within a global Paleoproterozoic supercontinent. Gondwana Research, 5: 85–94.
Wilde S A, Wu F, Zhang X (2003) Late Pan-African magmatism in Northeastern China: SHRIMP U-Pb zircon evidence from granitoids in the Jiamusi Massif. Precamb. Res., 122: 311–327.
Wilde S A, Cawood P A, Wang K et al. (2004) Determining Precambrian crustal evolution in China: a case study from Wutaishan, Shanxi Province, demonstrating the application of precise SHRIMP U-Pb geochronology. In: Malpas J et al. (eds.), Aspects of the Tectonic Evolution of China. London: The Geological Society. Special Publication, 226. 5–26.
Wu F Y, Xu Y G, Gao S et al. (2008) The main discussion in the lithosphere thinner and the craton destroy. Petrologica Acta Sinica, 24: 1145–1174 (in Chinese).
Wu G Y, Li Y J, Wang G L et al. (2006) The volcanic rocks of Jingning Period oceanic arc in Bachu area, Xinjiang. Geosciences, 20 (3): 361–369 (in Chinese with English abstract).
Wu G H, Li H W, Xu Y L et al. (2012) The tectonic-thermal event and its structure evolution in the Tarim craton basement ancient uplift. Acta Petrologica Sinica, 28 (8): 2435–2452 (in Chinese).
Wu J S, Gen Y S, Shen Q H et al. (1998) Archean Geological Character and Tectonic Evolution in Sino-Korea Ancient Continent. Beijing: Geological Publishing House, 1–212 (in Chinese with English abstract).
Wu Z, Xu H Z, Yang F L et al. (2014) Extension tectonic model restrain for NE part of South China Sea lithosphere. Geotectonica et Metallogenia, 38(1): 71–81 (in Chinese with English abstract).
Xia L X, Zhao F Q, Luo Z H et al. (1995) North Qilian tectonic-volcanic magma evolution dynamics. Northwest Geological Science, 16(1): 1–28 (in Chinese with English abstract).
Xia X P, Sun M, Zhao GC et al. (2006) LA-ICP-MS U-Pb geochronology of detrital zircons from the Jining Complex, North China Craton and its tectonic significance. Precambrian Research. 144: 199–212.
Xiao R G, Sui D C, Luo Z H et al. (1995) Discovery and its petrology research for Paleoproterozoic metamorphism system, in northern Inner Mongolia. Geosciences, (2):140–148 (in Chinese).
Xiao W J, Kröner A, Windley B F (2009a) Geodynamic evolution of Central Asia in the Paleozoic and Mesozoic. International Journal of Earth Sciences, 98: 1185–1188.
Xiao W J, Windley B F, Huang B C et al. (2009b) End–Permian to mid-Triassic termination of the accretionary processes of the southern Altaids: Implications for the geodynamic evolution, Phanerozoic continental growth, and metallogeny of Central Asia. International Journal of Earth Sciences, 98: 1189–1287.
Xiao W J, Santosh M (2014) The western Central Asian Orogenic Belt: A window to accretionary orogenesis and continental growth. Gondwana Research, 25:1429–1444.
Xiao W J, Sun M, Santosh M (2015) Continental reconstruction and metallogeny of the Circum-Junggar areas and termination of the southern Central Asian Orogenic Belt. Geoscience Frontiers, 6 (2): 137–140.
Xiao X C, Tang Y Q, Feng Y M et al. (1992) The Tectonic Evolution of Northern Xinjiang and Its Adjacent. Beijing: Geological Publishing House, 1–169 (in Chinese).
Xiao X C, He G Q, Xu X et al. (2010) Crust Structure and Geological Evolution in Xinjiang China. Beijing, Geological Publishing House, 1–317 (in Chinese).
Xin H T, Zhao F Q, Luo Z H et al. (2011) Paleoproterozoic foundation of fine chronology and its geological significance, at Aktashitag area, southeast boundary of Tarim Basin. Acta Geologica Sinica, (12):17–33 (in Chinese).
Xu Y H, Zhao T P, Peng P et al. (2007) Proterozoic Xiaoliangling Formation volcanic rock geochemistry characteristics and its geological significance in Lvliang, Shanxi. Acta Petrologica Sinica, 3(5): 1123–1132(in Chinese).
Xu Z M (2011) The Ordovician Petroleum Reservoir System and Its Resource Potential. PhD thesis of China University of Geosciences (Beijing), 1–126 (in Chinese with English abstract).

Xu Z Q, Xu H F, Zhang J X et al. (1994) Nanshan Caledonia subduction complex accretion terrain and its dynamic in North Qilian passage. Acta Geologica Sinica, 68(1): 1–15 (in Chinese).

Xu Z Q, Yang J S, Zhang J X (1999) Compare with two sides tectonic units and shear mechanism of lithosphere in Altun Fault. Acta Geologica Sinica, 73 (3): 193–205 (in Chinese).

Xu L M, Zhou L F, Zhang Y J et al. (2006) The characteristic of tectonic stress field and its tectonic background for the Ordos Basin. Geotectonica et Metallogenia, 30(4): 455–462 (in Chinese with English abstract).

Xu Z Q, Yang J S, Ji S C et al. (2010) Some recognitions for China continental tectonics and its dynamics. Acta Geologica Sinica, 84 (1):1–29 (in Chinese)

Xu Z Q, Yang J S, Li H B et al. (2011) India-Asia collision tectonics. Acta Geologica Sinica, 85 (1): 1–33 (in Chinese).

Xu Z Q, Wang Q, Cai Z H et al. (2015) Kinematics of the Tengchong Terrane in SE Tibet from Late Eocene to Early Miocene: Insights from coeval mid-crustal detachments and strike-slip shear zones. Tectonophysics, https://doi.org/10.1016/j.tecto.2015.09.033.

Yan F, Chen G, Hou B et al. (2014) Research on the zircon U-Pb dating, micro-elements and Hf isotopic in volcanic clasolite of drilling in Junggar Basin. Acta Geologica Sinica, 88(6): 1068–1080 (in Chinese).

Yang J S, Meng F C, Zhang J X et al. (2001) Reconsidered the age of Hongliuxia volcanic rock and its tectonic significance in the eastern Altun Fault. Science in China (D), 31 (supplement): 83–89 (in Chinese).

Yang J S, Liu F L, Wu C L et al. (2003) Two super high pressure metamorphism in the Central collision zone: The chronology evidence come from zircon with coesite. Acta Geologica Sinica, 77 (4): 463–477 (in Chinese).

Yang J S, Xu Z Q, Zhang J X et al. (2009) The discussion on tectonic background and subduction / return mechanics in China main high-super high pressure metamorphism zone. Acta Petrologica Sinica, (7):3–34 (in Chinese with English abstract).

Yang J S, Xu Z Q, Duan X D et al. (2012) Discovery on the Jurassic SSZ type ophiolite in Mykyina, Burma. Acta Petrologica Sinica, 28 (6): 1710–1730 (in Chinese).

Yang M G, Wang Y G, Li L et al. (1994) Regional geological characteristics of South China. In: Chen Y Q (eds.), Introduction of China Regional Geology. Beijing: Geological Publishing House. 313–384 (in Chinese).

Yang W R, Wang G C, Jian P (2000) The Dabieshan Orogenic Belt Tectono-Chronology. Wuhan: China University of Geosciences Press, 1–141 (in Chinese with English abstract).

Yang W R, Jian P, Han Y Q (2002) The super high pressure metamorphism determine and its significance in Dabie orogenic belt. Earth Science Frontier, 9(4): 273–283 (in Chinese with English abstract).

Yang X, Xu X H, Qiang Y X et al. (2014) Discussion the regional deference for the forming basement of Tarim basin. Geotectonica et Metallogenia, 38(3): 544–556 (in Chinese with English abstract).

Yin H F (Chief editor.) (1988) China Paleontology Geography. Wuhan: Press of China Geosciences Press, 1–329 (in Chinese).

Yoon S (2001) Tectonic history of the Japan Sea region and its implications for the formation of the Japan Sea. Journal of Himalayan Geology, 22 (1): 153–184.

Yoshikura S, Hada S, Isozaki Y (1990) Kurosegawa terrane. In: Ichikawa K, Mizutani S, Hara I (eds.), Pre-Cretaceous Terranes in Japan. Publication of IGCP Project No. 224, Osaka.

Zhai M G (2004) Decompose the North China Craton 2.1–1.7 Ga geological events and its significance. Acta Petrologica Sinica, 20 (6):1343–1354 (in Chinese).

Zhai M G (2007) North China craton Paleoproterozoic tectonic event. Acta Petrologica Sinica, 23(11): 2665–2682 (in Chinese with English abstract).

Zhai M G (2010) How to form the continental crust of Earth? – Mystical and interesting Pre Cambrian geology. Nature Magazine, 32 (3): 126–129 (in Chinese).

Zhai M G, Bian A G. (2000) The end of Neoarchean supercontinent cohere and Late Paleoproterozoic-Mesoproterozoic splitting action in North China Craton. Science in China (supplement), 129–137 (in Chinese).

Zhai M G, Guo J H, Li Z et al. (2007) The Su–Lu orogeny extension in Korean Peninsula: Orogeny, Precambrian basement and Paleozoic sedimentary evidences and restrict. Universities Acta Geologica, 13 (3): 415–428 (in Chinese with English abstract).

Zhang B R, Han Y W, Xu J F et al. (1998) The geochemistry evidences for North Qinling Neoproterozoic belonging to Yangzi plate. Universities Acta Geologica, 4(4): 369–382 (in Chinese with English abstract).

Zhang C L, Li H K, Wang H Y (2012) Discussion on the progress in Precambrian geology research for Tarim Block. Geological Review, 58(5): 923–936 (in Chinese).

Zhang F Q, Chen H L, Dong C W et al. (2008) Precambrian basement evidence in the Northern Song-Liao Basin. Geology in China, 35 (3): 421–428 (in Chinese).

Zhang G W, Zhang B R, Yuan X C (eds.) (1996) Orogenic Process of Qinling Orogenic Belt and Three Dimensional Structural Map Series of Lithosphere. Beijing: Science Press (in Chinese with English abstract).

Zhang G W, Zhang B R, Yuan X C et al. (2001) Qinling Orogenic Belt and Continental Dynamics. Beijing: Science Press, 1–855 (in Chinese with English abstract).

Zhang H F (2009) Interaction of peridotite-melt: The key of Craton type lithosphere mantle breakup. Science Bulletin, 54: 2008–2026 (in Chinese).

Zhang H F, Luo Z P, Zhou Z G et al. (2009) Paleoproterozoic collision orogenic time in middle-northern North China: Come from ultra-aluminum Granite and the constraint of ductile shear period. Mineral and Rock, (1): 62–69 (in Chinese).

Zhang J X, Xu Z Q, Chen W et al. (1997) Discussion on the period of subduction-accretion complex/volcanic arc in Middle part of North Qilian. Magazine of Petrology and Mineralogy, 16(2): 112–119 (in Chinese with English abstract).

Zhang J X, Zhang Z M, Xu Z Q et al. (1999) The discovery of khondalite and the preliminary research of petrology and isotopic chronology in west Altun. Science in China, D (4): 11–18 (in Chinese).

Zhang J X, Yang J S, Xu Z Q et al. (2000) U-Pb and Ar-Ar ages of eclogites from the northern margin of the Qaidam basin, northwestern China. Journal of the Geological Society of China (Taiwan), 43(1): 161–169.

Zhang J X, Mattinson C G, Meng F C et al. (2009) U-Pb geochronology of paragneisses and metabasite in the Xitieshan area, north Qaidam Mountains, western China: Constraints on the exhumation of HP/UHP metamorphic rocks. Journal of Asian Earth Sciences, 35: 245–258.

Zhang J X, Li H K, Meng F C et al. (2011) Multiple tectonic-thermal events recorded by "metamorphism basement" in SE boundary of Tarim (Altun): the constraint of zircon U-Pb age. Acta Petrologica Sinica, (1): 25–48 (in Chinese).

Zhang J, Li J Y et al. (2012) The relationship between Alxa and North China blocks in Early Paleozoic: Got the information from Middle Ordovician clastic zircon in east border of Alxa. Acta Petrologica Sinica, 28(9): 2912–2934 (in Chinese).

Zhang L F, Ai Y L, Li Q et al. (2005) The formation and evolution in super-high pressure metamorphic zone in Southwest Xinjiang. Acta Petrologica Sinica, 21 (4): 1029–1038 (in Chinese).

Zhang L G (1995) Block-Geology of East Asia Lithosphere: Isotope Geochemistry and Dynamics of Upper Mantle, Basement and Granite. Beijing: Science Press, 1–252 (in Chinese with English abstract).

Zhang Q, Chen Y, Zhou D J et al. (1998) Ophiolite geochemistry characteristics and its origin in Dachadaban, North Qilian. Science in China, 28(1): 30–34 (in Chinese).

Zhang Y W, Jin Z J, Liu G C et al. (2000) Main unconformity formation and erosion research in around Major areas, Tarim Basin. Earth Science Frontier, 7(4): 449–457 (in Chinese with English abstract).

Zhang Z C, Dong S Y, Huang H et al. (2009) The geology and geochemistry of Permian intermediate-acid intrusion in Southwest Tianshan: Rock origin and tectonic background. Geological Bulletin, 28 (12): 1827–1839 (in Chinese).

Zhang Z J, Zhan Z, Qin S X et al. (2003) On the basic tectonic framework and evolution for Pre-Neoproterozoic in South China. Journal of Earth, 24(3): 197–204 (in Chinese with English abstract).

Zhang Z M, Wang J L, Shen K et al. (2008) Paleozoic orogeny around the eastern Gondwana: Petrology and evidence of chronology in Namkagbawa Group, Eastern Himalaya. Acta Petrologica Sinica, 24 (7): 1627–1637 (in Chinese).

Zhao C J, Peng Y Q et al. (1996) The tectonics and crust evolution in the eastern Jilin and Heilongjiang. Shenyang: Liaoning University Press, 1–186.

Zhao D P, Liu L (2010) Deep structure and origin of active volcanoes in China. Geoscience Frontiers, 1 (1): 31–44.

Zhao G C (2001) Paleoproterozoic assembly of the North China craton. Geological Magazine, 138: 87–91.

Zhao G C (2007) When did plate tectonics begin on the North China craton? Insights from metamorphism. Earth Science Frontiers, 14 (1): 19–32.

Zhao G C (2014) Precambrian Evolution of the North China Craton. Amsterdam: Elsevier, 1–194.

Zhao G C, Wilde SA, Cawwod P A et al. (1998) Thermal evolution of Archaean basement rocks from the eastern part of the North China Craton and its bearing on tectonic setting. International Geological Review, 40: 706–721.

Zhao G C, Wilde S A, Cawood P A et al. (2002) SHRIMP U-Pb zircon ages of the Fuping Complex: Implications for accretion and assembly of the North China Craton. American Journal of Science, 302, 191–226.

Zhao G C, Sun M, Widle SA (2004) Late Archaean to Paleoproterozoic evolution of the Trans-North China Orogen: In slights from synthesis of existing data of the Hengshan-Wutai-Fuping belt. In: Malpas J et al. (eds.), Aspects of the Tectonic Evolution of China. London: The Geological Society. Special Publication, 226: 27–56.

Zhao G C, Sun M, Widle S A et al. (2005) Late Archean to Paleoproterozoic evolution of the North China Craton: Key issues revisited. Precambrian Research, 137: 149–172.

Zhao H, Meng W B, Tian J C et al. (2011) Paleogene sedimentary phase and sedimentary evolution characteristics in Kuche depression, Tarim Basin. Journal of Sichuan Geology, 31(2):137–141 (in Chinese with English abstract).

Zhao J M, Huang Y, Ma Z J et al. (2008) Discussion on the basement structure and property in northern Junggar Basin, Acta Geophysica Sinica, 51: 1767–1775 (in Chinese).

Zhao J Z, Li Q M, Wang Q H et al. (2004) The formation and distribution of great and middle types oil and gas fields. Journal of Northwest University (Nature Science), (2): 93–98 (in Chinese with English abstract).

Zhao T P, Zhai M G, Xia B et al. (2004) The Research on Xionger Group volcanic rock zircon SHRIMP age: The constraint for the original time to develop the cover of North China craton. Science Bulletin, 49 (22):2342–2349 (in Chinese).

Zhao W J, Nelson K D, Xu Z X et al. (1997) Double intracontinental subduction and the characteristics of partial melting layer. Acta Geophysica Sinica, 40 (3): 325–336.

Zhao W J, Xue G Q, Zhao S et al. (2004a) INDEPTH-3 seismic tomography: The evidences of North Xizang–Indian lithosphere subducted fault. Journal of Earth, 25(1): 1–10 (in Chinese).

Zhao W J, Zhao X, Shi D N et al. (2004b) Progress in the study of deep profiles of Tibet and the Himalayas (INDEPTH). Acta Geologica Sinca, 78 (4): 931–939.

Zhao W J, Wu Z H, Shi D N et al. (2014) Kunlunshan deep structure and orogenic mechanism. Geology in China, (1):5–22 (in Chinese with English abstract).

Zheng T Y, Zhu R X, Zhao L et al. (2012) Intra-lithospheric mantle structures recorded continental subduction. J. Geophys. Res., 117: B03308, https://doi.org/10.1029/2011.jb008873.

Zhong D L (1998) Paleo-Tethys orogenic belt in Western Yunnan-Sichuan. Beijing: Science Press, 1–231 (in Chinese with English abstract).

Zhou H Y, Mo X X, Li J J et al. (2007) The single zircon U-Pb age of mica plagioclase gneiss in Qingeletu, Alxa, Inner Mongolia. Journal of Mineral, Petrology and Geochemistry, 126 (3): 221–223 (in Chinese with English abstract).

Zhou J B, Zhang X Z, Wilde S A (2011a) Determine the khondalite series ~500 Ma Pan-African period and its significance. Acta Petrologica Sinica, (4): 345–355 (in Chinese).

Zhou J B, Zhang X Z, Wide S A (2011b) The determine and it means for the Pen-Africa period (500 Ma) in NE China. Acta Petrologica Sinica, 4: 345–355 (in Chinese with English abstract).

Zhou X M, Zhou H B, Yang J D et al. (1989) Sm-Nd isochron age of ophiolite suite and its geological significance in Fuchuan, Xixian, Anhui Province. Chinese Science Bulletin, 34 (16): 1243–1245 (in Chinese).

Zhu D C, Mo X X, Zhao Z D et al. (2010) Presence of Permian extension and arc type magmatism in Southern Tibet: Paleography implications. GSA Bulletin, 122: 979–993.

Zhu G, Hu Z Q, Chen Y et al. (2008) The development of Early Cretaceous extension basin to show the craton destroy. Bulletin of Geology, 27: 1594–1604 (in Chinese with English abstract).

Zhu G, Wang Y S, Liu G S et al. (2005) ^{40}Ar-^{39}Ar dating of strike-slip motion on the Tan-Lu fault zone, East China. Journal of Structural Geology, 27: 1379–1398.

Zhu J S, Cao J M, Cai X L et al. (2002) the high resolution ratio for S-wave seismic tomography in East Asian-Western Pacific boundary seas. Acta Geologica Geophysica Sinica, 45(5): 646–664 (in Chinese).

Zhu R X, Chen L, Wu F Y et al. (2011) The time, extent and mechanism for North China Craton. Science in China (Earth Science), 41: 583–592 (in Chinese).

Zhu R X, Xu Y G, Zhu G et al. (2012) The destroy of North China craton. Science in China, 42 (8): 1135–1159 (in Chinese).

Zhu W B, Zheng B H, Shu L S et al. (2011) Neoproterozoic tectonic evolution of the Precambrian Aksu blueschist terrane, northwestern Tarim, China: Insights from LA-ICP-MS zircon U-Pb ages and geochemical data. Precambrian Research, 185(3–4): 215–230.

Zhu Y F, Xu X, Wei S N et al. (2007) The geochemistry of OIB type pillow basalt and its geological significance in Karamay, western Junggar. Acta Petrologica Sinica, 23: 1739–1748 (in Chinese).

Zolin Y A, Zorina L D, Spiridonov A M (2001) Geodynamic setting of gold deposits in Eastern and Central Trans-Baikal (Chita Region, Russia). Ore Geology Reviews, 17: 215–232.

Zuo G C, Liu Y K, Zhang Z C et al. (2011) Tectonic evolution and its metallogene analysis for Middle-Southern Tianshan orogeny. Geosciences, 25 (1): 1–13 (in Chinese).

3 The Tectonic Evolution of Asian Continental Lithosphere

The Asian continental lithosphere underwent very complex formation processes. During its long-term tectonic evolution, multi-directional collisions and convergences had occurred among many blocks for many times, until its lithosphere was fixed and became the main part of Eurasian continental lithosphere plate. Even after the plate's convergence was completed, the Asian continental lithosphere plate further had undergone extremely widespread and rather strong intraplate deformations. These characteristics are very distinctive and complex in the global tectonic evolution. So to study in deep, the tectonic evolution characteristics of the Asian continental lithosphere plate are not only of great theoretical significance, but also of important practical value for the exploration of mineral resources, protection of the environment and mitigation of the natural disasters.

It is generally recognized that Archean–Paleoproterozoic is the period of inhomogeneous accretion of planetesimals and the formation period of proto-continental nuclei. It is considered that the continental crust began to form between 4.2 and 3.6 Ga. The periods of Archean (4.6–2.5 Ga) and Paleoproterozoic (2.5–1.6 Ga) account for more than 60% of the whole of the Earth history. Therefore, it is not surprising that the main part of the present continental crust, 70–80% or more, was formed during Neoarchean (2.8–2.5 Ga) and Paleoproterozoic (2.5–1.6 Ga). The formation process of the Asian continental crust is similar to that of other continents on the Earth. The rocks formed during the Archean and Paleoproterozoic are located mainly in the lower crust, and their outcrops are distributed between 5 and 10% of the surface area of the Asian continent. It is believed that the Earth was produced by the accumulation and accretion of planetesimals, including the heavier elements, which had previously condensed from the solar nebula (Safranov 1972; Dai 1979; Bai et al. 1996; Hofmann 1997; Ouyang 1995; Ouyang et al. 2002; Jayananda et al. 2000, 2008; Zhai 2007, 2010, 2014).

Ouyang (1995) and Ouyang et al. (2002) proposed that the process of accumulation of the Earth could be divided into two stages: One is the inhomogeneous accretion stage, and the other is the multi-period accumulation stage. (1) Proto-Earth was formed by the accumulation of "giant stars," with a diameter of more than 3000 km, accreting to about 70–90% of the present Earth mass. The giant stars are composed of M-group planetesimals (iron meteorites and chondrites), mainly iron, and L-group planetesimals (less ferrous chondrites) which composition is similar to that of the Moon. During the accretion and accumulation process, the heavier materials sank toward the center of the Earth under the influence of gravity. When material was continually added into the exterior, the pressure would be increased and the materials would be heated adiabatically, until the interior of the Earth became large or completed molts. Continual differentiation, with the sinking of denser materials toward the center and the rise of lighter materials toward the surface, has led to the development of the present spherical structure of the Earth. Due to its high density, iron was segregated toward the center of the Earth to form an iron- and nickel-rich core, which now accounts for about one-third of the mass of the Earth's interior. Toward the surface, lighter silicates formed the proto-crust and proto-mantle. Mg-rich silicates, with a perovskite crystal structure and a higher density, sank and condensed to form the lower mantle. Low-density silicates, with a composition similar to lunar rocks and rich in rare earth elements (REEs), potassium, phosphorus, uranium and thorium, moved upward to form a proto-crust, equivalent to the present transition layer of the mantle at a depth of 400–670 km. The crust, mantle and core of proto-Earth have subsequently undergone fractional melting, forming the layered structure. On the whole, the Earth has maintained a stable gravitational equilibrium and a spherical structure since it was formed at 4.5 Ga. (2) Because the upper mantle and crust are laterally inhomogeneous at the present time, it is proposed that smaller C-group (carbonaceous chondrites) or L-group planetesimals, with an average diameter of 400 km, were accumulated at the late period in the accretion process to

T. Wan, *The Tectonics and Metallogenesis of Asia*,
https://doi.org/10.1007/978-981-15-3032-6_3

form the outer layers of the Earth. These planetesimals were partially melted and differentiated after they were accumulated on the surface of the cooling proto-Earth. The proto-crust was formed from the partial melting of this proto-upper mantle layer, and the process of differentiation of the upper mantle and crust has continued since the period of 4.46 Ga.

In the 1970s, great progress was made in Paleoarchean–Neoarchean tectonic research by using unstable isotopic geochemistry to determine the ages of tectono-thermal events. This made it possible to identify groups of rocks affected by particular tectono-thermal events and to arrange them into a series of tectono-stratigraphic periods according to the isotopic age (Kröner and Greiling 1984).

Granitic gneiss is a main rock type in granulite–gneiss areas, including the Paleoarchean–Mesoarchean deep intrusions, metamorphosed supra-crustal rocks and the remains of greenstone belts, and tonalite–trondhjemite–granite (TTG) rock suites. The supra-crustal rocks are mainly amphibolite and pyroxene granulite from the metamorphism of basic volcanic rocks, gneiss and leptite, or mainly metamorphosed rocks from dacitic and rhyolitic protoliths, and biotite plagiogneiss, granulite and almandine (magnetite) quartzite (Bai et al. 1996).

In early studies of the Archean rocks, the continental nuclei were identified as uplifted "mantled gneiss domes" (Eskola 1949). Gneiss domes were described in Dabie Mountain area (Ma and Cai 1965; Cai 1965) and at Fuping in the Taihang Mountain area in China. Many of these gneiss domes are now interpreted as layered structures related to ductile shear deformation and crustal vertical accretion by underplating in zones of extension (Qian 1996).

The recognition that the East Siberian plate was composed of six continental nuclei (mantled gneiss domes) was acknowledged by the academics. As to the formation of gneiss domes, it is generally predicted that the meteorite impact induced the mantle material uplift and diapir (Bai et al. 1996). Now, most researchers consider that the continental nucleus's formation period before Paleoproterozoic does not belong to the plate development period (Zhai 2007, 2010, 2014).

Condie and Aster (2013) pointed out that the Hf, Nd and Sr isotopes could be used to better understand the circulation process of the early supercontinent in the early Earth evolution period. These indicate that there existed complex evolution process in different stages of the early Earth evolution, which needs further research. Due to limited data about the early Earth evolution, discussion on the Asian tectonic evolution in this monograph will start from the late period of Paleoproterozoic.

3.1 The Tectonic Evolution in Late Paleoproterozoic (1800–1600 Ma)

In the late period of Paleoproterozoic, there formed the Siberian plate (1600 Ma) [1], Song–Nen block [10], Junggar block [11], Sino–Korean plate (1800 Ma) [14], Alxa–Dunhuang block (1800 Ma) [16], Qaidam block (1800 Ma) [18], North and South Tarim blocks (1800 Ma) [20]. Almost meanwhile, there formed ten continental plates in other areas of the globe, including Baltic plate [70] (1600 Ma), North American plate [71] (1600 Ma), Australian plate [72] (1600 Ma), coastal East Antarctica plate, Greenland plate, Kalahari (Botswana) plate, Madagascar plate, South American plate (1600 Ma), West African plate and Zimbabwe plate (Rogers and Santosh 2002; Fig. 3.1). Most of these plates developed their united crystalline basements at the period of 1600 Ma, while the Sino–Korean plate developed its united crystalline basement at 1800 Ma.

The Late Paleoproterozoic (1600 Ma)–Early Mesoproterozoic was the main period for the formation of Columbia continental plates (Fig. 3.1). In the continental reconstruction sketch made by Rogers and Santosh (2002, 2004), however, positions of China continental blocks were not found. In this period, the rocks developed in each continental plate around the globe were basically combination of tonalite–trondhjemite–granite (TTG) and granulites; it shows that these continental plates had many similarities and there might exist a united paleo-continent (Bai et al. 1996; Rogers and Santosh 2002, 2004).

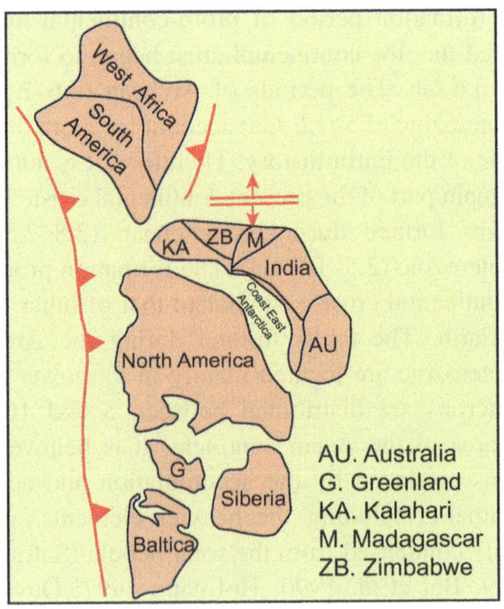

Fig. 3.1 Columbia continental reconstruction sketch at the end of Paleoproterozoic (~1600 Ma) (After Rogers and Santosh 2002)

Zhao et al. (2004) also proposed their viewpoints about the reconstruction of Columbia continental plates. However, till now the reconstruction of Columbia paleo-continent can only rely on the isotopic ages, orientations of tectonic lines (foliations) and lithological characteristics; therefore, the location of the paleo-continent still has many uncertainties, which is not appropriate to be overemphasized and overinterpreted. Piper (2013a, b) tried to reconstruct the positions and migration velocities of the paleo-continent by using the continental paleo-magnetism database for Pre-Cambrian (2.7–0.6 Ga) and the data of crustal accretion and global cooling periodicity. However, the related research is still in the process of exploration.

3.2 The Tectonic Evolution in Early–Middle Mesoproterozoic (1600–1200 Ma)

From the early to late periods of Mesoproterozoic, the Columbia supercontinent broke up gradually. In Mesoproterozoic, the development of Mackenzie dyke swarm in North America and basic dyke swarm in the Siberian plate, Sino–Korean and Indian plate (Ernst and Baragar 1992, 2001; Bai et al. 1996; Hou 2012; Figs. 3.2 and 3.3) indicated the breakup of the Columbia supercontinent. Depending on the distribution characteristics of the dyke swarm, the area equivalent to the recent Arctic area (North Pole) may be the center of radiate dyke swarm at that time; thus, many scholars predicted that this area may be the center of mantle activity, i.e., the center of mantle plume uplift. However, due to lack of relatively reliable paleo-magnetic data during these periods, the paleo-continent reconstruction is only a hypothesis based on the isotopic ages and similarity of tectonic–lithological characteristics.

Yakobchuk (2010), Zhang et al. (2012), Piper (2013a, b) and Kaur and Chaudhri (2013) all proposed totally different reconstruction proposals for the Columbia supercontinent; however, evidences for the reconstruction are not so sufficient.

3.3 The Tectonic Evolution in Late Mesoproterozoic (1200–1000 Ma)

In the late period of Mesoproterozoic, according to some similarities of tectonic–lithological characteristics, it can be inferred that many paleo-continents around the globe underwent widespread convergence and formed the Rodinia supercontinent (Fig. 3.4). Due to limitations of paleo-continent reconstruction which are only based on isotopic ages and tectonic–lithological characteristic comparison, opinions proposed by different scholars (Hoffman 1997; Dalziel 1997; Lu et al. 2005) are so various that a unified recognition could not be yet agreed upon for the time being. Lu et al. (2005) recognized that the Siberia, China continental blocks and Indian plate might not take part in the formation of Rodinia supercontinent, and instead they kept independence in the paleo-ocean (Fig. 3.4), while Hoffman (1997) and Dalziel (1997) never put the China blocks into their reconstruction sketches. It seems that it is reasonable to recognize that some Asian plates and blocks were still separated during the period when most of the global blocks converged into the Rodinia supercontinent at about 1100 Ma.

3.4 The Tectonic Evolution in Middle Neoproterozoic (~850 Ma)

In the middle period of Neoproterozoic, the Rodinia supercontinent broke up and most of the global continental blocks were dispersed. However, there happened partial collisions in the Asian continent and formed the Southern Anhui–Northeastern Jiangxi–Xuefeng Mountains–Eastern Yunnan collision zone (also called Jiangnan collision zone) [23] (JN, i.e., between the North and South Yangtze plates in Fig. 3.5; Wan 2011) and Central Tarim Neoproterozoic collision zone [21] (between TB and TN in Fig. 3.5; Wu et al. 2006). The above two collision zones caused the formation of the Yangtze plate and Tarim block. However, similar collisions have not been found in other blocks of the world. In this period, the Tarim block and Qaidam block had been separated from the Sino–Korean plate and possessed paleo-biogeographic and sedimentary characteristics of the Yangtze plate. When the main part of Rodinia supercontinent was broken up, the Tarim block and Yangtze plate underwent partial convergence and collision, indicating that the global tectonics was not evolving synchronously. It might be a reasonable phenomenon, from the global point of view, that some areas converged while others extended. This may be a reasonable result of maintaining global stability and balance of gravity.

3.5 The Tectonic Evolution in Late Neoproterozoic–Early Cambrian (635–510 Ma)

In the late period of Neoproterozoic–Early Cambrian, there happened the important Pan-African Tectonic Event, in which many blocks eventually converged into the united Gondwana supercontinent. These blocks include Southern Qiangtang–Sibumasu plate ($\sim$510 Ma) [34], Gangdise plate ($\sim$510 Ma) [36], Himalayan block ($\sim$510 Ma) [38], Indian plate ($\sim$510 Ma) [40], Turkey–Iran–Afghanistan plate ($\sim$510 Ma) [43], Arabian plate ($\sim$510 Ma) [46], Oman plate [47], Western Burma (Pegu Mountains–Rangoon) plate

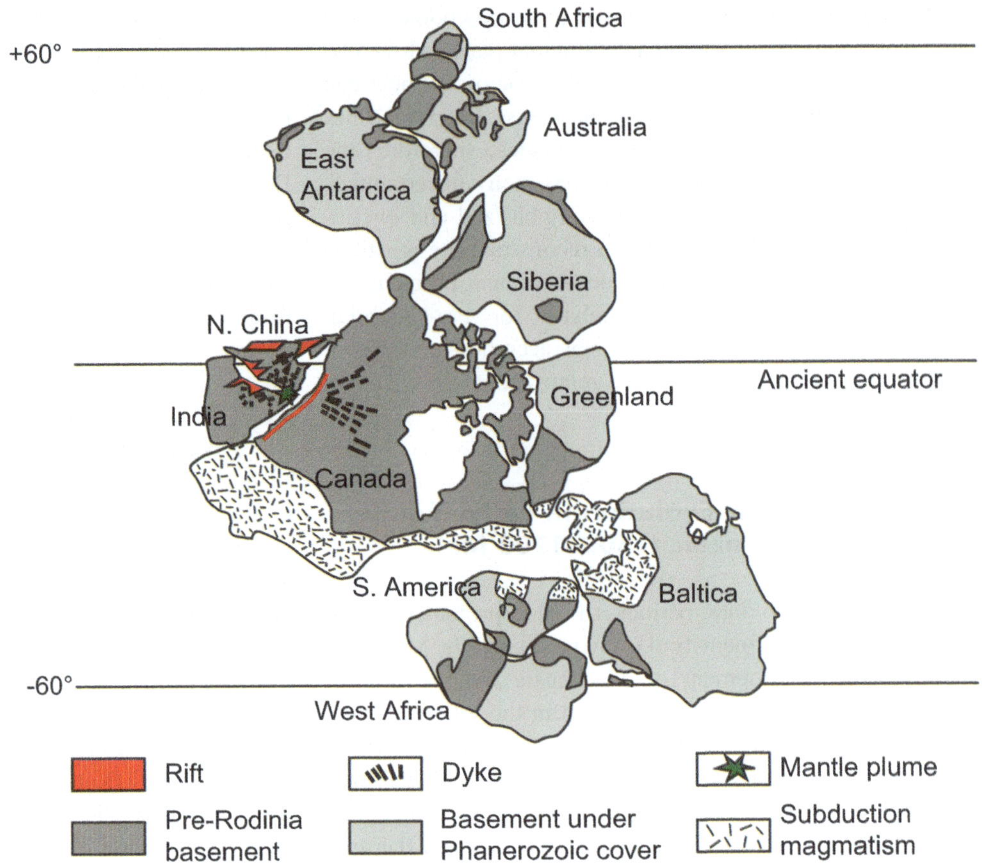

Fig. 3.2 Reconstruction sketch at the early period (Early Mesoproterozoic) of breakup for the Columbia supercontinent (After Hou 2012)

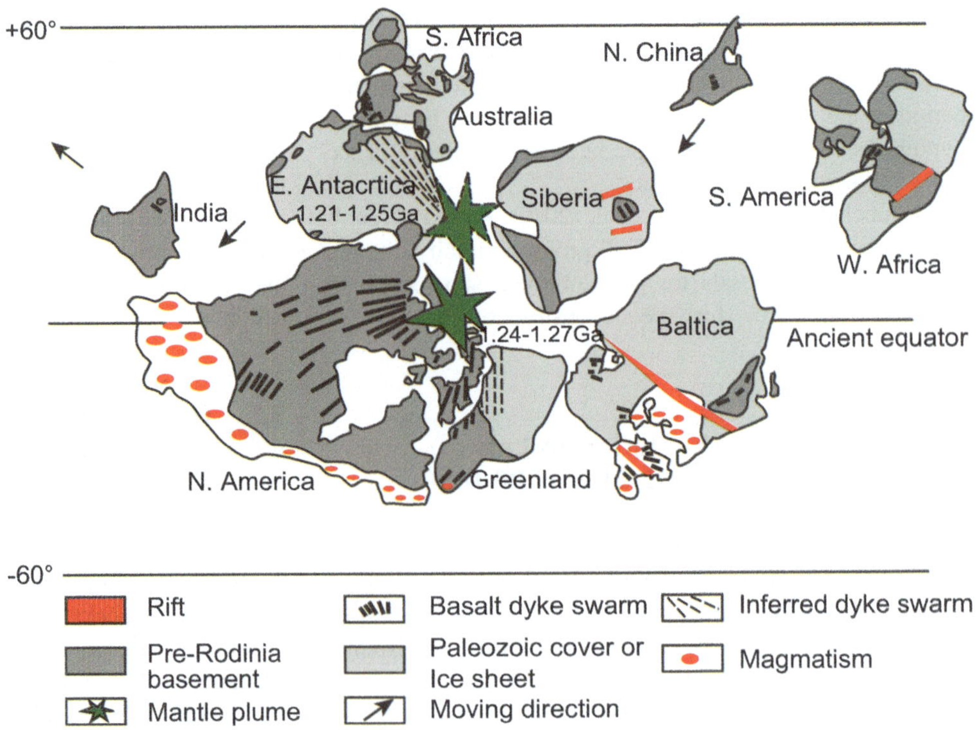

Fig. 3.3 Reconstruction sketch at the middle period (Middle Mesoproterozoic) of breakup for the Columbia supercontinent (After Hou 2012)

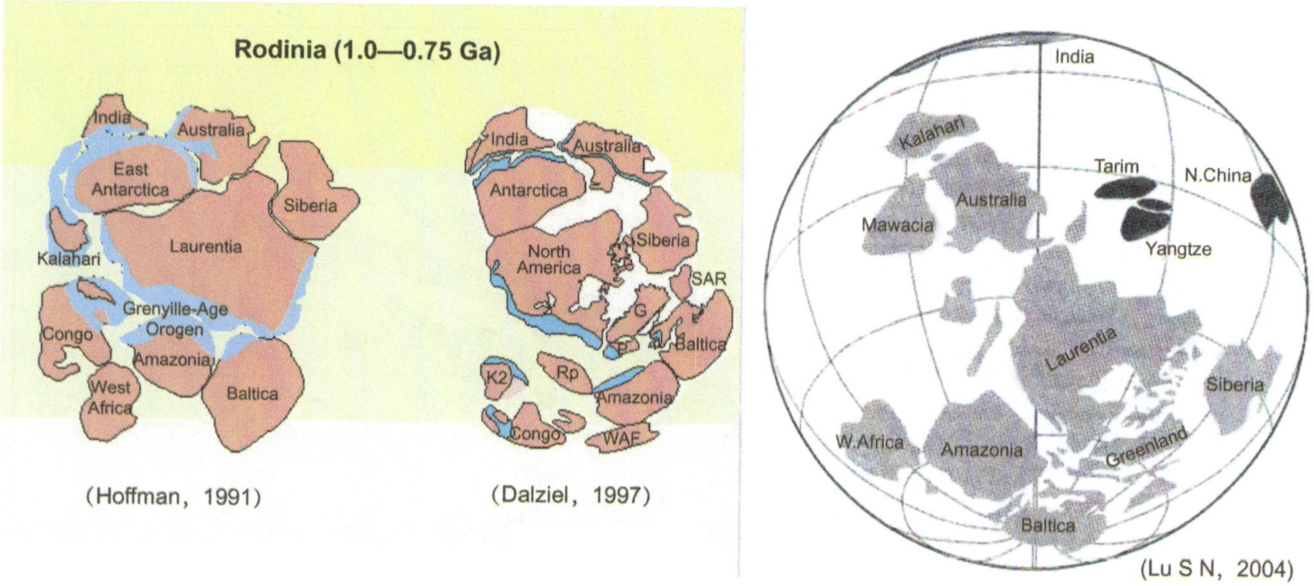

Fig. 3.4 Reconstruction sketch of Rodinia supercontinent in Late Mesoproterozoic

(~510 Ma) [49], Sunda plate (~500 Ma) [51], Sulawesi Sea block (~500 Ma) [53] and Eastern Argo block [54]. Besides the above blocks, Gondwana also includes the African plate, South American plate, Australian plate and Antarctica plate (Fig. 3.6; http://dictionary.reference.com/browse/Gondwana). At that time, the Gondwana supercontinent was located at the southern part of the Southern Hemisphere, developed amphibolite metamorphism and formed the united crystalline basement at about 510 Ma. This tectonic event was first confirmed in the study of African continent, so this tectonic period was often called Pan-African Tectonic Event (Kennedy 1964). However, most of the Asian blocks, European plate and North American plate maintained a discrete state in the Tethys Ocean in Proterozoic–Paleozoic, instead of becoming part of Gondwana.

As the Gondwana supercontinent was located at a high latitude in the Southern Hemisphere, in Carboniferous–Permian, there existed widespread glacial phenomenon on Gondwana and developed cold water fauna which was the Gondwana biota first discovered by Suess (1885) in North and Central India.

In addition, in the Asian continent, many small crystalline blocks formed in Early Cambrian (~510 Ma) and its previous periods were found to distribute in Bureya, Jiamusi, Song–Nen, Tuotuoshang–Xilingol, Hailar, Middle Mongolia, Altay, etc (Zhou et al. 2011; Wan 2011; Fig. 2.7). They may have been part of Gondwana continent in Early Paleozoic, and after the continent breaking up, they migrated to the middle latitude of the Northern Hemisphere with a large plate (possibly the Siberian plate) and then converged and collided into the Altay–Middle Mongolia–Hailar Early Paleozoic accretion collision zone [6] and Balkhash–Tianshan–Hingganling Late Paleozoic accretion–collision zone [10]. However for the time being, the accurate paleomagnetic data have not been obtained. Huang and Jiang (1962) first described this tectonic event in Northeast China as "Xingkai Movement."

Some scholars regard the Sino–Korean plate, Yangtze plate, Cathaysian plate and Indosinian plate which were south to the above small crystalline blocks as part of Gondwana supercontinent, and it is not appropriate obviously. Maybe they do not know the formation age of crystalline basement for those plates which was before the Pan-African Tectonic Event. However in the eastern, southwestern and southern sides of the Sino–Korean plate, Yangtze plate, Cathaysian plate and Indosinian plate, there existed crystalline basement blocks formed in the Pan-African Tectonic Event. It would be difficult to explain the above phenomenon if the plates had not undergone significant displacement. Because the paleo-magnetic data in the metamorphic rocks at this period are very controversial and not reliable, and the basis for the paleo-continental reconstruction of the whole area is insufficient, no researchers have been able to provide relatively convincing paleo-continental reconstruction of small blocks.

Based on Scotese's (1994, www.scotese.com Web site) paleo-continental reconstruction, Wan and Zhu (2011) revised and supplemented the paleo-magnetic data for many blocks in Asia and re-drew the global paleo-continental reconstruction map since Paleozoic (Figs. 3.6, 3.7, 3.8, 3.9, 3.10, 3.11 and 3.12). It was recognized that all the Asian continental blocks were scattered to the south of the equator and distributed along the latitude in Cambrian (Fig. 3.6).

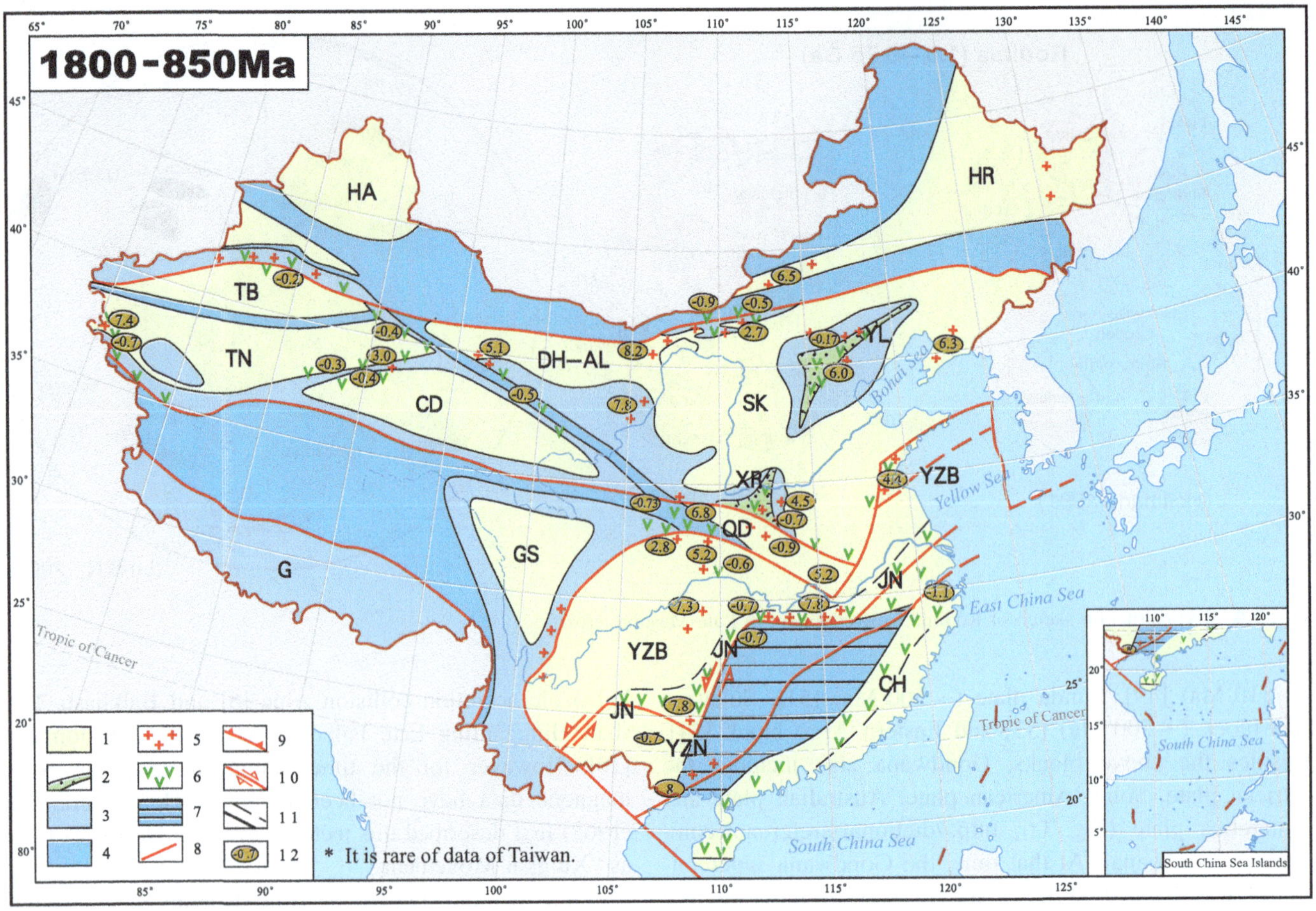

Fig. 3.5 Paleo-tectonic sketch of the China continent in Mesoproterozoic and Neoproterozoic (1800–850 Ma). **Legend**: (1) Area of erosion on paleo-continental plate; (2) rift system; (3) shallow seas with sedimentation; (4) oceanic crust; (5) granitic intrusions; (6) island arc with intermediate-acid volcanic rocks; (7) continental slope with turbidite sedimentation; (8) collision zone or boundary of plate; (9) subduction zone (including ophiolite belt); (10) strike-slip fault; (11) boundary of tectonic province; (12) rate of plate movement ("−" is rate of extension; "+" is rate of shortening (cm/y); see Appendix 5-1 of Wan 2011). **Tectonic domain**: Peri-Siberian tectonic domain (HA. Paleo-Kazakhstan plate, HR. Paleo-Harbin plate); Proto-Sino–Korean tectonic domain (TB. Paleo-Tarim plate, CD. Paleo-Qaidam plate, SK. Paleo-Sino–Korean plate including the North China area, the Korean Peninsula and the Alxa block); Peri-Yangtze tectonic domain (YZB and YZN. Paleo-Yangtze plate, including the Qinling–Dabie block, GS. Paleo-Songpan–Garze block, CH. Cathaysian paleo-plate, JN. the Jiangnan subduction and collision zone, formed at the period of about 850 Ma); Peri-Gondwanan tectonic domain (G). **Notes:** In this figure, the eastern part of Paleo-Yangtze plate is shown to be indented into the Sino–Korean plate in the southern Yellow Sea area, which tectonic event occurred in Triassic. In the tectonic reconstruction, part of the Yellow Sea in this map should be shown to belong to the Sino–Korean plate (Data from Bai et al. 1996; Liu et al. 1994; Wan 2011; completed and redrawn by the author)

During Early Paleozoic, the Siberian plate [1] rotated 12° clockwise with respect to a central reference point and the paleo-latitude changed from 31.4° S to 18.4° N (Khramov et al. 1981). Over this period, the position of the plate changed through latitude 49.8° (about 5000 km of displacement), and the average movement velocity reached 4.53 cm/y. The Siberian plate was the fastest moving plate in Early Paleozoic. However, most of the other Asian plates migrated northward relatively slowly, but till now the dynamic source has not been found. The Sino–Korean plate rotated 13.8° clockwise in Early Paleozoic and migrated from 20.2° S to 12.9° S with a rate of movement of slightly more than 0.8 cm/y. From Cambrian to Early Silurian, the Yangtze plate rotated 24.4° clockwise, during Middle–Late Silurian rotated 70° counterclockwise and migrated from 11.7° S to 2.8° N, with a rate of northward movement of 0.7 cm/y. The test data of other small blocks were not so enough reliable, so it needs further research (Wan 2011).

3.6 The Tectonic Evolution in the Late Period of Early Paleozoic (443–419 Ma)

In this period, the plate collisions were very strong in the Altay–Middle Mongolia–Hailar Early Paleozoic accretion–collision zone [6] and Karaganda–Kyrgyzstan (Qirghiz)

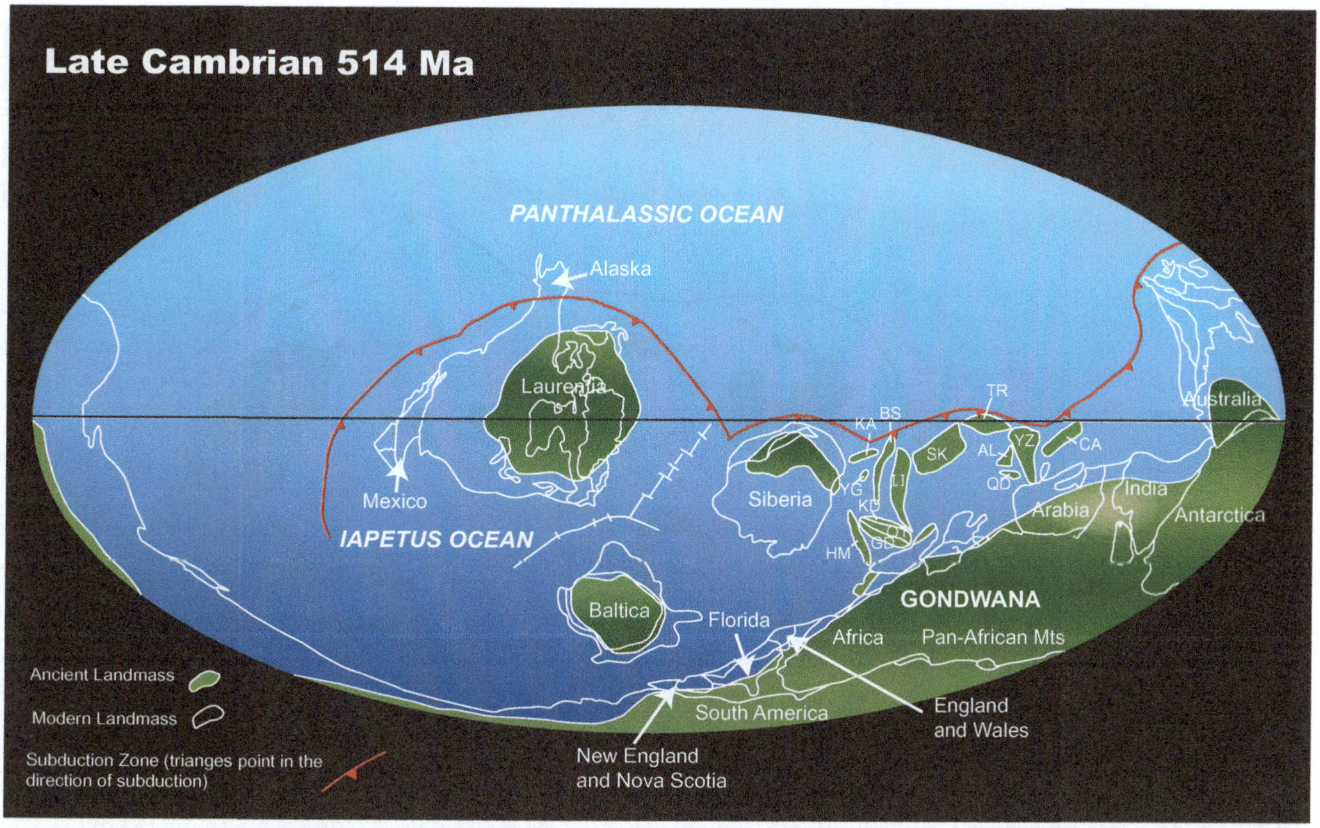

Fig. 3.6 Global paleo-continental reconstruction during Late Cambrian (514 Ma). AL. Alxa–Dunhuang block; BS. Baoshan–Sibumasu block; CA. Cathaysian plate; GD. Gangdise block; HM. Himalayan block; JG. Junggar block; KD. Eastern Kunlun block; KZ. Turan–Karakum plate; LI. Lanping–Simao–Indosinian plate; QD. Qaidam block; QT. Southern Qiangtang block; SK. Sino–Korean plate; TR. Tarim block; XM. Hingganling–Mongolia blocks; XY. Tarim–Dunhuang block; YG. Yagan block; YZ. Yangtze plate. These tectonic units' codes are also used in Figs. 3.7, 3.8, 3.10, 3.11, 3.12, 3.15, 3.20, 3.24, 3.27, 3.32. The data of paleo-magnetism and the central reference point positions in Asia and its adjacent areas are shown in Appendix 6 of Wan (2011) (After Scotese 1994; www.scotese.com Web site; revised by Wan and Zhu 2011)

Early Paleozoic accretion–collision zone (541–419 Ma) [7], which could also be regarded as the southwestern boundary accretion zones of the Siberian plate. There existed many small blocks (Fig. 2.7) in the above collision zones which had faster northward migration velocities than that of the Siberian plate, thus converging into the above two collisions on the southwestern side of Siberian plate. According to the recent magnetic orientation, the NW trending faults were dextral strike-slips, and the NE trending faults were sinistral strike-slips, while the nearly E–W trending faults were reverse faults or thrusts (Fig. 3.9; Xiao et al. 1992; Allen et al. 1993; Che et al. 1994; Li et al. 2002; Buslov et al. 2004; Charvet et al. 2007; Wang et al. 2008; Wan 2011; Wan and Zhu 2011).

The tectonic event happened in the late period of Early Paleozoic was rather special in the Sino–Korean tectonic domain and Yangtze tectonic domain. Most blocks in the above two domains remained separately in the Tethys Ocean, while only the Alxa–Dunhuang block [16], Qaidam block [18] and Tarim block [20] underwent convergence and collision, forming the Qilian Early Paleozoic accretion–collision zone [17] and Altun Early Paleozoic sinistral strike-slip–collision zone [19] (Fig. 3.9). After their amalgamation, the "Xiyu Plate" (which means western area in Chinese) was formed (Gao and Wu 1983; Wang and Chen 1987; Ge and Liu 2000; Wan 2011). The tectonic lines of the Xiyu plate are mainly WNW and ENE trending (based on present magnetic orientations). The fauna and flora of the Xiyu plate in Early Paleozoic were similar to those of the Yangtze tectonic domain. However, the independent Xiyu plate existed for a very short time, only for 140 Ma. In the late period of Late Paleozoic, the Xiyu plate moved northeastward together with the Sino–Korean plate and spliced with the Balkhash–Tianshan–Hingganling Late Paleozoic accretion–collision zone [10] and Sino–Korean plate, all incorporating into the Pangea supercontinent.

Almost at the same time, the Cathaysian plate [26] (Fig. 2.19) and Turan–Karakum plate [8] (Fig. 3.9) formed their united crystalline basements separately in the late period of Early Paleozoic (Brookfield 2000; Garzanti and

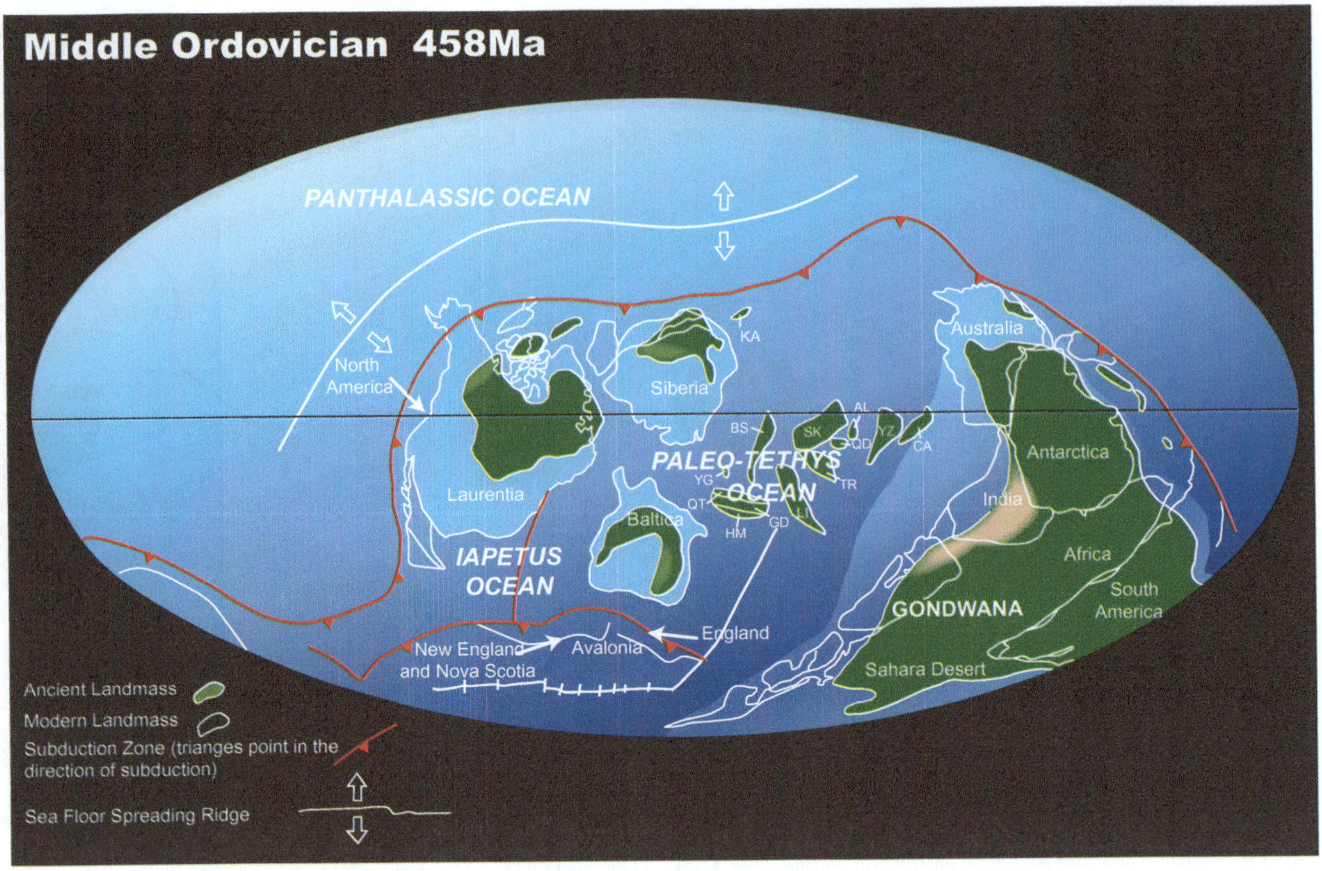

Fig. 3.7 Global paleo-continental reconstruction during Middle Ordovician (458 Ma). The legend and illustration are as same as those of Fig. 3.6. The data of paleo-magnetism and the central reference point positions in Asia and its adjacent areas are shown in Appendix 6 of Wan (2011) (After Scotese 1994; www.scotese.com Web site; revised by Wan and Zhu 2011)

Gaetani 2002; Luo et al. 2005). In the Cathaysian plate, there developed widespread low greenstone system, mingled with many small Archean–Proterozoic metamorphic systems. In the shallow metamorphic system within the crystalline basement of the Cathaysian plate, there developed more than 200 folding axes (not only on the border, but also distributed in the whole area, according to the 1:200,000 regional geological survey), which are mainly nearly N–S trending (including NNE or NNW trending, according to the present magnetic orientations; Fig. 2.19; Table 2.1; Wan 2011). This is quite different from the tectonic characteristics of the Yangtze plate.

In the late period of Early Paleozoic, at the Western Hemisphere the most important tectonic event was the formation of Caledonian collision zone (430–426 Ma; Brenchley and Rowson 2006), which put together the North American plate and Baltic plate, and led to the formation of the Laurasia supercontinent. However, no final conclusion has been reached on the dynamic mechanics of the collisions in the late period of Early Paleozoic (Wan 2011; Li et al. 2014).

To sum up, it can be seen that at the end of Early Paleozoic, the characteristics of tectonic events in the Asian continent were very different. It is not suitable to use one united tectonic event or local term (e.g., the term "Caledonian Movement or Event"; Huang et al. 1965) to refer to the tectonic events happened in Asia and China during the late period of Early Paleozoic, unless you insisted on Stille's proposition (1924) of "Universal Orogenic Episode." This term is one of the mistakes in China's geological academy for a long time.

3.7 The Tectonic Evolution in the Early Period of Late Paleozoic (419–323 Ma)

The Siberian plate [1], North American plate and Baltic plate continued to move northward, and together with the South American plate, African plate, Antarctica plate and Australian plate, were amalgamated to form the Pangea supercontinent with nearly N–S trending in the early period of Late Paleozoic (Figs. 3.10, 3.11 and 3.12). From Devonian

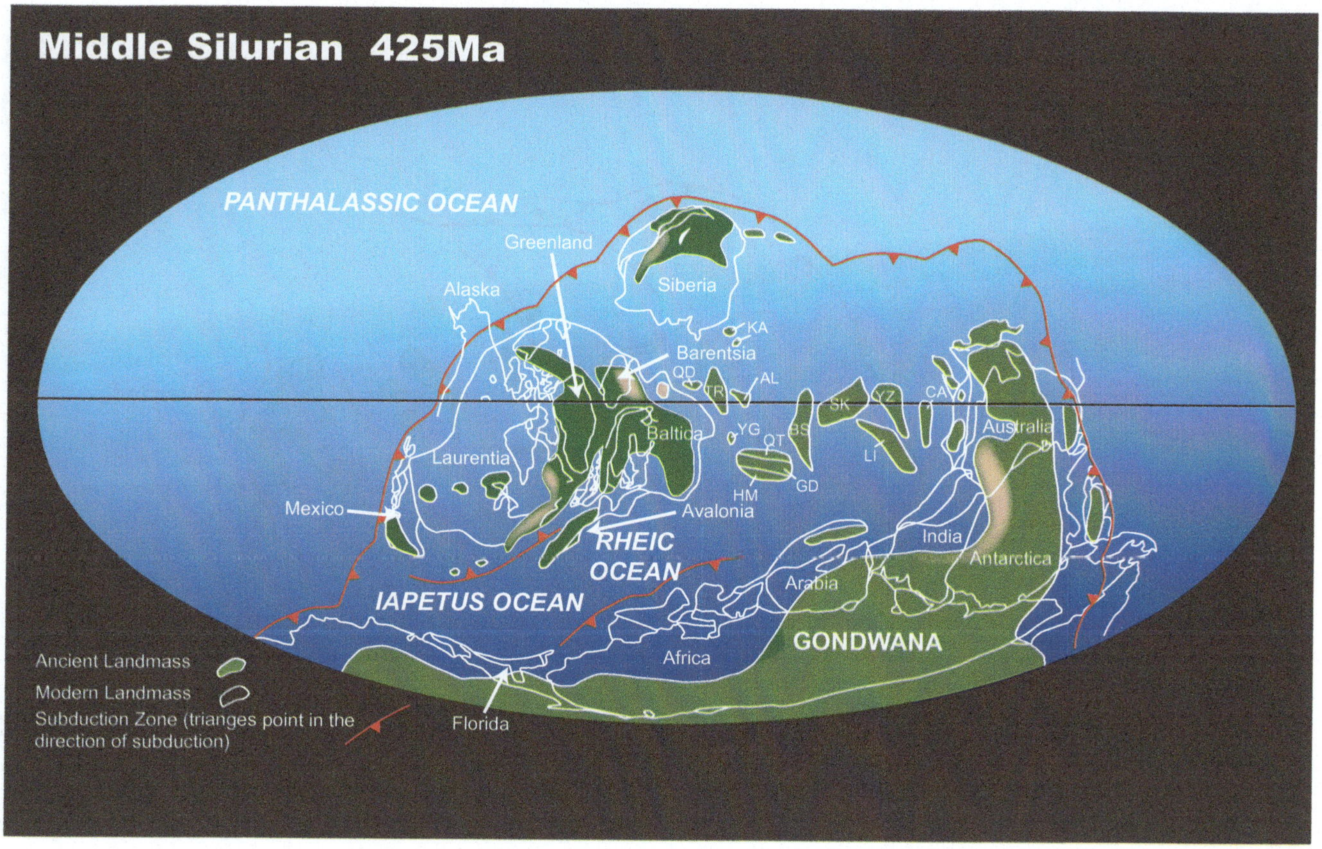

Fig. 3.8 Global paleo-continental reconstruction during Middle Silurian (425 Ma). The legend and illustration are as same as those of Fig. 3.6. The data of paleo-magnetism and the central reference point positions in Asia and its adjacent areas are shown in Appendix 6 of Wan (2011) (After Scotese 1994; www.scotese.com Web site; revised by Wan and Zhu 2011)

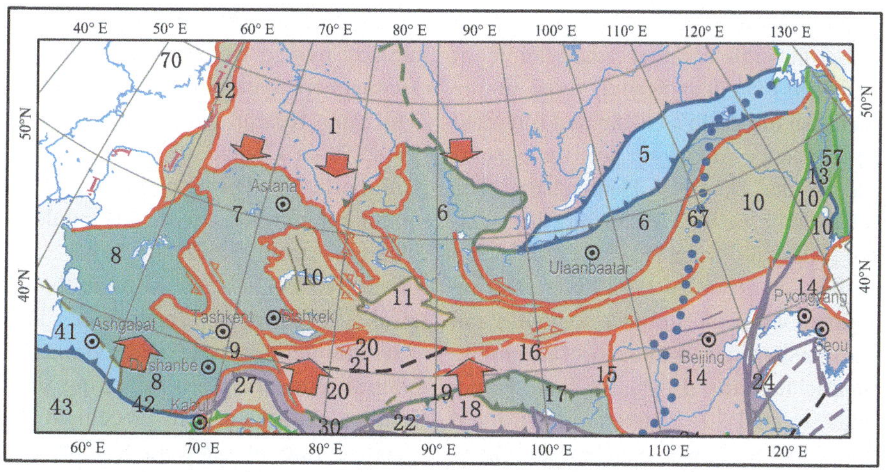

Fig. 3.9 Tectonic sketch of Central Asia–Mongolia tectonic domain [6, 7, 17, 19, etc.] in the late period of Early Paleozoic. The red line shows collision zone in the late period of Early Paleozoic. The big red arrow shows the block migration orientation, with the size showing the strength of stress. The small red arrow shows the moving trending of the faults. The numbers of tectonic unit are as same as those in the contents, Appendix and text

to Early Permian, using a central reference point to represent each of continental blocks, one can find the Siberian plate migrated from 33.4° N to 37.5° N latitude, a paleo-latitude change of 4.1°, and the rate of latitudinal displacement was only 0.34 cm/y, much lower than its rate of movement during Early Paleozoic. The measurements of

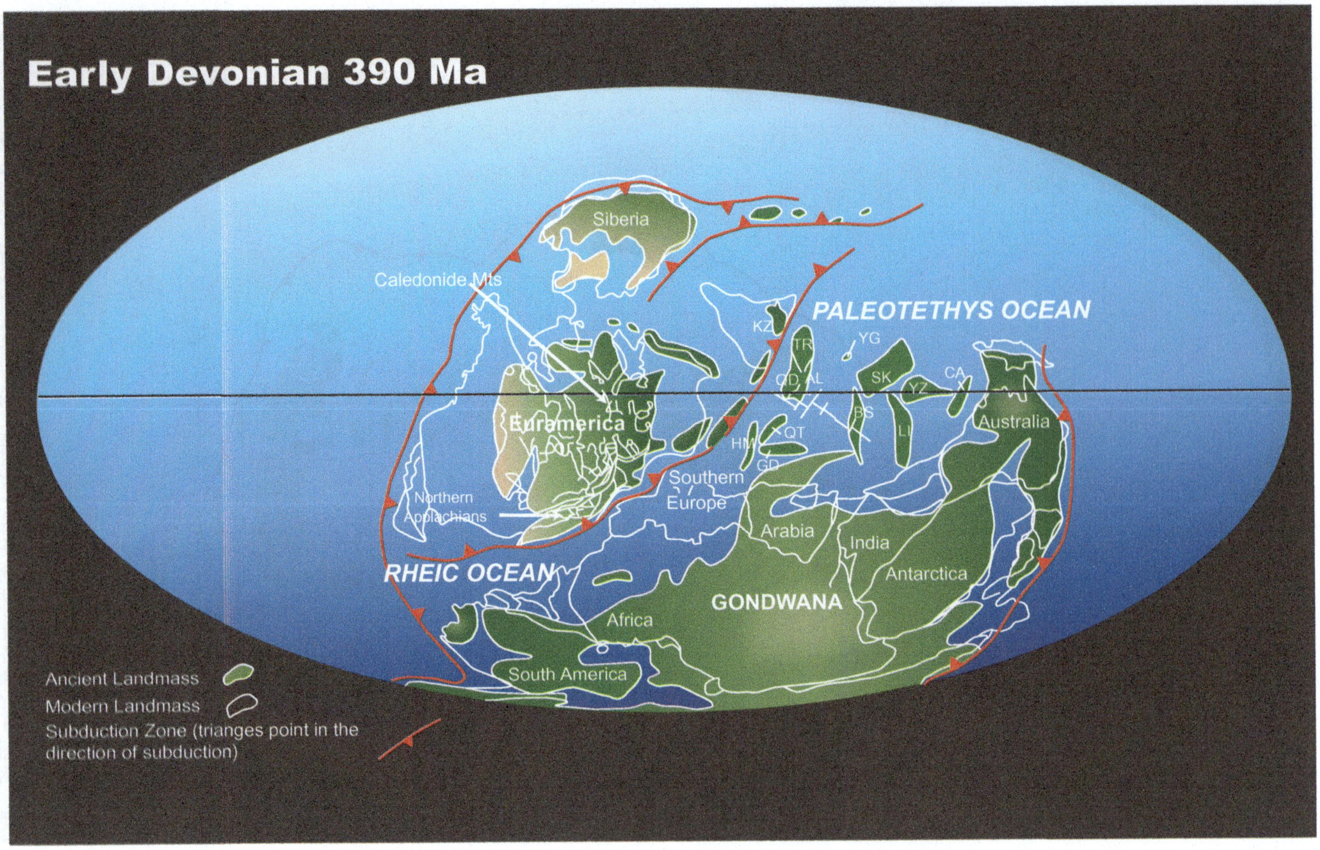

Fig. 3.10 Global paleo-continental reconstruction during Early Devonian (390 Ma). The legend and illustration are as same as those of Fig. 3.6. The data of paleo-magnetism and the central reference point positions in Asia and its adjacent areas are shown in Appendix 6 of Wan (2011) (After Scotese 1994; www.scotese.com Web site; revised by Wan and Zhu 2011)

paleo-magnetic declination show that the Siberian plate rotated counterclockwise through a rather small angle of 13.9° (Khramov et al. 1981). In that period, most of the Asian blocks were located in the Tethys Ocean and migrated northward with different distances, reaching near the equator or the middle latitude of the Northern Hemisphere. The latitude change of Indian plate in Late Paleozoic was relatively great. It migrated southward from 28.4° S to 37.3° S, a change of 8.9° in latitude, at a rate of 0.74 cm/y. During Devonian, the Indian plate rotated 40° counterclockwise and then during Carboniferous–Permian rotated 67° clockwise (Klootwijk and Radhakrichnamurty 1981). In Late Paleozoic, the Australian plate rotated 20° counterclockwise and migrated southward from 4.4° S to 56.3° S paleo-latitude, at a rate of 4.3 cm/y (Van der Voo 1993). The above data show that in Late Paleozoic, the Asian continental blocks migrated at different rates. The Siberian plate and Indian plate were relatively stable, while the Australian plate migrated southward at a rather fast rate, which may be related to the extension of the Eastern Tethys Ocean. As a result of the blocks' migration, the Pangea supercontinent was formed during Late Permian, with a nearly N–S trend (Fig. 3.12).

In the whole Late Paleozoic, the Sino–Korean plate had already moved from the Southern Hemisphere to Northern Hemisphere with little paleo-magnetic declination change. The Sino–Korean plate migrated northward from 12.9° S to 10.8° N, a change of 23.7° in paleo-latitude. In Middle Carboniferous–Permian, the Sino–Korean plate rotated 18.5° counterclockwise, with its paleo-magnetic declination changing from 338.2° to 319.7°, and migrated northward about 310 km at a rate of more than 1 cm/y in latitude (Wu et al. 1991; Ma and Yang 1993). The Yangtze plate rotated 7.6° counterclockwise in Devonian–Early Permian with its central reference point migrating from 6.9° S to 3.3° N latitude (a northward displacement about 1000 km) at an average rate of 0.84 cm/y, more slowly than that of the Sino–Korean plate (Zhang et al. 2001). In Devonian–Early Carboniferous, the Cathaysian plate remained at 11.8° S–10.3° S paleo-latitude (latitude migrated and revised by Wan and Zhu 2011), moving only 1.5° (~150 km), but rotated counterclockwise through 52.3° from 111.2° to 58.9° (Chen et al. 1991). The Junggar block remained in the Northern Hemisphere during Late Paleozoic, moving only from 29.7° N to 28.3° N latitude, but rotated clockwise through 77°

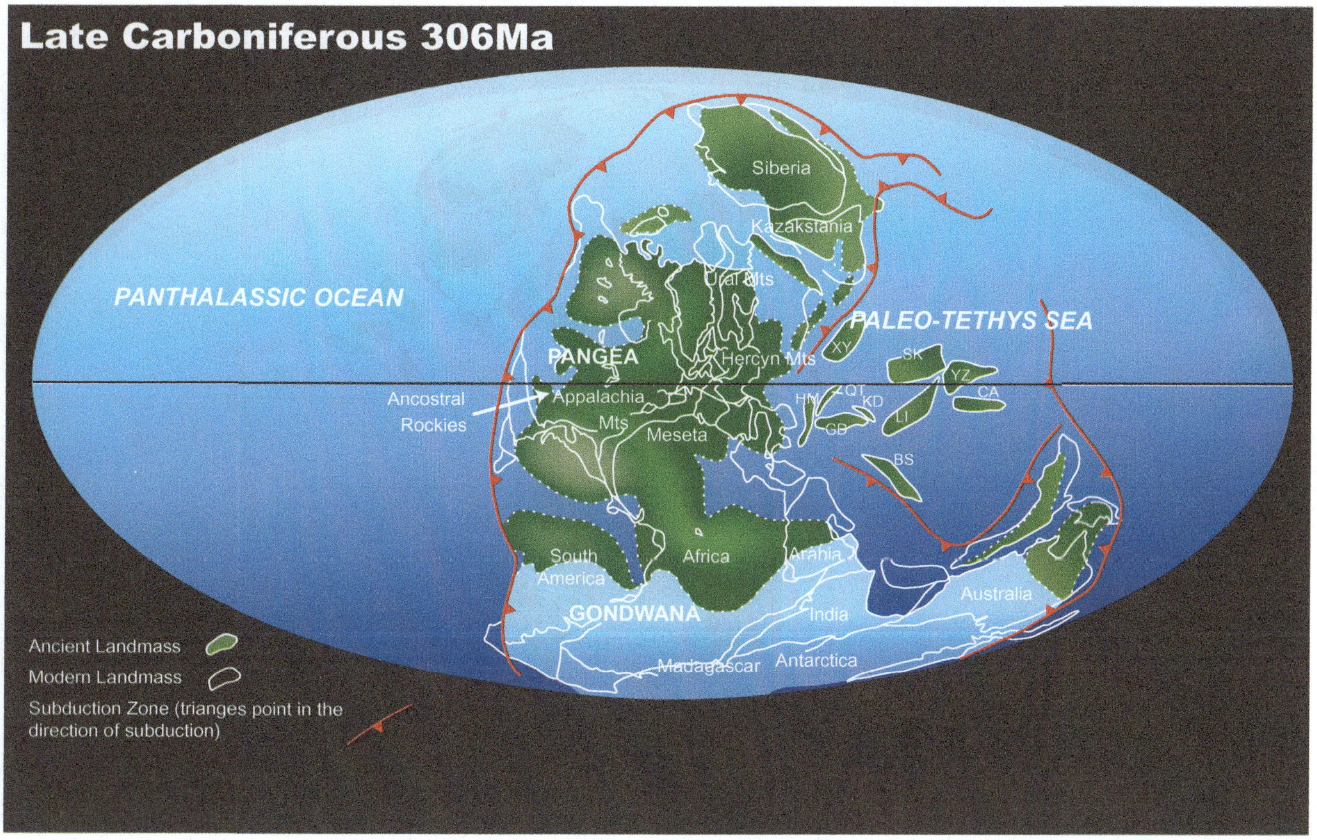

Fig. 3.11 Global paleo-continental reconstruction in Late Carboniferous (306 Ma). The legend and illustration are as same as those of Fig. 3.6. The data of paleo-magnetism and the central reference point positions in Asia and its adjacent areas are shown in Appendix 6 of Wan (2011) (After Scotese 1994; www.scotese.com Web site; revised by Wan and Zhu 2011)

from NW 342.4° to NE 59.4° (Li et al. 1992). The Tarim block, which formed a part of the Xiyu plate, moved through 10° in latitude (~1000 km) northward from 21.2° N to 31.3° N between Devonian and Permian, at a rate of more than 0.84 cm/y, and rotated counterclockwise through 73.4°, from 94.5° to 21.1° (Fang et al. 1992, 1996).

Many blocks in Southern Asia (such as India, Iran, Turkey, Himalaya, Gangdise and Southeast Asia), and Australian, African, South American and Antarctica plates which used to belong to Gondwana supercontinent, have continued to belong to the Gondwana supercontinent. These blocks and plates began to break up gradually during Late Paleozoic, but still remained in the middle latitude of the Southern Hemisphere with a little migration (Figs. 3.6, 3.7, 3.8, 3.9, 3.10, 3.11, 3.12; Van der Voo 1993).

Paleo-geomagnetic determinations show that in Paleozoic the three major plates surrounding the Asian continental blocks became dispersed (Figs. 3.6, 3.7, 3.8, 3.9, 3.10, 3.11, 3.12). In the whole Paleozoic, the Siberian plate moved directly northward across 71° of latitude (about 7000 km; Khramov et al. 1981); the Australian plate migrated southward across 42° of latitude (near 4100 km; Van der Voo 1993); at first, the Indian plate was relatively stable, but in Late Paleozoic moved rapidly southward over a total distance of 27° of latitude (about 2600 km; Klootwijk and Radhakrichnamurty 1981). The above data could explain the dispersed kinematic characteristics of the Asian continental blocks throughout the whole Paleozoic.

To sum up, it can be seen that the Siberian plate migrated a great distance northward during Paleozoic, while during Late Paleozoic the Gondwana and Indian plate migrated southward. The paleo-Asian continental blocks remained dispersed, migrated gradually northward by different degrees during the whole of Paleozoic. At particular times, blocks collided and amalgamated; for example, the Altay–Junggar–Ergun collision zone was formed in Early Paleozoic, the blocks forming the Xiyu plate were amalgamated together in the end of Early Paleozoic, and the Balkhash–Tianshan–Hingganling collision zone occurred during Carboniferous and Early Permian.

The Asian continental blocks were arranged along the equator in Early Paleozoic, however in the process of gradual movement arrived at a nearly N–S arrangement in Carboniferous, close to the arrangement of present continental blocks. This was an important turning point in the assembly of the continental blocks.

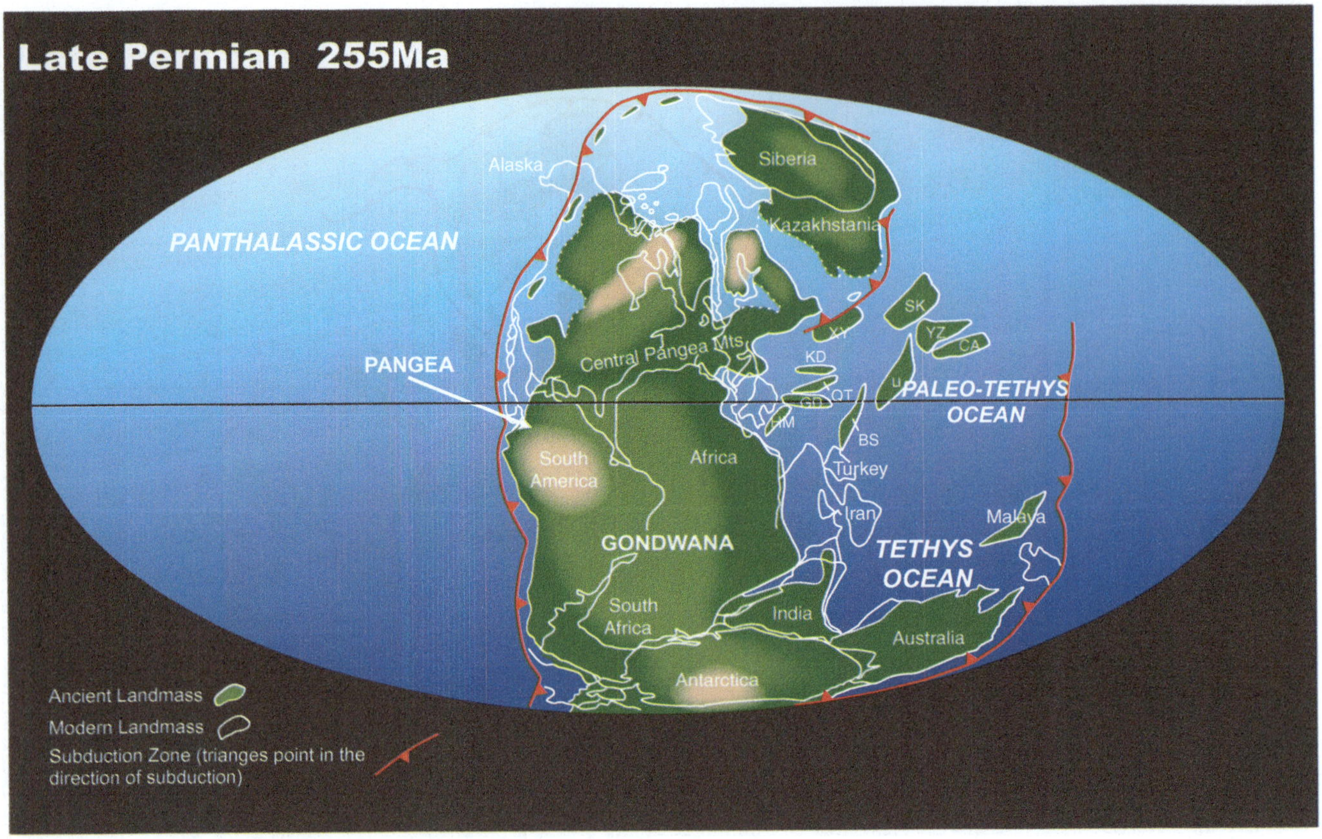

Fig. 3.12 Global paleo-continental reconstruction in Late Permian (255 Ma). The legend and illustration are as same as those of Fig. 3.6. The data of paleo-magnetism and the central reference point positions in Asia and its adjacent areas are shown in Appendix 6 of Wan (2011) (After Scotese 1994; www.scotese.com Web site; revised by Wan and Zhu 2011)

It seems that the Asian continental blocks may be separated and re-arranged by the Tethys Ocean extension. However, the conclusive evidence of plate migration dynamic mechanism has not been verified till now. About that dynamic mechanism, although the recognitions of the positions and kinematics for Paleozoic continental plates are almost same (Powell et al. 1993; Scotese 1994; Cavazza et al. 2004; Schettino and Scotese 2005), many hypotheses are rare in certain evidences, only a guess, such as the thesis of Torsvik et al. (2008, 2014).

In the early period of Late Paleozoic (Late Devonian–Early Carboniferous, 385–323 Ma), between the Siberian plate and its surrounding small blocks, Altay–Middle Mongolia–Hailar collision zone, Xiyu plate and Sino-Korean plate, there formed some great arc accretion–collision zones: the Western Tianshan Late Paleozoic (385–260 Ma) collision zone [9] and Balkhash–Tianshan–Hingganling Late Paleozoic (385–260 Ma) collision zone [10] (Figs. 3.9, 2.10). In that period, the collision was mainly N–S trending (according to the present magnetic orientation), and the collision of the western part happened in Late Devonian–Early Carboniferous (385–323 Ma, Fig. 3.13; Buslov et al. 2004; Xiao et al. 1992; Allen et al. 1993; Che et al. 1994; Charvet et al. 2007; Wan 2011; Wan et al. 2015), while the collision of the eastern part (northern side of the Sino–Korean plate) occurred mainly in Middle Permian (270–260 Ma, Fig. 3.14; Wan 2011). It shows that the western Balkhash–Tianshan area collided before the eastern Inner Mongolia–Hingganling area. The reason may be that there still reserved the paleo-ocean in north to Hailar, but the eastern part never collided with the Siberian plate. In other words, in Late Paleozoic, the paleo-ocean (which is part of the Tethys Ocean, so the term of "Paleo-Asian Ocean" is not appropriate) between the eastern Balkhash–Tianshan–Hingganling Late Paleozoic collision zone and Siberian plate was relatively narrow in its western part, while relatively wide in its eastern part.

The tectonic line in the western Balkhash–Tianshan–Hingganling Late Paleozoic collision zone was mainly WNW or nearly E–W trending. In Late Devonian–Early Carboniferous (385–323 Ma), the nearly E–W trending faults were the main faults (reverse fault or thrusts) caused by the collision, and the NW trending regional faults were dextral strike-slip faults (Fig. 3.13). However, in the eastern part of the collision zone, the tectonic line was mainly NE or nearly E–W trending, and the NE trending faults were sinistral strike-slip faults (Fig. 3.14).

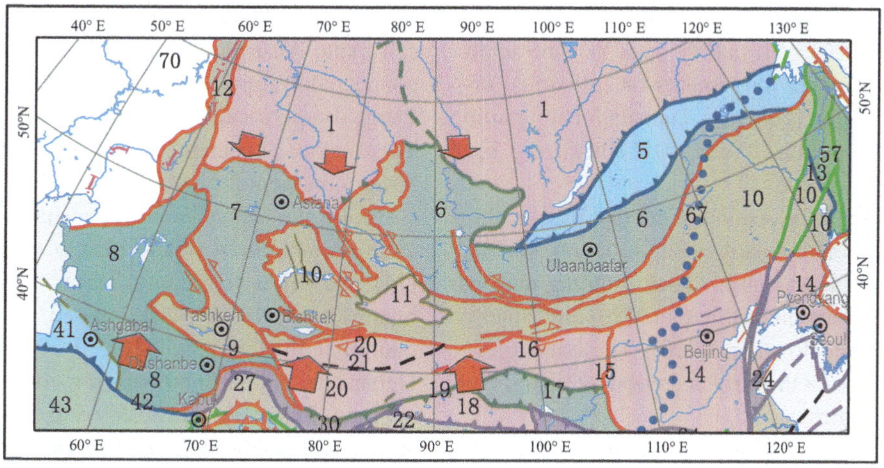

Fig. 3.13 Tectonic sketch of Central Asia–Mongolia tectonic domain in the early period of Late Paleozoic (Late Devonian–Early Carboniferous, 385–323 Ma). The big red arrow shows the block migration orientation, with the size showing the strength of stress. The small red arrow shows the moving trending of the faults. The numbers of tectonic unit are as same as those in the contents, Fig. 1.1 and text

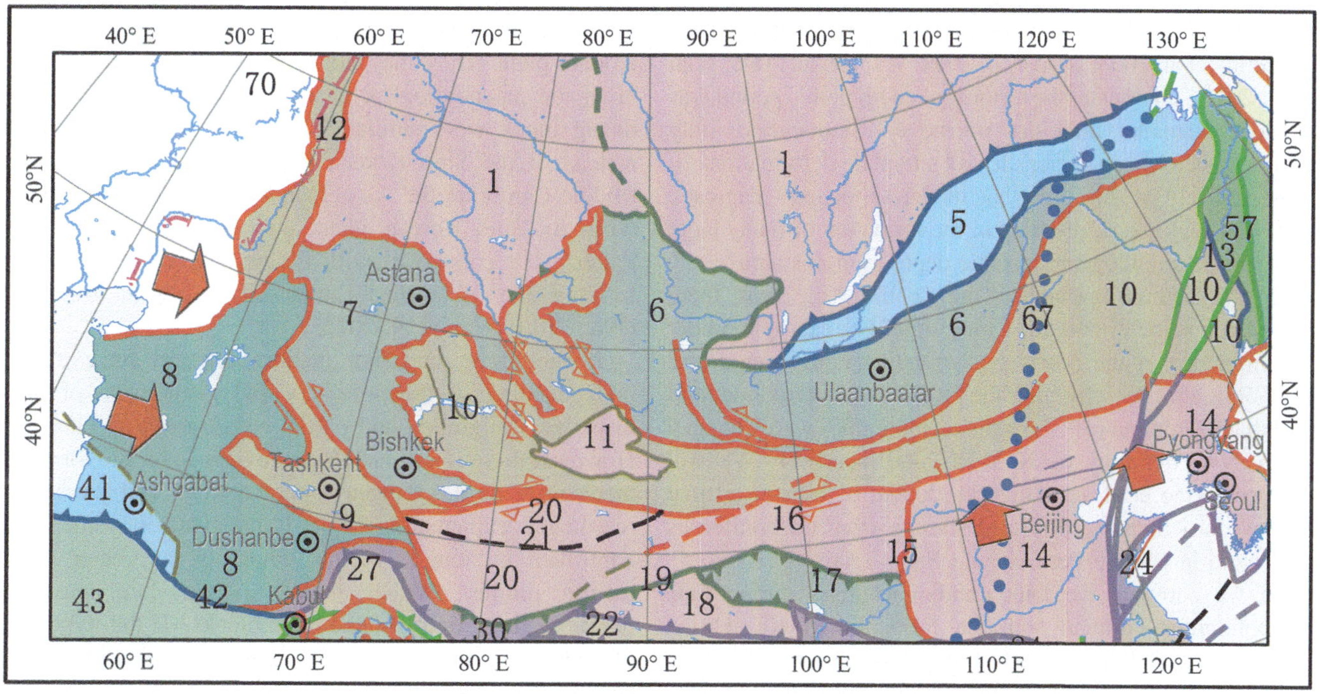

Fig. 3.14 Tectonic sketch of Central Asia–Mongolia in the late period of Late Paleozoic (Late Carboniferous–Middle Permian, 323–260 Ma). The red line shows the fault zone distribution areas for this period. The big red arrow shows the direction of compression, while the small red small arrow shows the moving trending of the faults (sinistral or dextral). The numbers of tectonic unit are same as those in the contents, Fig. 1.1 or text

3.8 The Tectonic Evolution in the Late Period of Late Paleozoic (323–260 Ma)

In recent years, researchers have discovered that in the late period of Late Paleozoic (Late Carboniferous–Middle Permian), the regional NW trending faults in the western part of the Balkhash–Tianshan–Hingganling Late Paleozoic collision zone [10] were sinistral strike-slip faults (Buslov et al. 2004; Zonenshain et al. 1990; Bazhenov et al. 2003; Fig. 3.14), which were totally different from the early period of Late Paleozoic. And the nearly E–W trending faults were dextral strike-slip faults (Shi et al. 1994; Li et al. 2002; Gao et al. 2006; Pickering et al. 2008; Xiao et al. 2008; Wang et al. 2008). However, in the eastern part of this collision zone near the Hingganling area during Late Carboniferous–

Middle Permian, the collision was mainly N–S trending (Fig. 3.14; Wan et al. 2015). It can be seen that the understanding of the two tectonic movements or collisional orogenic movements of the Balkhash–Tianshan–Hingganling Late Paleozoic collision zone (Wan 2011) has long been considered to be questionable.

On the western side of the Asian continental blocks, there formed the Ural accretion–collision zone (400–260 Ma) [12] in Late Paleozoic, which was nearly N–S trending and was the collision zone between the Baltic plate [70] and Siberian plate [1] (Brenchley and Rowson 2006). The convergence and movement direction of the Ural accretion–collision zone were nearly E–W trending.

The long-distance eastward compression effect of the Ural accretion–collision zone must have changed the fault characteristics of the Balkhash–Tianshan zone (Fig. 3.14), turning the NW trending regional faults into sinistral strike-slip faults and the pre-existing nearly E–W trending reverse faults into dextral strike-slip faults (Buslov et al. 2004; Zonenshain et al. 1990; Shi et al. 1994; Li et al. 2002; Bazhenov et al. 2003; Pickering et al. 2008; Xiao et al. 2008; Wang et al. 2008; Wan et al. 2015). Of course, the tectonic strength at this period was obviously smaller than that of the collision period, only showing tectonic characteristics of intraplate deformations. It was precisely at this time that moderate tectonic-magmatic activities took place, resulting in the formation of many large or super-large endogenic metallic deposits in this area.

In the tectonic research of the Central Asia areas, there have been some doubts for a long time. For example, under the influence of the nearly N–S trending convergence and collision effects, why did the Kazakhstan Orocline and Balkhash Orocline (of which the tectonic lines were changed from NW trending to N–S–NE and nearly E–W) still formed? Xiao et al. (2003, 2008, 2009a, b) considered that the two oroclines were formed starting from Carboniferous and fixed in Middle Permian and Early Triassic.

If we correlate the fault trending changes during Late Carboniferous–Permian with the formation of Kazakhstan Orocline and Balkhash Orocline, it will be very reasonable to explain these phenomena by the long-distance effect of eastward compression derived from the collision of Ural accretion–collision zone in Late Paleozoic. In the past, this long-distance eastward compression of Ural accretion–collision zone was overlooked. It was often assumed that there were only two orogenic episodes of nearly N–S shortening or that it was the product of multiple accretion and orogenesis.

Due to the development of the above Western Tianshan collision zone [9], Balkhash–Tianshan–Hingganling collision zone [10], Ural accretion–collision zone [12] and the Variscan (or Hercynian) collision zone (in Europe), the European plate (Baltic plate as its main part) and the Asian continental blocks with Siberian plate as its center were connected (Brenchley and Rowson 2006). They were also combined with the North and South American plates, African plate, Australian plate and Antarctica plate which were connected in Early Paleozoic, leading to the formation of the Pangea supercontinent in the end of Permian (Fig. 3.12). This tectonic event incorporated more than one half of the Asian continental blocks into the Pangea supercontinent.

As for the Helanshan–Liupanshan collision zone [15], the author infers that it was formed when the Sino–Korean plate [14] and Xiyu (including Tarim, Qaidam and Alxa–Dunhuang) plate [16, 18, 20] collided with the Tianshan–Hingganling collision zone [10]. It is quite probable that the Helanshan–Liupanshan collision zone was formed in the end of Late Paleozoic because of the development of many tectono-magmatic activities of this time near this collision zone (Wan 2011; Wan et al. 2015; Gen and Zhou 2012). However, because of the strong E–W trending compression in Jurassic, a very complex ramp overthrust system was formed, and the early structural traces were covered up, so the data were not sufficient. Some researchers insisted that this collision zone was just an "aulacogen" in the Sino–Korean plate. They may not know that after Middle Cambrian the paleontological characteristics of the Alxa–Dunhuang block were similar to those of the Yangtze plate, instead of the Sino–Korean plate. It is obvious that after Middle Cambrian the Alxa–Dunhuang block was separated with the Sino–Korean plate, and until the end of Late Paleozoic it converged with the Sino–Korean plate again, forming the Helanshan–Liupanshan collision zone [15] (Wan 2011; Wan et al. 2015).

It should be said that modern scholars have a close understanding of the processes and phenomena of block dispersion, migration and partial convergence in Paleozoic. However, there is still insufficient evidence for the formation mechanism of the Pangea supercontinent, and there is no reasonable explanation for it. If the plate movement was controlled by mantle plume, where is the center of the mantle plume? Which mantle plume can cause the re-convergence and dispersion of blocks? If the hypothesis of meteorite impact is correct, when and which plate is the center of meteorite impact? So far, no scholar has provided convincing evidences.

3.9 The Tectonic Evolution in Triassic (Indosinian Tectonic Event, 252–201 Ma)

In Mesozoic, the characteristics of migration, convergence and dispersion of global plates (Figs. 3.15, 3.16, 3.17) were extremely different from those in Paleozoic. In Triassic, large-scale collisions occurred in the central and southern parts of the Asian continent, which resulted in many East Asia blocks (such as South China Sea, Indochina, Southern Qiangtang blocks) merging northeastward into the Eurasian plate. Six collision zones were formed in the central-southern part of the Asian continent, including the Qinling–Dabie [24] (Maruyama et al. 1992; Li et al. 1996, 1997; Dong et al. 1999; Zhang et al. 2001), Shaoxing–Shiwandashan [25] (Wan 2011), Western Hindukush–Pamir–Kunlun [30] (Jin et al. 1999), Jinshajiang [31] (Zhong 1998), Shuanghu [32] (Li 1997; Li et al. 2006) and Changning–Menglian–Central Malay [33] (Liu et al. 1991; Zhong 1998; Hutchison and Tan 2009; Metcalfe 2011) collision zones (Figs. 3.16, 3.17). After the collision of Indosinian tectonic period, nearly two-thirds of the Asian continent has been amalgamated into the Pangaea supercontinent (Figs. 3.15, 3.16, 3.17).

The formation of the above six collision zones was related to the northward migration of the Asian continental blocks at different rates. Depending on the paleo-magnetic data (Appendix 6 in Wan 2011), in Triassic the Siberian plate migrated northward at a rate of about 3 cm/y (Khramov et al. 1981), while the Sino–Korean plate moved northward at an average rate of 1.76 cm/y (Ma and Yang 1993). So, the Sino–Korean plate did not collide with the main part of the Siberian plate; instead, it was connected with the Central Asia–Mongolia Paleozoic accretion–collision zone, and there still remained residual oceanic basin in the present Trans-Baikal or Mongolia–Okhotsk area. The average northward movement rate of the Yangtze plate in Triassic was 3.32 cm/y (Zhu et al. 1988; Opdyke et al. 1986), obviously more than that of the Sino–Korean plate, resulting in the formation of the Qinling–Dabie collision zone. Unfortunately, the paleo-geomagnetic data of Cathaysian plate were not sufficient enough to discuss its relationship with surrounding blocks for the time being. The average northward migration rate of Indosinian plate during Triassic was 1.22 cm/y (Van der Voo 1993; Yang and Besse 1996), while rate of the Baoshan–Sibumasu plate was 2.42 cm/y (Zhuang 1988). The above different northward migration rates may be the main reason for plate collisions in the central-southern Asian continental blocks. The average northward movement rate of the Indian plate was about 2 cm/y (Klootwijk and Radhakrichnamurty 1981) in Triassic, and rate of the Australian plate was 1.6 cm/y (Van der Voo 1993), but they were both still located in the Tethys Ocean.

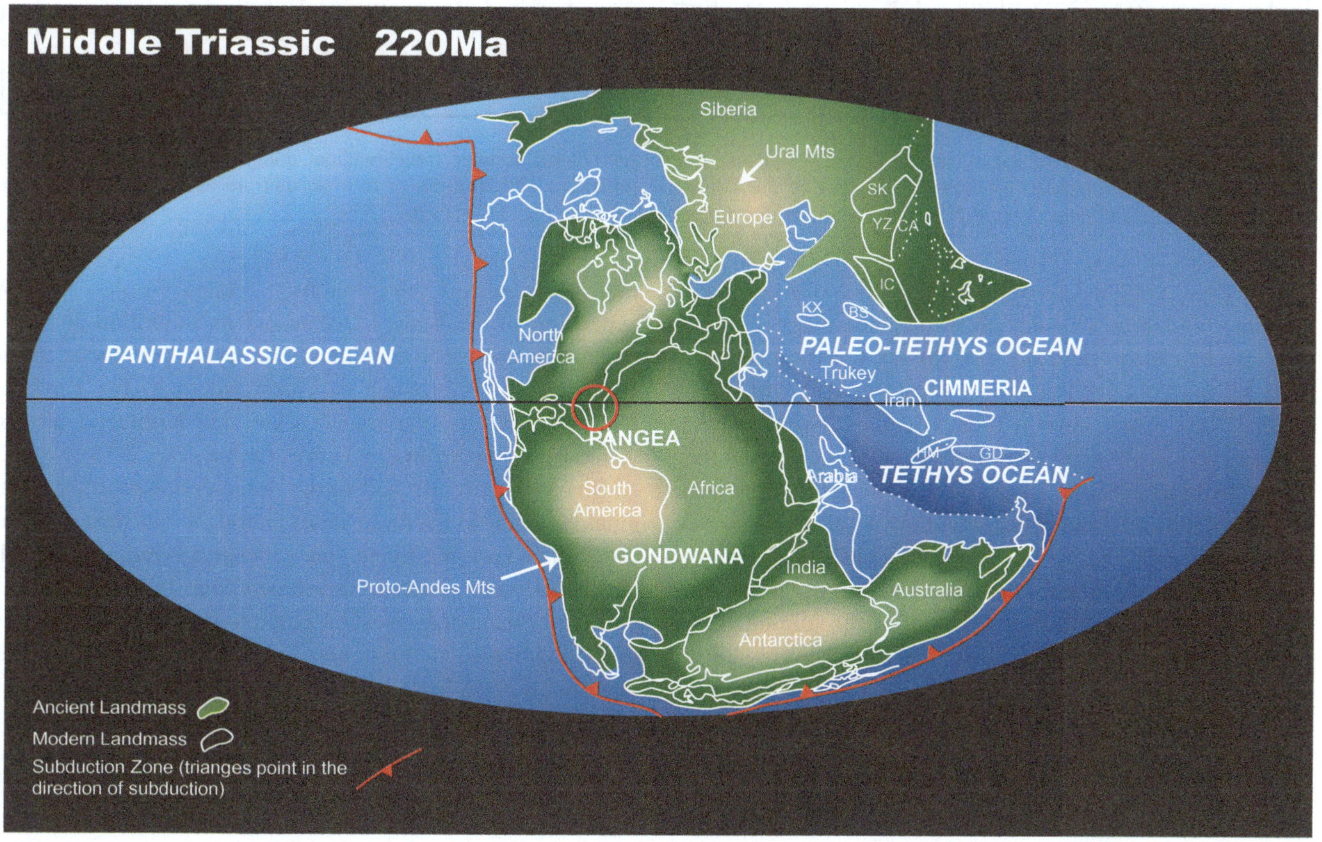

Fig. 3.15 Global paleo-continental reconstruction in Middle Triassic (220 Ma). BS. Baoshan–Sibumasu block; CA. Cathaysian plate; GD. Gangdise block; HM. Himalayan block; KX. Western Kunlun block; KZ. Turan–Karakum plate; IC. Lanping–Simao–Indosinian plate; QT. Southern Qiangtang block; SK. Sino–Korean plate; YZ. Yangtze plate. The red circle is the dissociation center of the Pangea supercontinent. The data of paleo-magnetism and the central reference point positions in Asia and its adjacent areas are shown in Appendix 6 of Wan (2011) (After Scotese 1994; www.scotese.com Web site; revised by Wan and Zhu 2011)

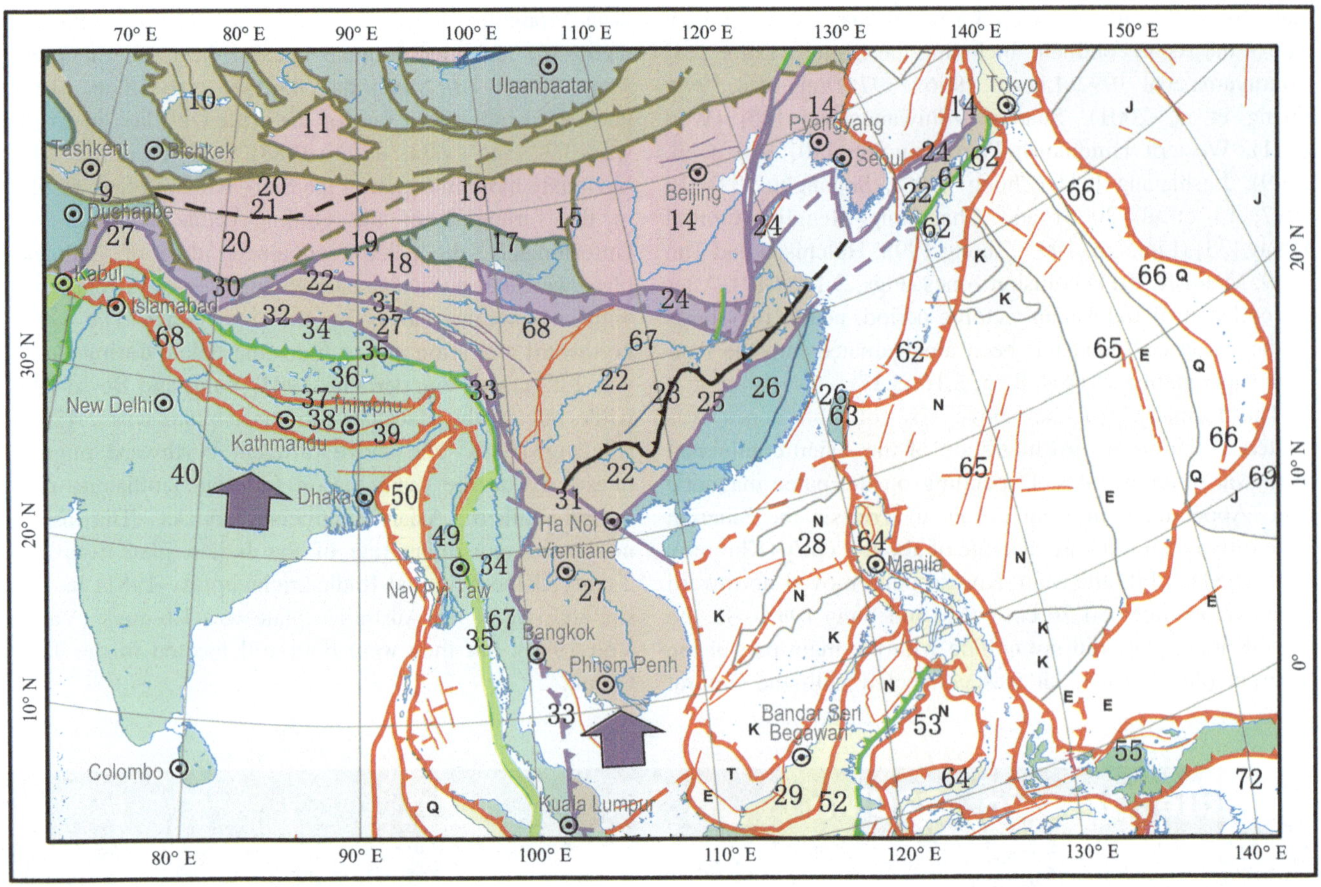

Fig. 3.16 Tectonic sketch of the Asian continent in Triassic (the Indosinian period). The purple line shows the location of the Triassic collision zones or main faults. The purple arrow shows the direction of the Triassic tectonic force (without doing the paleo-magnetic reconstruction). The numbers of tectonic unit are as same as those in the contents, Fig. 1.1 or text

However, some paleomagnetist thought that these blocks had not collided in Triassic, since the paleo-magnetic poles of many blocks on both sides of the six collision zones in Asia did not completely coincide (Zhao et al. 1996; Huang et al. 2008). They have withdrawn this improper understanding in recent years. The premise of their understanding is that the blocks can no longer move or undergo large-scale intraplate deformations after the convergence. However, many blocks in the Asian continent could also undergo uneven rotation and rather large tectonic deformations after they were joined together, so even after the joining of the plates, the paleo-magnetic poles of each block were still not necessarily identical.

In Asia, the tectonic event happened in Triassic was usually called the "Indosinian Tectonic Event" which was first named by French Geologist Fromaget (1934) in Vietnam. This tectonic event significantly influenced the central and southern parts of the Asian continent, i.e., from the Indochina, Cathaysian, Yangtze, Sino–Korean to the Kazakhstan and Mongolia areas. The many collision zones formed in central and southern part of Asia during the Indosinian period are obviously related to the NE–SW expansion of the Tethys Ocean and its migration and subduction to the Asian continent. However, the blocks of the Southwestern Asia, such as blocks south to the Kavkaz–Alborz areas, i.e., the Iran, Turkey and Kimmeria blocks, were still scattered in the Tethys Ocean (Fig. 3.15).

Since the northern half of the Asian continent finished its amalgamation in the Indosinian period, the influence of the Indosinian Tectonic Event was widespread. Indosinian folds and faults were generally developed from Indosinian plate–Northern Qiangtang–Western Kunlun area from the south to the Mongolia–Kazakhstan area in the north. And the Erdenet super-large porphyry copper–molybdenum deposits were formed in Northern Mongolia in Triassic (Jiang et al. 2010). According to the present magnetic orientation, the trending of tectonic line in the Indosinian period is nearly E–W trending (Fig. 3.17), but is NW–SE trending according to its paleo-magnetic orientation (Fig. 3.18). Indosinian intraplate deformations were so widespread because of the plates completing a large area of convergence in this period, so the tectonic stress could be transferred easily within the

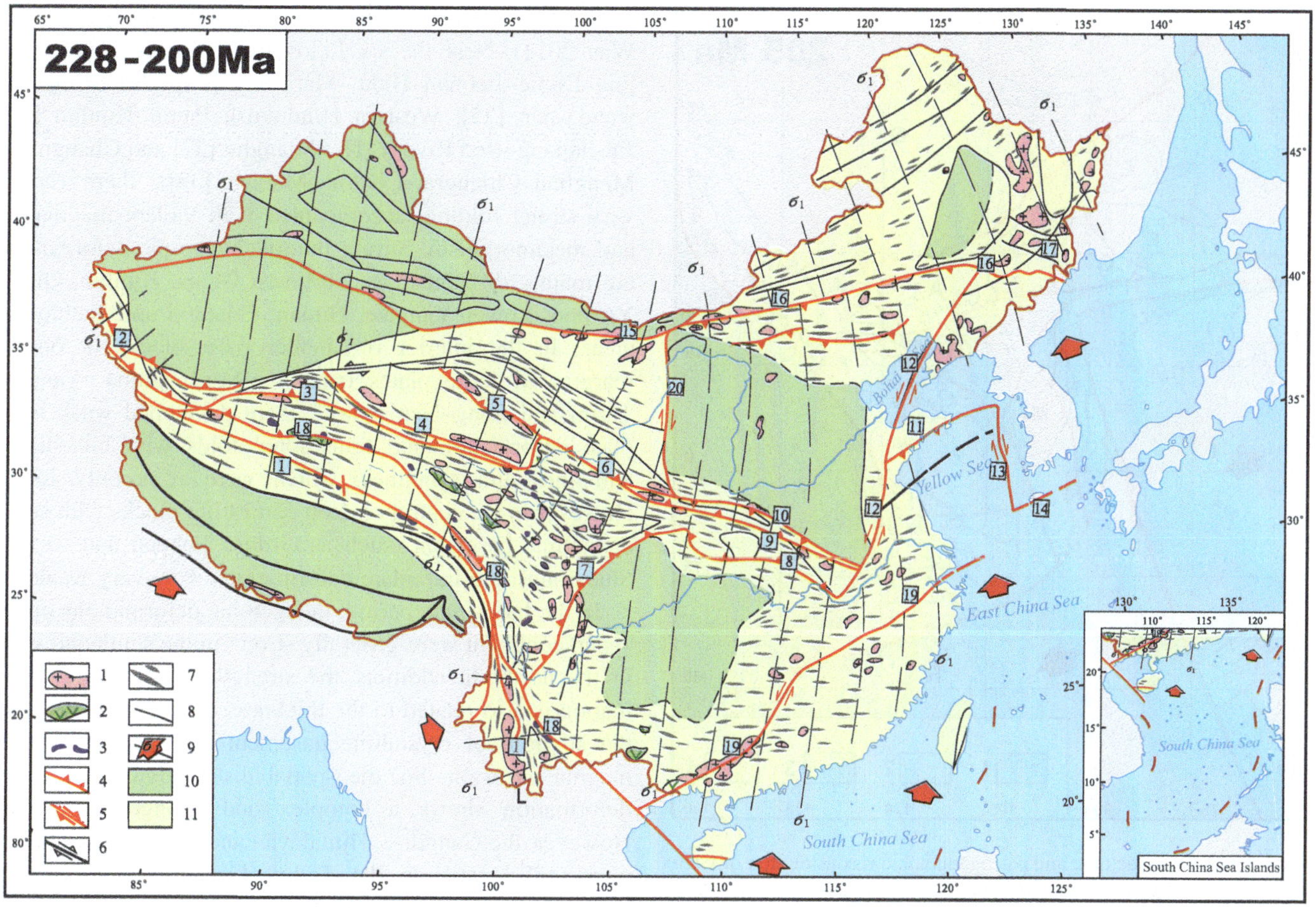

Fig. 3.17 Tectonic sketch of the China continent in the Indosinian period (228–200 Ma). **Legend**: (1) Indosinian granite; (2) Indosinian volcanic rocks; (3) Indosinian ophiolite and ultramafic rocks; (4) number of collision zones or thrust belts; (5) number of normal and strike-slip faults; (6) block boundaries or faults with weak activity, unnumbered; (7) anticline fold axes (data in Attachment Table 3.4 of Wan 2011); (8) trace of maximum principal compressive stress (σ_1); (9) direction of plate movement; (10) areas with conformity or disconformity; (11) areas with parallel disconformity. **No. of plate boundaries, collision zones and fault belts**: (1) Lazhulung–Shuanghu–Lancangjiang–Changning–Menglian–Dak, Thailand–Central Malay Peninsula collision zone; (2) Kangxiwa–southern border of Tarim thrust belt; (3) southern border of Kunlun; (4) Central East Kunlun thrust belt; (5) Junulshan–Qinghainanshan thrust belt; (6) Wushan–Baoji collision zone; (7) Longmenshan collision zone; (8) Mianxian–Lueyang–Dabashan–Fangxian–Xiangfan–Guangji (northern border of Yangtze plate) thrust belt; (9) Shandan–Tongbo collision zone; (10) Luonan–Fangcheng (southern border of Sino-Korean plate) collision zone; (11) Zhucheng–Qingdao–Rongcheng collision zone; (12) Tancheng–Lujiang sinistral strike-slip fault zone; (13) eastern border of Yellow Sea dextral strike-slip fault; (14) Cheju Do collision zone; (15) northern border of Alxa thrust belt; (16) Northern Yinshan–Xar Moron River (northern border of Sino-Korean plate) thrust belt; (17) Dunhua–Mishan sinistral strike-slip fault zone; (18) Jinshajiang–Red River collision zone; (19) Shaoxing–Shiwandashan collision zone (sinistral strike-slip fault zone); (20) Helanshan–Liupanshan (western border of Sino-Korean plate) dextral strike-slip fault zone

continental plate. Prior to this period, most of the Asian continental blocks were dispersed and the Mesoproterozoic–Paleozoic sedimentary systems in most of the blocks were basically horizontal with weak tectonic deformations without united tectonic line.

Although the collision zones in the Indosinian period had different directions, based on paleo-magnetic data, these collision zones were mainly concentrated in the northeastern direction. In other words, the regional folding axes and regional tectonic lines during Triassic were actually NW–SE trending according to paleo-magnetic orientation (Fig. 3.18), but they were nearly E–W trending according to present magnetic orientation. That is because that the paleo-magnetic declination of the eastern Asian continental blocks was about 30° to the east of their present magnetic declination. So if paleo-tectonic characteristics were restored, the Triassic folding axes and tectonic line were obviously NW 300°–SE 120° trending (Wan 2011). It means that the large number of E–W trending tectonic structures formed in the Indosinian period in the Asian continent is absolutely not meanings of the "latitudinal tectonic belt" named by Lee (1926, 1929, 1947). In fact, their formations

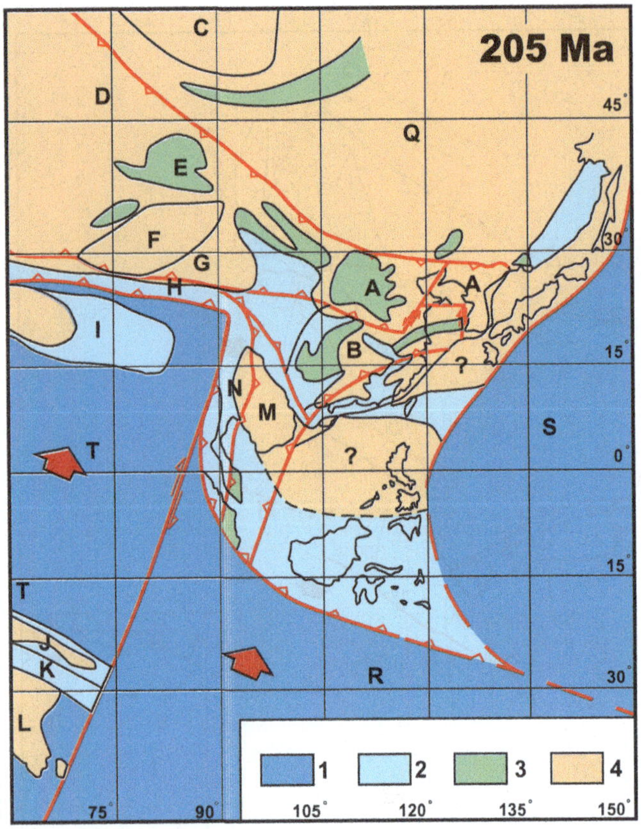

Fig. 3.18 Paleo-tectonic and paleogeographic reconstruction of China continent and its adjacent areas in the Indosinian period (Triassic, 205 Ma). **Legend**: (1) Ocean; (2) shallow sea; (3) continental sedimentary basin; (4) eroded continent. **The symbol for tectonic units**: (A) Sino–Korean plate; (B) South China plate (including Yangtze and Cathaysian plates); (C) Siberian plate; (D) Kazakhstan blocks; (E) Junggar block; (F) Tarim block; (G) Qaidam block; (H) Kunlun block; (I) Northern Qiangtang block; (J) Gangdise block; (K) Himalayan block; (L) Indian plate; (M) Simao–Indosinian plate; (N) Baoshan–Sibumasu block; (O) Pacific plate; (P) Philippine Sea plate; (Q) Tianshan–Mongolia–Hingganling Late Paleozoic collision zone; (R) Australian plate; (S) Izanagi plate; (T) Tethys oceanic plate. The paleo-magnetic data and their location of central reference point of the blocks are from Appendix 6 of Wan (2011) (After Wan 2011)

have nothing to do with the Earth's latitudinal deformation or rate change of rotation.

The author has got the 111 differential stress values of the Indosinian period, which are all 100–125 MPa (Wan 2011). The magnitude of differential stress was stronger in the southern part than in the northern part, indicating that the dynamic was originated from south of the Asian continent, i.e., from the long-distance effect of the expansion and northeastward subduction of the Tethys Ocean.

Based on data of regional geological survey of China, in Triassic there developed 2195 macro- and mesoscale folds (including 1109 anticlines and 1086 synclines) in China continent. From Inner Mongolia and Heilongjiang Province to Guangdong Province, there developed widespread E–W trending folds (in present magnetic orientation, Fig. 3.17; Wan 2011). Near the six Indosinian collision zones (Qinling–Dabie–Jiaonan–Hida Marginal [24], Shaoxing–Shiwandashan [25], Western Hindukush–Pamir–Kunlun [30], Jinshajiang–Red River [31], Shuanghu [32] and Changning–Menglian–Chiangrai–Central Malaya [33]), there formed very strong folding, accompanied with violent magmatism and metamorphism. Strong intraplate tectonic deformations are mainly distributed in Northeast China, Kunlun, Qilian, Yanliao, Lower Yangtze, Hunan, Guangxi and Cathaysian areas. In the interior of smaller paleo-blocks or on the margins of large plates (such as North China, Yangtze, Tarim and Song–Nen plates), there developed weak intraplate deformation and transitional folds with box-shaped anticlines and synclines, which were commonly found. However in the central or deep part of big blocks with stable crystalline basement (such as Ordos, Sichuan and western Shandong), the intraplate deformations were very weak.

In the Indosinian period, the tectonic deformations on the Asian continent were generally strong in the south and weak in the north. In addition, the strength of intraplate deformation is also related to the thickness of sedimentary strata, the strength of crystalline basement and the presence of magma intrusion. So, the spatial distribution of intraplate deformation shows a complex and changeable pattern. However, the Gangdise, Himalayan and Indian plates, which were still located in the Tethys Ocean in the Southern Hemisphere during Triassic, had not yet been amalgamated into the Asian continent, without any folding deformation at this time (Fig. 3.17).

In most areas of the Sino–Korean, Yangtze and Indosinian plates, the Indosinian Tectonic Event is rather important, for the earliest and most widespread folding of the sedimentary cover occurred during the Indosinian period. On the margins and eastern part of Sino–Korean plate, all the Mesoproterozoic, Neoproterozoic, Paleozoic and Triassic Systems were folded during this period, and the Indosinian Tectonic Event in this area happened in the end of Triassic (~200 Ma). In the northern and eastern areas of the Yangtze plate, rocks from the Jinning and Nanhua up to the Middle or Late Triassic Systems were also folded during this period. Detachment occurred in incompetent units, such as the shale of Silurian Fentou Formation, Longtai Coal Formation of Permian and the Middle Triassic gypsum salt layer, so that strata above and below showed very different styles of folding. In the Ordos block of the Sino–Korean plate and Sichuan basin of the Yangtze plate, due to their rather solid crystalline basement in depth, almost no strong Indosinian folding was formed.

In the southern part of the Yangtze plate (mainly in the Guangxi area), the E–W trending folds were formed in Early Paleozoic, and strata of the Upper Permian and Lower–Middle Triassic Systems were folded during the Indosinian period. The orientations of the fold axes and maximum principal stress were very similar during these two separate tectonic events, so that unconformity between the Triassic and Paleozoic Systems is not easily distinguished. Therefore, some scholars (such as Guo 1998) believed that there was no Indosinian folding at all, which might be a mistake.

However, in the Cathaysian plate, where new and old fold systems intersect each other at a large angle, unconformities can be distinguished easily. These relationships are demonstrated clearly by hundreds of folds, shown on 1:200,000 regional geological maps, distributed on the Cathaysian plate and its borders. In the Cathaysian plate, including the East and South China Sea areas, Paleozoic and Lower and Middle Triassic Systems were affected by folding during the Indosinian event. Unconformity between the Anyuan Coal Formation of Upper Triassic and its underlying strata is very obvious. The Indosinian Tectonic Event in the Southern Yangtze, Cathaysian and Indosinian plates all occurred in the end of Middle Triassic (~237 Ma) (Wan 2011).

However, at the Northeast China and Hingganling areas, only the strata of the Permian and Triassic Systems were folded in the Indosinian period, with the NE trending. So, some scholars (Zhao 1990; Cui 1999; Ge and Ma 2014; Ge et al. 2014) confused the Triassic folds with the Jurassic ones and thought that the tectonic lines in Triassic and Jurassic were both NE trending for the whole China continent. Of course, it is rather difficult to distinguish the Triassic and Jurassic folds in northeastern China.

An important characteristic of the Indosinian period was the formation of arcuate tectonic zones with linear folds and faults, for example, the famous Guangxi and Huaiyang (along the lower Yangtze River, from Hubei to Jiangsu). The existence of these two arcuate tectonic zones is beyond doubt. Strong crystalline blocks existed in the central part of the northern side of these two tectonic zones, and sedimentary strata with lower strength were nearby. It is natural to form the arcuate tectonic zones under the N–S trending compression. However, there are problems if we talk about the two areas as "ε" (epsilon)-type tectonic zones (Li 1926, 1929, 1947, 1962). In fact, the so-called backbone and spine found later were not formed in Triassic.

When folding occurs in areas with inhomogeneous lithological characters, arcuate folds and faults may occur at different scales. The development of Indosinian arcuate folds is related to the Indosinian folding event, which is the earliest and most widespread folding after the sedimentary cover in the Sino–Korean and Yangtze plates was formed, with obvious detachment and folds.

Along N–S longitudes (103.5° E, 112.5° E, 115° E and 124° E) in the eastern parts of China according to the present magnetic declination, the shortening ratio, shortening magnitude, shortening rate, deformation time and linear strain rate for intraplate folds have been calculated. It is found that the greatest shortening ratio (up to 50%) occurred in eastern South China (115° E), and the shortening ratios in Northeast China (124° E), central South China (103.5° E) and western South China (112.5° E) were 36.69%, 20.18% and 14.13%, respectively. In North China, folding was rather weak with a much smaller amount of shortening. The length of time over which deformation occurred, calculated from the four sections above, is between 2.1 and 8.6 Ma, meaning that strong deformation only occupied 5–17% of the total Indosinian period. Linear strain rates were very small (1.39×10^{-15} to 2.13×10^{-15}/s), indicating that deformation took place while the rocks were in a rheology state (Wan 2011).

Under the influence of the Indosinian Tectonic Event, many regional nearly N–S trending (including NNE and NNW) extension-shear faults or joints (in present magnetic declination) often became important parts of penetration and condensation of magma, supercritical fluid or hydrothermal ore deposits, resulting in magmatism and formation of many giant endogenic metallic deposits. In Triassic, there formed many collision zones and fractures; however, endogenic metallic deposits which were actually formed during the collision period within the collision zone were generally small in scale and few in quantity (according to the data of Mao et al. 2002). It may be related to the too strong tectonism, which made the metallogenic fluids not only easy to flow but also easy to lose.

In the end of Triassic (200 Ma), there formed radial extension and shoshonite dyke swarm between the North America, South America and Africa in the Western Hemisphere. The dyke swarm was NW trending in the North American plate, SW trending in the South American plate, and E–W or SE trending in West Africa (Fig. 3.19). The isotopic ages of shoshonite are about 200 Ma, with an error of only 1 Ma. Then, the original Atlantic Ocean began to appear, which means that the Pangea supercontinent began to extend and break up (Figs. 3.15, 3.19; Marzoli et al. 1999; Hames et al. 2000; Condie 2001).

The above data show that when the blocks in the Eastern Hemisphere were under convergence in Triassic, the blocks in the Western Hemisphere began to break up. At this time, the radial extension and dyke swarm in the Western Hemisphere may be derived by the mantle plume uplift from the core–mantle boundary (CMB). Of course, it is also possible that the giant meteorite impact caused huge mass deficit on the Earth surface, inducing the mantle material uplift to form a mantle diapir. The above two hypotheses are both possible. However, the evidences are so far not enough to reach a conclusion for the time being.

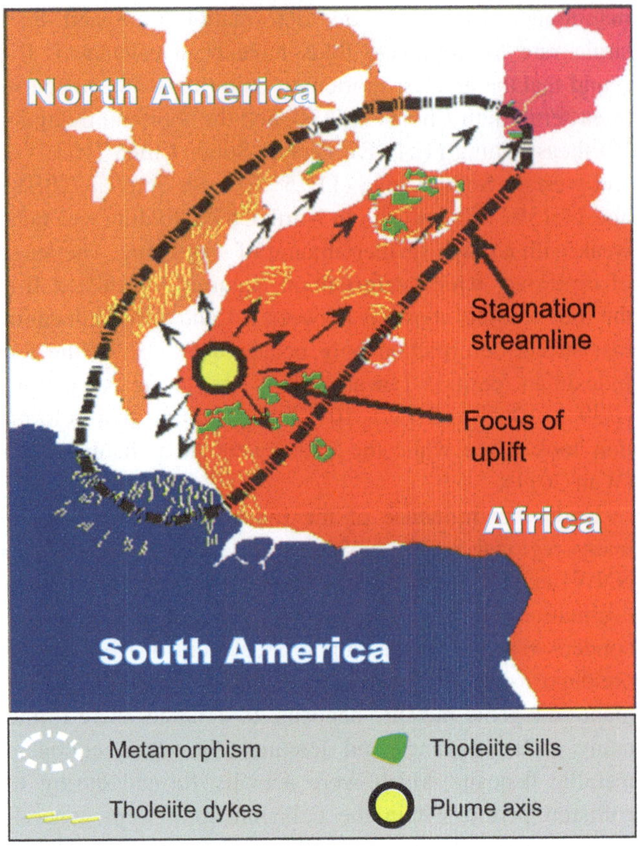

Fig. 3.19 Plume-related magmatism in the late period of Triassic (After Marzoli et al. 1999; Hames et al. 2000; Condie 2001)

3.10 The Tectonic Evolution in Jurassic–Early Period of Early Cretaceous (Yanshanian Tectonic Event, 200–135 Ma)

The Yanshanian Tectonic Event (in early time usually to be called Yanshan Movement) mainly occurred in Jurassic, which was first named by Weng (1927). In the Jurassic–early period of Early Cretaceous, under the influence of the North American plate moving and compressing to the WSW trending, the East Asian continental crust generally had a counterclockwise rotation of 30°–20°. It made the Siberian region turn to SW by 36.2° (Khramov et al. 1981) and made eastern part of Central Asia (including Junggar and Tarim) migrate southward by about 5° in latitude (Figs. 3.20, 3.21; Li et al. 1989, 1992; Wan 2011). The crust of East China and Korea Peninsula in East Asia counterclockwise rotated by 30°–20° (Ma and Yang 1993; Kim and Van der Voo 1990; Opdyke et al. 1986) and slipped in the direction of ESE, onto the ancient oceanic lithosphere mantle (Wan 2011; Wan et al. 2016; Wan and Zhao 2012; Wan and Lu 2014). Thus, since then, the magnetic north of the Asian continental crust has become almost identical to the modern magnetic north (Fig. 3.22). At this time, the Izanagi plate subducted and compressed to the WNW trending (Moore 1989; Wan 2011; upper left of Fig. 2.45), which played a role in hindering the rotation and migration of continental crust. This understanding has not been valued by scholars for a long time, and it is a new understanding of the author.

For the counterclockwise rotation of East Asian continental crust, there are not only the paleo-magnetic evidences, but also the reliable geological evidences. So, the distribution of tectono-magmatism and its migration are discussed first. In Northeast China, from east to west, the volcanic eruption zone of Early Jurassic was concentrated on the eastern border of Dunhua–Mishan fault zone and Yanbian area (Bureau of Geology and Mineral Resources of Jilin 1988); that of Middle Jurassic was concentrated near Laoyeling (Xu et al. 2008); and that of Late Jurassic mainly distributed from the deep of Daqing Oilfield to west of Harbin (Lei 2011, personal communication), and in the Dahingganling, the volcanic eruptions mainly occurred in the Cretaceous (Bureau of Geology and Mineral Resources of Inner Mongolia 1991). They are all characterized by the magmatism induced by crustal faults. The distribution of volcanic zones exhibited the phenomena of westward migration, i.e., the crust block showing the characteristic of counterclockwise rotation. In Northeast China, the Jurassic volcanic zones had the westward migration by 400 km with the maximum migration velocity of about 0.8 cm/y. However, the formation period of granite in Northeast China is generally late, mainly in Middle–Late Jurassic (160–135 Ma), so the zonation of their distribution is not obvious.

In South China, as a result of the erosion of the upper crust, the zoning of Jurassic volcanic rocks is not clear. However, the Jurassic granite intrusions in South China have obvious zoning. Based on the data of regional geological survey, Zhan (1994) proposed firstly: From Triassic to Jurassic, the granite zone in South China had the characteristic of gradual eastward migration. It means that the Triassic granite is mainly distributed in Shiwandashan, Guangxi–Changsha, Hunan, i.e., in the vicinity of southwestern part of Shaoxing–Shiwandashan Indosinian collision zone [25]. The Jurassic crust re-melting-type granite (S-type) is mostly penetrated along the low-angle thrust. The Early Jurassic granite is mainly distributed in the central and southwestern parts of Jiangxi; the Middle Jurassic granite is in the northern and eastern parts of Jiangxi; the Late Jurassic granite is in the western parts of Fujian and Zhejiang and the central and eastern parts of Guangdong (Wan et al. 2016; Wan and Zhao 2012). According to the above data, it can be seen that the Jurassic granite zones in South China gradually migrated eastward for about 180 km (the average migration velocity of 0.26 cm/y).

The origin of S-type granitic magma in South China is 15–22 km in depth; i.e., the source is near the layer of low seismic velocity and high electric conduction near the

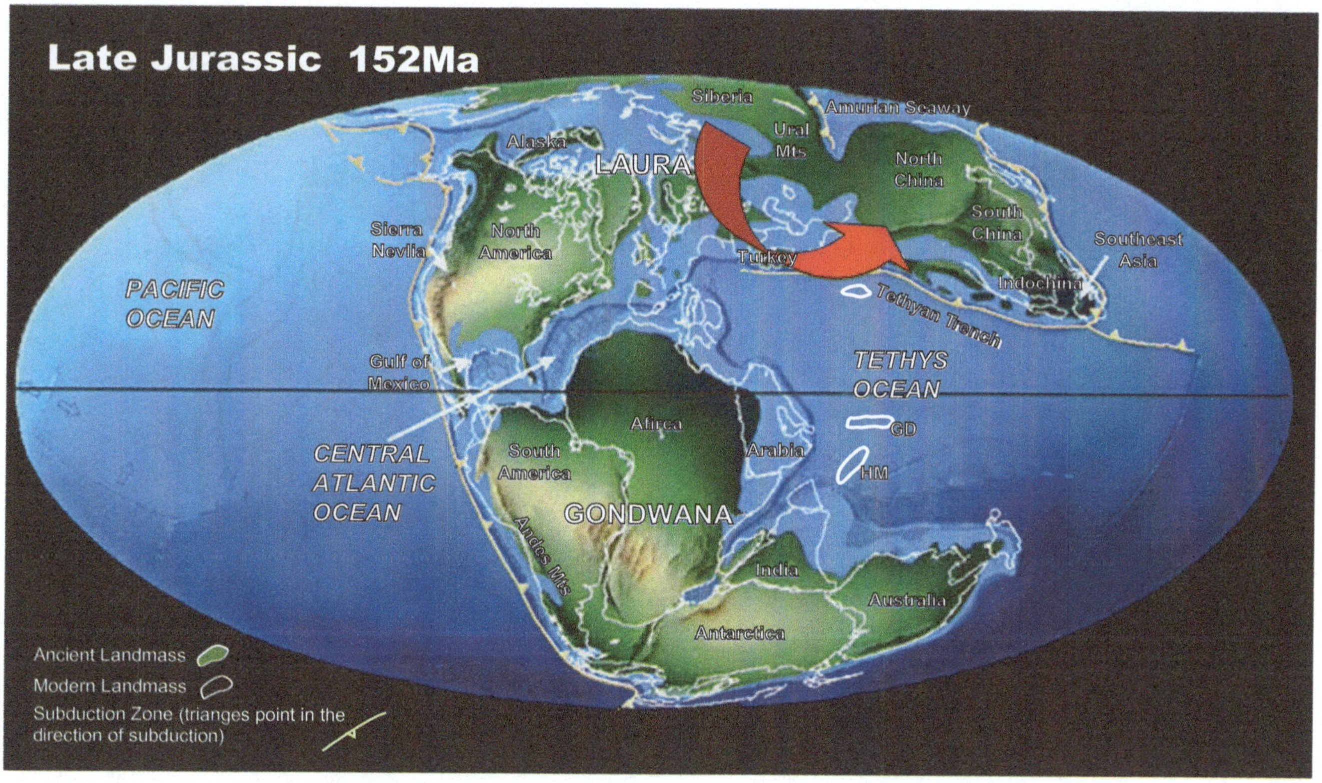

Fig. 3.20 Global paleo-continental reconstruction in Late Jurassic. The legend and instruction are as same as those of Fig. 3.6. The red arrow shows the direction of continental crustal rotation of East Asia. The data of paleo-magnetism and the positions of central reference points in Asia and its adjacent areas are listed in Appendix 6 of Wan (2011) (After Scotese 1994; www.scotese.com Web site; revised by Wan and Zhu 2011)

middle crust. The source of I-type or A-type granitic magma has a depth of 32–40 km, near the bottom of crust, i.e., the Moho discontinuity (Zeng et al. 2000). In the above tectono-magmatism, the continental crust suffered from the horizontal compression, producing a series of NNE trending folds and thrusts. The crust is properly thickened for 4–8 km (Wan et al. 2016; Wan and Zhao 2012).

In North China, near Beijing, the Jurassic tectono-magmatic zone also rotated counterclockwise, from the ENE direction in Early Jurassic, the NE direction in Middle Jurassic and to the NNE direction in Late Jurassic (Bureau of Geology and Mineral Resources of Hebei 1989). However, the migrations of tectono-magmatic zone are not clear which may be related to the fact that the North China was located near the rotation center of continental crust (Wan 2011; Wan et al. 2016; Wan and Zhao 2012).

Due to the strong tectono-magmatism in Jurassic and Cretaceous all being originated from the low-velocity and high-conductivity layer of the middle crust or the bottom of the crust (Moho discontinuity), it is speculated that the main detachments mainly occurred at the bottom of crust and in the middle crust (Wan et al. 2008). In addition, the petrologists and geochemists (Lu et al. 2006; Lu 2010; Zhou 2006) recognized that the mantle in the lower part of the lithosphere and the asthenosphere in East Asia was undisturbed or slightly disturbed. According to the data, many minor disturbances occurred in Archean and Proterozoic (Lu et al. 2006; Lu 2010). Till now, researchers have not found any reliable evidences for the great disturbance of Mesozoic–Cenozoic mantle magma, which means that in the depth of lithosphere there has been no tectonic activity since Mesozoic. In addition, the lower lithosphere of East Asia has the properties of oceanic crust or mantle (Xu et al. 2012; Yu et al. 2010). As a result, it seems that since Jurassic, beneath the East Asian crust, it should be the ancient, stable oceanic mantle with lower activity, rather than the hot and highly active continental lithosphere mantle with lower density. Due to fact that the East Asian continental crust migrated onto the oceanic lithosphere mantle along the Moho discontinuity in Jurassic, forming the thin lithosphere with continental crust and oceanic mantle which is distributed in the east of the blue dotted line (Fig. 3.21), the thickness of continental crust is 30–35 km, that of oceanic lithosphere mantle is more than 40 km, and the total lithosphere thickness is about 80 km.

Thus, the reduction of the lithosphere of East Asia is not due to the decrease of the lithosphere itself (in Jurassic, in fact, the crust has a moderate thickening), but is the result

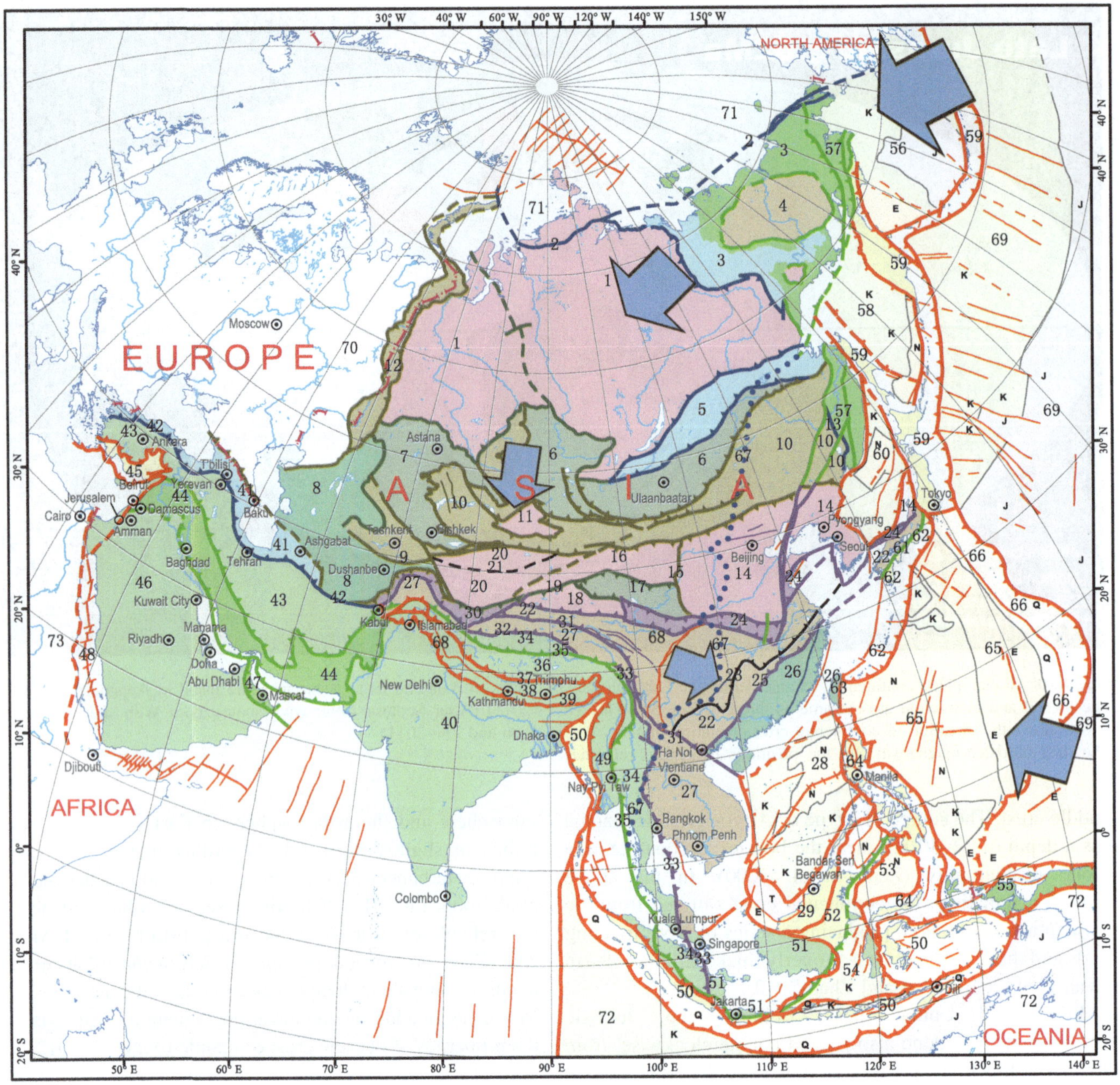

Fig. 3.21 Tectonic sketch of Asian continent in Jurassic–early period of Early Cretaceous. The blue arrow shows the orientation of tectonic maximum principal compressive stress. The purple line shows the location of Triassic main fault. The numbers of tectonic units are as same as those in the contents and text. The eastern area of the blue dotted line shows the distribution of East Asian continental crust and oceanic mantle lithosphere, i.e., the thinned lithosphere, and the western side of that line is normal continental lithosphere

formed by the continental crust moving onto a fairly thin oceanic mantle (Wan 2011; Wan et al. 2016; Wan and Zhao 2012; Wan and Lu 2014). The author believes that there is no "thinning" of the lithosphere in East Asia. It seems that the activity of inner detachments of continental lithosphere should be paid necessary attention in the study of continental tectonics.

However, in recent years, the popular hypothesis is that the stronger tectono-magmatism and "thinning" of lithosphere in East Asia since Jurassic–Cretaceous were caused by hot mantle uplift, crust underplating (Deng et al. 1992, 1996) or mantle uplift derived from the Pacific plate subduction (Wu et al. 2008; Zhu et al. 2008; Zhang 2009; Zhu et al. 2011, 2012). The key question is that they have never found any evidences of thermal activity or strong turbulent in the depth of East Asian lithosphere mantle since Mesozoic (Huang and Zhao 2006). It seems that there are many contradictions between these hypotheses and facts.

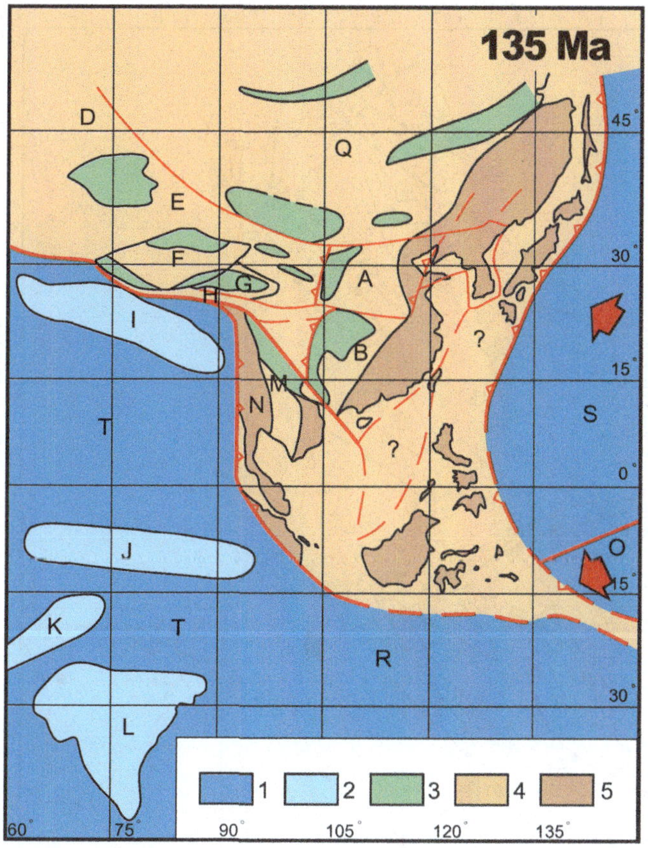

Fig. 3.22 Sketch of tectonic-paleogeographic reconstruction of East Asia in Late Jurassic. **Legend**: (1) Ocean; (2) shallow sea; (3) continental sedimentary basin; (4) eroded continent; (5) mountains. **Symbols of tectonic unit**: (A) Sino–Korean plate; (B) South China block; (D) Kazakhstan block; (E) Junggar block; (F) Tarim block; (G) Qaidam block; (H) Kunlun block; (I) Northern Qiangtang block; (J) Gangdise block; (K) Himalayan block; (L) Indian plate; (M) Simao–Indosinian plate; (N) Baoshan–Sibumasu block; (O) Pacific plate; (Q) Tianshan–Mongolia–Hingganling Late Paleozoic collision zone; (R) Australian plate; (S) Izanagi plate; (T) Tethys Ocean plate. The data of paleo-magnetism and the central reference point positions are from Appendix 6 of Wan (2011) (After Wan 2011)

In addition, the asthenosphere under the East Asian lithosphere has almost equal temperature (−1280 °C). Under such thinner lithosphere condition, the geothermal gradient should be higher, so when the faulting occurs it will be easy to appear the phenomenon such as local decompression, expansion and warming, resulting in the formation of the magma chamber. The magmatism or the migration of ore-bearing ultra-critical fluid results in forming a large number of endogenous metal deposits in East Asia.

In Jurassic, the East Asian continental crust (Figs. 3.21, 3.23) underwent the WNW trending compression and shortening, and NNE trending extension, forming a series of NNE–NE trending thrusts and folds (be called Neo-Cathaysian Tectonic System by Lee 1939) and WNW trending extension-shear tectono-magmatic zone (Dayishan Structure called by Lee 1939) and endogenic metallic ore deposits. The rock deformation climaxed mainly in the late epoch of Yanshanian period (Late Jurassic–the early period of Early Cretaceous) (Fig. 3.23).

The characteristics of intraplate deformation in the Yanshanian period are very clear and have long been recognized by Chinese geologists. Wong (1927) first considered the particularity of Yanshan Movement in East Asia. It is totally different from the Alps Movement, and it occurred only in Jurassic. However, thereafter many famous geologists (Huang 1945, 1960; Huang et al. 1965; Zhao 1959) extended the Yanshan Movement to the Jurassic and Cretaceous tectonic event, bringing the two tectonic episodes with completely different deformation characteristics and dynamic sources together to be the Yanshan Movement. But such an approach is very inappropriate. According to the significantly different characteristics of magmatic activities, the petrologists have long called the Jurassic period as "Early Yanshanian period" and the Cretaceous period as "Late Yanshanian period." However, the related isotopic ages of this period are lacking.

Almost at the same time when Weng (1927) proposed the term of Yanshan Movement, Lee (1929) proposed the "Neo-Cathaysian Tectonic System" (in the earlier time, some researchers called it Sinian System) featured by a series of NNE–NE trending folds, reverse faults in the east of China continent, and believed that the dynamic mechanism of the system was the regional counterclockwise rotation. Li's (1929) recognition was very advanced. It is extraordinary that he could exactly point out this dynamic mechanism almost 80 years ago. When he proposed the opinion, he believed that the system was formed from Late Mesozoic, but the time limit was not accurate. It should be noted that there was no isotopic dating method at that time. The proposal of the Neo-Cathaysian Tectonic System is of great significance for systematically studying the combination law of tectonic deformation in the eastern part of the Asian continent. Judging from the available data, the formation of the Neo-Cathaysian Tectonic System is exactly the same as that of the Yanshanian period. The Neo-Cathaysian Tectonic System is mainly distributed in the range of the continental crust–oceanic mantle-type lithosphere in East Asia. The tectonic system was formed under the control of the Yanshanian tectonic stress field (Fig. 3.23).

In the Yanshanian period, the thickness of continental crust for East Asia increased a little bit. According to the author's estimation method for the folding attitude recovery, the East Asian continental crust was shortened about 11–23.4% (Wan 2011). Thus, it is estimated that the thickness of crust increased about 4–8 km. The surface topography may be relatively elevated. According to the relationship between the current crust thickness and height of landform, in Late Jurassic the height of East China landform may reach about

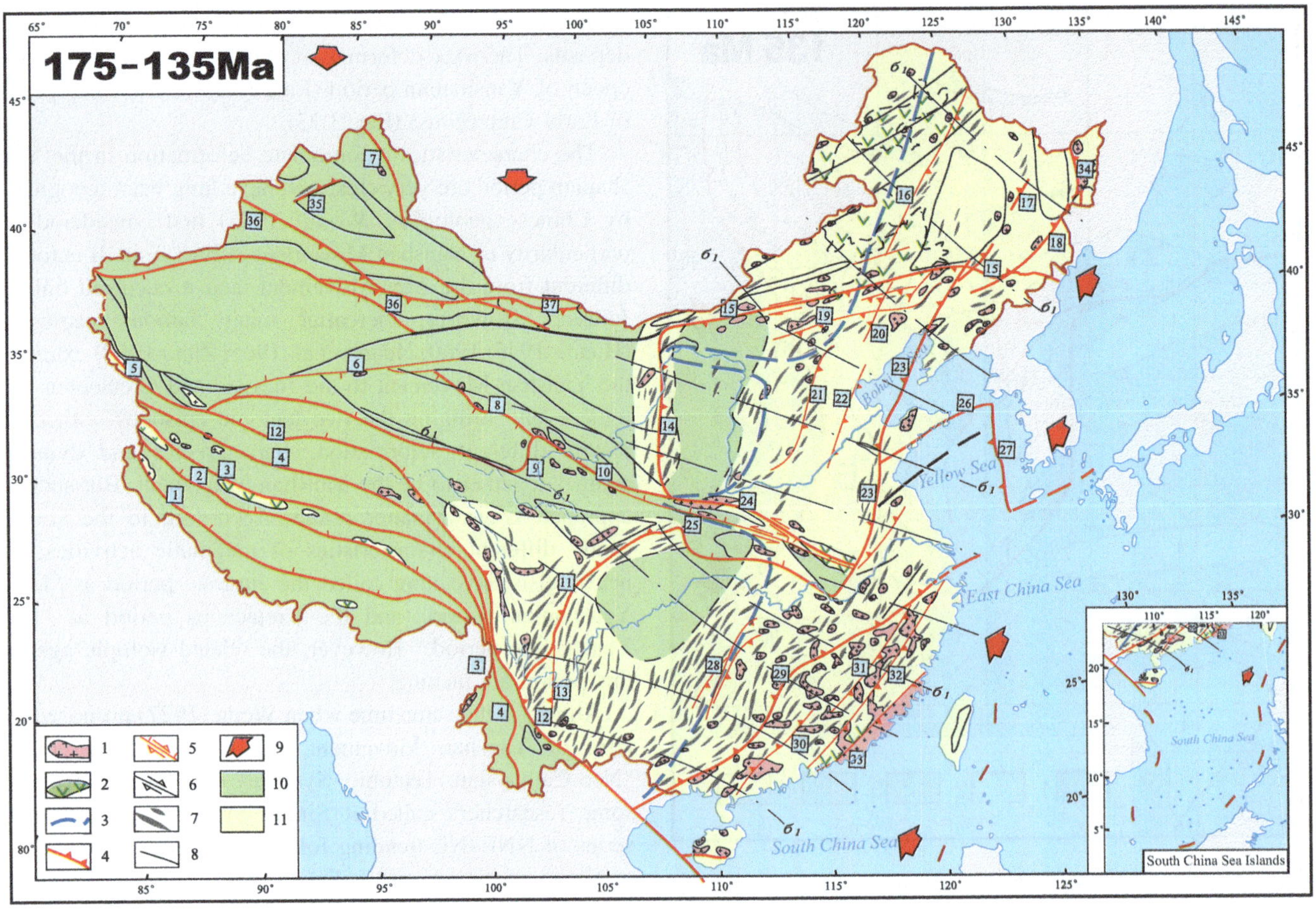

Fig. 3.23 Tectonic sketch of China in Yanshanian period (175–135 Ma). **Legend**: (1) Yanshanian granites; (2) Yanshanian volcanic rocks; (3) the transformation boundary between continental lithosphere (west) and the lithosphere with continental crust and oceanic mantle (east); (4) collision zones or thrust belts; (5) normal and strike-slip faults; (6) boundaries of blocks with weak activity or faults, without number; (7) fold axes, only showing anticlines (data in Attachment Table 3.5 of Wan 2011); (8) trace of maximum principal compressive stress (σ_1); (9) direction of plate movement; (10) distribution areas with conformity or disconformity; (11) distribution areas with angular unconformity. **No. of plate boundary, collision zone or fault belt**: (1) Boundary of plates along Yarlung Zangbo tectonic zone (with oceanic crust); (2) Gegyai–Nyainqentangulha fault; (3) boundary of plates along the Bangongco–Nujiang tectonic zone (with oceanic crust); (4) Shuanghu–Lancangjiang fault; (5) Kangxiwa–southern border of Tarim strike-slip fault; (6) Altun dextral strike-slip fault belt; (7) Kurt–Narmande strike-slip fault belt; (8) Junulshan–Qinghainanshan strike-slip fault belt; (9) southern border of Qaidam (Wenquan–Chaka) reverse fault belt; (10) Wushan–Baoji strike-slip fault belt; (11) Longmenshan reverse fault belt; (12) Jinshajiang–Red River sinistral strike-slip fault belt; (13) Panzhihua–Xichang reverse fault with strike-slip; (14) Liupanshan–Helanshan reverse fault belt; (15) Yinshan–Xar Moron River dextral strike-slip fault belt; (16) East of Dahingganling reverse fault belt; (17) Yilan–Yitong reverse fault belt; (18) Dunhua–Mishan reverse fault belt; (19) Shangyi–Gubeikou–Pingquan dextral strike-slip reverse fault belt; (20) Fuxin–Jinzhou (western Liaoning) reverse fault belt; (21) east of the Taihangshan reverse fault belt; (22) east of the Cangzhou–Liaocheng reverse fault belt; (23) Tancheng–Lujiang reverse fault belt; (24) Luonan–Fangcheng sinistral strike-slip fault belt (southern border of the Sino–Korean plate); (25) Shangdan–Tongbo sinistral strike-slip fault belt; (26) Zhucheng–Qingdao–Rongcheng dextral strike-slip reverse fault belt; (27) eastern border of Yellow Sea reverse fault belt; (28) Xuefengshan thrust belt; (29) Shaoxing–Shiwandashan reverse fault; (30) Wuchuan–Sihui reverse fault; (31) Chong'an–Heyuan reverse fault; (32) Lishui–Lianhuashan reverse fault; (33) Changle–Nan'ao reverse fault; (34) Wandashan collision zone; (35) Karamay buried thrust during Mid–Late Jurassic; (36) Tianshan main reverse fault belt; (37) North Yagan thrust during Early–Mid Jurassic (After Wan 2011)

1500–2000 m above sea level. Thus, the weather changed from the warm moisture in Early Jurassic to the rather drought arid in Late Jurassic, due to the landform uplift (Wan 2011).

Zhang et al. (2001) recognized that the Middle–Late Jurassic volcanic rocks of East China are all "adakite," deducing a "volcanic plateau," and explained that all the volcanic activities in North China were derived from the subduction of the Pacific plate. Unfortunately, his point of view is not conformed to the fact (recently, Zhang Q. had pointed out the volcanic rocks in East China not to be adakites). In fact, the high content of strontium in the Mesozoic volcanic rocks of North China is due to the high content of strontium in the marine sedimentary system of

Early Paleozoic. When the magma uplifted and intruded to assimilate those high strontium sedimentary strata, the content of strontium in volcanic rocks of Jurassic and Cretaceous would be generally higher. This is a reasonable explanation. It cannot be considered that these volcanic rocks are the "adakites" from the oceanic plate subducting below the North China continent. The modern seismic tomography data show that when the oceanic plate subducts underneath the North China continent, it has reached a depth of about 600 km, but the magma chamber just only in 30–40 km depth. It has never been found so far that there is any relationship between the oceanic plate subduction and formation of the magma chamber in the North China.

In the Yanshanian period, the typical characteristics of intraplate deformation in East Asia are shown as a series of NNE or NE trending folds, reverse faults, thrusts and WNW or NW trending extension faults or strike-slip faults (Fig. 3.23). According to the statistics of geological survey data at the scale of 1:200,000 in China, the mega-folds and meso-folds are composed of 1566 anticlines and 1603 synclines, and the dip angles of their limbs are mainly medium (30°–60°). They are all distributed in the east of the Ordos area and Sichuan basin (Wan 2011). Only in the vicinity of the weaker zone in the basement or the regions with thick sedimentary strata in East Asia, the linear folds can be formed, i.e., the East Asian thinner lithosphere with continental crust and oceanic mantle (Wan 2011; Wan et al. 2016; Wan and Zhao 2012; Wan and Lu 2014).

However, in the area with common continental lithosphere or the areas with stable basement and thinner sedimentation in Western and Central Asia, the folds are very gentle, forming open fold, and even the Jurassic folds are not developed, such as Sichuan, Ordos and its west area. In addition, in West Shandong where more crystal basements are exposed, there are neither folds nor magma intrusion or volcanic eruption.

The Yanshanian tectonic differential stress values in East Asia (the difference values between the maximum and minimum principal stresses) usually are about 100 MPa. In some parts of East Asia and South Tibet, the tectonic differential stress values may be more than 100 MPa, while obviously less than 100 MPa in other areas (Wan 2011).

According to the author's (2011) research in East China, it is found that the axial directions of the Yanshanian folds all changed with the time: For the end of Early Jurassic, the folding axes are commonly ENE trending; for Middle Jurassic, usually NE trending; and for the end of Late Jurassic, commonly NNE trending. It means that in chronological order from the early to the late, the tectonic line rotated counterclockwise by about 45°, and the rotation angle of continental block rotated by about 20°–30°, while the rest may be the accumulation of plastic deformation. This indicates that in the Yanshanian period the orientation of near horizontal maximum principal compressive stress was changed gradually, i.e., from NNW in the early epoch, through NW in the middle epoch and to WNW in the late epoch.

The Jurassic tectonic stress field in East Asia, at last, was characterized by the WNW–ESE trending shortening, i.e., the trending near horizontal maximum principal compressive stress, and NNE–SSW trending horizontal extension. Thus, the WNW–ESE trending extension in faults or other reduction surfaces (Dayishan-type Structure, named by Lee 1929) is easy to open, showing the extension or extension-shear features, which is a good site for ore-bearing fluid migration, intrusion or preservation. This determines the occurrence of common endogenic metal ore bodies or deposits in the Yanshanian period. As mentioned above, most of the (about 70–80%) endogenic deposits are mainly preserved in WNW–ESE trending fractures. Especially, in the east of the transformation line of East Asian lithosphere type [67], i.e., in the continental crust and oceanic mantle lithosphere area, the lithosphere is thinner, and the geothermal gradient is higher. Once structural deformation occurs, it is easy to form magma chambers and pneumatic–hydrothermal activities and easy to form endogenous metal deposits. It is the reason why one-fourth of the endogenic metallic mineral deposits in China were formed in such a short span of the Jurassic (lasting only about 40 million years).

In Late Jurassic–early period of Early Cretaceous (170–135 Ma), the plate collisions mainly occurred in the southern margin of East Siberian Sea Jurassic collision zone (200–135 Ma) [2], the Verkhojansk–Chersky Jurassic accretion–collision zone [3], the Transbaikalia (or Mongolia–Okhotsk) Jurassic accretion–collision zone [5] and Wandashan Jurassic collision zone [13] (Fig. 3.21), and the oceanic crust in this area is subducted underneath the continental crust (Figs. 2.8, 2.9). The Wandashan collision zone was the southern part of Verkhojansk–Chersky Jurassic collision zone at first. The above collision zones were obviously suffered from the integrated impact of collision and compression by the North American plate [71] in the direction of WSW and the NE trending subduction of Tethys Ocean plate, thus resulting in the counterclockwise rotation of the East Asian continental crust and the formation of the East Asian continental crust and oceanic mantle-type lithosphere (Wan 2011; Wan and Lu 2014; Wan et al. 2016).

In Jurassic, the orientation of the maximum principal compressive stress in most parts of the Asian continent and in the western area of China was dominated by near N–S direction, and the differential stress value is about 80 MPa. In general, the structural deformation is very weak, and only near the great faults, for example, there only formed some tight folds on the sides of Altun fault zone [19] (Wan 2011).

In the Southwest Asian continent, the tectonic activities in Jurassic were weak in general. Only in the Kavkaz–Alborz collision zone [41] in the Kimmeridgian epoch of Late Jurassic, further convergence and collision occurred, which was the main tectonic event in West Asia. This collision event was related to the continuous northeastward expansion and subduction of the Tethys Ocean (Fig. 3.20), forming some NW–SE-oriented folds. In the other areas, the tectonic activities were very weak at this time, with the strata basically showing the conformity relationship.

3.11 The Tectonic Evolution in the Middle Period of Early Cretaceous–Paleocene (135–56 Ma)

The middle period of Early Cretaceous–Paleocene (135–56 Ma) tectonic event was first named as "Sichuanian Movement" by Tan and Li (1948) when they finished the monograph of *Geology of Xikang*. However, at that time they thought that the tectonic event occurred at the end of the Cretaceous. The geologists of Bureau of Geology and Mineral Resources of Sichuan Province (1991) discovered that the unconformity was formed between Paleocene and Eocene, but not between the Cretaceous and Paleocene, and that the tectono-magmatism occurred in Early–Middle Cretaceous (135–99 Ma) in the main parts of China. From the middle period of Early Cretaceous to the end of Paleocene (135–56 Ma), all the continental and oceanic plates of the world were migrated northward (Moore 1989; Wan 2011). That is, the radical plate cracking and extension with the Weddell Sea near the Antarctica as the center (the upper right of Fig. 2.45) occurred, causing the Gondwana supercontinent to break up. The Indian plate, whose northern part was oceanic plate, migrated northward from 45° S to the vicinity of the equator at a very fast speed in Late Cretaceous, with a maximum velocity of 18 cm/y (Lee and Lawver 1995). The northward velocity of other plates' dissociation from the Gondwana was only a few centimeters per year. The Australian plate was basically stable at the area of 50° S–60° S. The North Africa reached the region of 20° N, and the South American plate was still located at the South Hemisphere (Fig. 3.24; Van der Voo 1993; Wan and Zhu 2012). The above phenomenon could be explained by the long-distance effect of the plate radical extension of Gondwana and the formation of the great basalt volcanic province with the Weddell Sea near the Antarctica as the center at the southern end of the southern Atlantic Ocean during Jurassic (Storey 1995; Storey and Kyle 1997; Wan 2011). That is to

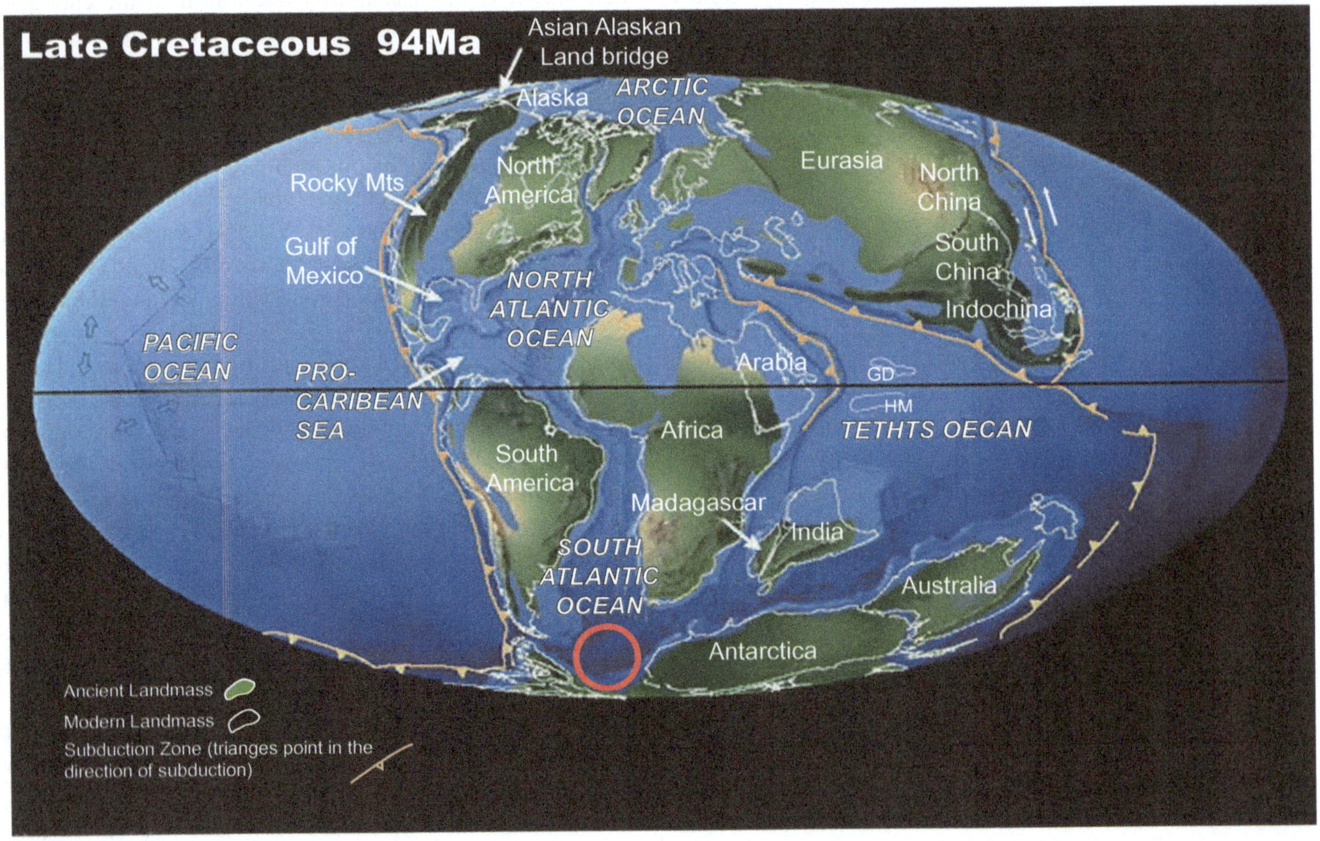

Fig. 3.24 Global paleo-continental reconstruction in Late Cretaceous. The red circle shows the center of radial plate extension. The data of paleo-magnetism and the central reference point positions in the Asia and its adjacent areas are from Appendix 6 of Wan (2011) (After Scotese 1994; www.scotese.com Web site; revised by Wan and Zhu 2011)

say, the radial plate expansion centered on the Weddell Sea mantle plume had gradually affected the whole world, making most of the world's plates in Cretaceous, which was basically characterized by moving northward. As for whether the large volcanic rock province of Weddell Sea is really a mantle plume or a plate expansion caused by the impact of meteorites, there are still not enough evidences. However considering the surprisingly fast migration of the Indian plate, it may be caused by the meteorite oblique impact, whose orientation was exactly the migration direction of Indian plate, so that the northward movement of the Indian plate is higher than that of other plates. That is just only an assumption or inference. Due to the northward expansion of the Tethys Ocean, the northward migration of the Indian plate [40] and Australian plate [72] (upper right in Figs. 2.45, 3.24), the whole Asian continent was suffered from the NNE trending horizontal compression. Significant near NNE–SSW direction shortening and near E–W extension occurred in the East Asian continental crust–oceanic mantle lithosphere, forming a series of NNE extension-shear fault zone, magmatic zone, metamorphic core complex and many endogenic metallic deposits (Wan 2011). In that period, the rock deformations were featured by forming the WNW trending wide folds and thrusts, NNE trending normal faults, NE or NW strike-slip faults (Fig. 3.25).

The tectonic system of the Sichuanian period was completely different from that of Jurassic period. The orientation of maximum principal compressive stress for the Sichuanian period was almost vertical to that of Yanshanian period. In the most areas of South and North China, the two limbs of fold are usually all with the low angles. In Northeast China, the two limbs of fold are only less than 5°. In the vicinity of the Bangongco–Nujiang collision zone on the Qinghai–Tibet Plateau, the local folds are tightly closed, and the axis of the fold is almost identical to the strikes of the collision zone or intersects at a small angle. In Late Cretaceous, due to the change of Indian plate moving orientation to the NE40° (Lee and Lawver 1995), the main thrust of Bangongco–Nujiang collision zone shows characteristics of the sinistral strike-slip, causing the separation of the Shuanghu Triassic collision zone [32] from Changning–Menglian–Chiangrai–Central Malaya Triassic collision zone [33] (Fig. 3.26). As the result of the change of Indian plate migration orientation from north to NE, the direction of maximum principal compressive stress in the Cretaceous System of China continent changed from NNE in Early Cretaceous, through NE in Middle Cretaceous, to ENE in Late Cretaceous. So, their fold axes appear as the WNW trending in Early Cretaceous, the NW trending in Middle Cretaceous and NNW trending in Late Cretaceous. According to the paleo-magnetic data, since Cretaceous, the magnetic in north of the Asian continental crust has been still stable, with only a few degrees of deviation from the modern magnetic north (Wan 2011). Thus, the above changes of maximum principal compression orientations were caused by the change of the direction of convergence and compression of the Indian plate, rather than the results of the rotation of Asian continental plate. After Middle Cretaceous, the tectonic forces became weaker, and the residual folds were mainly formed in Early Cretaceous; that is, the axis of the folds is still dominated by WNW. According to the statistics in the *Atlas of Petroleum Geology*, there are 2008 buckling folds, including 1032 anticlines and 976 synclines (in attachment table of Wan 2011). Some researchers recognized that after Cretaceous the Asian continent only experienced extension, without any compression. This is a misunderstanding. In fact, the compression and extension were formed at same time, and they could be derived from each other.

In the central and western parts of Yunnan and the Hengduan Mountains areas, the Jurassic and Cretaceous Systems are continually deposited without angular unconformity and sedimentary discontinuity between them. It means that the tectonics of Cretaceous was rather weak in the Yunnan areas. However, recently it can be found that the Jurassic and Cretaceous strata are formed to tight folds with intermediate and high angles (30°–70°), and usually the eastern limbs are steeper. They may be formed after Cretaceous, possibly in Eocene, Oligocene, or even in Early Miocene. In the past, the author (Wan 2011) ever considered they were all formed in the Sichuanian period (mainly Cretaceous), and now it seems to be incorrect.

The tectonics of East Asian continent in the Sichuanian period was strong in the southwest and weak in the northeast. It can be clearly seen that in the change of 176 tectonic (differential) stress values it can reach about 180 MPa in the Ali, Tibet; about 140 MPa in the central part of China; and only 100–90 MPa in the North and Northeast China (Fig. 3.37; Wan 2011). The areas with stronger deformation are distributed in Tibet–Southwest Sichuan, Gangdise, Altun–Qilian and the Hunan–Hubei–Guangxi areas (Fig. 3.25); the areas with weaker deformation are located at the eastern part of China, Yunnan and Northeast China. The distribution of stronger or weaker deformation areas was mainly controlled by the fast northward migration of the Indian plate (Schettino and Scotese 2005) and the slow northward migration of the Australian plate (Van der Voo 1993). The difference line was at the 90° E ridge in the Eastern Indian Ocean. It is also related to the distance between them, the thickness of Cretaceous strata, the coalescence degree of deep faults and so on.

In Middle Cretaceous (~100 Ma), Oceanic Anoxic Event, which may be derived from the meteorite impact, occurred in all the oceans around the world (Wan 2011), resulting in the extinction of organisms. The Middle East, Central Asia and North Africa used to have just shallow marine environment and now become the most important oil-gas fields in the world. In 1952–1965, general surveys for

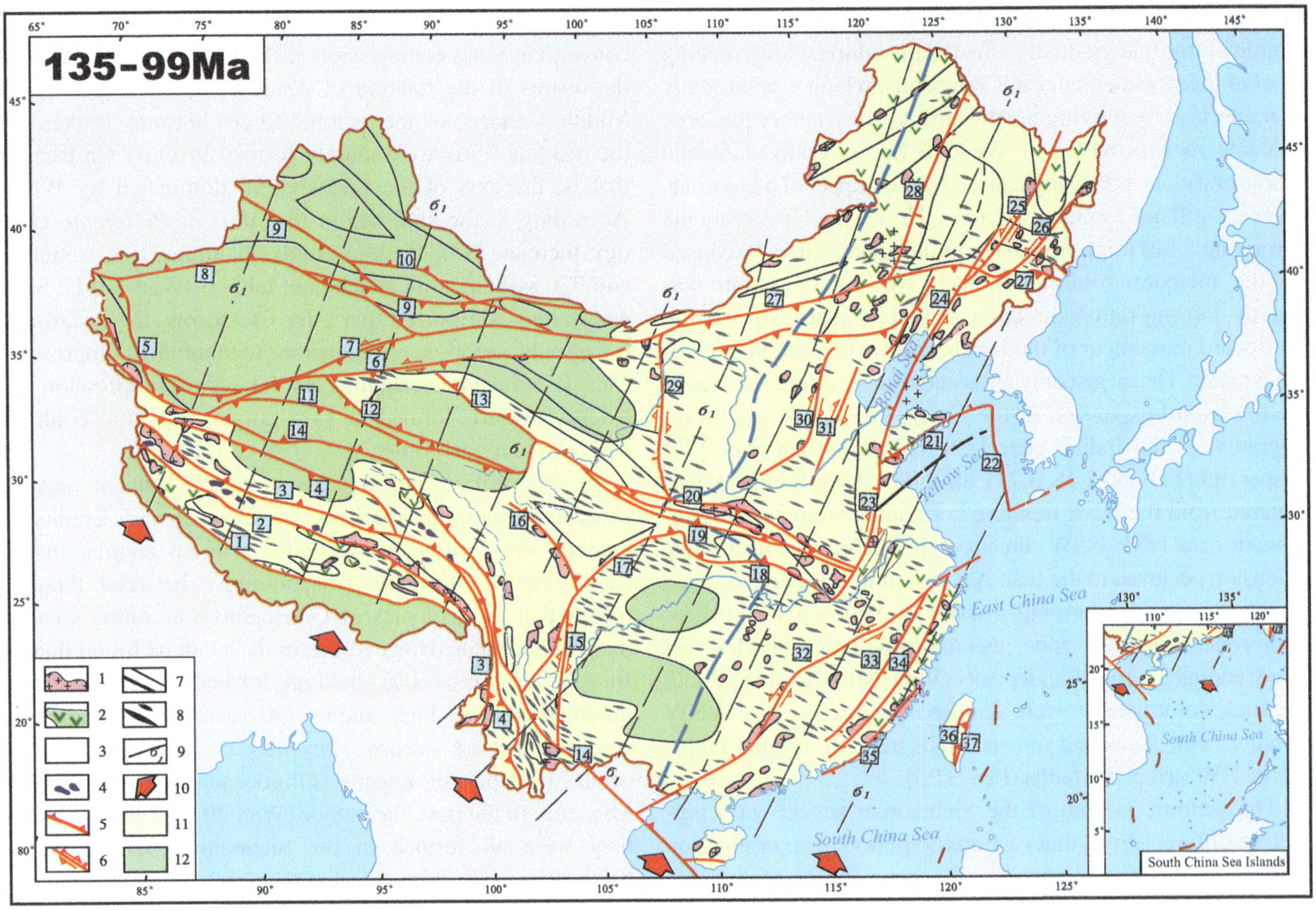

Fig. 3.25 Tectonic sketch of the China continent in early–middle period of Cretaceous (135–99 Ma). **Legend**: (1) Sichuanian granites; (2) Sichuanian volcanic rocks; (3) the transformation boundary between continental lithosphere (west) and the lithosphere with continental crust and oceanic mantle (east); (4) Sichuanian ophiolite and ultramafic rocks; (5) collision zones or thrust belts; (6) normal and strike-slip faults; (7) block boundaries or faults with weak activity, unnumbered; (8) anticline axis (data in Appendix 6 of Wan 2011); (9) trace of maximum principal compressive stress (σ_1); (10) direction of plate movement; (11) distribution areas with angular unconformity; (12) distribution areas with conformity or disconformity. **No. of plate boundaries, collision belts and fault zones**: (1) Yarlung Zangbo suture zone, with oceanic crust; (2) Garze–Nyainqentanglha reverse fault; (3) Bangongco–Nujiang collision zone; (4) Shuanghu–Lancangjiang reverse fault; (5) Kangxiwa–southernmost Tarim strike-slip and reverse fault zone; (6) Altun sinistral strike-slip fault zone; (7) Ruoqiang–Dunhuang reverse and sinistral strike-slip fault; (8) Korla–Wuqia overthrust zone; (9) Nilek–Yilinhabirga reverse fault zone; (10) Bogda reverse fault; (11) Eastern Kunlun reverse fault (or central Kunlun fault); (12) Northern Kunlun (southernmost Qaidam zone) reverse fault; (13) Junulshan–southernmost Qinghai Lake (northernmost Qaidam) reverse fault zone; (14) Jinshajiang–Red River dextral strike-slip and reverse fault zone; (15) Anninghe dextral strike-slip and reverse fault zone; (16) Daofu–Kangding dextral strike-slip and reverse fault zone; (17) Longmenshan sinistral strike-slip and normal fault zone; (18) Southern Dabashan–Fangxian–Guangji overthrust zone; (19) Shangdan–Tongbo overthrust zone; (20) Wushan–Baoji–Luonan–Fangcheng reverse fault zone; (21) Zhucheng–Rongcheng reverse fault zone; (22) east marginal Huanghai (Yellow Sea) dextral strike-slip fault zone; (23) southern section of Tancheng–Lujiang dextral strike-slip and normal fault zone; (24) middle section (Liaohe–Siping) of Tancheng–Lujiang sinistral strike-slip and normal fault zone; (25) Yilan–Yitong sinistral strike-slip and normal fault zone; (26) Dunhua–Mishan sinistral strike-slip fault zone; (27) Xar Moron reverse fault zone; (28) eastern edge of Dahingganling dextral strike-slip and normal fault zone; (29) Liupanshan–Helanshan dextral strike-slip and normal fault zone; (30) eastern edge of Taihangshan dextral strike-slip and normal fault zone; (31) eastern Cangzhou dextral strike-slip and normal fault zone; (32) Shiwandashan–Shaoxing sinistral strike-slip and normal fault; (33) Chong'an–Heyuan normal fault; (34) Lishui–Lianhuashan normal fault; (35) Changle–Nan'ao normal fault; (36) Shoufeng fault; (37) Yuli sinistral strike-slip fault zone (After Wan 2011)

petroleum were carried out in China (Petroleum Specialized Committee of Geological Society of China and Petroleum Geology Specialized Committee of Chinese Petroleum Society 1966), focusing on geological survey in the Cretaceous sedimentary basins. It is found that in most of the Cretaceous System in China there developed red sandstone and shale systems in arid and hot subtropical climate, which does not have conditions for hydrocarbon generation but it is propitious to the formation of the gypsum and salt deposits. Only in Northeast China, the Cretaceous System formed in the temperate zone with moist weather could preserve giant oil and gas fields, such as Daqing oil-gas field.

Fig. 3.26 Tectonic sketch of Asian continent in the middle period of Early Cretaceous–Paleocene (135–56 Ma). The green arrow shows the orientation of tectonic maximum principal compressive stress. The green line shows the Early Cretaceous–Paleocene collision zone and main reverse faults. The east of the blue dotted line areas shows the distribution of East Asian continental crust and oceanic mantle lithosphere. The numbers of tectonic unit are as same as those in the contents or text

In Jurassic, a series of NNE trending reverse faults or thrusts were developed in the East Asian continent. However, in the middle period of Early Cretaceous–Paleocene, those faults were transformed into normal faults, such as the Dahingganling–Taihangshan fault zone, Tancheng–Lujiang fault zone and southeast coast fault zone. They were all suffered from the NNE trending shortening and WNW–ESE trending extension, and a series of metamorphic core complexes were easy to form nearby (Liu et al. 2005, 2006). Someone thought that this is the "elastic rebound" after squeezing, which is not proper. Elastic rebound only occurs during the seismic wave broadcast, and after elastic deformation, it bounces back to its original position. However, all structural deformations that can be recorded during the geological history are not elastic deformations, but permanent plastic deformations, which are impossible to restore. In recent years, some researchers thought that the E–W trending extension in Cretaceous in the East Asia was caused by the "rollback" of the Pacific subduction zone. However, from the point of view of the author, it is impossible. Because the rollback is very limited to the trenches and island arcs in area near the subduction zone (Figs. 2.43, 2.45), it is impossible to affect the inner area of the continental plate thousands of kilometers away. Moreover, the affecting distance was much shorter in Cretaceous, and the Pacific plate migrated and subducted to the north (Osozawa 1998), but not to the west.

Many plates migrated northward in Cretaceous and Paleocene, resulting in the formation of the Bangongco–Nujiang–Mandalay–Phuket–Northern Barisan Cretaceous collision zone [35], Kavkaz–Alborz Late Paleozoic–Late Jurassic accretion–collision zone [41], Anatolia–Tehran Middle Cretaceous–Paleocene collision zone [42], Zagros–Kabul accretion–collision zone (since Cretaceous) [44], Oman Cretaceous accretion–collision zone [47] and Eastern Kalimantan–Southern Sulu Sea Cretaceous collision zone [52]. However, at this time, for the tectonic force of the northward subduction of the Indian plate was rather weaker, the structural deformation in Cretaceous in this area was weak. Due to the fact that the moving velocity of Indian plate [40] was obviously faster than that of Australian plate [72], the middle part of the Bangongco–Nujiang–Mandalay–Phuket–Northern Barisan Cretaceous collision zone [35] and the eastern area of 90° E ridge gradually rotated from near E–W trending to near N–S trending, causing the regional fault with the characteristic of dextral strike-slip. In the same time, the 90° E ridge controlled by the dextral strike-slip was formed in the ocean basin. The oceanic drilling data of the ridge show that the isotopic ages of oceanic basalts regularly get older from north to south (Fig. 2.38; Condie 2001), indicating the gradual northward movement of the Indian Ocean plate.

In the sedimentary system of West Asia (Iran–Arabian Peninsula–Turkey), a series of NW–WNW trending thrusts, folds with arc type and collision zones were formed, including the Anatolia–Tehran Middle Cretaceous–Paleocene collision zone [42] (Fig. 2.40), Zagros–Kabul accretion–collision zone (since Cretaceous) [44] and Oman Cretaceous accretion–collision zone [47] (Fig. 3.26). In the adjacent plates (Turkey–Iran–Afghanistan plate [43] and Arabian plate [46]), the intraplate deformations occurred, and the area was in a shallow sea environment, providing a very good structural condition for hydrocarbon to migrate and accumulate.

In the Sichuanian period (135–52 Ma), the climax of tectonic events was at the end of Paleocene. In the Asian continent, the strata between Cretaceous and Paleocene were all continuously deposited, without tectonic events, which was very different from the oceanic sedimentary strata. Till now, it has been discovered that the sedimentary discontinuities and micro-tektite impacts between Cretaceous and Paleocene Systems are mainly distributed in the depth of the Central Atlantic Ocean (Norris and Kroon 1998). The meteorite impact event at the end of Cretaceous (65 Ma) in the Yucatan Peninsula, Mexico, had great influences on the organism catastrophe, such as the extermination of dinosaur and most of the gymnosperm (Sharpton et al. 1992). But so far it has not found any significant influence on the migration of global plates (Moore 1989; Wan 2011).

3.12 The Tectonic Evolution in Eocene–End of Oligocene (56–23 Ma)

In Oligocene ($\sim$36 Ma), the migration orientation of the Pacific plate turned from NNW to WNW suddenly. In the Western Pacific Ocean, there formed a subduction zone (Fig. 3.27), which may be caused by the oblique impact of micro-tektites from Caribbean to East Asia (Yin and Wan 1996; Wan et al. 1997; Wan 2011) (Fig. 3.28). The micro-tektites are distributed in the strata beneath the ocean floor at the depth of 400 m, in the direction of WSW. The impact center is located near the Eastern Pacific Rise (Glass 1982; Wan 2011).

According to the prediction of Yin and Wan (1996), before 36 Ma, the Pacific plate moved toward the NNW orientation, with a velocity of 7.1 cm/y. Due to the oblique impact of micro-tektites to the Earth's surface, the Pacific plate was subjected to the force toward the WSW direction, with a possible migration velocity of about 8.7 cm/y. Thus, the direction of the join forces influenced the Pacific plate suddenly to change to WNW with a migration velocity 10.6 cm/y (the yellow arrow shown in Fig. 3.29).

As the Pacific plate suddenly turned to the WNW, the subduction zones and a series of trench–island arcs were formed between the East Asian plate and Pacific plate, such as the Aleutian–Kamchatka–Kurile–Northeast Japan Cenozoic subduction and island arc zone [59], South Honshu–South Shikoku Ryukyu subduction and island arc zone [62], and the Izu–

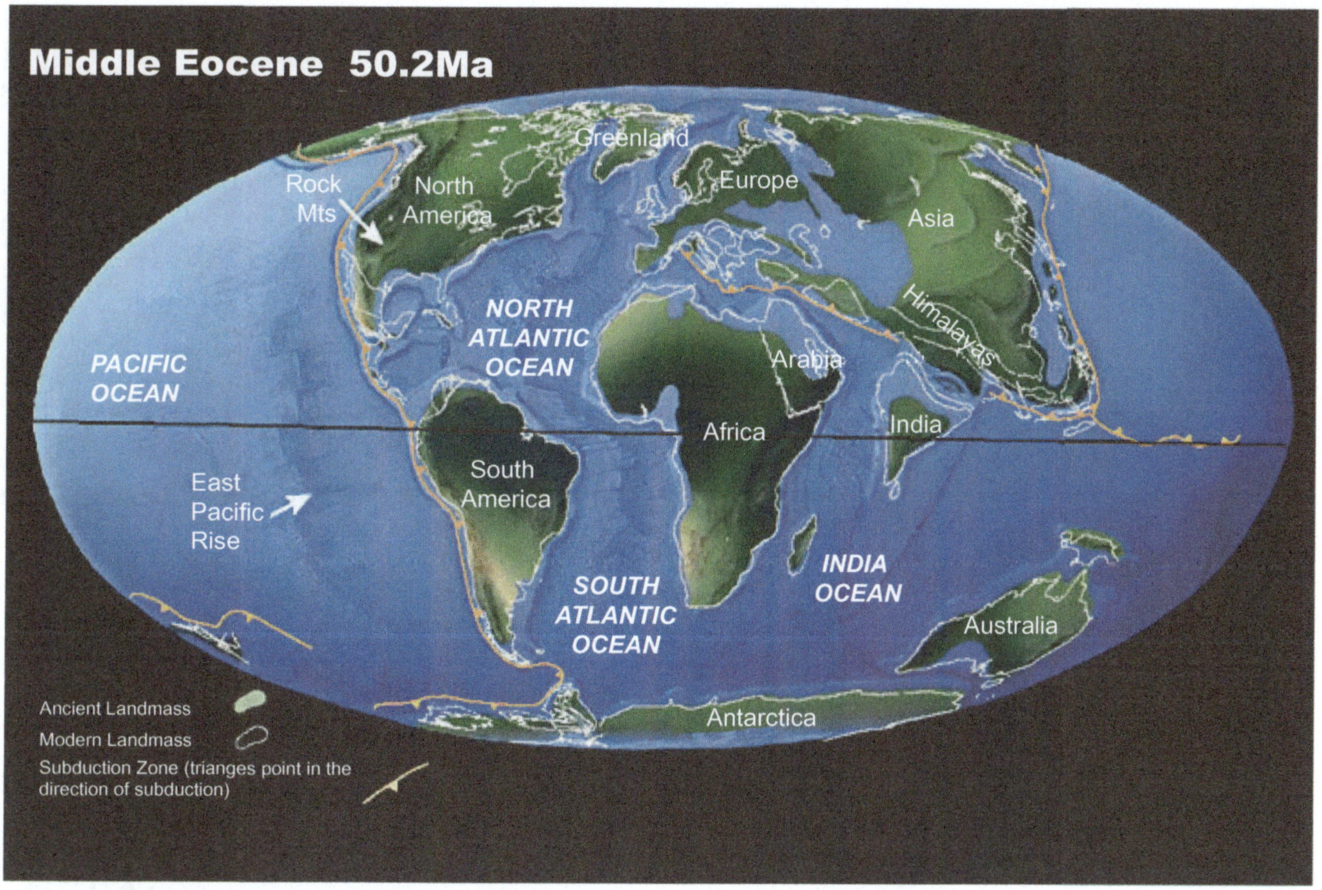

Fig. 3.27 Global paleo-continental reconstruction in Middle Eocene (50.2 Ma) (After Scotese 1994)

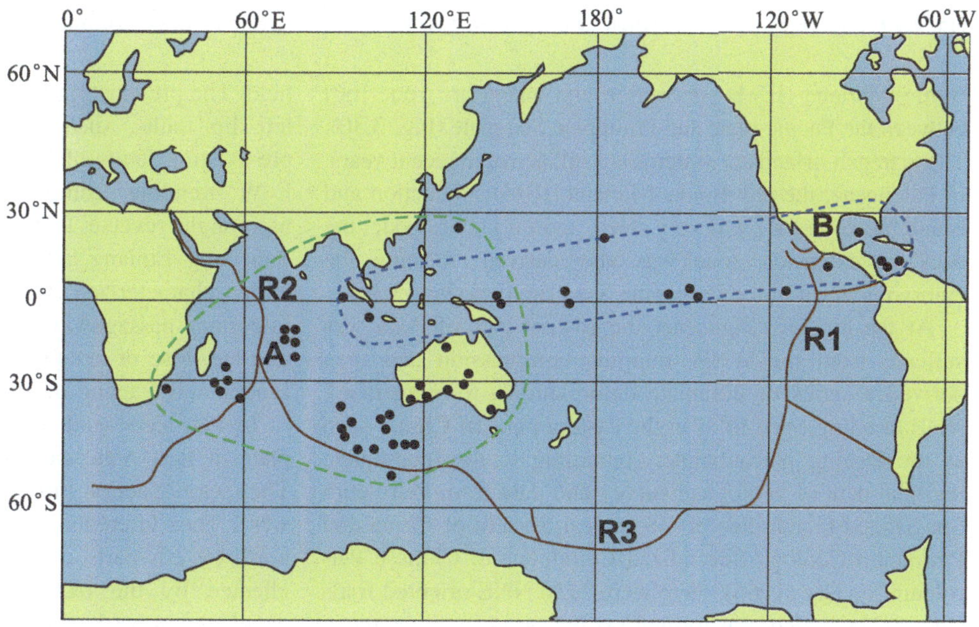

Fig. 3.28 Distribution of micro-tektites related to the meteor impact events of Asia, Australia and North America–Caribbean. (A) Area covered by micro-tektites from the Australasian event; (B) area covered by micro-tektites from the North America–Caribbean event; (R1) East Pacific Rise; (R2) Mid-Atlantic Ridge; (R3) Carlsberg Ridge; (R4) Circum-Antarctic Ridge. Small black circles are DSDP well locations, and the arcuate double lines near the offshore Baja California represent circular fractures in the ocean floor (Modified from Glass 1982)

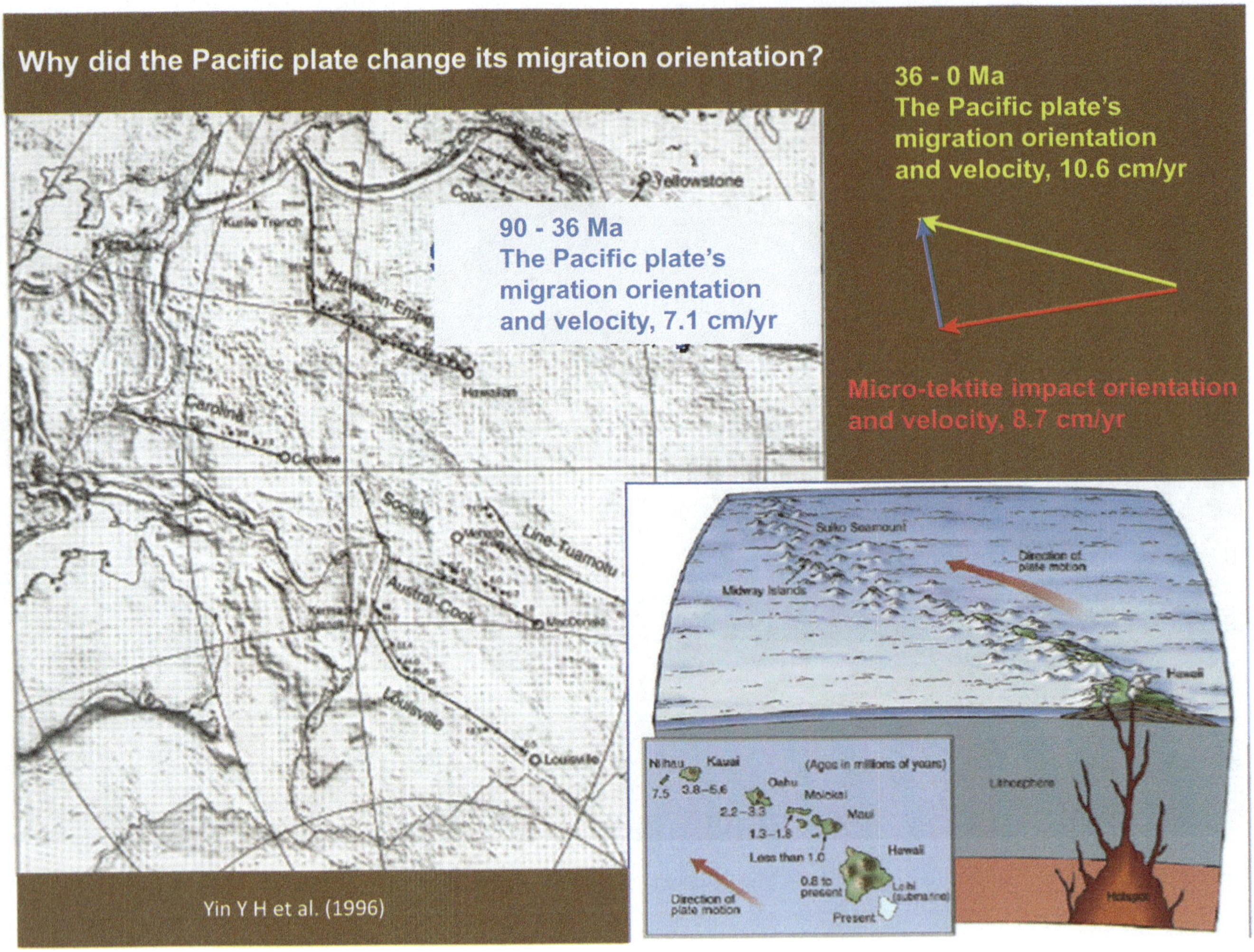

Fig. 3.29 Sudden change of migration trending in the Pacific areas during Eocene (Modified from Yin and Wan 1996; Wan et al. 1997)

Bonin–Mariana (IBM) subduction and island arc zone [66] between the Pacific plate and Philippine Sea plate (Fig. 3.30). Those trench–island arc systems are still active in recent years. In Paleogene, the Izu–Bonin–Mariana (IBM) subduction and island arc zone [66] was very steep, and on its west side, the back-arc extension zone was also derived, forming the embryonic form of the Philippine Sea plate [65] (Fig. 3.30).

At the end of Oligocene, the eastern part of Asia was suffered from the WNW trending compression and then derived a series of intraplate deformations, forming many NNE trending open folds in the eastern parts of China, such as the Daqing placanticline. According to the incomplete regional data of geological survey and data from Petroleum Specialized Committee of Geological Society of China and Petroleum Geology Specialized Committee of Chinese Petroleum Society (1966), there were 2126 NNE-oriented folds (Fig. 3.31), which transformed the NE trending faults (such as the northern section of Tancheng–Lujiang fault zone), to the dextral strike-slip fault, and controlled the formation of echelon anticline oil reservoirs in the Eastern Liaohe oil field. The pre-existing faults in the NW direction became the left-slip faults, such as the Red River fault zone. The pre-existing faults of N–S direction were suffered from the E–W trending compression and transformed into the high-angle reverse faults, such as the middle part of the Tancheng–Lujiang fault zone. The original WNW trending faults characterized by the extension-shear became the important passageway or position for the hydrocarbon to migrate, lose or accumulate, such as the Shengli, Western Liaohe and Dagang oil fields.

In Paleocene–end of Oligocene, the tectonics of Pacific plate in East Asia was stronger and weaker in West Asia. The tectonic action time was gradually lagging from east to west. The differential stress value is about 70 MPa. The northwestern parts of China (Xinjiang area) were weakly affected by the tectonic movement in this period and developed some broad open folds with axis of N–S in soft strata. Until Late Miocene–Neocene, the tectonic movement had begun to strongly affect the Xinjiang area (Figs. 3.31, 4.20; Wan 2011).

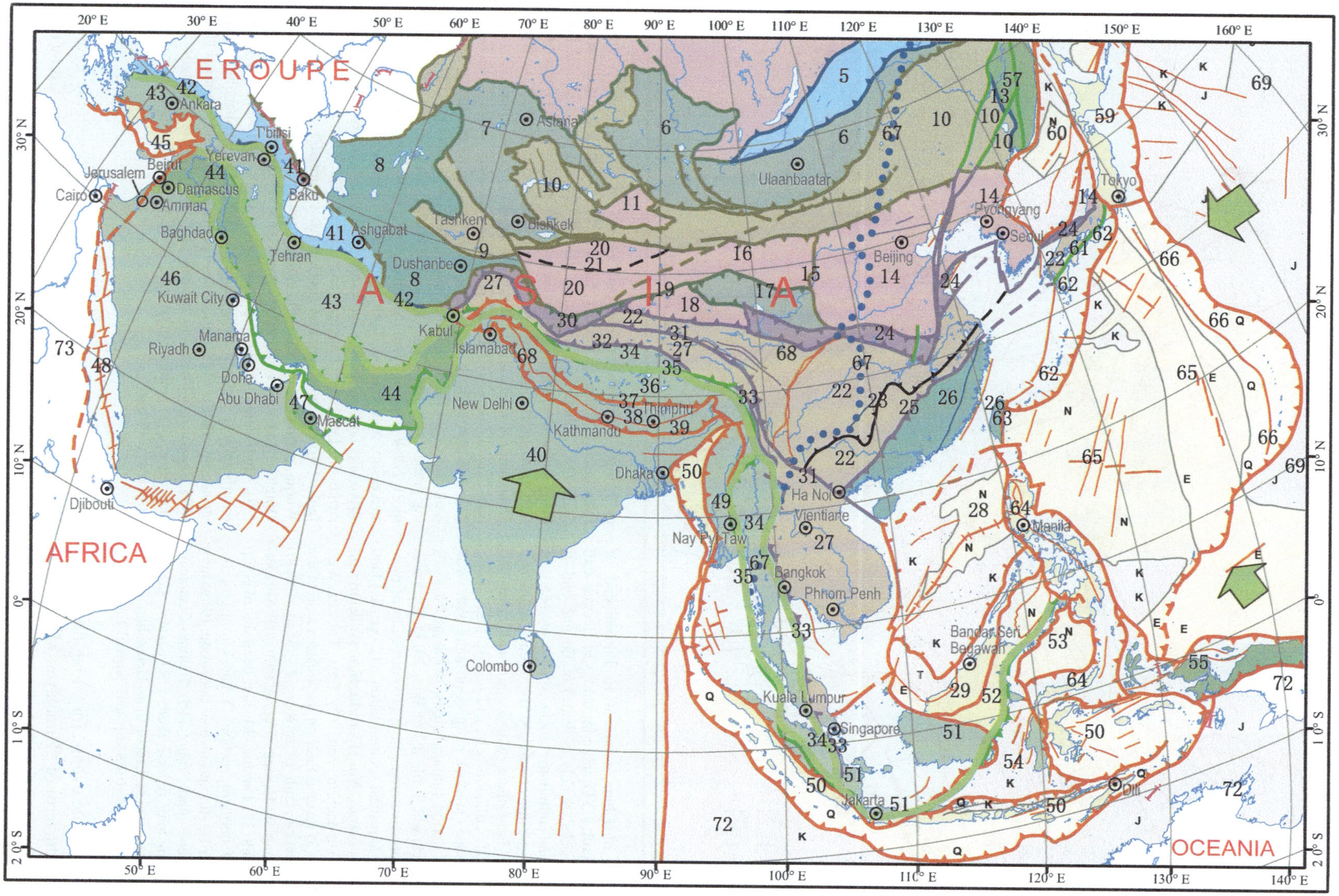

Fig. 3.30 Tectonic sketch of Asian continent in Eocene–end of Oligocene. The numbers of tectonic unit are as same as those in the contents or text. The east side of the blue dotted line shows the distribution of East Asian continental crust and oceanic mantle lithosphere, which west side is the area of the normal continental lithosphere. The green arrow shows the plate's moving direction

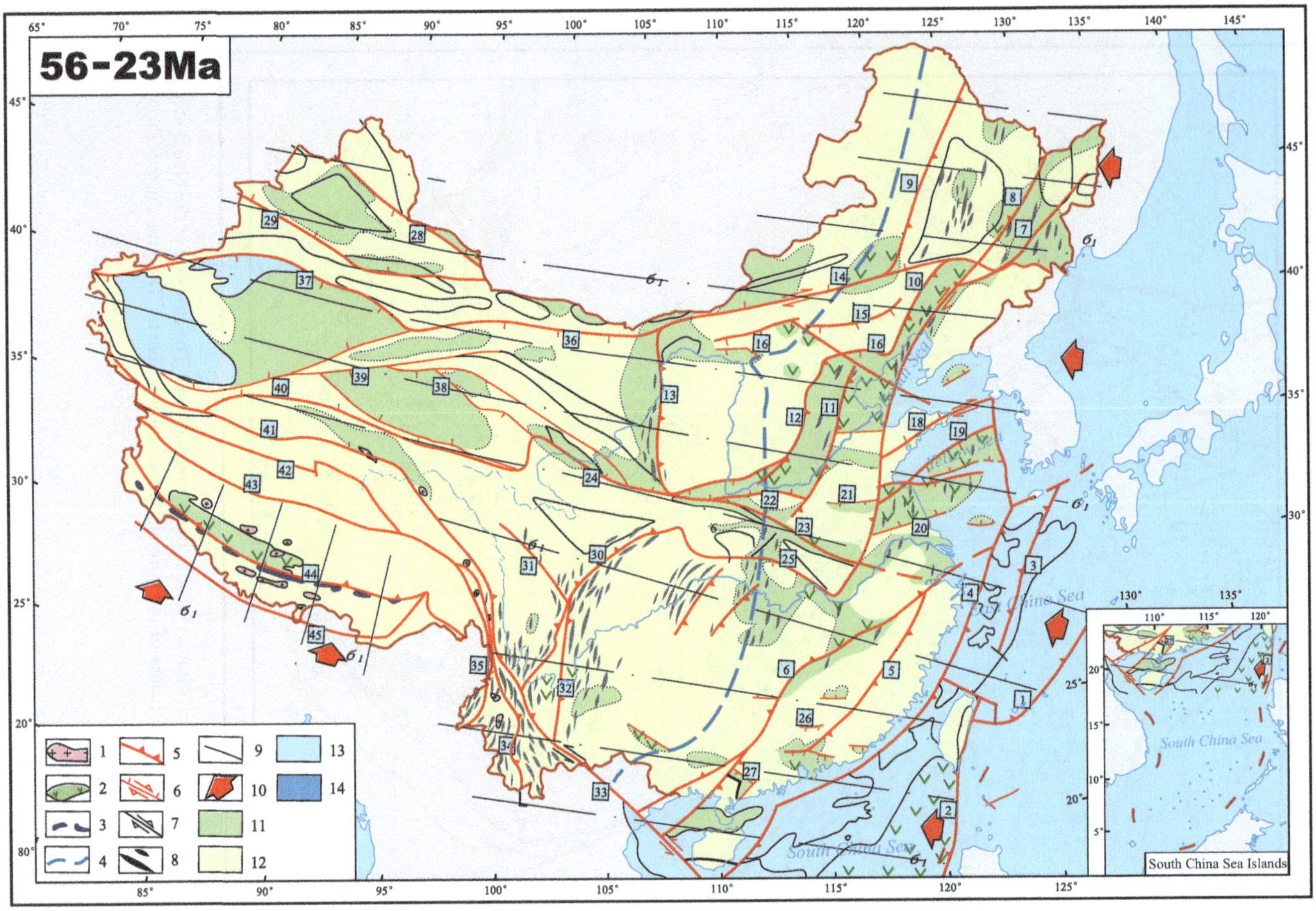

Fig. 3.31 Tectonic sketch of China continent in Eocene–end of Oligocene (56–23 Ma). **Legend**: (1) Granitic intrusions of Eocene–end of Oligocene; (2) volcanic rocks (mainly tholeiite) of Eocene–end of Oligocene; (3) ophiolitic and ultramafic rocks of Eocene–end of Oligocene; (4) the transformation boundary between continental lithosphere (west) and the lithosphere with continental crust and oceanic mantle (east); (5) plate collision belts and reverse fault zones; (6) normal and strike-slip faults; (7) block margins or faults with weak activity in the North Sinian period (unnumbered); (8) axial traces of anticlinal folds (data in Attachment Table 3.7 of Wan 2011); (9) traces of maximum principal compressive stress (σ_1); (10) direction of plate movement; (11) basins with terrestrial sedimentation; (12) eroded land areas; (13) shallow seas; (14) oceans. **No. of subduction, collision and fault zones**: (1) Ryukyu subduction belt; (2) East Taiwan–West Philippines subduction belt; (3) reverse fault on the western side of the Tiaoyu Island uplift; (4) reverse fault of offshore (~50 m isobath) Fujian–Guangdong; (5) Chong'an–Heyuan reverse fault; (6) Shiwandashan–Shaoxing reverse fault; (7) Dunhua–Mishan dextral strike-slip and reverse fault zone; (8) northern section (Yilan–Yitong) of the Tancheng–Lujiang dextral strike-slip and reverse fault zone; (9) eastern side of Dahingganling reverse fault zone; (10) Beipiao–Jianchang reverse fault zone; (11) Cangdong reverse fault; (12) eastern side of Taihangshan reverse fault zone; (13) Liupanshan–Helanshan reverse fault zone; (14) Xar Moron dextral strike-slip and normal fault zone; (15) Jining–Gubeikou normal fault zone; (16) Southernmost Yinshan–Daqingshan–Yanshan normal fault; (17) Guangrao–Jiyang normal fault; (18) Zhucheng–Rongcheng dextral strike-slip and normal fault; (19) Guanyun–southern Yellow Sea dextral strike-slip and normal fault; (20) Jiangdu–Hai'an dextral strike-slip and normal fault; (21) Wuhe–Huaiyuan normal fault; (22) Luoning–Luoyang normal fault; (23) Luonan–Fangcheng sinistral strike-slip and normal fault; (24) Baoji–Tianshui normal fault zone; (25) Fangxian–Xiangfan–Guangji normal fault zone; (26) E–W trending Nanling normal fault group; (27) Maoming graben; (28) Derbugan–Kelameili normal fault zone; (29) southernmost Nilek–Turpan–Hami normal fault; (30) Longmenshan dextral strike-slip and reverse fault zone; (31) Daofu–Kangding sinistral strike-slip fault; (32) Anninghe reverse fault; (33) Red River sinistral strike-slip fault; (34) Lancangjiang reverse fault; (35) Nujiang reverse fault; (36) the northernmost part of Alxa normal fault; (37) Korla–Wuqia normal fault; (38) the southernmost part of the Junulshan–Qinghai Lake normal fault zone (northernmost Qaidam zone); (39) Altun dextral strike-slip and normal fault; (40) Northern Kunlun (southernmost Qaidam zone) normal fault; (41) Jinshajiang sinistral strike-slip fault; (42) Kongkela–Tanggula–Wenquan normal fault; (43) Bangongco–Dongqiao normal fault; (44) Yarlung Zangbo collision zone; (45) southernmost relic of Himalayan (Tethys) Ocean, and now it is a thrust

As to the Cenozoic tectonic events, for a long time, many researchers used to use the term of "Himalayan Movement," which was first named by Huang (1945, 1960) and Huang et al. (1965). He referred to the division scheme of tectonic events in Alps area to call the Mesozoic tectonic event as "Old Alps Movement" and the Cenozoic tectonic event as "New Alps Movement." He called the Jurassic–Cretaceous tectonic event as "Yanshan Movement" and the Cenozoic tectonic event as "Himalayan Movement." In fact, in that time little was known about the structures of Himalayan areas. In 2004, when the author did field geological investigation in the Alps, he learned that French and Italian scholars have long recognized the "Old Alps Movement" not to exist at all, and both Triassic and Cretaceous Systems are continually deposited with conformable contact. The end of Paleocene with the strong tectonics is the true period of Alps tectonic deformation, which characteristics in Neogene have been basically as same as those of modern times, with NW trending migration and compression.

In fact, during Paleogene in the Asian continent, the Himalayan areas did not uplift to be mountains and had no strong tectonic or orogenic events. The naming principle of tectonic event is determined by tectonic event that causes the strongest rock deformation and forms the mountains. Therefore, it is obviously inappropriate to consider the Cenozoic tectonic event to be "The Himalayan Movement" (Huang 1945, 1960; Huang et al. 1965).

Due to the fact that in Paleogene the Indian plate, migrating toward the North, just began to subduct and then collided with the Asian continent, the rock deformations were rather limited and weaker (Fig. 3.31). The strong tectonic factor at this period was the westward subduction and compression of the Pacific plate. This tectonic event was firstly proposed by Petrologist Tang (1979). He called the Cretaceous–Paleocene rock deformation as "First Epoch of North China Event" and the Eocene–Oligocene rock deformation as "Second Epoch of North China Event." However, his "First Epoch of North China Event" is the same as the "Sichuanian Tectonic Event" named by Tan and Li (1948), so it is not necessary to re-name it. The "Second Epoch of North China Event" is a new idea, and it is very consistent with the fact. Thus, the author agrees to the viewpoint that the Eocene–Oligocene tectonic event is called "North China Tectonic Event."

In the East Asia, the strong tectonic deformation zone, formed by the migration and subduction of the Pacific plate, changed with time. The direction transition of Emperor–Hawaii Ridge occurred at the period of 36 Ma near the Midway Island (Raymond et al. 2000). That is, the Pacific plate began to migrate in the direction of WNW with a velocity about 10.6 cm/y at the time of 36 Ma. Tectonics affecting the Western Pacific island arc occurred at about 30 Ma, which is called "Takachiho Movement" in Japan (Maruyama and Seno 1986) and "Puli Movement" in Taiwan (Zhang 1984; Chen 1984; Stephan et al. 1986). The event (23 Ma) in the eastern part of China is called North China Tectonic Event (Tang 1979), in the western part of Yunnan at about 15 Ma (He and He 1993) and in the western part of Xizang and Xinjiang about 10 Ma (Guo et al. 1991; Wang et al. 2006). That means the stronger rock deformations in different areas are not formed at the same time, undergoing a process of gradually transferring to the west.

The compression of Pacific plate in the direction of WNW also drove a near N–S orientation extension in East Asia. On the basis of the old fault zones, there formed a series of high-angle normal faults in the WNW direction and three near E–W trending mountains, including the Yinshan–Yanshan Mountains, Qinling–Dabie Mountains and Nanling Mountains, as well as four drainage basins, including the Songliao and inland, Yellow River, Yangtze River and Zhujiang River drainage basins. But at that time they had not yet formed a complete water system.

In Eocene–Oligocene, the Asian continent was flattened, and thus the moist air masses of Pacific Ocean and Indian Ocean could be driven straightly in. Most parts of the Asian continent were warm and humid, with luxuriant organisms and propitious to organic matter accumulation. It is also an important period for hydrocarbon generation and accumulation in Asian continent.

In the Qinghai–Tibet region, some stronger deformations were developed in Eocene–Oligocene, which were caused by the northward migration of Indian plate with the velocity ~6.0 cm/y (Lee and Lawver 1995) and the differential stress value about 82–100 MPa (Wan 2011). The collision evidences between the Indian plate and Eurasian plate are mainly manifested in the Yarlung Zangbo–Myitkyina Paleogene collision zone [37] (Figs. 3.30, 3.31). According to the time of the oceanic crust finally disappearing in Middle and Late Oligocene at (~34 Ma) (Wang et al. 2002; Aitchison and Davis 2001; Aitchison et al. 2007), the onset of continent collisions may be after that period. Prior to that time, the Indian plate subducted beneath the Eurasian continent. Recently, some scholars have recognized that oceanic basin is just a residue oceanic basin, which means that the onset of collision may be earlier. However, the discussions about that problem are still going on and should be paid sufficient attention to.

The 90° E ridge fault between the Indian plate and Australian plate at that time was characterized by the dextral strike-slip (Fig. 2.38), due to the northward migration velocity (~5 cm/y) of Indian plate being faster than that of Australian plate (~2 cm/y). The 90° E ridge fault penetrated into the Asian continent in the direction of NNE. Although a coherent fault was not formed on the surface, strong deformation occurred on the west side of the fault Qinghai–Tibet Plateau and Northwest China, while the deformation on its east side was obviously weak. The main structural deformation in the eastern region was derived from the westward compression of the Pacific plate. As a result, the structural boundary of 90° E ridge fault–Hengduanshan–Helanshan–Liupanshan was formed. The performance of this boundary in modern tectonic activities is the formation of "North–South Seismic Zone" (Yong 1988). However, in the Asian continent, this N–S tectonic belt does not form a coherent fault zone, but appears as a reactivity of the pre-existing fault zone with a series of incoherent faults with slightly different directions. Therefore, after Paleogene, the deformation of

Asian continent shows characteristics—stronger in the west and weaker in the east, with rather different structural line directions; the geomorphology is characterized by the higher in the west and lower in the east.

3.13 The Tectonic Evolution in Neogene–Early Pleistocene (23–0.78 Ma)

The Neogene–Early Pleistocene was the main tectonic period of forming the Himalayan Mountains, so it lives up to its name to be called "Himalayan Tectonic Period" (Huang 1945, 1960; Huang et al. 1965). In this period, the collision zones or island arcs, formed in the Asian continent, include: the Southern Himalayan main boundary thrust [39], Toros accretion–collision zone (since Neocene) [45], Arakan–Sunda Cenozoic subduction and island arc zone [50] and Northern New Guinea island arc zone (since Neocene) [55]. Then, the Indian plate [40], Turkey–Iran–Afghanistan plate [43] and Arabian plate [46] migrated northward, and collided and cohered to the Asian continent. The collision zones were all formed on the basis of pre-existing faults and block boundary. Therefore, the strikes of collision zones and main faults changed variously, but the basic moving directions of blocks were relatively consistent, mainly influenced by the different velocities of northward migrations of Indian plate and Australian plate (Figs. 3.32, 3.34; Wan 2011).

In the southwestern area of China, the tectonic stress (average differential stress) value is about 92.6 MPa, while in the northern and eastern area of China that is about 21.5 MPa (Wan 2011), indicating a obviously weaker tectonic activity. At that time, the northward migration velocity of Indian plate [40] was about 5 cm/y (Lee and Lawver 1995) and that of Australian plate was only 2 cm/y (Hall and Blundell 1995; Hall et al. 2011). Thus, the Pamir–Qinghai–Tibet areas, to the north of the Indian plate, were suffered from stronger N–S shortening, resulting in strong tectonic magmatism and obvious thickening of the lithosphere (in the yellow dotted line area of Fig. 1.1). The thickness of lithosphere reaches 170–200 km, including the crust thickness of about 60–70 km and the lithosphere mantle thickness of about 130–150 km. Such a lithosphere is considered as the thickened continental lithosphere [68]. However, in the north of the Australian plate, the Southeast Asia and the eastern part of China, only weak deformations occurred, which were characterized by weak E–W trending extension near the pre-existing faults (such as along the Da Hingganling–Taihangshan, Tancheng–Lujiang or Fujian–Guangdong coastal

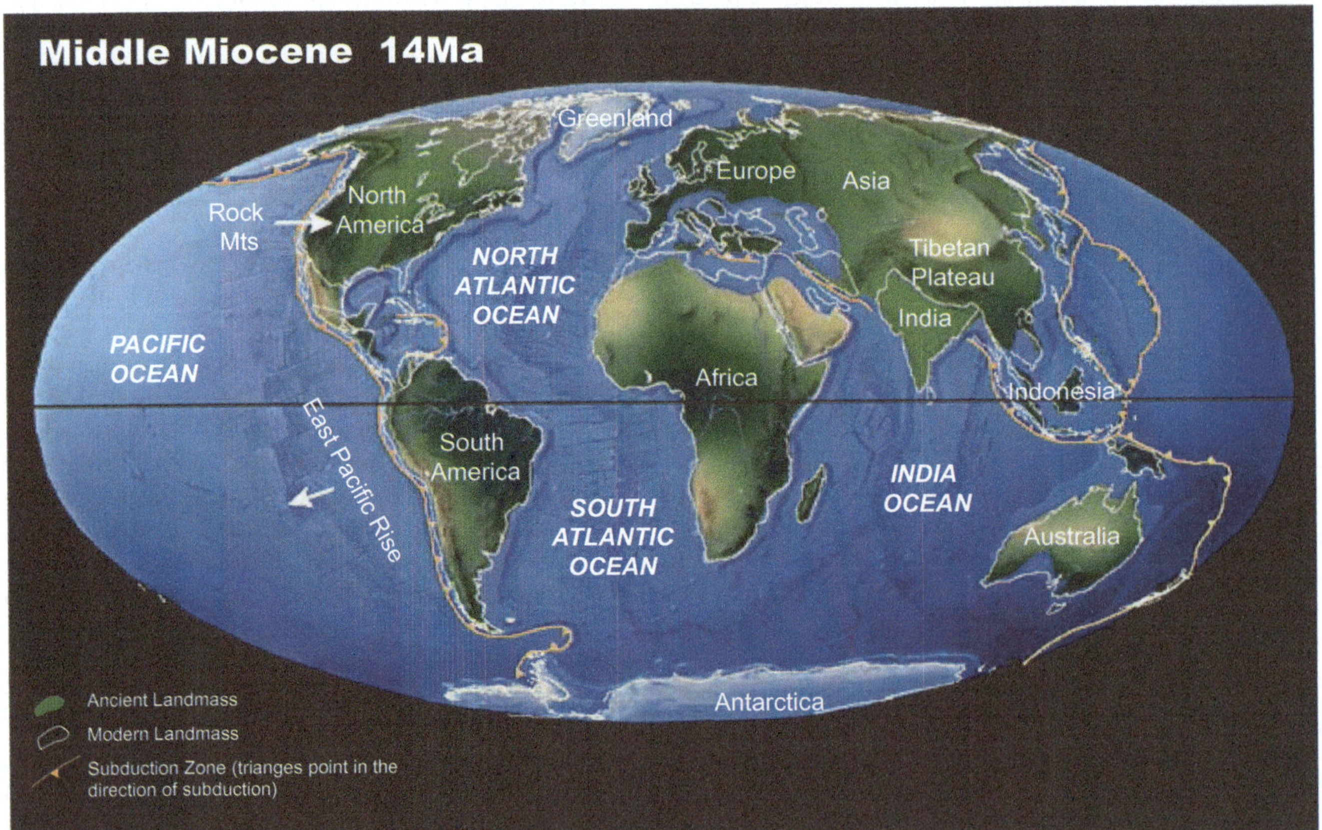

Fig. 3.32 Global paleo-continental reconstruction in Middle Miocene (After Scotese 1994)

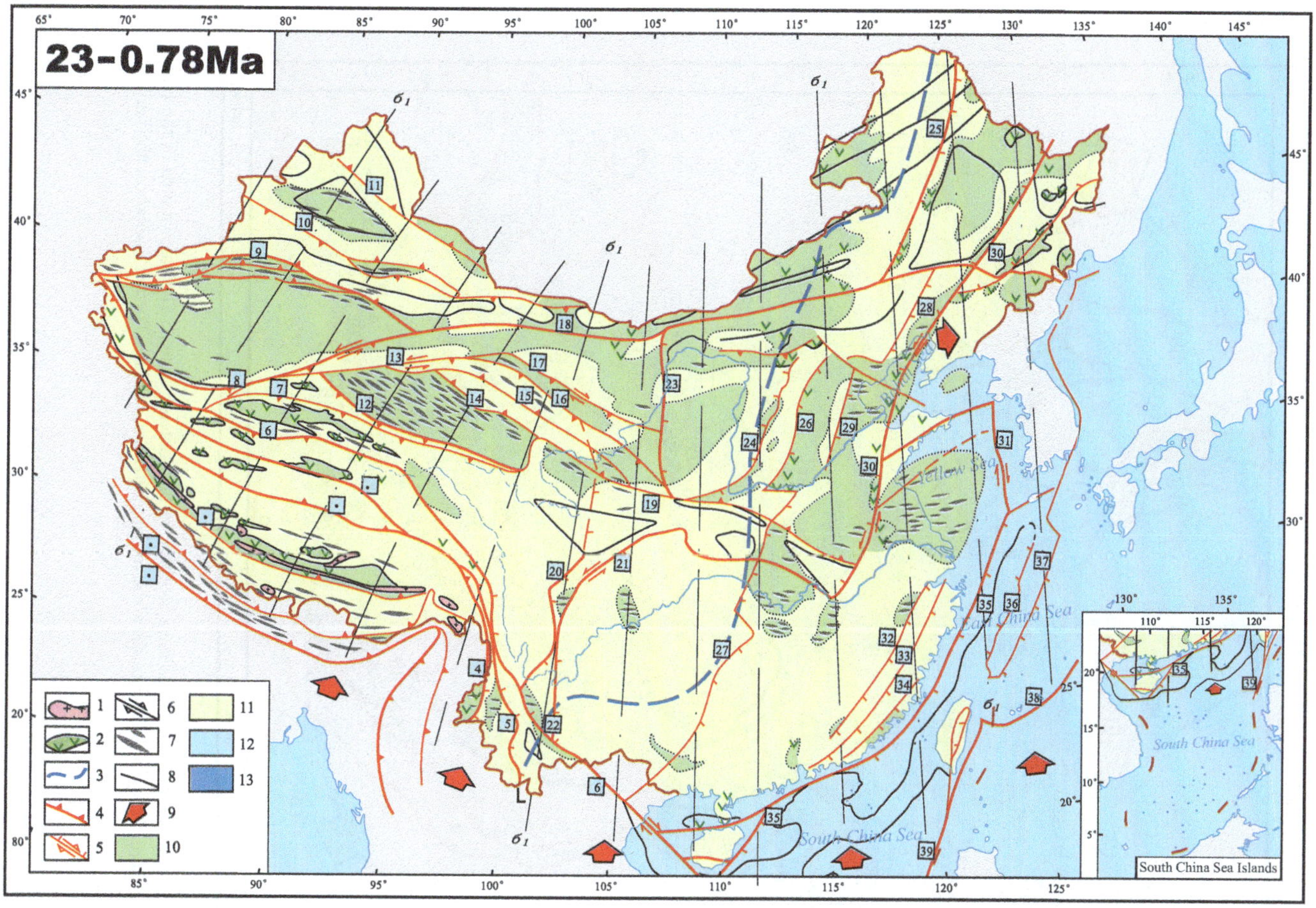

Fig. 3.33 Tectonic sketch of China continent in Neogene–Early Pleistocene (23–0.78 Ma). **Legend**: (1) Himalayan granite; (2) Himalayan volcanic rocks; (3) transformation boundary between continental lithosphere (west) and the lithosphere with continental crust and oceanic mantle (east); (4) collision zones or thrust belts; (5) normal and strike-slip faults; (6) boundaries of blocks or faults with weak activity during the Himalayan period; (7) fold axes, only anticlines are shown (in Attachment Table 3.8 of Wan 2011); (8) trace of axes of maximum principal compressive stress (σ_1); (9) direction of plate migration; (10) areas of terrestrial sedimentation; (11) land areas subjected to erosion; (12) areas of shallow sea; (13) oceanic areas. **No. of plate boundaries, collision belts and fault zones**: (1) Main boundary thrust of the Himalayas (MBT, boundary between the Indian and Eurasian plates); (2) main Himalaya thrust (MHT); (3) Yarlung Zangbo thrust; (4) Bangongco–Dongqiao collision zone; (5) Kongkela–Wenquan Tanggula thrust; (6) Jinshajiang–Red River thrust zone, with dextral strike-slip on the Red River section; (7) Kunlun thrust (i.e., central Kunlun thrust); (8) Kangxiwa–Ruoqiang–Dunhuang reverse fault zone with strike-slip; (9) Korla–Wuqia thrust zone; (10) Nile–Yilinhabirga–Yagan thrust zone; (11) Debuga–Kramali thrust zone; (12) southern margin of the Qaidam thrust; (13) Altun sinistral strike-slip fault zone; (14) Junulshan–southern margin of the Qinghaihu thrust; (15) southern margin of the central Qilianshan thrust; (16) northern margin of the north Qilianshan thrust; (17) Longshoushan thrust; (18) northern margin of the Alxa thrust; (19) Wushan–Baoji–Luonan–Fangcheng thrust zone; (20) eastern margin of the Daxueshan normal fault zone; (21) Longmenshan sinistral strike-slip fault zone; (22) Xiaojiang dextral strike-slip and normal fault zone; (23) Liupanshan–Helanshan normal fault zone; (24) Fenhe graben zone; (25) eastern edge of Dahingganling normal fault zone; (26) eastern edge of Taihangshan dextral strike-slip and listric fault zone; (27) Wulingshan–Damingshan normal fault; (28) Beipiao–Jianchang normal fault zone; (29) eastern Cangzhou normal fault zone; (30) Tancheng–Lujiang sinistral strike-slip and normal fault zone; (31) eastern marginal Huanghai (Yellow Sea) dextral strike-slip fault zone; (32) Chong'an–Heyuan normal fault; (33) Lishui–Lianhuashan normal fault; (34) Changle–Nan'ao normal fault; (35) Coastal fault zone of Fujian and Guangdong; (36) western marginal normal fault of Tiaoyu Island; (37) east marginal normal fault of Tiaoyu Island; (38) Ryukyu subduction zone; (39) West Philippine–East Taiwan longitudinal valley sinistral strike-slip fault zone (After Wan 2011)

fault zones), basalt intrusion and eruption, and extremely weak E–W folds (Figs. 3.33, 3.34; Wan 2011).

Since Neocene, the velocity of northward migration of the Asian continent had been different from that of its surrounding plates; therefore, the tectonic line of the Northern Arakan–Sunda Cenozoic subduction and island arc zone [50] and the Eastern Zagros–Kabul accretion–collision zone [44] transformed to near the N–S orientation and the Western Zagros collision zone (Nezafati 2006) to NW orientation. Similarly, between the African plate [73] and Arabian plate [46], due to the different migration velocities, there formed the Red Sea–Dead Sea rifts [48] with the NNW or NNE trending extension. Moreover, because the northward migration velocity of the African plate was greater than that

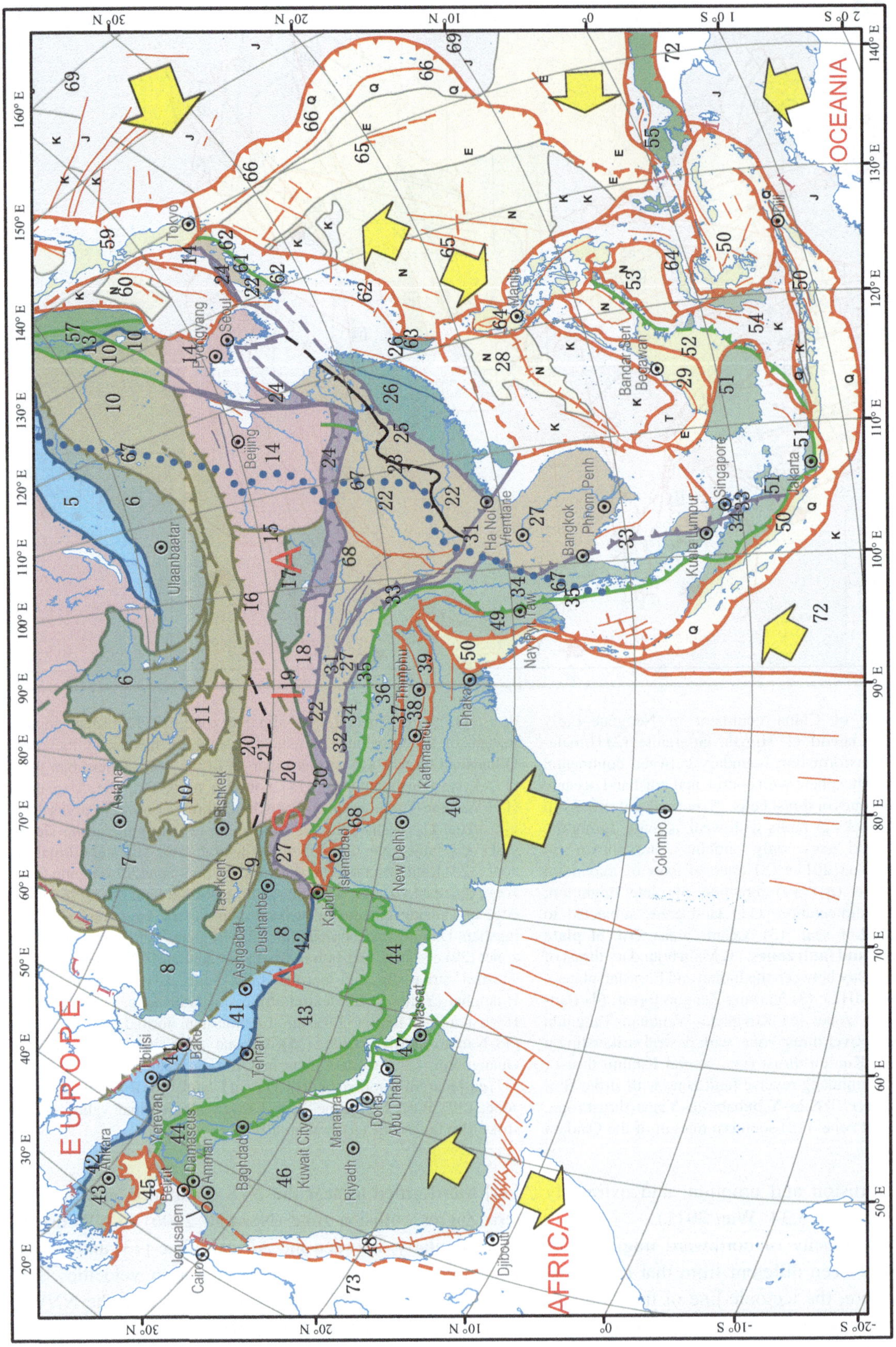

Fig. 3.34 Tectonic sketch of Asian continent in Neocene–Early Pleistocene. The numbers of tectonic unit are as same as those in the contents or text. The eastern area of the blue dotted line shows the distribution of East Asian continental crust and oceanic mantle lithosphere. The yellow arrows show the moving direction of plate. The red lines show the fault of Neocene–Early Pleistocene

of the Arabian plate, it caused the Dead Sea–Aqaba fault to show the dextral strike-slip features (Garson and Miroslav 1976; Nezafati 2006).

Based on the formation of the Pamir–Qingzang thick-type continental lithosphere, influenced by the gravitational equilibrium, there formed the "Roof of the World"—Pamir–Qingzang Plateau with an elevation more than 4000 m above sea level. It obstructs the moist air northward moving from the Indian Ocean, thus resulting in the arid desert landscapes and bad ecological environments in the Central Asia and Northwest China.

The Pamir–Qingzang Plateau was also formed by the crustal uplift. On the Asian continent, there are many longer river systems, which are all controlled by the tectono-geomorphology and shown as the unshaped or radial distribution. The higher mountains in Asia are mainly distributed at the Pamir–Qingzang Plateau and Armenia volcanic plateau. They are the source areas for many rivers. In the center of high mountains, surrounded by the Pamir–Qingzang Plateau, Altun, the eastern side of the Mongolia Plateau, Altay Mountains, Kazakhstan Hills, Turgan Plateau and southern boundary of Iran Plateau, there are widely inner continental river systems, for example the rivers of Syr, Amu, Darya, Tarim and Yili, which are all distributed in Central and Southwest Asia. Surrounded by the Verkhoyanskiy, Starov, Khr Dzhugoau, Sayan, Kazakhstan Hills and Ural Mountains, there are the Lena, Yenisey and Obe Rivers. They are all running into the Arctic Ocean. East to the Mongolia Plateau, it is the Heilongjiang System.

In Neocene–late period of Early Pleistocene, around the Pamir–Qingzang Plateau, the strong headward erosion occurred in the river systems to form the Yellow River and Yangtze River (Zhu 1989; Yang 1988; Yang and Lv 1992), Zhujiang, Red River and Lancangjiang–Mekong River systems, which belong to the Pacific Ocean drainage area. In the south of Himalaya, Hindukush and Toros Mountains, there is the Indus River, Ganga River, Yarlung Zangbo–Brahmaputra River, Nujiang–Salween River, Irrawaddy River, Euphrates River and Tigris River, which belong to Indian Ocean drainage area. Besides, there some short rivers are running into the Black Sea and Mediterranean Sea.

The density of river networks is closely related to the distribution of rainfall. In the Central and Southwest Asia, the weather is dry, and the rainfall is slight. So, the river network is sparse. However, in the East and Southeast Asia, especially in South China, the Indochina Peninsula and the Malay Islands, the weather is moist and the precipitation is abundant; the river networks are concentrated with the most abundant water.

Since Neogene, the Pacific plate had migrated further to the WNW orientation, moreover causing the Philippine Sea plate to derive the NNE–SSW trending extension under the WNW squeezing action and the South Honshu–South Shikoku–Ryukyu subduction and island arc zone [62] and Philippines–Moluccas Cenozoic subduction and island arc zone [64] to enhance their subductions (Hall and Blundell 1995).

The fault plane of the East Taiwan subduction was formed in Paleogene and became almost vertical in Neogene. Under the above-mentioned action, the East Taiwan longitudinal valley [63] is characterized by sinistral strike-slip fault and slight extension (Stephan et al. 1986). The extension of Japan Sea and South China Sea and occurrence of the oceanic crust are all products of this tectonic stage, instead of the "back-arc basin" (Tamaki et al. 1992; Yoon 2001; Liu 2002; Wan 2011) caused by the so-called subduction of the Pacific plate and the Philippine Sea plate by Hsu et al. (1988).

3.14 The Evolution in Neotectonic Period (0.78 Ma–)

The recent (since Middle Pleistocene) tectonics, or called neotectonics, was first proposed by Obruchef in 1948. He defined the "neotectonics" as the tectonic movement that caused modern terrain. Later, his student, Niglaief, revised his definition and considered that the beginning time of the neotectonics was Miocene or Oligocene. In the 1990s, Niglaief revised the definition again and recognized that it was not suitable to determine the beginning time of neotectonic period according to the formation of landform (Ding 2003, personal communication).

According to the periodic characteristics for tectonic evolution, between Neogene and Early Pleistocene Systems, there is no unconformity, yet without any tectonic event in the Asian continent. However, there are unconformities between Early and Middle Pleistocene strata, which are widely distributed in the Asian continent. Since Middle Pleistocene, the strata had kept the characteristics of continuous deposition. Thus, the author (Wan 2011) suggested that the neotectonic period should start from Middle Pleistocene (0.78 Ma).

At the neotectonic period, the oceanic plates mainly inherited the Neogene activity, and their basic migration patterns had not changed obviously, but the migration velocities and strength must change obviously, because just such small changes could not result in great influences to form the recent volcano, earthquake and other geological disasters (Figs. 3.35, 3.36). In the Asian continent, the maximum principal compression stress orientations and the stress values (usually to use the differential stress, i.e., the difference between the stress values of maximum and minimum), had undergone the long-distance effect of the surrounding oceanic plates continuing to subduct and compress.

The Russian scholars published the modern geodynamic system in the central and north Eurasia on the Internet. The green area in Fig. 3.35 shows the North Asian dynamic system, which indicates that the clockwise rotation of the crust centering on the north end of Siberia may be influenced

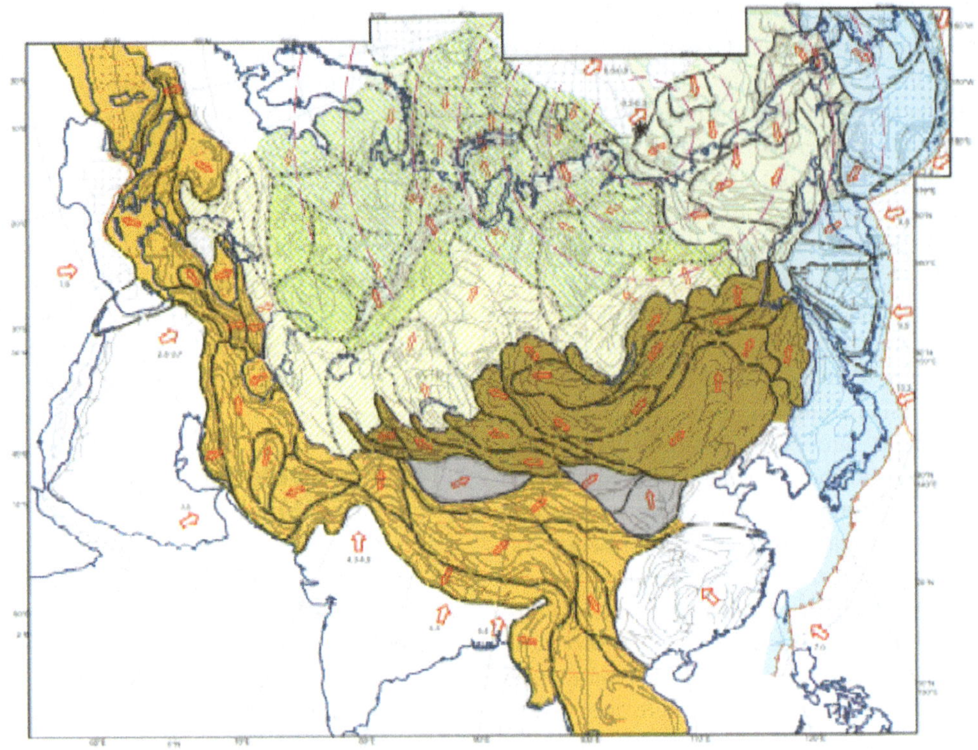

Fig. 3.35 Modern geodynamic system in the central and northern parts of Eurasia (From Russian Geodynamic Institute, on the Web). The red arrows show the recent compression or migration orientation; the red dotted lines show migration trace; the thicker red lines show the oceanic trench

by the North American plate migration. The counterclockwise and main southeastward compression occurred near the Ural Mountains and North Europe, which may be suffered by the influence of extension for North Atlantic Ocean. The shallow yellow area in Fig. 3.35 shows the main northward compression of Northern–Central Asia, and it could be influenced by the join forces for whole oceanic plates; in the northeastern part, there produced the crust's clockwise rotation and the westward moving, which could be influenced by the North American plate westward migration. The brown area in Fig. 3.35 is the Central Asian–Mongolian dynamic system, with the main northeastward compression in the eastern part and the westward compression in the western part—the phenomena may be resulted from the long-distance effect of the Indian plate northward collision. The shallow brown area in Fig. 3.35 is the South Asian dynamic system, which also must undergo the main northward migration. The white and gray regions are the relative stable areas, of which Southeast China block belongs to the northwestward migration. Due to a lot of granite intrusions, the crust is rather stable to be rare in earthquakes (Fig. 3.36).

In India, due to the widespread distribution of crystalline basement, there is rather stable and continual northward migration, but the moving velocity is as lower as only about 5 cm/y. The Arabian areas had continually migrated northeastward since Cretaceous, whose moving velocity was obviously lower than that of the Indian plate. The shallow blue region is the East Asian dynamic system, i.e., the Western Pacific trench–arc system, which had mainly westward compressed and subducted, with the velocity of 10 cm/y (Fig. 3.36). Taiwan and its adjacent areas are influenced by northwestward migration of Philippine Sea plate (Fig. 3.36), with a velocity of about 5–8 cm/y.

The central area of East Asia is suffered by the compression and subduction of the Pacific plate and Philippine Sea plate, in which the intraplate maximum principal compression stress orientation is near E–W trending. The differential stress values in the northern part of East Asia are usually between 12 and 22 MPa (Wan 2011), and in South China, South China Sea and Philippine areas, the compression orientations all change to NW trending (Fig. 3.35) and the differential stress values can reach 22–40 MPa, obviously more than that in North China (Wan 2011). The reason is that there are widespread granitic intrusions formed in the crust of South China. As a result, the greater differential stress could be accumulated in the crust. In the Malay Peninsula–Indonesian regions, influenced by NE or N–S trending continual subduction of Australian plate, the compression orientations are mainly in NNE of N–S trending.

The pre-existing faults and fractures in various directions in the Asian continent are very numerous, caused by the intraplate deformation. Thus, the trending of pre-existing fractures can be found to be similar to the orientation of

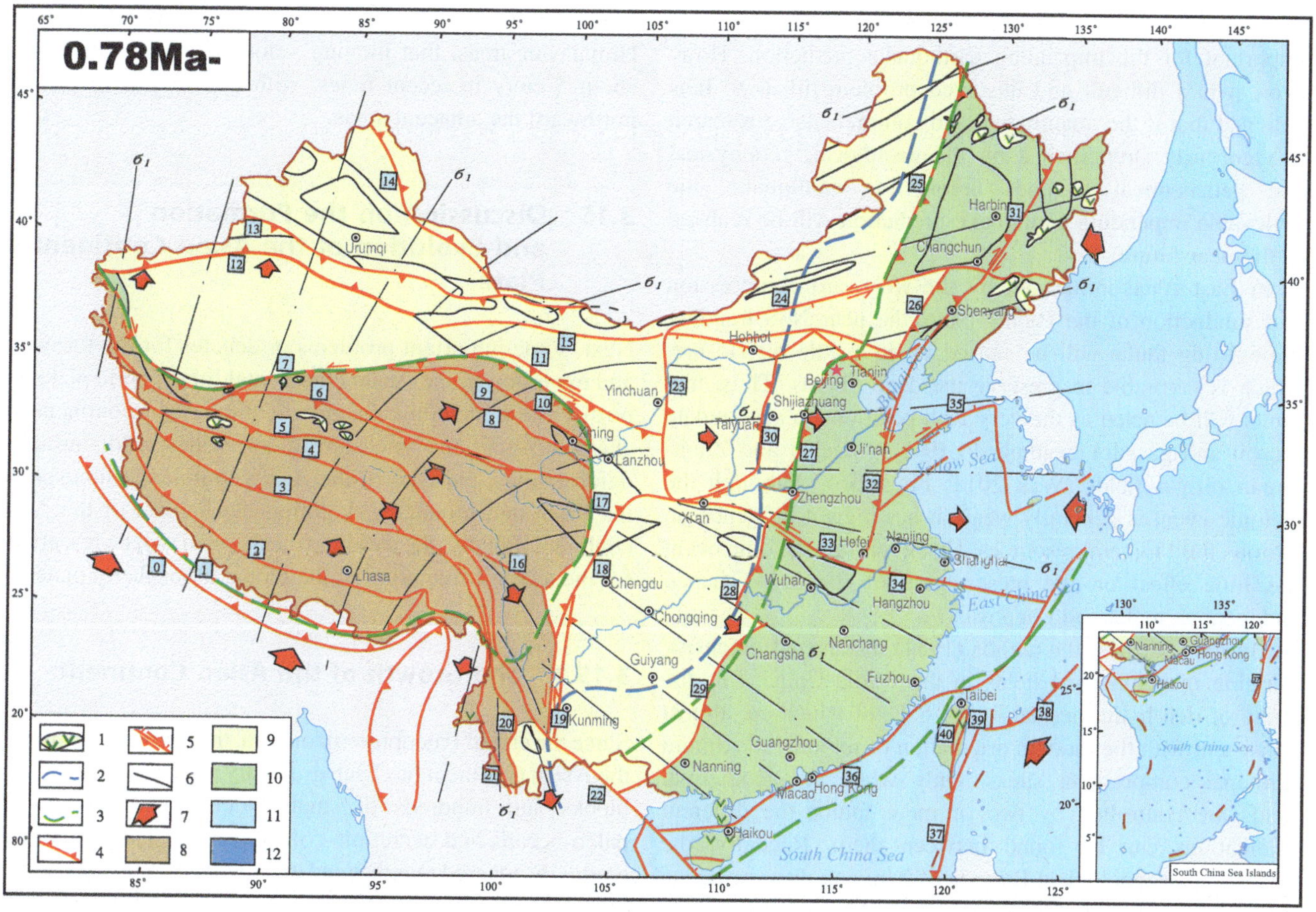

Fig. 3.36 Tectonic sketch of China continent in the neotectonic period (0.78 Ma–). **Legend**: (1) neotectonic volcanic rocks; (2) the boundary of lithosphere type, its western areas are the normal continental lithosphere, eastern areas are the continental crust and oceanic lithosphere mantle type; (3) the Bouguer gravity anomaly gradient zones (the normal lithosphere type in the middle; thicker type lithosphere near the Qingzang–Pamir areas; the thinner lithosphere with continental crust and oceanic mantle in the east); (4) modern plate boundaries or large-scale active reverse faults; (5) active strike-slip or normal faults; (6) maximum principal compressive stress trace (according to earthquake fault plane interpretation); (7) directions of plate movement; (8) Qinghai–Tibet Plateau, mean height of 4000 m above sea level; (9) Inner Mongolia–Loess plateau–Yunnan–Guizhou plateau, mean height of 2000 m above sea level; (10) lower mountains, hills and plains in East China, mean height of 1000–0 m above sea level; (11) shallow marine area; (12) oceans. **No. of boundary of plates, collision zones and fault belts**: (0) Himalayan main boundary thrust (MBT); (1) Himalayan main central thrust (MCT); (2) the Yarlung Zangbo thrust; (3) Bangongco–Nujiang thrust belt; (4) Shuanghu–Tanggula reversed fault zone; (5) Lazhulong–Jinshajiang reversed fault zone; (6) Kangxiwa–Kunlun mountain reversed fault zone; (7) southern edge of Tarim (Keziletao–Kuyake–Altun) thrust with sinistral displacement; (8) southern edge of Junulshan–Qinghaihu (northern edge of Qaidam) reversed fault; (9) southern edge of mid-Qilian overthrust; (10) north edge of north Qilian overthrust; (11) Longshoushan reversed fault; (12) Korla–Wuqia thrust; (13) Yilin Habirga–Yagan thrust; (14) Ertis–Kramali thrust; (15) northern edge of Alxa reversed fault; (16) Yidun–Litang thrust; (17) eastern edge of Daxueshan reversed fault; (18) Jiajinshan reversed fault; (19) Xiaojiang normal-sinistral strike-slip fault; (20) Lancang River normal fault zone; (21) Luxi–Menglian normal fault; (22) Red River sinistral strike-slip fault; (23) Liupanshan–Helanshan reversed fault; (24) Xar Moron sinistral strike-slip normal fault; (25) eastern edge of Dahingganling thrust; (26) Beipiao–Jianchang reversed fault; (27) eastern edge of Taihangshan thrust; (28) eastern edge of Funiushan–Wulingshan reversed fault; (29) Xuefengshan–Damingshan reversed fault; (30) Fenhe–Weihe reversed fault zone (dextral strike-slip at the southern end); (31) Yilan–Yitong dextral strike-slip reversed fault; (32) Tancheng–Lujiang dextral strike-slip reversed fault; (33) Baoji–Luonan–Fangcheng sinistral strike-slip normal fault; (34) Fanchang–Ningbo hidden active normal fault zone; (35) Zhucheng–Rongcheng sinistral strike-slip normal fault; (36) coastal Fujian–Guangdong reversed fault; (37) West Philippine–East Taiwan sinistral strike-slip fault zone; (38) Ryukyu subduction belt; (39) East Taiwan longitudinal valley (sinistral strike-slip); (40) West Yushan thrust

regional maximum principal compression stress, and those fractures show the extension-shear features with best permeability. It is useful for the fluids to migrate or preserve in the crust, including the oil-gas, basement fracture water, geothermal fluid, coalbed methane or shale gas, etc. Along and near those fractures, the strength of whole rocks is lower to induce easily the earthquakes, volcanoes and many geological disasters, such as landslide, rock burst, shaft wall failure, tube deformation and crack, gas blast, shaft submergence and hot damage under the shaft, etc.

The research on the modern tectonic stress field is very important for the impending earthquake prediction. However, it is a difficult and unsolved problem till now. It is believed that if the quantitative and comprehensive research is vigorously strengthened by the geological, geophysical and geochemical methods before the earthquake, the believable impending earthquake prediction will be realized in the near future.

In East Asia, influenced by the westward compression and subduction of the Pacific plate, the near N–S trending pre-existing faults will be caused to be a little bit closure, which is favorable to preserve the deep oil-gas. Thus, that faults will be acted as the very well positions to accumulate the oil and gas, for example PL19-3, Yingehai and Zhongyuan oil-gas fields (Wan 2011; Fig. 3.36). Although the tectonic event is generally weak, it has a great influence on various fluid mineral resources. However, this tectonic event has little effect on the preservation of the early shaped endogenous metal solid deposits.

By the statistics, the author (1984) discovered extensive trending of geothermal fields in the whole China, i.e., the strike of rich hydrothermal faults, 83% of which are almost consistent with the modern orientation of regional maximum principal compression stress. Only when the geothermal fields are controlled by two or more faults, the different orientations can be found between them. Recently, the Second Hydrogeological Brigade of Xinjiang discovered the high-temperature geothermal field in Tashkorgan on the east side of Pamir Plateau, whose water temperature in the bottom of well is higher than 160 °C. In that area, the ancient (at the end of Paleozoic) regional compression orientation was near E–W trending; however, the modern regional compression orientation is near N–S trending (Fig. 3.36). Controlled by the modern tectonic stress field, the fault zones are enriched with the high-temperature fluids and are all in the N–S direction, to be similar to the recent maximum principle compressive stress orientation.

The different characteristics of the various directions of the lithospheric plate in the Asian continent during the neotectonic period are obviously affected by the long-distance effects of subduction or collision for adjacent plates (Fig. 3.36). In East Asia, it is mainly suffered from the influences of westward subduction of the Pacific plate; in the southeastern part, it is influenced by the northwestward subduction and compression of the Philippine Sea plate; in the southern part, it is influenced by the northward collision of the Indian plate and slow subduction of the Australian plate; in the northwestern part, it is influenced by the Atlantic Ocean and Arctic Ocean extension. In the ancient geological periods, the migration velocities of the Asian continental lithospheric plate were about one-tenth of that of the surrounding oceanic plates, usually only several millimeters per year. Merely near the Himalayan areas, that moving velocities have been kept at about 5 cm/y in recent times, with the obvious decrement northward the adjacent areas.

3.15 Discussion on the Formation and Evolution of the Asian Continental Plate

Next, some important problems or debates for the formation and evolution of the Asian continental lithospheric plate will be discussed, including the growth of the Asian continent, the characteristics and mechanism of widespread intraplate deformations, the type of the Asian continental lithosphere, the basin and mountain evolution mechanism of the Asian continent. Finally, the dynamic mechanism will be researched for the tectonic evolution of the global lithospheric plates.

3.15.1 The Growth of the Asian Continent

Since the end of Paleoproterozoic Era, the lithospheric plate of the Asian continent has been gradually assembled by 28 large blocks and hundreds of small blocks distributed in the paleo-oceans and accretion–collision zone. They have been gradually cohered together and experienced 14 tectonic events with the different active characteristics, subductions, collisions and intraplate deformations. The dynamic mechanism of each tectonic event is different, as well as the migration orientations, moving velocities are different from its adjacent plates. It would result in different structural styles, strengths and differential stress values, so the Asian continental plate shows a very complex and diverse tectonic pattern. Those tectonic characteristics are very special and rare for the tectonic evolution of global continental plates. So, it is very difficult and interesting to research those special phenomena.

Since Late Paleoproterozoic, why have the Asian blocks experienced 14 tectonic events of subductions, collisions and intraplate deformations to form the main part of the biggest Eurasian plate, not to be broken or crushed yet? What is the reason?

At first, let us explore the process of ocean–continent subduction on how to affect the growth of continent. For the density of oceanic lithospheric plates being significantly higher than that of the continental ones, when they are converged, the oceanic plates must subduct under the continental plates, with the velocity of several or tens of centimeters per year. The strain rates are very slow, with the high-temperature and high-pressure rheological behavior in the deep, not to impact fast. The subduction between oceanic and continental plates could induce medium–deep source earthquakes and volcanic eruptions to form some fractures in a short time. But that main deformation patterns are still attributed as the rheological behavior, it is impossible to cause the overall destruction and cracking of the continental lithospheric plate; on the contrary, it is possible to increase its strength and stability. Some Chinese researchers (Deng et al. 1992, 1996; Wu et al. 2008; Zhu et al. 2008; Zhang 2009; Zhu et al. 2011, 2012) considered that the Pacific plate subduction could cause the destruction and fragmentation of the East Asian continent. It seems not to coincide with the facts.

According to the global results of deep seismic tomography, the African plate subducted underneath the European continental plate (Cavazza et al. 2004), the Indian–Australian plate subducted underneath the Asian plate (Hall et al. 2011), and the Pacific plate subducted underneath the Asian plate (Zhao et al. 2007; Zhao and Liu 2010; Fig. 2.50); their maximum subduction depths are all limited in the depth of 400–670 km, i.e., to transition layer of middle mantle. When the oceanic plates subduct to the deep, the temperature and density of the subduction oceanic lithosphere will be likely to reach as same as those of the middle mantle. It is difficult to distinguish the difference between the two by seismic data. But some researchers recognized that the Farallon plate could subduct into the depth of about 2885 km, near the boundary between mantle and core, or reached the lower mantle in the depth of about 1600 km. However, the above seismic tomography results are not sure, and there are some different opinions (Grand et al. 1997; Sigloch et al. 2008).

It seems that the oceanic plate, subducting underneath the continental plate, may be favorable to increase the opportunity for the continental lithosphere formation, but not to be destroyed and cracked. Furthermore, it has never found any one case of continent being destroyed and cracked by the oceanic plate subduction to the continental plate.

As to the continental–continental collision, could it make itself destroy and crack? Depending on the existing paleo-magnetic and rock deformation data, because the thickness of continent is obviously greater than that of the oceanic plate, their convergence will be more difficult without the oceanic water as the "lubricant." The convergent velocity is significantly less than that of subduction, usually less than 6 cm/y; for example, the collision speed of the Himalayan collision zone since Paleogene was 6–5 cm/y, and the convergence velocity of the Qinling–Dabie collision zone in Triassic was only between 2 and 1 cm/y. Its strain rate at the time of convergence was as low as $n \times 10^{-15}$/s to $n \times 10^{-16}$/s. It means that the ductile deformation occurs in the deep and is attributed to the rheological behavior, rather than the fast and violent collision, so the continent will never break and crack.

Of course, the rock block must produce many fractures (faults, fissures or joints) in the collision zone. In the collision zones and their sides, the rocks are located in the relative closed system. In the deep, at the Moho discontinuity or middle crust with the low seismic velocity and high electric conduction, it is easy to form ductile deformations, the partial pressure decreasing and temperature increasing. When the temperature increases beyond the rock's solidus, the rock will be partially melted into magma and many ultra-critical fluids, to migrate upward. In the process of uplifting and expanding its volume, the magma consumes energy and the temperature gradually decreases, so that it condenses into intrusive rocks in the Earth's crust (especially in the fault) or ejects to the surface to form volcanic rocks. In a word, it is preferentially filled into the structural fracture to consolidate the broken rock. In the fault zones, all the deep ultra-critical fluids will be gradually condensed and crystallized during the uplifting process, thus to get the crushed rocks as a consolidation body.

In addition, the rock deformations, beneath the depth of 5–10 km, are all of the ductility to form the different metamorphic rocks and almost to cause the break phenomena to be disappeared, thus to increase the degree of rock integration. To sum up, under the condition of very low strain rate, the compression, collision and tectonic break are just the local and temporary phenomena. The magma, fluid, ductile deformation and metamorphism will result in the rock to be consolidated, but not to be crushed or cracked. Thus, the continent–continent collisions not only can make the breaking, destroying or splitting, but also can accelerate many small blocks to be cohered. In short, the continent–continent collision is the main dynamic mechanism for the continental lithospheric growth.

It concludes that the Asian continental blocks have experienced 14 events of convergences, collisions and intraplate deformations (Table 3.1). The above actions are influenced by 14 tectonic events and the intraplate deformations, which result in the tectonics of Asian continental lithospheric plate to be more and more complex.

3.15.2 The Widespread Intraplate Deformation

On the research of Asian continental lithosphere, the formation mechanism of widespread intraplate deformation is also an important and very interesting project. In the Asian continental lithosphere, there exists the most widespread intraplate deformation, which extension length, no matter in N–S or E–W trending, is more than 3 km. Why is there such a large range of intraplate deformations in the Asian continent, accompanied by strong tectonic-magmatic activities? What is the formation mechanism of them?

Table 3.1 Convergent or compressional orientations in the Asian continent at each tectonic period

No.	Tectonic period	Convergent orientation (according to the modern magnetism)
1	Late Paleoproterozoic (1800–1600 Ma)	Forming the Columbia supercontinent, with different orientation convergences
2	Early–Middle Mesoproterozoic (1600–1200 Ma)	Cracking of the Columbia supercontinent in different orientations
3	Late Mesoproterozoic–Early Neoproterozoic (1200–850 Ma)	Multi-directional convergence of the Rodinia supercontinent
4	Middle Neoproterozoic (~850 Ma)	Convergence and collision for the North–South Tarim and North–South Yangtze plates
5	Late Neoproterozoic–Early Cambrian (635–510 Ma)	Multi-directional convergence for the Pan-African Tectonic Event on the Gondwana
6	Late period of Early Paleozoic (~397 Ma)	Near N–S trending collision in the Central Asia–Mongolia, forming the Xiyu plate, Cathaysian plate E–W trending migration
7	Early period of Late Paleozoic (385–323 Ma)	Near N–S trending convergence and collision in the Central Asia–Mongolia and Tianshan
8	Late period of Late Paleozoic (323–260 Ma)	E–W trending shortening for Ural collision; the Mongolia–Western Tianshan, Alxa and Sino–Korean plates' collision because of long-distance effect; the Hingganling and Sino–Korean plate collision in N–S trending
9	Late period of Triassic (220–200 Ma)	Near N–S trending shortening in the Central and South Asian continent (according to modern magnetic orientation)
10	Late Jurassic–early period of Early Cretaceous (170–135 Ma)	WSW trending compression of the Siberian plate, near N–S migration for the Central Asia, WNW–ESE trending shortening for the East Asian crust, the counterclockwise rotation for the Asian continental crust
11	Middle period of Early Cretaceous—the end of Paleocene (135–56 Ma)	NNE–NE–ENE trending shortening for the East Asian continent
12	The end of Oligocene (~23 Ma)	Near E–W trending shortening for the East Asian continent and near N–S trending compression for the South Asian continent
13	Neocene–Early Pleistocene (23–0.78 Ma)	N–S trending shortening in the whole Asian continent, especially strong in the Qinghai–Tibet–Pamir areas
14	Since Middle Pleistocene (0.78 Ma–)	Clockwise rotation in North Asia, near E–W trending with radial compression in East Asia and near N–S shortening in South Asia

During the oceanic plate subducting underneath the continental plate, how many influences could be produced in the continental lithosphere? It has been a controversial topic for all along. Globally, after the uniform crystalline basement was formed in the central part of many great plates, for example the North American plate, the intraplate deformation would be very weak, with the differential stress values less than 20 MPa (Ben et al. 1997; Wan 2011). So, the sedimentary cover strata are mainly shown at almost horizontal attitude. In some areas, for example in the South African blocks, the structural joints have been never formed (Qin et al. 2000), except for some diagenetic joints. These phenomena are easy to be found in the horizontal strata in the North American plate or South African plate. So when the Plate Tectonic Theory was first proposed in the 1960s and 1970s, many scholars considered that the plate must be a "rigid body," without so many intraplate deformations, only a few intraplate deformations and magmatism in the boundary area of the plate (Le Pichon et al. 1973; Press and Siever 1974; Turcotte and Schubet 1982; Pollard and Aydin 1988).

On the west side of the North American plate, since Mesozoic, the Farallon plate continued to eastward subduct underneath the North American plate at an intermediate angle, to form the intraplate deformation. In those areas, the intraplate deformation is only limited in the narrow zone with the 200 km width (Ben et al. 1997). In the widespread eastern area of North American plate, since Paleozoic, there were almost no intraplate deformations with the horizontal attitude (The Geological Society of America 1989; Schmidt et al. 1993). Thus, many researchers confused on the widespread intraplate deformation and related magmatism, and thought those phenomena were not related to the plate horizontal moving and expected to find some different reasons, for example the intra-continental orogeny or someone vertical dynamics (Hsu et al. 1988; Ge 1989; Zhao 1986; Cui 1999; Song 1999; Neves and Mariano 2004; Shu et al. 2006, 2008; Shao et al. 2007).

Faced with the characteristics of stronger intraplate deformation in the Asian continent, many Chinese geologists had named those special tectonic phenomena as: platform activation or multi-cycle structure (Huang 1945, 1984; Huang et al. 1965, 1977), "Diwa" (its meaning is depression block; Chen 1960, 1978, 1998), continental margin activation (Ren et al. 1980, 1990, 2000), platform folding zone (firstly named by Ma et al. 1961; Regional Geological Group, Beijing College of Geology 1963), intra-continental orogeny (Hsu et al. 1988; Ge 1989; Cui 1999; Song 1999; Wu and Zhang 1999), etc. From the literal understanding of these terms, it seems that there are different meanings; in fact, they all show the same geological phenomenon—intraplate deformation. They all discussed the strong intraplate deformations from different perspectives; i.e., since Mesoproterozoic, most of the ancient Asian blocks had experienced multiple and rather strong intraplate deformations with the different orientations.

The author summarizes the reasons for the intraplate deformation in China as follows:

(a) The China continental plate initially had not an uniform or great crystalline basement, which was composed of 28 small plates and tens of micro-blocks. It is just like a glass plate bonded by a lot of broken glass pieces; of course, its strength is always much lower than that of a whole glass plate. It is also like in the experiment of rock mechanics; i.e., the strength of jointed rocks will be obviously lower than that of the whole rock mass. Therefore, the total strength of the continent that is composed of many small blocks and undergone many tectonic events must be lower and easy to form the plastic deformations and then to produce the weak tectonic stability. Especially, it will be easy to form intraplate deformations by the tectonics in the late periods. Its characteristics are so different from those of the North American plate and South African plate.

As for those relative large blocks with the great and complete crystalline basement in the Asian continent, the intraplate deformations of their upper strata are notably weaker, such as in the central part of Siberian plate [1], Turan–Karakum plate [8], the Ordos block in the Sino–Korean plate [14] and the Sichuan basin in Yangtze plate [22].

(b) The thickness of sedimentary cover is not uniform, and the rock strength of upper part for the continental lithosphere is weaker. In general, the strength of metamorphic crystalline basement is stronger, but its sedimentary cover is weaker. The thicker the sedimentary cover is, the easier it is to deform. The reason is that the hardness grade of the sedimentary cover is lower with the near horizontal beddings, and the strengths of each layer are different.

Statistical results of the whole sedimentary cover for each period in the Asian continent show that the average thickness of sedimentary systems since Middle Proterozoic or Neoproterozoic is about 16,985 m. However in the stronger tectonic zones or collision zones, the average thickness of sedimentary system is 42,036 m; in the Altun–Kunlun zone, the maximum thickness of sedimentary system reaches 61,712 m; in the Western Qinling and Qilian zone, the thickness is between 55,000 and 58,000 m (Wan 2011). In the huge sedimentary stratum areas, it not only could cause the strong deformation at the collision period, but also could produce the later intraplate deformation. However in the widespread exposed metamorphic crystalline basement areas, the sedimentary cover is only several hundred meters. Because the crystalline basement could be suffered by the great tectonic stress, the rock deformation in the upper covered sedimentary systems is very weak, such as in the East Siberian plate, Indian plate, West Shandong and Ordos basin in the Sino–Korean plate. These phenomena are seldom found in the Asian continent, which areas are less than 10% of the total.

(c) Due to the multi-period, different collisions and convergences with different velocities and orientations, since Mesoproterozoic, the Asian continental lithosphere had undergone 14 tectonic events to form 38 collision zones, subduction zones or the strike-slip fault zones. For the long-distance effects, there formed multi-period and different directional intraplate deformations, with the complex appearances and styles, and the related tectonic magmatisms (Table 3.1).

The strong and multi-directional intraplate deformations in the Asian continental plate are the rare tectonic phenomenon in the global lithospheric plates. The reason is that the Asian continental blocks had been located in the

Paleo-Tethys Ocean for a long time and had suffered the migration, convergence and collision from many oceanic or continental plates. Their long-distance effects could result in multi-period and multi-directional intraplate deformations and related tectonic magmatism–metamorphism.

As to the Gondwana tectonic domain, from the perspective of modern magnetic orientation, since Triassic, the convergent and compressional orientations had been rather stable, mainly north toward and NNE toward migrating, and its strength had some changes. In other tectonic domains, the migration velocities and orientations had changed a lot. In terms of the strength of intraplate deformation, it is usually stronger near the plate boundary and gradually decreases toward the intra-continental plate (Fig. 3.37). The tectonic stress values and the span of action time also decrease from the boundary to the inside.

Depending on the intraplate stress value change or migration at the stronger deformation period, the dynamic mechanism of the intraplate deformation can be preliminarily judged. Based on the collected 176 differential tectonic stress values of the Early Cretaceous–Paleocene, they are about 180 MPa in Ali areas, Xizang; about 140 MPa at Central China (Qinling); and only 100–90 MPa in Northeast China. With the transformation of the tectonic stress field from SW to NE trending, the stress value becomes gradually weaker, and the differential stress value decreases by 1.8 MPa/100 km extending forward (Fig. 3.37; Wan 2011). Thus, the author considers that the intraplate deformation at the middle period of Early Cretaceous was caused by the long-distance effect from the Indian plate's NEE direction subduction and compression.

Another noteworthy fact concerning the long-distance effect of plate migration is that: After the migration orientation of the Pacific plate changed from NNW to WNW trending in Eocene–Oligocene, the stronger rock deformation zone caused by the tectonic action would gradually migrate westward, and the stronger deformation periods would gradually become younger (see Sect. 3.12). For the orientation changing period of Imperial–Hawaiian Islands, it was at 36 Ma near the Midway Island (Raymond et al. 2000); i.e., from this period on, the Pacific plate had begun to change to the WNW trending, with the migration velocity about 10.6 cm/y. The Western Pacific island arc begun to be influenced at about 30 Ma, which corresponding tectonic movement is called as "Takachiho Movement" in Japan (Maruyama and Seno 1986) and "Puli Movement" in Taiwan (Zhang 1984; Chen 1984; Stephan et al. 1986); for East China, it was at 23 Ma (called as North China Tectonic Event; Tang 1979); for West Yunnan, it was about 15 Ma (He and He 1993); for West Xizang and West Xinjiang, it was about 10 Ma (Guo et al. 1991; Zheng 1996, personal communication; Wang et al. 2006). It means that the stronger rock deformations in different areas would not be formed at the same time, with a gradual process of westward delivery. According to the above data, the author (Wan 2011) estimated that the distance of transformation of WNW trending strong deformation zone would be up to 13,000 km, and the migration of strong deformation zone would last for 20 million years, with the velocity about 65 cm/y. Of course, due to that the data on the isotopic age, tectonic stress and migration of structural deformations in various periods and regions are limited, and the long-distance effects of the plate-related effects have yet to be further demonstrated after the accumulation of more data.

Some scholars thought that the intraplate deformation and magmatism were far away from the plate boundary and were originated from the mantle plume uplift. In addition, they considered that there were six mantle plumes in East China in Cenozoic, which controlled six radial dyke swarms, and all the rocks belonged to the basalts with mantle-derived

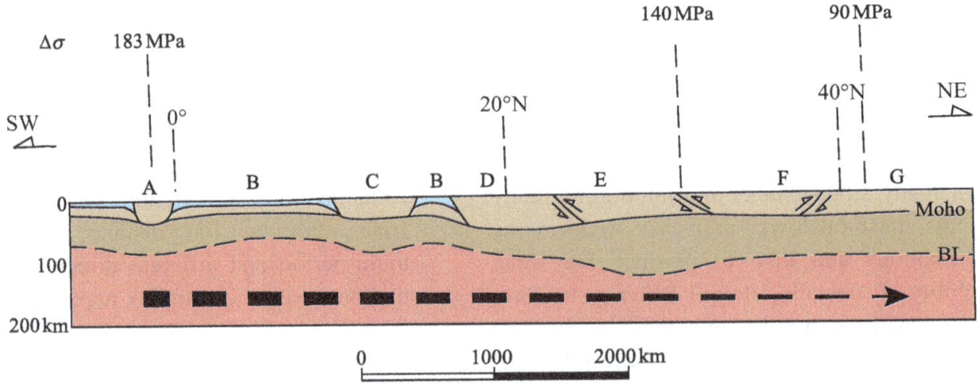

Fig. 3.37 Differential stress value changes resulted from the long-distance effect of collision in the Asian continental plate. (A) Ali area, Himalaya block; (B) Neo-Tethys Ocean; (C) Gangdise block; (D) former Southern Qiangtang and Indosinian block; (E) former Yangtze plate; (F) former Sino–Korean plate; (G) former Central Asia–Mongolia collision zone; (D–G) the part merged in the Eurasian plate in Cretaceous; BL. base of lithosphere; Δσ. differential stress value. The long black arrow shows the tectonic stress value decreasing from SW to NE (After Wan 2011)

xenoliths (Deng et al. 1992, 1996). It is a pity that they ignored the formation ages of those dyke swarms. In fact, the WNW or E–W trending dyke swarms are mainly the tholeiites, being formed in Paleogene (52–23 Ma), and the NNE or near N–S trending dyke swarms are mainly the alkali basalts, being formed in Neogene–Early Pleistocene (23–0.78 Ma). Their formation periods are obviously different, and the formation periods are differences as 20 million years between the tectonic and magmatism. The magmatism is controlled by the different intraplate tectonic stresses, without any evidences of mantle plume absolutely. Oppositely, in Paleogene and Neogene, there were the abundant evidences of long-distance effects of horizontal compression, subduction or collision that multi-stage tectonic magnetism has experienced (Wan 2011).

The intraplate deformations in the Asian continent are not only shown as the rock deformation, the stratum folding and faulting on the Earth surface, but also shown as the tectonic detachments in the lithosphere. Zhang (1984) pointed out that in the lithosphere there were many tectonic detachments, such as at the bottom of sedimentary covers, low seismic velocity and high electric conductivity layer near the middle crust and Moho discontinuity. Through the tectonic action, all the tectonic detachments will be produced (Fig. 3.38). The detachments in the lithosphere are shown obviously in the recent earthquakes. Ma (1989), Xue and Huang (1989) and Deng et al. (2007) recognized that more than 90% earthquakes at Ms.6.0 or >6.0 were located at the 10–25 km depths, especially in the low-velocity and high-conductivity layers of the middle crust and near the Moho discontinuity. On the Qinghai–Tibet Plateau, the earthquakes usually occur in the crust and its bottom, i.e., less than 70 km. Only in Pamir, some earthquakes are originated from the depth of 100–280 km.

The author and his colleagues (Wan et al. 2008) collected the data on the originated depths of magma source areas in East China. The results indicate that the distributions of magma source areas and tectonic detachments during the Yanshanian (200–135 Ma) and Sichuanian (135–70 Ma) periods are very similar. According to the collected and calculated pressure, temperature and depth for the magma source areas, most of them (764 magma source areas) are distributed at the intersections among the middle crust, Moho discontinuity and regional branch faults, and only a few (33 magma source areas) are scattered near the bottom of lithosphere.

Influenced by the Pacific plate subduction, on the east side of the Asian continent, there is a Benioff zone with the intermediate subduction angle, which trench is 200 km away from the east to Japan island arc. The China continent is influenced by that zone, but only in the boundary area of Northeast China, the hypocenters are found at the depth of 500–590 km; in most areas of East China, there is not any deep seismic focus found in the deep. Toward east to the Taiwan Island, the Philippine Sea plate subducted westward in Paleogene. However, since Neogene, the Benioff zone had changed to the high-angle attitude, with the sinistral strike-slip features, i.e., without any subduction. On the Taiwan Island, the hypocenter depths for most of earthquakes are less than 70 km, and in the northeastward and southeastward oceanic areas, there are the meso-hypocenters with the depth of 100–280 km. It means that the seismic faults occur in the mantle, beneath the lithosphere (Xiu et al. 1989). From the Taiwan Island to Central Fujian, the hypocenter depth becomes gradually shallower, from 70 to 5 km, which is a continental listric fault zone (Wan and Chu 1987).

To sum up, it can be seen that the tectonic detachments of middle crust and Moho discontinuity in the Asian continental lithosphere are very important. Near the southern boundary of the Asian continent—Pamir, the tectonic actions and earthquakes have been revealed in the lithospheric basement and asthenosphere.

Gao et al. (2011a, b, 2013) conducted researches on seismic reflection profiles of the Kunlun Mountains in the eastern margin of the Qinghai–Tibet Plateau and Gao et al. (2013) in the Western Qinling Mountains and the eastern

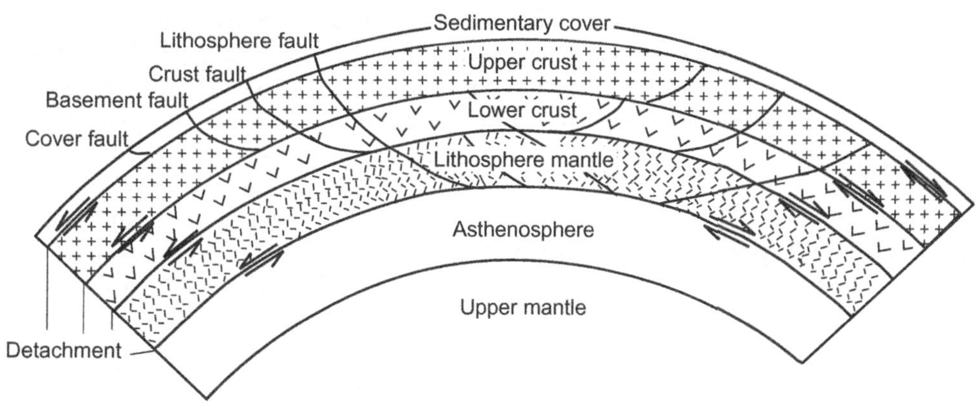

Fig. 3.38 Distribution of tectonic detachments and faults in the continental lithosphere (After Zhang 1984)

margin of Qinghai–Tibet Plateau. Dong et al. (2014) completed the study of deep seismic reflection profiles in the Sichuan basin and completed a three-dimensional visual display of seismic reflection profiles of Lujiang–Zongyang areas. Zhang et al. (2009) studied deep seismic reflection profiles in North China. Their results manifested clearly that the structural deformation styles up and below the Moho discontinuity are extremely different in China and that the Moho discontinuity is an important tectonic detachment in the continental lithosphere.

According to the Gravitational Equilibrium Theory, it infers that "mountain root" would exist in the mountain area. However, due to the tectonic detachments near the Moho discontinuity (about 40–50 km depth) in the Asian continental lithosphere, the "mountain roots" are all broken. Zhao et al. (2014; Fig. 2.20) researched the deep structures of the East Kunlun and also discovered that there was no longer "mountain root," but formed the tectonic detachments near the Moho discontinuity. Wang et al. (2011) researched the thrust tectonic styles in the depth of the Xuefeng Mountains by the data of deep seismic reflex sections and found the obvious tectonic detachments near the Moho discontinuity, as well as the extreme rock deformations.

According to the above data, the rock deformations in the Asian continent are not only shown as the faults and folds near the Earth surface, but also shown as the inner tectonic detachments in the lithosphere. So according to the recent data of original depth of magmatism, seismic reflection sections and the intraplate earthquakes, the tectonic detachments in the continental plate are mainly formed near the Moho discontinuity and middle crust. The detachments at the bottom of lithosphere are not very obvious, but relatively weak. This is an important characteristic for intraplate deformations in the Asian continental plate.

At the initial developing period of Plate Tectonic Theory, the main detachments were thought to occur only beneath the bottom of the whole oceanic and continental lithosphere plates, i.e., in the asthenosphere. Till now, the whole Asian continental plate has not been surveyed by the deep seismic reflex sections; therefore, the distributions of tectonic detachments in the Asian continental plate are unknown. It can be believed that the detailed and believable results on the tectonic detachments at the bottom of the continental lithosphere or inner continental lithosphere must be obtained by the deep research.

3.15.3 The Types of Asian Continental Lithosphere

In the research on the Asian continental lithosphere, its type is another special problem to be paid attention to. According to Wan et al. (2016), besides the common continental lithospheric type (with the thickness of 100–150 km) in the Asian continent, there are two special lithospheric types, i.e., the continental crust and oceanic mantle–continental lithosphere, with a total thickness of 70–80 km, and thicker type of continental lithosphere, with a thickness of 160–200 km (Cai et al. 2002). The continental crust and oceanic mantle–continental lithosphere are actually a transitional type between continental lithosphere and oceanic lithosphere.

In the recent geological research, most of researchers always recognize that the East China block is neither a typical platform (i.e., orogenic belt) nor a typical continental lithospheric plate (i.e., collision zone). The East China block is characterized by the uniform crystallization basement, which was formed at the end of Paleoproterozoic for the Sino–Korean plate, in Neoproterozoic for the Yangtze plate, at the end of Early Paleozoic for the Cathaysian plate (Wan 2011). Those blocks became the typical stable plates before the end of Paleozoic, but the strong tectono-magmatism occurred since Jurassic. Thus, based on the characteristics of tectonic unit, the East China block had been called as many names by some famous researchers, for example, the active or para-platform (Huang 1945, 1960), "Diwa" (Chen 1960, 1998), active continental margin (Ren et al. 1980, 1990, 2000), folding zone in platform (Ma et al. 1961), intraplate orogenic belt or intra-continental orogenic zone (Ge 1989; Zhao 1986; Cui 1999; Song 1999). In recent years, many researchers have attempted to study "the destruction of North China Craton" (Wu et al. 2008; Xu et al. 2008; Zhang et al. 2009; Zhu et al. 2011). Due to lack of the data on deep geology, geochemistry and geophysics, many geological phenomena have not been fully understood in the past sixty years, thus to result in a lot of debates and many hypotheses.

The special lithospheric tectonic was first recognized by Melcher et al. (2002), and they discovered that the continental crust could thrust and move onto the oceanic mantle between the European continent and African continent during Paleogene, and called the mantle as "pre-oceanic subcontinent mantle." Their results have greatly inspired the author.

This monography will discuss the thickness change, tectonic characteristics and transitional periods of lithospheric type and their mechanisms. In viewpoints that above problems have become the discussion hot spots for geosciences, the author just proposes his own opinion for discussion.

As to the thinner lithosphere in East China, by the seismic tomography, Xin Jisan et al. (Institute of Geophysics and Geochemistry, Bureau of Shanxi Geology and Exploration) first discovered the lithospheric thickness to be obviously thinner (about 80 km) to the east of Taihang Mountains (unpublished paper); their map has been quoted by many researchers. Fan and Menzies (1992) first published the paper and pointed out clearly that based on the formation

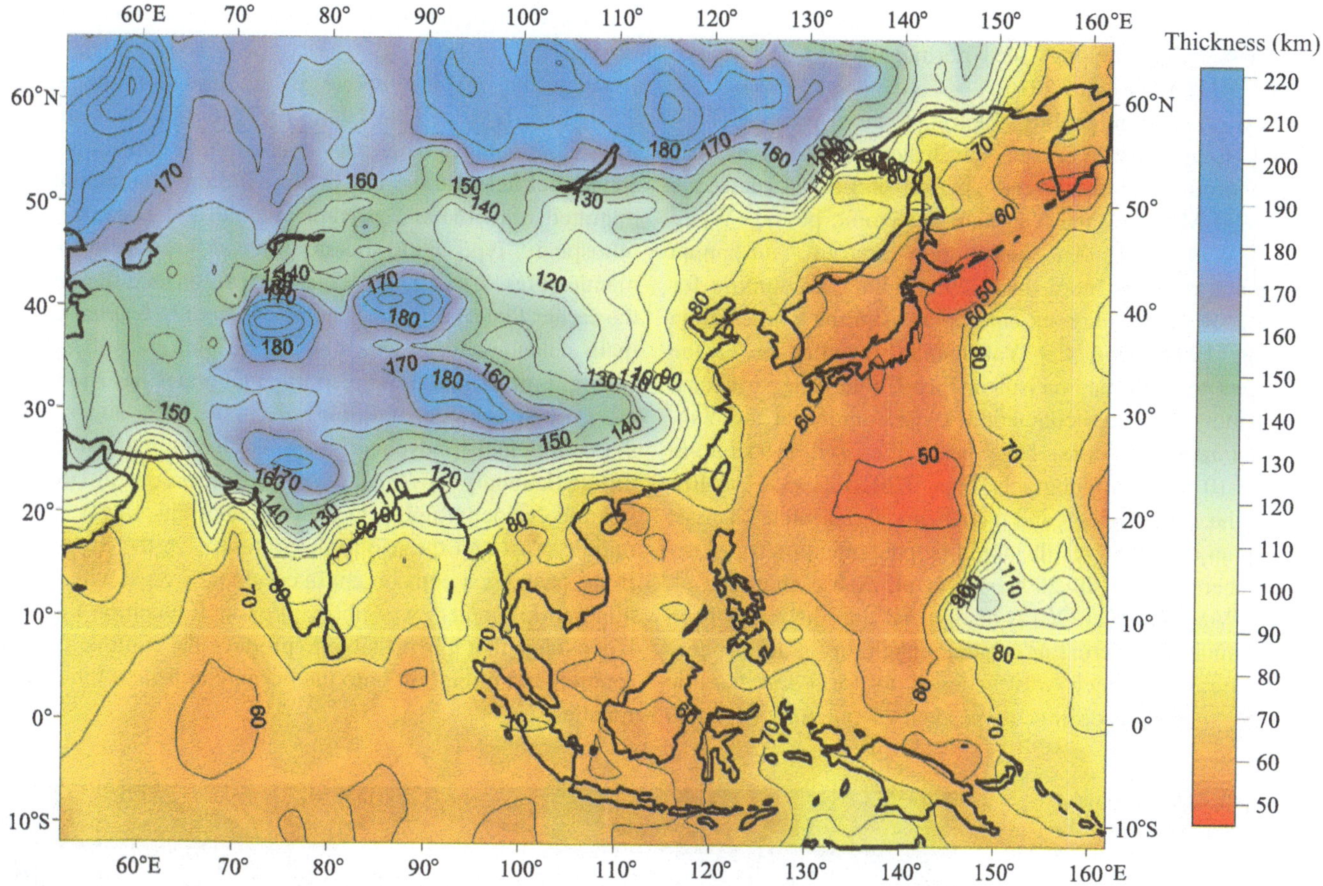

Fig. 3.39 Lithospheric thickness of East Asia–Western Pacific Ocean (After Cai et al. 2002)

depth of kimberlite, the thickness of lithosphere was about 200 km in Paleozoic or previous periods, but thinning to 120 km since Mesozoic. Later, the above conclusion has been confirmed by data of original formation depth of a lot of mantle source xenoliths and also verified by the data of two-pyroxene geothermal barometer and more geophysical exploration results (Wan et al. 1996, 2008; Cai et al. 2002; Lu et al. 2005; Li 2010) (Fig. 3.39).

Wan et al. (2008) firstly drew the border [67] between eastern strong magmatic area and western weak magmatic area (Figs. 2.53, 3.25, 3.30). Some petrologists called that boundary as "magmatic line" in China (Xue 2011, personal communication). This border is located between China and Mongolia, through the Western Okhotsk Sea, westward across the Dahingganling, through Central Shanxi, Western Hubei, Western Hunan, to Eastern Yunnan, even southwestward to West Thailand. From north to south, it becomes the western border of East China thinned lithosphere. In the eastern areas of the border, the lithospheric thickness is about 70–80 km; only in Wuhan and its surrounding areas, it becomes thicker, about 90 km. In the western areas of the border, the continental lithospheric thickness is 100–160 km, belonging to normal-type continental lithosphere with a crustal thickness about 40–60 km and lithospheric mantle thickness about 60–100 km (Wan 2011), which constitutes the main part of Asian continental lithosphere (Fig. 3.39). The border of thickness change of East China lithosphere is located west to present gravity anomaly gradient zone (Figs. 2.53, 3.21, 3.23, 3.25, 3.26, 3.30, 3.31, 3.34, 3.36, 3.39). The above borders are rather near.

The current eastern border of East China thinned lithosphere is the trench–subduction zone among the Asian plate, the Pacific plate and Philippine Sea plate (Figs. 3.21, 3.26). Thus, the "East China thinned lithosphere" also called as "East Asian thinned lithosphere" includes areas such as East China, Far East of Russia, Korean Peninsula, Japan Islands, most of Indochina Peninsula, Philippine Islands, Indonesia (Sunda) Islands and their surrounding seas. The thickness of East Asian thin lithosphere is about 70–80 km at continental regions and generally is 70–60 km beneath islands and sea areas. Only in the nearest areas of the recent subduction zone, the thickness of continental crust was a little bit thickened in the arc island zone by subduction rollback and compression. The thickness of lithosphere in the Pacific plate and Philippine Sea plate is less than 60 km, usually about 50–60 km; the thickness of oceanic crust is less than

10 km and the oceanic mantle lithosphere of 40–50 km thickness, which belongs to a normal oceanic lithosphere (Wan 2011; Wan et al. 2016).

East to Dahingganling–Central Shanxi–West Hubei–West Hunan–East Yunnan, the tectono-magmatism was rather strong in Jurassic–early period of Early Cretaceous (200–135 Ma), i.e., the Yanshanian tectonic period (Weng 1927; Wan 1994). The distribution area of Yanshanian magmatic rocks is more than 229,000 km^2, accounting for about 25% of magma rock outcrop areas in the whole China (Cheng 1994), and the Yanshanian magmatism is the strongest tectono-magmatism in East China. The structures and continental crust counterclockwise rotation at the Yanshanian tectonic period have been discussed in detail in Sect. 3.10 of this monograph (Figs. 3.20, 3.21, 3.23, 3.40).

In the petrological and geochemical research of East China, an important result should be paid attention to. Many researchers (Lu et al. 2004; Yu et al. 2010; Zhang et al. 2001; Wan and Zhao 2012) recognized that there had been the oceanic lower crust or oceanic mantle in the lithosphere since Jurassic, which was believed to form the basaltic magma source according to the data of magma-originated depth and characteristic of chemical compositions of basic magma.

According to the above recognition for the special lithosphere type of East China or East Asia, it means that East China lithosphere was a common continental lithosphere before Jurassic; but in Jurassic, the continental crust migrated onto the old oceanic mantle and became a new lithosphere type with the continental crust and oceanic mantle (Figs. 2.53, 3.40). This recognition is different from the earlier Plate Tectonic Theory, which only emphasizes the main detachments at the lithospheric bottom and asthenosphere. Further research on continental plates has shown that there are detachments (sliding) or ductile shear zones within the continental lithosphere, such as discontinuities between the sedimentary cover and crystalline basement, the low seismic velocity/high electric conductivity layer in the middle crust and the Moho discontinuity in the lithosphere. In the past, great detachment in continental crust with several hundred kilometers was not paid enough attention to. Now, it considers that it is possible to produce the continental crust rotation and migration onto the oceanic crust and lithosphere mantle.

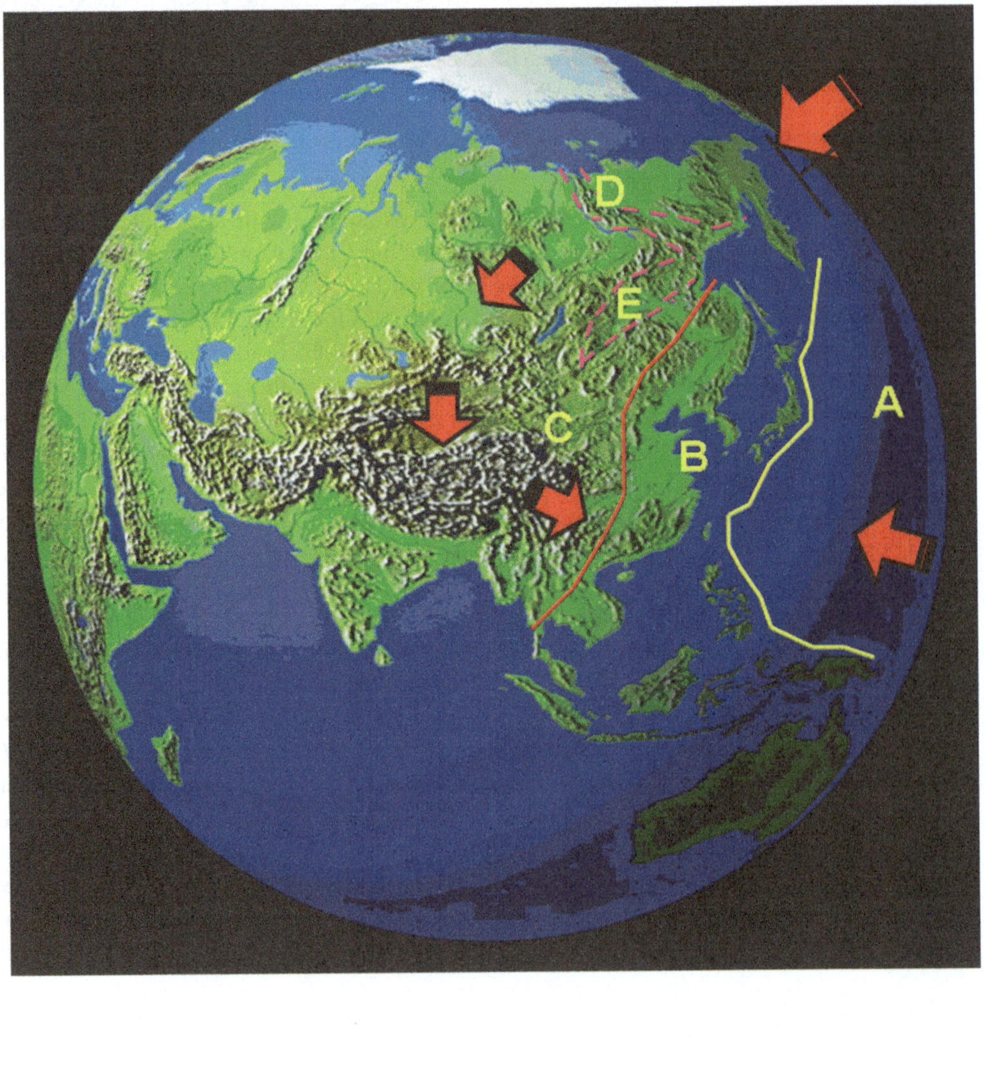

Fig. 3.40 Distribution and formation mechanism of East Asian lithosphere in Jurassic. The biggest red arrow shows the compression and migration trending of North American plate; other red arrows show the migration orientation of blocks and exhibit the counterclockwise rotation for East Asian continent. Red line is the boundary between Asian normal-type continental lithosphere and East Asian continental crust and oceanic mantle lithosphere. Light yellow line is the border between East Asian continental crust and oceanic mantle lithosphere, and the oceanic lithosphere. (A) Izanagi oceanic lithosphere plate; (B) East Asian continental crust–oceanic mantle lithosphere; (C) Asian continental lithosphere; (D) Verkhojansk–Chersky Jurassic (200–135 Ma) collision accretion zone; (E) Trans-Baikal (or Mongolia–Okhotsk) Jurassic (~145 Ma) collision accretion zone (After Wan and Lu 2014)

In view of many new data and results on geology, geochemistry and geophysics, it is more and more difficult to utilize the hot mantle hypotheses to explain the lithosphere thinning (Cai et al. 2002; Lu et al. 2004, 2006; Zhang et al. 2009) and some oceanic plate (such as the Pacific plate) subduction resulting in the continental lithosphere thinning (Zhu et al. 2008, 2011).

It must be considered that, firstly, the extremely strong tectono-magmatism in East Asia began in Jurassic (Cheng 1994; Wan et al. 2008), instead of Cretaceous. Some researchers have paid much attention to the Cretaceous magmatism and got many measurement data. Next, in Cretaceous, the Pacific plate migrated northward (Figs. 2.45, 2.47) to form the great sinistral strike-slip fault between the Pacific plate and Asian continental plate (Yoshikura et al. 1990; Fig. 2.46), but not to influence North China continent. It was not until Late Cretaceous (80 Ma) that the Pacific plate started to subduct toward NNW orientation. Once more, till now, it has never found the believable evidences on the thermal activity or tectonic disturbance in Meso–Cenozoic for East Asian lithosphere mantle (Xu et al. 2012; Wang et al. 2012). In recent years, all the isotopic ages of mantle-derived xenoliths are at Archean or Paleoproterozoic, only with some slight disturbance (Zhou 2006; Lu 2010). If the hot mantle was caused by the Pacific plate subduction beneath the East Asian continent, there would be the Meso–Cenozoic isotopic ages and the stronger disturbance. In the geochemical research of East China, many scholars found that the lower crust or lithospheric mantle (basaltic magma source) had the characteristics of oceanic crust or mantle, without hot and tectonic disturbance in them (Lu et al. 2006; Zhang 2009; Zhang et al. 2012; Yu et al. 2010).

At last, obviously, it is really reasonable and correct to utilize the horizontal subduction and compression of the surrounding oceanic plates to explain the intraplate deformations and magmatism. But it is far-fetched to interpret such a large area of strong tectono-magmatism in East Asia as the hot mantle uplift derived from plate subduction. The subduction zone migrated underneath the Asian continental plate to about 600 km depth, but most of the Jurassic–Cretaceous magmatic activities in East Asia are originated from the Moho discontinuity and middle crust low seismic velocity/high electric conductivity layer, and the Cenozoic basaltic magma with limited distribution also originated in the depth of 60–100 km, that is, near the bottom of the lithosphere (Wan et al. 2008; Wan and Zhao 2012). In recent, through the gravity and seismic tomography methods, Wang X. S. et al. have discovered that the mantle underneath North China is rather cold with the high density at the 100 km depth, and never found any hot and low-density mantle at about 100 km depth. Till now, there are not any evidences to prove the relationship between the subduction zone with about 600 km depth and the magmatism or mantle plume from the shallow mantle and crust.

The other type is the thicker continental lithosphere, which boundary is shown in Figs. 1.1, 2.30 and 3.34. The thicker continental lithosphere was caused by the Indian plate northward subduction and collision, thus to make the thickness of lithosphere increase greatly. The thickness of Pamir–Qinghai–Tibet Plateau continental lithosphere is about 160–200 km (Cai et al. 2002), in which the crust is 56–74 km thick, and the lithospheric mantle is about 100–120 km thick (Figs. 2.35, 2.36, 3.39). The area of the thicker continental lithosphere of the Pamir–Qinghai–Tibet Plateau is almost as same as that of today's Pamir–Qinghai–Tibet Plateau. There is little controversy about this thickening lithosphere.

To sum up, for the Asian continent, there are three types of continental lithosphere. It is extremely distinguishing in the globe and greatly affects the evolution of Asian continental lithosphere.

3.15.4 The Mechanism of Basin and Range Tectonics in Asian Continent

As well known,on the Asian continent, there are the widespread mountains, basins, plateaus, grasslands, deserts, lakes, plains and marshlands with different heights.

In 1854–1855, J. H. Pratt and G. B. Airy, respectively, proposed their hypotheses of gravitational isostatic compensation. They both recognized that the thicker the crust was, the lower the density, the higher the mountains would be (Zeng 2005). It seemed that the change of the topography on the continent would be mainly caused by the vertical migration of crust materials.

Now, it is well known that, for the Pamir–Qinghai–Tibet Plateau with the height of more than 4000 m above sea level, its thickness of lower-density crust is more than 60 km; for the Mongolia Plateau, Loess Plateau, Yunnan–Guizhou Plateau, Shan State Plateau, with the average height of about 2000 m above sea level, the thickness of lower-density crust is about 40–50 km; for the East Asian plains and basins with the height of about 100–50 m above sea level, the crust thickness is only 35–30 km (Li and Mooney 1998; Teng et al. 2003).

Because the mountains are composed of lower-density granitic rocks and sedimentary rocks,after the stronger weathering and erosion, the height will be decreased continuously; however, due to the compensation of gravity balance, the mountains will be uplifted. According to the calculation data, the average uplifting velocity for the planation surface of Pamir–Qinghai–Tibet Plateau varies from 1.6 to 1.0 mm/y (some researchers predicted the average uplifting velocity for the Himalayan Mountains as more than 80 mm/y, but it may be not sure); the average value for the Yunnan–Guizhou Plateau and Loess Plateau is between 0.8 and 0.4 mm/y; the average value of the hill areas in East Asia is only between 0.1 and 0.04 mm/y (Tian and Chen 2010, personal communication). As to the plains, the basins and marginal seas of East Asia, they are all subsided at different velocities, about 1.6–0.3 mm/y (Yang 1999). Although the surface of plains, basins and marginal seas are all covered by lower-density sediments and water, these areas are really subsided by the compensation of gravity balance, apparently due to the presence of many high-density basic or ultrabasic rocks underneath the above areas.

In the collision zones, the rather strong horizontal tectonic stress can not only result in the rock deformations and landscape uplifts, but also form many faults and fractures, and produce the intermediate and acid magma intrusion and eruption. After cooling, those rocks will be formed as low-density rocks (2.5–2.6 g/cm^3, lower than the crustal average density of 2.7 g/cm^3). So after the main collision, it is easy for the mountains to uplift. Thus, many geologists prefer to call the collision zone as "orogeny belt," which is a term that inherits the Hypothesis of Geosynclines and Platforms.

However, the collision zone is not certain to form the mountains. For example, the Shaoxing–Shiwandashan Triassic collision zone [25], the Jinshajiang–Red River Triassic collision zone [31] and the Yarlung Zangbo–Myitkyina Paleogene collision zone [37] never developed mountains. The reason is that there are no large-scale, low-density intermediate-acid intrusive rocks in the above collision zones; thus, mountains are not formed.

To sum up, it is an objective fact that under gravity balance, mountains are rising continuously, while low-lying plains, basins and marginal seas are sinking (Zeng and Wan 1999). The idea of the vertical movement has a profound impact on continental geologists, so the Hypothesis of Geosyncline and Platform has been popular for one hundred years; till now, some scholars have persisted in this hypothesis. It should be concerned that the velocities of uplift or depression are mainly several millimeters per year; the oceanic plate horizontal migration velocities may reach several centimeters to ten centimeters per year. In the South Asian continent, the horizontal migration velocities reach 4–5 cm/y, and in other areas they are gradually decreased to several millimeters per year. In general, in the Pamir–Qinghai–Tibet area with strong collisions, the recent horizontal migration velocity is usually 10–6 times as that of the vertical, even about 30 times for some areas. In plains, basins and marginal seas, the horizontal migration velocity is 3–2 times as that of the vertical, while in other areas it is between the above two values (Wan 2011).

Generally speaking, the horizontal migration velocities are obviously greater than those of the vertical on the Earth surface. This important conclusion must be paid attention to. However, Li (1962, 1976) pointed out that the horizontal stress value could be greater 600 or 2000 times than those of the vertical. His recognition is confirmed to be wrong, for his calculation is based on the hollow "arc, thin-crust construction." If the calculation was based on the solid Earth, the results would be confirmed not to get so great horizontal stress value. This conclusion is clear and definite in mechanics. However, many geologists keep silent about the matter.

In the continental tectonic research, if the vertical migration is emphasized excessively, it will not be proper, and also the influences of horizontal tectonic stress should be paid attention to. In the past, basin tectonic was studied only by seismic sections, which can only detect vertical changes in physical properties, so much attention has been paid to the changes of sedimentary thickness for a long time, especially to regression or transgression. It is believed that the formation, migration, accumulation and loss of the hydrocarbon are mainly controlled by the vertical migration of plates. In recent years, three-dimensional seismic exploration has been carried out in many areas, and it is gradually realized that the horizontal stress is in fact dominant. For example, in the beginning, the researchers did not understand what tectonic stress to control the oil-bearing structures in the east of Liaohe oil field, after utilizing the three dimension seismic exploration, they reinterpreted those structures and discovered a series of small anticlines enriched with hydrocarbon along the NE-trending faults, which were actually derived from the dextral horizontal shearing of NE trending faults under the action of regional nearly E–W trending compression and shortening in Late Paleogene (Li and Xu 2001; Wan 2011).

On the continent, there are many low-lying basins, which are the main location to preserve the oil and gas. Just in those basins, there are the normal and high-gravity equilibrium anomalies. So many geophysical researchers assumed that there exists the "mirror symmetry"; i.e., on the surface, there is a depression, and in the deep there is a rise of Moho discontinuity. This recognition has ever been believed correctly for a long time. Actually, the so-called rise of Moho discontinuity was assumed by the geophysical researchers, not results of measurement.

In recent years, the results of seismic reflection profiles (Zheng et al. 2009, 2012; Qin et al. 2011; Zhang et al. 2009) have shown underneath the basins in North China and South China Sea, and the Moho discontinuity is almost distributed flat (Fig. 3.41) without any rise.

How did the basin come into being? According to research in recent years, the author considers that the formation of basins is mainly controlled by the gravitational

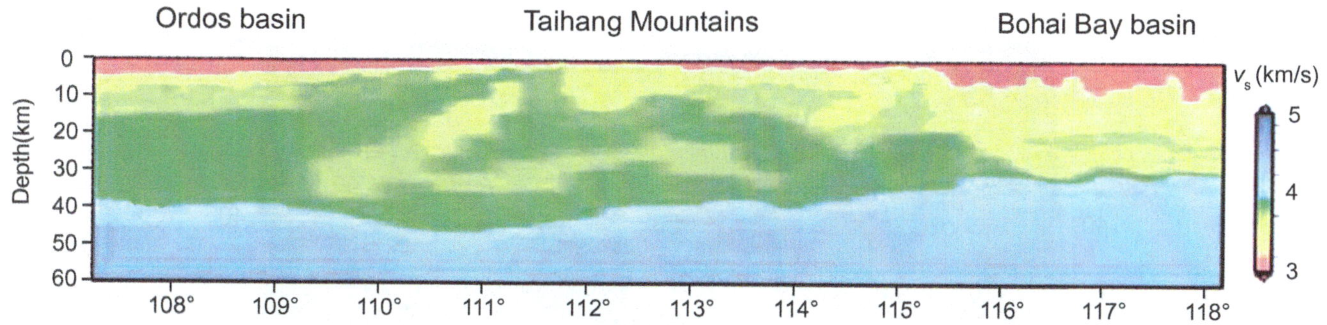

Fig. 3.41 S-wave velocity structure on the near E–W trending seismic tomography section in North China. S-wave velocity value is marked with the right color scale bar. In the figure, the depth of Moho discontinuity, from west to east, is a little bit rising leisurely. At the Bohai Bay basin, the depth of Moho discontinuity is about 30 km, and at the western North China (Ordos basin), it is more than 40 km. Because the vertical scale of this figure is enlarged as 4 times, the Moho discontinuity to the east is ascended rather clearly, and its dip angle is about 10°. If the same horizontal and vertical scale is used to draw the figure, the rise of dip angle will be rather small, about 2°–3°. It is obvious not to form any rise of Moho discontinuity under the Bohai Bay and North China areas. In addition, it must be noted that: For the deep crust, at the depth of Central Shanxi (111.5° E), there exists a low-velocity zone (yellow color) dipping to the west, which may be the manifestation of crust fault zone (ancient subduction zone) in North China block formed in the end of Paleoproterozoic and also is probably the boundary of paleo-continent and paleo-ocean in the end of Paleoproterozoic. However, at the depth of central Hebei (115.4° E), the fault zone only penetrates into the upper crust (yellow color), but never penetrates into the lower crust. The author infers that it may be caused by the different velocities; i.e., the moving velocity of upper crust is greater than that of the lower part. The above two fault zones show the upthrust characteristics (After Zheng et al. 2009)

isostatic compensation, and the shape of basins is confined by the normal faults or strike-slip faults.

Zhang (2005) made an outstanding research of the Bohai Bay basin. He used a lot of data from the Institute of Geology and Geophysics of the Chinese Academy of Sciences which accurately re-verified the epicenter of historical earthquakes and carried out seismic tomography with relatively high accuracy (Figs. 3.42, 3.43).

In Fig. 3.42, it shows the seismic velocity interference under the Moho discontinuity (depth more than 35 km), in which there are four distinct high-velocity interference areas (in blue color), indicating the existence of high-density bodies (density > 3.3 g/cm^3). Form NE to SW, they are distributed at the Central Bohai, Huanghua depression (Dagang oil field) and Puyang depression (Zhongyuan oil field) and Laizhou Bay (near the Shengli oil field), which are parts with relatively large sedimentary depressions in the Bohai Bay area. And in the depths of the Taihang and Yanshan Mountains, they are all shown as the lower-velocity interference areas (in red color), which indicates that the rock density in the deep part of those areas may be rather low, and there may be the lower-density acid-intermediate intrusions and sedimentary systems.

Figure 3.43 shows a velocity structure profile, which just crosses the Bozhong depression and exhibits a positive high-velocity interference area under the Moho discontinuity (in blue color); i.e., it may be the high-density body. According to geological data, the basins around Bohai Bay were initially formed in Cretaceous, so the author infers that the above high-velocity body (i.e., high density with strong velocity interference) may be ultrabasic intrusions developed under Moho discontinuity in Early Cretaceous. It is precisely the formation of these ultrabasic intrusions that significantly increase the overall density of the lithosphere in this area, resulting in excessive compensation, so that the basin will subside under gravity balance, which may be the main reason for the formation of basins. However, in the low-velocity interference areas of the Taihang–Yanshan Mountains (in red color), there are many lower-density intermediate-acid intrusive rocks or sedimentary rocks (2.5–2.7 g/cm^3), resulting in mass deficit. Thus under the compensation of gravity balance, those areas will uplift to form mountains, just like the woods floating on the water surface.

In the lithosphere, some areas will be uplifted and some areas will be subsided, so between them, it usually is easy to make the pre-existing fault reactivity. Thus, the NNE trending faults in front of the Taihangshan Mountains and the nearly E–W trending faults in the southern foot of the Yanshan Mountains naturally become the western and northern boundaries of the Bohai Bay basin, and the Tancheng–Lujiang fault zone (Fig. 3.25) was formed as its eastern boundary. For the maximum principal compression stress orientation of Early Cretaceous at NNE–SSW trending, the above fault zones would transfer the Jurassic reverse fault into the normal fault in Cretaceous, to produce a series of near E–W trending extension in the Bohai Bay basin. The author considers that the above faults mainly control the boundaries in the basin formation process, but not to play a leading role, also the basin is affected by the extension of fault zone. These phenomena are often found in the evolution of the Asian continent.

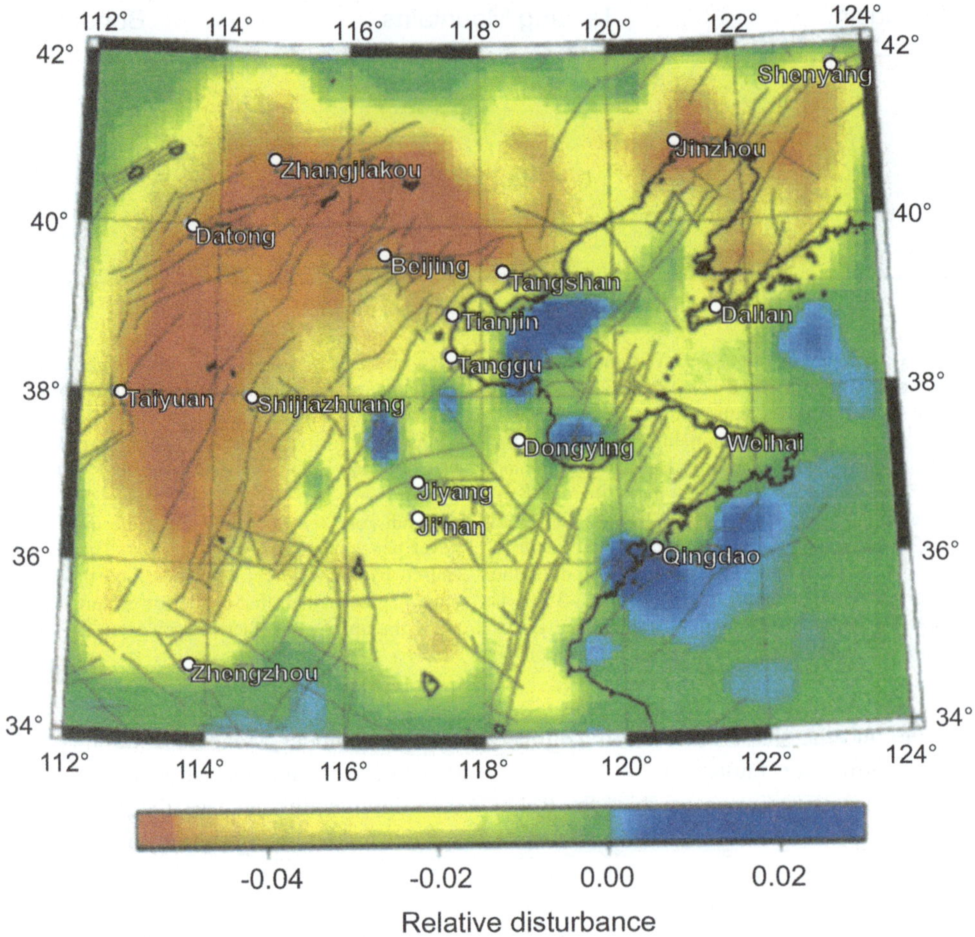

Fig. 3.42 Seismic velocity disturbance around Bohai Bay in the depth more than 35 km (After Zhang 2005)

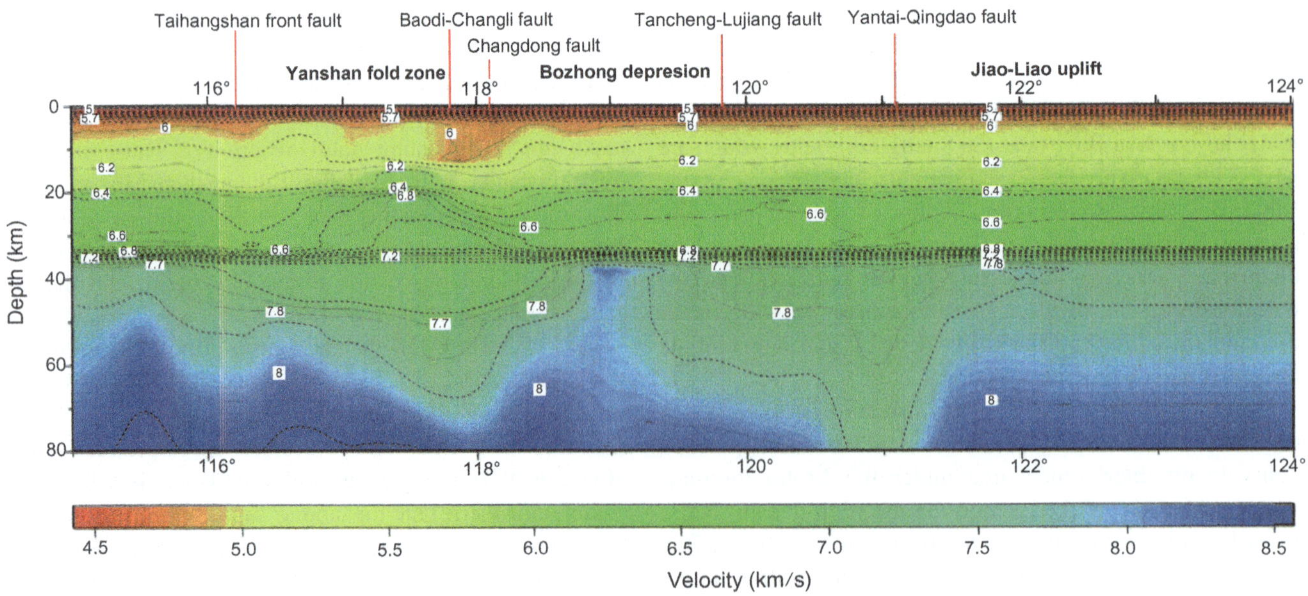

Fig. 3.43 BH-01 velocity structure map of the crust–upper mantle section (After Zhang 2005)

In Cretaceous–Paleocene, being controlled by the NNE trending maximum principal compression stress, the NNE trending faults were all transformed as normal faults in East Asia, for example, in the Dahingganling–Taihangshan–Wuling, Tancheng–Lujiang, Okcheon–Honam zone (Chough et al. 2000; Kim et al. 2005)—the coastal fault zone of Zhejiang–Fujian–Guangdong, there formed a series of NNE trending strong-folded and granitic rock belts (Figs. 3.25, 3.26; Wan et al. 2008), at their nearby areas (without granitic rock distribution), there formed fault-depression basins, such as the Songliao, North China (Bohai Bay) and South Yellow Sea, and on the eastern side of the Qinglai–Dafu–Central Malaya granitic belt, there formed the Chao Phraya basin in Thailand and Eastern Malaya Plain.

In Eocene–Oligocene, for being affected by the WNW toward subduction and compression of the Pacific plate (Figs. 3.30, 3.31), in the Asian continent, the N–S trending extension occurred to form three fault-block mountains with abundant granitic rocks, i.e., Yinshan—Yanshan, Qinling—Dabie and Nanling; besides, the E–W trending high-angle normal faults were formed. Among them, there formed the basements of four drainage basins (Song–Liao and Inner Mongolia slope basins, Yellow River, Yangtze River and Zhujiang River basins); at that time, the above river systems had not been linked up together (Zhu 1989; Yang 1988). However, in West Asia, the Turkey–Iran–Afghanistan Plateau (with an average altitude of 2000 m above sea level) was formed by the northward compression of the Indian plate.

In Neogene–Early Pleistocene, due to the northward collision of the Indian plate (Figs. 3.33, 3.34), the thickness of lithosphere and crust was significantly thickened, forming the Pamir–Qinghai–Tibet Plateau (with an average altitude of more than 4000 m above sea level), and in its surrounding areas there developed the Central Asia–Mongolia–Yunnan–Guizhou–Shan State Plateaus and it made the Turkey–Iran–Afghanistan Plateau continue to uplift to 2000–3000 m above sea level. At the more peripheral areas, such as the western part of Central Asia, Siberia and East Asia lower mountains, hills, plains and marginal seas, they were all lower than 200 m above sea level. The Neogene–Early Pleistocene is the shaped period for the modern Asian landscapes.

To sum up, for the formation of basins and mountains, at the initial research stage, it seemed to be caused only by the vertical uplift and subsidence, and to be dominated by the compensation of gravity balance. However, with the further research, it considered that the rocks could be distributed not to be homogeneous, which should be influenced by the continental lithosphere plate intraplate deformations, and controlled by the different regional horizontal orientations of the tectonic stress and the tectono-magmatism. If there were not the different densities for rock blocks, the subsidence or uplift would not occur. The regional morphologic characteristics are controlled obviously by the tectonic background.

In a word, the formation of continental topography or the formation of basin–mountain systems is the result of combined action for the intraplate horizontal stress and compensation of gravity balance, but the regional horizontal tectonics stress dominates. It indicates that it is not enough or improper only to apply gravity balance for studying the formation of basin–mountain systems. This is the important difference between the Platform–Geosyncline Hypothesis and Plate Tectonic Theory. This point is worth keeping in our minds.

3.15.5 Dynamic Mechanism for the Global Lithospheric Plate Tectonics

After studying the various issues on the Asian continental tectonics, it is necessary to focus on the dynamic mechanism for the global lithospheric plates. It is an unavoidable and difficult problem in geosciences.

Next, the author will discuss three problems on the dynamics of the lithospheric plates:

(1) **The driving action of plate tectonic migration**

The migration of oceanic and continental lithospheric plates is obviously driven by the horizontal tectonic stress of regional plates (Bott and Kusznir 1984, 1991; Zoback and Magee 1991), which was recognized from the time when the Plate Tectonics Theory was found, and the main detachment layer is located in the asthenosphere beneath the bottom of the lithosphere (Le Pichon et al. 1973; Press and Siever 1974; Hilde et al. 1977).

However, as to the dynamic mechanism of plate horizontal migration, there were some great debates in 1970–1990. At first, most of the geophysical scholars emphasized the importance of subduction (Forsyth and Uyeda 1975; Turcotte and Schubert 1982) and recognized that in the common condition, the tectonic stress values of plate extension, drag and subduction could be up to 20–30 MPa and up to 1000 MPa on plate boundaries. The negative buoyant force of a subducting plate would be very great, which is the major driving force for plate movements.

The above values obviously seem too great. Later, many researchers concluded more factors, such as the various physical parameters changing with depths, different subduction velocities, the thickness of plate and different subduction processes, and then calculated the negative buoyant force not to be so great, only about 40–290 MPa (Zang and Ning 1994). It is not in accordance with the "negative buoyant force" defined by some scholars (Forsyth and Uyeda 1975; Turcotte and Schubert 1982). In fact, although the density of oceanic lithosphere (2.9–3.0 g/cm^3) is greater than that of the continental lithosphere (2.6–2.7 g/cm^3), it is absolutely impossible to exceed that of the mantle (3.3–5.56 g/cm^3). How can the downward drag of the oceanic plate "negative buoyancy" become the dominance driving force for plate migration due to its so-called high density with locally formed eclogites?

The subduction of oceanic plate must meet great obstruction of high-density mantle. Due to the influences of the mantle temperature and confining pressure, when the oceanic plate subducts into the deep, its temperature and density will be increased, and the subducting plate will be shown as same as the surrounding mantle rocks. In recent years, quite a lot of seismic tomographic data have shown that plate subduction often stops or suspends near the depth of 670 km. For example, when the Indian plate downward inserts into the depth of 600–800 km, it will result in winding and rotation, and some fragments will be left in the lower mantle (Bigwaard et al. 1998; Van der Voo et al. 1999; Replumaz 2004). The depth of the Australian plate downward thrusting into the Asian continent is also more than 600 km (Replumaz 2004). In Saudi Arabia (Maruyama 1994) and South Europe (Cavazza et al. 2004), it is found that the subducting plate can hardly insert more than 600 km deep into the lower mantle with perovskite-structured silicates, and the material densities of the subducting plate and the middle mantle tend to be the same.

The recent research shows that the migration velocity of lithospheric plate is faster than that of the mantle. According to the estimated migration velocity of lithosphere plate, the hot spot is commonly assumed to be fixed. Later, Minster and Jordan (1978) found that the hot spot can also migrate a little, usually about several millimeters per year, only a few up to 2 cm/y. Its velocity is obviously less than that of lithospheric plate, and the downward subducting velocity of the lithosphere plate is about 1–1.5 cm/y in the mantle (Grand et al. 1997). Compared with the faster migration velocity of lithosphere plate, most of scholars recognized the migration velocity of mantle convection and the hot spots to be rather slow (Liboutry 1982). Based on many data, Bott and Kusznir (1984, The Royal Institute of Geophysics) described it as a famous sentence: "The plate driving the mantle convection, rather than the mantle convection driving the plate migration." It means that the migration velocity of plate is obviously faster than that of mantle. Till now, some scholars (Domeier and Torsvik 2014; Torsvik et al. 2008, 2014) have still persisted that the mantle moves faster than plates, which is debatable.

While the plate continues to subduct and penetrate into the mantle, some hot mantle materials from the Gutenberg discontinuity between core and mantle will be formed as the mantle plume to uplift. On the one hand, the cold lithospheric plate will downward penetrate into the middle mantle; on the other hand, the deep hot mantle materials (mainly supercritical fluids) will result in the extension of upper lithosphere, then to form a huge mantle convection system (Mattauer 1999; Condie 2001; Xu et al. 2003).

In the later twentieth century, there was a prevalent recognition that it was impossible to form the whole mantle convection. Nowadays, that recognition must be abandoned. Although convection in the mantle has been demonstrated, movements of the mantle cannot be held responsible for transporting the lithospheric plates like a conveyor belt, because there is no evidence of movement rates in the mantle exceeding 2 cm/y. A "conveyor belt" moving at a slow speed is not able to transport the "goods" at a faster rate. And it should be remembered that the "mantle convection hypothesis" was proposed by Griggs in 1939 and by Holmes in 1944, before the foundation of the Plate Tectonics Theory. Thus, the above fact negates the hypothesis that the mantle convection drives the plate migration by "conveyor belt." Only some American scholars are not willing to face this fact, but to express tactfully "the problem of establishing the dynamic mechanism driving plate movements would be solved in twenty or thirty years" (speech on IUGG, 1990). However, in fact, it has not been solved so far.

To sum up, according to recent research, the basic migration model for the oceanic and continental plates is mainly horizontal and paralleled to the Earth's surface. The migration velocity of the oceanic and continental plates is obviously faster than that of the mantle. It shows that the above recognition may be correct and consistent with the fact.

(2) **Breakup of paleo-continent and the activities of hot mantle plume**

In this monography, the author mainly discusses the accretion and formation of the Asian continent. However, it is necessary to discuss why the paleo-continent was broken and dispersed.

In recent geological data, the first acknowledged tectonic event was the gradual breakup of the Columbia supercontinent during the early–middle periods of Mesoproterozoic (1600–1200 Ma). The Mackenzie dyke swarms in North America, the basic dyke swarms in Siberian plate, Sino–Korean plate and Indian plate (Figs. 3.2, 3.3, 2.3, 2.37; Hou 2012) are all manifestations of the breakup of the Columbia supercontinent. According to the distribution and compressional orientation of the North American Mackenzie and Siberian dyke swarms, most scholars recognized that since the early period of Mesoproterozoic, a great mantle plume had existed at recent Arctic area. However, it must point out that the Siberian plate was located at the Southern

Hemisphere at that time, but not at the recent location. As a result of many supercritical hot fluids uplifting to form magma chamber, the crust would be extended radially to form the radial dyke swarms. It means that the continent dissociation, dispersion and the radial extension were caused by the hot mantle material upwelling and controlled by the vertical tectonic actions. Finally, it resulted in the horizontal migration of lithospheric plate near the Earth's surface.

As to the breakup mechanism of Rodinia supercontinent in the late period of Mesoproterozoic, till now it seems not to obtain the final conclusion. In Late Permian, at West Sichuan of China, there developed the Emeishan basalt province, which was originated from the hot spot between the Yangtze plate and Southern Qiangtang plate. Maybe there formed the upwelling hot mantle to cause the partial extensions in the Yangtze plate, but never to cause the Yangtze plate to break up. It seems that only the mantle diapir occurred, because the seismic tomography data show that there is no lower-velocity zone below 200–300 km, i.e., without thermal anomaly (Fig. 2.23; Zheng et al. 2012) or mantle plume. Shellnut (2014) recognized that the Emeishan igneous province was formed by the rise of mantle plume, but he also admitted that the rise of the hot mantle occurred only in a very short time, only confined in the Capitanian period of late Middle Permian ($\sim$260 Ma), and it could not result in overall destruction and breakup of the Yangtze plate.

Many researchers agreed that the breakup of the Pangea supercontinent was related to the formation of radiating dyke swarms centered in West Africa (Fig. 3.19) at the end of Triassic (200 Ma). That is to say, it is possible that the rise of mantle plume and upward intrusion of magma caused radial extensions on the Earth's surface, and then the Pangea supercontinent was broken up, destroyed and separated into the North American, South American and African continental plates (Marzoli et al. 1999; Hames et al. 2000; Condie 2001). Those magmatic rocks are all the tholeiites, which errors of $^{40}Ar/^{39}Ar$ isotopic age are only in 1 Ma. The volume of eruption magma reaches 7×10^6 km^3 (Marzoli et al. 1999; Hames et al. 2000). However, as to the Triassic super-mantle plume, till now its deep status has not been obtained for that period; neither it can be determined as a "mantle plume" or a "mantle diaper," nor can it be determined whether there is the influence from a giant meteorite impact.

Other examples of radial extensions near the Earth's surface mainly include the Middle Jurassic Karoo–Ferrar–Chon Alke flood basalt provinces among the African, Antarctic and South American continents (Storey 1995; Storey and Kyle 1997; Fig. 3.44). The Karoo volcanic activities in South Africa occurred at 195–177 Ma, with the areas of 3×10^6 km^2. The Ferrar basalt province is mainly shown as the Antarctic mafic flood basalt, sill and dyke swarm, which isotopic ages are between 193 and 170 Ma, as well as the Chon Alke flood basalt province in South America. Storey (1995) and Storey and Kyle (1997) considered these three basaltic provinces centering on the Weddell Sea near the Falkland Islands. A super-plume seemed to occupy this location in Middle Jurassic. Three flood basalt provinces with a rifted triple junction could be formed by the mantle plume upwelling, which led to extension and breakup of Gondwana. The deep structure of this proposed mantle plume has not yet been explored by the geophysical exploration.

The radial migration of the plate made South America, Africa, India, Australia and Antarctica move northward, but their rates of migration were extremely different. The Indian plate moved northward at a rate of 18 cm/y in Cretaceous (Lee and Lawver 1995), only 0.8 cm/y for the Australian plate (Van der Voo 1993), several centimeters per year for the African and South American plates. Why did the Indian plate move northward so fast and others rather slow? It seems not very good to explain the above phenomena only using the vertical rise of mantle plumes. Is it possible that it was caused by the meteorite oblique impacting directly the Indian plate? It needs studying deeply in the future.

According to the third oceanic floor magnetic anomaly, Moore (1989) firstly proposed the radial migration model of each plates in the Pacific area in Late Jurassic ($\sim$138 Ma; upper left of Fig. 2.45); i.e., the Izanagi plate migrated toward NW trending (Maruyama and Seno 1986; Moore 1989); the Pacific plate at the Southern Hemisphere migrated and subducted toward SW trending underneath the Australian plate; the Farallon plate migrated and subducted toward NE trending underneath the North American plate; the Phoenix plate migrated and subducted toward SE trending underneath the South American plate. It has been suggested that the radial movement of these four plates was due to the uplift of a super-plume (Pavoni 1997; Condie 2001). However, there is no satisfactory explanation of why uplift of a super-plume occurred in this particular area.

From contrasting swells in the geoid and relief on the core–mantle boundary (D″ layer) estimated from seismic waves (Hager et al. 1985), and the distribution zones of oceanic crustal accretion indicated by magnetic anomalies on the oceanic floor, Pavoni (1997) proposed that there existed super-plumes below the African and Pacific areas in the past 180 Ma (since Middle Jurassic). He proposed a "geotectonic bipolarity" of two mantle plumes. The rise of these two mantle plumes caused the horizontal displacement of the lithospheric plates and the upper mantle. The anomalous swells in the geoid, with a maximum uplift of 800 m in Central Africa and 1200 m in the Central Pacific, are corresponded to elevations in the core–mantle boundary

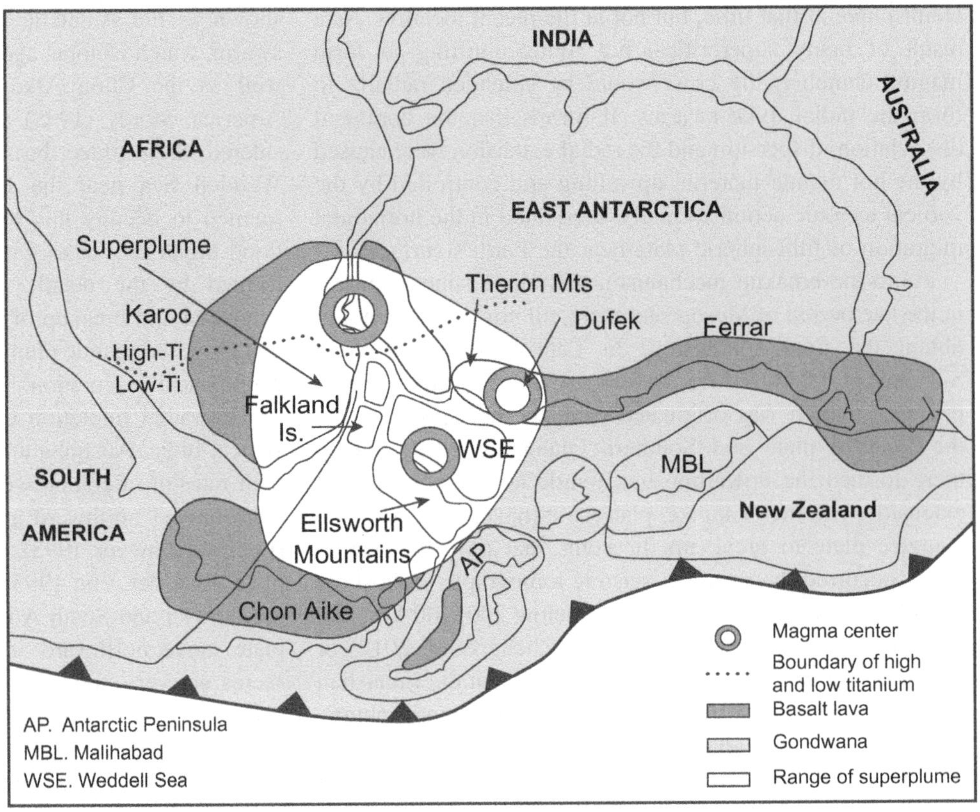

Fig. 3.44 Middle Jurassic mantle plume in South Antarctica and breakup of Gondwana (After Storey 1995; Storey and Kyle 1997; from Wan 2011)

of 3.5 km below Africa and 3 km below the Pacific. It is generally accepted that mantle plumes arise from core–mantle boundary, and may be caused by differences in the rate of movement of the liquid outer core and the solid mantle, causing the core–mantle boundary swell and the formation of the mantle plumes.

In recent years, the uplifting supercritical hot fluid from the core–mantle boundary is commonly called as **mantle plume** (Fig. 3.45; Condie 2001; Brandon and Walker 2005). In the whole mantle, there only exists supercritical thermal fluid with irregular shapes rising along the crystal lattice cracks, and there is not any columnar hot mantle composed mainly with silicates. The supercritical hot fluid only developed in the upper and middle mantle should be called **mantle diapir** (Condie 2001), which is the uplift of hot mantle materials without root. There are hundreds of hot spots (including the Hawaii hot spot) in the depth near the middle mantle all over the world, which seismic velocities are not different between the hot mantle material and its surrounding parts. It means that most hot spots belong to the mantle diapir. The worldwide distribution is very irregular; most of them are scattered in the interior of the plate, only a few near the oceanic ridge. Thus, many scholars inferred that the mantle diapir might be caused by giant meteorite impact (Wan et al. 1997; Wan 2011).

In the twentieth century, there were some prevalent tectonic hypotheses, such as the pulse expansion and contraction hypothesis (Bucher 1933; Grabau 1940; Umbgrove 1947; Zhang 1959; Milanovsky 1980), the great expansion hypothesis (Glikson 1980), the finite (15–20%) expansion hypothesis (Owen 1992; Wang et al. 1997), the isostatic un-equilibrium hypothesis (Ma and Jiang 1987), the hypothesis of velocity change of the Earth rotation (Li 1947, 1962; Scheidegger 1963, 1982; Wang and He 1979), the hypothesis of mantle convection driving the plate displacement, namely the "conveyor belt model" (Holmes 1944; Griggs 1939; Wilson 1970; Le Pichon et al. 1973) and surge tectonics hypothesis (Meyerhoff et al. 1996). After decades of detailed research, there have obtained enough data to negate the above hypotheses. Here, it is unnecessary to repeat them in this monograph.

From the view of the global tectonics, the vertical migration of mantle plume or mantle diapir results in the radial horizontal migration and extension of the lithospheric plates and causes the breakup and radial horizontal migration of paleo-continents and oceanic plates.

To sum up, the plate's horizontal subduction and collision can produce the lithosphere accretion, and the uplifting mantle plume or mantle diapir can cause the horizontal migration, breakup or convergence of plates. The intraplate deformation, originated from the lithospheric plate horizontal migration and the inhomogeneous material distribution, can form basins and mountains by the gravity equilibrium and the vertical displacement. In the different

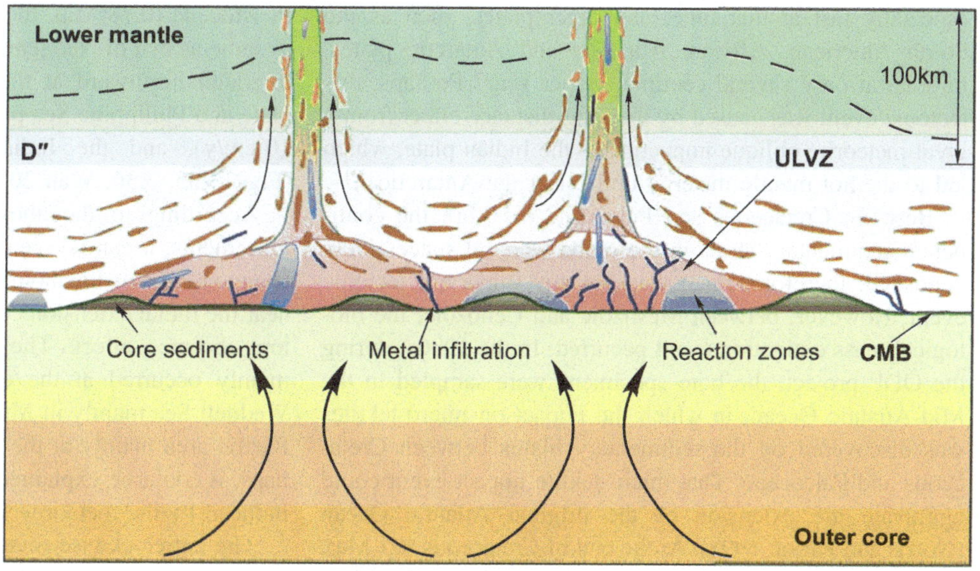

Fig. 3.45 Model of mantle plume formation in the D″ layer originated from core disturbance. ULVZ. Upper seismic low-velocity zone; CMB. core–mantle boundary (After Brandon and Walker 2005; Wan 2011)

scale tectonic researches, it seems that the horizontal and vertical motions for solid Earth materials can be driven from each other. Of course, it is necessary to make clear the primary and secondary actions.

(3) **The possible tectonic mechanism of global Mesozoic and Cenozoic lithospheric plate**

In the oceanic lithosphere plates, the rocks could be recorded only in about last 200 Ma. The earlier oceanic plate had subducted into the continental plates. So, the continual accretion of continental plate could remain the imperfect geological records about last 2000 Ma. There are few paleo-geological records remained near the Earth surface and most being hidden in the continental lithosphere. That is why till now, nobody can effectively solve the dynamics mechanism of global lithospheric plate tectonics in the whole geological history.

As for the global plate migration and its mechanism of the last 200 Ma (i.e., since Mesozoic), it has made some progress. As discussed earlier in this book, the mode of global plate movement in Mesozoic and Cenozoic is now outlined below:

At the end of Triassic (200 Ma), the mantle plume uplift or the giant meteorite impact resulted in the Pangea super-continent splitting (Figs. 3.15, 3.19; Marzoli et al. 1999; Hames et al. 2000; Condie 2001), and the original Atlantic Ocean began to form; however in the Eastern Hemisphere, due to the extension and the northward subduction of Tethys Ocean, the Asian continental blocks produced the obvious convergence and accretion.

In Middle Jurassic (195–177 Ma), among the African, Antarctic and South American continents, the extension center was located at the Weddell Sea near the South Pole, the hot mantle material uplifted to form the large volcanic province (Fig. 3.44; Storey 1995; Storey and Kyle 1997), and then the Gondwana began to radially break and dissociate. Thus, that long-distance effect caused the surrounding plates to move northward at different velocities in Late Jurassic to Cretaceous.

In Late Jurassic (~138 Ma), due to the hot mantle material uplift, the four oceanic plates radially migrated and subducted to the surrounding continental plates (upper left of Fig. 2.45; Maruyama and Seno 1986, Moore 1989): The Izanagi plate migrated, compressed and subducted toward northwest (Maruyama and Seno 1986, 1992; Moore 1989); the Pacific plate in Southern Hemisphere migrated, compressed and subducted toward the Australian plate at the SW trending; the Farallon plate migrated and subducted toward the North American plate at the NE trending; the Phoenix plate migrated and subducted toward the South American plate at the SE trending. According to the above four-plate radial migration model, it may be inferred that there existed the mantle plume uplift or mantle diapir to cause them to migrate from the centers to the surrounding areas (Pavoni 1997; Condie 2001). At the same time of Jurassic, the North American plate suffered expansion of the Atlantic Ocean, its northern part migrated and compressed toward the Northeast Asia at the WSW trending, and then the anticlockwise rotation took place in the eastern part of Asian continental crust (Figs. 3.20, 3.39).

In the middle period of Cretaceous (100 Ma), the whole continental and oceanic plates showed the characteristics of northward displacement to form some near N–S trending extension rift and strike-slip fault zones (upper right of Figs. 2.45, 3.26; Moore 1989). Among them, the Indian plate had the fastest northward velocity, up to 18 cm/y (Lee and Lawver 1995). Why did the Indian plate migrate especially fast at that time, and other plates, such as the South American, African, Australia and Antarctic plates, migrate at only several centimeters per year? Perhaps, this tectonic event was caused by the long-distance effect from a great meteorite oblique impacting to the Indian plate, which led to the hot mantle material uplift near the Antarctica.

Between Cretaceous and Paleocene (65 Ma), the continental sedimentary strata in the globe were all successively deposited; therefore, there was not any important tectonic event. However, between Mesozoic and Cenozoic, the biological mass extinction event occurred. In the 1990s, during the ODP project, the bore specimens were sampled in the Mid-Atlantic Ocean, in which the impact on micro-tektites was discovered on the sedimentary hiatus between Cretaceous and Paleocene. That micro-tektite impact event could aggravate the extension of the original Atlantic Ocean (Norris and Kroon 1998). At the end of Cretaceous (65 Ma), the meteorite impact event in the Yucatan Peninsula, Mexico, may have great influences on the organisms, such as the extinction of dinosaur and gymnosperm (Sharpton et al. 1992). However, till now, it has never found any important influences on the migration of global plate tectonics (Moore 1989; Wan 2011).

During the late period of Eocene (36 Ma), the Pacific plate suddenly turned to migrate from the NNW trending to the WNW trending with a velocity of 10 cm/y and subducted into the Asian continental plate (lower left of Figs. 2.45, 2.46, 3.29, 3.30; Glass 1982; Yin and Wan 1996; Wan et al. 1997), which may be caused by the giant micro-tektite oblique impact to Middle America–SE Asia. However, the Indian plate, African plate and South American plate still slowly were migrating northward with a velocity of 5–2 cm/y, and the Atlantic Ocean continued to extend at E–W trending.

In Miocene–Early Pleistocene (23–0.78 Ma), the Indian plate, Australian plate, African plate and South American plate were migrating northward to form the recent Eurasian continental plate finally; the Pacific plate continued to migrate westward; the Philippine Sea plate migrated toward northwest and resulted in the NNE–SSW extension. The Atlantic Ocean continued to extend slightly at the EW trending (Figs. 3.32, 3.34; Lee and Lawver 1995; Hall and Blundell 1995; Hall et al. 2011; Scotese 1994; Wan 2011).

Between Early and Middle Pleistocene (0.78 Ma), the Australasian micro-tektite impact event occurred (Glass 1982). The micro-tektites were strewn over a wide area with the layer up to 10 m thick in Australia, Indian Ocean and nearby Southeast Asia. The age of formation of this tektite field was determined by K–Ar and fission track methods as 0.9–0.7 Ma, showing two separated meteorite impact events. The impacting site was located near the triple junctions of three lithospheric plates in the Indian Ocean, which could extend to the deep of the Indian Ocean. After Middle Pleistocene (0.78 Ma), the main plate migration model was as same as that of Neogene–Early Pleistocene, and the plates migrated northward at the lower velocity, but the Pacific plate and Philippine Sea plate migrated relatively fast (about 10 cm/y), and the Indian Ocean continued to extend (Figs. 3.35, 3.36; Wan 2011).

According to the above model changes of plate displacements, it can be seen that the plate migration model is changed obviously about every 33 Ma, some are located near the radial extension center, and others are caused by the long-distance effect. The radial cracking of West Africa mainly occurred at the end of Triassic ($\sim$200 Ma), the Weddell Sea mainly in Middle Jurassic ($\sim$177 Ma) and the Pacific area mainly at the end of Jurassic ($\sim$140 Ma). Perhaps, it could be explained by the uplift of mantle diapir or induced by the meteorite vertical impact.

The anticlockwise rotation of the East Asian continental crust may be caused by the Atlantic Ocean extension to form the long-distance effect of the North American plate toward WSW compression in Jurassic. In Middle Cretaceous, most of the oceanic and continental plates migrated northward, which may be caused by the mantle diapir near the Antarctica; also, it may be a result from the giant meteorite impacting toward the Indian plate.

However, in Late Eocene–the end of Oligocene (36–23 Ma), the sudden change of the migration orientation of the Pacific plate could not be well explained by the uplift of mantle diapir or its long-distance effect. It reasonably seems to be explained by a low-angle oblique impact of the giant meteorite (Figs. 3.28, 3.29). Since Neogene, the changes of plate migration velocity and tectonic stress could be explained by the long-distance effect of the migration of surrounding plates.

Recently, in the hypothesis of mantle plume uplift or mantle diapir causing the migration of lithospheric plates, it emphasizes the uplift of a large number hot fluids and the topping action of the mantle head, which causes the partial melting of the basement lithosphere to form the magma upward intrusion, to develop the radial extension center and then to be separated into several plates, for example, forming the triple junction among the plates and the great-scale magmatism. If the magma erupts on the Earth surface, it will be formed as the wide basalt areas. If the magma intrudes into the horizontal strata, it will be developed as the great-scale sill. If the magma intrudes into the vertical and radial fractures, it will be formed as the radial dyke swarm. The whole magmatism commonly occurs simultaneously, and the differences of isotopic ages from magma rocks should be within the span of 1 Ma (Condie 2001).

According to the recognition that the mantle plume is originated from the core–mantle boundary, the mantle plume may be easily formed in the deep position near the equator. Due to the lower migration velocity of supercritical fluid in the mantle, it will last several hundred million years for its forming and reaching the Earth surface.

It seems that it is not easy to form the fast migration of the lithospheric plate in only one direction by the mantle plume uplift. Why could the plate migration model suddenly change every 33 Ma? So, it is difficult to explain the lithospheric plate migration model only by the mantle plume or diapir. These problems need to research deeply.

In addition, there are more than hundreds of irregular mantle diapirs in the lithospheric plates, but only a few are formed on the plate boundaries. It means the mantle diapir is not certain to cause the plate extension. Till now, it is not very clear about the mantle diapir formation. Perhaps, the mantle diapir is caused by not too great meteorite impact.

Some researchers supposed that the meteorite impact would make the surface rocks explode and eject far away, and cause the serious surface mass deficit and the pressure decrease of upper crust rocks (Fig. 3.46). Thus, the deep mantle hot fluids would migrate upward, i.e., forming the mantle diapir. The outline of this hypothesis is that macro-meteorite hits the Earth with a periodicity of 33 ± 3 Ma (Rampino and Stothers 1984, 1988) and forms huge craters (Fig. 3.46).

When a meteorite, with a diameter greater than 1 km, impacts the Earth's surface, it will produce a giant crater with the diameter of hundreds of kilometers and the depth of more than 20 km (Fig. 3.46). The circular and radial fracture areas extend over thousands of kilometers (Grieve 1990).

① Meteorite impact will produce fractures on the surface layers and result in a huge mass deficit. When the mantle rises to maintain isostatic equilibrium, it will form a mantle diapir in the upper part of the mantle and cause the large-scale flood basalts to erupt (the covering areas up to about 100 km^2). The vertical impact can produce circular and radial extension fractures, which may be filled with magma and evolved into triple junctions within the lithospheric plates. The separated plate fragments will be pushed apart horizontally to migrate far away from the impact site at different directions (Wan et al. 1997).

② A violent impact with the velocity of 30–50 km/s can cause the meteorite and its adjacent rocks immediately to vaporize, while the rocks far away will be melted or shattered. When the powdered rock and gas form a plume into the atmosphere and spread out to form the black clouds, they will block the sun to decrease temperature over a large part of the Earth, perhaps lasting several months. The greenhouse gases—CO_2, H_2S and H_2O—will be emitted from the target rocks, including limestones or evaporites. A period of low temperature is followed by a period of high temperature, perhaps resulting in wildfires (Alvarez et al. 1980; Wolbach et al. 1985; Rampino and Volk 1988). The hydrogen sulfide mixed with water will form the acid rain.

③ High temperature near the impact site can ignite fires to kill off all the plants and animals in the surroundings. If the dust clouds cover the sunlight for a long time, the plants will be unable to photosynthesize and then to wither or die. First, the herbivores will die of starvation and then the carnivores. If these subsequent impacting effects cover a very large area and last a long time, many species of plants and animals will become extinct. Acid rain can also kill the vegetation, contaminate freshwater and inhibit the formation of calcareous shells in marine and freshwater environments, so many microorganisms will become extinct, and food chains will be destroyed. Many scientists believe that all the mass extinctions recorded in the fossil may be caused by meteorite impacts (Alvarez et al. 1980; Prinn and Fegley 1987).

④ If the impact occurs in oceans, a giant tsunami (seismic sea wave) will attack the adjacent coasts or low-lying areas to form the far inland-deposited chaotic tempestites (tsunamites), and the trace elements will also change a lot (Erickson and Dickson 1987).

⑤ It is possible for a major impact event to cause the polarity reversal of the Earth's magnetic field. Although so many reversals are recorded in the rocks during the Earth's history, neither has been linked to a specific impact event.

⑥ When a macro-meteorite hits the fracture rocks over a large area, it will produce a mass deficit around the crater. The gravitational isostatic compensation not only can induce to form the shallow mantle plumes (i.e., mantle diapir) and the large-scale magmatism, but also can break the lithospheric plates with radial or circular extension fractures, form the triple junctions and slowly push the segments apart. Evidently, the major meteorite impact events will affect the atmosphere, hydrosphere, biosphere, lithosphere and the magnetic field.

According to data from Rampino and Stothers (1984, 1988) and Yin et al. (1988), there is a close periodicity relationship between the tectonic evolution of the Earth and the occurrence of giant meteorite impacts. There are two kinds of periodicity in the movement of the solar system

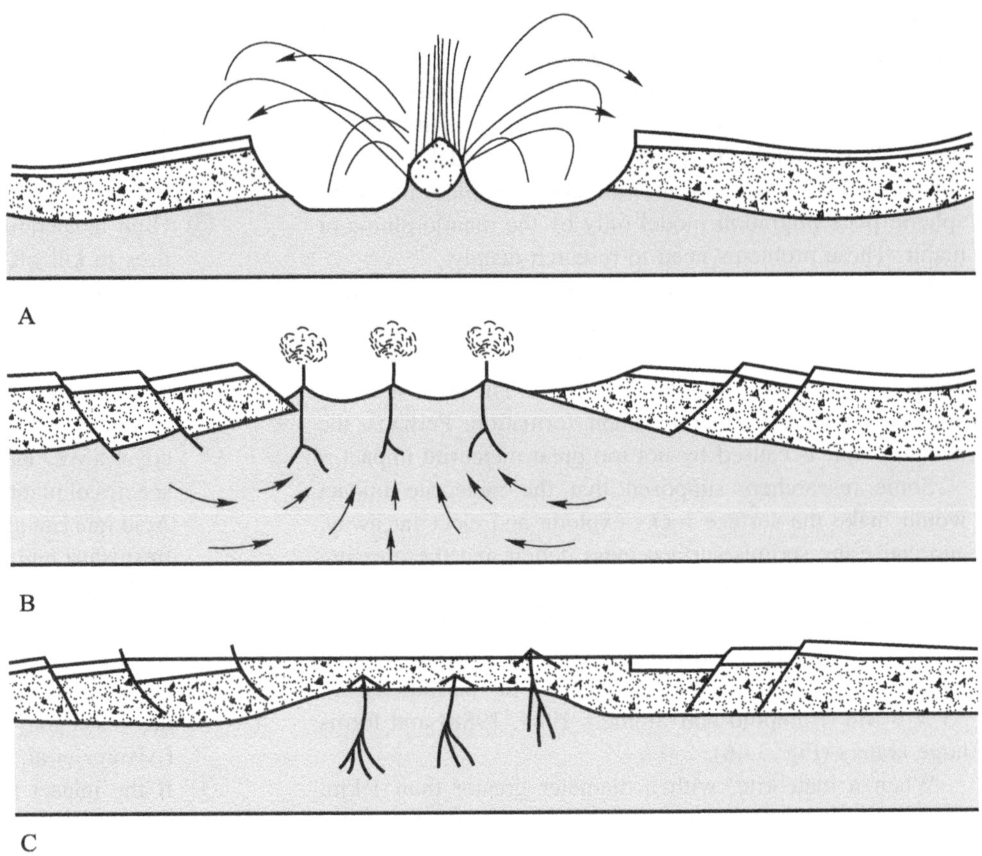

Fig. 3.46 Sketch illustrating the formation of a meteorite impacting crater. (A) A giant meteorite impacts the Earth surface to form a crater, and the rock materials (mixed and crushed meteorites and rocks) are ejected over a wide area; (B) the impact results in high temperature, crushed rock and mass deficit near the surface, the deep materials are upwelling over the surroundings, and volcanism may be induced, then to form fractures and more crushed rocks; (C) the ejected impact materials and lava may completely cover the meteorite crater. This impact process imitates the formation of craters on the lunar surface (After Press and Siever 1974)

around the galaxy: one of 33 ± 3 Ma and another of 265 ± 60 Ma. The former is related to the solar system crossing the galactic disk, where the interstellar matter is dense. The changes of the gravitational field may eject asteroids from their orbits to impact the Earth. The latter periodicity is the interval of the galactic year, i.e., the time of the solar system traveling around the galaxy. In a galactic year, the solar system crosses the galactic disk for eight times.

As to the periodicity of 33 ± 3 Ma, when the solar system crosses the galactic disk, it will be easy for a giant meteorite to impact on the Earth surface. If the meteorite impacts at a high angle, i.e., almost vertical to the Earth surface, it will be easy for the lithosphere plate to form the radiate extension. When the meteorite impacts at the low angle, the Earth will be suffered toward WSW tangential acting force; just like the Oligocene North American–Southeast Asian micro-tektite event (Fig. 3.28), the migration direction of the Pacific plate changed from the original NNW trending to WNW trending (Fig. 3.29; Yin and Wan 1996; Wan et al. 1997).

During the splitting process of Gondwana in Late Cretaceous, most of plates in the world were northward migrating; however, the Indian plate migrated at the fastest velocity (18 cm/y), and other plates migrated very slowly, just several centimeters per year. It had better explained this phenomenon by the giant meteorite oblique impact toward the Indian plate. If the meteorite impacted at a high angle or almost vertically, it would result in the lithospheric plate radiate extension, and the migration velocities would be almost same for the surrounding plates.

Perhaps, it is a better hypothesis to explain the change of plate migration model in Mesozoic–Cenozoic by some thoughts, such as the giant meteorite impact (Fig. 3.46) inducing the mantle diapir, forming the triple junction of plates and radial extension, sudden change of plate migration trending.

Due to the lack of data on change of Paleozoic plate migration model before Triassic (Figs. 3.6, 3.7, 3.8, 3.10, 3.11, 3.12), it has not obtained any believable clues till now. As to the long-period (about 200 Ma) formation and breakup models of supercontinents, such as the Columbia, Rodinia and Gondwana, some scholars predicted that they may be produced by the deep mantle plumes.

Torsvik et al. (2008, 2014) attempted to explain the mechanism of Paleozoic plate migration using the hypothesis that the uplift of mantle plume caused the horizontal migration of plate, and gave an illustration of the formation of the Pangea supercontinent. It is easy to understand the Western African mantle plume causing the Pangea supercontinent to split and break. Meanwhile, they thought that the deep mantle plume could make the plates converge into the Pangea supercontinent. It is difficult to understand, because there are too many conjectures and lack of data and facts.

To sum up, as for the lithosphere, as a very thin sphere on the Earth, its average thickness is about 1/60 of the Earth's radius. For its obvious migrations, differential stresses and rock deformations, it must be resulted from the comprehensive actions, such as the evolution of inner Earth, celestial movement and meteorite impact. The super-mantle plume, rising from the discontinuity between core and mantle, could cause the long-period (several thousand million years) and slow plate migration. As to some rootless hot spots and mantle diapirs, they could be caused by the giant meteorite impact, and they could be used to explain the short period (about 33 million years) and sudden change of the plate migration, as well as multi-orientations of the plate displacements.

It seems that there is no contradiction between two hypotheses of mantle plume and meteorite impact. These two hypotheses can be complementary to each other. In some day in the future, based on the deep research and systematically obtained data on the Earth's deep tectonics, it is believed that the dynamic mechanism of global lithospheric plate tectonics will make a breakthrough progress. The geodynamics and mechanism of lithospheric plate tectonics are indeed an extremely attractive and difficult frontier project. Only in the course of serving the human society, solving the actual problems of social development and accumulating all kinds of geological, geochemical, geophysical and astronomical data, can this difficult problem be perfectly solved. In the absence of a large amount of actual data, it is not feasible to establish a new doctrine or a new theory, and the benefits to scientific development are limited.

References

Aitchison J C, Davis A M (2001) When did the India–Asia collision really happen? In: International Symposium and Field Workshop on the Assembly and Breakup of Rodinia, Gondwana and Growth of Asia. Osaka City University, Japan. Gondwana Research, 4: 560–561.

Aitchison J C, Jason R A, Davis A M (2007) When and where did India and Asia collide? Journal of Geophysical Research, 112: B05423.

Allen M, Windley B, Zhang C (1993) Paleozoic collisional tectonics and magmatism of the Chinese Tianshan, Central Asia. Tectonophysics, 220: 89–115.

Alvarez L W, Alvarez W, Asaro F et al. (1980) Extraterrestrial cause for the Cretaceous-Tertiary extinction. Science 208: 1095–1108.

Bazhenov M L, Collins A Q, Degtyarev K E (2003) Paleozoic northward drift of the North Tian Shan (Central Asia) as revealed Ordovician and Carboniferous paleomagnetism. Tectonophysics, 366: 113–141.

Bai J, Huang X G, Wang H C et al. (1996) The Precambrian Crustal Evolution of China (2nd Edition). Beijing: Geological Publishing House, 1–223(in Chinese with English abstract).

Ben A, Van der Pluijm, John P et al. (1997) Paleostress in Cratonic North America: Implications for continental interiors. Science, 277(8): 794–796.

Bigwaard H, Spakman W, Engdahl E (1998) Closing the gap between regional and global travel tomography. Journal of Geophysical Research (B), 103: 30055–30078.

Bott M H P, Kusznir N J (1984) The origin of tectonic stress in the lithosphere. Tectonophysics, 105: 1–13.

Bott M H P, Kusznir N J (1991) Sublithospheric loading and plate-boundary forces. Phil. Trans. R. Soc. Lond., A 337: 83–93.

Brandon A D, Walker R J (2005) The debate over core-mantle interaction. Earth Planet. Sci. Lett. Frontiers, 232: 211–225.

Brenchley P J, Rowson P E (2006) The Geology of England and Wales (2nd Edition). The Geological Society of London.

Brookfield M E (2000) Geological development and Phanerozoic crustal accretion in the western segment of the south Tien Shan (Kyrgyzstan, Uzbekistan and Tajikistan). Tectonophysics, 328: 1–14.

Bucher W H (1933) The Deformation of the Earth's Crust. Princeton: Princeton University Press, 1–518.

Bureau of Geology and Mineral Resources of Jilin Province (1988) Regional Geology of Jilin Province. Beijing: Geological Publishing House, 1–698 (in Chinese with English abstract).

Bureau of Geology and Mineral Resources of Hebei Province, Beijing and Tianjin (1989) Regional Geology of Hebei Province. Beijing: Geological Publishing House, 1–741 (in Chinese with English abstract).

Bureau of Geology and Mineral Resources of Sichuan Province (1991) Regional Geology of Sichuan Province. Beijing: Geological Publishing House, 1–730 (in Chinese with English abstract).

Bureau of Geology and Mineral Resources of Inner Mongolia (1991) Regional Geology of Inner Mongolia. Beijing: Geological Publishing House, 1–725 (in Chinese with English abstract).

Buslov M M, Watanabe T, Fujiwara Y et al. (2004) Late Paleozoic faults of the Altai region, Central Asia: Tectonic pattern and model of formation. Journal of Asian Earth Sciences, 23: 655–671.

Cai X L (1965) The geological structure of Pre-Sinian in Dabieshan area: Also discussing the structural features of gneiss arch. In: Beijing College of Geology, Collection on Scientific Research, No.4, Special of Structure Geology and Regional Geology, 15–26 (in Chinese).

Cai X L, Zhu J S, Cao J M et al. (2002) Structure and dynamics of lithosphere and asthenosphere in the gigantic East Asian–West Pacific rift system. Geology in China, 29 (3): 234–245 (in Chinese with English abstract).

Cavazza W, Roure F M, Spakman W et al. (2004) The Transmed Atlas: The Mediterranean Region, From Crust to Mantle. Berlin, Heidelberg: Springer, 1–141.

Charvet J, Shu L S, Laurent-Charvet S (2007) Paleozoic structural and geodynamic evolution of eastern Tianshan (NW China): Welding of the Tarim and Junggar plates. Episodes, 30: 163–186.

Che Z C, Liu H F, Liu L (1994) The Formation and Evolution for Middle Tianshan Orogeny. Beijing: Geological Publishing House, 1–135 (In Chinese with English abstract).

Chen G D (1960) The Activation of Platforms and Their Significances in Deposits Investigation. Beijing: Geological Publishing House, 1–408 (in Chinese).

Chen G D (1978) Methods of Metallotectonic Research. Beijing: Geological Publishing House, 1–413 (in Chinese).

Chen G D (ed.) (1998) Crustal body Geotectonics of Asian Continent and Adjacent Seas. Changsha: Education Publishing House of Hunan, 1–322 (in Chinese).

Chen H H, Sun S, Li J L et al. (1991) New results of paleo-geomagnetism of South China plate. In: Annual Reports (1989–1990) of Open Laboratory of Lithosphere Tectonic

Evolution, Institute of Geology, China Academy of Sciences. Beijing: Science Press, 102–106 (in Chinese).

Chen Z (1984) The preliminary analysis of terrane characteristics in Taiwan, China. Oceanic Science and Technology of South China Sea, 6: 7–15 (in Chinese).

Cheng Y Q (1994) Introduction to the China Regional Geology. Beijing: Geological Publishing House, 1–517 (in Chinese with English abstract).

China Geological Society Petrological Division and China Petrological Society Geological Division (1966) The Atlas of Petrological Geology in China. Beijing: Geological Publishing House (in Chinese).

Chough S K, Kwon S T, Ree J H et al. (2000) Tectonic and sedimentary evolution of the Korean peninsula: A review and new view. Earth-Science Reviews, (52): 175–235.

Condie K C (2001) Mantle Plumes and Their Record in Earth History. Cambridge University Press, 1–306.

Condie K C, Aster R C (2013) Refinement of the supercontinent cycle with Hf, Nd and Sr isotopes. Geoscience Frontiers, 4 (6): 667–680.

Cui S Q (1999) On global Meso-Cenozoic intracontinental orogenesis and orogenic belts. Earth Science Frontiers, 6 (4): 283–293 (in Chinese with English abstract).

Dalziel I W D (1997) Overview: Neoproterozoic–Paleozoic geography and tectonics: Review, hypothesis, environmental speculation. Geological Society of America Bulletin, 109 (1): 16–42.

Deng J F, Zhao H L, Wu Z X et al. (1992) A mantle plume beneath the north part of China continent and lithosphere motion. Geoscience, 6 (3): 267–274 (in Chinese with English abstract).

Deng J F, Zhao H L, Mo X X et al. (1996) Continental Roots: Plume Tectonics of China. Beijing: Geological Publishing House, 1–105 (in Chinese).

Deng Q D, Ran Y K, Yang X P (2007) The Active Tectonics Map of China (1:4 000 000). Beijing: Seismological Press (in Chinese).

Domeier M, Torsvik T H (2014) Plate tectonics in the Late Paleozoic. Geoscience Frontiers, 5(3): 303–350.

Dong S W, Li T D, Chen X H et al. (2014) Deep Exploration showing the China crust structure, deep process and metallogene. Earth Science Frontiers, 21(3): 201–225.

Dong Y P, Zhang G W, Lai S C et al. (1999) An ophiolitic tectonic mélange first discovered in Huashan area, south margin of Qinling Orogenic Belt and its tectonic implications. Science in China (D), 29(3): 222–231.

Erickson D J, Dickson S M (1987) Global trace element biogeochemistry at the K/T boundary: Oceanic and biotic response to a hypothetical meteorite impact. Geology, 15: 1014–1017.

Ernst R E, Baragar W R A (1992) Evidence from magnetic fabric for the flow pattern of magma in the Mackenzie giant radiating dyke swarm. Nature, 356: 511–513.

Ernst R E, Grosfils E B, Mege D (2001) Giant dike swarms: Earth, Venus and Mars. Ann. Rev. Earth Planet. Sci., 29: 489–534.

Eskola P E (1949) The problem of mantled gneiss domes. Geol. Soc. London Quart. Jour., 104: 461–476.

Fan W M, Menzies M A (1992) Destruction of aged lower lithosphere and accretion of asthenosphere mantle beneath eastern China. Geotectonica et Metallogenia, 16(3–4): 171–180.

Fang D J, Jin G H, Chen H l et al. (1992) Preliminary approach to Late Paleozoic–Mesozoic geomagnetism and tectonic evolution in north margin of Tarim plate. In: Jia Runxu (chief editor), Research of Petroleum Geology of Northern Tarim Basin in China: Structural and Petroleum Geology, (2), 96–104. Wuhan: China University of Geosciences Press (in Chinese).

Fang D J, Jin G H, Jiang L P et al. (1996) Paleozoic paleo-magnetic results and the tectonic significance of Tarim Plate. Acta Geophysica Sinica, 39 (4): 522–532 (in Chinese with English abstract).

Forsyth D, Uyeda S (1975) On the relative importance of the driving forces of plate motion. Geophys. J. R. Astron. Soc., 43: 163–200.

Fromaget J I (1934) Observations et reflexions Sur La geologic stratigraphique et Indochina. Bull. Soc. Geol. France, Ve str. t. 4.

Gao J, Long L L, Qian Q et al. (2006) South Tianshan: A Late Paleozoic or Triassic orogen. Acta Petrologica Sinica, 22(5): 1049–1061 (in Chinese with English abstract).

Gao R, Wang H Y, Zhang Z J et al. (2011a) Penetrated crust-upper mantle, discovery continental deep structure and the effect for the resources and environment (Sino Probe-02) introduction of project and the science key problem. Journal of Earth, 32 (S1): 34–48 (in Chinese with English abstract).

Gao R, Wang H Y, Wang C S et al. (2011b) The lithosphere shortening in Northeast side Qingzang Plateau: The evidences by deep reflection section. Journal of Earth, 32 (5): 513–520 (in Chinese with English abstract).

Gao R, Wang H Y, Yin A et al. (2013) Tectonic development of the northeastern Tibetan Plateau as constrained by high-resolution deep seismic reflection data. Lithosphere, 5(6): 555–574.

Gao X Y, Gao R, Keller G R et al. (2013) Imaging the crustal structure beneath the eastern Tibetan Plateau and implications for the uplift of the Longmen Shan range. Earth and Planetary Science Letters, 379: 72–80.

Gao Z J, Wu S Z. 1983. Tectonic evolution of Tarim paleo-continent in Xinjiang. Chinese Science Bulletin, 28 (23): 1448–1450.

Garson M S, Miroslav K R S (1976) Geophysical and geological evidence of the relationship of Red Sea transverse tectonics to ancient fractures. GSA Bulletin, 87(2): 169–181.

Garzanti E, Gaetani M (2002) Uprooting history of Late Paleozoic magmatic arc within the "Turan Plate" (Tuarkyr, Turkmenistan). Sedimentary Geology, 151: 67–87.

Ge X H (1989) The history of formation of intraplate orogenic belts in the north China paleo-plate. Geological Review, 35 (3): 254–261 (in Chinese with English abstract).

Ge X H, Liu J L. (2000) Be broken "Western China Craton". Acta Petrologica Sinica, 16 (1): 59–66 (in Chinese with English abstract).

Ge X H, Ma W P (2014) The Course of China Regional Tectonics. Beijing: Geological Publishing House, 1–466 (in Chinese).

Ge X H, Liu J L, Ren S M (2014) The formation and evolution of Meso-Cenozoic tectonic in Eastern China. Geology in China, (1):19–28 (in Chinese and with English abstract).

Gen Y S, Zhou X W (2012) Early Permian magma-thermal event in Alxa metamorphic basement: Evidences from isotopic ages. Acta Petrologica Sinica, (9): 3–21.

Glass B P (1982) Tektites. In: Introduction to Planetary Geology. Cambridge University Press, 145–172.

Glikson A (1980) Precambrian sial-sima relation, evidence for Earth expansion. Tectonophysics, 6 (3): 193–234.

Grand S P, Van der Hilst R D, Widiyantoro S (1997) Global seismic tomography: A snapshot of convection in the Earth. GSA Today, 7 (4): 1–7.

Grabau A W (1940) The Rhythum of the Ages. Beijing: H. Wetch, 1–561.

Grieve R A F (1990) Terrestrial impact: the record in the rocks. Meteoritics, 26: 175–194.

Griggs D T (1939) Creep of rock. J. Geol., 47: 225–251.

Group of Regional Geology, Beijing College of Geology (1963) Regional Geology in China. Beijing: Industry Publishing House of China, 1–404 (in Chinese).

Guo F X (1998) Meso-Cenozoic Nanhua (South China) orogenic belt: Subaerial tri-directional orogen. Acta Geologica Sinica, 72 (1): 22–33 (in Chinese with English abstract).

Guo T Y, Liang D Y, Zhang Y Z et al. (1991) Geology of Ngari Tibet (Xizang). Wuhan: China University of Geosciences Press, 1–464 (in Chinese with English abstract).

Hager B H, Clayton R W, Richards M A et al. (1985) Lower mantle heterogeneity, dynamic topography and the geoid. Nature, 313: 541–545.

Hall R, Blundell D J (1995) Reconstructing Cenozoic SE Asia. Geological Society Special Publications, 106: 153–184.

Hall R, Cottam M A, Wilson M E J (eds.) (2011) The SE Asian Gateway: History and Tectonics of Australia–Asia Collision. Geological Society of London, Special Publication, 355: 1–381.

Hames W E, Renne P R, Ruppel C (2000) New evidence for geologically instantaneous emplacement of earliest Jurassic Central Atlantic magmatic province basalts on the North American margin. Geology, 28: 859–862.

He H S, He K Z (1993) The planation surface formation and its Quaternary tectonics. Geosciences, (1): 33–40.

Hilde T W, Uyeda C S, Kroenke L (1977) Evolution of the western Pacific and its margin. Tectonophysics, 38: 145–165.

Hofmann A W (1997) Early evolution of continents. Science, 275: 498–499.

Holmes A (1944) Principles of Physical Geology. London: Thomas Nelson and Sons, 1–532.

Hou G T (2012) Basic Dyke Swarm in North China. Beijing: Science Press, 1–177.

Hsu K J, Sun S, Li J L et al. (1988) Mesozoic overthrust tectonics in South China. Geology, 16: 418–821.

Huang B C, Zhou Y X, Zhu R X (2008) Based on the paleomagnetism to research the formation and evolution of China continent. Earth Science Frontier, 15(3): 348–359 (in Chinese with English abstract).

Huang T K (Jiqing) (1945) On the Major Structural Forms of China. Geological Memoirs, A (20):1–165(in English with Chinese abstract).

Huang T K (Jiqing) (1960) The main characteristics of the geological structure of China: Preliminary conclusions. Acta Geologica Sinica, 40(1): 1–37 (in Chinese).

Huang J Q (1984) New Research on tectonic features of China. Bulletin of the Chinese Academy of Geological Sciences, (9): 5–18. Beijing: Geological Publishing House (in Chinese with English abstract).

Huang J Q, Jiang C F (1962) From the view point of multiply tectonics cycle to preliminary discuss the crust evolution. Acta Geologica Sinica, (2): 3–50 (in Chinese).

Huang J Q, Zhang Z K, Zhang Z M et al. (1965) The Eugeosynclines and Miogeosynclines of China and their development of multi-cycle. In: Professional Papers of Academy of Geological Sciences, Section C: Regional Geology and Structure Geology. Beijing: Industry Publishing House of China, No. 1: 1–71 (in Chinese).

Huang J Q, Ren J S, Jiang C F et al. (1977) An outline of the tectonic characteristics of China. Acta Geologica Sinica, 51 (2): 117–135 (in Chinese).

Huang J L, Zhao D P (2006) High-resolution mantle tomography of China and surrounding regions. Journal Geophysical Research, 111 (B9), https://doi.org/10.1029/2005jb004066.

Hutchison C S, Tan D N K (eds.) (2009) Geology of Peninsular Malaysia. Kuala Lumpur: Murphy. The University of Malaya and the Geological Society of Malaysia, 1–479.

Jayananda M, Moyen J F, Martin H et al. (2000) Late Archaean (2550–2500 Ma) Juvenile magmatism in the Eastern Dharwar craton: Constraints from geochronology of Nd-Sr isotopes and whole rock geochemistry. Precambrian Research, 99: 225–254.

Jayananda M,Kano T, Peucat J J et al. (2008) 3.35 Ga Komatiite volcanism in the western Dharwar craton: Constraints from Nd isotopes and whole rock geochemistry. Precambrian Research, Elsevier. https://doi.org/10.1016/j.precamres.2007.07.010, v.162: 160–179.

Jiang S H, Nie F J, Su Y H et al. (2010) Chronology and mechanism of Erdenet super-large copper-molybdenum ore deposit in Mongolia. Acta Geoscientica Sinica, 31(3): 289–306.

Jin X C, Wang J, Ren L D et al. (1999) Problems concerning the tectonics of the west Kunlun orogen. In: Ma Z J et al. (eds.), Advances in Structural Geology and Geophysics in China—in Honor of Academician Ma Xingyuan's 80th Birthday. Beijing: Seismological Press, 105–113 (in Chinese with English abstract).

Kaur P, Chaudhri N (2013) Metallogeny associated with the Paleo–Mesoproterozoic Columbia supercontinent cycle: A synthesis of major metallic deposits. Ore Geology Review, http://dx.doi.org/10.1016/j.oregeorev.2013.3.005.

Kennedy W Q (1964) The structural differentiation of Africa in the Pan-African (500 m.y.) tectonic episode. Res. Inst. Ar. Geol., University of Leeds, 8th Ann. Rep., 48–49.

Khramov A N, Petrova G N, Peckersky D M (1981) Paleomagnetism of Soviet Union. In: McElhinny M W, Valencio D A (eds.), Paleoconstruction of the Continents, Geodynamic Series, Boulder. Colorado: Geological Society of America, 177–194.

Kim K H, Van der Voo R (1990) Jurassic and Triassic paleomagnetism of South Korea. Tectonics, 9 (4): 699–717.

Kim S W, Oh C W, Hyodo H et al. (2005) Metamorphic evolution of the Southwest Okcheon Metamorphic belt in South Korea and its regional tectonic implications. International Geology Review, 47: 344–370.

Klootwijk C T, Radhakrichnamurty (1981) Phanerozoic paleomagnetism of the Indian plate and India–Asia collision. In: McElhinny M W and Valencio D A (eds.), Paleoconstruction of the Continents. Geodynamics Series. Boulder, Colorado: Geological Society of America, 93–105.

Kröner A, Greiling R (1984) Precambrian Tectonics Illustrated. Stuttgart, 1–419.

Le Pichon S, Francheteau J, Bonin J (1973) Plate Tectonics. New York: Elsevier Publishing Company, 1–300.

Lee J S (Li Siguang) (1939) The Geology of China. London: Thomas Murby & Co. 1–528.

Lee T Y, Lawver L A (1995) Cenozoic plate reconstruction of Southeast Asia. Tectonophysics, 251(1–4): 85–138.

Li C (1997) The $^{40}Ar/^{39}Ar$ age and its significance of the crosstie from blueschists in the mid-Qiangtang Area, Tibet. Chinese Science Bulletin, 42 (4): 488 (in Chinese).

Li C, Zhai Q G, Chen W et al. (2006) The eclogites Ar-Ar dating in Middle Qiangtang, Qingzang Plateau. Acta Petrologica Sinica, 22 (12): 2843–2849 (in Chinese with English abstract).

Li H W, Xu K (2001) The dextral strike-slip faulting of Tan-Lu fault zone and the structural oil fields distribution in Liaohe basin. Earth Science Frontiers, 8 (4): 467–470 (in Chinese with English abstract).

Li J H, Li W B, Wang H H et al. (2014) Global tectonic setting the Late Paleozoic convergence of Pangea: Discussion of plate vortex trajectory. Acta Geologica of Sinica, 88(6): 980–991 (in Chinese).

Li J Y, Wang K Z, Li W Q et al. (2002) Eastern Tianshan tectonic and metallogeny. Geology of Xinjiang, 20(4): 295–301 (in Chinese with English abstract).

Li S G (Lee J S) (1926) The main genesis of visualization variation of Earth surface. Journal of Geological Society of China, 5 (3–4): 209–262.

Li S G (Lee J S) (1929) Some characteristic structural types in Eastern Asia and their bearing upon the problem of continental movement. Geological Magazine, Vol. 96, No. 782, 358–375; No. 784, 457–473; No. 785, 501–526 (Reprinted in 1976, Geological Mechanics Methods, pp. 65–112. Beijing: Science Press, in Chinese).

Li S G (Lee J S) (1947) Fundaments and methods of geological mechanics. In: Geological Mechanics Methods, (reprinted in 1976). Beijing: Science Press, 131–194 (in Chinese).

Li S G (1962) An Introduction to Geomechanics. Beijing: Geological Publishing House, 1–131 (Reprinted in 1973, in Chinese).

Li S G (1976) Orogenic history and structural contours of China. In: Geological Mechanics Methods. Beijing: Science Press, 215–228 (in Chinese) (Originally published in Contribution to the 7th Pacific Science Conference in New Zealand. 1949, 2: 26–44).

Li S G, Jagoutz E, Xiao Y L et al. (1996) Ultrahigh pressure metamorphic chronology in Dabieshan–Sulu terrane: I. Sm-Nd isotopic system. Science in China (D), 26 (3): 249–257 (in Chinese).

Li S G, Li H M, Chen Y Z et al. (1997) Ultrahigh pressure metamorphic chronology in Dabieshan–Sulu terrane: II. Zircon U-Pb isotopic system. Science in China (D), 27 (3): 200–206 (in Chinese).

Li S L, Mooney W D (1998) Crustal structure of China form deep seismic sounding profiles. Tectonophysics, 288: 105–113.

Li Y A, Li Q, Zhang Z K et al. (1989) Study on paleomagnetism of Tarim craton since Late Paleozoic. Xinjiang Geology, 7 (3): 1–77 (in Chinese with English abstract).

Li Y A, Cheng G L, Li Q et al. (1992) Paleomagnetic study of the eastern Xinjiang. Xinjiang Geology, 10(4): 329–371 (in Chinese with English abstract).

Liboutry L (1982) Tectonophysics and Geodynamics. Paris: Masson, 1–270 (in French).

Liu B J, Xu X S, Xia W J et al. (1994) Atlas of Petrological Phase Paleogeography of South China (Sinian to Triassic). Beijing: Science Press (in Chinese).

Liu B P, Feng Q L, Fang N Q (1991) Tectonic evolution of the Paleo-Tethys in Changning–Menglian and Lancangjiang belts, Western Yunnan. In: Proceedings of 1st International Symposium on Gondwana Dispersion and Asian Accretion (IGCP Project 321), Kunming, China. Beijing: Geological Publishing House, 189–192.

Liu J L, Davis G A, Lin Z Y et al. (2005) The Liaonan metamorphic core complex, southeastern Liaoning province, North China: A likely contributor to Cretaceous rotation of eastern Liaoning, Korea and contiguous areas. Tectonophysics, 407: 65–80.

Liu J L, Guan H M, Ji M et al. (2006) Late Mesozoic metamorphic core complex and its restrain for lithosphere thickness in North China. Natural Science Progress, 16 (1): 21–26 (in Chinese).

Liu Z S (2002) Geology of South China Sea. Beijing: Science Press, 1–502.

Lu F X (2010) Multiple geological events in ancient lithosphere mantle in North China Craton: Inspiration from peridotite xenolith in the Kimberlite. Acta Petrologica Sinica, 26 (11): 3177–3188 (in Chinese with English abstract).

Lu F X, Zheng J P, Shao J A et al. (2006) Asthenosphere upwelling and lithosphere thinning in late Cretaceous–Cenozoic in eastern North China. Earth Science Frontiers, 13(2): 86–92 (in Chinese with English abstract).

Lu S N, Chen Z H,Li H K et al. (2004) The Meso–Neoproterozoic geological evolution in Qinling orogenic belt. Bulletin of Geology, (2):11–16 (in Chinese with English abstract).

Lu S N, Li H K, Xian Z Q (2005) The research of Middle Proterozoic isotopic geological age. China Geology, 37(4): 1002–1013 (in Chinese)

Luo J H, Zhou X Y, Qiu B et al. (2005) The oil-gas geological characteristics and regional evolution in Karakum area. Geological Review, (4): 409–415 (in Chinese with English abstract).

Ma X H, Yang Z Y (1993) The collision and suturing of the three major blocks in China and the reconstruction of the Paleo-Eurasian Continent. Acta Geophysica Sinica, 36 (4): 476–488 (in Chinese with English abstract).

Ma X Y (chief editor, Editorial Board for Lithospheric Dynamics Atlas of China, State Seismological Bureau) (1989) Lithospheric Dynamics Atlas of China. Beijing: China Cartography Publishing House (in Chinese with English abstract).

Ma X Y, Cai X L (1965) Features of structure changes in Early Archean in east China: Problems about Tectonics of China. Beijing: Science Press, 141–150 (in Chinese).

Ma X Y, You Z D, Tan Y J et al. (1961) Some basic problems in China Tectonics. Acta Geologica Sinica, 41(1): 30–44 (in Chinese).

Ma Z J, Jiang M (1987) Strong earthquake periods and episodes in China. Earthquake Research in China, 3 (1): 47–51 (in Chinese with English abstract).

Mao J R, Tao K Y, Li J Y et al. (2002) Mesozoic grano-diorite characteristics and its tectonic evolution in Southwest Fujian. Journal of Petrology and Mineralogy, 21(2): 135-142 (in Chinese with English Abstract).

Maruyama S (1994) Plume tectonics. Journal of the Geological Society of Japan, 100(1): 24–49.

Maruyama S, Seno T (1986) Orogeny and relative plate motions: Example of the Japanese Islands. Tectonophysics, 127: 305–329.

Maruyama S, Liou J G, Zhang R (1992) Tectonic evolution of the ultrahigh-pressure (UHP) and high-pressure (HP) metamorphic belts from central China. The Island Arc, 3 (2): 112–121.

Marzoli A, Renne P R, Piccirillo E M et al. (1999) Extensive 200-million-year-old continental flood basalts of the Central Atlantic magmatic province. Science, 284 (5414): 616–618.

Mattauer M (1999) Seismique et tectonique. Pour la Science, (265): 28–31.

Metcalfe I (2011) Paleozoic–Mesozoic History of SE Asia. In: Hall R et al. (ed.), The SE Asian Gateway: History and Tectonics of the Australia–Asia Collision. London: The Geological Society of London Special Publication, 355: 7–36.

Meyerhoff A A, Taner I, Morris A E L et al. (1996) Surge tectonics: A new hypothesis of global geodynamics. Dordrecht, Boston, London: Kluwer Academic Publishers, 1–323.

Milanovsky E E (1980) Problems in the tectonic development of the Earth in the light of concepts on its pulsation and expansion. Rev. Geol. Dynam. Geol. Physique, 22 (1): 15–27.

Minster J B, Jordan T H (1978) Present-day plate motions. J. Geophys. Res., A, Space Physics, 83 (B11): 5331–5354.

Moore G W (1989) Mesozoic and Cenozoic paleogeographic development of the Pacific region (Abstract). In: 28th International Geological Congress. Washington D C, USA, 2–455.

Neves S P, Mariano G (2004) Heat-producing elements-enriched continental mantle lithosphere and Proterozoic Intracontinental orogens: Insights from Brasiliano/Pan-African Belts. Gondwana Research, 7(2): 427–436.

Nezafati N (2006) Au-Sn-W-Cu-Mineralization in the Astaneh–Sarband Area, West Central Iran (including a comparison of the ores with ancient bronze artifacts from Western Asia). der Geowissenschaftlichen Fakultät, der Eberhard–Karls–Universität Tübingen, 1–116.

Norris R D, Kroon D et al. (1998) Proceedings of the Ocean Drilling Program, Initial Report, vol. 171 B, College Station, TX (Ocean Drilling Program).

Opdyke N D, Huang K, Xu G et al. (1986) Paleomagnetic results from the Triassic of the Yangtze Platform. J. Geophys. Res. (B), 91(9): 9553–9568.

Osozawa S (1998) Major transform duplexing along the eastern margin of Cretaceous Eurasia. In: Flower M F J et al. (eds.), Mantle Dynamics and Plate Interactions in East Asia, Geodynamics Series. Washington, D C: AGU, 27: 245–257.

Ouyang Z Y (chief editor) (1995) Formation and evolution of Planetary Earth. Geology and Geochemistry, (5): 1–105 (in Chinese).

Ouyang Z Y, Liu J Z, Zhang F Q et al. (2002) A preliminary study on the origin and evolution of the planetary Earth′s heterogeneity. Earth Science Frontiers, 9 (3): 23–30 (in Chinese with English abstract).

Owen H G (1992) Has the Earth increased in size? In: Chatterjee S, Horton N III (eds.), New Concepts in Global Tectonics. Lubbock: Texas Tech. University Press.
Pavoni N (1997) Geotectonic bipolarity: Evidence of bi-cellular convection in the Earth's mantle. S. Afr. J. Geol., 100 (4): 291–299.
Pickering K T, Koren T N, Lytochkin V N (2008) Silurian–Devonian active–margin deep marine systems and paleogeography, Alai Range, Southern Tian Shan, Central Asia. Journal of the Geological Society of London, 165: 189–210.
Piper J D A (2013a) Continental velocity through Precambrian times: The link to magmatism, crustal accretion and episodes of global cooling. Geoscience Frontiers, 4(1): 7–36.
Piper J D A (2013b) A planetary perspective on Earth evolution: Lid tectonics before plate tectonics. Tectonophysics, 589: 191–223.
Pollard D D, Aydin (1988) Progress in understanding jointing over the past century. DOI:10.1130/SPE253-p313.
Powell C M, Li Z X, McElhinny M W et al. (1993) Paleomagnetic constraints on timing of the Neoproterozoic breakup of Rodinia and the Cambrian formation of Gondwana. Geology, 21: 889–892.
Press F, Siever R (1974) Earth. W H Freeman and Company, 1–613.
Prinn R G, Fegley B (1987) Bolide impacts, acid rain, and biospheric traumas at the Cretaceous-Tertiary boundary. Earth and Planetary Science Letters, 83: 1–15.
Qian X L (1996) The character and tectonic evolution in the Early Pre-Cambrian continental crust, Acta Petrologica, 12(2): 169–178 (in Chinese)
Qin D X, Yan Y F, Tian Y L et al. (2000) The geological characteristics and meneralloge evolution in the Dahongshan copper deposit. Geologica Sinica, 35(2): 129–130.
Qin J X, Hao T Y, Xu Y et al. (2011) The depth distribution of Moho discontinuity in South China Sea and the relationship with the tectonic unit. Acta Geophysica Sinica, 54 (12): 3171–3183.
Rampino M R, Volk T (1988) DMS and the K/T boundary: Phytoplankton extinctions, reduction in cloud albedo, and climatic warming. Nature, 322: 63–65.
Rampino M R, Stothers R B (1984) Terrestrial mass extinctions: Cometary impacts and the Sun's motion perpendicular to the galactic plane. Nature, 308(5961): 709–712.
Rampino M R, Stothers R B (1988) Flood basalt volcanism during the past 250 million years. Science, 241(4866): 663–668.
Raymond C A, Stock J M, Cande S C (2000) Fast Paleogene motion of the Pacific hotspots from revised global plate circuit constraints: The history and dynamics of global plate motions. Geophysical Monographys, 121: 359–375.
Ren J S, Jiang C F, Zhang Z K et al. (1980) The Geotectonic Evolution of China (explanation of 1: 4 M scale China Tectonic Map) Beijing: Science Press, 1–124 (in Chinese).
Ren J S, Chen T Y, Niu B G et al. (1990) Tectonic Evolution of the Continental Lithosphere and Metallogeny in Eastern China and Adjacent Areas. Beijing: Science Press, 1–205 (in Chinese).
Ren J S, Wang Z X, Chen B W et al. (2000) China Tectonics, Outlook from the Globe: Brief Direction of Tectonic Map of China and its Adjacent Areas. Beijing: Geological Publishing House, 1–50 (in Chinese).
Replumaz A et al. (2004) 4-D evolution of SE Asia's mantle from geological reconstructions and seismic tomography. Earth and Planetary Science Letters, 221: 103–115.
Rogers J J W, Santosh M (2002) Configuration of Columbia: A Mesoproterozoic Supercontinent. Gondwana Research, 5(1): 5–22.
Rogers J J W, Santosh M (2004) Continents and Supercontinents. New York: Oxford Press, 1–289.
Safranov V S (1972) Evolution of the Protoplanetary Cloud and Formation of the Earth and Planets. Moscow: Nauka (Translated by the Israel Program for scientific translation).
Scheidegger A E (1963) Principles of Geodynamics (2nd edition). Springer-Verlag, Berlin, 1–254.
Scheidegger A E (1982) Principles of Geodynamics (3rd edition). Springer-Verlag, Berlin, 1–395.
Schettino A, Scotese C R (2005) Apparent polar wander paths for the major continents (200 Ma to the present day): A paleomagnetic reference frame for global plate tectonic reconstructions. Geophys. J. Int., 163: 727–759.
Schmidt C J. Chase R B, Erslev E A (1993) Laramide basement deformation in the Rocky Mountain foreland of the western United States. Geological Society of America, Special Paper, 280: 1–372.
Scotese C R (1994) Continental Drift (6th edition). The PALEOMAP Project. University of Texas at Arlington.
Shao J A, He G, Zhang L (2007) Deep crustal structures of the Yanshan intracontinental orogeny: A comparison with pericontinental and intercontinental orogenics. Geological Society, London, Special Publications, 280: 189–200. https://doi.org/10.1144/sp280.9.
Sharpton V L, Dalrymple G B, Marin L E et al. (1992) New links between the Chicxulub impact structure and the Cretaceous/Tertiary boundary. Nature, 359, (6398): 819–821.
Shellnut J G (2014) The Emeishan large igneous province: A synthesis. Geoscience Frontiers, 5(3): 369–394.
Shi Y S, Lu H F, Jia D et al. (1994) Paleozoic plate tectonic evolution of Tarim and Western Tianshan region, Western China. International Geological Review, 36: 1058–1066.
Shu L S, Michel F, Jiang S Y et al. (2006) SHRIMP zircon U-Pb age, litho- and bio-stratigraphy analyses of the Huaiyu Domain in South China: Evidence for a Neoproterozoic orogen, not Late Paleozoic-Early Mesozoic collision. Episodes, 9(4): 244–252.
Shu L S, Michel F et al. (2008) Late Paleozoic-Early Mesozoic geological feature of South China: Response to the Indosinian collision events in Southeast Asia. Geoscience (340):151–165.
Sigloch K, McQuarrie N, Nolet G (2008) Two-stage subduction history under North America inferred from multiple-frequency tomography. Published online 29 June, https://doi.org/10.1038/ngeo231.
Song H L. 1999. Characteristic of Yanshan type intraplate orogenic belts and a discussion on its dynamics. Earth Science Frontiers, 6 (4): 309–316 (in Chinese with English abstract).
Stephan J F, Blanchet R, Rangin C et al. (1986) Geodynamic evolution of the Taiwan–Luzong–Mindoro belt since the Late Eocene. Tectonophysics, 125 (1–3): 245–268.
Stille H (1924) Grundfragen der Vergleichenden Tektonik. Berlin: Borntraeger, 1–443.
Storey B C (1995) The role of mantle plumes in continental breakup: Case histories from Gondwanaland. Nature, 377: 301–308.
Storey B C, Kyle P R (1997) An active mantle mechanism for Gondwana breakup. S. African J. Geology, 100 (4): 283–290.
Suess (1885) Das Antlitz der Erde (3 Volumes). Prague and Leipzig, F. Tempsky and G. Freytag.
Tamaki K, Suyehiro K, Allan J et al. (1992) Tectonic synthesis and implications of Japan Sea ODP drilling. Proc. Ocean Drill, Program Sci. Results, 127–128: 1333–1348.
Tan X C, Li C Y (1948) Geology of Xikang (Sichuan Province) In: Central Geological Survey, Geological Memoirs. A. No. 15 (reprinted by Geological Publishing House in 1959, in Chinese).
Tang Z (1979) Structural features of oil-gas-bearing basins of eastern China. Petroleum Exploration and Exploitation (1): 30–37 (in Chinese).

Teng J W, Zeng R S, Yan Y F et al. (2003) Depth distribution of Moho and tectonic framework in eastern Asian continent and its adjacent ocean. Science in China, D 46 (5): 428–446.
The Geological Society of America (1989) The Geology of North America, vol. A, The Geology of North America: An overview.
Torsvik T H, Muller R D, Van der Voo R et al. (2008), Global plate motion frames: Toward a unified model. Rev. Geophys., 46, RG3004, https://doi.org/10.1029/2007RG000227.
Torsvik T H, Van der Voo R, Doubrovinea P V et al. (2014) Deep mantle structure as a reference frame for movements in and on the Earth. www.pnas.org/lookup/suppl/, https://doi.org/10.1073/pnas.1318135111/-/dcsupple-mental.
Turcotte D L, Schubet G (1982) Geodynamics: Applications of Continuum Physics to Geophysical Problems. New York: John Wiley & Sons, 1–303.
Umbgrove J H F (1947) The pulse of the Earth. Nihoff: The Hague, 1–358.
Van der Voo R (1993) Paleomagnetism of the Atlantic, Tethys and Iapetus Oceans. Cambridge University Press, 1–273.
Van der Voo R, Spakman W, Bigwaard H (1999) Tethyan subducted slabs under India. Earth Planet. Sci. Lett.,171 (1): 7–20.
Wan T F (2011) The Tectonics of China: Data, Maps and Evolution. Beijing: Higher Education Press; Heidelberg, Dordrecht, London, New York: Springer, 1–501.
Wan T F, Chu M J (1987) The active listric fault in Fujian and Taiwan. Earth Science, 12(1): 21–29 (in Chinese with English abstract).
Wan T F (1994) Intraplate Deformation, Tectonic Stress Field and their Application for Eastern China in Meso-Cenozoic. Wuhan: Press of China University of Geosciences, 1–156.
Wan T F, Lu H F (2014) The Eastern China continental crust-oceanic mantle lithosphere and its forming mechanism. Geotectonica et Metallogenia, 38(3): 495–511 (in Chinese).
Wan T F, Zhu H (2011) Chinese continental blocks in global paleocontinental reconstructions, during the Paleozoic and Mesozoic. Acta Geologica Sinica, 85 (3): 581–597.
Wan T F, Zhao Q L (2012) The genesis of tectono-magmatism in eastern China. Science China, Earth Sciences, 55 (3): 347–354.
Wan T F, Yin Y H, Zhang C H (1997) On the extraterrestrial impact and plate tectonic dynamics: A possible interpretation. Proceedings of 30th International Geological Congress, VSP, 26: 87–95.
Wan T F, Wang Y M, Liu J L (2008) Detachment and magmatic source depth in lithosphere of Eastern China during Yanshanian and Sichuanian stages. Earth Science Frontiers, 15(3):1–35 (in Chinese with English abstract).
Wan T F, Zhao Q L, Wang Q Q (2015) Paleozoic Tectono-Metallogeny in Tianshan–Altay Region, Central Asia. Acta Geologica Sinica, 89 (4):1120–1132.
Wan T F, Zhao Q L, Lu H F et al. (2016) Discussion on the special lithosphere type in Eastern China. Earth Sciences, 5(1): 1–12.
Wan T F, Zhu H, Zhao L et al. (1996) Formation and Evolution of the Tancheng-Lujiang Fault Zone. Wuhan: Press of China University of Geosciences. 1–85.
Wang C L, Yang D P, Zhang B et al. (2011) Western Xuefengshan Mesozoic thrust style and deformation characteristic, Recent Geology (6): 3–13 (in Chinese with English abstract).
Wang C S, Li X H, Hu X M et al. (2002) Latest marine horizon north of Qomolangma (Mt. Everest): Implications for closure of Tethys seaway and collision tectonics. Terra Nova, 14: 114–120.
Wang H Z,Li X, Mei S L et al. (1997) Pangea cycles, earth rhythms and possible earth expansion. In: Wang H Z, Jahn Borming, Mei S H (eds.), Origin and History of the Earth. Proc. 30th Intern. Geol. Congr., 1: 111–128. VSP, Utrecht, The Netherland.
Wang R, He G Q (1979) Rate change of Earth and the global stress field caused by its tide-drawing force under the condition of axial symmetry. In: Astronomical Observatory of Shanghai. Astronomical Geodynamics Memoirs, Beijing: Science Press, 8–21 (in Chinese).
Wang X S, Fang J, Xu H Z et al. (2012) The lithosphere three-dimension density structure in North China Craton. Acta Geophysica Sinica, (4): 90–96 (in Chinese with English abstract).
Wang Y, Chen Z L, Liu J et al. (2006) Neotectonic characteristics and controlled the sandstone uranium deposit in southern Yili basin. Geotectonica et Metallogenia, 30 (4): 486–494 (in Chinese with English abstract).
Wang Y, Li J Y, Sun G H (2008) Post-collision eastward extrusion and tectonic exhumation along the eastern Tianshan orogen, Central Asia: Constraints from dextral strike-slip motion and $^{40}Ar/^{39}Ar$ geochronological evidence. Journal of Geology, 116: 599–618.
Wang Y S, Chen J N (1987) Metamorphic belts and metamorphism of Qinghai province and adjacent areas, Geological Memoirs, No. 6. Beijing: Geological Publishing House (in Chinese).
Weng W H (1927) Mesozoic orogenic movement of east China. Journal of Geological Society of China, 8(1): 33–44.
Wilson J T (ed.) (1970) Continents Adrift: Readings from Scientific American. San Francisco: W. H. Freeman and Company.
Wolbach W S, Lewis R S, Anders E (1985) Cretaceous extinctions: Evidence for wildfires and search for meteoric material. Science, 230(4722): 167–170.
Wu F Y, Xu Y G, Gao S et al. (2008) The main controversy of thinner and the break for North China lithosphere. Acta Petrologica Sinica, 24: 1145–1174 (in Chinese).
Wu G Y, Li Y J, Wang S L et al. (2006) Neoproterozoic oceanic island volcanic rock in Bazu area, Western Xinjiang. Recent Geology, 20 (3): 361–369 (in Chinese with English abstract).
Wu H N, Chang C F, Liu C et al. (1991) Evolution of the Qinling fold belt and the movement of the North and South China blocks: The evidence of geology and paleomagnetism. Scientia Geologica Sinica (Northwest University, science edition), (3): 201–214 (in Chinese with English abstract).
Wu Z W, Zhang C H (1999) On the Chinese orogenic belt theory. Earth Science Frontier, 6(3): 21–29 (in Chinese with English abstract).
Xiao W J, Windley B F, Hao J et al. (2003) Accretion leading to collision and the Permian Solonker suture, Inner Mongolia, China. Tectonics, 22(6): 1069. https://doi.org/10.1029/2002tc001484.
Xiao W J, Han C M, Yuan C et al. (2008) Middle Cambrian to Permian subduction-related accretionary orogenesis of Northern Xijiang, NW China: Implications for the tectonic evolution of Central Asia. Journal of Asian Earth Sciences, 32:102–117.
Xiao W J,Kröner A, Windley B F (2009a) Geodynamic evolution of Central Asia in the Paleozoic and Mesozoic. International Journal of Earth Sciences, 98: 1185–1188.
Xiao W J, Windley B F, Huang B C et al. (2009b) End-Permian to mid-Triassic termination of the accretionary processes of the southern Altaids: Implications for the geodynamic evolution, Phanerozoic continental growth, and metallogeny of Central Asia. International Journal of Earth Sciences, 98: 1189–1287.
Xiao X C, Tang Y Q, Feng Y M et al. (1992) Tectonic Evolution of Northern Xinjiang and Its Adjacent Regions, Beijing: Geological Publishing House, 1–169 (in Chinese with English abstract).
Xu W L, Ge W C, Pei F P et al. (2008) Chronology of Mesozoic volcanism and its tectonic significance in Northeast China. Proceedings of China Geochemistry Conference, Guiyang (unpublished paper; in Chinese).
Xu Y G, Zhang H H, Qiu H N (2012) Oceanic crust components in continental basalts from Songliao, Northeastern China: Derived from the mantle transition zone? Chemical Geology. https://doi.org/10.1016/j.chemgeo.2012.01.027.
Xu Z Q, Zhao Z X, Yang J S et al. (2003) Tectonics beneath plates and mantle dynamics. Geological Bulletin of China, 22 (3): 149–159 (in Chinese with English abstract).

Xue F, Huang J W (1989) Distribution of earthquake focal depths. In: Ma X Y (chief editor), Lithospheric Dynamics Atlas of China, Fig. 25. Beijing: China Cartography Publishing House (in Chinese).

Yakobchuk A (2010) Restoring the supercontinent Columbia and tracing its fragments after its breakup: A new configuration and a Super-Horde hypothesis. Journal of Geodynamics, 50(3–4): 166–175.

Yang D Y (1988) Genetic mechanism of the alluvial terranes along the Three-Gorges course of the Yangtze River. Acta Geologica Sinica, 43 (2): 120–126 (in Chinese with English abstract).

Yang D Y, Lv G N (1992) Study on the through time and its geological implication of the Three Gorges along the Yangtze River, In: Liu D S, An Z S (chief editors), Loess, Quaternary Geology and Global Change. Volume 3, 140–143. Beijing: Science Press (in Chinese).

Yang H T (1999) Sea level rise and coastal disasters in China's coast. Quaternary Sciences, (5): 456–465 (in Chinese with English abstract).

Yang Z Y, Besse J (1996) Paleomagnetic study of Permian and Mesozoic sedimentary rocks from North Thailand supper Report. In: The Laboratory of Lithosphere Tectonics and its Dynamics (MGMR China). Beijing: Geological Publishing House, 122–132.

Yin H F, Xu D Y, Wu R T (1988) Catastrophism of Geological Evolution: Wuhan: China University of Geosciences Press, 1–201 (in Chinese).

Yin Y H, Wan T F (1996) The possibility and dynamics of a microtektite impacted the Pacific plate and caused the change of its moving direction in the end of Eocene. In: 1995 Annual Report, The Laboratory of Lithosphere Tectonics and its Dynamics (MGMR China). Beijing: Geological Publishing House, 122–132.

Yong Y Z (1988) The division of N–S seismic zone in China and its significance. Sichuan Seismic, (1): 33–36.

Yoon S (2001) Tectonic history of the Japan Sea region and its implications for the formation of the Japan Sea. Journal of Himalayan Geology, 22 (1): 153–184.

Yoshikura S, Hada S, Isozaki Y (1990) Kurosegawa terrane. In: Ichikawa K, Mizutani S, Hara I (eds.), Pre- Cretaceous Terranes in Japan. Publication of IGCP Project, No. 224, Osaka.

Yu S Y, Xu Y G, Ma J L et al. (2010) Remnants of oceanic lower crust in the subcontinental lithospheric mantle: Trace element and Sr-Nd-O isotopic evidence from aluminous garnet pyroxenite xenoliths from Jiaohe Northeast China. Earth Planet. Sci. Lett., 297: 413–422.

Zang S X, Ning Y J (1994) Advanced and problems of global dynamics research. In: Contribution to Symposium of Modern Geodynamics. Beijing: Seismic Publishing House, 13–18 (in Chinese).

Zeng H L (2005) Gravity Field and Gravity Exploration. Beijing: Geological Publishing House, 1–273 (in Chinese).

Zeng H L, Wan T F (1999) Gross differences between two isostatic gravity anomaly maps of China. Tectonophysics, 306 (2): 253–257.

Zeng H L, Wan T F, Teyssier C et al. (2000) The 3-D geometry of the Linglong granitic complex from 2-D gravity forward modeling, Shandong Province, East China. Geophysics, 65 (2): 421–425.

Zhai M G (2007) North China craton Paleoproterozoic tectonic event. Acta Petrologica Sinica, 23(11): 2665–2682 (in Chinese with English abstract).

Zhai M G (2010) How the earth crust to be formed?—Mystical and interesting Pre-Cambrian geology. Nature Journal, 32 (3): 126–129 (in Chinese with English abstract).

Zhai M G (2014) Multi-stage crustal growth and cratonization of the North China Craton. Geoscience Frontiers, 5(4): 457–469.

Zhan M G (1994) South China Mesozoic tectono-magmatism-metallogeny research. Ph. D Thesis of China University of Geosciences, 1–102 (in Chinese with English abstract).

Zhang G W, Zhang B R, Yuan X C et al. (2001) Qinling Orogenic Belt and Continental Dynamic. Beijing: Science Press, 1–855 (in Chinese with English abstract).

Zhang H F (2009) Peridotite-melt intersection: The key of the craton type lithosphere mantle breakup. Science Bulletin, 54: 2008–2026 (in Chinese).

Zhang H F, Yang Y H, Santosh M (2012) Evolution of the Archean and Paleoproterozoic lower crust beneath the Trans-North China Orogen and the Western Block of the North China Craton. Gondwana Research., 22 (1): 73–85.

Zhang J J, Zheng Y F, Zhao Z F (2009) Geochemical evidence for interaction between oceanic crust and lithospheric mantle in the origin of Cenozoic continental basalts in east-central China. Lithos, 110: 305–326.

Zhang L (2005) Seismic tomography research for crust and upper mantle around Bohai Bay areas. Ph D Thesis of China Academy Institute, 1–79 (in Chinese with English abstract).

Zhang Q, Qian Q, Wang E Q et al. (2001) The Eastern China Plateau in Middle-Late Yanshan Period: Inspiration of adakite. Geological Science, 36(2): 248–255 (in Chinese with English abstract).

Zhang S H, Zhu H, Meng X H (2001) New paleomagnetic results and its paleogeography significant in Devonian–Carboniferous for the Yangtze block. Acta Geologica Sinica, 75(3): 303–313 (in Chinese with English abstract).

Zhang S H, Li Z X, Evans D A D et al. (2012) Pre-Rodinia supercontinent Nuna shaping up: A global synthesis with new paleomagnetic results from North China. Earth and Planetary Science Letters, 353–354: 145–155.

Zhang W Y (ed.) (1959) An Outline of Tectonics of China. Beijing: Science Press, 1–320 (in Chinese).

Zhang W Y (1984) An Introduction to Fault-block Tectonics. Beijing: Petroleum Industry Press, 1–385 (in Chinese with English abstract).

Zhao D P, Liu L (2010) Deep structure and origin of active volcanoes in China. Geoscience Frontiers, 1 (1): 31–44.

Zhao D P, Maruyama S, Omori S (2007) Mantle dynamics of Western Pacific and East Asia: Insight from seismic tomography and mineral physics. Gondwana Research, 11: 120–131.

Zhao G C, Sun M, Wide S A et al. (2004) Late Archean in Paleoproterozoic evolution of the Trans-North China orogen. In: Aspects of the Tectonic Evolution of China. London, The Geological Society, Special Publication, 226: 27–56.

Zhao W J, Wu Z H, Shi D N et al. (2014) The deep structure and orogeny mechanism for Kunlun Mountains. Geology in China, (1): 5–2 (in Chinese with English abstract).

Zhao X X, Coe R S, Gilder S A et al. (1996) Paleomagnetic constraints on the paleogeography of China: Implications for Gondwanaland. Australian Journal of Earth Sciences, 43 (6): 643–672.

Zhao Y (1990) The Mesozoic orogeny and tectonic evolution in Yanshan area. Geological Review, (1): 3–15 (in Chinese with English abstract).

Zhao Z F (1959) On the Yanshan movement. Geological Review, 19 (8): 339–346 (in Chinese with English abstract).

Zhao Z F (1986) Review of the 50th anniversary of the Indosinian Movement. Scientia Geologica Sinica, (1): 7–15 (in Chinese with English abstract).

Zheng T Y, Zhao L, Zhu R X (2009) New evidence from seismic imaging for subduction during assembly of the North China craton. Geology, 37: 395–398.

Zheng T Y, Zhu R X, Zhao L et al. (2012) Intra-lithospheric mantle structures recorded continental subduction. Journal of Geophysical Research, 117: B03308, https://doi.org/10.1029/2011.jb008873.

Zhong D L (1998) Paleo-Tethys orogenic belt in Western Yunnan-Sichuan. Beijing: Science Press, 1–231 (in Chinese with English abstract).

Zhou J B, Zhang X Z, Wilde S A (2011) Determine the Pan-African event (~500 Ma) Khondalite in Northeast China and its significance. Acta Petrologica Sinica, (4): 345–355 (in Chinese with English abstract).

Zhou X H (2006) On the lithosphere transform type and thinner research during Meso-Cenozoic, in Eastern China. Earth Science Frontier, 13 (2): 50–64 (in Chinese with English abstract).

Zhu G, Hu Z Q, Chen Y et al. (2008) The significance of Early Cretaceous extension basin development for the craton breakup in Eastern China Craton. Geological Bulletin, 27: 1594–1604 (in Chinese).

Zhu R X, Chen L, Wu F Y et al. (2011) The time, scope and mechanism for the North China craton breakup. Science in China (D), 41: 583–592 (in Chinese).

Zhu R X, Xu Y G, Zhu G et al. (2012) North China craton breakup. Science in China (D), 42 (8):1135–1159 (in Chinese).

Zhu Z Y (1989) The river terrace formation and river system evolution in the middle reaches of Yellow River. Acta Geographic Sinica, 44 (4): 429–440 (in Chinese with English abstract).

Zhu Z W, Hao T Y, Zhao H S (1988) The paleomagnetism of Mesozoic strata in Panxi region and its tectonic significance. In: The Collections of Panxi Rift in China (3). Beijing: Geological Publishing House, 199–211 (in Chinese).

Zhuang Z H (1988) The paleo-magnetism research in Cretaceous-Paleogene, from Ya′an to Tianquan in Shichuan Basin. Geophysical and Geochemical Exploration, 12(3): 224–228 (in Chinese).

Zoback M L, Magee M (1991) Stress magnitudes in the crust: Constraints from stress orientation and relative magnitude data. Phil. Trans. R. Soc. Lond., A 337 (1645): 181–194.

Zonenshain L P, Kuzmin M L, Natapov L M (1990) Geology of the USSR: A Plate Tectonic Synthesis. American Geophysical Union, Geodynamics Series, Volume 21: 1–242, Washington D C.

4 Tectono-Metallogenesis in Asian Continent

4.1 The Giant Ore Fields and Deposits in Tectonic Domains

In this part, the main characteristics of the giant ore fields and deposits formed in every tectonic domain will be discussed. This part mainly aims to research the giant ore-field structures for the endogenic metallic deposits. The main characteristics of the super-large, giant ore fields and deposits are listed in the Appendix. This monograph pays more attention to the ore-field structures of endogenic metallic deposits and their ore-controlling tectonics, which may be very useful to explore the ore fields and deposits in the future.

4.1.1 The Giant Ore Fields and Deposits in Siberian Tectonic Domain

4.1.1.1 The Giant Ore Fields and Deposits in Siberian Plate [1]

There are many kimberlite pipes preserved in the Siberian plate, 15% of which are related to the diamond deposits. According to the seismic exploration, known diamond-bearing kimberlite pipes are mostly located in high-density rocks, i.e., ultra-mafic rocks in Archean blocks. The kimberlite pipes were finally formed in Late Paleozoic (about 360 Ma) or Cretaceous (127–90 Ma), which were controlled by the intraplate rift magmatism. The kimberlite-type diamond deposits are distributed at Malobatuoba, Alakit, Daldyn, Mun, Nakyn (Fig. 2.2). In the most of situations, the isotopic ages of ultra-mafic rocks with the diamond-bearing kimberlite pipes and eclogite xenolith are Archean–Paleoproterozoic (3.5–3.2 Ga and even to 2.0 Ga). Generally speaking, the high-density blocks of lower crust and upper mantle are the remains after partial melting, which are the major sources of the diamond-bearing kimberlite pipes. As seen from the space distribution of ore fields, they are controlled by the boundary or radiate faults in continental nucleus. In other words, they mostly are the ancient tectonic fragments of continental crust. It seems that the "continental nucleus model" in the Pre-Cambrian Earth dynamic mechanism is best one to explain the space distribution of the Siberian kimberlite-type diamond deposits (Moralev and Glukhovsky 2000; Fig. 2.2).

Masaitis (2002) and Yelisseyeva et al. (2013) pronounced that a small planet impact area with the diameter about 100 km existed at the **Popigai Astroblem** area in North Siberia (111° 11′ E; 71° 39′ N), and thought that the small planet hit the Earth 35 Ma ago. In the meteorite crater, they discovered a large-scale diamond deposit. According to their estimation, the diamond reserves would be about 10^{12} karats, which is the ten times of the global reserves, and could satisfy the future requirements for 3000 years. The hardness of impact diamonds is two times than that of common diamonds. However, the above estimation should be checked by deep exploration.

After Neoproterozoic Era, in the West Siberia, the violent basalt eruption occurred and the fault-depression formed, which are covered by the Paleozoic and Mesozoic–Cenozoic basic volcanic and sedimentary rocks. The high-density basic volcanic rocks by gravity balance have caused the depression areas to sink for long times (Fig. 2.3). In Late Permian (about 252 Ma), the large-scale continental basalt flow erupted to form the giant Siberian traps from the **Noril'sk, Taimyr and Tunguska** of the North Siberia to the east side of Ural Mountains (Fig. 2.3; Saunders and Reichow 2009). Till now the basalt flow keeps the horizontal attitude. The volume of traps is estimated to 2×10^6–$5\ 10^6$ km^3 (Dobretsov et al. 2008).

In **Noril'sk** (88° 24′ E, 69° 18′ N), North Siberia, there is the **Pechenga copper–nickel sulfide-type ore deposit** enriched with platinum and palladium elements. There may be the mantle plume activation near the center position. The Siberian traps and mantle plume were formed at 1900 Ma. The ore deposit in the lower part of basic volcanic system of Late Permian belongs to magma liquation deposit (Figs. 4.1 and 4.2). The ore reserves are up to 339 million tons. The nickel average grade is 1.18% (Naldrett 1989, 1999).

T. Wan, *The Tectonics and Metallogenesis of Asia*,
https://doi.org/10.1007/978-981-15-3032-6_4

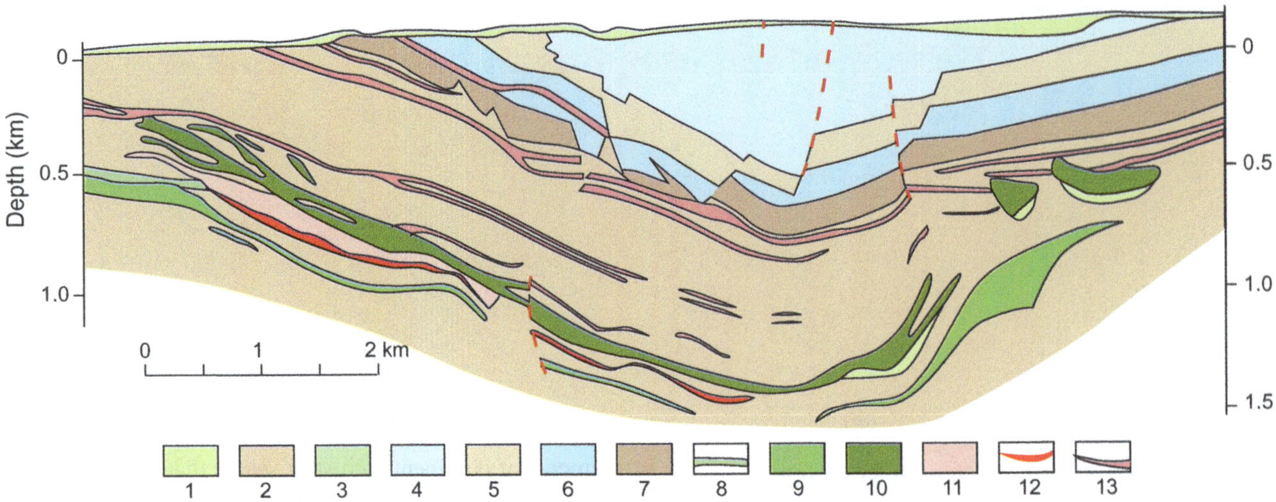

Fig. 4.1 Geological section of Tarnah intrusion in the Noril'sk ore deposit. 1. Quaternary; 2. sedimentary cover; 3. Molongov non-differentiation volcanic rock; 4. Najiskin strong differentiation volcanic rock rare in nickel; 5. Gudqi new volcanic rock including enriched nickel picrite; 6. server epoch volcanic rock rare in nickel; 7. Ewajin alkaline volcanic rock rare in nickel or chromium; 8. Tarnah non-differentiation sill; 9. lower Tarnah intrusion sill; 10. upper Tarnah intrusion sill, gabbro–diorite; 11. picrite and gabbro–diorite with dissemination mineralization; 12. massive sulfide ore; 13. trachyte–lamprophyre with ore (Modified from Naldrett 1999)

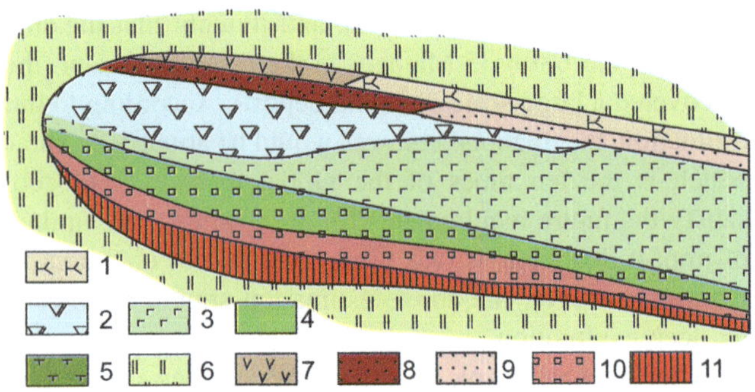

Fig. 4.2 Geological section of upper and lower ore beds of Tarnah intrusion, in the Noril'sk ore deposit. 1. Contact zone of gabbro–diabase; 2. diorite; 3. olivine gabbro–lamprophyre; 4. picritic gabbro–diabase; 5. porphyritic gabbro–diabase; 6. outer contact zone, metasomatic rocks; 7. metasomatic rocks; 8. skarn platinum (Pt) ore; 9. upper porphyritic rock with some sulfide Pt ore; 10. sulfide Pt ore in picrite and lower porphyritic rock; 11. compact sulfide Pt ore (Modified from Naldrett 1999)

The Russia has reserved 17.4 million tons for the nickel metal recoverable reserves, ranking the first in the world. The ore bodies, with the main orientation of NNW or NNE trending, are all controlled by the regional extension faults.

In the Irkutsk, East Siberia, there is the **Nepa supergiant potassium salt deposit** (108° 00′ E, 59° 02′ N), which belongs to the carnallite deposit formed in Early Cambrian. This deposit is divided into nine layers, in which total thickness is up to 35 m, each layer's thickness about 1.5–5 m, the burial depth about 600–900 m. The total area is more than 10,000 km^2; the reserves are estimated to more than 70 billion tons, accounting for 31% of the global reserves. The thickness of most huge ore layer reaches 18.6 m, and the areas are also rather great. In the potassium salt ore layer, the content of KCl is about 30–40% (Petrychenko et al. 2005).

The potassium salt deposit commonly is formed in the middle latitude subtropical zone with extremely dry climates. So the occurrence of the potassium salt deposit is a powerful evidence for great migration of Siberian plate. It means that the Irkutsk area, East Siberia, had been located at the middle latitude subtropics with extremely dry climates (Goncharenko 2006).

Tunguska (its center at 105° 00′ E, 62° 42′ N) is a **super-large and biggest coal field** in Russia. The geological reserves are 1.745 × 10^{12} tons, accounting for 16% of the global reserves with the depth of 600 m. The maximum thickness of coal-bearing strata reaches 500 m, about 100–300 m in the southern areas. The anthracite is dominant, and the others are the brown coal and meager coal. They are located at West Siberia, at Krasnoyarsk and Irkutsk areas between the Yenisey River and Lena River, in which areas reach 1.045 million km^2. Those coal fields are distributed in the synclines of Noril'sk, Tunguska and Western Angara. They are all separated by the hills and uplifts. In Late Paleozoic–Early Mesozoic, the magma intruded into the coal-bearing strata along the normal faults, which broken the completeness of those strata. From the sections to observe, the magmatic rocks account for 10–75% of the coal-bearing strata. In the Northern and Central Tunguska syncline, the volcanic tuff and lava cover on the coal-bearing strata. The occurrence of those could prove that the Siberian plate had been located in the warm, wet and moist areas at that time (Volkov 2003; the free dictionary by Farlex on Web).

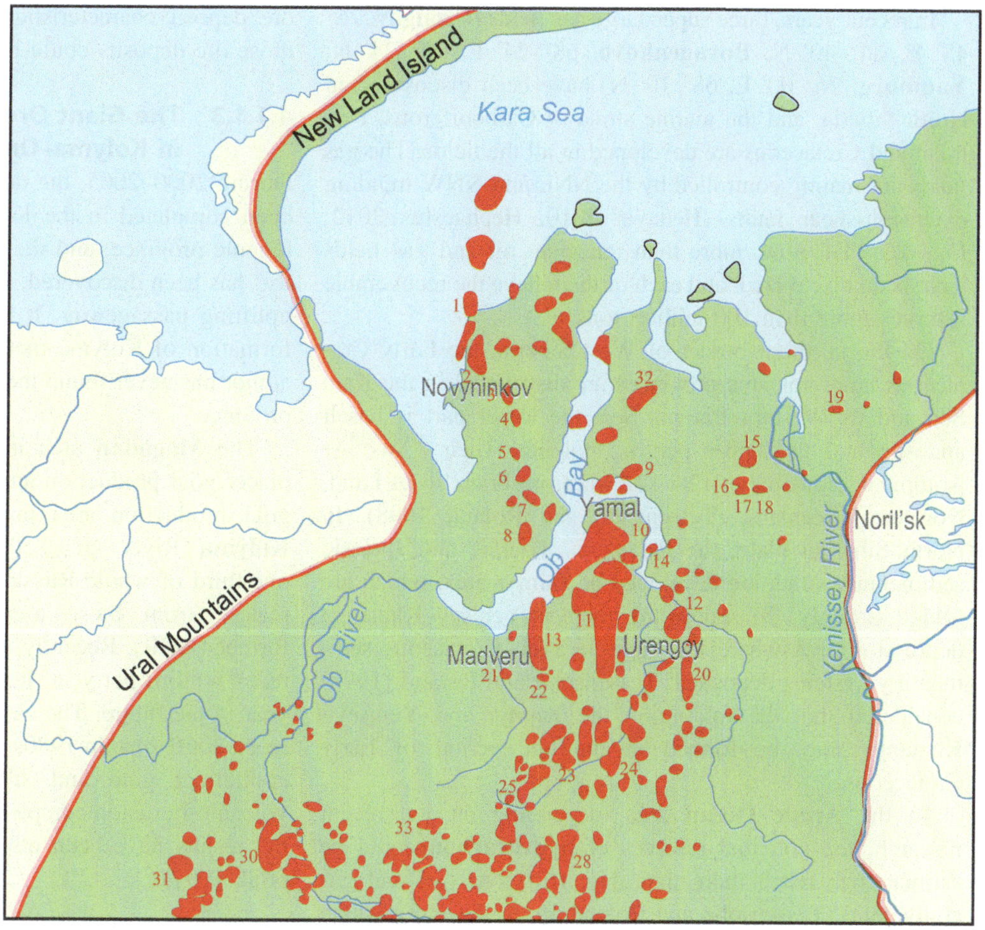

Fig. 4.3 Distribution sketch of oil and gas fields in West Siberia. The red areas are the oil and gas fields, the numbers show the oil and gas fields (the names omitted) (Modified from Li et al. 2005)

In Jurassic, influenced by SW trending migration and compression of Kolyma–Omolon plate [4] and North American plate [71], on the eastern side of Siberian plate, Verkhojansk–Chersky Jurassic accretion–collision zone [3] was formed (Oxman 2003), and on its south boundary, Transbaikalia Jurassic accretion–collision zone [5] was formed (Zolin et al. 2001). Thus, the strong folding and thrusting occurred in East and South Siberian plate. The folded and thrusted strata are the Early Paleozoic and Carboniferous–Middle Jurassic clastic systems, and the Neoproterozoic terrigenous carbonate rocks (Parfenov et al. 1995).

The **Sarylakh super-large gold–antimony ore deposit** (128° 24′ E, 61° 59′ N) is located at the Yakutsk area, East Siberian plate. The ore deposit was formed by the tectono-thermal action, which was controlled by the Verkhojansk–Chersky Jurassic accretion–collision zone [3]. That ore deposit was directly controlled by the Adycha–Taryn fault zone in Jurassic. All the faults, lineations and folds are distributed along the NW trending and controlled by the deep tectono-thermal events.

The boundary of antimonite body can be delineated by the sulfide–quartz–sericite alteration and the secondary dickite and illite alteration zone near the antimonite ore veins, which width reaches 100 m (Bortnikov et al. 2010). On the south boundary of Siberian plate, there is the **Golev lead–zinc ore deposit** in Krasnoyarsk area. Its lead–zinc reserves account for more than 40% of total reserves in the Russia, which are the high-grade (Pb 7%) ores.

After Neoproterozoic, due to the basic magma intruding and erupting in the West Siberia, the crustal density increased to form the fault-depression and Paleozoic and Mesozoic–Cenozoic sedimentary strata enriched with hydrocarbon (the thickness more than 4000 m), then to form the famous and biggest **Turmin oil and gas field** (its main field is the Samotlor) in Russia. Those oil and gas fields are located at the middle of the Ob River and the northeastern areas of Turmin oil and gas field. Till now, the reserves of 874 oil fields in West Siberia are 218×10^{12} tons, the outputs about 104×10^{12} tons; the gas reserves are 49.5×10^{12} m^3, the outputs about 15×10^{12} m^3 (Fig. 4.3; Liang et al. 2014).

In recent years, three super-large gas fields (**Urengoy**, 76° 45′ E, 66° 40′ N; **Bovanenkovo**, 68° 24′ E, 70° 24′ N; **Yamburg**, 76° 00′ E, 68° 10′ N) have been discovered in North Siberia, and the marine strata of Carboniferous, Permian and Cretaceous are developed in all the fields. The gas fields are mainly controlled by the NNE and NNW trending extension-shear faults (Hedayat 2010; Hephaestus 2012; Fig. 4.3). Till now, more than ten giant oil and gas fields have been discovered, and each of them have the recoverable reserves more than 100 million tons.

On the northern border of West Siberia, the Early Cretaceous large volcanic provinces are distributed in the Kara Sea and its Western Barents Sea, the lower part is basalt andesite and the upper part is tholeiite. Their ^{40}Ar-^{39}Ar isotopic ages are about 128–132 Ma from Franz Josef Land, Northern Greenland (Tessensohn and Roland 1998). In North Siberian plate, the Paleozoic, Triassic and Jurassic sedimentary strata are developed at Taimyr area, which are folded strongly. To southward, there is Yenisey–Khatanga depression area where the huge Jurassic–Cretaceous sedimentary systems deposited. Tessensohn and Roland (1998) considered that the folding in the Taimyr and Yenisey–Khatanga may be formed in the late period of Early Cretaceous.

In the **Arctic Ocean** area, depending on the recent research, the potential reserves of **hydrocarbon** could be estimated to reach more than 1×10^{12} tons (oil and gas equivalent) (Tessensohn and Roland 1998; Fig. 4.3), which are mainly reserved in Triassic, Jurassic and Cretaceous sedimentary strata. The United States Geological Survey proposed a new estimate in the July 2008. They estimated that there are 9×10^{10} barrels for oil reserves and 47×10^{16} m^3 for gas reserves. It means that they account for 13% unproved oil reserves and 30% unproved gas reserves in the world. In addition, on the Arctic Ocean continental shelf, northern of Siberian plate, the abundant gas hydrates are reserved, with the important potential resources, in the nearby there are many mud volcanos. On the Arctic Ocean continental shelf, there are also enriched with the sedimentary lead–zinc, mercury, tin ores and quartz veins with the gold.

4.1.1.2 The Giant Ore Fields and Deposits in Verkhojansk–Chersky Jurassic Accretion–Collision Zone [3]

In the Chersky Peninsula, the tin reserves of the primary tin-sulfide-type ore deposits are about 350 thousand tons; however, till now they have been never exploited (Zhang 2009). After checking the mercury ore deposit in the Chersky volcanic zone, Guo et al. (2005) considered that ore deposit had some comparable and similar in the regional tectonic background, volcanic rocks, wall rock alteration and ore deposit characteristics. The part mineral resources of those ore deposits could be exploited for bloodstone.

4.1.1.3 The Giant Ore Fields and Deposits in Kolyma–Omolon Plate [4]

During 2000–2005, the deep reflex seismic exploration has been completed in the Kolyma gold, silver and tin metallogenic province, and the "Sky Light of Moho Discontinuity" has been discovered, which may be the mantle hot fluid uplifting passageway. It is used to reasonably explain the formation of Kolyma metallogenic province. However, the author has never found the detailed data for the metallogenic province.

The **Magadan** area in Far East is the most important placer gold production area in Russia, also is one of bigger gold production area in the world. Only in the **Yana–Kolyma** River areas, the placer gold output occupies one-third of whole Russia's output (a large number of "inmates" from those areas to exploit and excavate in former-USSR). Recently, the gold minerals are exploited to reach 60 tons per year. It is expected to produce 88 tons per year in the future. The estimated reserves of the placer gold reach 3750 tons. In 1980s, the Russian geologists discovered epithermal gold and silver deposits in Late Paleozoic volcanic-intrusion complexes of the Omolon plate; however, the results have been never published in public (Khomich et al. 1997).

4.1.1.4 The Giant Ore Fields and Deposits in Transbaikalia (or Mongolia–Okhotsk) Jurassic Accretion–Collision Zone [5]

In the central and eastern regions of Transbaikalia gold zone (Figs. 2.4 and 2.5), there are three types of gold deposits: (1) Middle–Late Jurassic high-sulfide-type deposits with some porphyry-type deposits; (2) Middle–Late Jurassic low-sulfide-type deposits; (3) Cretaceous low-sulfide Au-Ag epithermal deposits, only a few.

The orientation of Jurassic gold deposit zone is as same as the trending of regional fault, with all NE trending (Fig. 4.4); and the gold quartz-sulfide veins are distributed as NW-SE trending (Fig. 4.5; Kulikova et al. 1996). These veins show a little bit extension, mainly WNW trending, as same as the trending of East Asian maximum principle compression stress (Wan 2011). It means they are formed by the same regional tectonic stress field. The **Shakhtama** porphyry-type copper ore deposit was formed at the period of 160–175 Ma, and Wulugetu porphyry-type copper ore deposit was formed at the period of 177 Ma (Mao et al. 2014). Both were formed in Jurassic. The above data show that the NE orientation regional faults control the metallogenic transmission zone (Fig. 4.4) and the maximum principle compression stress control the WNW trending of ore

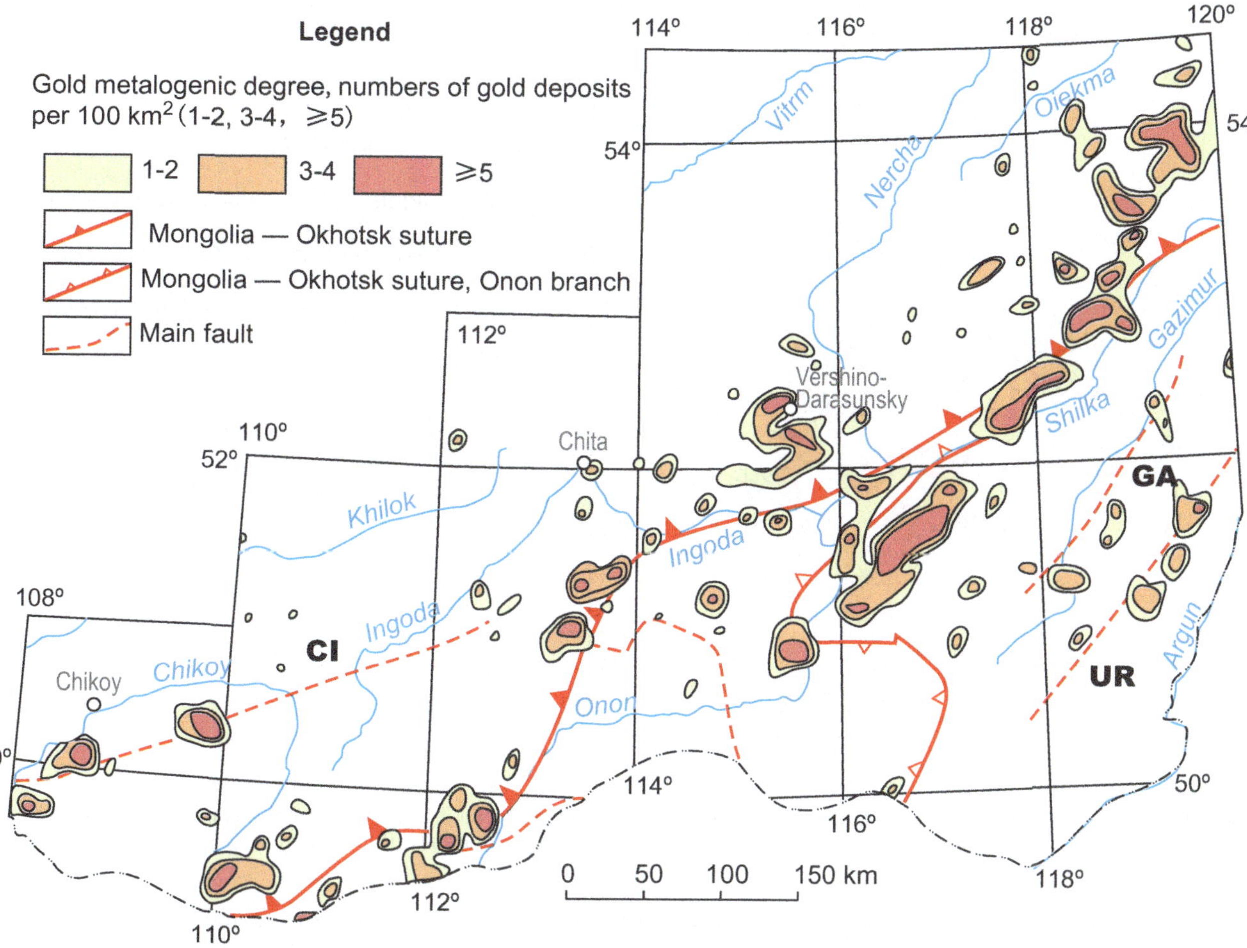

Fig. 4.4 Distribution of Middle and Eastern Transbaikalia gold deposit zone. CI—Chikoy-Ingoda; GA—Gazimer; UR—Urulunguy. The locations of two sutures are shown in Figs. 2.4 and 2.5 (Modified from Zolin et al. 2001)

deposits (Fig. 4.5). However, some Cretaceous gold veins are located around the porphyry body, which are controlled by the cyclic active faults (Fig. 4.5; Zolin et al. 2001).

4.1.2 The Giant Ore Fields and Deposits in Central Asia–Mongolia Tectonic Domain

4.1.2.1 The Giant Ore Fields and Deposits in Altay–Middle Mongolia–Hailar Early Paleozoic Accretion–Collision Zone [6]

The western part of Altay–Middle Mongolia–Hailar (Figs. 2.6 and 2.8) tectono-metallogenic zone is also called as Altay–Zaysan metallogenic province (Wu et al. 1993; He and Zhu 2006; Fig. 4.6), which is an important copper, gold, polymetallic and rare-element metallogenic province (80°–85° E; 48°–52° N), near the boundaries among Russia, Kazakhstan, China and Mongolia.

In those areas, there are more than 1800 ore deposits. From NE to SW, they are divided into five sub-ore zones: Horz–Salemsakta (I), Ore–Altay (II), Erjix (III), Changba–Naram (IV) and Sikakaba (V), in which the Northeastern **Ore–Altay** sub-ore zone shows the strongest metallogenesis (Fig. 4.6, II), with the length of 500 km and width of 10–90 km. It is evaluated as extremely abundant ore deposit, accounting for about 80% ore reserves for all the deposits. The representative ore deposits are the marine volcanic massive sulfide polymetallic type, such as the **Nikolaev** VMS copper and zinc deposit in Kazakhstan, the discovered metal reserves: Cu 1.042 million tons (grade of 2.52%), Zn 1.527 million tons (grade of 3.83%), Pb 0.197 million tons (grade of 0.49%), Au 20.1 tons (grade of 0.5 g/t), Ag 1446 tons (grade of 33×10^{-6}). That tectono-metallogenic zone was formed in Early Paleozoic. The basement rocks are Ordovician–Silurian strata, with the green schist facies metamorphism. However, the tectono-metallogenesis mainly occurred in the Devonian–Lower Carboniferous volcanic systems. Most of them were formed in the volcanic-depressions of the late period of Early Devonian–Middle Devonian, not formed in Early Paleozoic (Mao et al. 2012a, b). It means those ore deposits were formed

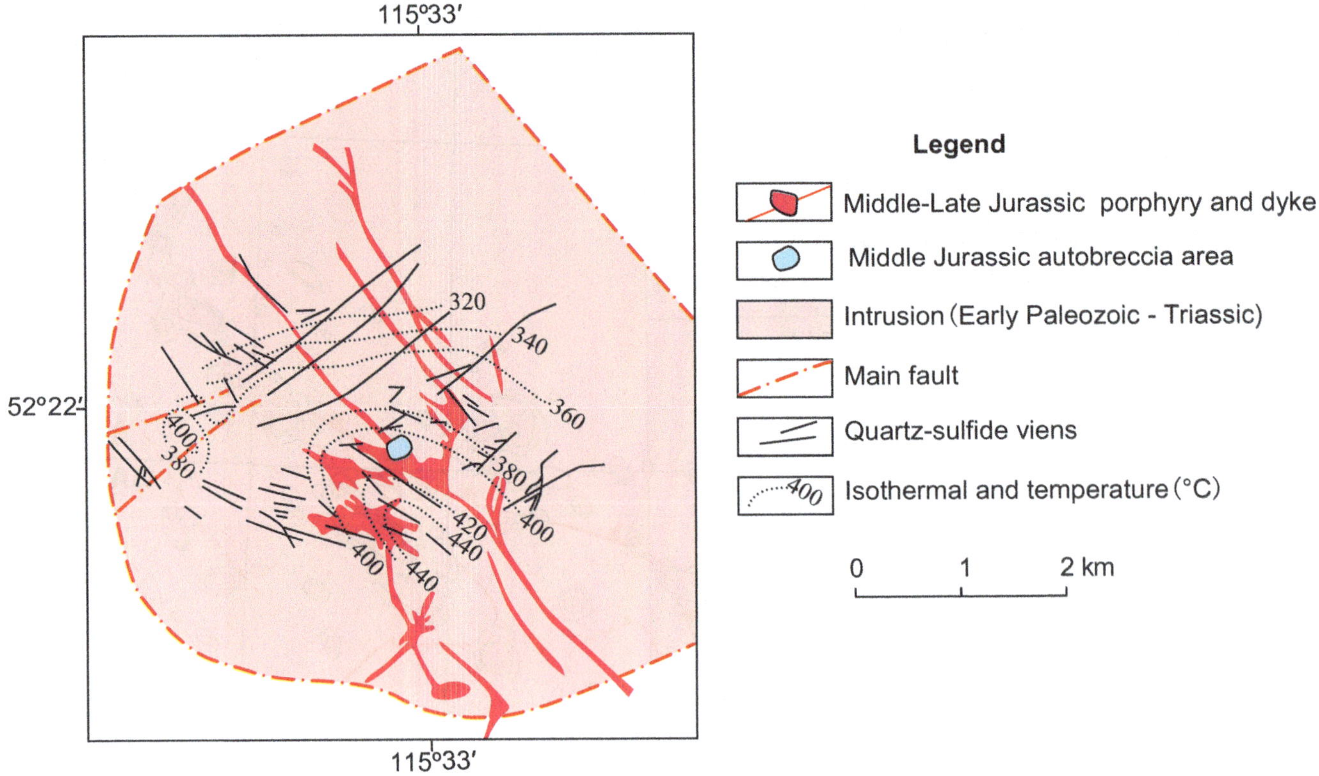

Fig. 4.5 Distribution of dykes and quartz–sulfide veins with gold for the Darasun gold deposit. The NW–SE orientations of dykes and quartz-sulfide veins with gold are as same as the trending of East Asian maximum principle compression stress (Modified from Kulikova et al. 1996)

after the main collision period. The wall rocks of ore bodies are characterized by the bimodal volcanic rocks, clastic systems and continental rift (Gritsuk et al. 1995). The giant ore provinces of Kazakhstan are located at the intersection of NW and near W–E trending, the ore fields and deposits are controlled by the Late Paleozoic volcanic systems and their volcanic tectonic assemblages.

As similar as that, there is the **Ziliangnovsk** supergiant lead–zinc polymetallic deposit in the Ore–Altay, which Pb-Zn reserves reach more than 5 million tons, with the average lead grade of 4%, that of zinc of 7%. In addition, there are the silver reserves of 12 thousand tons (grade at 73.33×10^{-6}) and the gold reserves of 1.1 tons (grade at 0.15×10^{-6}). They were all formed in Devonian (Mao et al. 2012a, b).

According to the deep seismic data, the depth of middle crust with low seismic velocity and high electric conduction layer of Ore–Altay is about 22–24 km; in the surrounding areas, that layer's depths are about 26–28 km (Chen et al. 1999). It is obviously that the depth of middle crust with low seismic velocity and high electric conduction layer of Ore–Altay is located at higher position, so the geothermal gradient is rather higher, and the supercritical fluid will be richer. It may be important to easily concentrate the ores. Only a few of ore provinces were mineralized in the collision stage, most were formed after the collision, i.e., Late Paleozoic (Devonian–Carboniferous) (Wu et al. 1993).

The **Fuyuan** (89° 49′ E, 47° 12′ N) contains violent granitic pegmatite-type rare metal and rare earth deposit in Altay–Middle Mongolia–Hailar tectono-metallogenic zone of China, in which the representative deposit is **Kokotohai** granitic pegmatite-type Li (lithium), Be (Bismuth), Nb (niobium), Ta (tantalum), Rb (rubidium), Se (cesium) and Hf (hafnium) deposit (Fig. 4.7; Zou and Li 2006). Till the 1999, the proved reserves have been acquired as BeO 61, 373 tons, Li_2O 52, 451 tons, Nb_2O_5 657 tons, Ta_2O_3 825 tons. That ore deposit at the northern contact of Aral granite body was formed in the period of 330–250.3 Ma, i.e., Carboniferous–Permian (the isotopic age). It means that ore deposit was formed after Altay Early Paleozoic collision zone forming, mainly controlled by Late Paleozoic continental crust remelted granite. The ore deposit is located at the outer contact of that granite body. Some northern deposits were formed in Triassic–Jurassic. In the Altay area, there are 100 thousand pegmatite veins. In Kokotohai ore areas, there are 2100 pegmatite veins, averaging 10 veins per 1000 m^2, in the most concentrated areas, averaging 50 veins per 1000 m^2, in which No. 3 vein is the best ore-bearing one

Fig. 4.6 Metallogenic structure in Altay–Zaysan metallogenic province. **Legend**: 1. Pre-Cambrian mineralization; 2. Early Paleozoic mineralization; 3. Late Paleozoic mineralization; 4. pyrrhotitization; 5. iron ore; 6. pyrite–polymetallic ore zone; 7. pyrite-polymetallic ore calculus; 8. gold zone; 9. gold calculus; 10. rare-element mineralization area; 11. polymetallic rare-element mineralization area; 12. rare metal mineralization area; 13. rare metal ore calculus; 14. tectonic boundary; 15–18. main regional faults. **Sub-mineralization zones**: I. Horz–Salemsakta; II. Ore–Altay;III. Erjix; IV. Changba–Naram; V. Sikakaba (Modified from Wu et al. 1993)

(Fig. 4.7). In the ore deposit area, the trending of most ore veins is in according with regional tectonic line, showing the NW trending with the distribution of brush structure. The veins show the step-like alternation with the steep and gradual trends, which are controlled by the stratifications and joints, all dips toward the southwest direction. Till now its deep boundaries have been never delineated. The attitudes of that ore deposit are extremely favorable to concentrate the volatility ore fluid.

From the plane distribution of brush structure for the Kokotohai ore veins, they converge generally NW–SE toward Late Paleozoic crustal remelted granite, diverge toward NW trending, and to the tail change to E–W and NE trending (Fig. 4.7; Zou and Li 2006). According to the character of Late Paleozoic regional tectonic stress field, it shows that stress field changes in the Balkhash–Tianshan–Hingganling Late Paleozoic accretion–collision zone [10]. In Early Carboniferous (345–325 Ma), it was controlled by the NW trending dextral strike-slip, the joints in the ore deposit appeared the extension-shear (left of Fig. 4.7) to form the brush joint system, thus to make the good condition for ore fluid injection. And during Early Permian (280–270 Ma), the regional NW direction faults changed to sinistral strike-slip faults, the above brush joint system turned to compression-shear (right of Fig. 4.7); thus, it is favorable to seal up and condense the ore fluids and then to form this supergiant granitic pegmatite-type rare metal and rare earth element ore deposit.

Changing with the buried depth, the ore bodies appear vertical mineralized zoning: The deep part is enriched with

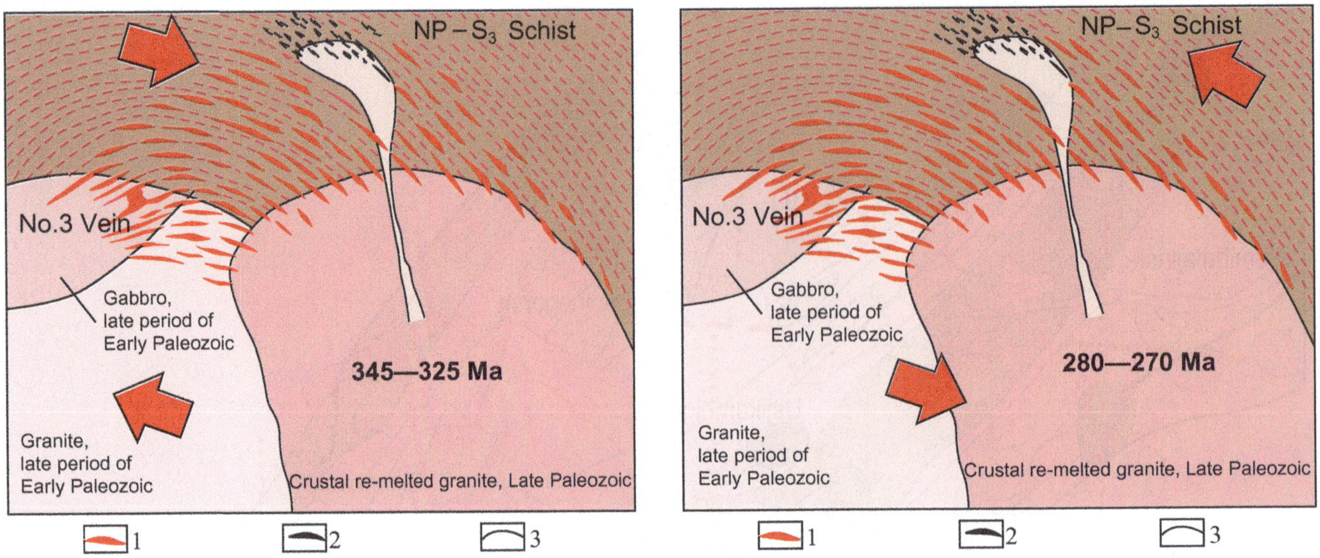

Fig. 4.7 Sketch of forming model for Kokotohai granitic pegmatite-type rare metal and rare earth ore deposit in Xinjiang. 1. Li, Be, Nb, Ta, Rb, Se and Hf ore granitic pegmatite veins formed in late Paleozoic; 2. Be, Nb and Ta ore mica granitic pegmatite veins formed in Triassic; 3. geological boundary. The red arrows show the migration orientation of block at different periods (Modified from Zou and Li 2006)

the Be muscovitization, up to the second zone that is Be, Nb and Ta mineralized zone, the third zone is Li, Be, Nb, Ta and Hf mineralized zone, the fourth zone is Ta, Nb and Hf mineralized zone, and at the top zone is Ta, Cs, Li, Rb and Hf mineralized zone. Near to the center or the boundaries, it shows the above similar mineralized zone. After that ore deposit forming, it has also suffered partial influences of Triassic and Cretaceous magma crystalline differentiation (Zou and Li 2006). Wang et al. (2007b) got the SHRIMP zircon U–Pb ages of 220–198 Ma. So it seems the pegmatite veins may be at last fixed in Triassic. To sum up, although the above ore deposit is located in the Early Paleozoic collision zone, it was not formed in Early Paleozoic, but formed after collision and in the intraplate deformation periods, namely from Late Paleozoic to Triassic.

On the southern border of Altay, China, there is the NW trending precious and ferrous metal metallogenic zone, represented by **Ashle** massive sulfide copper and zinc ore deposit (Fig. 4.8). Its southern border is Ertix fault in Malkaku fault zone, and the northeast side is Bisala fault in Wuqia–Abagong fault zone. In Early Paleozoic, that zone belonged to the southwestern of Altay–Middle Mongolia–Hailar Early Paleozoic accretion–collision zone, and after collision, the strong Late Paleozoic tectono-magmatism occurred to form the Ashle volcanic zone and massive sulfide copper and zinc ore deposit.

The Ashle volcanic rocks are characterized by rich sodium, which are the marine spilite–keratophyre formation. The isotopic ages of the Ashle volcanic rocks are 352.3–386 Ma, and ages of hydrothermal mineralization are mainly

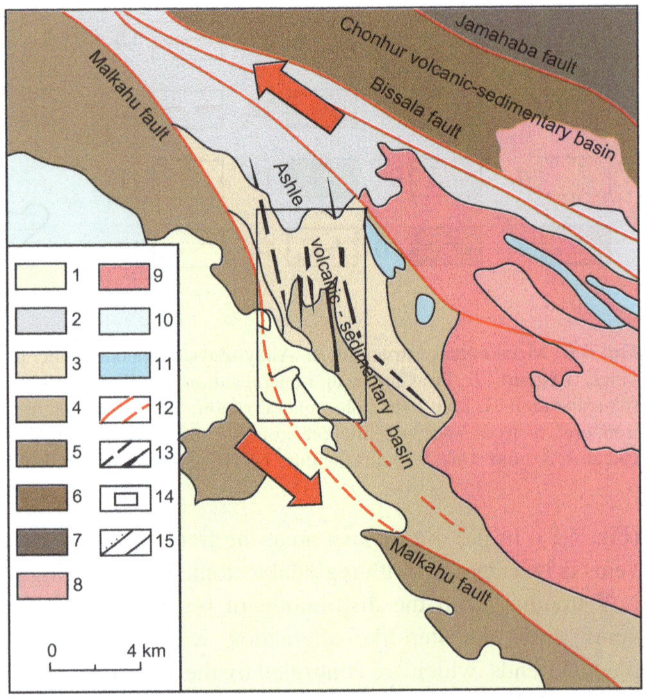

Fig. 4.8 Geological sketch of Ashle volcano-sedimentary basin, in Altay Permian metallogenic zone. 1. Quaternary sediments; 2. lower carboniferous series; 3. middle and upper Devonian series; 4. Ashle Formation of middle–lower Devonian series; 5. Tuokesalei Formation of middle–lower Devonian series; 6. Altay Formation of middle Devonian series; 7. lower Devonian series; 8. Late Paleozoic monzonite; 9. Late Paleozoic plagioclase; 10. Late Paleozoic tonalite; 11. Late Paleozoic gabbro–diorite; 12. fault; 13. anticline and syncline; 14. Ashle mining area; 15. geological boundary. The red arrows show Permian blocks affected by sinistral strike-slip (Modified from Chen et al. 1996)

262–242 Ma, i.e., at the end of Permian (Chen et al. 1996). According to the regional tectonic stress state (the detail change shown in Balkhash–Tianshan–Hingganling Late Paleozoic accretion–collision zone [10]), in Early Carboniferous (345–325 Ma), being influenced by NW trending dextral strike-slip faults, the near N–S trending faults in Ashle area showed the extension; however, in Early Permian (280–270 Ma), the near NW trending faults turned to sinistral strike-slip, the ore deposit underwent the near E–W partial compression, and then the strata were folded to form a series of near NS trending folds (Fig. 4.8), the near N–S trending fractures showed the compression-shear and a little bit closing, favorable to reserve the massive sulfide copper and zinc ore deposit.

In the central and northern parts of Altay–Middle Mongolia–Hailar Early Paleozoic accretion–collision zone [6], there reserved the **Erdenet** (104° 08′ E, 49° 02′ N) **super-large porphyry copper (molybdenum) ore deposit**, which is the second largest ore deposit in Mongolia (Fig. 4.9), with the copper reserves more than 10 million tons (Zhang 2009). The dating ages of zircon SHRIMP and LA-MC-ICP-MS and molybdenite Re–Os isochronism in ore porphyry (quartz diorite) are acquired to be about 240 Ma, i.e., formed in Early Triassic (Jiang et al. 2010). The metallogenic age was too late as for the collision zone age. The ore-bearing Erdenet system is controlled by the fault (Fig. 4.10), which is distributed in NW 330° with extension-shear. The ore-bearing Erdenet system was influenced by Triassic (Indosinian tectonic event) NS toward compression (from recent paleomagnetism evidences). In the ore area, the ore body extends scope of 2 km × 1 km, with the maximum vertical depth of 560 m, including 100–300 m for the secondary enrichment zone of copper. The main ore body is obviously controlled by the NW trending fault, which intersects as the small angle with the Triassic near NS trending maximum principal compression stress. Thus, it could be predicted that the ore body and its related intrusion body were intruded along the extension-shear fault, which occurred after the formation of Early Paleozoic collision zone, and were controlled by the Triassic intraplate transverse extension fault. The ore was not formed in Early Paleozoic collision period, but in the intraplate deformation period.

In Mongolia, more than 130 gold ore deposits have been discovered, with the prospective reserves of 3000 tons, now the demonstrated reserves of 160 tons (Zhang 2009). In the Bayanhongol areas of the Central-Southern Mongolia, there are enriched with gold resources, for example, the Chagansublga porphyry copper deposit with the reserves about 118 tons.

Near the border of China, Russia and Mongolia, the porphyry copper–molybdenum deposit is located in the Wunugtu Mount, south to the Manzhouli about 22 km. The ore reserves reach 849.7 million tons, with the copper average grade of 0.46% and the molybdenum average grade of 0.053%. According to the Rb–Sr whole rock isotopic dating, the ore ages are about 130–140 Ma; it seems to be formed in Late Jurassic–Early Cretaceous. However, the U–Pb zircon ages and K–Ar isotopic ages are all 180–190 Ma.

4.1.2.2 The Giant Ore Fields and Deposits in Karaganda–Kyrgyzstan Early Paleozoic Accretion–Collision Zone [7]

In the Early Paleozoic accretion–collision zone [7], the Northeastern Tajikistan, there is the second largest silver ore deposit in the world, the Great Kanimansur silver field

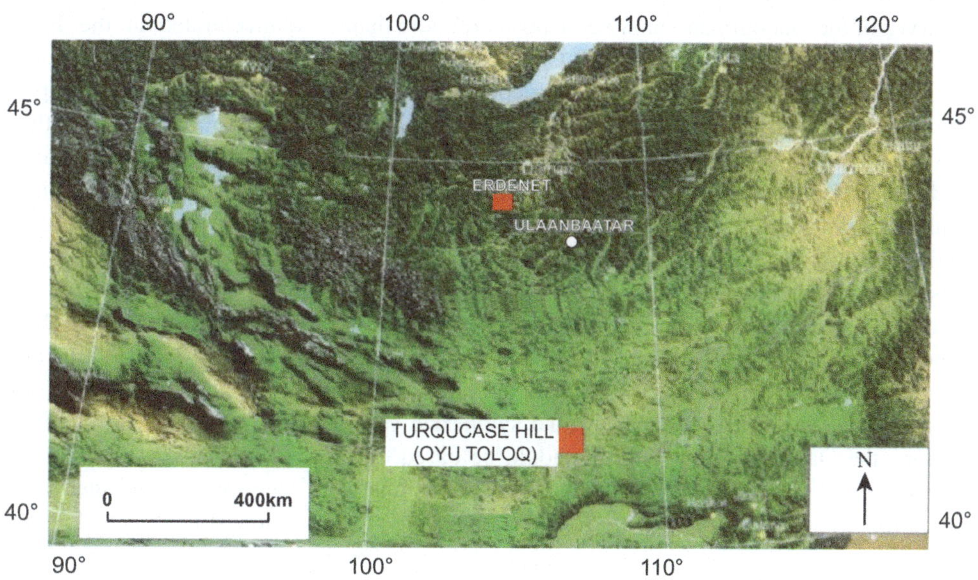

Fig. 4.9 Location of Erdenet copper–molybdenum and Oyu Tolgoi supergiant copper gold ore deposits. Turquoise Hill (Oyu Tolgoi) is Oyu Tolgoi supergiant copper gold ore deposits, details to see Figs. 4.22, 4.23 and 4.24 (After Nie et al. 2011; personal communication)

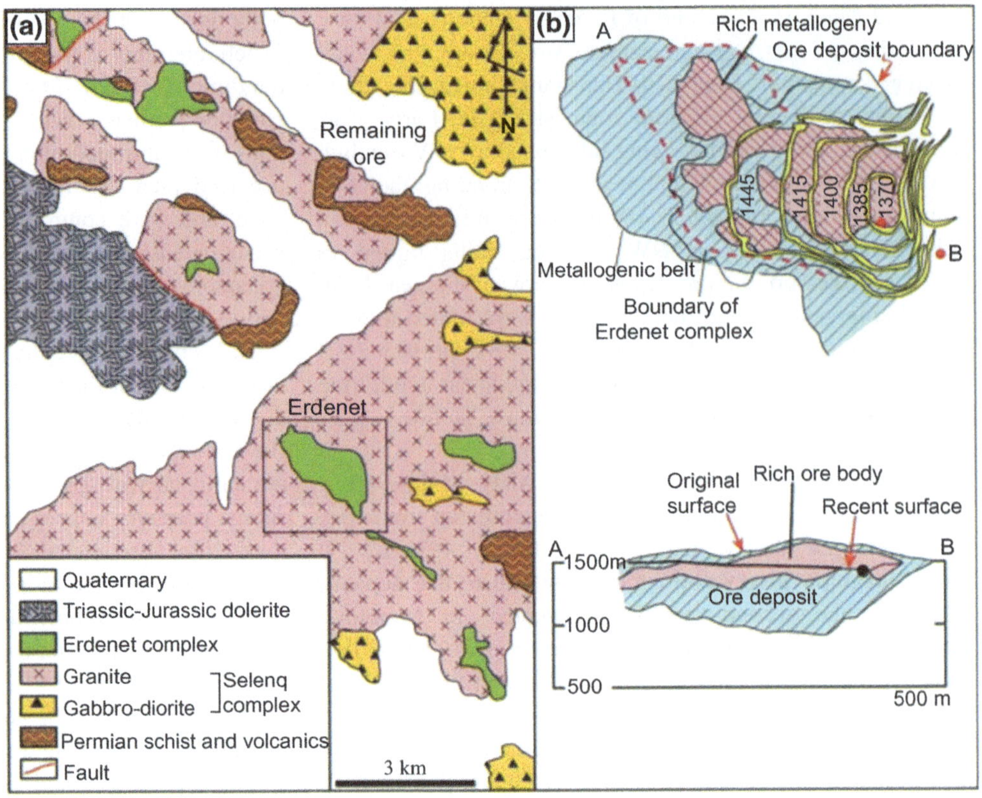

Fig. 4.10 Geological sketch of Erdenet copper–molybdenum deposit (Nie et al. 2011; personal communication)

(about 70.0° E, 40° 39′ N), near Hudrad City in Ashle area, the Western Tianshan reserves the silver–polymetallic ore deposit, and contains fluorite, gold, antimony, bismuth, etc. The ore total reserves are estimated to be about 1 billion tons; now its silver proved reserves are more than 60,000 tons. That ore deposit is located in the Late Paleozoic intermediate-acid volcanic system with the thickness of 6 km and makes up a 2000 m × 800 m × 500 m mesh-vein ore deposit, with the silver average grade of about 50 g/t, the richest part up to 500 g/t. The lead and zinc total reserves of the Dakoniman Sur silver deposit are more than 8 million tons (Wang and Wang 2010; Chen et al. 2012a and on the Web). The Dakoniman Sur silver deposit is located at the Northeastern Adlasmask volcanic depression zone; there are the dacite, trachyandesite, trachyte, rhyolite and their tuff and ignimbrite in the Devonian–Cretaceous volcanic system with the total thickness of 6 km, which cover the Triassic and Cretaceous sedimentary strata. The fault zone is composed of three fault systems: near EW–NWW, NE and NW trendings. The main ore minerals include native silver, argentite, proustite, galena, sphalerite, chalcopyrite, pyrite, tetrahedrite and haematite. The main veinstone minerals are quartz, carbonate, barite and fluorite. To synthesize various factors, the Dakoniman Sur deposit should be based on the volcanic rocks to suffer the hydrothermal-metasomatism alteration, then to form the silver–polymetallic porphyry ore deposit. However till now, it has not got the isotopic age (Wang and Wang 2010).

The **Uzbekistan–Kyrgyzstan gold ore belt** and the southern Middle Tianshan Late Paleozoic volcanic activity zone (for Central Asia) are located at the association between Karaganda–Kyrgyzstan Early Paleozoic accretion–collision zone [7] and Western Tianshan Late Paleozoic accretion–collision zone [9] (Figs. 2.6 and 4.11). In Kyrgyzstan, the gold and copper ore resources are rather rich, the forecast ore reserves are up to 1.5 million tons, mainly concentrated at the **Kumtor** supergiant gold deposit and **Pojimoqiak** skarn copper and gold ore deposit. The gold industry output occupies the half of Kyrgyzstan total industry outputs, and the exploited gold reserves are about 1200 tons. The main ore deposit types belong to hydrothermal, skarn and porphyry types (Chen et al. 2010).

The **Kumtor** supergiant gold deposit (~77.5° E, 42° N; Fig. 4.11) is located on the southern border of Karaganda–Kyrgyzstan Early Paleozoic accretion–collision zone (541–419 Ma) [7], i.e., the Eastern Uzbekistan–Kyrgyzstan metallogenic belt. Those gold reserves are about 1100 tons (Xie et al. 2015).

That ore deposit is located near the Isak Lake, Eastern Kyrgyzstan, with the high topography, 3200–4150 m above sea level, and located at the front of glacial zone. The ore deposit is distributed as belts, 15 km long and 0.1–0.4 km wide (Fig. 4.12). The main ore bodies are filling in the

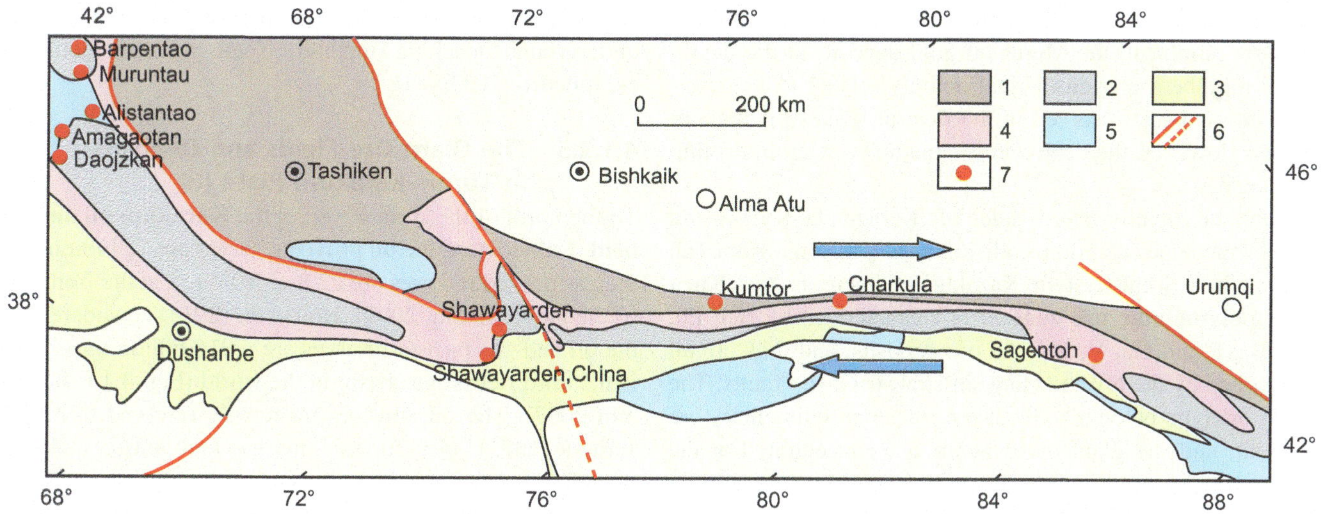

Fig. 4.11 Sketch of Uzbekistan–Kyrgyzstan gold ore belt, Western Tianshan. 1. Turgai–middle Tianshan Early Paleozoic collision zone; 2. southern Middle Tianshan Late Paleozoic collision zone; 3. the boundary zone of Karakum and Tarim blocks; 4. Late Paleozoic volcanic activity zone in the southern Middle Tianshan; 5. the middle-small blocks in the collision zone; 6. fault and predicted fault; 7. gold deposit. The blue arrows show the dextral strike-slip direction of fault in Early Permian (Modified from Chen et al. 2010)

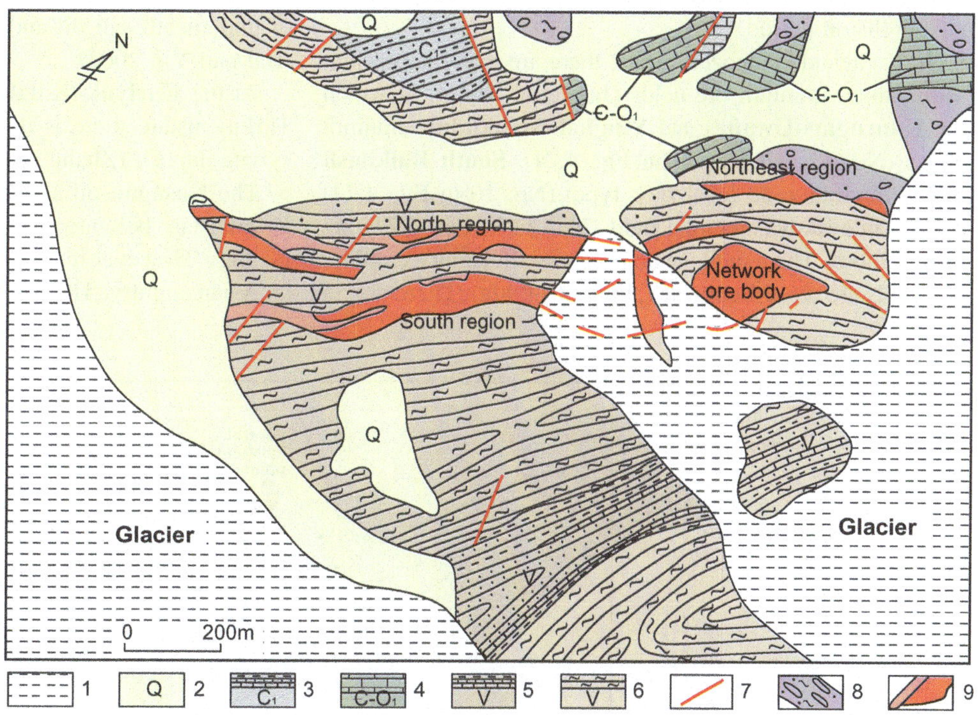

Fig. 4.12 Geological sketch of the middle part in Kumtor gold supergiant deposit. 1. Glacier; 2. Quaternary system; 3. Early Carboniferous sandstone, siltstone and shell; 4. Cambrian–Ordovician marble; 5. Vendian period shale and phyllite with limestone of Neoproterozoic; 6. Vendian period carbon-mud shell of Neoproterozoic; 7. fault; 8. tectonic melange; 9. ore body and mineralization zone (Modified from Chen et al. 2010)

Vendian period shallow metamorphic carbon-bearing shale system of Neoproterozoic. The trending of ore bodies and the foliation of strata are all ENE direction. Depending on the appearance of ore body and the ore structures, it can be predicted that the ore hydrothermal could be injected into compression-shear fracture system (Mao et al. 2012a, b). Based on the Ar–Ar plateau ages of beresite, the gold ore whole rock with sericite and the ages of sericite mineral are all between 284.3 ± 3.0 and 288.4 ± 0.6 Ma, so it can be determined that the ore deposit should be formed in Early Permian. The Kumtor gold deposit was formed by the long-distance effect from the eastward compression of the Ural collision zone (detailed illustration to see Fig. 2.5), and ENE trending regional faults changed to dextral strike-slip, then Neoproterozoic shallow metamorphic carbon-bearing shale system deposited, finally the shear migration occurred almost along the stratification, thus the violent small veins and mesh-veins gold ore were formed.

This ore deposit should be to say that is hydrothermal ore deposit, similar to the Muruntau gold deposit. In the depth, there may be the hydrothermal activity related to the magmatism, and just reserved in the dextral strike-slip fracture zones. However, the deep status is unknown without drilling evidence.

On the southwestern side of Karaganda–Kyrgyzstan Early Paleozoic accretion–collision zone [7], in the Kara Tau Mountains, Southwestern Kazakhstan, there is the **Kara Tau polymetallic ore zone**. It is characterized by rich Pb, Zn, Ba, P and Cu, V, Mo, Au ore deposits, and they are all controlled by the NW trending sinistral strike-slip faults. The main ore type belongs to the layer volcanic–sedimentary ore deposit, and the giant ore deposits are concentrated at the Central Kara Tau region. The ore deposits were mainly concentrated in the dolomite or dolomitic limestone of Famennian period of Late Devonian. The metal elements are reserved in the bedding with small veins or disseminated areas. In Late Devonian, the NW trending faults were not the thrust or reverse fault caused by collision, but had the characteristics of sinistral strike-slip. It is the result of post-collision period.

In Karaganda Paleozoic strata, there are many supergiant sedimentary uranium ore fields (Jiao et al. 2015), in which the **Kanrugan–Uwanas** and **Yingkai–Menkuduk** uranium zones (No. 12 and No. 13 in Fig. 4.74), **South Balkhash Lake** uranium zone (coal rock type) (No. 10 in Fig. 4.74) and **Lazalevskeya** uranium field (No. 22 in Fig. 4.74) are most famous. The metallogenic time of sedimentary uranium ore fields are all in Cenozoic (about 30–1 Ma). There are 55 uranium ore deposits, and 70% deposits are exploited by underground leaching method. Their reserves could be exploited for 100 years.

4.1.2.3 The Giant Ore Fields and Deposits in Turan–Karakum Plate [8]

To the north of the Amu River, in the **Karakum oil and gas field, Uzbekistan**, the oil proved reserves are 12 billion tons, the gas proved reserves are 22.8 × 10^{12} m^3, at the front rank of the globe (Zhang 2009). Hou et al. (2014) considered that the oil and gas proved reserves are 4.016 billion tons, and it is the third richest gas basin in the world (Fig. 4.13; Bai and Yin 2007). The oil and gas are mainly reserved in Middle Jurassic and Upper Jurassic marine carbonites, including 68.0% oil reserves, 84.0% condensate oil reserves and 44.2% gas reserves, some gas reserves (36.4%) in Lower Cretaceous sandstones. The distribution of Upper Jurassic underneath salt oil-gas fields is mainly controlled by the favorable reservoir and ancient uplift structure. The biogenic reef and ancient structure are mainly developed at the northeastern basin—North More sub-basin, and the reserves of underneath salt oil and gas are concentrated in this area (Bai and Yin 2007).

At the **Karlyuk–Kalaber** (66° 23′ E, 37° 29′ N), in the Turkmenistan, there is the supergiant Triassic sedimentary sylvite deposit (Zhang et al. 2005b).

The **Kasagan** oil and gas field (50° 10′ E, 46° 15′ N) in Kazakhstan belongs to the Baltic plate (Europe) in the geology. However in the administration district, it belongs to an Asian country. The Kasagan oil and gas field is located on

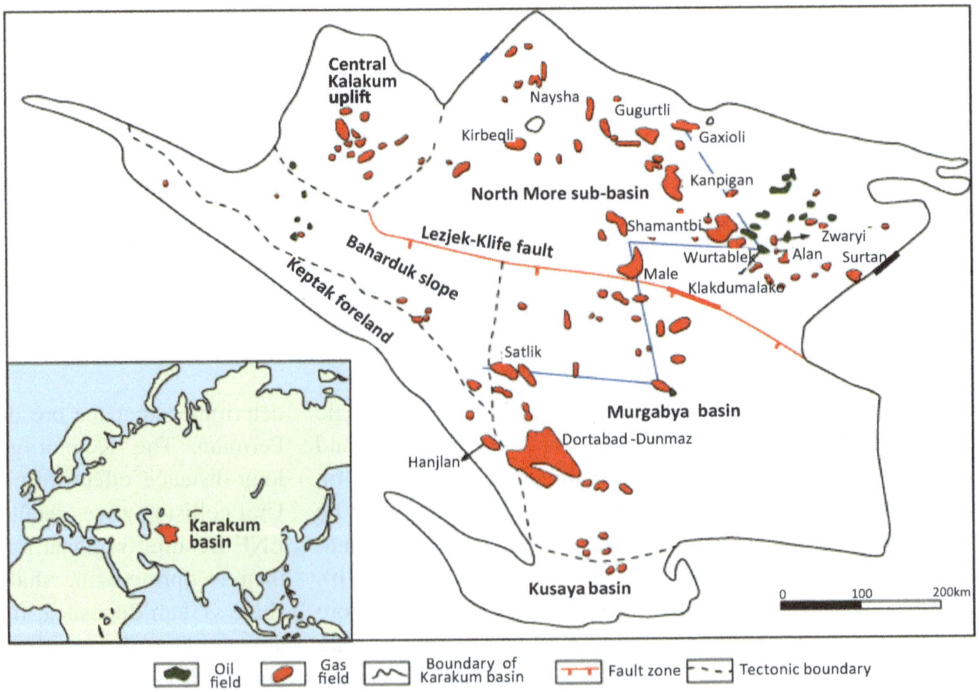

Fig. 4.13 Distribution of Karakum oil and gas field (Modified from Bai and Yin 2007)

the north side of the Caspian Sea, which reserves are ranked fifth in the world with the oil and gas reserves of 4.739 billion tons (Hou et al. 2014). But the exploration project has been delayed again and again. In recent, China and other international companies will jointly explore it, possibly in the near future. The reserve layers are concentrated in the Paleozoic sedimentary strata. Most of Caspian Sea beaches belong to Eastern Kazakhstan, Atlao, Aktobi States in the Kazakhstan. Their areas are about 40×10^4 km^2. In the center of that basin, the depth of Pre-Cambrian basement is about 22 km. In the sedimentary cover, there are two oil-gas-bearing associations: cover salt association and underneath salt association, their thickness more than 10 km. The Lower Permian Kongu Epoch–Kashan Epoch salt layer separates the cover and underneath salt associations, to form more than 1500 salt hills with the uplift ranges about 8–10 km. The oil-gas reserves of underneath salt association are distributed at the boundaries of oil-gas-bearing basin, and Middle–Upper Devonian and Lower Carboniferous clastic rocks, Middle–Upper Carboniferous and Lower Permian carbonites are all the oil-gas layers with the industry value. More than 90% proved reserves are distributed in 29 underneath salt association oil-gas fields, in which the preservation conditions of the Tianjiz, Kalazaganak and Zananol oil and gas fields were strongly deformed by the dissolution of fissures and carbonates. The whole giant oil and gas fields (such as Tianjiz, Kalazaganak) are all located in the underneath salt association. Due to the geological structure of basin rather complex, the oil-gas reservoirs are buried underneath the deep position and the dissolution of salt layers, so for open or pit mining, it must prepare the high strength and special equipment.

In Central Asia–Mongolia tectonic domain, there are a lot of **sedimentary sandstone uranium ore fields** (Fig. 4.74), which are the famous uranium ore province in the world (Jiao et al. 2015). Why are the sandstone uranium ore fields easy to form in that tectonic domain? It may be related to a lot of rich uranium granitic intrusions. After they were weathered and eroded, the weathering materials would be reserved in the basin. Since Neogene, that region has been located at the middle latitude dry weather belt, so that few uranium materials could be migrated off, and remnants were concentrated to form the ore deposits.

4.1.2.4 The Giant Ore Fields and Deposits in Western Tianshan Late Paleozoic Accretion–Collision Zone [9]

In the eastern regions of Western Tianshan Late Paleozoic collision zone (Fig. 2.6), i.e., the **Kulama–Fergana** metallogenic zone (Wu et al. 1993), the most important deposit is porphyry copper deposit in the Central Asia; the second one is lead–zinc ore field.

The porphyry copper deposits are mainly related to the stock or dyke of Carboniferous granodiorite porphyry. The metallogenic zone is mainly distributed in NE trending, which is intersected as a small angle with the maximum principal compression stress orientation of Late Carboniferous. It means that regional metallogenic faults have undergone extension-shear process. The ore bodies are often distributed along the faults or at the intersection point of faults (Fig. 4.14). The metal metallogenesis is concentrated in the strong alteration zone of greisen or beresite.

At the Eastern Uzbekistan, southeastward 65 km to Tashkent, there is the **Almalyk porphyry copper and gold field** (Fig. 4.14), in which **Kalmakyr** (69° 37′ E, 41° 03′ N) **supergiant porphyry copper and gold deposit** has copper reserves of 2.802 million tons, with grade of 0.38% of copper, gold reserves of 2604 tons with grade of 0.5 g/t; Besides, there are associated elements Mo, Ag, Se, Te, Re, Bi and In. The formation of ore deposit is closely related to Carboniferous–Permian granodiorite porphyry or quartz syenite porphyry, in which U–Pb zircon, K–Ar and Re–Os isotopic ages are between 320–290 Ma, i.e., in the end of Carboniferous. The ore deposit is reserved at the top of porphyry body and also is controlled by E–W trending regional fault and NW 300° trending secondary fault or breccia zone. The 65–75% of ore bodies are located in dykes and veins, and their thickness is only between several millimeters to 4 cm, length about several centimeters to tens centimeters; 30–35% of ore bodies are distributed as disseminated state. The ore veins at the top of intrusion body are distributed as lamp bulb shape. The intrusion body zone is distributed in the NW trending, which maximum planar area is about 3520 m × 1430 m, the greatest depth up to 1240 m (Mao et al. 2012a, b; Xie et al. 2015). The most concentrated faults and the ore bodies with highest grade are all trending in E–W and NW direction, which are as same as the orientation of main regional faults. The author predicts that ore deposit may be controlled by the regional EW orientation compression and sinistral strike-slip of NW trending fault in Late Carboniferous–Early Permian (Figs. 4.14 and 2.11). It means those ore deposits should be all formed in the intraplate deformation periods.

In the Southwestern Tianshan area, there is the famous gold–mercury–antimony–rare metal metallogenic zone. In the Eastern Uzbekistan, the gold resources are extremely abundant, the largest gold deposit is **Muruntau deposit** (64° 32′ E, 41° 22′ N; Figs. 4.11 and 4.15; Mao et al. 2002a, b). The length of that gold metallogenic zone is about 12 km, located at the hinterland of Qirkum basin. The gold output is 21 tons per year (however, some one estimated the output up to 80 tons per year). In recent, the proved gold reserves are 6137 tons, with average grade of 2–11 g/t, and the gold deposit could be estimated as reserves of 1830 tons in the depth of 1500 m. In Muruntau ore deposit (Fig. 4.15), the

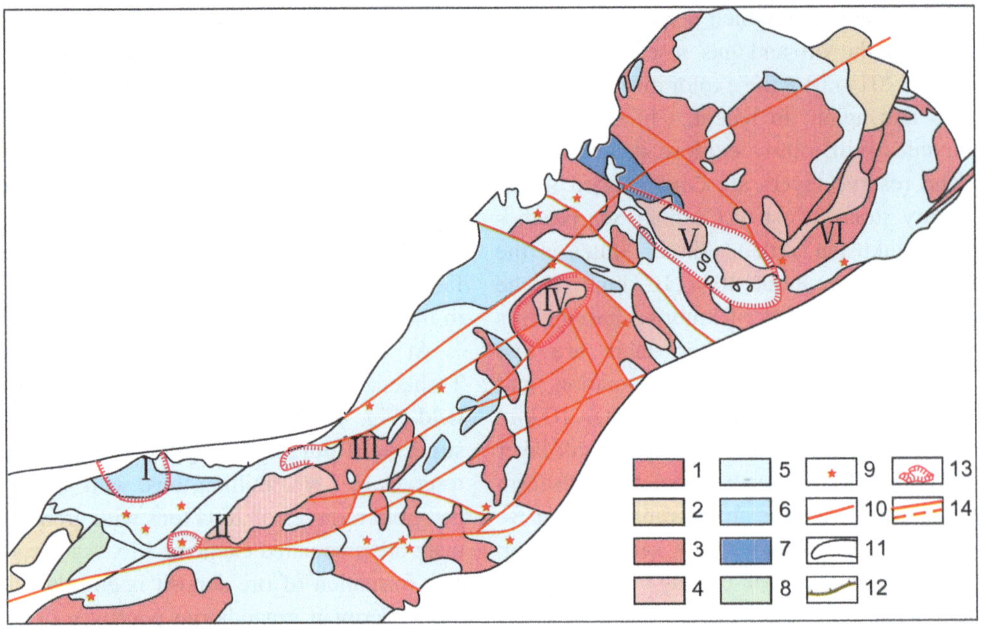

Fig. 4.14 Sketch of Kulama–Fergana metallogenic zone. **Legend**: 1. Late period of Early Paleozoic granite; 2. Late Paleozoic carbonite and dolomite; 3. Late Permian and Early Triassic granite; 4. Early Permian granite; 5. Late Permian–Early Cretaceous volcanic rocks; 6. Late Permian–Early Cretaceous sub-volcanic rocks and volcanic rocks; 7. distribution area of volcanic neck; 8. Cretaceous–Paleogene sedimentary strata; 9. eruption center; 10. fault; 11. boundary of ore field; 12. boundary of tectono-metallogenic zone; 13. outcrop of granite; 14. regional fault. **Numbers of ore field**: I. Almalyk porphyry copper field; II. Saokblak ore field; III. Kandal ore field; IV. Grand ore field; V. Kandagan ore field; VI. Bojimhak ore field (Modified from Wu et al. 1993)

main fold is Tazgan anticlinorium, which axis is near EW trending, with the eastward plunging crown and 15°–30° dip angle. The secondary fold, i.e., the Muruntau anticline, makes up its southern part. In the ore deposit, there are the well-developed strata from Cambrian to Ordovician, about 5 km thick, which are called Besapen Formation, belonging to a series of metamorphic siltstone, sandstone and mudstone. At the bottom of ore strata, there are a series of carbonate–continental volcanic–sedimentary strata; on the upper, there is the terrigenous flysch formation. According to the isotopic chronology and paleontology research, the Besapen Formation is a tectono-mixed system in fact. On the boundary, there are the vein zones with light color components, and two granodiorite stocks are distributed at the southeast part of ore areas. In the central ore areas, there are the well-developed NW trending schists and flow cleavage fracture zone, more than 10 km long and about 1 km wide, the rocks have all undergone strong faulting to form the crushed-breccia and mylonite, some researchers call that "phyllonite" zone. In that zone, there are such strong alterations as silicification, biotitization, potassium feldspathization, and the gold ore body mainly occurs in the strong silicification. The attitudes of gold body are controlled by sinistral shear zone and its ductile–ductile–brittle fracture system strictly. The ore body was formed by a lot of netted veins. The ore veins are usually the pyrite with gold–arsenopyrite–quartz veins about 15–20 cm wide. The ore deposit generally is a large-scale, complex structure rised steeply styloid with a little bit eastward dipping. The average content of sulfide in the ore vein is about 0.5–1.5%. The gold element shows the characteristics of multiple separation and re-distribution, with the silver, copper, lead, bismuth, arsenic, iron, etc. The silver average grade is about 100–300 g/t. Based on the Rb–Sr, Sm–Nd and Re–Os isotopic chronology for the synchronism ore mineral, the metallogenic ages of Muruntau gold deposit are 270–290 Ma, i.e., in Early Permian. In that period, the near EW trending faults showed the dextral strike-slip, and the NW trending faults, derived sinistral extension-shear fracture systems, then could form the ore deposit. In the Uzbekistan gold zone, the isotopic age of tectono-magmatism is 310 Ma, and the ore-forming ages are 271–261 Ma (Groves et al. 1998).

In the gold deposit, the ENE trending faults show the characteristics of compression-shear. It is not favorable to reserve the ore. It is obviously that formation of gold deposit is not caused by the black sedimentary strata. According to the data of two super-deep boreholes (depth more than 6000 m), it can be found that the source of metallogenic materials is related to the deep granites, but not been proved from the variegated Besapan Formation or only related to the black strata (Mao et al. 2002a, b). Due to the Baltic plate and Ural Late Paleozoic accretion–collision zone [12] migrating

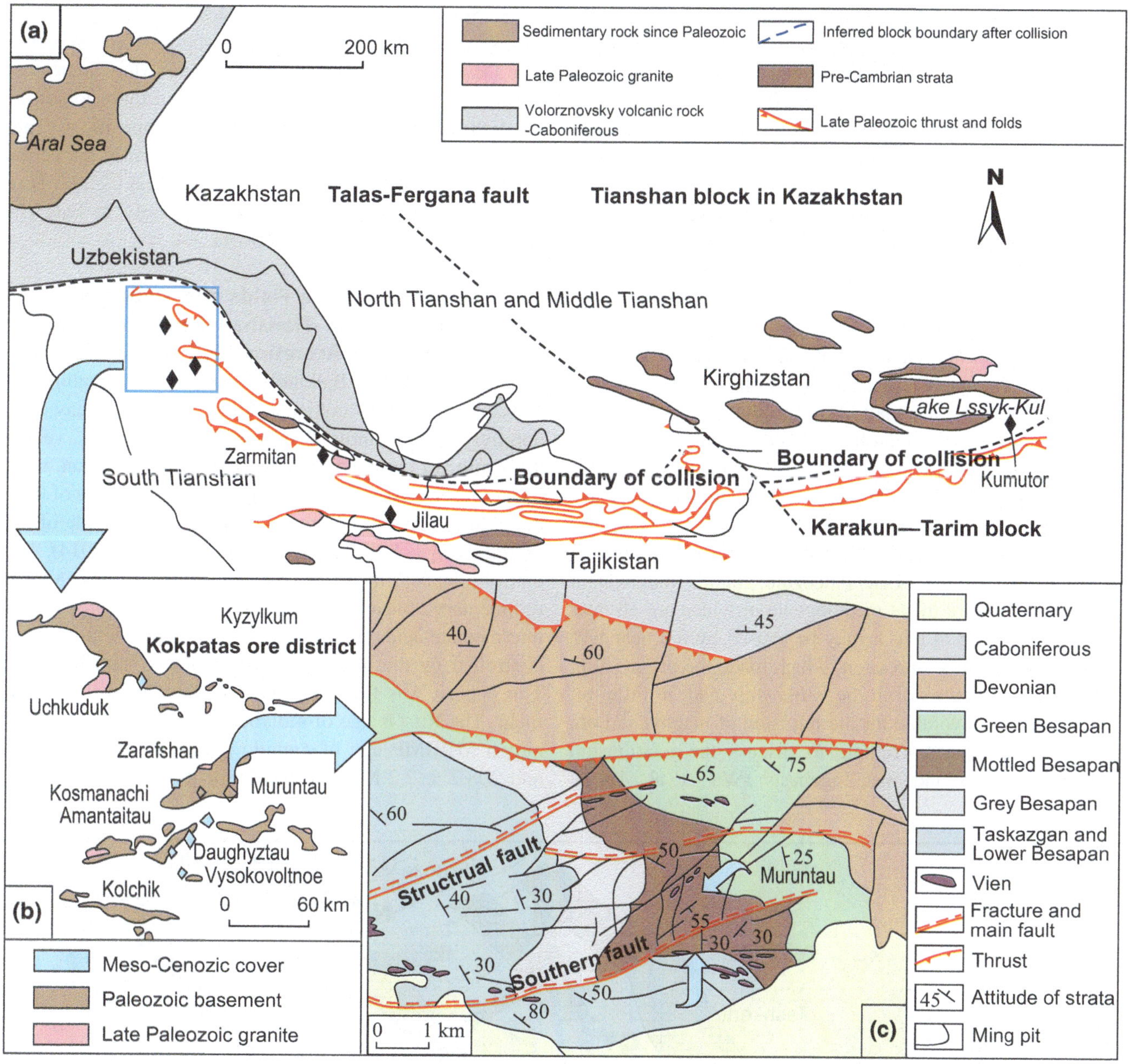

Fig. 4.15 Sketch of regional geology and ore areas in Muruntau gold deposit in Southwestern Tianshan (Modified from Graupner et al. 2006 and Mao et al. 2012a, b)

and compressing eastward, it caused the near EW and NW trending foliations showing appropriate extension, and at the crossing point of faults, the useful elements were concentrated highly to form netted veins and vertical steeply styloid generally. Formerly, some Chinese geologists tried to find the Muruntau-type gold deposit only in the black strata; in recent, it seems that is indecent obviously.

At Northern Alay Mountains of Kyrgyzstan, there is the **Dzhikrut** (68° 55′ E, 39° 14′ N) giant hydrothermal mercury and antimony deposit, in which the ore layer is rather huge to easy for exploitation. In ores, there are not only rich antimony, but also associated mercury, gold, thallium and tellurium, so it makes Kyrgyzstan be the third largest mercury production country in the world. The proved reserves of mercury are 20,900 tons, which are all concentrated in Southern Fergana mercury and antimony zone. The ore-controlling faults are mainly EW trending with high angle (Chen 1999).

In addition, in Kyrgyzstan there is the famous **Kadamse** (72° 02′ E, 40° 08′ N) **giant hydrothermal antimony deposit** and **Khaydarkan** (71° 00′ E, 39° 52′ N) **hydrothermal mercury and antimony deposit**. They are all reserved in Silurian, Devonian and Carboniferous Systems formed in Late Paleozoic accretion–collision zone. On the surface, it cannot find the relationship with the magmatism. The ore bodies are mainly reserved in the interstratum detachments. The regional faults are mainly the EW trending high-angle reverse faults, and the orientation of folds is also EW trending. However, the tectono-metallogenesis may occur in Triassic (Chen 1999). The tungsten and tin deposits are also enriched in Kyrgyzstan. The giant tungsten deposits are mainly distributed at Trudowaya and Kensu areas. The giant tin deposits are located at Trudowaya, Wqikoshgon and Saleblac (Zhang 2009).

In the Central Asian countries, Tianshan (Almaato, Tashken, Fergana, Dushangbe, etc.) areas, there is the important rare metal metallogenic province, which can be divided into North Tianshan, Middle Tianshan, South Tianshan and Southwest Tianshan (Fig. 4.16). They belong to the Western Balkhash–Tianshan–Hingganling Late Paleozoic accretion–collision zone [10] (Fig. 2.6); the orientation of collision zone is near EW trending (with ENE or WNW); however, the Late Paleozoic granitic magmatism and its related rare metal metallogenic province are all near NE trending (Fig. 4.16). It may be caused by the near EW trending regional compression, which made the pre-existing NE trending faults form the strike-slip and a little bit extension, and favorable for the intrusion of magma and ore fluids. It is a good example to discuss the relationship between magma and metallogenesis (Wu et al. 1993). Depending on the regional tectonics to predict, the main ore-forming period of that metallogenic province may be in Late Devonian–Permian. In the Tianshan–Altay areas, there are a lot of A-type granite intrusions, and their normal ε_{Nd} value shows that they have the mantle compositions. There were two strong tectonic events that occurred in the periods of 385–323 and 323–260 Ma (Figs. 2.10 and 2.11). However, the metallogenesis was formed mainly in Late Carboniferous–Permian at 318–260 Ma.

4.1.2.5 The Giant Ore Fields and Deposits in Balkhash–Tianshan–Hingganling Late Paleozoic Accretion–Collision Zone [10]

The western end of Balkhash–Tianshan–Hingganling Late Paleozoic accretion–collision zone reserves many giant porphyry copper, molybdenum and gold deposits, such as the **Kounrad copper and gold deposit**, Kazakhstan, which copper reserves are 7.9 million tons, with the grade of 0.9%; the gold reserves are more than 600 ton, with grade of 0.1–0.76 g/t (Figs. 4.17 and 4.18; Seltmann et al. 2014). Near that ore deposit, there are exposed for Early Carboniferous sedimentary-volcanic strata and Middle Carboniferous intermediate-acid granite stocks and veins, which are all controlled by the intersection point of NW and NE faults. The deposit was formed in Late Carboniferous–Early Permian. The Re–Os age of molybdenite is 284 Ma, the U–Pb zircon SHRIMP age of granitic porphyrite is 327.3 ± 2.1 and 308.7 ± 2.2 Ma, and the formation age of copper and

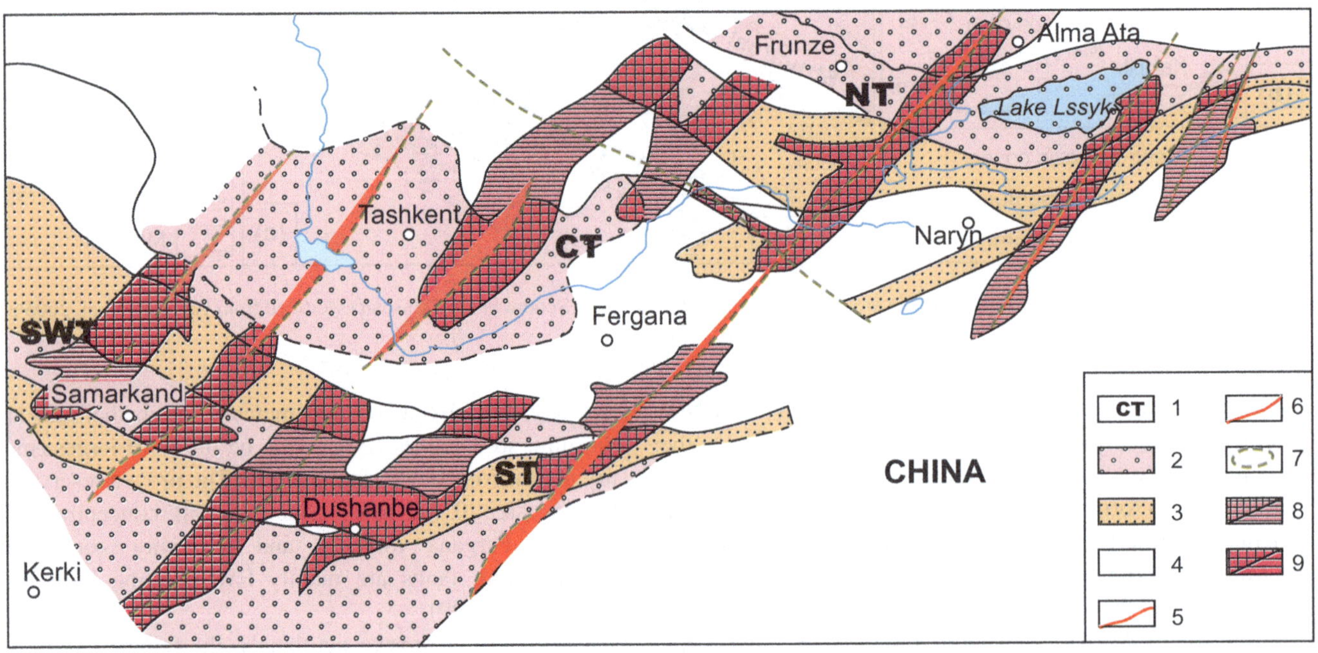

Fig. 4.16 Western Tianshan rare metal tectono-metallogenic province. 1. The code name of fold system (NT. North Tianshan, CT. Central Tianshan, ST. South Tianshan, SWT. Southwest Tianshan); 2. middle block; 3. uplift area; 4. fault-depression; 5. regional fault; 6. fault zone, to control the magmatism; 7. Late Paleozoic granite area with tin; 8. ore field with non-ferrous metals (tungsten and tin); 9 ore province with rare metals (tungsten and tin) (Modified from Wu et al. 1993)

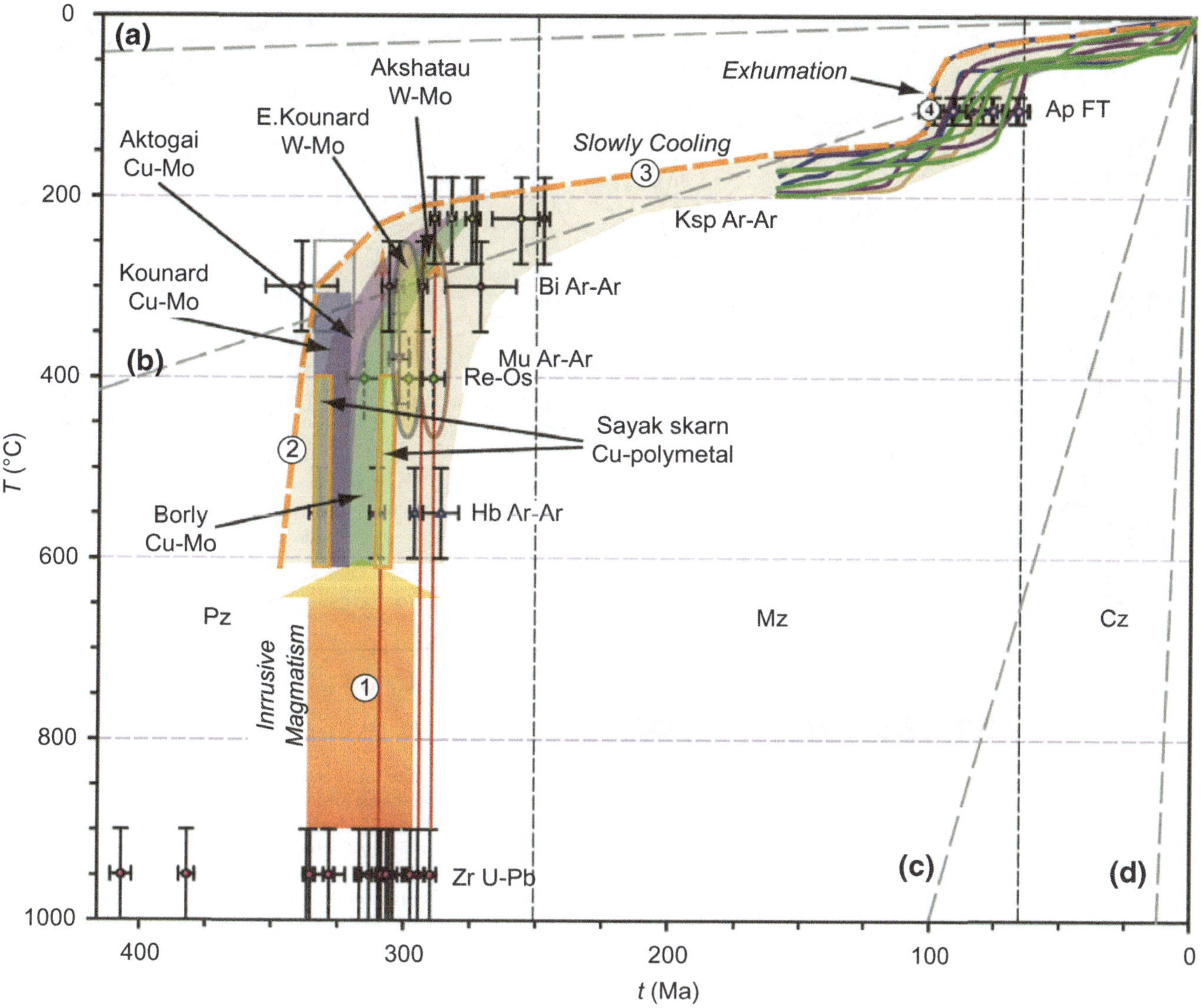

Fig. 4.17 Time and temperature change for main ore deposits (show the name and ore) metallogenesis in the Balkhash–Tianshan areas (After Nie et al. 2011; personal communication)

molybdenum mineralization is about 327 Ma (Chen et al. 2014). The controlled depth of that ore deposit is 500 m; it needs to further explore in deep.

In **Akshatau copper deposit** in the Balkhash metallogenic province, there are the copper reserves of 5.88 million tons, with the grade of 0.38%, as well associated W and Mo elements. The Re–Os ages are between 285 and 289 Ma. In the **Zhanet molybdenum deposit**, its Re–Os age is 295 Ma (Chen et al. 2014). The zircon U–Pb SHRIMP age is 316.3 and 305 ± 3 Ma for the mineralization in the **Borly copper and molybdenum deposit**. The Re–Os age is 284 Ma for mineralization in **East Kounrad copper deposit** (Nie et al. 2011; personal communication) (Figs. 4.17 and 4.18).

The **Aktogai copper field** is the biggest one in that area. Its copper reserves reach 17.20 million tons, with the grade of 0.34%, and the gold reserves reach 68 tons, average grade at 0.04 g/t (Seltmann et al. 2014). That ore deposit is located on the northeast side more than 60 km. That ore deposit was discovered in 1974. The ore deposit is developed in Carboniferous–Permian volcanic-intrusion complex zone, connecting with the granitic diorite intrusion. Near the intersection point of WNW and NE trending faults, there are crushed breccias and metal minerals developed in the quartz–feldspar small veins, to form the mesh-vein ores, and the metallogenesis was formed in Permian. The U–Pb SHRIMP ages for amphibole, biotite and potash feldspar

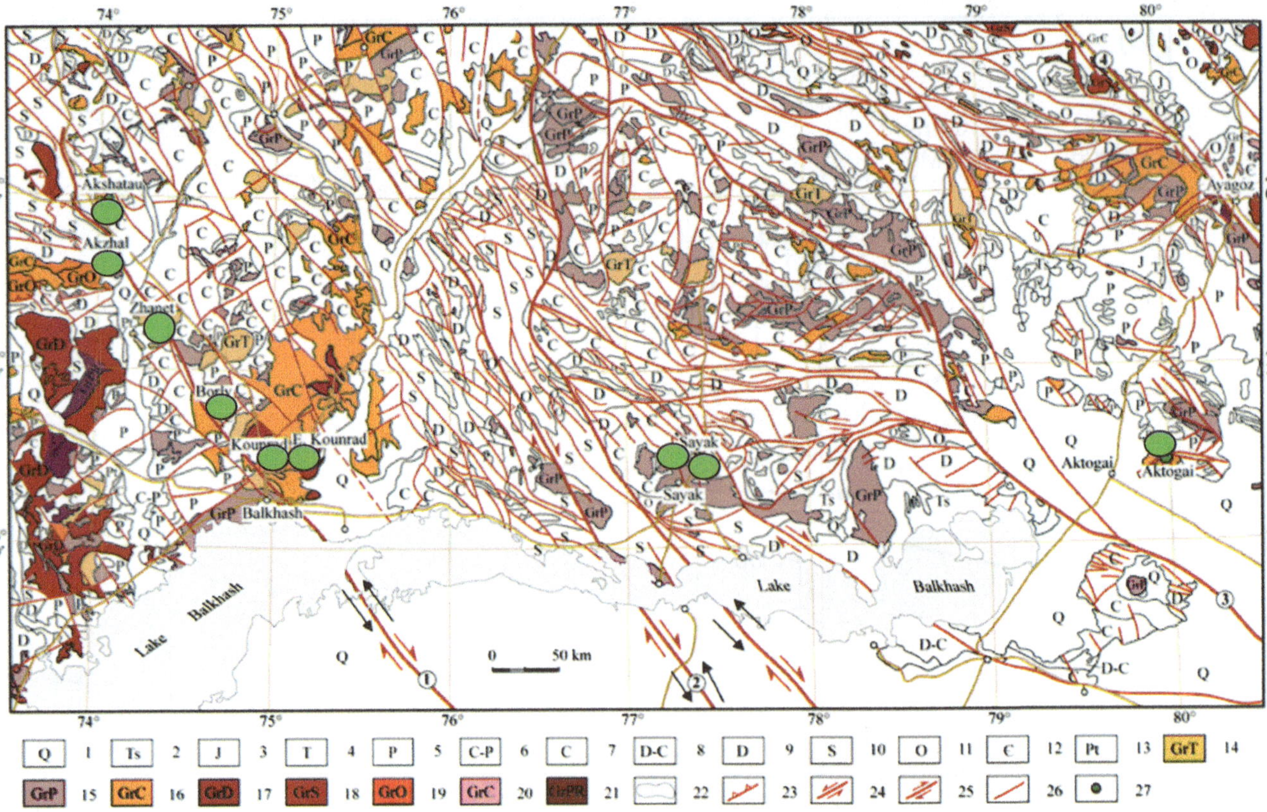

Fig. 4.18 Geological sketch of Balkhash metallogenic province. 1. Quaternary system; 2. Paleogene system; 3. Jurassic system; 4. Triassic system; 5. Permian system; 6. Carboniferous–Permian systems; 7. Carboniferous system; 8. Carboniferous–Devonian systems; 9. Devonian system; 10. Silurian system; 11. Ordovician system; 12. Cambrian system; 13. Proterozoic Erathem; 14. Triassic granite; 15. Permian granite. 16. Carboniferous granite; 17. Devonian granite; 18. Silurian granite; 19. Ordovician granite; 20. Cambrian granite; 21. Pre-Cambrian granite; 22. lake; 23. reverse fault; 24. sinistral strike-slip fault (the black arrows show the Late Carboniferous–Early Permian slip trending); 25. dextral strike-slip fault (the red arrows show the Late Devonian–Early Carboniferous slip trending); 26. the fault, activity unknown; 27. the giant metallogenic ore deposits (Akshatau, Akzhal, Zhanet, Borly, Kounrad, E. Kounrad, Sayak, and Aktogai etc.) (After Nie et al. 2011; personal communication)

in the quartz diorite and granodiorite are 335.7 ± 1.3 and 327.5 ± 1.9 Ma (Chen et al. 2014). That ore deposit is located at a little bit uplift of mantle, and the crust thickness is about 40–45 km (Figs. 4.17 and 4.18; Chen et al. 1999; Golovanov and Zhenodarova 2005; Cook et al. 2005).

The U–Pb SHRIMP ages for the syenite and granodiorite of **Sayak** ore deposit are 335 ± 2, 308 ± 10 and 297 ± 3 Ma. However, Ar–Ar cooling ages of the amphibole, biotite and potash feldspar are 287.3 ± 2.8 Ma, 307.9 ± 1.8 Ma and 249.8 ± 1.6 Ma, respectively. The formation ages of Sayak skarn ore deposit are 335 and 308 Ma (Chen et al. 2014).

The above whole ore deposits are all developed near the intersection point between regional NW trending fault and NE trending secondary faults, and mainly distributed along the NW trending faults, and formed in Late Carboniferous–Permian (about 307–257 Ma) (Chen et al. 2014). In the Balkhash ore province, there reserves the **Jezkazgan sandstone copper deposit** (copper reserves of 3.5 million tons, grade of 1.6%).

It is necessary to note that the whole supergiant ore deposits were not formed in the main collision period of Balkhash–Tianshan Late Paleozoic collision zone (Late Devonian–Early Carboniferous), but formed after the collision period, i.e., Late Carboniferous–Early Permian intraplate deformation period, controlled by the NW trending sinistral strike-slip faults or near EW trending dextral strike-slip faults (Fig. 4.17; Mao et al. 2012a, b; Chen et al. 1999).

In addition, in Northwestern Balkhash Lake areas, there are many ring-shaped structures, which appearance is so different to the intrusion bodies on the surface. It may be the surface manifestation of hidden intrusion bodies. So it is the important information to deeply explore the hidden ore deposits.

According to the above characteristics, the author predicts that in Balkhash–Tianshan areas, the Late Devonian–Early

Carboniferous magma intruded mainly along the NNE or near NS trending, as well the hydrothermal ore deposits. Because the regional maximum principal compression stress orientations at that time were mainly NNE or NS trendings, those faults showed the extension-shear, to get the magmas or ore fluids easy to inject cooling and reserve. However, in Late Carboniferous–Early Permian, the magma intrusion or hydrothermal fluids were reserved along the near EW trending or NW trending faults. Due to the regional maximum principal compression stress orientation was mainly WNW orientation at that time, the near EW or NW trending faults appeared extension-shear with sinistral or dextral strike-slip. It is favorable for magma or hydrothermal fluid to inject, cooling and reserve. The above recognition is coincided with the distribution of recent ore deposits very well, so it must be paid attention in the exploration for hidden ore deposits.

In west of Balkhash–Tianshan Late Paleozoic accretion–collision zone, at the Kazakhstan, there is the **Harik copper, nickel and gold metallogenic province**, **Jurutag gold, copper and iron metallogenic province**, Western Tianshan **Boruholo copper and gold metallogenic province, Nalati copper and nickel metallogenic province,** etc. (Karsakov et al. 2008).

The middle part of Tianshan is the important metallogenic belt in Central Asia and West China, where the copper, lead, zinc, nickel, gold and iron deposits are mainly reserved. In China, there is **Bogutu copper deposit** on northwest side of the Junggar basin. Its ore body is oriented in NE trending, formed on the side of regional fault. The Early Carboniferous porphyry deposit was mainly formed in the period of 310–322 Ma (Mao et al. 2014), namely the syn-collision period.

The **Axi epithermal gold metallogenic zone**, Western Tianshan of China, is distributed near EW trending Tulasu fault-depression-volcanic basin, in Yili basin. That fault-depression-volcanic basin is on the Early Paleozoic basement, where the Early Carboniferous continental facies intermediate-acid–intermediate volcanic systems are deposited. It is resulted from intraplate tectonic magmatism, not to belong to the collision, island arc or back island arc types. The epithermal gold metallogenic zone was mainly formed in late epoch of Early Carboniferous–Late Carboniferous (327–300 Ma; Yang et al. 2015), controlled by NW trending steep and sinistral strike–slip and extension-shear fault zone with more than 800 km. Along that zone there distribute **Axi, Abiyindi, Tawurbik, Yilmand and Jingxi gold deposits**. The ore bodies are usually reserved in WNW or NW trending en echelon extension-shear fractures derived by the dextral strike-slip fault, such as Axi gold deposit (Dong and Sha 2005).

The **Sawarden gold deposit** was found in Western Tianshan, by the Bureau of Geology and Mineral Resources of Xinjiang, China, which is the largest gold deposit in Xinjiang. The Xinjiang Second Geological Brigade found that deposit in Wuqia County, through 20-year exploration. They discovered 21 gold ore zones, which average grade is 2.45 g/t, at the height of 3, 100–4, 300 m above sea level. Up to June 2014, they got the gold reserves of 127 tons and estimated the future reserves to more than 200 tons. The No. 4 metallogenic zone is 85 m wide and 4000 m long, with the average grade of 2.57 g/t, in some parts the highest grade up to 63.88 g/t, the estimated gold reserves may reach 98.33 tons. In recent, the Xinjiang Second Geological Brigade have explored mainly the No. 4 metallogenic zone and found 11 gold mineralization veins, and the average thickness of ore bodies is about 25 m, with vertical dipping and NE 30° trending. That ore deposit contains the quartz fine network and impregnation deposits in the dolomitic carbonate clastic rocks, mudstone flysch formation, which genetic mechanism is as same as the **Muruntau gold deposit** (Liu et al. 2000). The intrusion bodies closely related to the metallogenesis are the alkali feldspar granite, which zircon U–Pb age is 261.5 ± 2.7 Ma (Yang et al. 2005a). It is obviously that granite intrusion was developed after the Tianshan collision (Late Devonian–Early Carboniferous) and formed in the intraplate deformation period. However, the metallogenic process occurred in Triassic (^{40}Ar-^{39}Ar age in the quartz gold-bearing veins dated at 210.59 ± 0.99 Ma; Liu et al. 2002). The antimony ore quartz veins were formed in Cretaceous (^{40}Ar-^{39}Ar age of 131.7 ± 1.8 Ma, Hu et al. 2000; 125 ± 17 Ma, Li et al. 2002b).

In recent, the Bureau of Geology and Mineral Resources of Xinjiang, China, have discovered a new **supergiant gold deposit—Katbasu**, in Xinyuan, Yili of Western Tianshan, which gold reserves were calculated in advance as 53 tons, the perspective gold reserves more than 100 tons. In recent, the gold proved reserves are 9.8 tons, the copper reserves up to 422 tons. The ore bodies are located in the Carboniferous adamellite and granodiorite with pyritization, yellow potassium jarosite. The ore bodies are controlled by the secondary faults, the ore alteration zone 2500 m long and 60–300 m wide. According to the Au grade >2.5 × 10^{-6}, it defines to circle 9 gold bodies with industrial value on the surface, 6 hidden gold bodies, 5 hidden copper bodies, 46 reserved ore bodies. The ore bodies generally show the ENE or ESE trending, dipping to south, with 20°–72° dip angle. The ore bodies appear stratiform-like lens, veins. The average gold grade varies from 2.56 × 10^{-6} to 9.98 × 10^{-6}, the average grade of hidden copper bodies at 0.19–0.42%. The ores are mainly the disseminated structure and fine veins. The gold is reserved in the pyrite and shows the disseminated and irregular granular (Zhao et al. 2012).

The **Yili giant sandstone uranium ore deposit** (Fig. 4.19) is located in the Yili fault-depression basin (81.9°

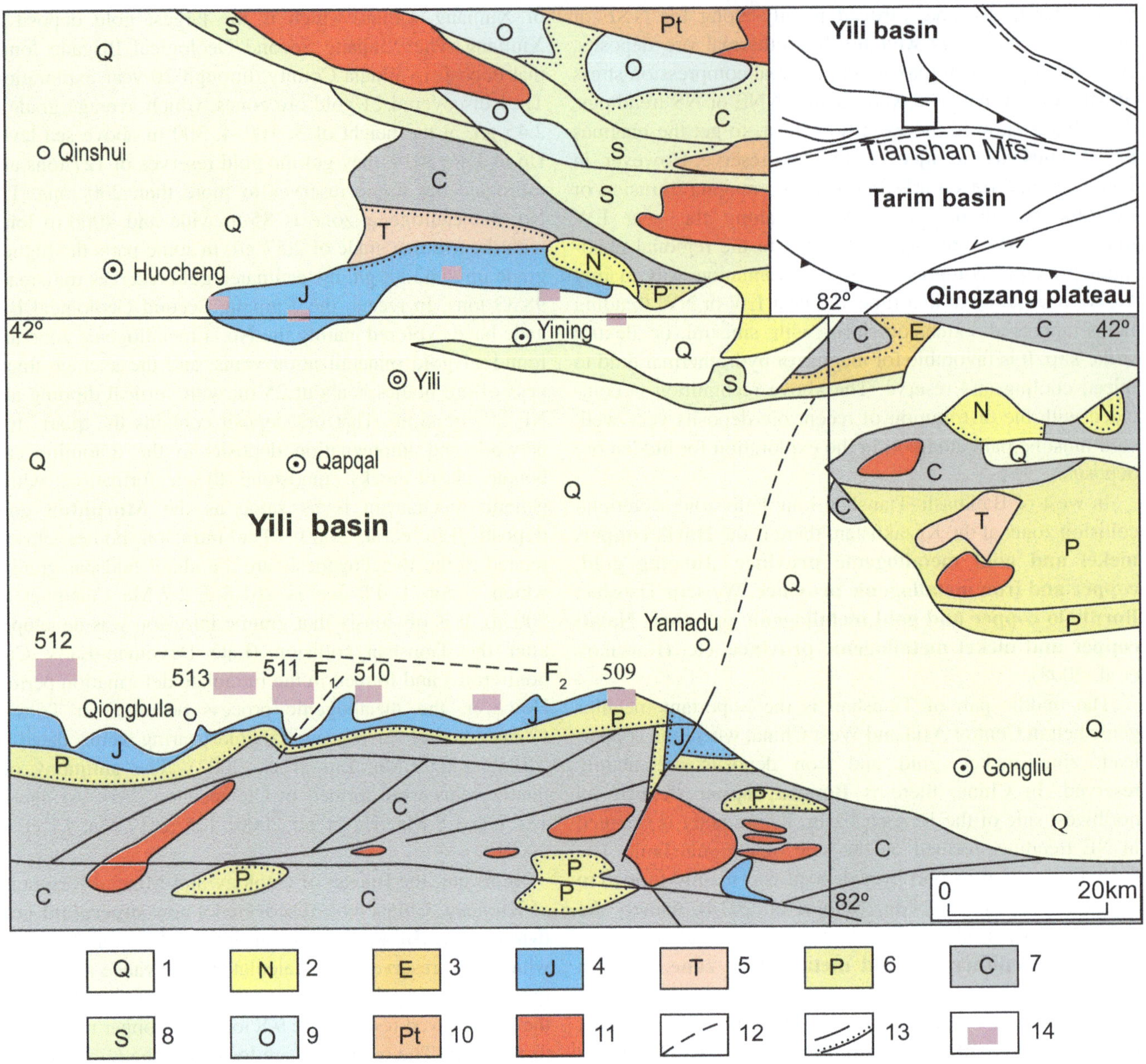

Fig. 4.19 Geological sketch of Yili sandstone uranium deposit. 1. Quaternary sediments; 2. Pliocene mudstone and sandstone; 3. Eocene sand-mudstone; 4. Jurassic sand-mudstone and sandstone; 5. Triassic sand-mudstone: 6. Permian tuff and volcanic breccia; 7. Carboniferous limestone and tuff; 8. Silurian sandstone and tuff-sandstone; 9. Ordovician limestone and slate; 10. Proterozoic limestone and dolomite; 11. intrusion body; 12. fault and inferred fault; 13. unconformity boundary; 14. uranium distribution area (Modified from Wang et al. 2006)

E, 41.9° N) in the Tianshan collision zone of China, but formed in Cenozoic. In that basin, the Early Jurassic lacustrine facies coal-bearing clastic systems are deposited, in which the high soil organic matters are concentrated and the uranium reserved. The ore body could be concentrated in the strata of Jurassic, Eocene, Pliocene and Pliocene. The sandstone uranium deposits occur mainly in the western side of Southern Yili fault-depression basin, where the late period tectonic movement is rather weak, so it is favorable to reserve the ore bodies. The main metallogenesis occurred in the period of 12–2 Ma, i.e., the late epoch of Miocene–Pliocene (Wang et al. 2006). There are the well-developed near NS trending open folds, which could be formed by the long-distance effect of the Pacific plate westward subduction since the late epoch of Paleogene, to cause the intraplate deformations. These folds greatly influenced the underground water to migrate. It is the most important influence factor for the in situ uranium mining. At the southern end of

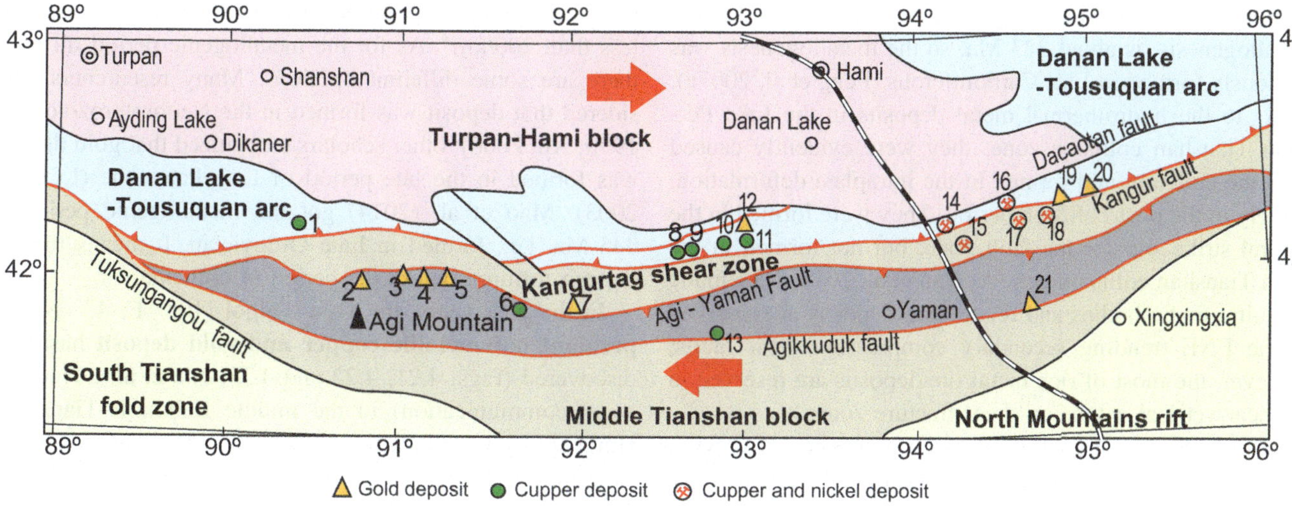

Fig. 4.20 Geological sketch of Tianshan (south of Turpan–Hami block) precious metal and polymetallic metallogenic belt. Numbers and names of ore deposit: 1. Xiaorequanzi copper deposit; 2. Shiyingtan gold deposit; 3. Kanxi gold deposit; 4. Kangur gold deposit; 5. Matoutan gold deposit; 6. Weiquan gold deposit; 7. Jiabaishan molybdenum and rhenium deposit; 8. Yandong gold deposit; 9. Tuwu copper deposit; 10. Linglong copper deposit; 11. Chihu copper deposit; 12. Xiaohongshan gold deposit; 13. Lubaishan copper deposit; 14. Tudun copper and nickel deposit; 15. Erhongwa copper and nickel deposit; 16. Xiangshan copper and nickel deposit; 17. Huangshan copper and nickel deposit; 18. Huangshandong copper and nickel deposit; 19. 148 gold deposit; 20. South Wutongwozi gold deposit; 21. Baishigou gold deposit. The red arrows show the strike-slip orientation in the metallogenic period (Modified from Mao et al. 2002a)

the Junggar block [11], there is the Kungurtag collision zone with near EW trending in the eastern Middle Tianshan (the main fault surface dipping to south).

In **Tuwu giant porphyry copper deposit** (No. 9 deposit in Fig. 4.20; Mao et al. 2012a, b), the ore intrusion bodies are mainly composed of diorite porphyry and plagioclase granite porphyry, which zircon U–Pb age is about 356 Ma, Rb-Sr isochron method age about 369 Ma, and the molybdenite forming age is 323 Ma (Mao et al. 2014). They were all resulted from Late Devonian to Early Carboniferous, i.e., the period of Tianshan main collision. It could be inferred that the bimodal volcanic system, surrounding of intrusion body, may be also formed in Devonian. Due to the main fault surfaces of collision zone dipping to south, the attitudes of Tuwu porphyry copper deposit are also EW trending and dipping to south. The ore deposit is reserved in the secondary fault zone, almost parallel to the main fault surface, thus to form the bigger ore body. The No. 1 ore body has the length of 1400 m, maximum width of 84 m, burial depth more than 600 m, with the veinlets' infection mineralization. The copper mineralization of that ore body is rather homogeneous, with the lower grade of 0.7%.

Near Hami, Eastern Tianshan, there is a precious metal and polymetallic metallogenic belt in the Kangurtag dextral strike-slip shear zone, forming a series of Kangurtag shear zone gold deposits (No. 2–8 and 19–21 deposits in Fig. 4.20); the **Xiangshan copper and nickel deposit** (Ni 0.6%; No. 16 deposit in Fig. 4.20) and **West Xiangshan** and **East Huangshan copper and nickel deposits** (No. 13–18 deposits in Fig. 4.20). To continue eastward, there is **Baishan porphyry molybdenum and rhenium deposit**, which is related to the plagioclase granite porphyry; in the ore body, the molybdenite Re–Os isochron ages are 224 ± 4.5 Ma (Zhang et al. 2005a, b) and 231.0 ± 6.5 Ma. They were all formed after the collision, i.e., in the intraplate deformation period. In **East Huangshan** copper and nickel deposit, its ore reserves are 6.92 million tons, with the nickel grade of 0.52% (No.18 deposit in Fig. 4.20; Mao et al. 2012a, b). Along that belt, the formation ages of volcano and magmatic ore deposits are mainly in 300–282 Ma, and the ages of mineralization and alternation are all in 261–252 Ma, i.e., after the collision period. According to the regional tectonics, the formation of all the volcanic rocks, mineralization and alternation is developed in the dextral strike-slip fault zone.

The **Caixiashan giant lead and zinc deposit** is located at Kawabla–Xingxingxia small block in the Eastern Tianshan Late Paleozoic collision zone, which is developed in Mesoproterozoic Xingxingxia Group. The ores are reserved in the pyritization dolomite marble, only a few in the carbonaceous siltstone. The ore bodies are controlled by the fracture zone, formed by the secondary fault. The shapes of deposit bodies are of vein-like, lenticular and lamellar. The wall rock alterations are rather strong; there are the silicification, pyritization, tremolitization and carbonatization. The origin type of that deposit is the sedimentary metamorphism–intermediate-low temperature hydrothermal reformed. The inferred reserves of lead and zinc are 3.42

million tons. The Rb-Sr isotopic age of diorite, related to the metallogenesis, is about 323 Ma, so the metallogenesis was obviously formed in Late Carboniferous (Peng et al. 2007a).

As to the hydrothermal metal deposits in the Late Permian Tianshan collision zone, they were evidently caused after the collision and formed in the intraplate deformation, namely in the post-collision period. They were formed in the dextral strike-slip for the fault zone, but not formed in the main Tianshan collision period (Wan et al. 2015). The mafic and ultra-mafic bodies and related ore deposits are reserved in the ENE trending secondary compression-shear faults; however, the most of rich metal ore deposits are reserved in the near vertical extension-shear fracture zones.

Those ore deposits have very well anomalies of gravity, magnetism, electricity and the dispersion flow, which are also favorable to adopt geophysical and geochemical methods to explore the ore bodies.

In Siziwang Banner, central–northern areas of Inner Mongolia, there is **Bainaimiao hydrothermal polymetallic deposit** (111.6° E, 41.6° N), which was formed in Proterozoic metamorphic system and Late Paleozoic granite. The ore deposit located in the Late Paleozoic accretion–collision zone, metal minerals came from the mantle mainly. It was usually accompanied with the mica or amphibole granite intrusion to form ore deposit. The ore bodies have the characteristics of volcanic massive sulfide type. The ore deposit is composed of quartz veins and quartz veins with alteration rocks. The trend of discovered ore deposit is mainly in accordance with the orientation of regional fault; however, the distribution density of ore veins is rather rare, less than 200/km^2. As for the metallogenic period till now, there are some different opinions. Many researchers considered that deposit was formed in the Neoproterozoic (Nie 1990; Xin 2006); other scholars recognized that gold deposit was formed in the late period of Late Paleozoic (Li et al. 2003). Mao et al. (2014) got the metallogenic period at 445 Ma, i.e., formed in Late Ordovician. It means the ore deposit forming before the period of collision.

In the past years, the **Oyu Tolgoi** (108° E, 43° N) **supergiant polymetallic copper and gold deposit** has been discovered (Figs. 4.21, 4.22 and 4.23; Nie et al. 2011, personal communication) in the middle Balkhash–Tianshan–Hingganling Late Paleozoic (385–260 Ma) accretion–collision zone [10], in Mongolia and near the border of China. The copper reserves of that ore deposit reach 37.58 million tons, with the copper grade of 0.98%; the gold reserves are 1425 tons, with the grade of 0.38 g/t (Seltmann et al. 2014), and other results are a little bit different (Fig. 4.21). The Oyu Tolgoi ore deposit is reserved in the Late Devonian porphyry body, and the wall rocks are mainly the acid-intermediate volcanics and volcanic clastic rocks, which were intruded by Early Carboniferous–Early Permian granitic bodies. Depending on the molybdenite of Re–Os age dating, the metallogenic ages of that deposit are 372–370 Ma, i.e., formed in Late Devonian (Zhang et al. 2010; Mao et al. 2014; Fig. 4.22). The ore deposit is resulted from the syn-collision. The deposit is controlled by the NNE trending fault obviously, but not along the near EW trending regional

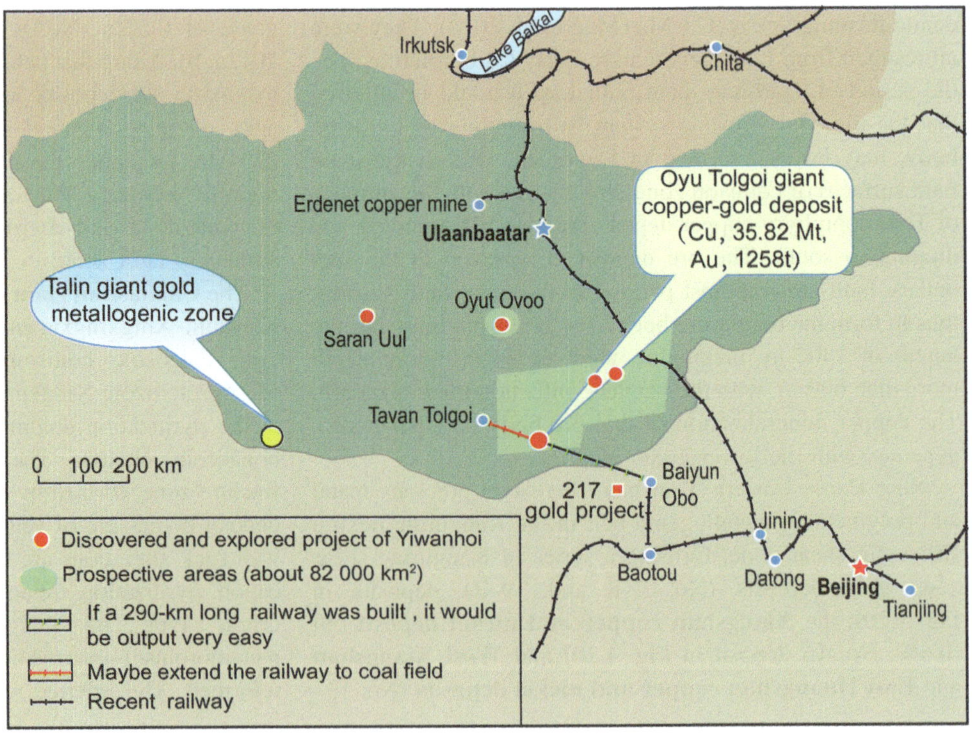

Fig. 4.21 Location of Oyu Tolgoi copper and gold deposit, in Southern Mongolia (After Nie et al. 2011; personal communication)

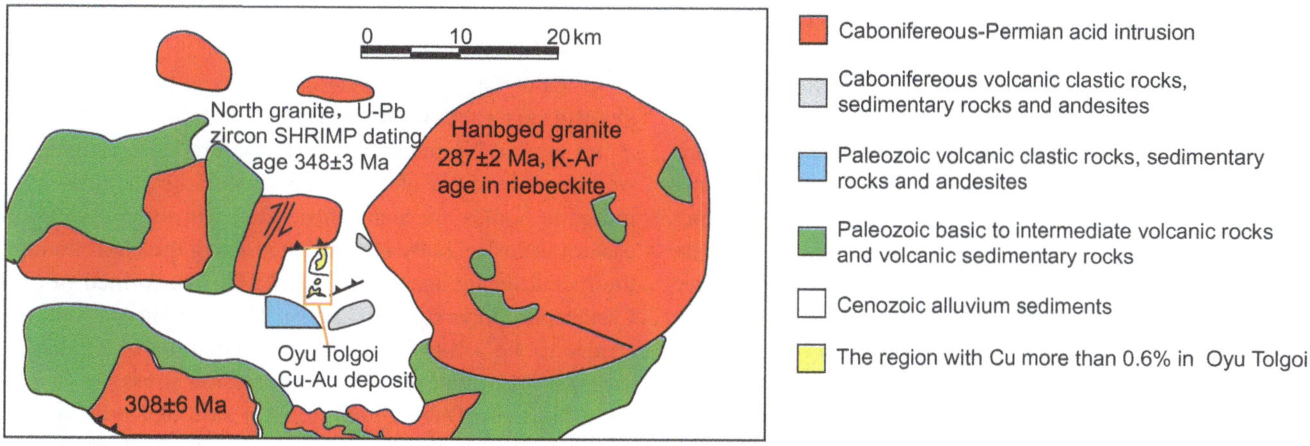

Fig. 4.22 Regional geological sketch of Oyu Tolgoi copper and gold deposit (After Nie et al. 2011; personal communication)

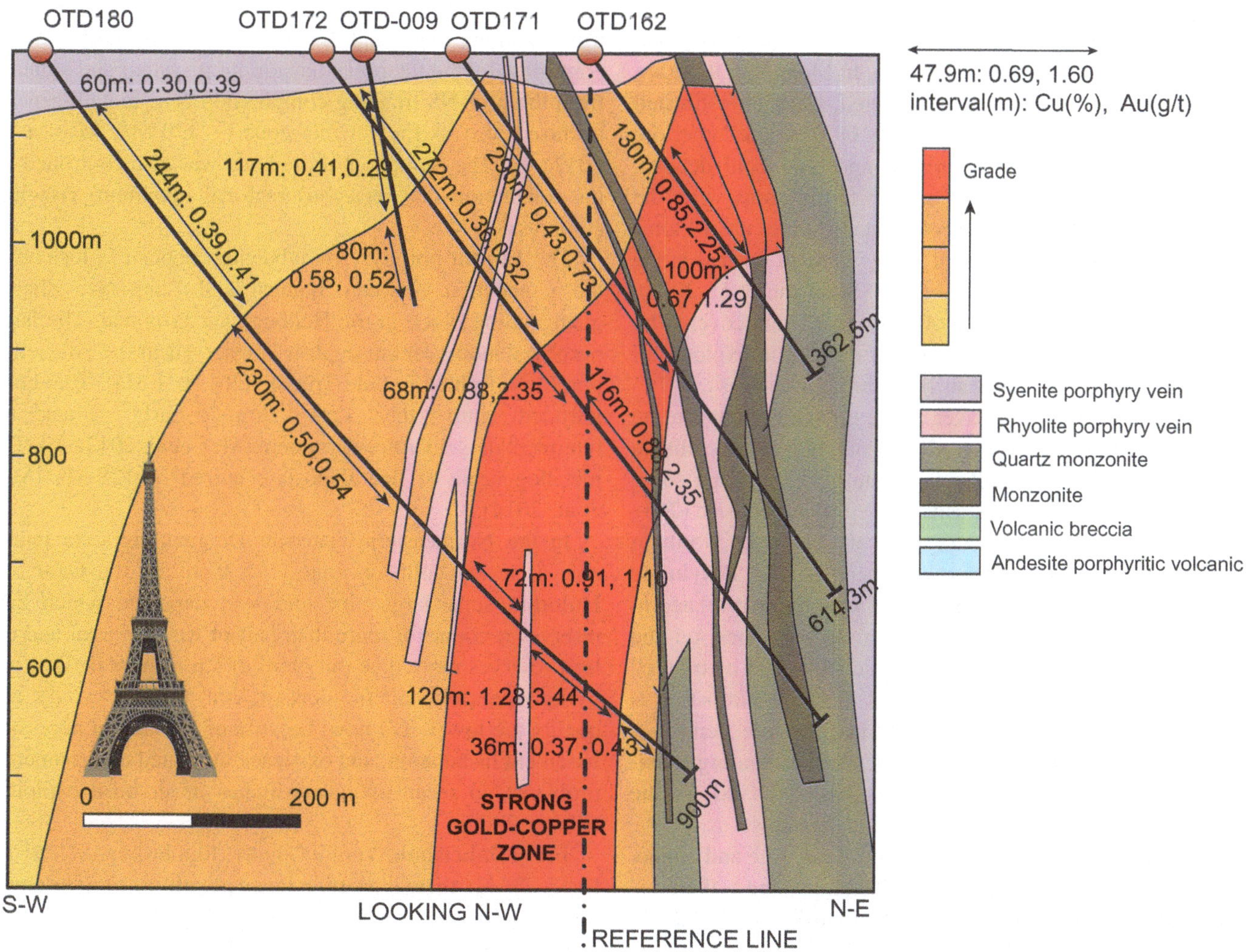

Fig. 4.23 Main geological section of Oyu Tolgoi copper and gold deposit (After Nie et al. 2011; personal communication)

faults of collision zone. The ore bodies are 400–1200 m long, 90–225 m wide, with the dip depths of 450–950 m (Fig. 4.23). In the SW toward and middle ore areas, there is the polymetallic ore deposit; in the north, there are the massive sulfide ore bodies. The ore deposits now are all covered by the Quaternary alluvium sediments. It needs only to build a long railway of 290 km for the Oyu Tolgoi copper and gold deposit to connect with the Chinese railway net, so it has the convenient communication. In the middle of Balkhash–Tianshan–Hingganling Late Paleozoic accretion–collision zone [10], the orientations of collision and compression are mainly NS direction, and the related extension-shear faults and fractures will be near NS trending.

Due to this area is far away to the Ural collision zone, the collision influence is rather weak. According to the understandings, in Mongolia near the China and Mongolia border, there are 56 polymetallic copper deposits, in which 40 ore deposits are related to the Late Paleozoic granites, 3 deposits were formed in Early Paleozoic and 13 deposits were developed in Jurassic–Cretaceous. The supergiant copper deposits may be mainly related to the Late Paleozoic granites, others were all formed after collision, and related to intraplate deep faulting. That metallogenic zone is developed in Mongolia, located at the rocky desert, with very good outcrop, so it is favorable to explore. In Mongolia, there are more than 130 gold deposits discovered, and their estimated reserves are 3000 tons (Zhang 2009). Of them the larger ore deposit is the Tsagaan Suvarga copper and gold deposit, which copper metal reserves are 1.3 million tons, with the copper grade of 0.53%; and gold metal reserves are 19 tons, with the gold grade of 0.08 g/t. They were all formed in the period of between 325 and 365 Ma (Seltmann et al. 2014). However, in Inner Mongolia of China, for being covered by the desert and grasslands, it is rather difficult to explore, till now it has not any important breakthrough progress.

Near the Xar Moron River at the boundary between the Tianshan–Hingganling Late Paleozoic accretion–collision zone [10] and the Sino–Korean plate [14], there are some porphyry molybdenum deposits formed in the Triassic (245–220 Ma; Nie et al. 2011; Zeng et al. 2013). In Northern Bairin Zuoqi, Inner Mongolia, there is the **Baiyinnor** (119.3° E, 44.1° N) **skarn lead and zinc deposit** related to the intermediate and acid intrusion. It is located on the boundary of Permian epimetamorphic strata and distributed along the near EW orientation. That metallogenic period is Middle–Late Jurassic at 171–140 Ma. That ore deposit is obviously controlled by the Jurassic WNW trending maximum principal compression stress and reserved in the heterogeneous lithology and extension fractures.

In the Dahingganling–Uran hot tin, lead, zinc and copper metallogenic zone, south of Dahingganling, there is **Huanggan skarn tin and iron deposit** (117° 22′ E, 43° ± 35′ N), in Hexigten Qi, which is the second largest iron deposit of Inner Mongolia, with rich tin, tungsten, zinc and copper. That deposit is distributed in NNE trending, the metallogenic periods at 137–122 Ma, i.e., Early Cretaceous (Mao et al. 2012a, b). That ore deposit was controlled by the NNE trending Early Cretaceous regional maximum principal compression stress, which led to the NNE trending fractures open appropriately, then favorable for the mineral-bearing fluids to intrude and reserve.

In Zalut Qi, Inner Mongolia, there is the **Barze alkali granite rare earth metal deposit** (121° E, 44.7° N), which is a rare type in Northeast China. In the ore area, there are the well-developed EW and NNE trending faults; the bearing-ore intrusion bodies are along those faults. In the ore areas, the eastern intrusion body is distributed along the NNE trending, the metallogenic ages at 125–127 Ma, i.e., formed in Early Cretaceous (Mao et al. 2012a, b). The formation of ore body seems to be controlled by the orientation of Cretaceous regional maximum principal compression stress.

In Northern Dahingganling, west side of Derbugan fault zone, there is the **Erentolegai epithermal silver deposit**. That deposit is controlled by the NNW trending Barjiger fault and NE trending Erentolegai fault, and the ore bodies are located in the quartz porphyry. The ore bodies are distributed along NNE and NNW trendings, which seems be controlled by the tracing extension joint system and suffered from the near NS trending compression. The metallogenesis occurred also in Early Cretaceous (~120 Ma, Mao et al. 2012a, b). The formation of ore body was also controlled by the orientation of Cretaceous regional maximum principal compression stress.

The **Lumin** porphyry molybdenum deposit is located at the connection between Jiamusi and Song–Nen–Zhangguangcailing block in the Heilongjiang Province. The large porphyry molybdenum deposit is reserved in the adamellite porphyry, formed in Late Triassic (201 ± 4 Ma). That body intruded into the Late Jurassic–Early Cretaceous intermediate-acid volcanic system (Mao et al. 2012a, b). The metallogenesis of that deposit occurred at 178 Ma (Mao et al. 2014).

In the Northeastern Tianshan–Hingganling Late Paleozoic accretion–collision zone [10], there are the **Ozer** and **Hulodna** super-large zinc and lead deposits, which zinc reserves account for more than half of Russian total and the lead reserves more than one-third of Russian total. That is a pyrite-type polymetallic deposit, but the ore grade and quality are lower. Till now, because of the lack of necessary basement installation, and existence of related environmental problems, in recent that deposit has never been exploited (Zhang 2009).

In the **Daheishan**, Yongji County, Jilin Province (126° 16′ E, 43° 29′ N), there is a **super-large porphyry molybdenum deposit** (Wang et al. 2012a), which is located at the southern Jiamusi block and in the eastern of Tianshan–Hingganling Late Paleozoic accretion–collision zone [10]. That deposit is controlled by the cross-location of NE and EW trending faults (Fig. 4.24), which belongs to a giant scale and low-grade ore body in the upper part of the granite–diorite porphyry. The ore body is bigger on the upper and smaller in the deep. Its surface areas are 2.7 km^2. Although that deposit is located in the Late

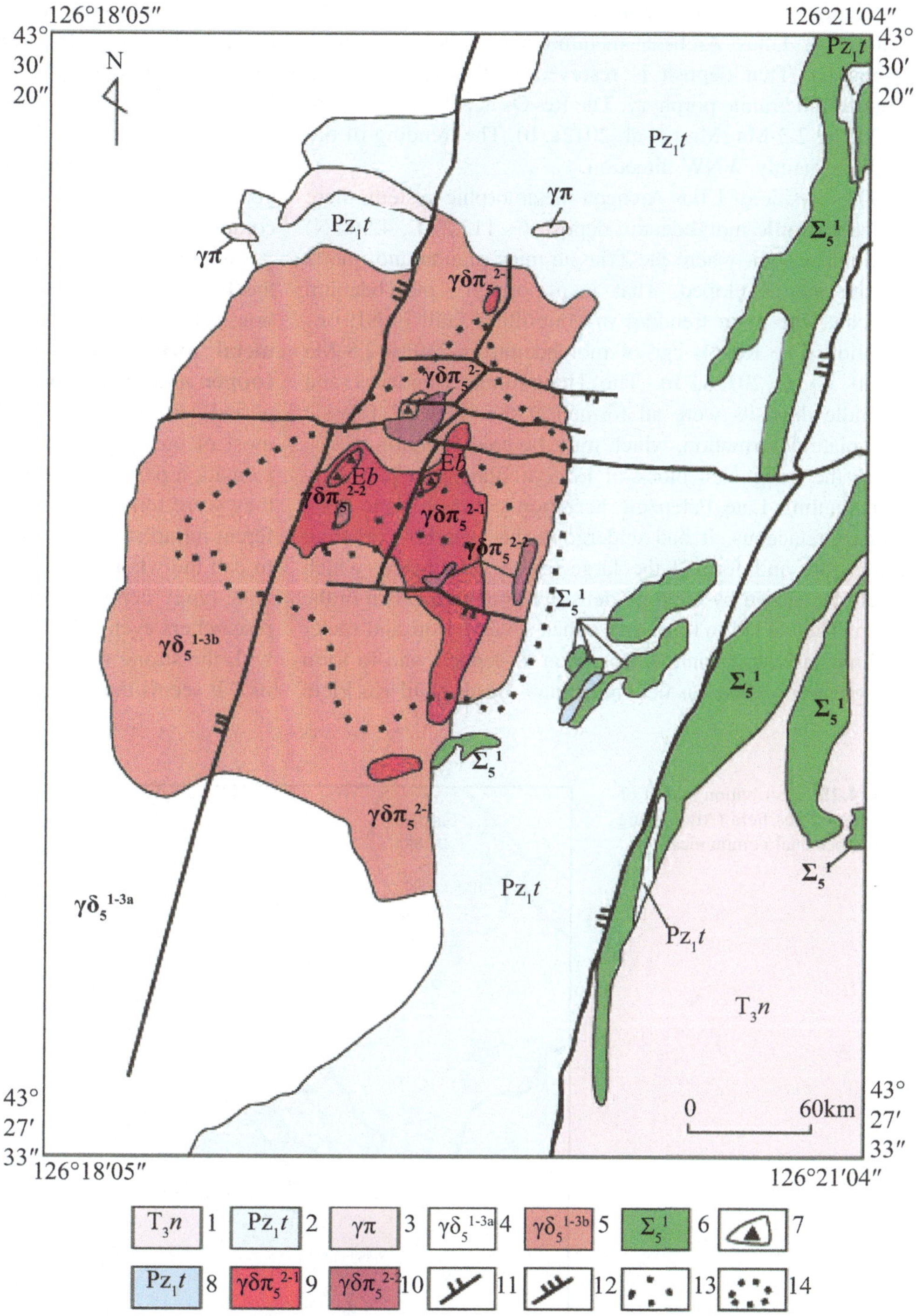

Fig. 4.24 Geological sketch of the porphyry molybdenum deposit in Daheishan, Yongji County, Jilin Province. 1. Nanloushan Formation of Late Triassic; 2. Toudaogou Formation of Early Paleozoic; 3. granite porphyry vein; 4. Changgangling mica granodiorite; 5. Qianzuiluo mica granodiorite; 6. ultra-basic vein; 7. breccia stack; 8. diorite vein; 9. Qianzuiluo mica granodiorite porphyry; 10. Qianzuiluo fine-grained mica granodiorite porphyry; 11. extension fault; 12. compression-shear fault; 13. rich molybdenum ore body; 14. poor molybdenum ore body (Modified from Wang et al. 2012a)

Paleozoic collision zone, the metallogenic period is of Jurassic at 168 Ma (Mao et al. 2014), i.e., that is controlled by the intraplate faulting.

In South Jilin, there is the **Hongqiling** copper and nickel deposit, which is a basic magma liquation deposit. That was formed in the Early Paleozoic hornblende gabbro and hornblende pyroxenite bodies, which intruded into the amphibolite and plagioclase–gneiss metamorphic system. The ^{40}Ar-^{39}Ar age of hornblende in the pyroxenite bodies is 228 ± 3.0 Ma, and the Re–Os isochron age of pyrrhotite is 208 ± 21 Ma (Mao et al. 2012a, b). It is obviously that the intrusion and metallogenic periods are all in Triassic. In that deposit, the main orientation of fault as passway for ore fluid is NW trending and the secondary is NE trending.

There is the **Bogdaura** molybdenum deposit in the middle of Ulias Archean metamorphic system, in Inner Mongolia. That deposit is reserved in the Triassic mica granite and granite porphyry. The Re–Os age of molybdenite is 235 ± 2.3 Ma (Mao et al. 2012a, b). The trending of ore body is mainly WNW direction.

On the side of Ulias Archean metamorphic system, there is the **Urandle** molybdenum deposit (~ 112.2° E, 42.6° N) in Sonid Zuoqi, where the Triassic mica granite and quartz diorite are developed. That is a porphyry molybdenum deposit. The main trending of controlling fault is NE orientation. The Re–Os age of molybdenite is 239 ± 2.8 Ma (Mao et al. 2012a, b). The Hongqiling, Bogdaura and Urandle deposits were all formed at the stage of Triassic intraplate deformation, which must be paid attention to.

In the Song–Nen block of Eastern Balkhash–Tianshan–Hingganling Late Paleozoic accretion–collision zone [10], since Cretaceous, it had undergone the near EW trending extension, and formed the large-scaled subsidence, which could be caused by the high-density basic magmatism in the deep. Thus, it led to huge, more than several-thousand-meter sedimentary strata-rich hydrocarbon to deposit, and to form the greatest oil and gas field in China—**Daqing** Oil-gas Field (Petroleum Geology Editorial Board of Daqing Oil-gas Field, 1993; Fig. 4.25). That oil field has been exploited for more than 50 years. The main source rock is the Qinshan Formation sandstone system of Early Cretaceous. The main reservoir structures are NNE trending open folds and related fractures (joints), which were formed at the end of Paleogene, caused by the Pacific plate's westward subduction and compression.

According to the recent data, in the formation process of the Balkhash–Tianshan–Hingganling Late Paleozoic accretion–collision zone [10], the **Balkhash–Tianshan rare metal metallogenic zone** and **super-large Oyu Tolgoi copper and gold deposit** were formed in the syn-collision period, i.e., in Devonian–Early Carboniferous. Although the most of ore deposits are located in that collision zone, the formation periods are all after Late Carboniferous. It means they were formed after the collision ending, i.e., in the different intraplate deformation periods. Some researchers used to call the whole deposits in the collision zone as "collision zone-type deposit" or "orogenic-type deposit"; some researchers even called the ore deposits in the intraplate and with the strong rock deformations as "collision-type deposits." It seems that their recognitions should be corrected. As

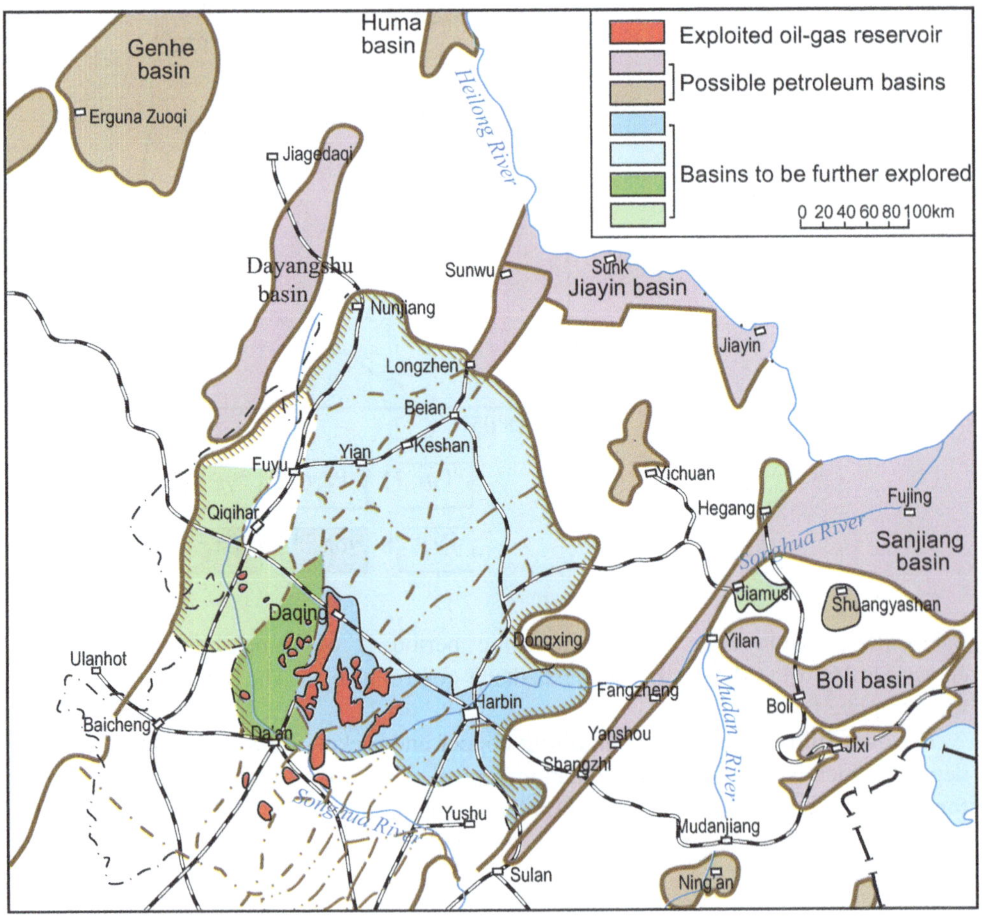

Fig. 4.25 Distribution sketch of Daqing oil-gas field (After Wang 2003, personal communication)

to the tectonic characteristics of metallogenic area, whether the periods of metallogenesis and tectonic events are formed in same time or not, it needs to be detailed research. Otherwise, it will be easy to make the wrong way for the exploration. In the main collision period, the strongest tectonic events were very favorable for the magma and ore fluids to move; on the other hand, they were easy to scatter and disappear into the hydrosphere and atmosphere. Of course, the ore deposits forming after the collision period will get more opportunity, the reason is that in the old collision zone, the secondary intraplate deformations were not so strong to get the useful elements to migrate and reserve, not easy to scatter and disappear into the hydrosphere and atmosphere, and then to reserve in the crust rocks.

4.1.2.6 The Giant Ore Fields and Deposits in Junggar Block [11]

In the Junggar block, there is the **Karamay** oil field (84.6° E, 46.2° N; The Petroleum Geology Editorial Board of Xinjiang 1993). The source rocks and reservoirs are located in the Permian, Triassic, Early and Middle Jurassic strata (Fig. 4.26). There are a series of NW trending high-angle reverse faults caused by the compression from north to south, which cut off Middle Jurassic strata and their beneath strata covered by Upper Jurassic strata. So it is obviously that the reversed structure was formed at the end of Middle Jurassic.

The **Karatongk** mafic magma-type copper and nickel sulfide deposit in Fuyun County (88.1° E, 47° N) is located on the northern border of the Junggar block. There are the well-developed Devonian marine facies intermediate-basic volcanic rocks and clastic formation, which is closer relationship with the Karatongk mafic magma-type copper and nickel sulfide deposit. In Early Carboniferous, there developed intermediate-acid volcanic lava, upper parts mainly the volcanic clastic rocks with the turbidite fan facies deposits. The directions of regional strata and faults are all NW trending. According to the isotopic dating results, the ages of rock formation and metallogenesis are all of 274–314 Ma, and the main metallogenic period is at 285 ± 17 Ma, i.e., in Early Permian (Mao et al. 2012a). In that metallogenic period, being suffered from the near EW trending compression, the NE trending fault zone showed the sinistral strike-slip characteristics and got the ore bodies have the characteristics of sinistral and en echelon range. That deposit was also formed in the intraplate deformation period.

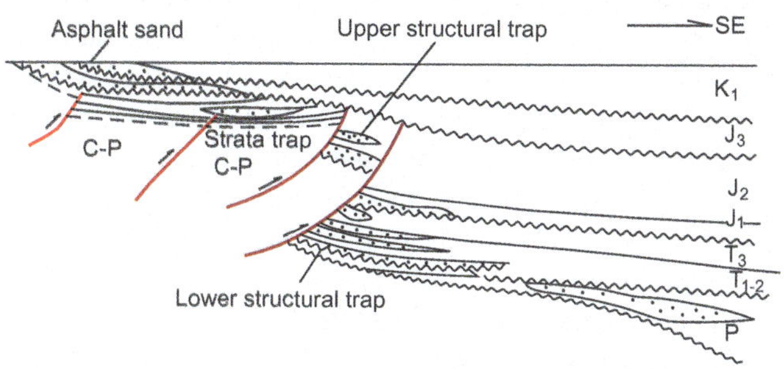

Fig. 4.26 Middle Jurassic reversed faults and the oil-gas trap (Modified from Zhai 2002)

4.1.2.7 The Giant Ore Fields and Deposits in Ural Late Paleozoic Accretion–Collision Zone [12]

In the Ural Late Paleozoic accretion–collision zone (Fig. 2.6), there are a lot of ophiolite suits, which are the main developed areas for the chromite and coulsonite and titanomagnetite, with the coulsonite reserves about 25 million tons, ranking the first in the world. At the southern end, the **Aqtobi** (~60° E, 50° N) in Kazakhstan reserves the best-quality chromite ore in the ultra-basic intrusions, in which the **Kempirsal** (58° 35′ E, 50° 15′ N) super-large ophiolite chromite deposit is best one, with the Cr_2O_3 ore grade of 20–59%, and the reserves of 3200 million tons. There are 120 ore bodies, which are distributed in the ore zone with 24-km length and 7-km width (Zhang 2009). The whole ore deposits were formed in Late Paleozoic syn-collision period. In the east, there is **Turgy** (61° 45′ E, 49° 35′ N) volcanic-type iron deposit, which was also formed in Late Paleozoic syn-collision period. In the Ural, there is also Shameika porphyry molybdenum deposit, and its Re–Os ages of molybdenum are 273 ± 5 and 282 ± 6 Ma (Mao et al. 2003).

4.1.3 The Giant Ore Fields and Deposits in Sino–Korean Domain

4.1.3.1 The Giant Ore Fields and Deposits in Sino–Korean Plate [14]

In the Archean distribution areas of Sino–Korea plate (Fig. 2.14), the **Anshan–Benxi** (122° 57′ E, 41° 02′ N) in Liaoning, **Qian'an–Qianxi** in Hebei (~118° E, 40° N), Central Inner Mongolia (~100° E, 41.5° N), Northern Shanxi (~112.3° E, 40° N), Western Shandong (~116.5° E, 35.5° N) and **Maoshan** in Korea (~129.2° E, 42.1° N), there are many giant banded iron formations (BIFs). They mainly were formed in Archean (3.1–2.9 and 2.7–2.5 Ga). Those ores were formed in the original continental nucleus. At that time, it was rare of the oxygen atmosphere, and the crust was rather thin with strong volcanic activity to form the intermediate-acid volcanic rocks and sedimentary systems, through the regional metamorphism and violent deformations. Now they are all located in the hinge zone or limbs of complex synclines and have undergone the deeply reformation and destruction by the metamorphism and migmatization. Those rich iron deposits have been almost extracted out, so now those deposits are mainly exploited for the low-grade ores (Mao et al. 2012b).

In recent, Bureau of Geology and Mineral Resources of Liaoning Province has discovered an iron deposit, which forecasting reserves are about 1 billion tons. The upper part of ore body is haematite with the iron grade of 50%, and in the lower part of ore body with the iron grade of about 30%.

Along the **Yingkou** of Liaoning (122.4° E, 40.6° N)–**Ji'an** of Jilin (126° E, 41.1° N), there are the sedimentary systems in the Paleoproterozoic ancient rift to form the boron metallogenic zone, with the NE trending, 300-km length and about 100-km width. The main boron deposits are the Houxianyu and Zuanmiao deposits in Liaoning. Those deposits are located in the Paleoproterozoic rift between two Archean continental nuclei. Besides enriching with the boron, the rock system in the Liaohe Group is associated with the iron, rare earth elements and uranium. There are a series of metamorphic-rich magnesium evaporite systems. The boron mainly is caused by the sea bottom hydrothermal fluids. All the wall rocks of deposits are the rich magnesium marble. The boron deposits were mainly formed in the period of 1852–1923 Ma, i.e., in the late period of Paleoproterozoic (Zhang 1984; Mao et al. 2012b).

The **Wengquangou,** Liaoning (~124° E, 41° N) super-large boron deposit is developed in the serpentinite on the side of the sericite amphibole leptynite, which belongs to the Liryu Formation of Proterozoic Liaohe Group. The distribution area of that is about 5 km^2. The 9 ore bodies are of the stratified, convex lens-like and hyacinth bean-like, in which the No. I and II are large. No. 1 body is EW trending, 2800 m long, 1500 m wide, with the most thickness of 150 m, the average thickness of 45 m. The average boron grade is 7.23%; the content of total iron is 30.65%, with the large coexisted deposit of iron and uranium. The ore minerals are mainly the magnetite and ludwigite, and the secondary ones are the ascharite and uraninite. The veinstone minerals are the serpentine, phlogopite and humite. The ore types are the ascharite–magnetite, colbranite–magnetite, ascharite–colbranite–magnetite and magnetite, of which the first two types are mainly in the deposit. The boron proved reserves and resources are 2185 tons, and the iron ore resources are 2800 million tons.

From **Haicheng** to **Yingkou Dashiqiao**, in the Southern Liaoning (the central area at 122.4° E, 42.6° N), there distributes the magnesite deposit with the length about 40 km, available reserves of 2.677 billion tons, occupying 85% in China, and 20% in the globe. In that magnesite deposit, the ore grade is rather high, with a little impurity. The ore bodies are reserved in the shallow, so it is easy to exploit (Zhao 2009). In recent, six large and super-large good-quality magnesite deposits (reserves more than 0.1–0.5 billion tons), one large talc deposit have been discovered. On the east side of those deposits, there is a near EW trending rift trough. At the north of that rift, it forms the Dashiqiao giant talc and magnesite deposit in the rich magnesium carbonite. The talc and magnesite deposits are controlled by the magnesium carbonite in secondary wedge basin of ancient rift. The thickness of whole wedges is more than thousand meters on its western side, and toward the east only several hundred meters. According to the research, the magnesite sediment was formed in the weather of dry tropical zone or subtropical zone, may be located at the latitude of 17°–28°. The talc ore bodies are mainly reserved in the compression-shear zone between the magnesite marbles or magnesites (Zhu 1984). Similarly, there are also the large magnesite deposits in the Korean Peninsula.

In the Korean Peninsula, eastern Sino–Korean plate, there are many lead, zinc, copper and gold deposits, which could be delegated by the **Komdok** (128° 47′ E, 40° 57′ N) super-large lead and zinc deposit. That deposit is located in the Paleoproterozoic metamorphic systems of near EW trending rift (Zhang et al. 2012).

The **Huadian** (~126° E, 42° N), Jilin Province, reserves a famous large gold deposit, forming in the quartz veins, developing in the Neoproterozoic gneiss system, of them the NW orientation Dalazi–Jiabigou structure zone with more than 40-km length and 5-km width. In that zone, there are a lot of fractures, foliations and mylonites, accompanied with multiple granite intrusions. In those structural bodies and the upper walls, the secondary fractures are filled with rock veins and gold quartz veins. That great structure zone controls more than hundreds gold deposits. The other structure zone is the NE–NNE trending fault zone distributed at the Sidaocha, Wudaocha, Jiabigou and Bajiazi. The scale of that structure zone is bigger, such as the **Jiabigou** main alteration zone up to 5 km length and 50–120 m width. In the rocks, alterations of silicification, chloritization, sericitization, etc., are all developed. That zone is filled with many quartz and other mineral veins (Hou et al. 2009). The Rb–Sr dating age is 244 ± 9 Ma, i.e., the main metallogenic process occurred in Triassic (Mao et al. 2012b); however, some deposits were formed in Jurassic (feldspar K–Ar method, 132.6–150 Ma).

In Huadian, Jilin Province, there is the **Bajiazi** large gold deposit, which is located in the Gaoyuzhuang Formation of Mesoproterozoic, dolomitic carbonite with some clastic, to form the quartz vein-type gold deposit. The metallogenic period is Triassic (Mao et al. 2012b) at 204 ± 0.53 Ma with Ar–Ar method. That deposit is reserved in the NE trending fault system, with the syenite porphyry (Hu et al. 2011).

In Liaoning Province, there is the **Qingchengzi** large gold, silver, lead and zinc field, which belongs to the important non-ferrous metal deposits, including many large and intermediate lead and zinc fields, such as Beilizi, Mapao, Nanshan, Domhyanggou and Zenzigou. They were formed in the Paleoproterozoic–Archean metamorphic system and developed for the quartz veins with NE and NW trending.

Their forming age, by the Ar–Ar method, is dated at 238 ± 0.6 Ma (Mao et al. 2012b; Hao et al. 2011). In recent years, at the outside of Qingchengzi ore field, a series of gold and silver deposits have been discovered, such as the Gaojiapuzi silver deposit and Xiaotongjiapuzi gold deposit.

In the middle Nulurhu Mountains, Inner Mongolia Autonomous Region, the **Jinchanggouliang** super-gold and molybdenum deposit is reserved in the granite and diorite bodies intruding into the Paleoproterozoic–Archean metamorphic system, which belongs to a porphyry-type deposit. Its molybdenite Re–Os age is 244.7 ± 2.5 Ma, namely Triassic (Mao et al. 2012b). The ore veins are distributed in the arc type with NNW trending. Li et al. (2010) recognized that the poor ore fluid formed in the early metallogenic period, may come from the Mesozoic mantle basaltic magma; the rich gold and copper fluid formed in the late metallogenic period, i.e., in the late period of basaltic magma evolution, was mixed with the crust material to form the supercritical fluid and magma.

The **Hadamengou** of Inner Mongolia large gold and molybdenum deposit is located on the north side of Sino–Korean plate, northwest to Baotou City. That deposit occurs in the Paleoproterozoic–Archean metamorphic system, where the potassium feldspathization is developed. It was determined to form in Triassic, according to the sericite Ar–Ar age of 239 ± 3 Ma (Mao et al. 2012b; Niu et al. 2015). The above five ore deposits were all formed in the Triassic intraplate deformation period, but not related to the plate collision or orogeny.

In the east to Tieling of Liaoning Province, there is the **Guanmenshan** lead and zinc deposit (124.8° E, 42.1° N) (Mao et al. 2012b). That deposit is located in the Meso–Neoproterozoic sedimentary basin, may be the result of thermal bittern metallogenesis. Depending on the recent isotopic chronology research, the ore material maybe come from source area in late period of Paleoproterozoic (1890 Ma), the main metallogenic period is at 467 Ma, i.e., the late period of Early Paleozoic, so it should be formed by the intraplate deformation. In that time the Tan–Lu fault zone had ever formed, but finally formed in Triassic. Some researchers thought that the Guangmenshan deposit was formed nearby the Tan–Lu fault zone and controlled by that fault zone; that opinion is not correct.

The **Baiyun-Obo** deposit (110° E, 41.8° N) far away 150 km north to Baotou City, Inner Mongolia, is a world-class supergiant REE ore deposit (Fig. 4.27) and is also the largest niobium (Nb) deposit in China. That deposit is located in the northern extension fracture zone of Sino–Korean plate [14], near to the Tianshan–Hingganling Late Paleozoic accretion–collision zone [10]. The Baiyun-Obo

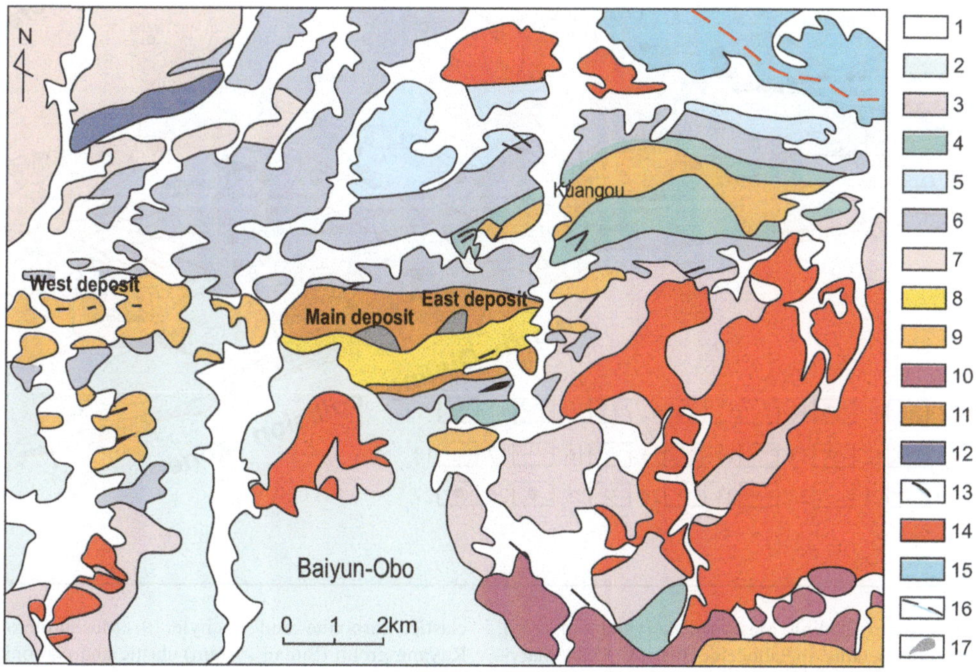

Fig. 4.27 Geological sketch of Baiyun-Obo super-large REE-Fe-Nb deposit. 1. Quaternary sediments; 2. Cretaceous system; 3. Late Paleozoic adamellite; 4. Baiyun-Obo group of Mesoproterozoic sandstone and quartzite; 5. Baiyun-Obo group of Mesoproterozoic charcoal slate; 6. Baiyun-Obo group of Mesoproterozoic charcoal and sandy phyllite; 7. Baiyun-Obo group Mesoproterozoic quartz schist; 8. Baiyun-Obo group of Mesoproterozoic, rich in potassium slate; 9. Paleoproterozoic complex; 10. Archean crystalline basement; 11. ore-bearing dolomite; 12. ultra-basic rocks; 13. carbonate vein; 14. Late Paleozoic alkali feldspar granite; 15. Late Paleozoic andesite; 16. fault; 17. ore body (After Yang et al. 2005a and Mao et al. 2012b)

ore region is extended for about 20 km long at EW trending, and the height difference is about 200 m. That was controlled by the EW trending continental boundary rift and underwent great-scale mantle-sourced carbonate alkaline magmatism and accompanied with a large number mafic dyke swarm intrusion, at that time the rare earth elements were formed in the early period. After that, in Neoproterozoic (1.1–0.8 Ga) through the elements' enrichment, then that supergiant REE-Fe-Nb deposit was formed (Fig. 4.27; Mao et al. 2012b). Now it is explored in the Baiyun-Obo bodies with more than 160 minerals, more than 70 elements, which are mainly composed of iron, niobium and rare earth minerals. Of them, the iron reserves reach 0.95 billion tons, the rare earth mineral reserves are 36 million tons, occupying 36% of the world reserves. So it is called as "Home of Rare Earth Mineral." In addition, there are so many minerals as copper, quartz, fluorite, apatite and pyrolusite.

Toward the west of Baiyun-Obo, the **Langshan** (107° E,41.5° N) areas, i.e., at the northwest end of Sino–Korean plate, there are massive sulfide-type lead and zinc deposits in the Mesoproterozoic continental boundary depression and the **Dongshenmiao**, **Hogqi** and **Tanyaokou** ore deposits, which ore bodies have the similar layer shape. The metallogenesis for those deposits was closely connected with the activity of hydrothermal and hot brine. After the main metallogenesis, they could be metamorphosed and deformed deeply.

On the southern boundary of Sino–Korean plate [10], i.e., the NE side of Qinling–Dabie Triassic accretion–collision zone [24], there is the **Eastern Qinling** Mesozoic molybdenum, gold, lead, zinc and silver. polymetallic metallogenic belt (Fig. 4.28). That belt shows the WNW trending. From the Xiaoqinling, south to the Tongguan of Shaanxi and Linbao of Henan, across the Luoning–Luanchuan, to the north of Fangcheng, it formed an important metallogenic belt with 400 km length and 100 km width. That belt was controlled by two regional faults: the southern border of Teiluzhi–Luonan–Luanchuan reverse fault zone and northern border of Qinling collision zone (containing Sanmenxia–Lushan fault zone).

In that belt, there is **Jinduicheng** giant porphyry molybdenum deposit (109.9° E, 34.3° N). The molybdenite isotopic ages are between 141 ± 4 and 127 ± 7 Ma. The main trending of ore body and ore-controlling fault is WNW

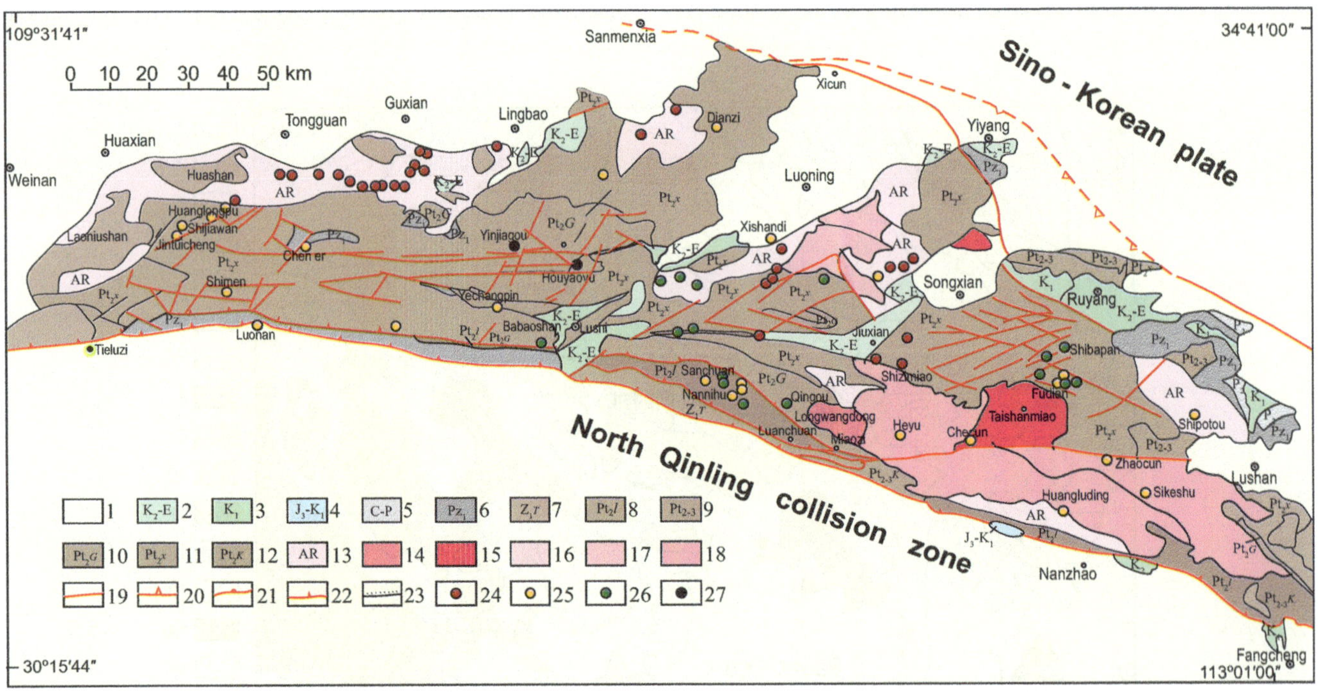

Fig. 4.28 Sketch of Eastern Qinling Mesozoic molybdenum, gold, lead, zinc and silver–polymetallic metallogenic belt. 1. Quaternary–Neogene strata; 2. Paleogene–Upper Cretaceous continental clastic mudstone; 3. Lower Cretaceous volcanic clastic rock; 4. Jurassic–Cretaceous continental clastic mudstone; 5. Carboniferous–Permian coal and aluminum-bearing system; 6. Cambrian–Ordovician marine carbonite and clastic rocks; 7. Sinian (Neoproterozoic) Taiwan carbonite and clastic rocks; 8. Neoproterozoic Luanchuan Group clastic, carbonite and trachyte; 9. Mesoproterozoic–Neoproterozoic Ruyang group (Sinian system) clastic and carbonite; 10–12. Mesoproterozoic Xionger group volcanic rock and Kuanping group marble; 13. Neoarchean granitic-greenstone; 14–15. Early Cretaceous granite; 16–18. Proterozoic granite; 19. fault; 20. extension fault; 21. high-angle normal fault; 22. overthrust; 23. unconformity; 24. gold deposit; 25. molybdenum deposit; 26. lead and zinc deposit; 27. lead, zinc and silver–polymetallic deposit (Modified from Ye et al. 2006)

trending. In Jurassic, the fault showed the extension-shear features, so it was favorable for ore fluid to migrate, and in Cretaceous that fault changed to compression-shear characteristics, so it must be related closely to or controlled by the regional tectonic stress fields.

In **Nannihu–Sandaozhuang**, Luanchuan, Henan Province, there is a super-large tungsten and molybdenum deposit (111.4° E, 33° 54′ N), and in the WNW trending ore veins, the molybdenite isotopic ages are 145–141 Ma. There is the Mo–Cu–Fe skarn-type deposit in **Yinjiagou**, **Huanglongpu** carbonate-type molybdenite lead and zinc deposit, and **Donggou** giant porphyry molybdenite deposit in Shangfanggou porphyry in Henan Province. In those deposits, the metallogenic ages (Re–Os and ^{40}Ar-^{39}Ar) are all about 143 Ma (Wu et al. 2014). However, at the outer contact zone of porphyry bodies, there are the skarn-type polymetallic sulfide iron deposits (in Luotuoshan and Yinhegou etc.). And in the fault, far away from the intrusion body, there are hydrothermal vein-type ore deposits, such as the **Xiao Qinling** giant vein-type gold field or silver, lead and zinc field, including **Fudian** molybdenum, lead, zinc and silver deposit, and **Lengshuibeigou**, **Hedaoca** and **Yindonggou** ore deposits. They are all distributed in the central area of geochemical anomalies. The zircon U–Pb ages are of 115–112 Ma, namely in Cretaceous. The ore veins are distributed along the EW, NW and NE trending faults or fracture zones (Ye et al. 2006) in the **Donggou** giant porphyry molybdenite deposit. In the **Yuchiling** giant porphyry molybdenite deposit, the ore body is reserved at the top of intrusion body, showing the similar layer, with near horizontal attitude. Its molybdenite Re–Os isotopic age is 131.2 ± 1.4 Ma.

To sum up, in that metallogenic belt, most of the ore deposits were controlled by Jurassic and the early period of Early Cretaceous WNW trending extension-shear faults. They were caused by the WNW orientation of regional maximum principal compression stress; or controlled by the middle period of Early Cretaceous–Late Cretaceous NNE trending extension-shear faults, which was caused by the NNE orientation of regional maximum principal compression stress (Wan 2011). Only the Huanglongpu carbonate-type molybdenite, lead and zinc deposit was formed in Triassic at 222–216 Ma (Mao et al. 2012b). The ore bodies were distributed in NWN orientation, which were influenced by the strong tectonics of Qinling–Dabie collision zone [24]. There are rather convenient traffic conditions in the Eastern Qinling metallogenic belt; nearby there are many old mines, and rather good prospects for deep exploration.

In the mountainous areas of south to Lingbao of Henan–Tongguan of Shaanxi, i.e., the **Xiaoqinling** areas, there are a lot of gold deposits (NW part of Fig. 4.28), which are the third largest gold deposit location in China. More than 1000 intermediate-low hydrothermal gold-bearing quartz veins are distributed in the Archean gneiss area with near EW length of 100 km, NS width of 50 km. The ore veins are rather narrow, and most of them are more than 100 m long, with higher grade, so it is easy for exploiting and dressing. In the recent twenty years, all the estimated K–Ar isotopic ages for gold veins have varied from 182 to 148 Ma (Mao et al. 2002a, b), so most geologists considered gold deposits to form in Jurassic. They may be influenced by the WNW trending maximum principal compression stress to form in the early period foliations or fractures with the extension-shear properties, then the hydrothermal fluids were filling into those fractures and cooling to form the gold deposits. However, in recent some researchers have used the Ar–Ar dating for the quartz in the gold-bearing veins and got some Triassic isotopic ages. The author recognizes that dating results are affected by the residue argon from quartz in gold veins and surrounding rocks. Those ages are not to believe. As for the Ar–Ar dating, if we want to get the believable ages, only the laser could be used to determine the gas–liquid inclusions. If only the powders of quartz mineral were used to determine the ages, it would get the mistake results. The author has ever had the failure experiences.

In the Songxian of Eastern Qinling Mesozoic molybdenum, gold, lead, zinc and silver–polymetallic metallogenic belt, there is the **Qiyugou** breccia pipe-type gold deposit (112° E, 34.2° N). There are more metamorphic breccias in its upper part. The gold is rich in pyrite ores. The metallogenesis occurred at the period of 115–130 Ma, namely Early Cretaceous. Under the regional compression near NS trending, the intersection of NE and NW trending fault may be the main passway of ore-bearing magma, to form the high-grade gold deposit (Qi et al. 2004a, b).

As for the gold outputs, Shandong is the largest province in China. In the Northern Shandong Peninsula, there is **Sanshandao–Jiaojia–Linglong** gold field (Fig. 4.29). That gold field, located in the Archean–Paleoproterozoic metamorphic system or intrusion, was formed in Jurassic and Early Cretaceous, and the injection and reserved period of the hydrothermal gold-bearing fluid was mainly in Middle Cretaceous (120–114 Ma). When the intrusion bodies were cooling, the ore-bearing fluid would penetrate and intrude into the boundary ductile shear zone between the intrusion and metamorphic rocks, to form the "Jiaojia-type" gold deposits. This type of deposits commonly has the characteristics of fine veins and dissemination, super-great scale and low grade, such as **Jiaojia of Laizhou** (Li et al. 2002c), **Cangshan, Sanshandao** (Zhao et al. 2013), **Xincheng, Penjiakuang of Rushan** (Yang et al. 1999) and **Guilaizhuang of Pingyi** (Chen et al. 1999). When the ore fluids penetrated into conjugate shear joint system with the NNE and ENE trending, it would form the "Linglong-type" gold deposits. This type of deposits commonly has the characteristics of bigger veins, smaller scale and high grade, such

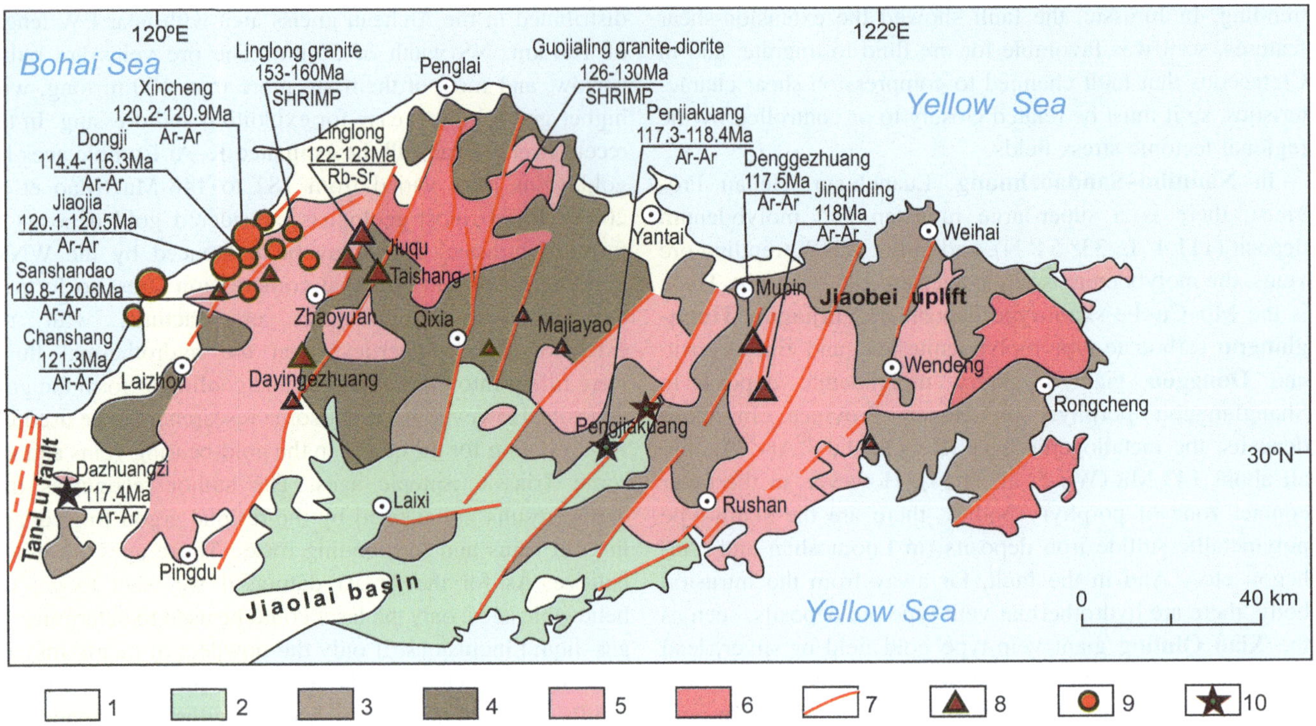

Fig. 4.29 Distribution of gold deposits in Eastern Shandong Province. 1. Quaternary sediments; 2. Cretaceous volcanic rock; 3. Proterozoic metamorphic rock; 4. Archean metamorphic rock; 5. Jurassic granite; 6. Cretaceous granitic diorite; 7. fault; 8. quartz vein-type gold deposit; 9. alternation-type gold deposit; 10. breccia gold deposit (Modified from Mao et al. 2008, 2012b)

as the **Linglong of Zhaoyuan** (Du et al. 2003), **Jinqinting of Rushan** (Gao et al. 2010), **Denggezhuang.**

They were all controlled by Middle Cretaceous regional tectonic stress field. At that time, the maximum principal compression stress orientation was NE trending, and the NNE orientation soft zones or the boundaries of intrusion changed to dextral strike-slip ductile shear zone, then formed the fine veins and dissemination, great-scale and low-grade alternation deposits; if the ore fluid penetrated into NNE and ENE trending conjugate shear joint system with extension-shear characteristics, it would form the big vein-type rich gold deposit. The above two types of gold deposits comprise great-scale gold metallogenic province. In addition, if the shear joint system was greater, it would form the fracture zone with the breccia zone, called as breccia gold deposit (such as the Penjiakuang deposit; Fig. 4.29).

The above gold deposits are the intermediate hydrothermal gold deposits, the most depth of exploration up to 1000 m. Till now, it may be continued to discover deep gold bodies. Depending on the recent research data, with the depth increasing, the metallogenic temperature will also increase and the grade of gold also will become higher. In the last years, the demonstrated gold reserves have been obtained as 51.83 tons at the Shizhuang, Laizhou City in the Shandong Province.

Goldfarb and Santosh (2014) expressed new opinion for the origin of Shandong gold fields. They thought those gold fields were caused by the asthenosphere uplift, and the oceanic plate subduction resulted in the dehydration and decarbonization, upward to form the metasomatic mantle wedge, and the gold metallogenesis was controlled by the Tan–Lu and other fault zones. That is the common thought for the researchers. When they research the magmatism and metallogenesis in the Eastern Asian continent, they have only focused on the controlling of the oceanic plate subduction, but they have not got any evidences for their ideas. They have never considered when the oceanic plate subducted underneath the Shandong Peninsula; their depth would reach 500–600 km. However, the strong tectonic magmatism only occurs near the Moho discontinuity or in the crust; their depth must be less than 35 km. Till now nobody has found any evidences for the oceanic subduction to cause the "asthenosphere uplift" and the shallow part magmatism and metallogenesis. Their ideas are not in conformity with facts, just their own idle dream. They do not understand the importance of very strong intraplate deformations that can also cause the magmatism and metallogenesis. That is a very special phenomenon in the globe.

In Shandong Province, there are more than several hundred ring-shaped structures not to be researched. Those

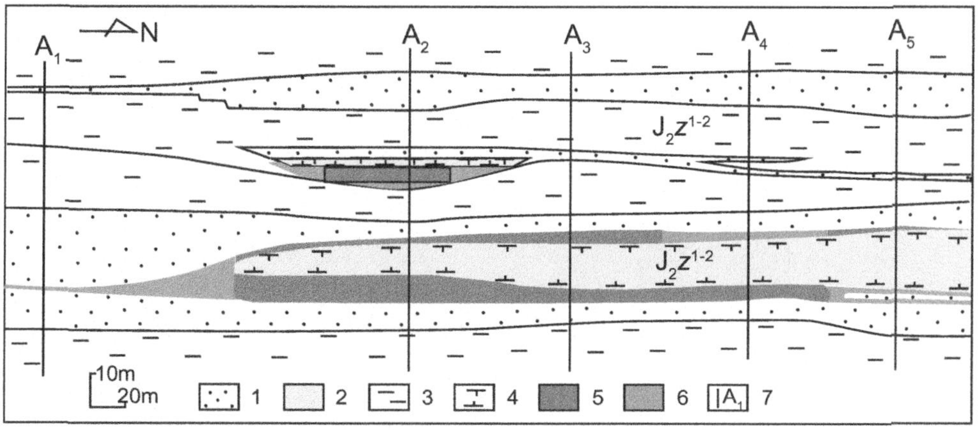

Fig. 4.30 Sketch geological section of No. 147 exploration line in Dongsheng sandstone uranium deposit (After Peng et al. 2007a, b). 1. Gray sandstone; 2. gray–green sandstone; 3. mudstone; 4. the front of interlayer oxidation zone; 5. industrial ore body; 6. mineralization body; 7. drilling locations and their numbers

boundaries of ring-shaped structures are completely different from any surface geological boundary lines. They may be the surface exhibitions for the deep hidden intrusion bodies. So it needs to find further the upper part of hidden intrusion bodies and their metallogenesis up to the 2000 m depth, which is a very important and formidable project. However, if we continue to explore the deep, other hypothermal deposits will be discovered. It may find other mineral resources. According to the element compositions of molybdenite which are commonly rich in Sino–Korean plate, we can guess that there may be the molybdenite or other hypothermal deposits reserved.

In Sino–Korean plate, there are two bases of energy resources: the Ordos basin and circum Bohai basin. The **Ordos** basin is the best one of China rich oil, gas, shale gas, coal and uranium comprehensive energy. The famous one is the Changqing oil and gas field (Hu et al. 2000; Editorial Board of Petrological Geology of Changqing Oil Field 1992). The oil and gas are mainly reserved in the Jurassic coal system, Triassic, the unconformity between Carboniferous–Permian and Early Ordovician, and some Early Paleozoic marine strata. The reservoir structures are mainly the Jurassic NS trending open anticline and its related fractures. The coal layers were mainly formed in Jurassic, especially to the northern of that basin. In the **Dongsheng–Shengfu** (109° 45′ E, 39° 50′ N) super-large coal field, their proved coal reserves reach 223.6 billion tons, which is the biggest coal field explored in China. The burial depths of those coal layers are very shallow, with very weak deformation, which is favorable to exploit with great-scale mechanized open-pit mining.

In some low-quality coal systems, there are great-scale uranium deposits. In recent years, the **Dongsheng** giant sandstone uranium deposit has been discovered (Peng et al. 2007b). Those ore bodies are reserved in the sand bodies of pigtail river sedimentary system belonging to Lower Zhiluo Formation of Middle Jurassic in Northern Ordos basin. In Late Jurassic–early period of Early Cretaceous, tectono-thermal event led to the hydrocarbon-bearing hydrothermal fluids to mineralize. They could not only provide the favorable condition for uranium activity, migration and concentration, but also make the uranium deposits completely occur in the reducing environment (Fig. 4.30). The ore bodies are generally distributed along NW–SE trending extension-shear fracture zones. The ore fluids were migrated and controlled by the Jurassic tectonic stress field; at that time, the orientation of maximum principal compression stress was WNW trending (Fig. 3.23). The highest contents of UO_2 in ore reach 53.75–74.60%. The main forming periods of ore were approved to be in Cretaceous (120–80 Ma) and Neogene (20–8 Ma) (Jiao et al. 2015).

The circum Bohai oil-gas-bearing basin (Fig. 4.31) is developed important great oil and gas province, in which there are the **Shengli** oil field (Editorial Board of Petrological Geology of Shengli Oil Field (Zhai chief ed.) 1993), **Liaohe** oil field (Editorial Board of Petrological Geology of Liaohe Oil Field (Zhai chief ed.) 1993), **Dagang** oil field (Editorial Board of Petrological Geology of Dagang Oil Field (Zhai chief ed.) 1991), **Jidong** oil field (The Company of Jidong, PCCL), Penglai 19-3 oil field (Guo et al. 2011) and **Zhongyuan** oil field of Henan. The petroleum source beds of those oil-gas fields are the Shahejie and Dongying Formations, which were deposited basically in the lacustrine environments under the Eocene and Oligocene wet weather. However, due to the different basement structural conditions, the reservoir structures are all different.

The **Shengli** and **Jidong** oil fields were controlled by the Eocene and Oligocene WNW trending extension-shear contemporaneous fault; in Miocene–Early Pleistocene, they underwent near NS trending compression to make those

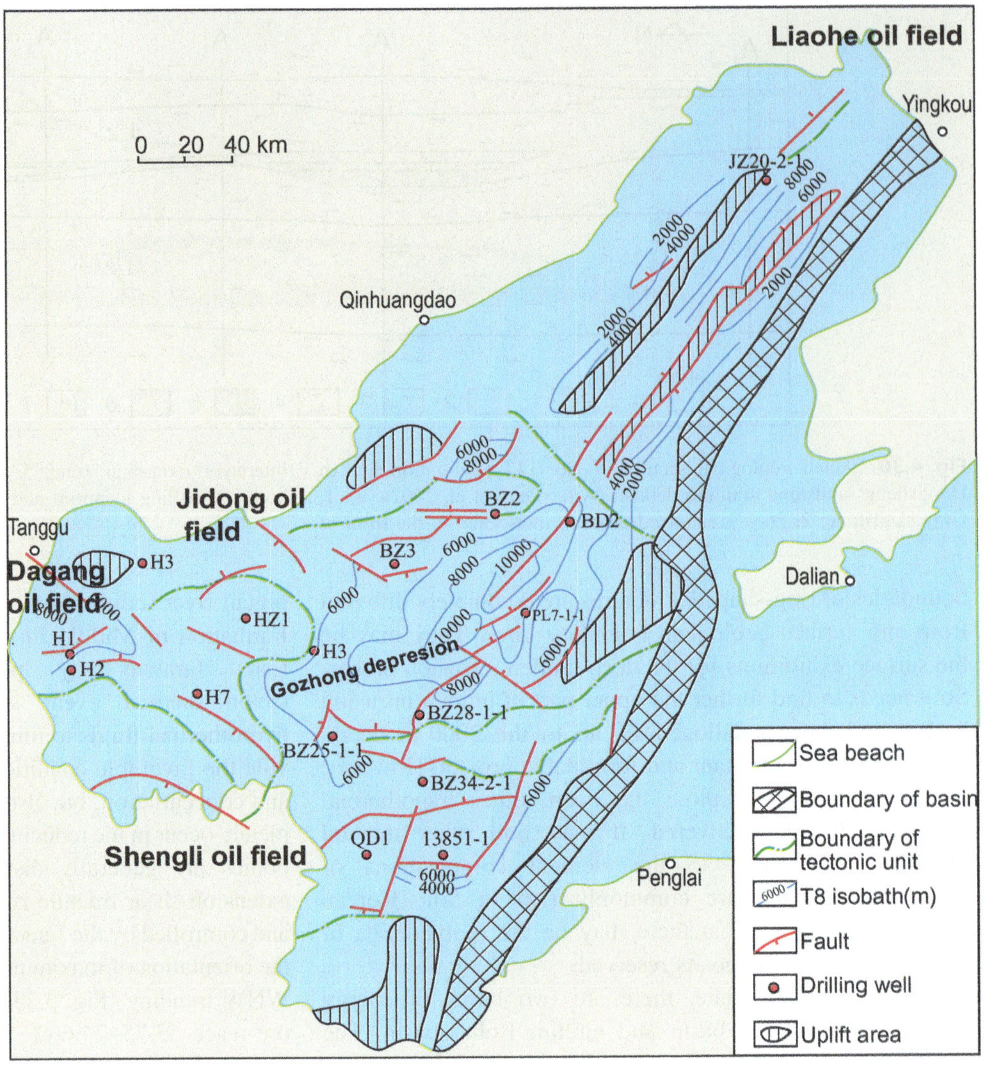

Fig. 4.31 Tectonic sketch of Bohai petroliferous basin (Modified from Wang 2004)

faults close, which was favorable to preserve the hydrocarbon. The Eastern Shengli oil field was reformed by Tan–Lu fault zone, in which the near NS trending faults suffered the ENE trending compression of recent tectonic stress field (Fig. 3.36), and it was also favorable to preserve the hydrocarbon.

However on the east side of Bohai Bay, the **Penglai 19-3** oil field is located in the No. 1 and No. 2 of Tan–Lu fault zone. During Paleogene, near EW trending compression caused the hydrocarbon close in the deep; during Neogene, NS trending compression (Fig. 3.33) caused the faults extension and open and got the hydrocarbon to migrate onto the near surface (about 1 km deep) and concentrate or form partial out leakage. At the recent period, by the recent ENE trending compression (Fig. 3.36), the faults were all closed to make the hydrocarbon be preserved at the rather shallow position and form a "shuttle" giant oil and gas field. However, due to the recent tectonic stress is rather weaker (the differential stress value is only 10–30 MPa), once the water was exceeded to inject in the depth and the pressure was increased, it would be easy to form the oil and gas out leakage accident, especially along the ENE trending secondary faults, nearly parallel to the orientation of recent maximum principal compression stress (also the extension easy to occur). This is the reason that the great oil and gas out leakage accident occurred in 2011.

In the **Zhongyuan** oil field, Puyang, Henan Province, the petroleum sources are as same as the above oil fields; however, the oil and gas reservoirs are mainly developed near the NS trending fault, and the principle of migration and concentration is as same as Penglai 19-3 oil and gas field. The petroleum sources were from Paleogene, Neogene–Early Pleistocene oil and gas migration and enrichment in the upper part, since Middle Pleistocene, the near NS trending fault had been rather close, so the oil and gas could be preserved and concentrated.

In the **Dagang** oil and gas field, the abundant oil and gas parts were controlled by the near EW trending faults.

However influenced by the ENE trending recent maximum principal compression stress of tectonic stress field (Fig. 3.36), it caused many ENE trending extension to make the oil and gas leakage so strong.

Located at the north end of Bohai Bay, the **Liaohe** oil field was controlled by the northern part of Tan–Lu fault zone. In the east of that field under the Late Paleogene EW trending compression, the NE trending faults showed the dextral strike-slip to form the near NS trending en echelon small anticlines controlling the oil and gas reserves. However, at the west of that oil field, it was controlled by the action of EW trending compression to form the EW trending extension fractures and the large oil and gas reservoir, and the re-deformations in the later periods were not very strong (Wan 2011).

As for the many important oil and gas fields in the circum Bohai Bay, the author has ever discussed deeply with geologists from the oil and gas field, on the combination research of the regional tectonics and the petroleum source, reservoir, migration and preservation conditions. The above recognitions of the author are expected to be referenced for the researchers. It is a great pity that some researchers from oil and fields have not paid attention to the multiplicity and variability the tectonic stress field; in other words, the horizontal principal stress can influence the petroleum source, reservoir, migration and preservation conditions. The structural research of Liaohe oil field is in the front rank (Li and Xu 2001).

The metallogenesis of Sino–Korean plate, except the Archean banded iron formation (BIF), is caused by the partial extension zone of intraplate deformation in basically. They were all formed after the formation of plate and going to rather stable. The metallogenesis usually is based on the early faults, boundary depressions or rifts, and controlled by the later tectonic stress fields and then to form the ore deposits. The ore materials in most of metallogenesis come from the Moho discontinuity or the middle crust low-velocity and high-conductivity layer. Due to the special geochemistry characteristics for the plate formation, it will form themselves special properties for the ore deposits.

In Sino–Korean plate, the metallogenesis is easy to concentrate the elements including molybdenum, iron, gold, magnesium, niobium, rare earth, lead, zinc and silver. The sedimentary deposits are controlled by the weather, geographic environment and tectonic background. After the formation of plate, the metallogenesis was controlled by different-period intraplate multiple extensions. These phenomena and recognitions must be paid attention to.

4.1.3.2 The Giant Ore Fields and Deposits in Alxa–Dunhuang Block [16]

In the Beishan area of Dunhuang block (Fig. 2.14), there are rare metal deposits such as Au, W, Sn, Mo and rare earth. They are connected with the granite and controlled by the Late Paleozoic basement rifts, with near NS trending in some degree. The basement structures have the characteristics of NS trending zone and with the EW trending range. There are Cu, Ni, W, Sn, Pb, Zn and Ag porphyry or skarn-type ore deposits and multi-epoch hydrothermal gold deposits. However, there are rare of the giant endogenous deposits, it is possible that many ore deposits are the hidden deposits. It shows the deep exploration has the well prospects. It should continue to do more deep exploration, in order to get the more important results.

On the southern boundary of Alxa–Dunhuang block, there is the **Jinchuan** (102° E, 38.3° N) copper and nickel (contained platinum) sulfide supergiant deposit in Gansu Province (Fig. 4.32). The ore reserves reach 0.515 billion tons, and the nickel grade is 1.06% (Naldrett 1999). The ore body and rock body were all formed in Mesoproterozoic (1508–1043 Ma) by the deep crust liquation–intrusion and were closely related to the intrusion of picritic tholeiite magma, which are usually formed on the boundary of extension zone of the stable block. The deposit was controlled by the NW trending fault and the ultra-mafic dykes. The ore-bearing intrusion body was located in the thrust sheet, then was uplifted near the surface by the later thrust nappe action. The intrusion rock body has the 6.5 km length, 20–527 m width, and the greatest depth reaches 1100 m. Of them, the richest ore body, No. 1, has the sideronitic texture;

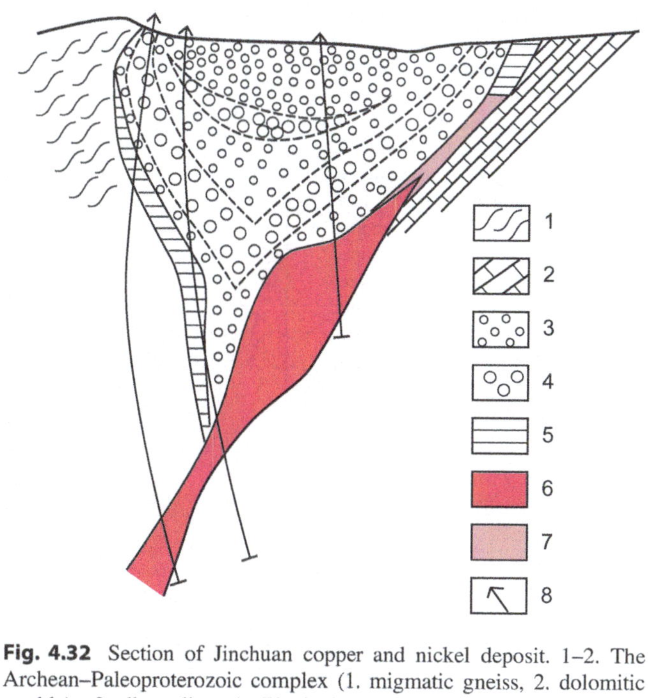

Fig. 4.32 Section of Jinchuan copper and nickel deposit. 1–2. The Archean–Paleoproterozoic complex (1. migmatic gneiss, 2. dolomitic marble); 3. lherzolite; 4. Plagioclase lherzolite; 5. websterite; 6. Iron-rich ore body; 7. copper and nickel ore body; 8. drilling hole (Modified from Tang and Li 1995)

the ore grade is rather high (Ni = 1–4%), and the distribution is stable (Tang and Li 1995).

4.1.3.3 The Giant Ore Fields and Deposits in Qilian Early Paleozoic Accretion–Collision Zone [17]

In the north of Qilian Early Paleozoic accretion–collision zone (Fig. 2.14), there is the **Baiyinchang** of Gansu (104.1° E, 36.6° N) volcanic massive sulfide (VMS)-type copper, lead and zinc deposit, which ores are developed in the bimodal keratophyre acid volcanic rocks (Fig. 4.33). The metallogenic period was at the later stage of the volcanic eruption, the metallogenic ages of 460–420 Ma (Late Ordovician–Silurian), which was formed at the syn-collision period. The most important Zheyaoshan deposit near the crater was controlled by the cross-point of NE and NW trending faults (Peng et al. 1996). Depending on the many different ring-shaped structures to determine (Fig. 4.33), there exist hidden intrusions in the deep. It may be developed for polymetallic deposits connected with the porphyry intrusions. As shown in Fig. 4.33, it needs to further explore some small ring-shaped structures in the deep, especially westward and southward deposits.

Similar to above ore deposit, in Northern Qilian Mountains, **Shijuligou**, Sunan County, Gansu Province (99.6° E, 38.9° N; Li et al. 1999), there is the massive sulfide-type copper and zinc deposit in Middle Ordovician basic volcanics, which was also resulted from syn-collision, but not too great. In Xiaoliugou, Sunan County, there is a giant skarn-type tungsten, molybdenum, iron, copper polymetallic deposit. The wall rocks are composed of Mesoproterozoic shallow metamorphic systems. The intrusions such as adamellite and granodiorite related to the ores, occurred at the Late period of Early Paleozoic (462 Ma), which were controlled by the WNW trending fault.

In the west of Northern Qilian Mountains, there exists the great-scale **Jingtieshan** ore deposit (98° E, 39.3° N, Fig. 4.34). Tens middle-small-scale iron deposits in the Proterozoic continental marginal intermediate-basic volcanic depression trough belong to the hydrothermal-sedimentary iron deposit. The SHRIMP zircon U–Pb age of ophiolite suit and volcanic system is 1777 Ma. However in the phyllite and carbonite rocks of Mesoproterozoic, there are the Sedex-type iron deposits (Fig. 4.34); those belong to the sedimentary deposits formed before the Paleozoic collision. The thickness of ore layers reaches 50–150 m, the average iron content in ore reaches 36.14%, and the proved reserves of iron ore are obtained as 0.484 billion tons (Mao et al. 2012b).

At **Ta'ergou**, northern Middle Qilian block (Fig. 4.34; Mao et al. 2012b), at the late period of Early Paleozoic there formed skarn–quartz vein-type tungsten deposits, connected with the Yeniugou granodiorite, which was formed at the Qilian collision period. The main ore deposits are the Baiyingchang and Targou ore deposits. The Jingtieshan Sedex-type iron deposit was formed before the collision, and other small ore deposits were formed after collision period.

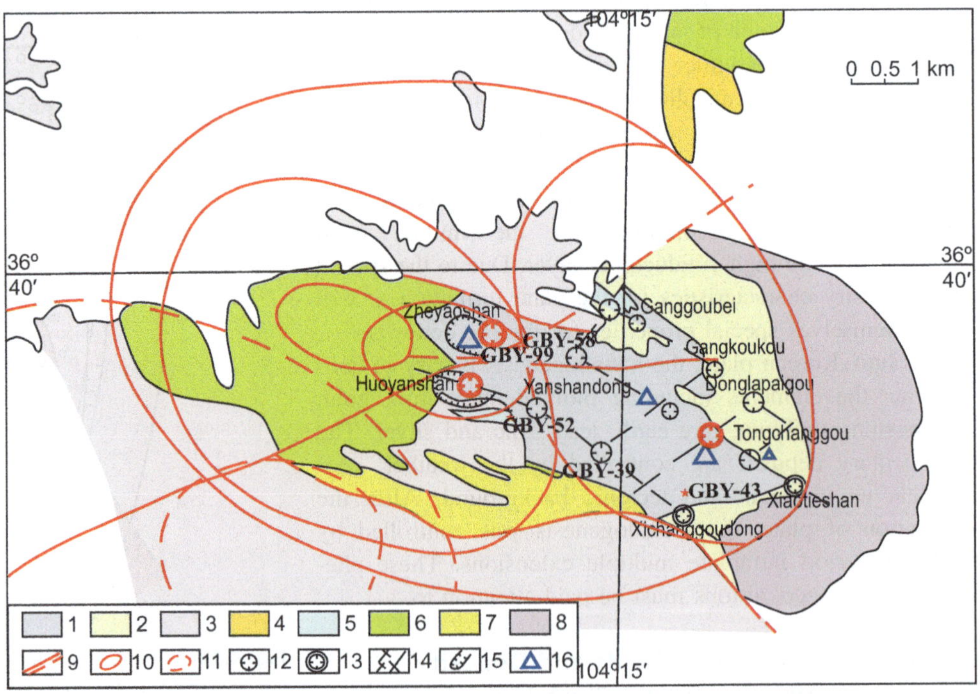

Fig. 4.33 Sketch of paleo-volcanic structure in the Baiyinchang ore field, Gansu Province. 1. Acid volcanic rock; 2. intermediate volcanic rock; 3. basic volcanic rock; 4. trachyte; 5. diabase; 6. phyllite; 7. chlorite-schist with carbonate; 8. clastic sedimentary rock; 9. fault and inferred fault; 10. ring-shaped structure (Beiyin); 11. ring-shaped structure (Heishishan); 12. paleo-crater; 13. paleo-crater determined by predecessors; 14. actual measurement fault system; 15. boundary of mining pit; 16. industrial ore deposit (Modified from Peng et al. 1996)

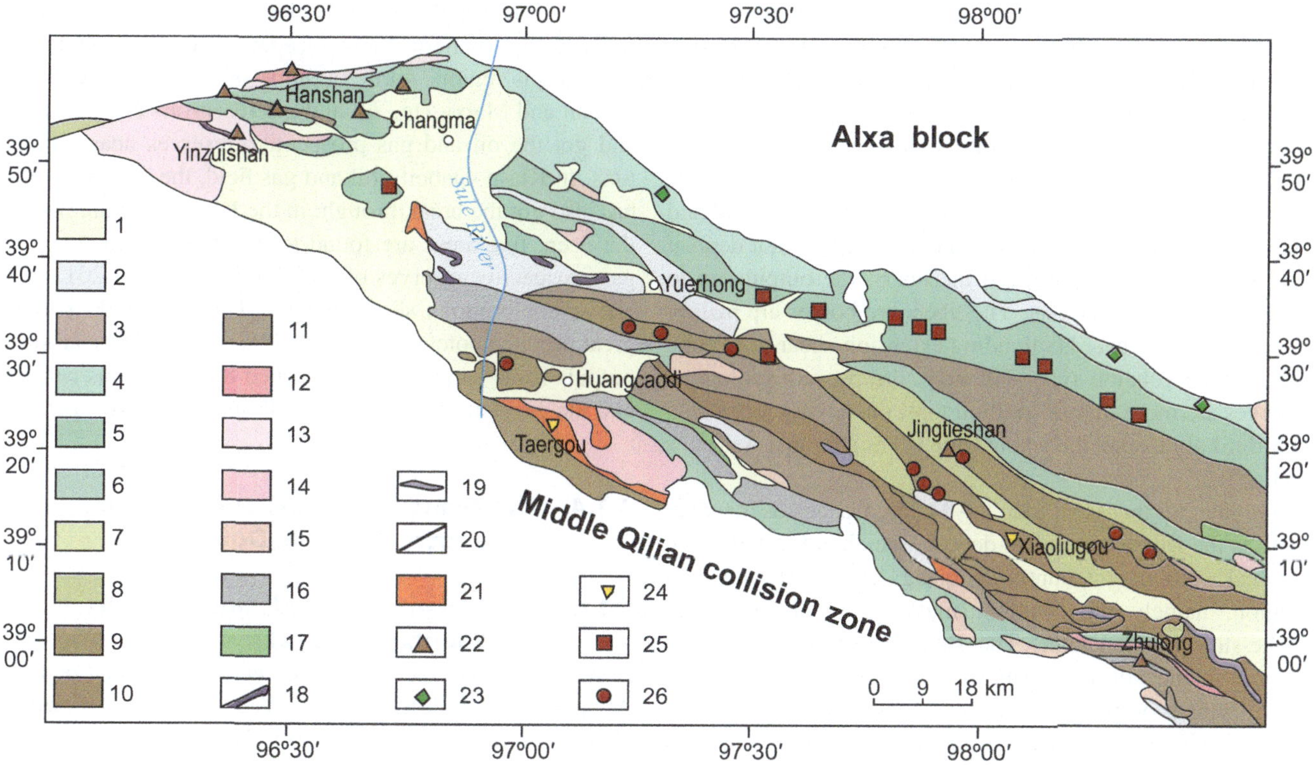

Fig. 4.34 Distribution sketch of geology and ore deposits in western part of Northern Qilian Mountains. 1. Quaternary sediments; 2. Carboniferous–Permian clastic rock; 3. Devonian continental sandstone; 4. Silurian marine sandstone; 5. Ordovician basic volcanic rock with carbonite; 6. Cambrian metamorphic clastic rock; 7. Neoproterozoic (Sinian) conglomerate; 8. Neoproterozoic carbonite; 9. Mesoproterozoic phyllite and carbonite; 10. Mesoproterozoic volcanic, carbonite and clastic rock; 11. Paleoproterozoic schist and amphibolite; 12. Archean gneiss; 13. Late Paleozoic granite; 14. Ordovician granodiorite; 15. Ordovician mica granite; 16. Ordovician–Silurian alkaline rock; 17. Ordovician basic rock; 18. Ordovician ultra-basic rock; 19. Mesoproterozoic ophiolite; 20. fault; 21. shear zone; 22. gold deposit and spot; 23. sandstone-type copper deposit; 24. skarn-type quartz vein-type tungsten; 25. Jingtieshan-type iron and copper deposit; 26. sedex-type ore deposit (Modified from Mao et al. 2011)

4.1.3.4 The Giant Ore Fields and Deposits in Qaidam Block [18]

In the Qaidam basin, in Pleistocene–Holocene (Fig. 3.36), there preserved the biggest continental salt sedimentary deposit in China, which is rich in the lithium, boron, salts, potassium and magnesium in the salt lake. The areas of salt lake reach 11,000 km^2. In the deep of eastern lake areas, there is the oil shale, methyl hydrate or shale gas deposit.

The **Qinghai** oil field is explored in three main areas: Huatugou, Shizigou and Qigequan, which total reserves of oil and gas reach 500 million tons (Yang 2009). Now the main exploration objects are focused on the fracture-type oil and gas reservoir; however, the output per year is less than 5 million tons of oil-gas equivalents. That resource potentiality needs deeply exploiting.

The Western Qaidam basin is the greatest strontium (celestine or strontium sulfate) field in China, which is the richest strontium ore area. There are more than ten strontium deposits or ore spots, of which the representative deposits are the **Dafengshan** and **Jiantingshan** ore deposits (Xie 1999; Sun et al. 2013). The reserves graded at B + C in Dafengshan (92.3° E, 36.8° N) celestine super-large deposit reach 10 million tons, with $SrSO_4$ grade of 31%. They are preserved in the Neogene Shizigou Formation (N_2s^2)—lacustrine facies carbonite-sulfate formation, with the slight fold deformation. The fold is a great open anticlinorium in the whole deposit area, and the axis is NW 290°, plunging toward two limbs, with the length about 10 km. The near NS trending extension joints were formed in Neogene (Fig. 3.33). By the water dissolution and concentration, it would form the celestine-rich ore veins.

4.1.3.5 The Giant Ore Fields and Deposits in Altun Early Paleozoic Sinistral Strike-Slip Collision Zone [19]

Along the Altun strike-slip collision zone (Fig. 2.14), there are a series of late period of Early Paleozoic ophiolite suite, and the strong serpentinization and **Mangya** giant asbestos deposit (90°–90° 20′ E, 38° 14′–38° 22.5′ N) (Web site of China Metal 2008-04-29). The ore areas are located at altitude of 3100 m above sea level. In the Mangya area, there is the biggest super-mafic intrusion-bearing serpentine asbestos. The ore belt is near EW trending extension, with the length more than 14 km. The Mangya asbestos deposit is located at the east end of that intrusion; the scale of ore body is the great, with rather high grade (2–5%). The asbestos fibers can be used to produce cement, in which physical and chemical characters are rather good. Since 1958, that deposit has been exploited to produce the fibers, the outputs up to about 60,000 tons per year. The fibers ore bodies are mainly developed in the near NS trending extension-shear joint zone with the en echelon type. Till now, the proved geological reserves are more than 21 million tons, occupying more than one-third of those in the whole China (The Mangya Asbestos Net, 2012).

On the northeastern side of Mangya asbestos deposit, there is the **Bazhou** asbestos deposit, which is located at the Itunblak, Noqiang County, Xinjiang Uygur Autonomous Region, namely the southern middle part of the Altun strike-slip collision zone. The Bazhou asbestos deposit is located at the northeast part of ultra-mafic intrusion body zone, which origin is as same as the Mangya asbestos deposit. Till now the proved geological reserves are 4.84 million tons, the fibers outputs near 60,000 tons per year (Bazhou Asbestos Deposit Net, 2012).

4.1.3.6 The Giant Ore Fields and Deposits in Tarim Block [20]

In the Tarim basin, the most famous oil field is the **Tabei** oil and gas field. In the whole Paleozoic, the Tarim basin had the peri-Yangtze plate biota and marine sedimentary characters, in Jurassic and Paleogene it still was located in the wet climate and continental basin, so there were the favorable conditions to produce the hydrocarbon. In Neogene, being influenced by Indian plate long-distance effect of northward subduction and compression (Fig. 3.33), near the surface, the Tarim block converged and subducted, and on the north boundary a series of near EW trending thrust-folding system were formed; under the thrust, the "triple-angle zone" became the well oil-gas reservoir to form the Tabei oil field. The oil reservoir could be the Cambrian–Ordovician, Carboniferous–Permian, Jurassic and Paleogene strata, etc.; in some parts, there formed the condensate gas reservoir. The last formation period for the above oil and gas reservoir structures was mainly in Neogene (Wan 2011). The above hydrocarbon-bearing systems are also to explore shale gas mainly, which have rather high research significance. However, the major problem is how to get the injection water. In short, the best prospective areas of giant oil and gas fields in Tarim basin are focused on the paleo-uplift and hidden foreland thrust zone (Zhao et al. 2004).

The Tarim Oil Company has advanced a project to research the southern slope of Tabei and Lunnan–Yinmailifu oil and gas region, northern slope of Tazhong and Bachu, Manan and Maixi, etc., oil and gas reservoirs and rich areas, and got the oil and gas prospective reserves near 4 billion tons. In Klasu–Dabeifu oil and gas field, the gas exploration has also got the breakthrough; in the Klasu structure zone, in the deep, five areas are found to preserve the gas, with the gas prospective reserves of 40 billion m^3 (Zhang 2011). The above explorations show that the Tarim oil field still has great produce potentiality (Kan 2000). In 2010, the crude oil outputs from Tarim oil field reached 8 million tons per year, and the gas outputs reached 20 billion cubic meters per year.

4.1.4 The Giant Ore Fields and Deposits in Yangtze Tectonic Domain

4.1.4.1 The Giant Ore Fields and Deposits in Yangtze Plate [22]

In Yangtze plate, there are many endogenic metal deposits. At the north-middle part of Yangtze plate—**Southeast Hubei** (about 115° E, 30° 10′ N), there are a lot of iron and copper deposits exploited earliest in China, which is an important ore, copper and gold reserved province (Figs. 4.35 and 4.36). The ore deposits are controlled by the diorite intrusions along the faults. The Jurassic–Early Cretaceous (200–135 Ma) intrusions are distributed along WNW trending and controlled by the maximum principal compression stress orientation (σ_1) of that period (Figs. 4.37 and 4.38; Wan 2011). However, the middle epoch of Early Cretaceous–Paleocene (135–56 Ma) intrusions is distributed along NNE trending and controlled by the maximum principal compression stress orientation (σ_1) of that period (Figs. 4.37 and 4.39; Wan 2011). Two-period magmatism and ore fluids were the main reasons for the rich ores and ore fluids to migrate, condensate and reserve in the extension-shear faults or joints (Figs. 4.37, 4.38 and 4.39). Thus, the intrusion was contacted with wall rocks at different attitudes, depths and lithology to form many-typed skarn iron-, copper- and gold-rich ore bodies (Fig. 4.36), such as **Tongling Anhui** and **Southeast Hubei** iron-, copper- and gold-rich ore provinces. Liu and Chang (1988) had vividly proposed the ore zone of Southeast Hubei ore province distribution as "T" shape; and that of Tongguanshan ore province distribution as "Π" shape. The author, according to the distributions and forming ages for above two ore provinces, recognizes that those horizontal lines, representing the WNW trending metallogenic zone, were formed in Jurassic–Early Cretaceous (200–135 Ma; Fig. 4.38); and the vertical lines, representing the NNE trending metallogenic

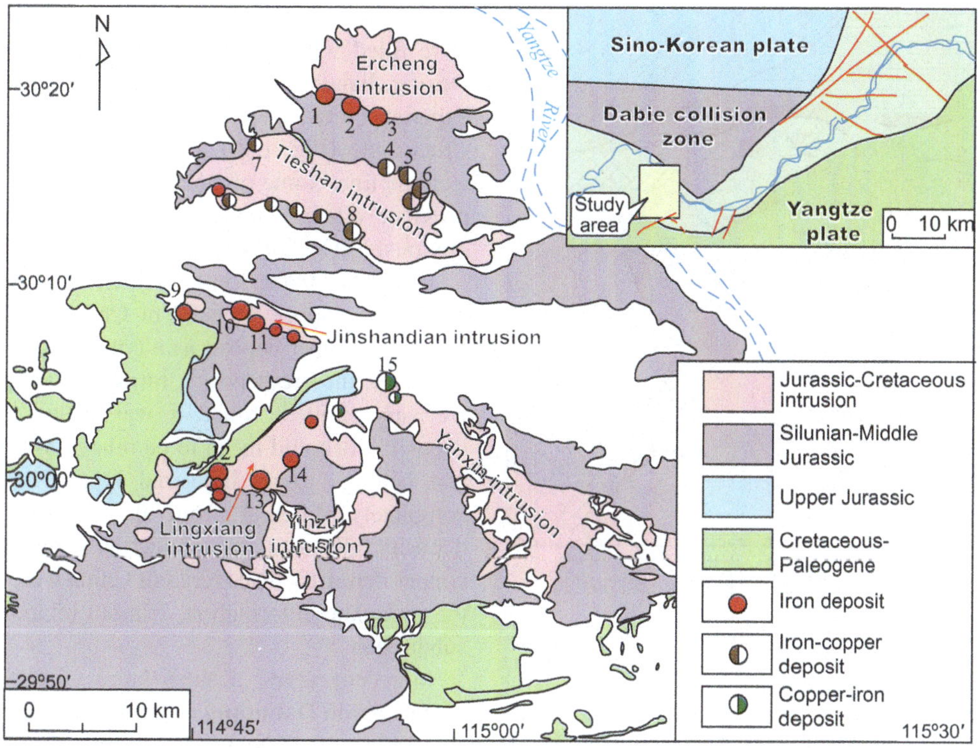

Fig. 4.35 Distribution of Daye-type skarn iron, copper and gold deposits in Southeast Hubei ore province. No. of ore deposits: 1. Guangshan; 2. Chengchao; 3. Licun; 4. Hehuachi; 5. Datongken; 6. Guangmiannao; 7. Dahongshan; 8. Tieshan; 9. Wangbaoshan; 10. Yuhuasi; 11. Zhangfushan; 12. Naojiao; 13. Liujiaban; 14. Daguangshan; 15. Tonglvshan (Modified from Zhai 1995)

zone, were formed in Early and Middle Cretaceous (135–56 Ma; Fig. 4.39). The metallogenic fluids were mainly caused from near the Moho discontinuity (Wan 2011).

The **Southeast Hubei** iron, copper and gold ore province (about 115° E, 30° N), **Chengmenshan** (copper)–**Yangchuling** (tungsten) of Jiangxi ore province (about 115.8° E, 29.7° N) and **Tongling** of Anhui iron, copper and gold ore province (about 117.9° E, 30.9° N) are composed the main parts of Middle and Lower Yangtze metallogenic belt. Continuing to the NNE, they extend to **Fanchang** of Anhui (118.2° E, 31.1° N)–**Wuhu–Ma'anshan–Nanjing** (118.5° E, 31.7° N). In fact, the Middle and Lower Yangtze iron, copper and gold metallogenic belt is composed of the network intersection of WNW (mainly formed in Jurassic; Fig. 4.38) and NNE (mainly formed in Cretaceous; Fig. 4.39) trending fault systems. Those two responding magmatic intrusions and ore fluids were reserved, and at the contact zones between intrusions and wall rocks, a series of iron, copper and gold deposits in different depths were formed. That is the reason for a lot of rich metal ore deposits to form (Wan 2011). The middle and lower Yangtze metallogenic province was controlled by regional NNE and WNW trending fault systems and the strata, but not to have the "middle and lower Yangtze great fault zone" recognized by Wu et al. (2008). Till now, by many geophysical exploration methods, yet nobody has discovered any great fault zone along the middle and lower Yangtze region.

That metallogenic zone is located near the Yangtze River and many industrial cities, so the traffic condition is very convenient. It is also the earliest exploitation area in China. Nowadays, in the deep (within 2000 m), there are new hidden ore deposits discovered in different positions. From the Chengmenshan (copper)–Yanchuling (tungsten) deposits extension to the ESE orientation, across the Poyang Lake, there is the **Dexing** copper polymetallic ore province (central position is 117.6° E, 28.9° N; Li et al. 2002a, b, c; Fig. 4.40). That ore province is located at the intersection of the WNW trending Jiujiang–Dexing fault and NE trending Northeast Jiangxi fault (Figs. 4.38 and 4.39). In porphyries formed in Jurassic (~171 Ma) and controlled by the WNW trending faults, there are such as **Tongchang**, **Zhushahong** and **Fujiawu** porphyry copper deposits, and **Jinshan** shear zone-type hydrothermal gold deposit. However, to be controlled by small intrusions and hydrothermal veins, many ore deposits, formed in Cretaceous (Figs. 4.38, 4.39 and 4.40) are mainly located along the NE trending faults. The **Yinshan** porphyry—epithermal silver–polymetallic deposit had twice enrichment and superposition. In the northwest of Yinshan deposit, some copper, lead and zinc ore bodies are distributed near EW trending, which related the dacite

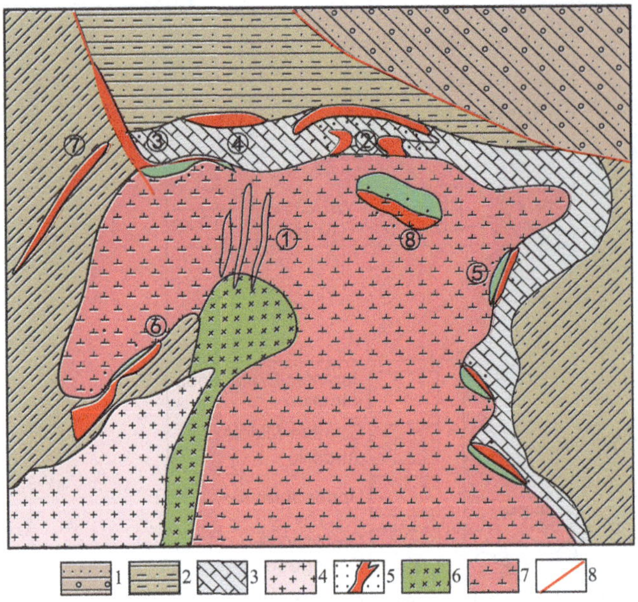

Fig. 4.36 Metallogenic model of Daye-type skarn iron, copper and gold deposits in Southeast Hubei ore province. **Legend**: 1. Sandstone and conglomerate; 2. siltstone and mud-carbonite; 3. carbonite; 4. granite; 5. ore body and alternation zone; 6. Gabbro or gabbro–diorite; 7. diorite–monzonite–granodiorite; 8. fault. **Ore body type**: ① ore magma type (Tieshan); ② skarn type at the top of intrusion; ③ skarn type controlled by the fault (Jinshandian); ④ ore body on the boundary of lithological change; ⑤ skarn type at contact zone (Tieshan); ⑥ skarn type reserved in the quartz diorite and granite (Chengchao); ⑦ intra-layer detachment body; ⑧ xenolith type in the marble (Tonglvshan) (Modified from Zhai 1995)

porphyrite was formed in the Early Period of Jurassic (~175–178 Ma). However, being influenced by the Cretaceous NE trending fault zone, the sinistral strike-slip extension-shear fractures were formed, and those fractures were filled with later lead and zinc veins filled. Thus, the Dexing, Jiangxi polymetallic metallogenic province was also experienced twice tectono-metallogenesis in the Jurassic and Cretaceous (Fig. 4.40) and controlled by WNW and NE tectonic line. Some researchers considered that metallogenic province was only controlled by NE trending tectonic zone; obviously, it is not reasonable.

In 1980s–1990s, some researchers propagated to find the "porphyry copper deposit" in China and took the Dexing porphyry copper deposit as a typical case (Zhu et al. 1983). However, the ore deposits, formed in the porphyry directly, are very limited. Since it was called "porphyry copper deposit," it should mean to be related to the porphyry, but it just belongs to an industrial type. That genetic type is completely different from the Andes porphyry copper deposits related to the subduction zone. But the "porphyry copper deposits" discovered in China were mainly formed in the intraplate deformation, without relationship to the plate subduction.

In recent years, a super-large tungsten field has been discovered in **Dahutang** area, Wuning County, Jiujiang City (~115° E, 29.1° N; Lin et al. 2006; Fig. 4.41), the proved tungsten reserves of which are 0.93 million tons with associated tin, copper, silver and molybdenum. They are distributed in 15 ore deposits. That ore field is located in the intersection of the east–west trending Jiuling–Zhanggongshan uplift and Wuning–Yifeng N–NE trending fault zone, which belongs to north of Dahutang–Tong'an tungsten (tin), tantalum, niobium poly-metallogenic zone, near the intersection of EW and NNE trending faults. The ore magmatic intrusion and the main metallogenesis were formed at the

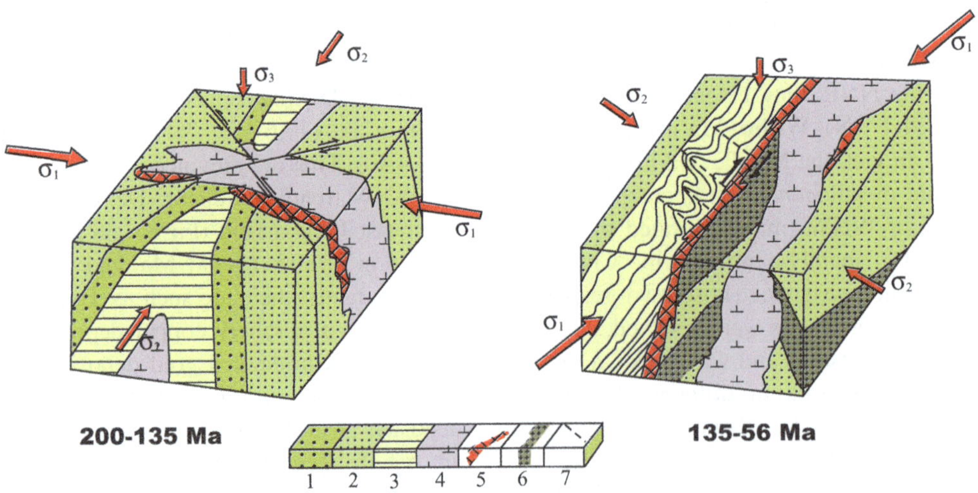

Fig. 4.37 Stress field, deformation and the influence to the metallogenesis in Jurassic–Cretaceous. 1. Conglomerate; 2. sandstone; 3. siltstone and shale; 4. diorite intrusion; 5. iron, copper and gold body; 6. breccia zone with ore; 7. axis surface of micro-fold. σ_1. Maximum principal compression stress orientation; σ_2. intermediate principal compression stress orientation; σ_3. minimum principal compression stress orientation

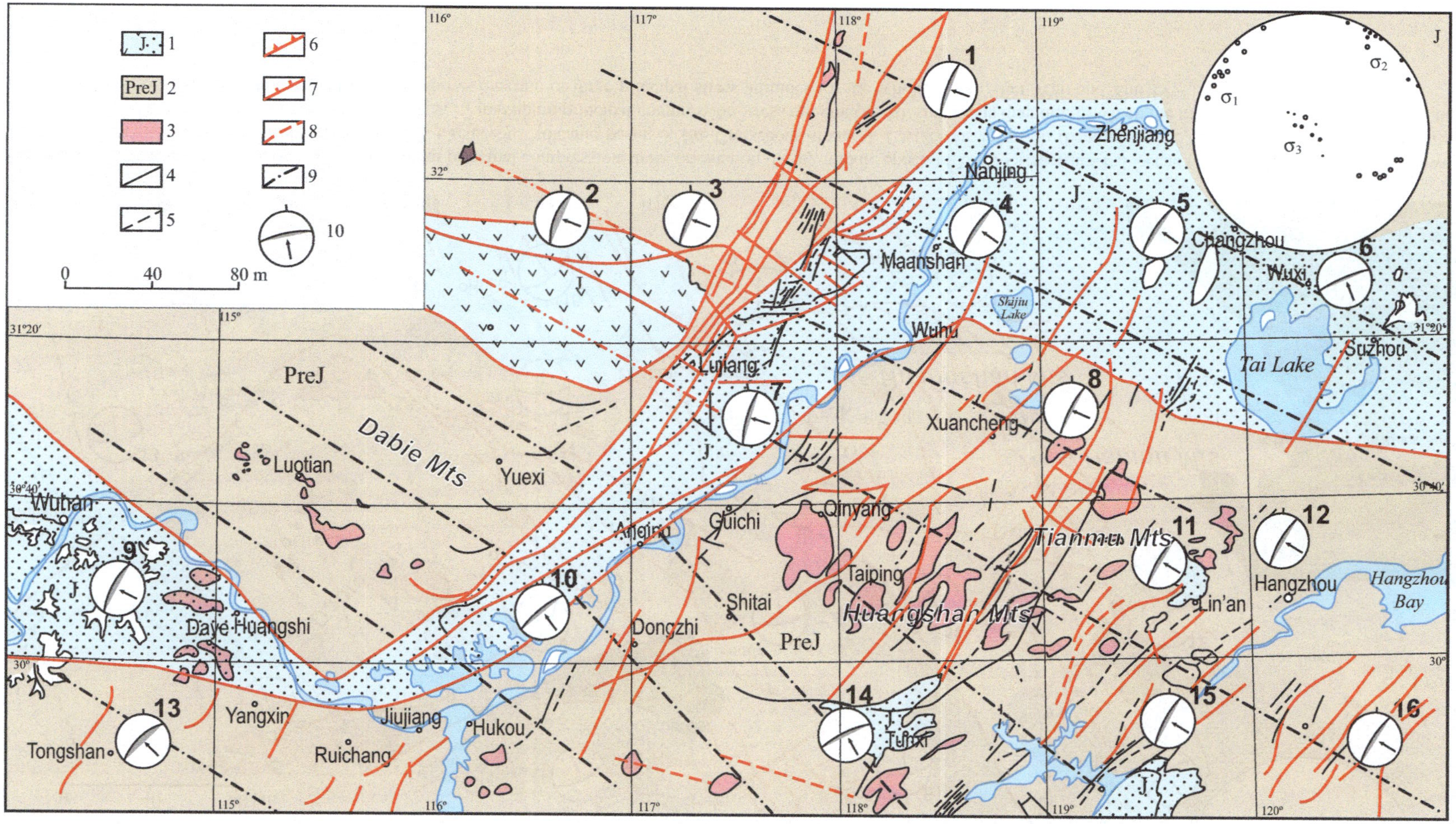

Fig. 4.38 Ore-bearing diorite body and the maximum principal compression stress orientation during Jurassic–Early Cretaceous (200–135 Ma) in middle and lower Yangtze region. 1. Jurassic sedimentary and volcanic basin; 2. Pre-Jurassic strata; 3. Jurassic diorite intrusion; 4. geological boundary; 5. inferred geological boundary; 6. normal fault; 7. reverse fault; 8. inferred fault; 9. σ_1 maximum principal compression stress orientation; 10. stress axis diagram, in which the line is σ_1, circular arc surface is the space projection of fold axis. On the right upper, it shows diagram of three principal stress attitudes for whole areas (After Wan and Zhu 1991)

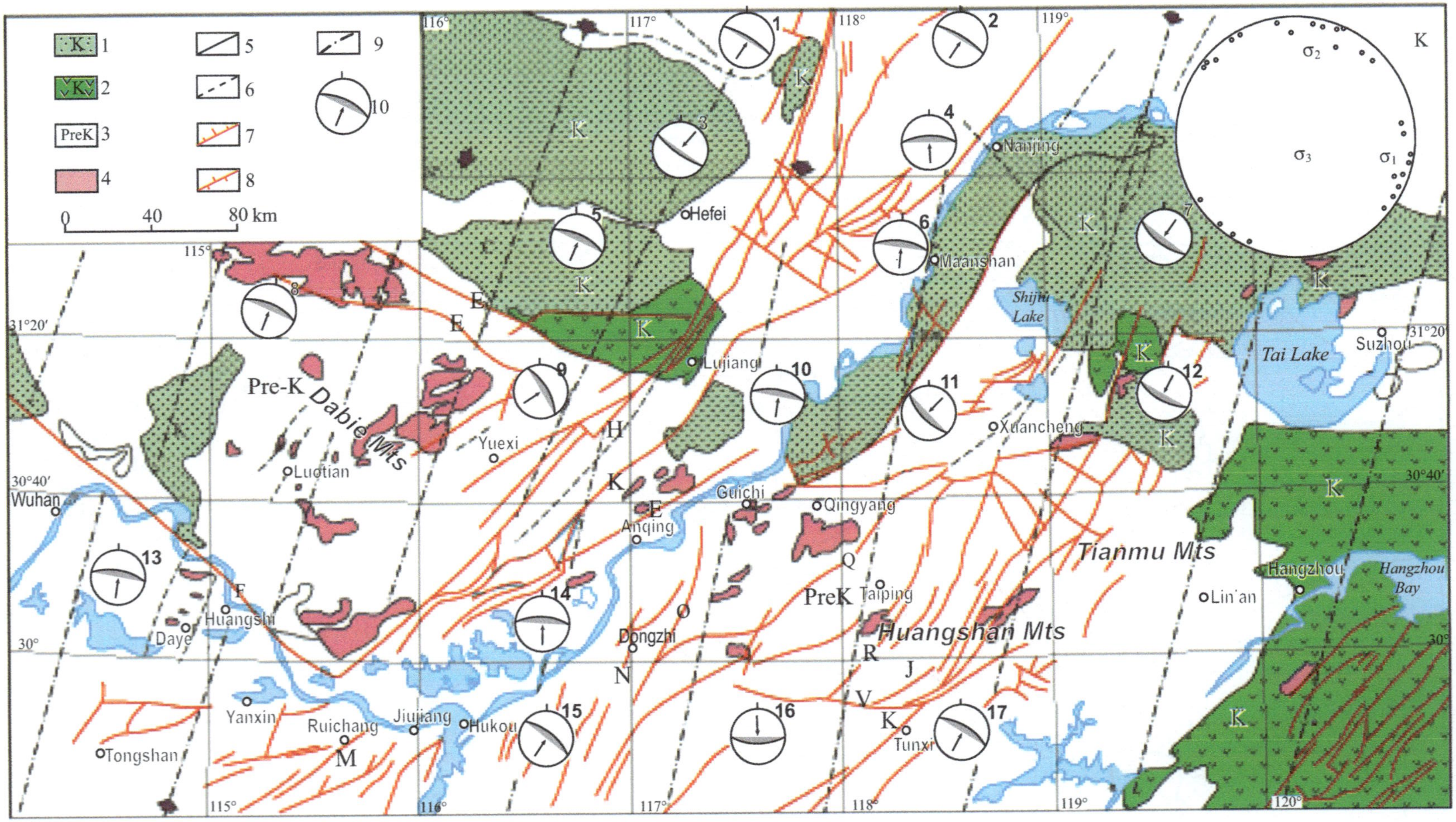

Fig. 4.39 Ore-bearing diorite body and the maximum principal compression stress orientation during middle epoch of Early Cretaceous–Paleocene (135–56 Ma) in middle and lower Yangtze region. 1. Cretaceous sedimentary basin; 2. Cretaceous volcanic basin; 3. Pre-Cretaceous strata; 4. Cretaceous intrusion (bearing ore); 5. geological boundary; 6. inferred geological boundary; 7. normal fault; 8. reverse fault; 9. σ_1 maximum principal compression stress orientation; 10. stress axis diagram, in which the line is σ_1, circular arc surface is the space projection of fold axis. On the right upper, it shows diagram of three principal stress attitudes for the whole areas (After Wan and Zhu 1991)

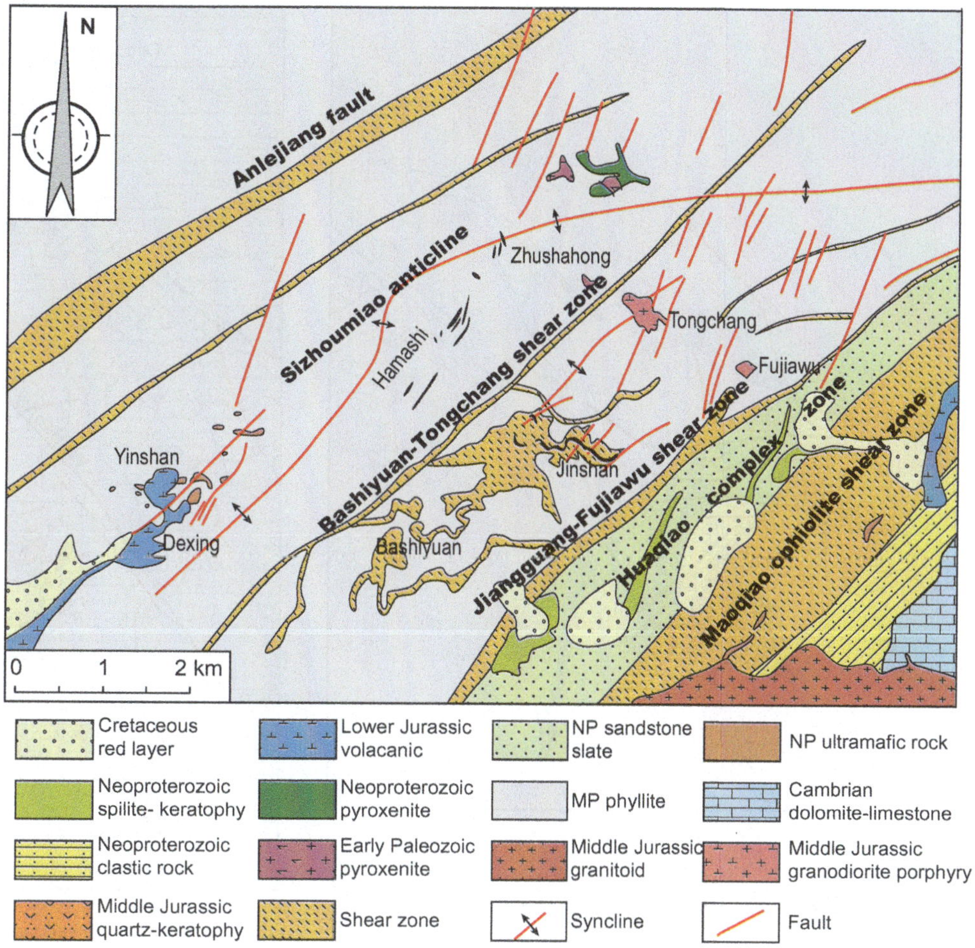

Fig. 4.40 Distribution of Dexing copper, gold and polymetallic ore province, Jiangxi (Modified from Mao et al. 2012a)

period of 149.9–134 Ma; and the Late Jurassic–Early Cretaceous was the main metallogenic epochs. In the ore field, the ore deposits are mainly distributed in the NE–NNE trending, the intervals of deposits are about 2.5–3.5 km. The ore veins are mainly distributed in EW or WNW trending. Till now, more than 60 wolframite quartz veins have been explored, as well as some quartz fine vein, fine-vein dissemination, greisen and the breccia (at the top of intrusion)-type wolframite, scheelite (with copper) deposits (Lin et al. 2006). That tungsten field is the biggest one discovered recently in the southeast of South Yangtze plate, which is near the Cathaysian plate.

In **Xikuangshan** (111.5° E, 27.7° N), Lengshuijiang City, Central Hunan of South Yangtze plate, there is a super-large hydrothermal antimony deposit. The antimony reserves reach 2.11 million tons, accounting for about half of the global reserves. The wall rocks are mainly Devonian carbonite system, and the wall rock alterations are mainly silication. The ore body is distributed along the stratification and shows the stratified structure. The metallogenic period can be divided as two epochs, the early one of which is the Jurassic (with the isotopic age being 156–155 Ma), controlled by the regional W–NW trending maximum principal compression stress and resulted in the N–NE axis open anticline, with the ore bodies being concentrated at the hinge zone of anticline; while the late one of which is the Cretaceous (with the isotopic age being ~124 Ma), influenced by the N–NE shortening process and resulted in the W–NW trending stretching process crossing the whole pre-existing anticline. At the top of the anticline overlapping twice, i.e., the structural detachment position, it is easy to form the strongest hydrothermal mineralization (Mao et al. 2012a). That ore deposit may be related to the deep magmatism that resulted in concentration of the ore-bearing hydrothermal fluid.

In the central Yangtze plate, Northeastern Guizhou, there is the **Wanshan** epithermal mercury deposit (109.2° E, 27.5° N), which is an integral mercury deposit formed in the carbonite system of Cambrian (Fig. 4.42). Similar to the Xikuangshan deposit, that deposit had also experienced twice important tectonic events: In Jurassic, under the action of WNW trending maximum principal compression stress, it formed the NNE trending open anticline (main part), with ore bodies being concentrated at the hinge zone of anticline;

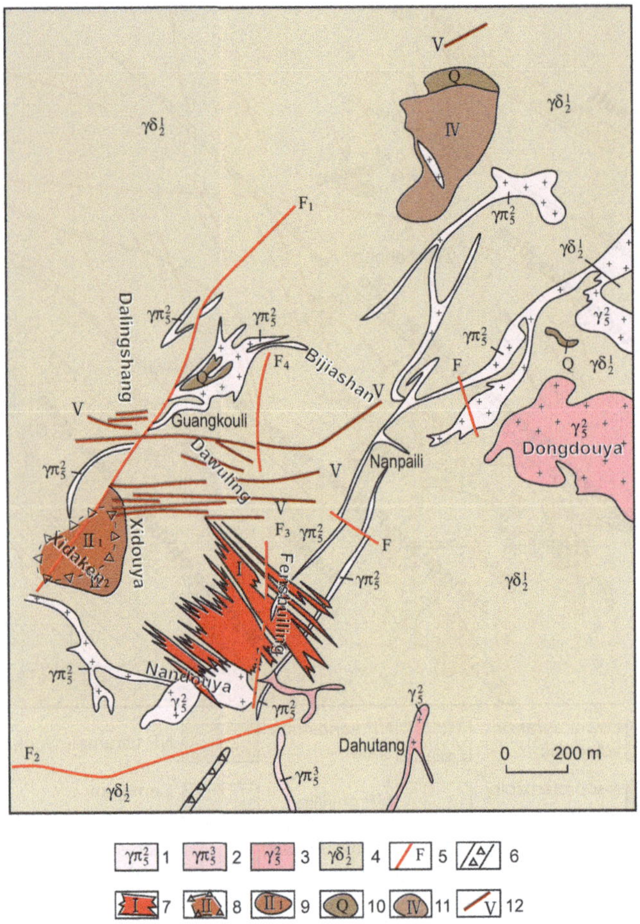

Fig. 4.41 Geological sketch of Dahutang area, Wuning County, Jiangxi tungsten field. 1. Late Yanshanian granite porphyry; 2. Middle Yanshanian granite; 3. Yanshanian mica granite; 4. Neoproterozoic mica granodiorite; 5. fault; 6. siliceous crush zone; 7. fine-quartz-vein tungsten deposit; 8. breccia-type tungsten body; 9. quartz net-vein-type tungsten body; 10. quartz pegmatite-type tungsten body; 11. greisen-type tungsten body; 12. quartz big vein-type tungsten body (width more than 0.1 m) (Modified from the original data of Northwest Jiangxi Geological Brigade and Lin 2006)

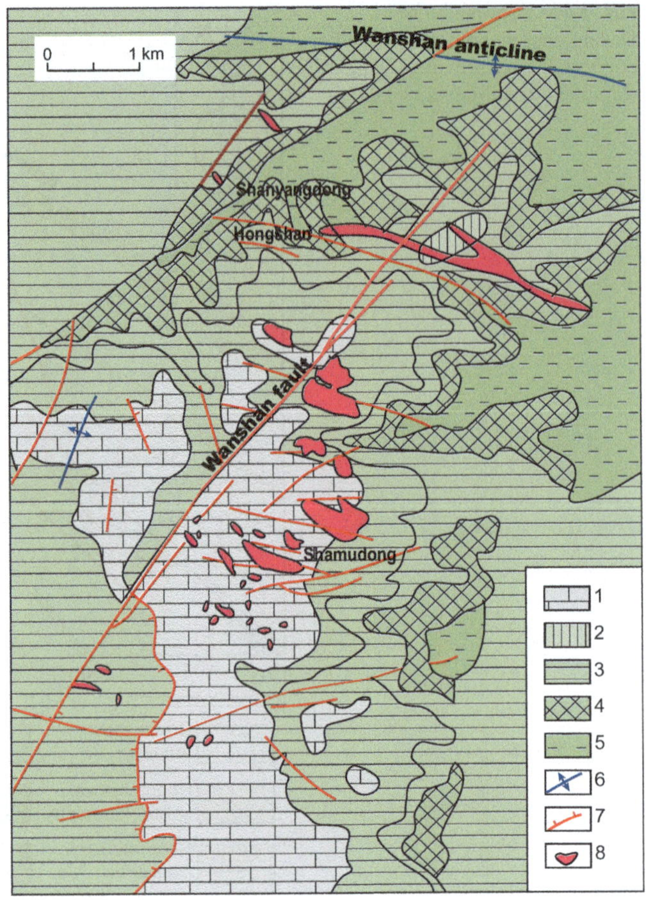

Fig. 4.42 Geological sketch of Wanshan, Guizhou mercury deposit. 1. Carbonite, Huaqiao Formation, Middle Cambrian; 2. dolomite, middle and upper members of Aoxi Formation, Middle Cambrian; 3. shale, middle member of Aoxi Formation, Middle Cambrian; 4. dolomite and black hydromica shale, lower member of Aoxi Formation and Qingxudong Formation, Middle Cambrian; 5. shale, Balang and Niutitang Formations, lower Cambrian; 6. anticline; 7. fault; 8. ore body (Modified from Chen and Zhu 1993)

while in Cretaceous (isotopic age is about 124 Ma), influenced by the NNE shortening, the pre-existing NNE trending anticline resulted in the WNW trending cross fractures and refolds. At the refolded position, it formed strongest structural prostration and best metallogenesis (after exploiting, it shows a great dome-type cave, just like an underground "hall," I visited it in 1966). That is why the main fold axis and faults are all distributed along NNE trending, but the ore bodies are preserved in the secondary anticline hinge zone with WNW trending and its fracture zone (Fig. 4.42; Chen and Zhu 1993). This example illustrates that it is so important to make clear the change of tectonic stress field for discussing the metallogenesis.

On the northern side of Yangtze plate, **Mayuan**, Hanzhong, Shaanxi Province (107.1° E, 32.8° N), there is the Mississippi Valley-type epithermal hydrothermal lead and zinc deposit. More than 20 ore deposits and spots have been discovered. Those lead and zinc deposits are controlled by strata obviously, which are located in the Neoproterozoic breccia dolomite on the boundary of paleo-continent. According to the sphalerite Rb-Sr method dating method, the isotopic age is 486 ± 12 Ma (Mao et al. 2012a), the ore-forming age at Early Ordovician, which belongs to pre-collision metallogenesis. The metallogenic temperatures are about 100–200 °C, with the characteristics of hot brine. The metals in ore came from the basement and cover metamorphic and sedimentary strata, and the sulfide belongs to the evaporated sulfide with reduction characteristics.

In **Dachang,** Guangxi (107° 35′ E, 24° 51′ N), there is a super-large layer metasomatism-type tin and antimony sulfide deposit. That deposit is located at South Yangtze plate (Fig. 4.43; Huang et al. 1985), which metallogenesis is

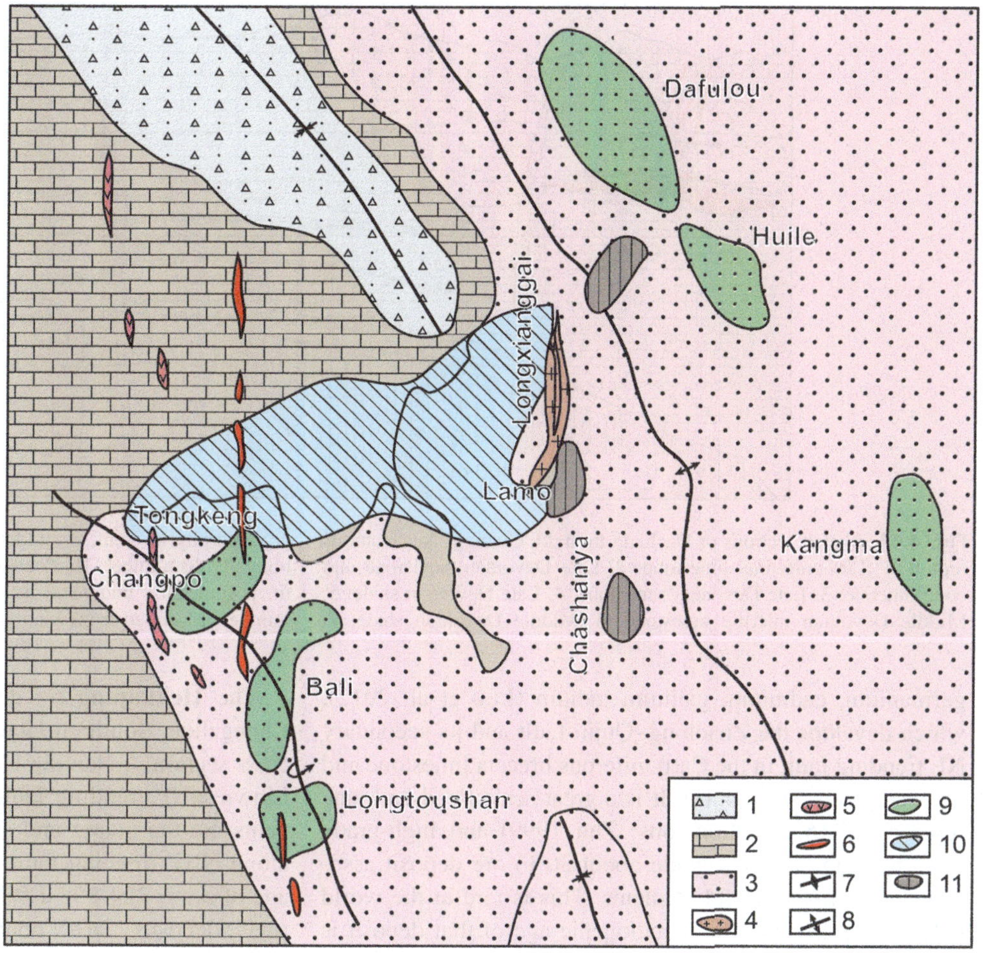

Fig. 4.43 Distribution sketch of Dachang tin field, Guangxi Autonomous region. 1. Carboniferous limestone; 2. Devonian carbonite, siliceous rock and siltstone; 3. Cretaceous granite; 4. quartz diorite porphyry; 5. granite porphyry vein; 6. tin polymetallic mineralization area; 7. anticline axis; 8. syncline axis; 9. zinc and copper mineralization area; 10. tungsten and antimony mineralization area; 11. skarn distribution area (Modified from Mao et al. 2012a)

closely related to the Cretaceous (94–91 Ma) hidden granite porphyry intrusion. Influenced by the Cretaceous near NS trending maximum principal compression stress, the near NS trending fractures show the extension-shear process, which are filled with ore fluids to form the tin bodies. The great skarn tin deposits are preserved near the outer contact zone of granite porphyry. Far from the intrusion body, there are the layer metasomatism-type tin-sulfide deposits, and at the top there are the cassiterite–sulfide deposits (Fig. 4.44).

In the intrusion body, there are characteristics of the rich tourmaline and fluorite. From the center of intrusion body to the outsides, it formed the proper sequence as zinc and copper (with tin)-zinc, lead, antimony and silver–tin polymetallic zone. At the top of intrusion body, there is the tungsten and antimony mineralization; in the more outside of intrusion body, it is the mercury mineralization.

Similar to Dachang of Guangxi tin deposit, at the Southern Xiaojiang fault zone, i.e., Southwest Yangtze plate, there is a supergiant tin (copper) polymetallic field of **Gejiu** of Eastern Yunnan (103.2° E, 23.3° N) (Fig. 4.45). The tin ore is closely related to equigranular granite and porphyritic granite. The porphyritic granite (isotopic ages of 81.2–82.8 Ma) and equigranular granite (81–77.4 Ma) are all intrusions formed in Late Cretaceous (Mao et al. 2012a). The ore zone was mainly controlled by NS and NNE trending faults, and the near EW trending secondary faults and folds controlled the ore bodies (right-lower of Fig. 4.45, near Gejiu). All the intrusions, faults and folds were all controlled by the Cretaceous tectonic stress field. Influenced by the NNE trending maximum principal compression stress, the near NS and NNE trending main faults showed extension-shear features, and a series of near EW trending secondary faults and folds controlled the intrusions and tin ore bodies. Being accompanied with the attitude of intrusion bodies and wall rocks, the tin ore was formed as different types: at the contact zone between intrusion and wall rocks, to form the skarn-type deposit; outside of intrusion between the layer detachment or soft zone, to form the stratiform-like ore body; along the fault or joints, to form the vein-like ore body (Fig. 4.46).

Near the Gejiu tin ore field, on the side of NS trending Zhaotong–Qujin fault, there is a super-large lead and zinc hydrothermal deposit of **Qilinchang** of Huize, Yunnan Province (103.3° E, 26.4° N), associated with silver,

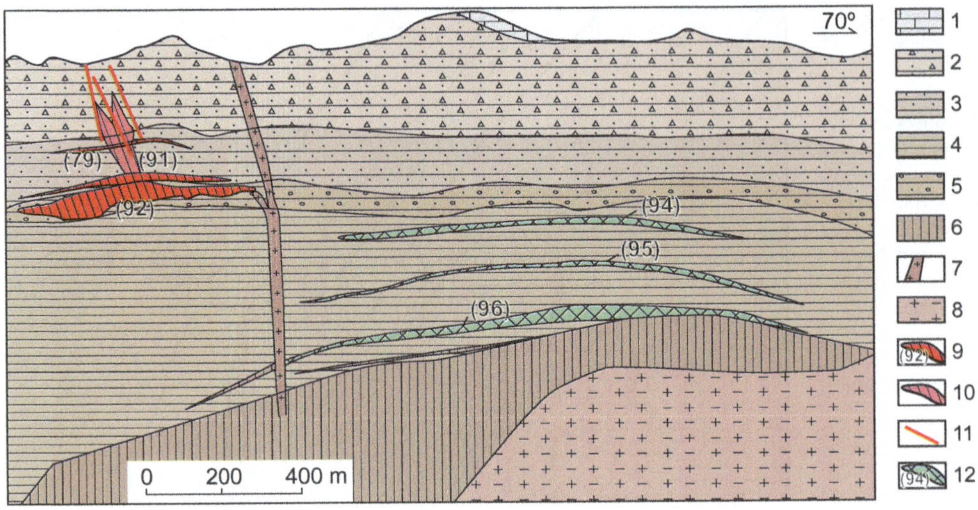

Fig. 4.44 Sketch sections of Dachang tin field, Guangxi Autonomous region. 1. Carboniferous limestone: 2. Late Devonian sandstone and conglomerate; 3. Late Devonian sandstone; 4. Late Devonian shale; 5. Middle Devonian micritic limestone; 6. Middle Devonian shale; 7. granoporphyrite vein; 8. hidden granitic intrusion; 9. tin polymetallic body and its number; 10. fine-vein-type polymetallic body; 11. big vein tin polymetallic body and its number; 12. zinc copper body and its number (Modified from Mao et al. 2012a)

germanium, cadmium, gallium, indium (Luo et al. 2012), which develops the Zhaotong–Qujin fault and its secondary NE trending fault in the Carboniferous breccia limestone and its nearly karst (Fig. 4.47). It is a great-scale (lead and zinc reserves of 15.2 million tons; Zhu 1999) and high-grade (lead and zinc average grade about 30%) ore deposit associated with many useful elements. This is rare in the world (Mao et al. 2012a). The metallogenic age of that deposit is determined by the sphalerite Rb–Sr isochron method as 224.8–226 Ma. Influenced by the Triassic near NS trending maximum principal compression stress, it resulted in near NS fault or NE trending secondary extension-shear faults. That deposit was controlled by the intraplate deformation, and the fault zone was filled with ore hydrothermal fluids to form the Qilinchang, Kuangshanchang and Yinchangpo deposits (Fig. 4.47). Some researchers considered those deposits may be related to the Emeishan Permian basalts. This understanding must not be suitable. Their element characteristics are different from each other, so the forming periods should be different.

Near that ore field, along the NS trending Anninghe fault, there is the Huili lead and zinc ore deposit. Along the near NS trending Zhaotong–Qujin fault in Yunnan, Xiaojiang fault and NW trending Ziyun–Yadu fault in Southwestern Guizhou, there are more than ten large and intermediate scale lead and zinc deposits (left-upper of Fig. 4.47).

In **Gongcheng**, Guangxi, there is the Limu tin, niobium and tantalum deposit, which is located in the Devonian and Carboniferous strata. The ore veins are quartz vein and pegmatite vein controlled by NS and EW trending faults. The muscovite Ar–Ar method age of that deposit is 214.1 ± 1.8 Ma (Wang et al. 2008).

The **Hehuaping** tin deposit is located at the south of Chengzhou, Southern Hunan. That deposit is located at the cross section of Nanling EW trending fault and Chengzhou–Linwu NNE trending fault, near the granite body. The ore veins are the quartz and skarn veins, which are distributed along NNE trending fault. The molybdenite Re–Os method age of ore is 224.1 ± 1.9 Ma (Mao et al. 2012a). The above two deposits are controlled by Triassic intraplate deformation.

At the westward of Xiaojiang fault, east to Anninghe fault, i.e., at the Dongchuan, Huili and Yimen areas, in the ancient fault-depression zone, there is a sedimentary-reformation-type bedding copper deposit; the famous one is **Dongchuan** copper deposit (Xiao et al. 2012). The Dongchuan deposit is preserved in the Mesoproterozoic Dongchuan Group Yinming and Luoxue Formations; the rocks are mainly amaranth breccia, sandstone, with the characteristics of evaporites. Those strata had covered on the Dahongshan Group marine volcanics with the iron and copper; after the erosion, it resulted in the metal elements to concentrate in the sedimentary strata. The ore minerals are mainly the chalcopyrite and bornite with the wall alternation of silication. The sedimentary period is Mesoproterozoic (1716–1607 Ma), and the hydrothermal reformation period is Neoproterozoic (794–712 Ma). According to the hydrothermal reformation, the copper element could be suitable to upward migrate and concentrate at interbedding detachment between the Yinming Formation slate and Luoxue Formation dolomite.

In the recent years, a giant **Yangla** copper deposit has been discovered in the Jinshajiang tectonic zone, at Deqing of Yunnan Province, which is reserved in the Devonian

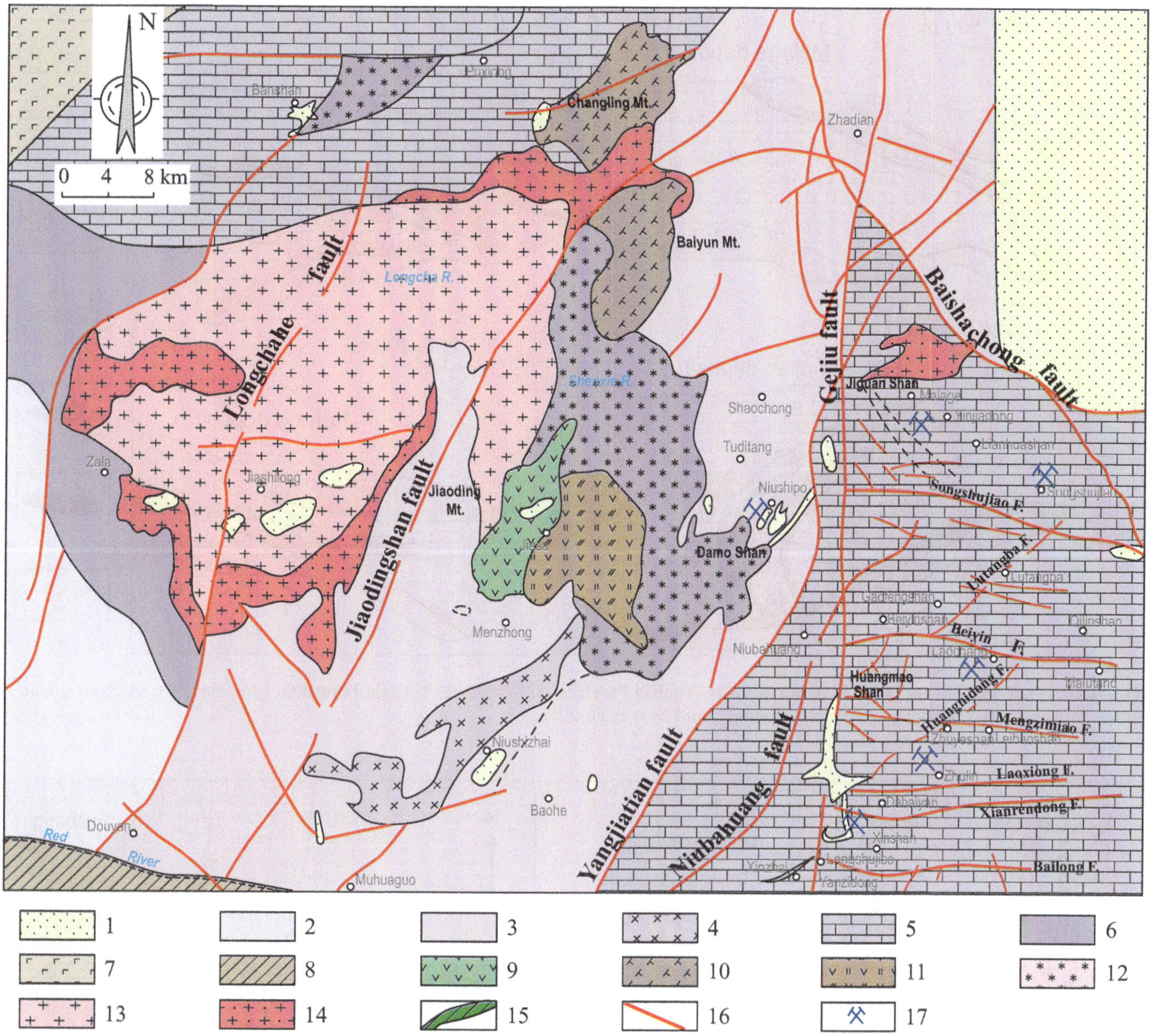

Fig. 4.45 Geological sketch of Gejiu tin field, Yunnan Province. 1. Quaternary sediments; 2. Late Triassic slate; 3. Middle Triassic sandstone, shale and basalt; 4. Middle Triassic basalt; 5. Middle Triassic carbonite; 6. Early Triassic sandstone and green sandstone; 7. Emeishan basalt; 8. Ailaoshan metamorphic zone; 9. gabbro; 10. nepheline syenite; 11. monzonite; 12. alkali granite; 13. porphyritic mica granite; 14. equigranular mica granite; 15. diabase dyke; 16. fault; 17. ore body (Modified from Mao et al. 2012a)

sandstone and shale with granodiorite intrusions. The ore deposit is the hydrothermal type, and the main trending of that is NW and NNE trending (Yang et al. 2012c), in which the molybdenite Re–Os isochron method age is 232 ± 2.9 Ma (Mao et al. 2012a). It should be controlled by Triassic intraplate deformation.

Recently, in **Xingren–Anlong**, Southwestern Guizhou, the deep seismic exploration has been done for Carlin-type super-large gold deposit (using the petroleum exploration data; Fig. 4.48, B–B′ line of seismic exploration), then gold bodies have been discovered to preserve in the rather soft strata to be easy detach (such as the Longtan coal system in Late Permian), and unconformity surface (between Upper and Lower Permian, or Silurian and Middle Devonian). The epithermal gold-bearing fluids were filled into low angle thrust or the nearby fracture zones; thus, it formed a gold deposit that was controlled by the typical epithermal intraplate fault and the bedding (in Zimudang, Huilong, Nipu, Dayaokou, Getang and Lannigou; Fig. 4.48). In that areas, the regional faults are mainly NE and NW trendings, the gold bodies are developed in the depth of 1500–3200 m. The Carlin-type gold deposits were formed in Jurassic and

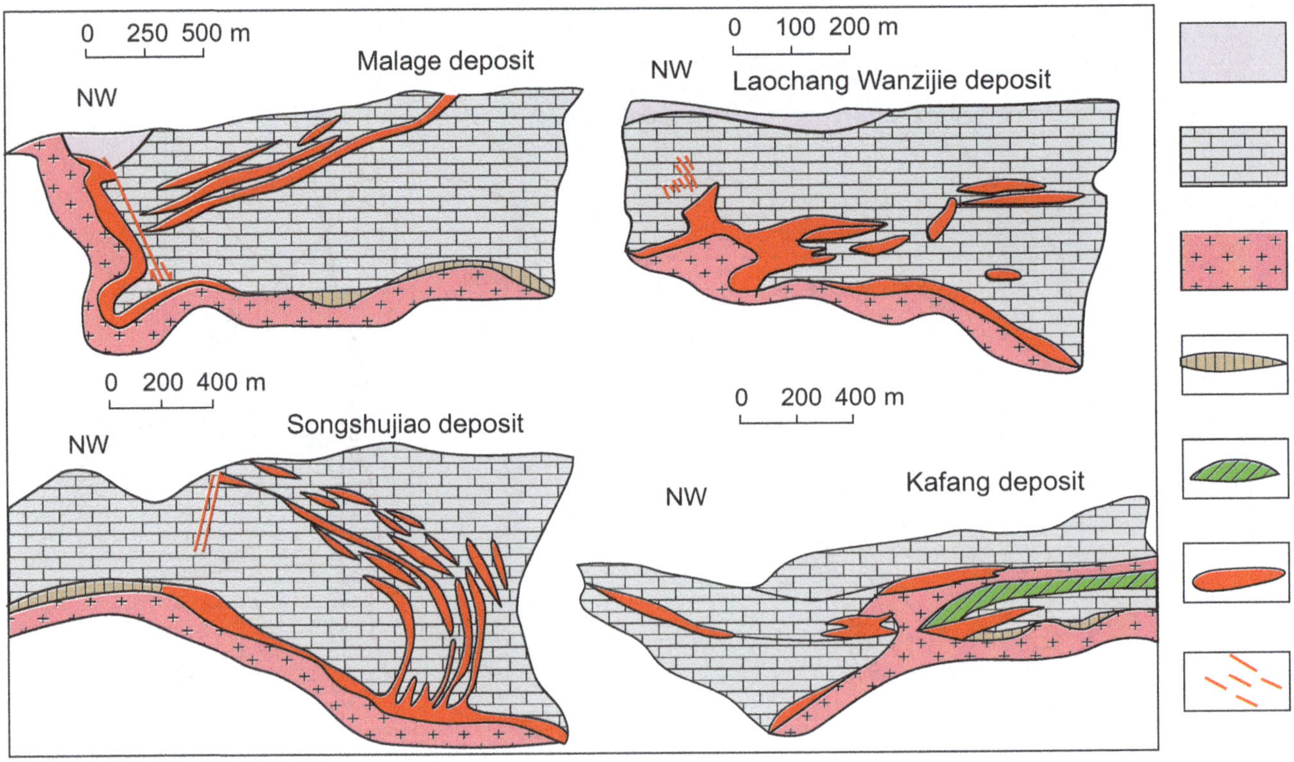

Fig. 4.46 Some sections for Gejiu main ore deposits, Yunnan Province. 1. Sandstone; 2. Gejiu Formation carbonite; 3. Cretaceous granite; 4. skarn; 5. diabase; 6. ore body; 7. ore vein (Modified from Mao et al. 2012a)

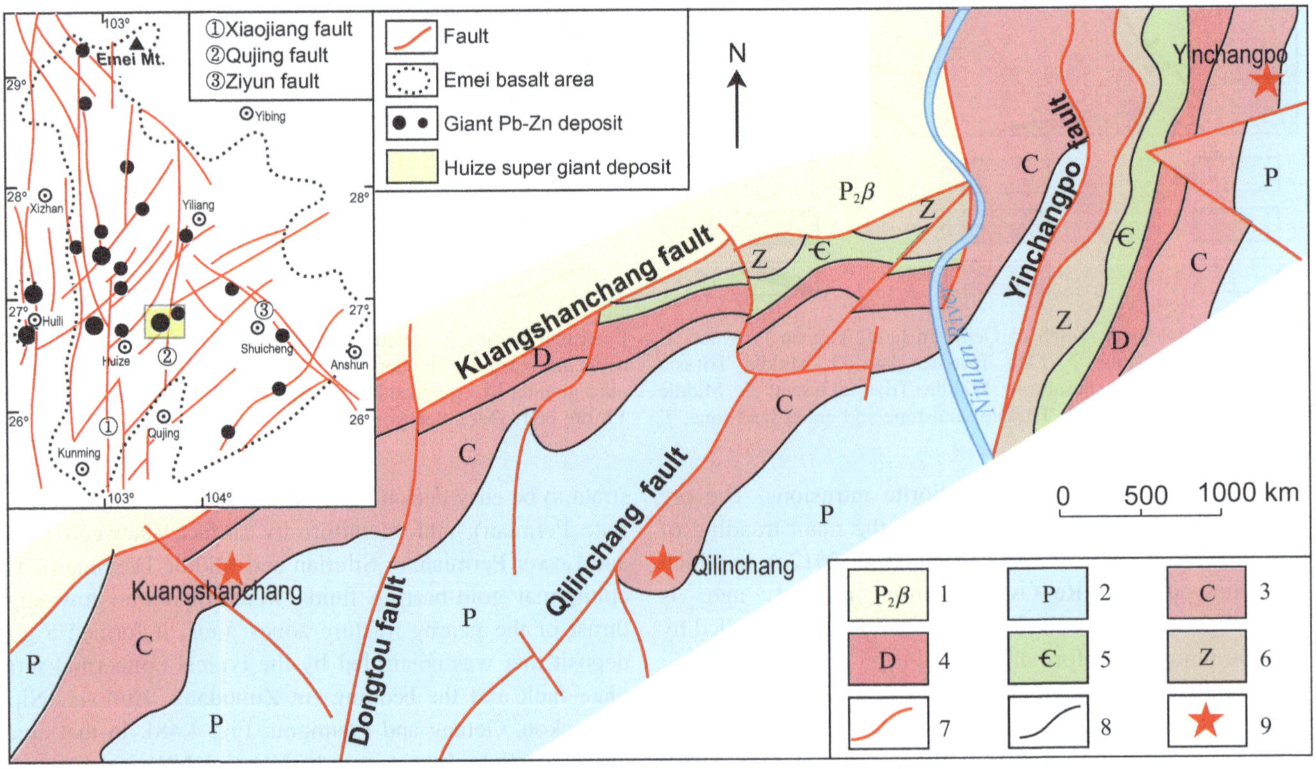

Fig. 4.47 Geological sketch of Huizex super-large lead and zinc hydrothermal deposit, Yunnan Province. 1. Permian Emeishan basalt; 2. Permian limestone and dolomite; 3. Carboniferous breccia limestone, oolite limestone dolomite; 4. Devonian limestone and dolomite; 5. Cambrian mud shale and sand shale; 6. Neoproterozoic siliceous dolomite; 7. fault; 8. strata boundary; 9. lead and zinc deposit (Modified from Mao et al. 2012a)

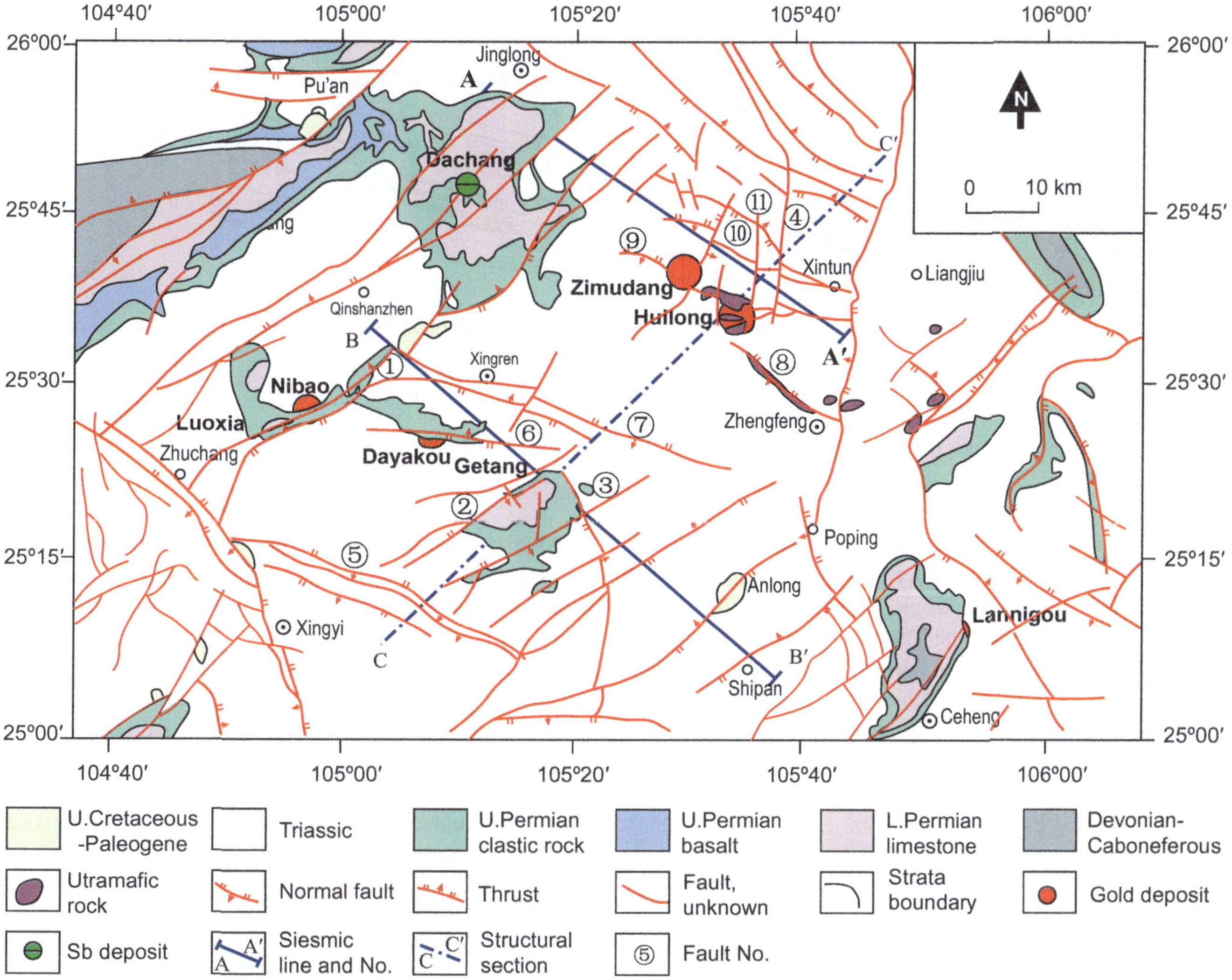

Fig. 4.48 Geological structure sketch in Southwestern Guizhou and the distribution of Carlin-type gold and antimony deposits. Fault names: ① Machang; ② Haimagu; ③ Shangheba; ④ Yongningzhen; ⑤ Zhuchang–Shangzhai; ⑥ Dayakou;⑦ Xingren; ⑧ Hetaoshu; ⑨ Dashan–Zhexiang; ⑩ Paodongwan; Donggang (Modified from Hu et al. 2012)

Cretaceous (the isotopic ages, by many methods, between 193–60 Ma in different deposits). The recent exploration geological and geophysics sections in **Huijiapu** gold field (such as Fig. 4.49) have well demonstrated the soft strata and the faults to control the metallogenesis (Hu et al. 2012).

The **Zimudang** gold deposit is preserved in the Permian Yelang Formation mud limestone, and its calcite Sm–Nd method age is 250.0 ± 1.4 Ma (Mao et al. 2012a). However, the above results cannot be determined for the metallogenic age. It needs to do further researches to get the correct ore period.

At **Jinlong**, Western Guizhou (105° 10′ E, 25° 41′ N; top of Fig. 4.48), a super-large hydrothermal antimony deposit is reserved in the Paleozoic System (Shen et al. 2010), which forming mechanism is as same as that of the gold deposit. The antimony bodies are mainly concentrated near the two limbs of NE trending open anticline and syncline. In Eastern Yunnan, also there are the similar deposits, such as **Muli** (105° 22′ E, 23° 59′ N), Guangnan County of Yunnan Province, large hydrothermal antimony deposit, which is located at the northeast end of Wenshan arc tectonic zone (the intersection site of Funing NW trending fault and Guangnan EW trending hidden basement fault), and the ore body is characterized by bedding. That is a sedimentary—reformation-type ore deposit, the antimony proved reserves are 0.17 million tons (Gong and Chen 2006).

Wenshan arc tectonic zone (the intersection site of Funing NW trending fault and Guangnan EW trending hidden basement fault) and the ore body are characteristized by bedding. That is a sedimentary-reformation-type ore deposit; the antimony proved reserves are 0.17 million tons (Gong and Chen 2006).

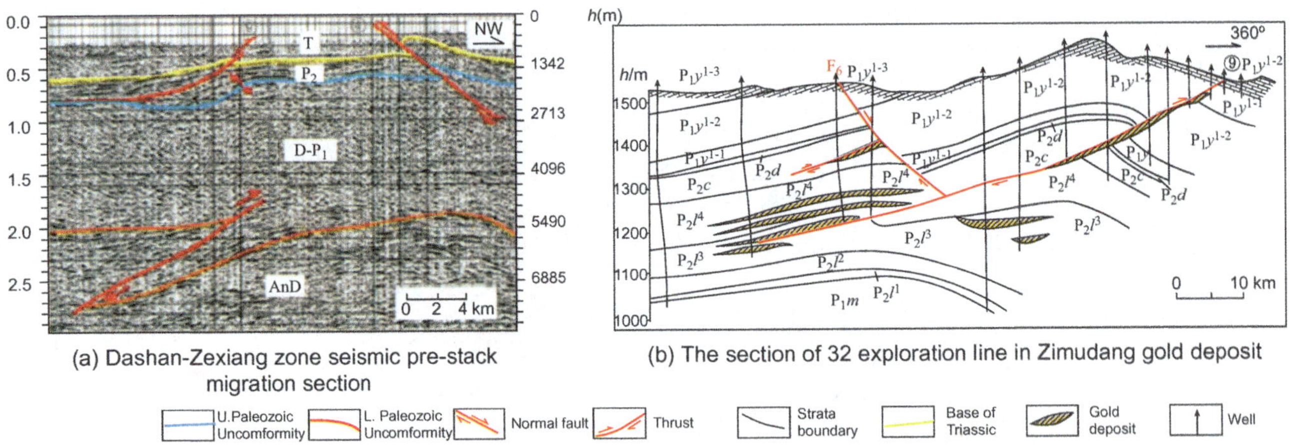

Fig. 4.49 Comparison between typical geological and seismic sections in Huijiapu gold field, Southwestern Guizhou Province (Modified from Hu et al. 2012)

In **Zhunyi–Qianxi–Shuicheng**, Northwestern Guizhou, there are the marine sedimentary carbonate-type manganese deposits formed under the tropic climates (Gao 2011). The ore layers occur in the Late Permian Longtang Formation and Early Permian Maokou Formation, and the ore layers commonly are 2–6 m thick. Under the later tectonic deformation, the strata and ore layer could be penetrated or folded. Generally speaking, the source of manganese could come from the weathering and migration of Emeishan basalts (Han and Pan 2009).

In Southwestern Guangxi (~107° E, ~23° N), also there are the similar marine carbonite sedimentary manganese deposits, such as the **Xialei–Hushun,** Tiandeng County manganese deposits (106° 42′ E, 22° 54′ N) (Chen and Zeng 1989). The ore layers are mainly developed in the Upper Devonian, the thickness of ore layer up to 12–20 m. The grades of manganese are about 20–40%.

The **Sichuan basin,** also called as "the Upper Yangtze basin," is located at the western-middle of Yangtze plate, which overlays on the rather stable Neoproterozoic crystalline basement. From the Paleozoic to Cretaceous, its subsidence had lasted for a long time. The thickness of sedimentary cover is about 4–12 km (Fig. 4.50); thus, it is favorable to reserve the gas and oil.

Now Sichuan basin becomes the greatest gas industrial base in China (Fig. 4.51; The Editorial Board of Petroleum Geology in Sichuan Oil and Gas Region 1989). The basin mainly produces the gas. In recent, from 125 gas fields and 12 oil fields, the oil outputs are about 150,000 tons per year, and the gas outputs are about 16 billion cubic meters per year. In the basin, the gas resources are 72,000 billion cubic meters. It was reported that China National Petroleum Corporation (CNPC) and Shell Corporation cooperated to discover and exploit the first one Shell gas field in China (China Economic Web site, 2011-12-8).

After comparing with the Paleozoic marine oil and gas fields in globe, Qian et al. (2003) considered that there should be the marine oil and gas layers in the Sichuan basin. Their suggestions have been proved by the recent exploration. According to the geophysical, geochemical and geological data for the depth of Sichuan basin, Liu et al. (2015) recognized that the distribution of hydrocarbon layers must mainly be controlled by the late period of Neoproterozoic–Early Cambrian near NS trending Mianyang–Changning aulacogen (the Chuanzhong Gaoshiti–Muoxi gas field, its gas reserves up to 1000 billion cubic meters), the Late Devonian–Early Triassic NW trending Guangyuan–Liangping aulacogen and other regional intra-continent fault-depressions (Fig. 4.51).

On the side of aulacogen and near the paleo-uplift areas, it is easy to form the great oil and gas fields, such as the famous Puguang gas field (Ma et al. 2005). The Puguang gas field is located at Yuanhan County, Dazhou City, in which exploration areas are 1118 km^2. The reserves are estimated as 8.916 billion cubic meters, which is of the large output and super-large-scale marine gas field. The "Science and Technology Daliay" (2014-3-21), according to the Southwest Oil and Gas Company (CNPC) information, reported the new gas proved reserves up to 4.40385 billion cubic meters including recoverable reserves of 1.875 billion cubic meters. It has the biggest marine facies carbonite gas reservoir in China.

In the East Sichuan basin, the storage structures of the gas were mainly formed in Late Paleogene, which were all controlled by the WNW trending regional shortening. So the reservoir structures in the East Sichuan basin are mainly distributed in the NNE trending folds formed at the end of Oligocene. However, in the West Sichuan basin the reservoir structures are mainly distributed near EW trending folds formed in Late Cretaceous and Paleocene. It must be

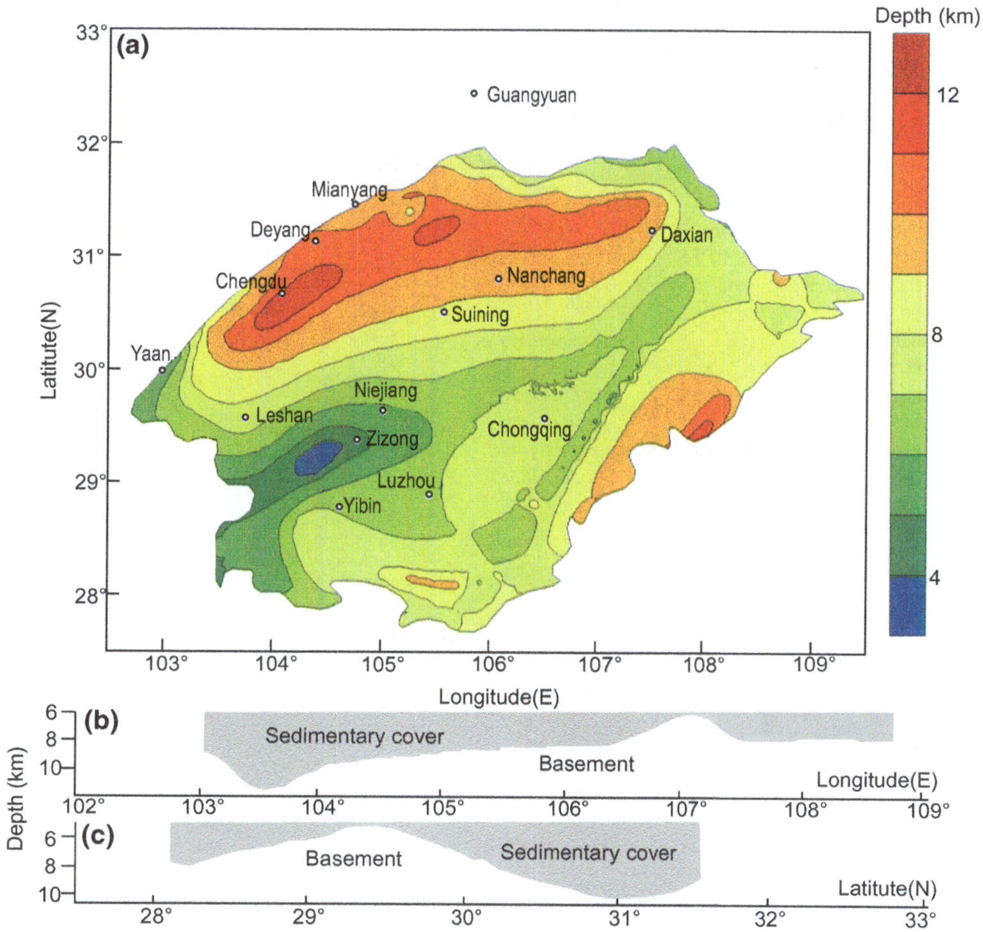

Fig. 4.50 Thickness contours of sedimentary cover in Sichuan basin (After Xiong et al. 2015)

considered that the recent maximum principal compression stress is trending NW–SE in the Sichuan basin. So the pre-existing fractures along the above orientations show the extension nowadays, and the fluid permeability will be so high that it is suitable for the gas or shale gas to migrate and concentrate (Wan 2011).

In the Ganze areas (among the Kangding, Yajiang and Daofu counties) of Sichuan Province, there develops the **Kiajika** pegmatite rare metal deposit (Wang et al. 2005a), which is reserved in the Triassic Xikang Group sandstone and shale system, and controlled by the NS trending mica granite intrusion near its inner and outer contractions, and associated with a series of rich in Li, Be, Nb and Ta pegmatite veins. In the ore veins, the muscovite Ar–Ar isochron ages are 199.4 ± 2.3 and 195.4 ± 2.2 Ma (Mao et al. 2012a). The above age data show that ore deposit should be formed in Early Jurassic.

In recent, at the Eastern Kunlun, **Dachang** gold deposit, in Qumalai County, Qinghai Province, there have obtained the gold reserves of 115 tons. This deposit was formed in Triassic. The metallogenesis occurred in the East Kunlun collision zone on the northwest side of Yangtze plate. In the Dachang deposit, there are 26 gold bodies explored at certain scale. They are distributed at the two limbs of EW trending anticline, in which 20 gold bodies are in the balance sheet of reserves, with the length of 80–3800 m, the average width between 1.0–9.65 m, the average grades between 3.45–10.53 g/t (Chen 2005).

At the Southwest Yangtze plate, there appeared obviously extension in Late Permian to form the Emeishan large-scale basalt eruption (Wan 2011), and at same time, a lot of metal deposits were reserved to form a super-large ore province (Liu et al. 2004; Mao et al. 2012a), such as **Panzhihua** coulsonite-titanomagnetite deposit (101° 45′ E, 26° 38′ N; Luo et al. 2002; Zhou et al. 2013; Fig. 4.52). That metallogenic zone commonly is near NS trending; however, the ore basic intrusion bodies and deposits are reserved in S-type extension-shear fault at the NNE orientation (Pecher et al. 2013).

The **Jinbaoshan** of Yunnan platinum deposit (Tao et al. 2003; Wang et al. 2007a), **Yangliuping** copper, nickel and platinum deposit (Song et al. 2004), and a lot of basalt-type copper deposits (Fig. 4.52) are all developed in that area. The NS or NE trending faults are the ore-conducting structures, and the ore-hosting structures are distributed at the Permian soft rock system and near its anticline hinge. The

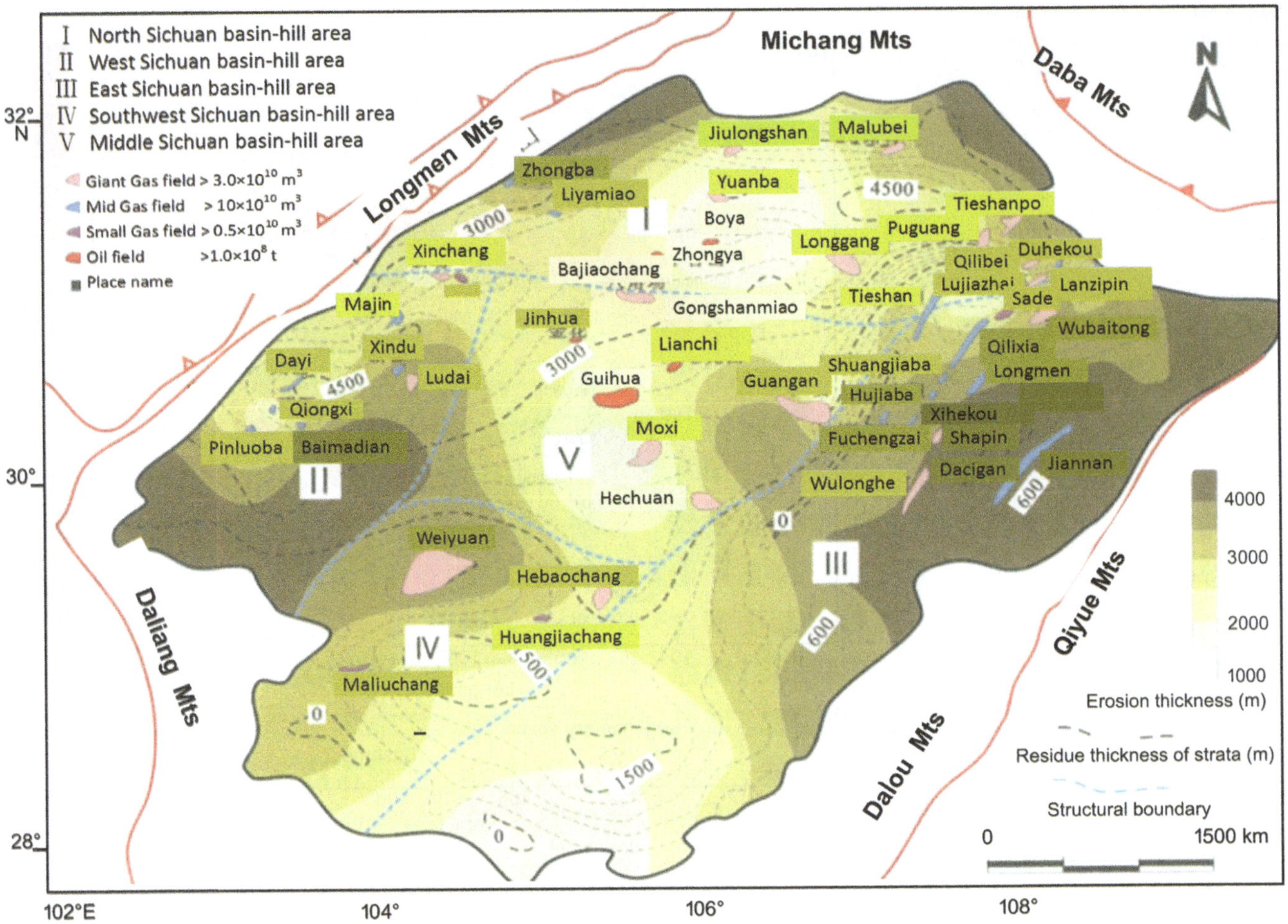

Fig. 4.51 Distribution of oil and gas fields in Sichuan basin (After Liu et al. 2015)

intrusion and eruption ages of basalts are between 262 and 251 Ma, and the metallogenic periods are mainly between 238–226 Ma, i.e., in Triassic, the periods of preservation of ore fluids were later than rock formation periods about 30 Ma (Mao et al. 2012a). Influenced by Triassic NS trending compression, the near NS trending faults showed extension-shear process; thus, it created the favorable condition to form the iron, vanadium, platinum and copper metallogenic zone.

Based on the seismic tomography, the three dimension velocity structural map shows that at the different depths (20–300 km, Fig. 2.23; Zheng et al. 2012) of Southwest Yangtze plate, there appears great-area low-velocity anomalies or arranges in crisscross pattern with low and high velocities, which is resulted from multiply and complex low-velocity mantle diapir (may contain supercritical ore-bearing fluids). The deep low-velocity mantle diapir is only distributed in the areas of Emeishan basalt. It may be the reason to form the Southwest Yangtze super-large non-ferrous metal and rare metal metallogenic province (Liu et al. 2004).

In 1980s, some researchers called this zone as "Panzhihua–Xichang Rift Zone," now that term seems not to be suitable. In addition, according to the recent seismic tomography, the Emeishan large basalt province has been regarded to originate from a mantle diapir (left-upper red area of Fig. 2.23), but there are not any evidences to illustrate the mantle plume (from the Gutenberg discontinuity, i.e., the boundary between core and mantle) existing in the depth of that area.

On the western side of Yalongjiang, in the Garze–Litang areas, from Middle Devonian to Late Triassic, the oceanic crust had subducted to the west, and formed high Sr/Yi porphyry intrusions, and enriched with Cu–Mo–Au porphyry–skarn ore deposits. They could extend to the Jinshajiang–Ailaoshan area (Deng et al. 2014a). On the eastern side of Jinshajiang collision zone, at the **Yitang (Deda) Jiacun** (99.3° E, 31.3° N) Triassic rift zone, there reserves the volcanic massive sulfide-type silver, lead, zinc polymetallic deposits (VMS; Fu and Xu 1989). That ore deposits are related to the bimodal volcanic eruptions (tholeiite–felsic volcanic rocks). The results of determined age of volcanic

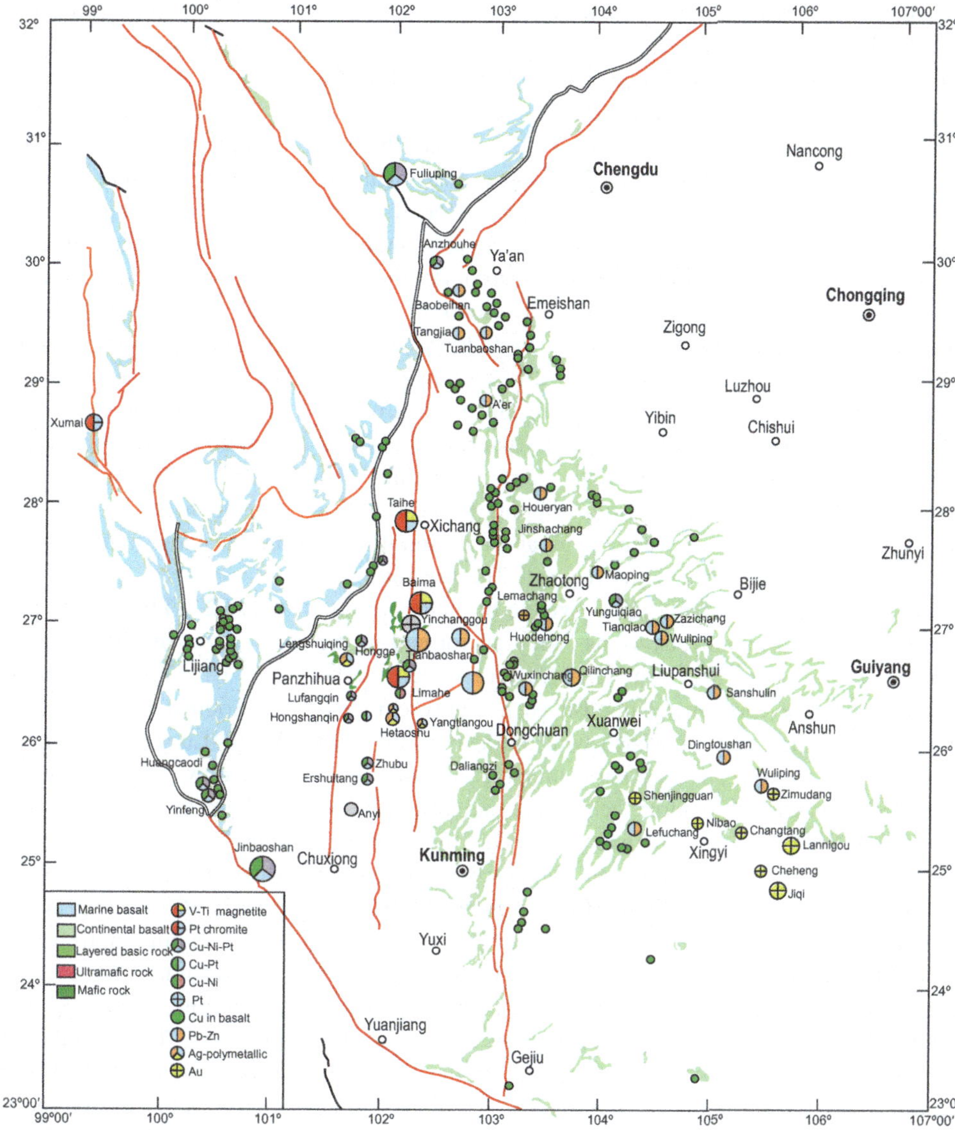

Fig. 4.52 Distribution of basic rock system metallogenic series in Southwest Yangtze plate. The red lines are the regional faults (Modified from Mao et al. 2012a)

rocks are as follows: The lower basalt is 217 Ma; the upper ore layer is 213 Ma. The Jiacun ore deposit was formed in the Late Triassic plate boundary volcanic arc and controlled by the near NS trending secondary extension fault zone. The metallogenesis would continue to Jurassic (Deng et al. 2014a). The massive sulfide ore bodies are all reserved in the acid volcanic systems, mainly associated with the barite, secondary with the siliceous rock and jasper. The ore body has the bedding ore sheets and net-vein structures (Mao et al. 2012a).

The **Pulang of Xianggelila** super-large copper deposit was discovered in the Yidun–Zhongdian block of the Western Sichuan Province, which copper reserves reach 6.5 million tons (Ren and Li 2007). That is a porphyry-type deposit. That deposit is developed in the NW trending mica quartz monzonite intruding the Late Triassic Tumugou Formation. According to the molybdenite Re–Os isochron age, the metallogenesis occurred at the period of 213 ± 3.8 Ma, which was resulted by Triassic intraplate deformation. After that it also had some metallogenic actions.

On the western boundary of Yangtze plate, i.e., east side of Jinshajiang–Red River collision zone, in last few years, a giant gold polymetallic deposit has been explored in **Beiya**, Heqing, Yunnan. That hydrothermal deposit is related to the Cenozoic-rich alkali porphyry (quartz albitophyre, quartz syenite porphyry and breccia) intrusion. The deposit is the porphyry, skarn and hydrothermal vein-type deposit and located in the NS trending syncline. The major ore body (KT 52) has the length of 1360 m, depth extension of 540 m, width of 0.86–103 m, and the gold grade of 2.68 g/t. The whole deposit is rich of gold (the average grade of 2.45 g/t) and associated with iron, copper, silver, lead, zinc and so on, which gold reserves reach 200 tons. The forming periods of

Cenozoic-rich alkali porphyry bodies are 36–32 and 26–24 Ma. The metallogenic period was a little later than that of the intrusion bodies. So this deposit should be formed in the Cenozoic intraplate deformation (He et al. 2013).

At the Anninghe, western side of Yangtze plate, there is the **Mianning** (~102.2° E, ~28.3° N) rare earth metallogenic zone (Xie et al. 2005), with the length about 150 km and width of 15 km, which is distributed along the Anninghe fault zone. The **Maoniuping** alkali rock–carbonite-type rare earth deposit is the most famous one. The rock formation and metallogenic periods are mainly in Oligocene–Early Miocene (31.7–15.28 Ma; Mao et al. 2012a). That deposit was obviously controlled by the long-distance effect of the Pacific plate westward subduction and compression, which led to the near NS trending fault showing the compression-shear, the fault surface dipping to west with a high angle, thus caused the deep ore alkali fluid migrating to the upper crust and forming the deposit. The ore intrusion bodies are mainly the quartz-alkali syenite or quartz syenite and their derived alkali rock veins. They all formed in the similar compression-shear fault and fracture systems. The Maoniuping rare earth metal deposit is, not only enriched with the light rare earth, but also with the heavy rare earth (europium, yttrium). The rare earth minerals are so simple that the ore is easy to exploit, beneficiate and smelt. In the ore, about 90% rare earth elements occur as the existence form of rare earth minerals, in which the bastnaesite accounts for 80–97%, the light REEs up to 97.48–98.81% (Shi 1992).

In **Shimian** County of West Sichuan Province, there is a Neoproterozoic (1000–800 Ma) as best deposit (Wan et al. 1988). The mafic and ultra-mafic intrusion bodies are distributed in NE trending with a high-dipping angle. They may be influenced by the Jiangnan Neoproterozoic collision zone [23]. However, it can be confirmed that deposit is not controlled by the later-period Anninghe or Daduhe faults.

On the northern side of Michangshan–Anjiamen fault, Southern Shaanxi and northern side of Yangtze plate, Zhenan–Xunyang basin, there is the **Jinlongshan** gold deposit, which belongs to a Carlin-type hydrothermal gold deposit (Yang et al. 2012b). The gold deposit is located in the EW trending anticline composed of the Devonian clastic rocks and carbonite, in which 26 ore veins have been discovered. The Ar–Ar isotopic age in the ore vein is 233 ± 7 Ma, indicating the gold ores to form in the Triassic intraplate deformation of the Yangtze plate. The gold geological reserves are about 80 tons.

On the northwestern border of Yangtze plate, Eastern Kunlun, there is the **Hutouya** polymetallic deposit, which is a large skarn-type copper, lead and zinc deposit (Shu et al. 2012). That deposit is closely related to the adamellite intrusion that intrudes into the Paleozoic sedimentary system. The main ore zone is distributed in the EW trending. The Re–Os isochron age of molybdenite is 230.1 ± 4.7 Ma, and that deposit was formed in the Triassic syn-collision period.

On the Western Yangtze plate, south from Kunyang and Huaning of Yunnan Province, north to the Mianxian of Shaanxi Province, Shennongjia of Hubei Province, west from the Xichang of Sichuan Province, east to the Shimen of Sichuan Province, in the Late Neoproterozoic–Early Cambrian, there had been a shallow sea carbonite platform, and in its surroundings, there had been a shallow sea or ocean. The phosphorus is concentrated in the rich phosphorus (biota) sea basin of continental boundary and at the continental slope, to form the rich phosphorus black shale–carbonite system. Of them, the **Kunyang** of Yunnan (102° 34′ E, 24° 44′ N; Yang et al. 2011) and **Dongshanfeng**, Shimen of Hunan (E110° 30′, N29° 53′) giant phosphorus deposits are the most famous ones.

4.1.4.2 The Giant Ore Fields and Deposits in Qinling–Dabie–Jiaonan–Hida Marginal Triassic Accretion–Collision Zone [24]

At Southwestern Qinling collision zone [24], besides the Maqu–Lueyang great fault in the **Yangshan** areas of Wenxian, Gansu Province (Fig. 4.53), there is a supergiant carlin-type gold deposit, with the proved reserves of 308 tons. The Yangshan gold deposit is located in the Anchang River–Guanyinba fault zone, and its secondary fault in ore area is 30 km long and several hundred meters wide, with ENE trending, northward dipping and dip angles at 50°–70°. In the fault zone, there are the secondary folds and shear zone (Fig. 4.53). The wall rocks are mainly composed of phyllites. The metallogenic period is determined by the zircon SHRIMP U–Pb method as 197.6 ± 2.2 Ma in the quartz veins of micro-dissemination-type ore, namely the Early Jurassic (Qi et al. 2003, 2004a). The Qinling collision zone was formed during the post-collision period in Triassic. Although the gold deposit is located in the collision zone, in Early Jurassic that area was changed into intraplate setting. Due to the regional WNW trending compression and shortening, the WNW trending faults and fractures showed the extension-shear; thus, it could be very suitable for the ore fluids to migrate and reserve.

Similar to the above deposit, nowadays the **Liba** giant gold deposit has been discovered in Lixian, Gansu Province, which is developed in Devonian System, nearby the granite intrusion. That is a hydrothermal vein-type deposit, and the orientations of ore veins are mainly WNW trending. The isochron ages of mica and quartz in ore veins are 216.4–210.6 Ma (Feng et al. 2003), i.e., formed in syn–collision period. In Fengxian, Shaanxi Province, there is the **Erlihe** giant gold deposit, which is also reserved in Devonian System, the granoporphyry and dioriteporphyry veins are intruded in the surrounding of ore bodies. The deposit is of hydrothermal

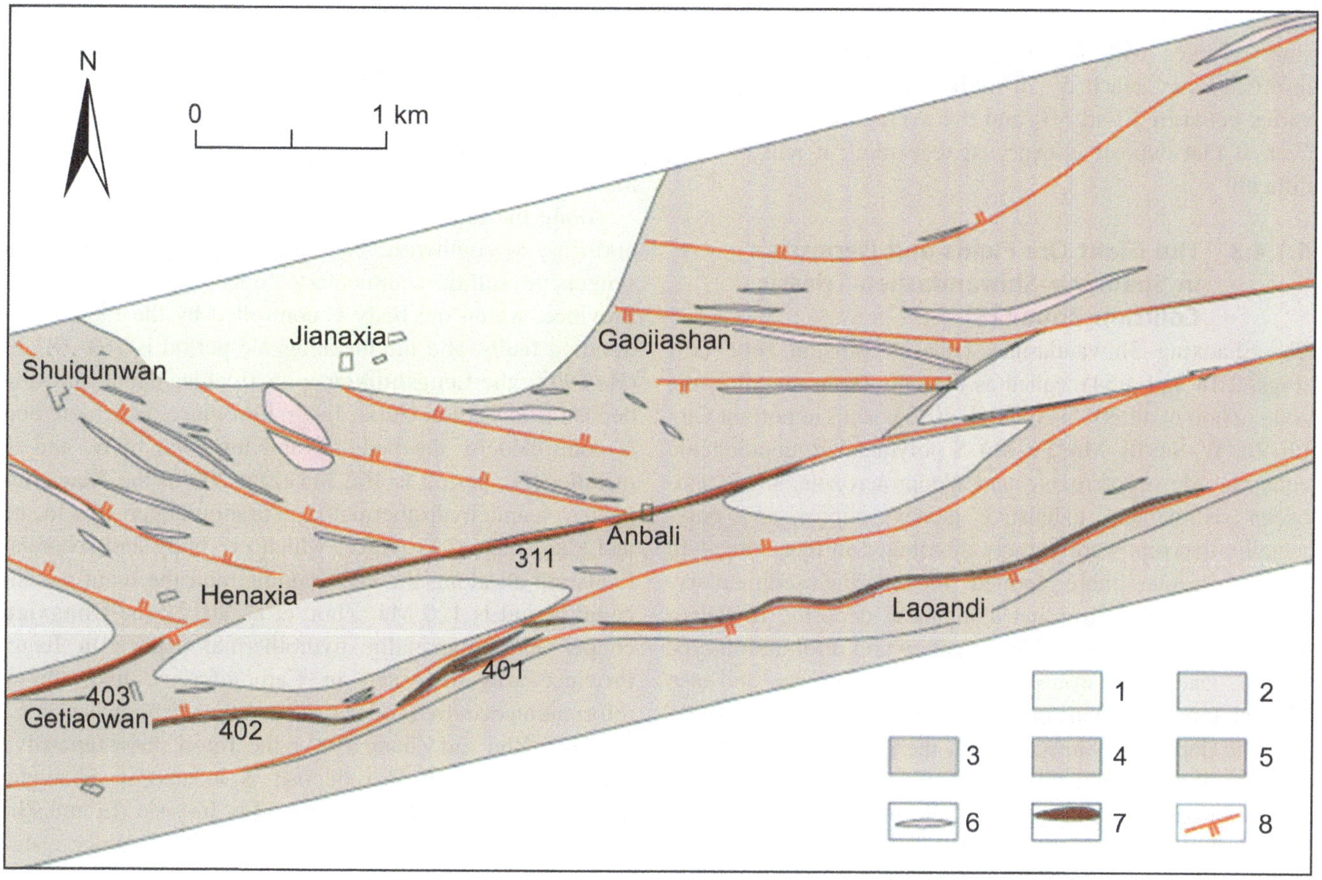

Fig. 4.53 Geological sketch of Getiaowan–Anba, Yangshan gold deposit. 1. Quaternary loess; 2. Middle Devonian carbonate rock; 3. Middle Devonian sandstone and phyllite; 4. Middle Devonian phyllite; 5. Middle Devonian carbonate and siliceous rocks; 6. granoporphyry; 7. gold body and its number; 8. fault (Modified from Qi et al. 2004b)

type, and the ore veins are mainly NE trending. The Rb–Sr isochron age of sphalerite in the ore vein is 220.7 ± 7.3 Ma. The **Wenquan** molybdenum deposit in Wushan, Gansu Province, is reserved in NS trending adamellite porphyry body. The molybdenite Re–Os age in the ore vein is 214.1 ± 1.1 Ma (Zhu et al. 2009). The **Baguamiao** gold deposit in Fengxian, Shaanxi Province, is a super-large deposit (Zhang et al. 2001) belonging to a hydrothermal quartz vein-type deposit, which NE trending ore body is very rich in golds. Its Ar–Ar method isotopic age in the quartz is 232.58 ± 1.59 Ma. The **Liziyuan** gold deposit in Tianshui, Gansu Province, is a porphyry-type deposit (Wei 2011), in which ore bodies are mainly in the EW trending. The K–Ar age of sericite is 206.8 ± 1.63 Ma. The **Manaoqiao** gold deposit, in Nanpin, Northwestern Sichuan Province, is a Carlin-type gold deposit. The hydrothermal veins are mainly in WNW and NNE–NE trendings and controlled by compression-shear fractures. In the quartz gas–liquid inclusions of ore vein, the Rb-Sr isochron age is 210 ± 11 Ma (Mao et al. 2012a). All the above six giant ore deposits were formed in the formation period of Western Qinling collision zone, and they were the results of syn-collision metallogenesis.

Recently, in Qinling areas many ^{40}Ar-^{39}Ar isotopic ages have been got from the metallogenic quartz veins in the Triassic intrusions or strata. Thus, some researchers recognized that many deposits were formed in the syn-collision period. For dating the age with ^{40}Ar-^{39}Ar isotopic method, it must be noted: If the gas–liquid inclusions are not taken aim by the laser, the isotopic ratio will be the reserved argons, so the age results will only show the wall rock-forming period, but not the true formation ages of the ore-bearing quartz veins. The author had determined more than ten quartz veins in the Ordovician–Silurian Systems in Qilian Mountains and just determined the ^{40}Ar-^{39}Ar ages for the quartz, all of which were as same as the wall rock's, i.e., about 400–500 Ma. This is the important issues for determining the ^{40}Ar-^{39}Ar isotopic age.

In Fangchang, Henan Province, the **Baishugang** super-large rutile deposit was discovered (Xu and Li 2003). That ore deposit zone has the length of 30 km and the area about 60 km^2. The C + D reserves of weathering-type rutile

(TiO_2) are obtained as about 1.2 million tons, and the possible reserves will be up to 12.39 million tons; the ore resources may reach 57.26 million tons. The rutile grade varies between 2 and 5%, and the average is 2.22%. However, if that deposit is expected to exploit, it will be very difficult.

4.1.4.3 The Giant Ore Fields and Deposits in Shaoxing–Shiwandashan Triassic Collision Zone [25]

The Shaoxing–Shiwandashan Triassic collision zone [25] (Figs. 2.18 and 4.54), called as the Qin–Hang metallogenic zone (Zhou et al. 2015; Xu et al. 2015), is an important Cu–Pb–Zn, W–Sn–Bi–Mo, Fe–Mn–S polymetallic metallogenic zone. In Mesoproterozoic and Neoproterozoic, the paleo-ocean sedimentary exhalative (sedex) copper and polymetallic deposits were densely distributed in those areas. In Neoproterozoic, there formed the marine sedimentary-metamorphic-type iron and manganese deposits. In Paleozoic, there formed marine sedimentary and reformation-type copper, lead, zinc, iron and manganese deposits, and also formed some tungsten, molybdenum, gold, silver and polymetallic deposits, connected with the granitoids. In Middle Triassic, the collision occurred. However, for the tungsten, tin, niobium, tantalum, uranium and polymetallic deposits; they were mostly formed in the post-collision extension period. In Jurassic–Cretaceous, the porphyry–skarn–hydrothermal vein-type copper, lead, zinc, gold, tungsten, tin and polymetallic deposits were formed, which were related to the partial extension of lithosphere and deep basalt diapir. They usually are reserved on the boundary of Mesozoic basins. Most of the ore deposits are distributed on the sides of main collision zone or paleo-continent.

Along the Shaoxing–Shiwandashan collision zone, from northeast to southwest, there is the **Yongping** skarn-type syngenetic sulfide sedimentary copper deposit in Jiangxi Province, which ore body is controlled by the EW and NE trending faults, and the metallogenic period is 163–183 Ma (He 1993); the **Lengshuiken** epithermal lead, zinc and silver deposit is located in Guixi, Jinagxi Province, which ore body is controlled by the bedding and intrusion body, and the metallogenic period is 162 Ma (Zuo 2008); the **Xiangshan** giant volcanic hydrothermal-type uranium deposit is located in Le'an, Jiangxi Province, which ore body and secondary faults are all along the NE trending, and the main metallogenic period is 120 Ma (Zhou et al. 2012); the **Dongxiang** copper and polymetallic hydrothermal deposit in Jiangxi Province was deposited in Carboniferous, hydrothermal reformation occurred in Jurassic, and the ore body is near EW trending (Zhu and Zhang 1981); the **Jiaoli** skarn-type silver, lead, zinc and tungsten deposit is located in Shangyou, Jiangxi, which metallogenic period is Jurassic (Li and Zhao 2004); the **Huken** tungsten deposit is located in Wugongshan, Jiangxi Province, which metallogenic period is Jurassic, and

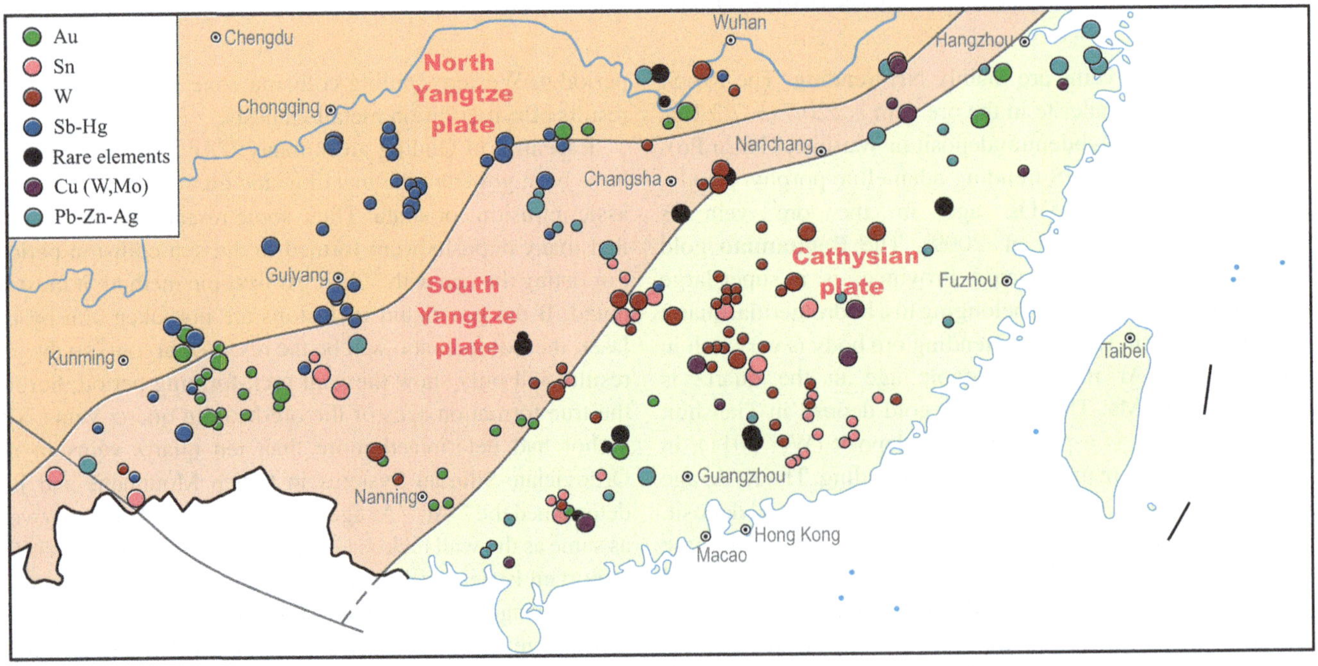

Fig. 4.54 Distribution of main deposits in South China. The collision zone between the South Yangtze plate and the Cathaysian plate is the Shaoxing–Shiwandashan Triassic collision zone, i.e., the Yongping–Shaoguan copper, polymetallic and rare metal metallogenic zone. The fault zone between South and North Yangtze plates is the Jiangnan (Southern Anhui–Northeastern Jiangxi–Xuefeng Mountains–Eastern Yunnan) Neoproterozoic collision zone (Original data from Mao J. W., personal communication)

the ore veins are distributed in NE and NW trending (Zhong et al. 2010); the **Songshugang** super-large tungsten, tin, niobium and tantalum deposit is located Geyuan (414), Hengfeng County, Jiangxi Province, which metallogenic period is Cretaceous, with the K-Ar ages of ore-bearing granite about 124–131 Ma, and the ore veins are mainly NE trending (Huang 1999); the **Qibaoshan** porphyry copper deposit is located in Liuyang, Hunan Province, in which ore body is controlled by the porphyry, mainly near EW trending, and the metallogenic period is 250–227 Ma (Hu and Yang 2003); the **Western Baoshan** porphyry copper deposit is located in Guiyang, Hunan Province (Wang 2005; Liao 2009; Zhou 2011); the **Shuikoushan** large lead and zinc deposit is located in Changning, Hunan Province, which ore body is controlled by EW trending fault, and the U–Pb age of granodiorite is 153.0 ± 0.9 Ma (Zhao et al. 2013; Lu 2013); the **Tongshanling** porphyry copper deposit is located in Jiangyong, Hunan Province, which metallogenic period is Jurassic–Cretaceous, with the ages of porphyry of 139–119 Ma, the deposit is reserved in the NE and SW trending intrusion bodies or the skarn contact zone (Tan 1983); The **Shizuyuan** super-large tungsten, tin and polymetallic ore field is located in Chengzhou, Hunan Province, which metallogenic period is Jurassic (Wang et al. 1987; Fig. 4.55); the **Dabaoshan** giant skarn copper and tungsten deposit is located in Shaoguan, Guangdong Province, which ore granite porphyry and ore body are distributed along the NNW trending, and the metallogenic period is 143–101 Ma (Zhuang 1986); the **Fankou** lead, zinc and silver–polymetallic giant ore deposit is located in Guangdong Province, which metallogenic period is 266–271 Ma, and mainly controlled by near NS trending fault (Sun et al. 2002).

The above ore deposits together constitute the important Yongping–Shaoguan copper, polymetallic and rare metal metallogenic zone (more than 1000 km long) (Fig. 4.54), which are located between the South Yangtze and Cathaysian plates, i.e., near the Shaoxing–Shiwandashan Triassic collision zone [25]. They are related to the granodiorite intrusions intruding the higher position and belong to the magnetic-type granite series with the high-degreed oxidation. The orientation of ore bodies is mainly WNW and NE trending or around the porphyry bodies. Although near the Triassic collision zone, they were mainly formed in Jurassic–Cretaceous, i.e., the post-collision period. Only the Qibaoshan copper deposit in Liuyang Hunan was formed in Late Triassic, i.e., the syn-collision period.

Near the Shaoxing–Shiwandashan collision zone [25], the rather crushed rocks resulted in the magma and supercritical fluids to be formed at the crust–mantle transition zone, which was easy to upward migrate to form the Yongping–Shaoguan copper, polymetallic and rare metal metallogenic zone. The transmit ore faults of above metallogenic zone may be the crustal faults of the Shaoxing–Shiwandashan Triassic collision; however, most of the ore deposits were formed in the Jurassic–Cretaceous, i.e., the post-collision period, which were controlled by the intraplate deformation (Figs. 3.23 and 3.25) with the partial extension. Only a few ore deposits were formed in the syn-collision period.

The **Shizhuyuan** (113° 08′ E, 25° 45′ N) super-large tungsten, tin and polymetallic ore field (Fig. 4.55) is located in Chengzhou, Hunan Province, which is a good example for concentrating the dominant elements to form the deposit in the South Yangtze and Cathaysian plates. That ore field just occurs at the boundary between above two plates, i.e., near the Shaoxing–Shiwandashan Triassic collision zone [25]. However, it must be considered that the Shizhuyuan ore field was also formed after the Triassic collision period, i.e., the Jurassic intraplate deformation period, at the same time the NE trending ore zone was formed. In that ore field, the formation period of Qianlishan ore-bearing granite is about 151–160 Ma, as well as the polymetallic deposit at 150–157 Ma (Mao et al. 2012a). Those are believed to be the production of Jurassic. Most ore types of that ore field belong to the skarn type, and some belong to ore types of massive greisen, vein or net vein. The metallogenesis of Shizhuyuan tungsten, tin, molybdenum and bismuth deposit has the zonation distribution from the intrusion to the outside. The orders are: ① the tungsten and tin in the massive greisen; ② the skarn dense big vein–greisen-type tungsten, tin, molybdenum and bismuth ore; ③ the skarn thin big net-vein tungsten, bismuth and tin ore; ④ the fine-net-vein–greisen-type bismuth and tin ore in marble; ⑤ the fine-net-vein–greisen-type tin and copper ore in mica granite. That ore deposit was related to the crust remelting near the collision zone, and the rich metal elements were easy to concentrate in the Southern Yangtze and Cathaysian plates, then to form a supergiant rare metal ore field.

4.1.4.4 The Giant Ore Fields and Deposits in Cathaysian Plate [26]

In the Central Cathaysian plate (Figs. 2.18 and 4.54), at Southern Hunan, Southern Jiangxi, Western Fujian and Northern Guangdong Provinces, there is the most important tungsten metallogenic zone in the world. At the Eastern Zhuguangshan, Chongyi–Dayu–Shangyou areas, with the area of 7800 km^2, there are 185 tungsten deposits and spots. In the Pangushan–Shangpin–Tieshanlong areas (1100 km^2), there are 111 tungsten deposits or spots. In the above tungsten metallogenic zone, there are many super-large and large tungsten deposits, such as **Xihuashan** super-large tungsten deposit (114° 15′ E, 25° 23′ N; the metallogenic period at 155–140 Ma, Mao et al. 2012a; Fig. 4.56).

The **Dajishan** super-large tungsten deposit is located in Quannan, Jiangxi Province (114° 21′ E, 24° 35′ N; the metallogenic period at 167–159 Ma, Chen et al. 1990; Re–Os method age of 161.0 ± 1.3 Ma, Zhang et al. 2011). For

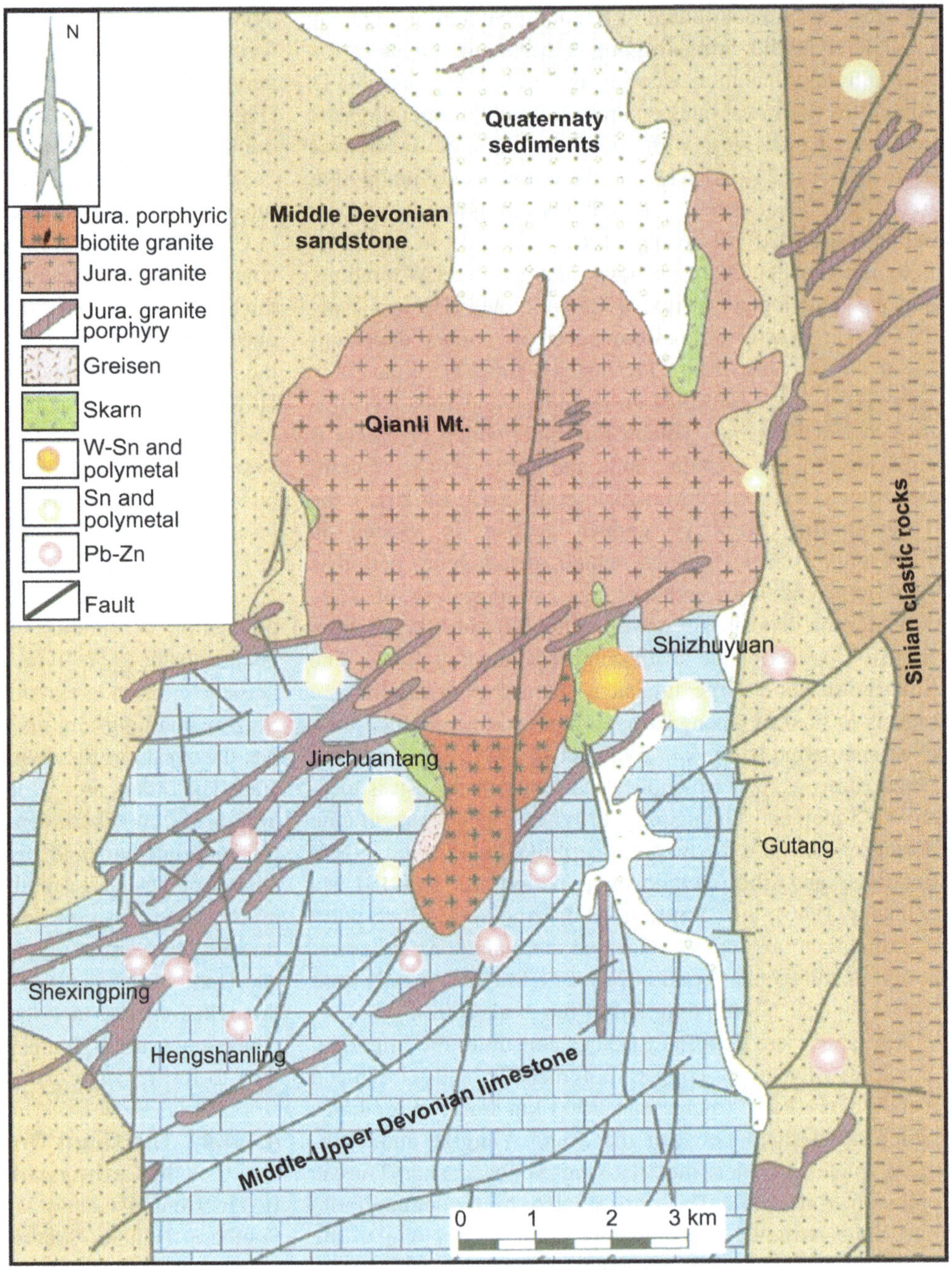

Fig. 4.55 Geological sketch at Shizhuyuan near Qianlishan granite, in Chengzhou, Hunan (Modified from Wang et al. 1987)

that crisis mine, the professors and students at China University of Geosciences have done further researches on the quantitative forecast, analyzed the tectonic stress field method and established the mathematics model to show that ore deposit with the great potential in the deep. According to the recent exploitation, the reserves of Dajishan deposit could be guaranteed to exploit in future 30 years (Zhou 2009).

In Jiangxi Province, there is the **Piaotang–Muziyuan** deposit (the metallogenic period at 155–150 Ma; Zhang et al. 2007b), the **Dalongshan** deposit (the metallogenic period at 156 Ma; Li and Shen 1990), the **Dawangshan**, **Huameishan** and **Pangushan** deposits in Yudu County (the metallogenic period at 157–158 Ma; Zeng 2013), the **Tieshanlong–Huangsha** deposit (the metallogenic period at Jurassic, Feng et al. 1989), **Xiaolong** and **Shangpin** quartz vein-type tungsten deposits in Taihe County (Zhang et al. 1998; Zeng 2013). In Hunan Province, there is the **Xintianling** super-large skarn tungsten deposit (112° 56′ E, 25° 40′ N; at the Northern Qitianling, the metallogenic period at 159–187 Ma; Yin and Wang 1994) and the **Yaogangxian** quartz vein-type tungsten deposit (the metallogenic period at 55–160 Ma; Guo et al. 2010). In Fujian Province, there is the **Xingluoken** super-large tungsten deposit (116° 55′ E,

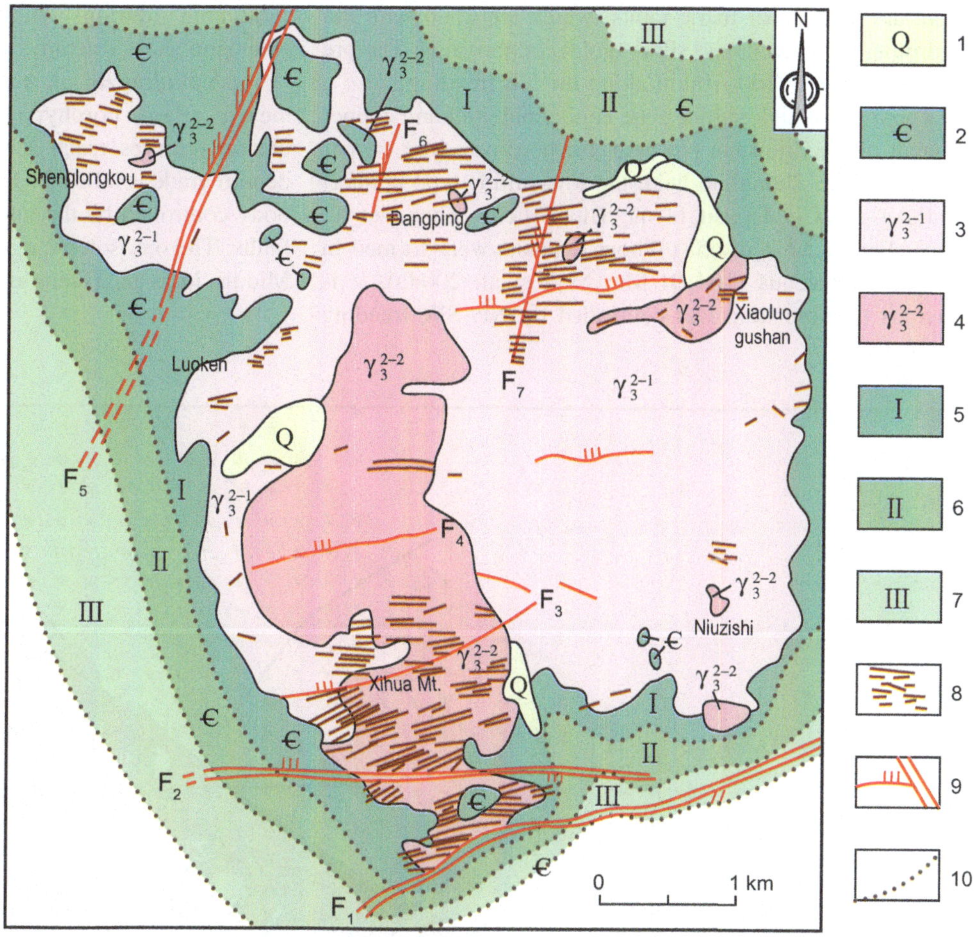

Fig. 4.56 Geological sketch of Xihuashan tungsten deposit in Dayu County, Jiangxi Province. 1. Quaternary sediments; 2. Cambrian system; 3. Cretaceous middle-fine biotite granite; 4. Jurassic middle-fine porphyritic biotite granite; 5. biotite plagioclase hornfel mineralization zone; 6. biotite-muscovite quartz hornfel mineralization zone; 7. the mineralization spot slate zone; 8. tungsten quartz vein; 9. fault; 10. rock facies boundary (After Huang et al. 2005, personal communication)

26° 21′ N; the metallogenic period at 147–156 Ma; Zhang et al. 2008).

All the above tungsten deposits are related to the granite intrusion closely and developed at the top zone or outside contact zone of intrusions. All the deposits belong to the hypothermal type, which tungsten quartz veins are distributed along mainly WNW trending, and developed at the upper part of granite bodies. From the lower to the upper, the ore bodies are ranged from big vein type, fine-vein type to net-vein type. Sometimes, at the contact zone of the intrusions, there formed the skarn-type tungsten deposits. The metallogenic periods of above giant tungsten deposits are the Jurassic, i.e., the Yanshanian tectonic period. At that period, influenced by Jurassic tectonic stress field, the maximum principal compression stress was along WNW trending, and the near WNW trending there existed joints and fractures in the intrusions and their contact zones extended to result in convenient passage ways and reserved positions for the ore fluids to form the tungsten deposits. The Xihuashan tungsten deposit in Dayu County, Jiangxi Province, is a typical example (Fig. 4.56).

The different-scale tungsten deposits or spots could be formed in Asia and China; however, near the boundaries among the Hunan, Jiangxi, Fujian and Guangdong Provinces, i.e., the Nanling metallogenic zone, there are the largest scale and best preservation condition for the tungsten deposits and tungsten metallogenic zone. The formation of Nanling metallogenic zone may be connected by the near EW trending hidden lithosphere fault (Wan et al. 2008; Wan 2011) to resulted in a series of near EW trending fractures and hypothermal action, which are favorable for many rich volatile component (specially the fluorine) granitic bodies and ore fluids to migrate upward, and to form the enrichment of tungsten. The tungsten enrichment is the very important and special characteristics of Cathaysian plate. It may be the reason why not to find similar metallogenic zone in the other areas in globe.

On the southwestern side of Shanghang volcanic basin, Southwest Fujian, there is the **Zijinshan** (116.4° E, 25.1° N) giant copper and gold field. The Cretaceous volcanic and clastic rocks covered on the surface. Between those systems and their underlying Neoproterozoic crystalline basement, there are a series of listric faults, which are filled with the hydrothermal breccia and silver–gold–copper veins. The ore bodies' length is 100–700 m along the NE trending.

There is the high-sulfide-type epithermal gold and copper deposit in the Zijinshan ore field, the porphyry-type copper deposit in the Zhongliao field, the low-sulfide-type epithermal Ag–Au–Cu deposit in the Bitian field (Zhang et al. 2003; Figs. 4.57 and 4.58). Those deposits were formed in Middle Cretaceous (104–91 Ma; Mao et al. 2004a). It is obvious that ore field was controlled by NE–SW trending shortening during Middle Cretaceous to form the extension-shear fracture system.

At **Yuanlingzai** of Anyuan, Southern Jiangxi Province, there is a large porphyry molybdenum deposit. The molybdenum reserves reach 199 thousand tons, and the molybdenum grade is 0.061%. That deposit occurs in the porphyry body controlled by the intersection of NE and NW trending faults. The ore isotopic age is 160–162.7 Ma, i.e., formed in Middle Jurassic (Huang et al. 2012).

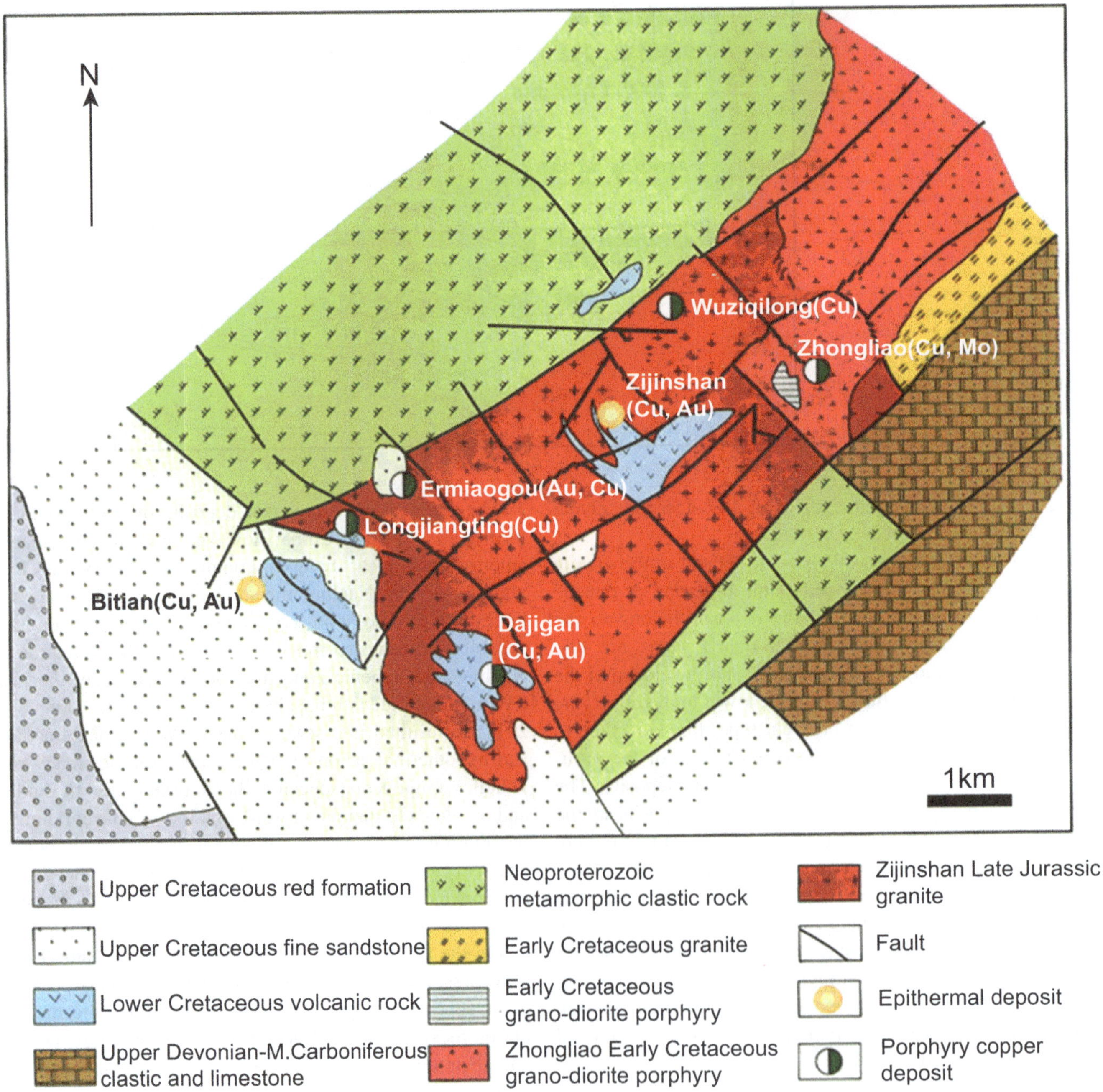

Fig. 4.57 Geological sketch of Zijinshan ore field in Southwestern Fujian (After Zhang et al. 2003)

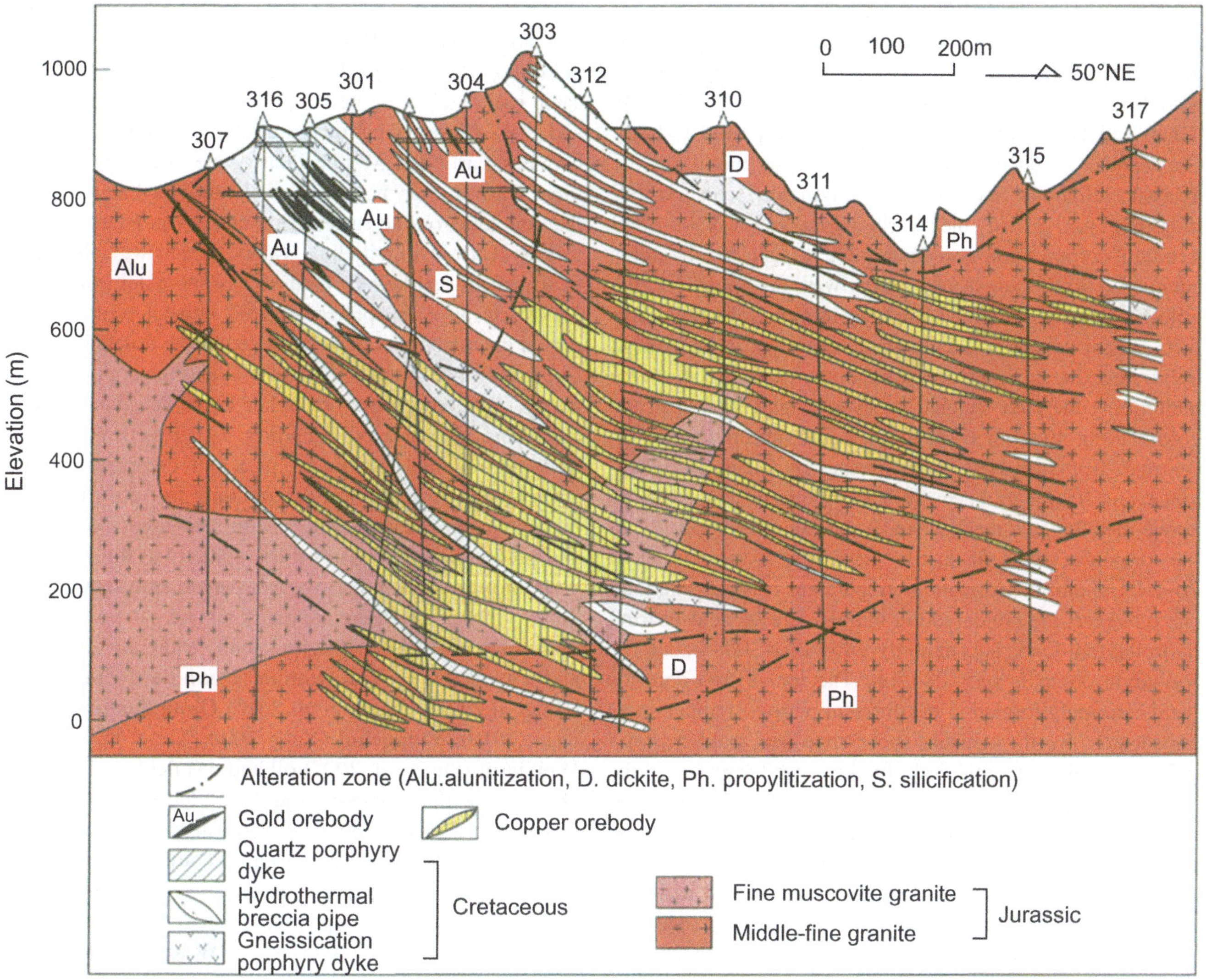

Fig. 4.58 Sketch of geological section of Zijinshan copper and gold deposit in Southwestern Fujian (After Zhang et al. 2003)

At **Dabaoshan** of Shaoguan, Guangdong Province, there is a giant porphyry-type molybdenum, rhenium, beryllium and tungsten deposit. The metal reserves reach 216 thousand tons, and the metal grade is 0.126–0.028%. The ore isotopic age is 164.7 Ma. The metallogenic period is Middle Jurassic (Mao et al. 2004b).

At the contact zones of **Yuanzhuding** porphyry body, Fengkai, Guangdong Province, there is a large porphyry-type copper, molybdenum, sulfur and silver deposit. The metal reserves reach 259 thousand tons, and the average metal grade is about 0.045%. The Re–Os isochron isotopic age is 155.6 Ma, i.e., the metallogenic period is Late Jurassic (Zhong et al. 2010).

In **Jilongshan**, Gaoya County, Zhaoqing City, Guangdong Province, there is a giant porphyry-type molybdenum deposit. The ore body is located at the outside contact zone of the porphyry body. The metal reserves are 183.7 thousand tons, and the average metal grade is 0.03–0.1%. The metallogenic period is Late Jurassic–Early Cretaceous (Huang et al. 2014).

In the Southwestern Cathaysian plate, Gaoyao County, Guangdong Province, there is the giant **Hetai** gold deposit (112.2° E, 23.3° N), which is a great and rich gold deposit in South China. The biggest ore body is No.VII in Gaoyao, which length is 1440 m, with the dip depth of 675 m and the average thickness of 2.98 m. In the past, many researchers usually called that deposit type as "ductile shear zone type." In fact, that deposit is located in the Neoproterozoic migmatite schist; the wall rock alterations are mainly silicification. Near the quartz veins, the silicification is relative strong. The above vein-type ores and silicification-type ores change gradually, not easy to distinguish. It means that they were reserved at the same time, and the same ore-bearing hydrothermal fluids intruded into the micro-fissures to form the deposit. The approximate forming age of migmatite with almost vertical attitude is Triassic (278–224 Ma); however, in the alteration mylonite and the quartz fluid inclusion in the gold body, the Rb–Sr isochron ages are 121.9–129.6 Ma (Chen and Li 1991), i.e., the metallogenic period is Early Cretaceous. It means that ore deposit was based on the Triassic foliations and joints of migmatite schist at NE trending, formed in the Early Cretaceous NNE–SSW trending compression and shortening to get the fractures change to extension-shear, which was favorable for the gold-bearing hydrothermal fluids to intrude and reserve. In fact, almost so-called ductile shear zone-type gold deposit only refers the earlier-period foliations of ductile shear zone undergoing the later tectonic stress to extend, and then, the gold-bearing hydrothermal fluids could intrude and concentrate. Both the Hetai gold deposit of Guangdong Province, and Jiaojia gold deposit of Shandong Province has the above metallogenic mechanism. It should be considered that the formation depth of ductile shear zone usually is about 10–20 km, and the formation temperature is about 500–600 °C; however, the condensation temperature of gold-bearing hydrothermal fluids is commonly lower than 300 °C, and the formation depth is only 2–4 km. The above two deposits were extremely formed in the different periods, temperatures and depths. So the "ductile shear zone-type gold deposit" only shows that the gold body finally is well reserved in the micro-fissures of ductile shear zone, and the ductile shear zone is just a nice wall rock for the gold metallogenesis. In other words, the gold deposit is not formed in the process of ductile shear zone. Do not confuse the periods of rock formation and metallogenesis. If the "ductile shear zone-type gold deposit" only meant that "the gold deposit is reserved in the ductile shear zone" or "the wall rock is ductile shear zone," it would be used properly.

The Late Jurassic–Early Cretaceous volcanic areas of Eastern Cathaysian plate are mainly distributed at Central Zhejiang Province and Central Fujian Province, where a lot of **epithermal fluorite deposits** are reserved. The fluorite veins could penetrate into the volcanic rock system and the upper cover of Cretaceous red clastic system. The forming period of fluorite veins is Late Cretaceous (90–70 Ma), younger than that of volcanic rocks about 40 Ma. The ratios of $^{143}Nd/^{144}Nd$ and $^{87}Sr/^{86}Sr$ from fluorites show that the characteristics are similar to the basement metamorphic system, and some fluorites come from the volcanic and sedimentary rocks. The fluorite veins are oriented in NE or EW trendings. It means that the formation of fluorite deposits was related to the deep recycle (about 5 km) of underground water. During the epithermal fluid migrating upward, they were condensed near the Earth surface to fill into the extension-shear fractures (Chao 1994; Wang et al. 2012b).

At the eastern end of Cathaysian plate, near Jilong of Taiwan Island, the **Jinguashi** high-sulfide-type epithermal gold and copper deposit is reserved. According to the volcanic rock age dating near that deposit, all the metallogenic and rock-forming periods are about 1 Ma, i.e., almost between Early and Middle Pleistocene (Tao 1997). At that time, Taiwan was not the arc-island. The Paleogene subduction zone, on the eastern side of Taiwan Island, changed to East Taiwan longitudinal valley, which was almost vertical, sinistral strike-slip fault with some extension. So that deposit is also an epithermal gold deposit with intraplate partial extension and closely related to the volcanic activity. The gold grade reaches 9.5×10^{-6}, and the copper grade is about 1%; both grades are rather high. The distribution of ore veins are near NS trending. Controlled by the NW trending regional maximum principal compression stress since Quaternary, the near NS trending fractures had shown the extension-shear features, being the favorable passway and preservation position for ore epithermal fluids.

In the East China Sea, Southeastern Cathaysian plate (Xu 2001; Jiang 2014), there are many oil and gas fields, such as **Chunxiao** (Gu 1996), **Pinghu** (Zheng and Gao 2004; Zhang and Zhang 2015), **Tianwaitian**, **Duanqiao**, **Zhujiangkou** (He et al. 2007; Zhong et al. 2010) **Qiongdongnan** (Zhang et al. 2010), **Yinggehai** (Yang 2000; Xie et al. 2012) and **Weixinan** (Sun et al. 2007) oil and gas fields. Although the above oil and gas fields nowadays are located in the sea, the source rocks were formed in Neogene or Paleogene lacustrine or fluvial sedimentary system. Their reservoir structures were mainly controlled by the end of Oligocene near EW trending shortening compression, the Neogene NS trending compression, or the nowadays NW or WNW trending shortening. So in those areas, the faults usually show the features of inversion structures. Due to the differences of basement faults and properties of rock matter, reservoir structures show the different characteristics on the shallow surface.

In summary, it shows that different plates can concentrate different useful elements and form special ore deposits, regardless of their forming times, positions and structures (see the Appendix).

In the North Yangtze plate, it is easy to reserve the iron, copper, vanadium, mercury, gold and rare metals deposits; in the South Yangtze plate, it is easy to reserve the tin, copper, lead, zinc, antimony and tungsten deposits; and in the Cathaysian plate, it is easy to reserve the tungsten, silver, lead, zinc, copper, uranium, gold, rare earth, tin and fluorine. Between the North and South Yangtze plates, there is the Jiangnan collision zone [23], and between the Cathaysian and South Yangtze plates, there is the Shaoxing–Shiwandashan Triassic collision zone [25], so the two zones have mixed the two plates' characteristics. Here, the author would like to summarize it as the following simple sentences: the South Yangtze plate and Indosinian plate may be called as "Rich Tin Plate," the Cathaysian plate called as "Rich Tungsten Plate."

On the boundary of two plates, the super-large tungsten and tin deposit are located Shizhuyuan of Hunan Province. Because the two kinds of elements of tungsten and tin have many similar characteristics, near the boundary of two plates, it is easy to form the tin deposit in the "Rich Tungsten Plate," or to form the tungsten deposit in the "Rich Tin Plate."

4.1.4.5 The Giant Ore Fields and Deposits in Eastern Hindukush–Northern Qiangtang–Indosinian Plate [27]

In the Yushu–Tuotuo River of Qinghai Province, i.e., in the eastern Northern Qiangtang block, the block extends at WNW trending. The Paleogene original WNW trending thrust changed to the normal fault to form a series of WNW trending fault-depression basins, where Paleogene Tuotuohe Formation and Yaxicuo Formation were deposited, and the hydrothermal fluids intruded into the normal fault nearly. In recent years, a large epithermal lead, zinc and silver metallogenic zone (Fig. 4.59; Yang et al. 2011, personal communication), including the **Dongmuzazhua, Muhailahen** and **Chaqupacha** giant lead and zinc deposits have been discovered. The ore isotopic ages are between 33–31 Ma. The giant epithermal lead, zinc and silver metallogenic zone may be caused by the long-distance effect of the Pacific plate westward compression and subduction, which resulted in the pre-existing EW trending faults to be the NS trending extension, then to form the fault-depressions. In addition, the epithermal rich lead, zinc and silver fluids could intrude into the extension fracture zone (Zhang et al. 2013; Fig. 4.59). If only the subduction and compression of Indian plate northward migration were used to explain the formation of those deposits (Deng et al. 2014b), it would be impossible and unconformity to the tectonic stress mechanism. The Indian plate northward compression could not form the EW trending normal fault zone with extension in Northern Qiangtang area.

On the eastern side of Northern Qiangtang block (East Tibet, North Ningjing Mountains, east to the Changdu, between the Jinsha River and Lanchang River), there is a giant porphyry copper and molybdenum ore zone—**Yulong** copper and molybdenum ore zone (Fig. 4.60), which copper geological reserves reach more than 10 million tons. The ore zone is distributed in NNW trending, with the length of 300 km and width of 15–30 km. From north to south, there are the Qinidong, Hengxinpu, Yulong, Manzon, Duoxiasongdo and Malasongdo deposits.

The Yulong giant porphyry copper and molybdenum deposit, in East Tibet (97.7° E, 31.3° N), is located at the plunging crown of Hengxinpu secondary fold, reserved in the adamellite body formed by the partial melting of mantle, which zircon U–Pb age of adamellite is 43.6 Ma. In the rock body, there develops the porphyry deposit, and at the contact zone between rock body and carbonate rock develops the skarn-type ore deposit (Fig. 4.60). The molybdenite Re–Os age is about 40.1 Ma in the copper ore (Ma 1990; Tang and Li 1995; Liang et al. 2008). The age difference between intrusion and metallogenesis is not obvious.

The Yulong copper and molybdenum metallogenic zone (Fig. 4.60) is located at the eastern part of Northern Qiangtang block (the crystalline basement may be formed at 800 Ma). Between the Jinshajiang–Red River [31] and Shuanghu [32] Triassic collision zones, those two collision zones are dipping to northeast, which should be the result of subduction and compression from the Tethys Oceanic plate in Triassic at first (Deng et al. 2014a, b). However, for the Yulong copper and molybdenum metallogenic zone in the eastern part of Northern Qiangtang plate, its metallogenesis was resulted from the Paleogene intraplate deformation, which was formed at the 200 million years later. It is so different from the porphyry copper of Andes Cenozoic subduction zone. The Yulong metallogenic zone is formed in extremely different tectonic background. In Paleogene, although that area began to be influenced by Indian plate northward subduction, the Indian plate subduction could not affect the Northeastern Qiangtang block. The Yulong areas could not be located on the Indian plate at that time.

It is more important that the Pacific plate had undergone the long-distance effect of westward subduction during Paleogene and got the NNW trending Jinshajiang fault zone to show the sinistral strike-slip in Paleogene (Zhong 1998; Wan 2011), then to form many WNW trending extension-shear crust faults (Fig. 4.60). Those contractions of WNW trending faults and NNW trending host faults caused the favorable positions for the porphyries forming. Some researchers guessed the metallogenic materials of Yulong metallogenic zone maybe come from oceanic plate and mantle, but till now nobody has found any believable evidences. In addition, there are a series of NNW trending dextral range en echelon folds (at middle part of Fig. 4.60), which were formed in Neogene (Fig. 4.32), as the result of Indian plate subduction and collision. At that time, the faults of Jinsha River and Lanchang River areas all showed the dextral strike-slip features (Zhong 1998).

The Yushu of Qinghai–Jinding of Yunnan epithermal lead, zinc and silver metallogenic zone is located in the Northeastern Qiangtang block. From the Northern Qiangtang extending southward, between the Jinsha River and

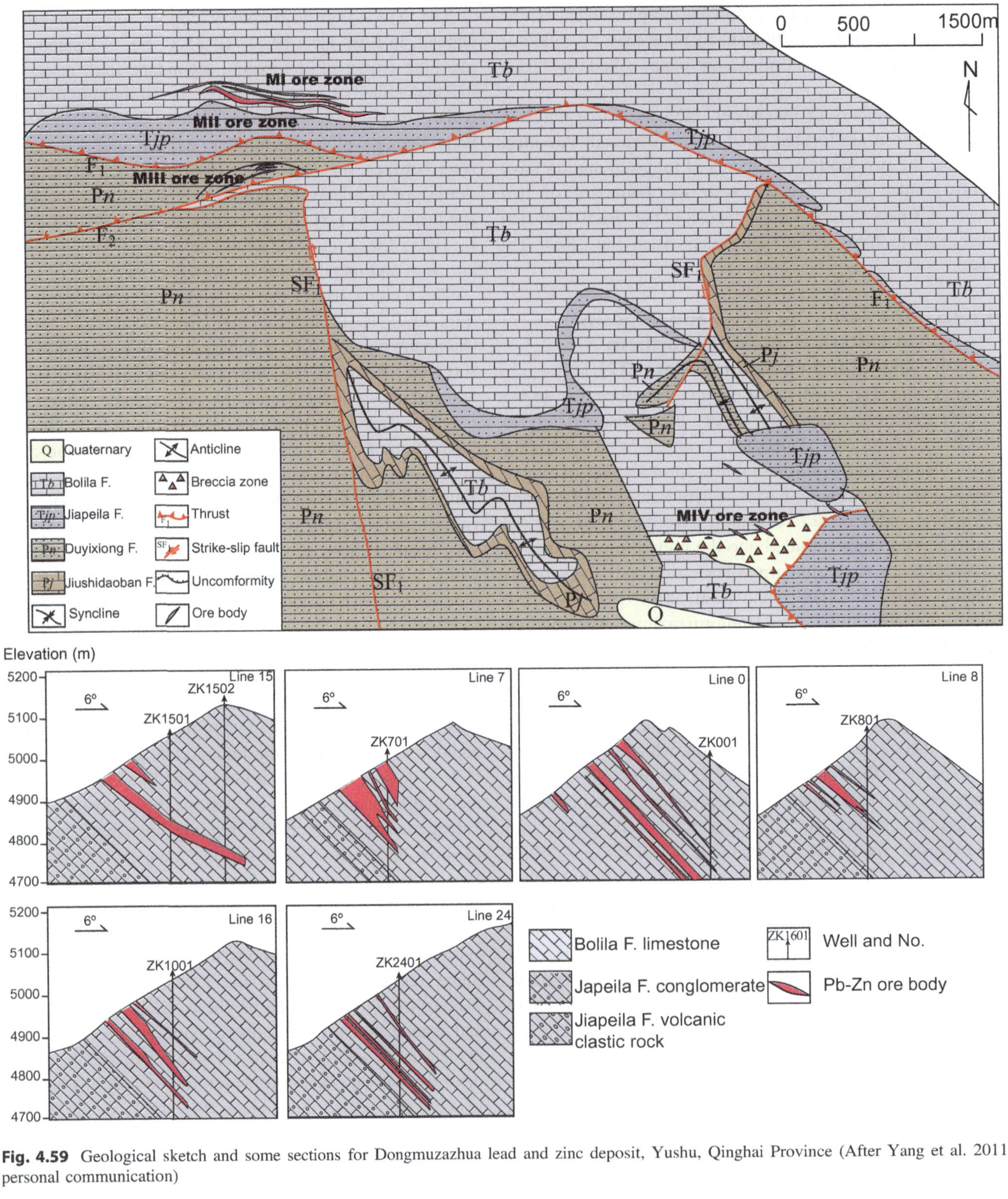

Fig. 4.59 Geological sketch and some sections for Dongmuzazhua lead and zinc deposit, Yushu, Qinghai Province (After Yang et al. 2011; personal communication)

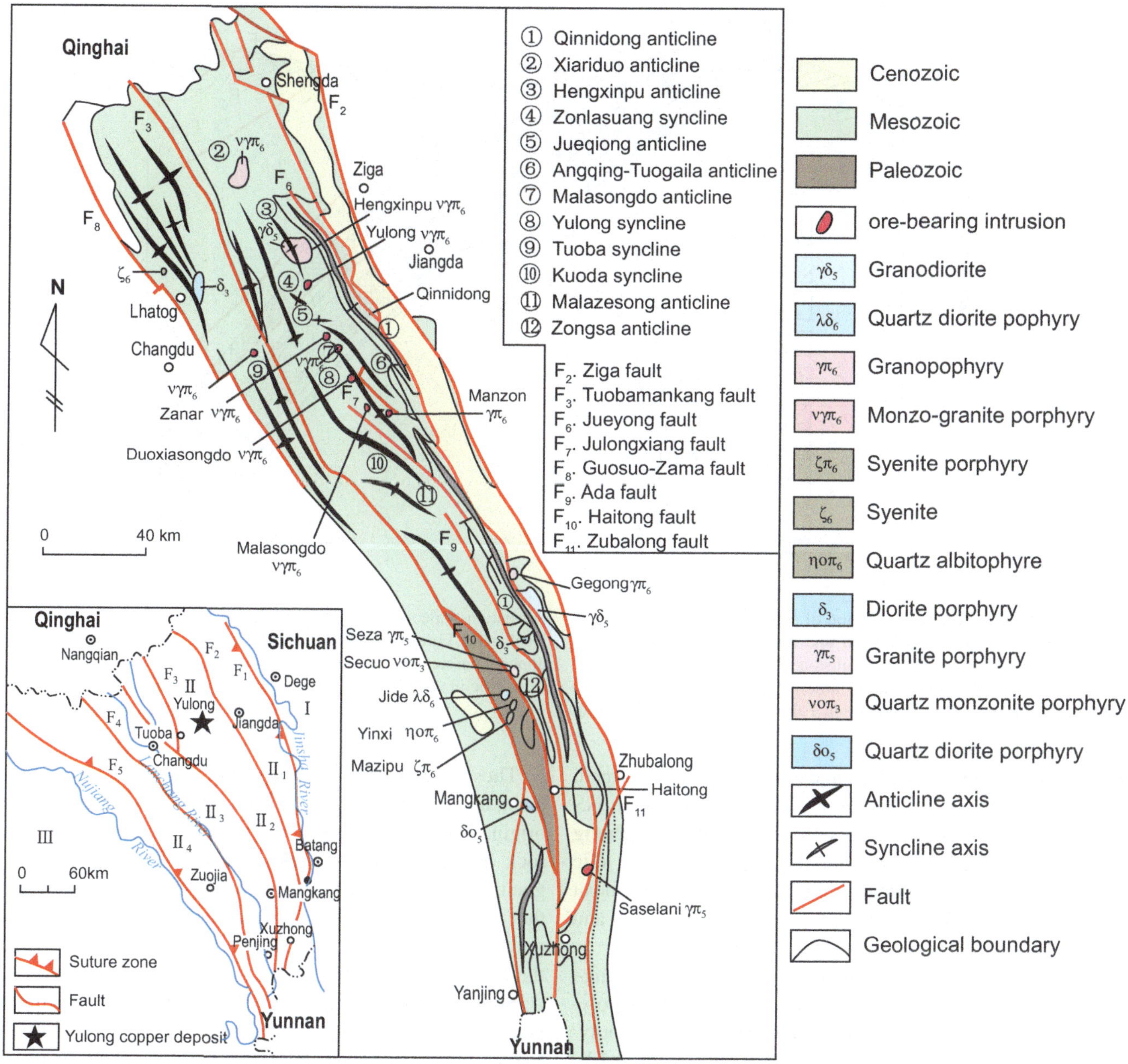

Fig. 4.60 Geological sketch of Yulong porphyry copper metallogenic zone (Modified from ang and Li 1995)

Lanchang River areas, across Eastern Xizang Autonomous Region to Lanping–Simao areas of Western Yunnan Province, there is the **Jinding** super-large lead, zinc and polymetallic deposit near NS trending in Lanping, Yunnan Province (Fig. 4.61).

The lead and zinc reserves reach 16 million tons; the lead and zinc grades are 8.44% (Zhu 1999). Its metallogenic period is Paleogene at 57–23 Ma (Xue et al. 2004, 2007). That deposit was formed almost at the same time with the Yushu–Tuotuohe areas. The reservoir strata are the Paleocene, Cretaceous and Jurassic Systems. Suffering Paleogene epithermal process, the celestine and metal sulfide could be turned into the calcite cementing and then concentrated into the ore. The Jinding is located in the south extension areas of Qiangtang block. The block extension orientation and the tectonic line were changed as NS trending, in Paleogene, influenced by the long-distance effect of the Pacific plate westward compression and subduction; those deposits were controlled by the overturn strata and thrusts, and the ore bodies were reserved on the sides of thrust zone (Fig. 4.61).

Near the Jinding deposit, there is the **Baiyangpin** large copper, cobalt and silver deposit. At the end of Paleocene–Early Eocene, the copper ores mainly formed, at the end of

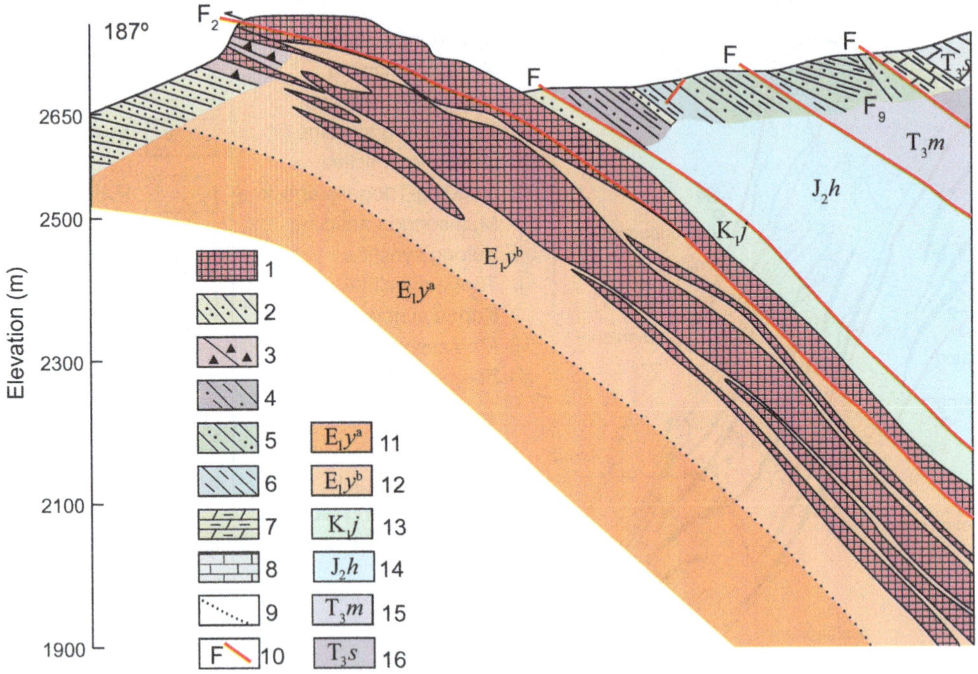

Fig. 4.61 No. 12 exploration section of Jinding lead, zinc and silver deposit, Yunnan Province. 1. Lead and zinc body; 2. sandstone; 3. conglomerate; 4. mudstone with siltstone; 5. siltstone–mudstone; 6. mudstone; 7. micritic limestone; 8. limestone; 9. strata boundary; 10. thrust; 11. lower Yunlong Formation, Paleocene; 12. upper Yunlong Formation, Paleocene; 13. Jingxing Formation, lower Cretaceous; 14. Huakaizuo Formation, Middle Jurassic; 15. Maichuqing Formation, upper Triassic; 16. Sanhedong Formation, upper Triassic (Modified from Xue al. 2004)

Eocene–Early Oligocene, lead and zinc ores mainly formed. The ore bodies are mainly reserved in the thrust with NS orientation (Tian 1997). Besides, there is the **Baiyangchang** middle-scale copper, silver and polymetallic deposit (He 1987; Zhao 2000), **Jinman** middle-scale copper deposit (reserved in the NS trending fault zone; He et al. 1998) and so on.

To sum up, the Eastern Hindukush–Northern Qiangtang–Lanping–Simao block [27] is really a great-scale polymetallic metallogenic zone. Why are there so many polymetallic deposits? Originally, that block was a near EW trending block, due the Indian plate northward subduction and the long-distance effect of the Pacific plate westward subduction and compression, the Eastern Hindukush–Northern Qiangtang–Lanping–Simao block turned to near NS trending, and that block changed into a shallow and long block, and resulted in strong intraplate deformations and the rocks to be crushed, thus to form a very favorable condition for deep magma and ore fluids to migrate upward to the shallow, finally to form the Yulong, Yushu, Jinding and Baiyangchang ore deposits. (Deng et al. 2014a). Those ore deposits were formed in Paleogene and Early Neogene; the forming depth of ore deposits and the uplifting ranges (about 3000–4000 m) are almost same, so those deposits could be exposed near the Earth surface.

Those ore materials were originated from the deep of crust, and that block may be rich in those ore materials in the origin. This is a very good case for the formation of stronger rock deformations and great-scale migrations to form the giant metallogenic belt. However, it should be considered that those ore deposits are located and reserved in the secondary faults and fractures, and no deposit is developed in the regional main faults.

Some researchers considered that ore deposits occurred near the Qingzang Plateau, they may be only caused by the Indian plate northward migration. However, at that time (Paleogene), the Indian plate had never collided with the Asian continent. Based on the structural characteristics of main ore deposits in the Northeastern Tibet, Northern Qiangtang, Lanping–Simao blocks, those deposits may be mainly influenced by the long-distance effect of the Pacific plate westward migration, compression and subduction. The author considers this point view rather reasonable. The Indian plate colliding to the Asian lithosphere mainly occurred at the end of Paleogene and Neogene.

In the Indosinian plate, the Northern and Central Vietnam, there are the large skarn-type iron deposits (proved reserves of iron about 0.86 billion tons), in which the **Shixi** iron deposit is located in Shi River County of Hejing (Wu 2009). In that deposit, the granite intruding period is the

Jurassic, the wall rocks are the Devonian–Permian carbonate rocks and terrestrial clastic rocks, the ore body is distributed along the strata bedding. In that deposit, the average iron content is reached more than 61% and the ore reserves are 320 million ton, and the total reserves reach 544 million ton, which is a hidden deposit. The deposit is located near the sea beach, with the good communication condition. In addition, in the Vietnam there are two volcanic–sedimentary–metamorphic iron deposits with more than 100-million-ton reserves.

In Vietnam, there are 119 tin deposits or spots and 194 tin placers surveyed, explored or exploited (the tin deposits in the basement rock have mainly been mined out; Chen et al. 2010). The highest grade of tin placers deposit reaches 1 kg/m^3, the reserves of SnO_2 are 23 thousand tons, and the WO_3 reserves are 1500 tons. In the **Haut Chhlong** of the Southern Cambodia and at **Duole** of Southwestern Vietnam, there is a giant bauxite deposit with high quality, which is located in the red clay weathering crust of the Neogene–Early Pleistocene tholeiite. The areas are more than 20,000 km^2, the depth of weathering zone is 60 m, and the proved reserves are 405 million tons (Zhang 2009).

In the **Siam** Plateau, Northeastern Thailand (Udon and Nong Khai) and Southern Laos, there are the salt deposits (including the carnallite, rock salt and sylvine). The distribution areas of salt layers reach 24,900 km^2, the geological reserves of salt deposits are about 240 billion tons (Chen et al. 2010). There is the **Sukhothai** kainite deposit, in Central Thailand, which is deposited in the hot and dry climates. It could be explained that Sukhothai block belonged to Asian continental block, but not to the Gondwana tectonic domain at that period.

In Late Permian–Early Triassic volcanic systems, the Northern Thailand, there are the skarn- and hydrothermal-type gold deposits, nearby the fault system, mainly NNE trending. The gold metallogenesis is connected with the porphyry bodies. Of them, the **Huai Kam On** gold deposit is the biggest one, with the proved reserves more than 200 tons (Wu 2011).

In the Central and Western Thailand, there are the tin, tungsten and antimony deposits. The primary tin deposits are mainly cassiterite (SnO_2)–quartz type developed in the inside or outside contact zones of granite bodies, and the tungsten is the associated mineral. The tin metallogenic zone is from the South Yangtze plate through Western Thailand, Eastern Burma and the Malay Peninsula to the Sumatra. That is the longest tin metallogenic zone in the world. The tin reserves account for 42% of the world total reserves. In that area, there are the antimony deposits else, which are mainly the pyrometasomatic type, with the geological reserves about 0.42 million tons (Chen et al. 2010).

The original hypothermal tin deposits (cassiterite–sulfate–quartz vein type) of Malay are developed at the Eastern Malay or the Central Malay collision zone, i.e., in the Indosinian plate. The tin deposits in **Malay Peninsula** are the placers (Fig. 4.62). The reserves and output of placers tin are more than 95% of Malaysia total, the placers tin are mainly from the collision zone and the basement rocks of eastward Indosinian plate (Hutchison and Tan 2009). From Western Yunnan, China, through Western Thailand, Eastern Burma and the Malay Peninsula to the Sumatra, and from the South Yangtze plate and Indosinian plate to the Malay Peninsula–North Sumatra, there distributes the main tin metallogenic zone in the globe. The South Yangtze plate and Indosinian plate are the "Rich Tin Plate." However, the tin and tungsten deposits concentrating in the bedrock occurred in the Jurassic–Cretaceous intraplate deformation period, and especially in the metallogenic period of the partial intraplate extension, for example, the **Sungai Lembing** cassiterite–quartz vein deposit (No. 6 in Fig. 4.62) is the biggest one in the Malay Peninsula. Because the exploiting and mining costs are too high, and the market demands are not enough, the deposit has been closed. The tin placers on the Malay Peninsula are majorly reserved in the Pliocene and Early Pleistocene alluvial and fluvial fan facies sediments. Some tin placers are intergrowth with the tungsten placers. The tin placers are distributed in six mining areas: **Dinding, Bangka Island, Kuala Lumpur, Phuket Phangnga Takuapa, Pahang–South Terengganu** fields (including Sungai Lembing field) (Fig. 4.62). In 2004, the output of tin reached 2578 tons (Hutchison and Tan 2009). The Dinding, Kuala Lumpur and Phuket Phangnga Takuapa mining fields are located at the Western Malay Peninsula, i.e., in the Sibumasu plate [34]. The Pahang–South Terengganu fields (including the Sungai Lembing deposit) are located at the Southern Indosinian plate [27]. The Bangka Island and Billiton fields are located on the eastern side of Sumatra Island, which belongs to the Sunda plate [51].

Besides, according to the statistics of 2004, on the Malay Peninsula there produced the golds of 43.48 tons, iron ores of 510,000 tons, kaolines of 220,000 tons and micas of 136,000 tons (Hutchison and Tan 2009)

To sum up, based on the very similar ore deposits and same ages of crystalline basement shown in the Yangtze plate, Eastern Hindukush–Northern Qiangtang block, and Indosinian plate, they could be a same plate in the origin. Later, they had undergone the different tectonic movements and different structural deformations, forming the different styles, positions and characteristics.

4.1.4.6 The Giant Ore Fields and Deposits in South China Sea Cenozoic Fault-Depression Basin [28]

On the shallow sea around the South China Sea fault-depression basin, the abundant oil and gas resources have been discovered since 1970s (Wei et al. 2005; Zhou

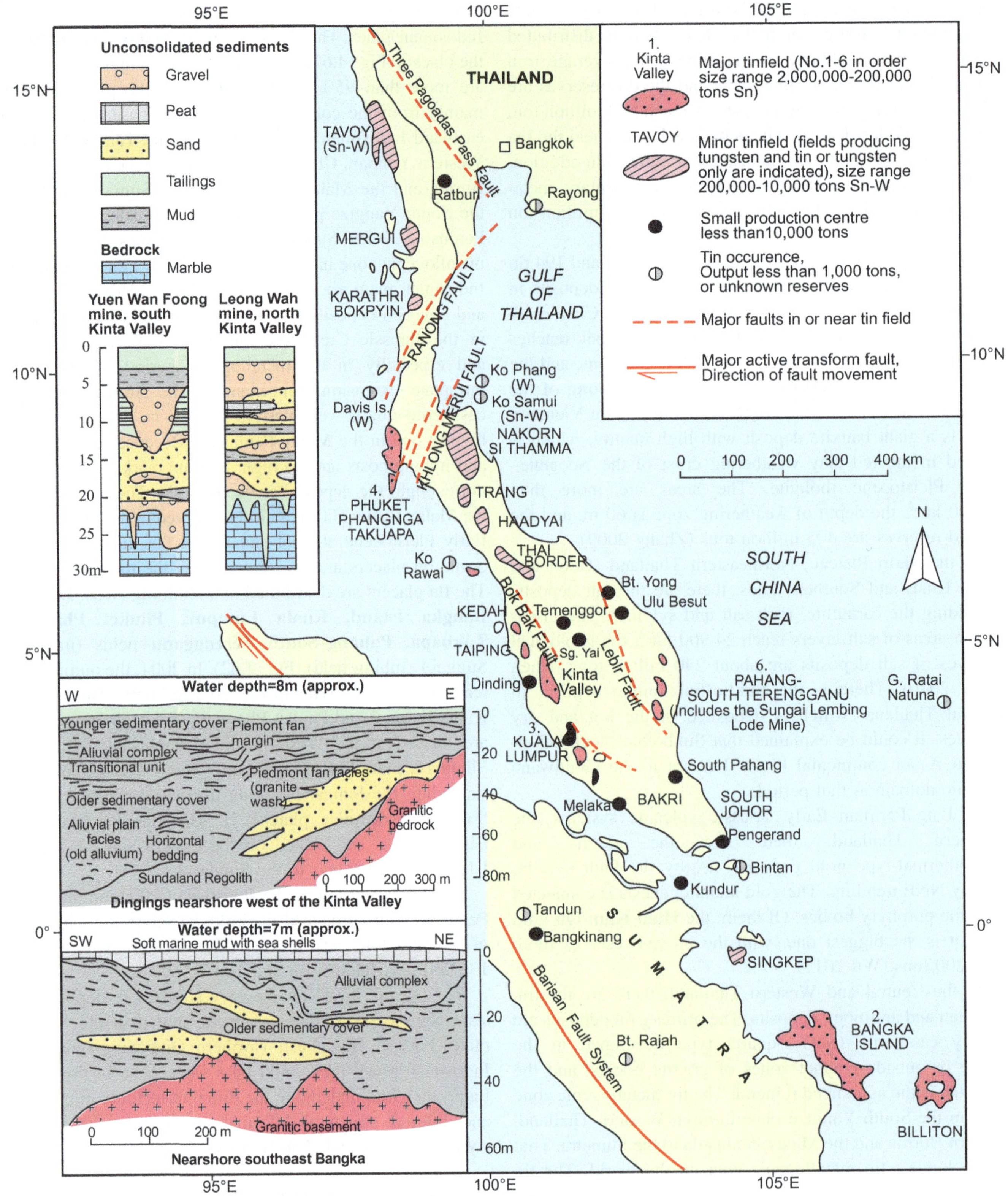

Fig. 4.62 Tin deposit distribution in the Malay Peninsula (After Hutchison and Tan 2009)

et al. 2005; Chen et al. 2009). The being exploited **Xisha** oil and gas field is a good example. At the Liyuetan, Wan'an, Zhongyue, and Zhenghe basins, there are the plentiful oil and gas resources.

In the South China Sea Cenozoic fault-depression basin, the total areas of bearing oil and gas basins are 410,000 km^2. On the sea boundary, those areas are about 260,000 km^2, the estimated oil and gas resource reserves are about 34.97 billion tons. Their main petroleum source beds are the Eocene and Middle Miocene mudstone, clay layer and siltstone. The bulk of reservoir rocks are mainly the Late Eocene, Oligocene and Miocene sandstone, turbidite and reef limestone, and the cover rocks usually are the Miocene and Pliocene mudstone (Wang et al. 2005b). In the June 2006, the **Liwan** oil and gas field was discovered, its exploited gas reserves are about 4000–6000 billion ft^3. That is the biggest gas field discovered on the sea in China. In the bathyal zones of the South China Sea, there are abundant gas hydrates ("combustible ice"), the research and exploration are in progress (Liu et al. 2005; Sa et al. 2005). In the recent, more than 200 petroleum companies of the West countries have explored here.

4.1.4.7 The Giant Ore Fields and Deposits in the Palawan–Sarawak–Zengmuansha Block [29]

In the marginal sea of the northern side of the Palawan–Sarawak–Zengmuansha block [29], there are the plentiful oil and gas resources. The Zengmu, Negara Brunei Darussalam–Sabah, Malaysia basin, Northwest Parawan and Andutan basins all reserve important bearing oil and gas fields. The Negara Brunei Darussalam–Sabah Malaysia have explored and exploited those oil and gas fields.

4.1.4.8 The Giant Ore Fields and Deposits in the Western Hindukush–Pamir–Kunlun Late Paleozoic–Triassic Accretion–Collision Zone [30]

At the western part of the Western Hindukush, Northwestern Pamir Plateau, north to the Pyandzh River, Tajikistan (72°–75° E, 37°–38.5° N), there are abundant metal deposits (Fig. 4.63). The antimony, lead and uranium reserves are ranked at the first in the CIS (Commonwealth of Independent State). The reserves of gold resources are estimated to more than 500 tons.

The **Great Kanimansur** (Большой Канимансур) silver field is the famous one, located in East Tajikistan, which part extends into the Karaganda–Kyrgyzstan Early Paleozoic accretion–collision zone [7]. There is a silver, gold and fluorite deposit, in which the silver ore reserves are 39 thousand tons, the silver reserves about 7.5 thousand tons, and the gold reserves about 429.3 tons (Zhang 2009). The ore body is net-vein, with the size of 2200 m × 800 m

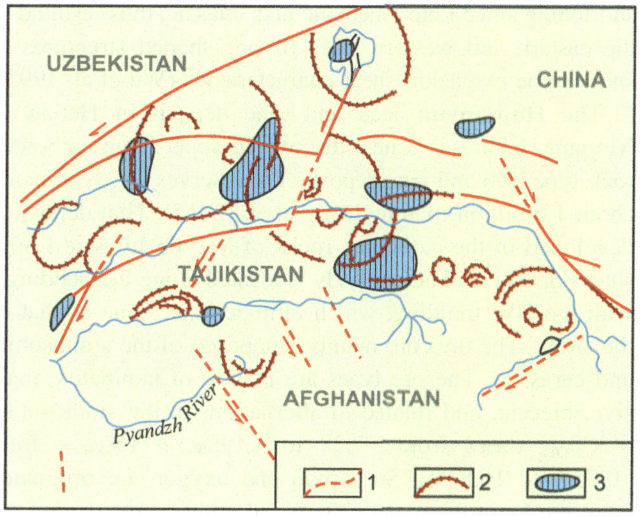

Fig. 4.63 Sketch of Tajikistan regional fault zone, the ring-shaped structure and metallogenic area. 1. Regional fault zone; 2. ring-shaped structure; 3. metal metallogenic areas (After Wu et al. 1993)

200 m. The silver deposit is located at the Northeastern Adlasmansk fault-depression zone and the volcanic sedimentary system that is controlled by the WNW–NW or NE trending ancient faults. They could be formed in the NS trending compression and shortening since Triassic. The total thickness of the Late Paleozoic intermediate-acid volcanic rocks and volcanic tuffs reaches 6 km. That deposit has the vertical differential characters, so during exploiting, it could be mined at the different layers to get the different ores. The ores are composed of the silver, lead, zinc, fluorite, copper and gold minerals. At the deeper position, there is the bismuth. The Great Kanimansur deposit was formed after the volcanic eruption and suffered the hydrothermal alteration, then to develop a silver–porphyry-type deposit.

The antimony reserves are ranked behind China and Thailand, at the third in the world. The **Zelafshang–Jisar** mercury and antimony metallogenic zone, in the middle of Tajikistan, are enriched with the antimony ore, with the antimony reserves about 300 thousand tons. The antimony deposits are concentrated in the **Senko–Majian, Jiriklut** and **Konjorge**. There are three mining areas in the Western Zelafshang–Jisar mercury and antimony metallogenic zone (Chen et al. 2012a). The biggest antimony deposit in the Tajikistan is **Skalinoya** deposit, which antimony reserves account for more 50% of the CIS, up to 36 thousand tons (Wang and Wang 2010).

There are many ring-shaped structures in Tajikistan discovered by remote sensing, which may be related to the hidden intrusion bodies in the deep. The metal ore fields are mainly distributed on the eastern and western sides of ring-shaped structures or the intersection points of regional faults (Fig. 4.63). It is obvious that the ring-shaped structures had been connected with the regional near NS trending shortening since Late Paleozoic and Triassic, thus resulted in the eastern and western sides of ring-shaped structures to show some extension-shear characteristics (Wu et al. 1993).

The **Huosaoyun** lead and zinc deposit in Hetian of Xinjiang, China, is a new discovered super-large carbonate rock-type lead and zinc deposit. The reserves were estimated about 1.6 billion tons by Dong et al. (2015). That deposit is developed in the carbonate rocks of Middle Jurassic Longshan Formation. The ore body is located along the beddings, with the NW trending, which attitude is as same as that of the strata. The ores are mainly composed of the smithsonite and cerussite. The ore types are mainly of laminated, massive, breccia, and related to alternation. In the smithsonite, $\delta^{13}C_{PDB}$ varies from −7.28 to 1.19‰, $\delta^{18}O_{SMOW}$ from 10.78% to 16.81‰. So carbon and oxygen are originated from the hydrothermal fluids mixed with magma water and marine water. The sphalerite Rb–Sr isochron age is 186 ± 6 Ma, i.e., Jurassic. So the Huosaoyun lead and zinc deposit is a new sedex type, in recent that deposit is continued to explore.

4.1.5 The Giant Ore Fields and Deposits in Gondwana Tectonic Domain

4.1.5.1 The Giant Ore Fields and Deposits in Southern Qiangtang–Sibumasu Plate [34]

The northern area of Southern Qiangtang–Sibumasu plate, during Late Permian–Middle Cretaceous and Late Cretaceous–Paleocene (~50 Ma), had undergone the Tethys Ocean plate subduction northeastward (i.e., to form the Bangongco–Nujiang collision zone). In the Baoshan–Tengchong block, there are many Late Cretaceous–Paleocene S-type granite intrusion-related skarn-type Pb–Zn, Sn–Fe deposits, and greisen-type Sn–W deposits. The formation of W, Mo, Ag and Au hydrothermal deposits is related to the granite intrusion (formed at 105–81 Ma). The formation of all those deposits is related to the subduction zone (Deng et al. 2014a, b).

In Western Thailand, during Triassic (250–200 Ma) and Cretaceous (82–77 Ma), there reserved many middle-scale iron, copper, lead, zinc, tungsten and tin deposits (Ridd et al. 2011). The Shan State, Eastern Burma, is the area enriching with metal reserves, at the **Laibiao–Ruimi**, Mandalay, the gold ore reserves reach 3.68 million tons, the average grade of 0.33–4.8 g/t. At the Western Shan State, there is a near NS trending lead, zinc and silver metallogenic zone. That zone could extend north to the Yunnan Province, China, and south to Thailand. That metallogenic zone is about 2000 km long, 300 km wide. The ore reserves from the **Bodwen** polymetallic deposit are 10 million tons, with the lead grade about 5.1%, zinc grade at 4%, silver grade about 93 g/t. In addition, there are many tin, tungsten and antimony deposits in that area (Zhang 2009; Chen et al. 2010). The northern area is enriched with tungsten, the southern area with tin. The primary tin and tungsten deposits are related to the hydrothermal processes near the granite bodies. As to the Western Malay Peninsula, the tin placers have been discussed in the above, here not to repeat.

4.1.5.2 The Giant Ore Fields and Deposits in Gangdise Plate [36]

In the eastern section of the Gangdise plate, there is the **Qulong** super-large porphyry and skarn-type copper deposit (Fig. 4.64). The ore porphyry body and copper deposit are controlled by the WNW trending branch-fault zone and its secondary NNE or NNW trending faults. The ore intrusion porphyry body is the biotite monzogranite porphyry, and the ore bodies are reserved in the porphyry body and its contact zone. The ore porphyry body belongs to calc-alkaline series and with high oxygen fugacity. The isotopic ages of ore porphyry and copper bodies are 16.4–17.58 Ma, i.e., they were formed in Miocene, namely in post-collision period, but not in the collision period. In that deposit, the copper ore reserves reach 10.36 million tons (Mao et al. 2012b).

Along the fault of Yarlung Zangbo Paleogene collision zone [37], there are many similar ore rock bodies, such as **Jiama** (Zheng et al. 2010), **Chongjiang**, **Bairong**, **Tinggong** ore porphyry bodies (Huang and Li 2004). The outcrop areas of porphyry bodies usually are about 1 km^2. The big outcrop areas of porphyry, due to the deep erosion, and the top parts of ore porphyry bodies have been eroded (Mao et al. 2012b). Those deposits were formed in the Neogene intraplate deformation period after the Yarlung Zangbo Paleogene collision. Although the above deposits are located near the Yarlung Zangbo Paleogene collision zone, it is not suitable to express them as the "orogenic zone type." Their forming setting is extremely different from the Andes copper zone that is resulted from the oceanic plate subduction to form the porphyry copper deposits.

At **Xiementong** area, the southern part of East Gangdise plate, there is the Xiongcun Jurassic porphyry-type copper, gold and polymetallic deposit, with the gold proved reserves of 120 tons (Tang et al. 2009; Lang et al. 2014a). The isotopic age of ore porphyry and copper bodies corresponds to Jurassic (molybdenite Re–Os model age of 172.6–161.5 Ma). The ore deposit is developed in the fine-vein zones of the quartz diorite porphyry body or alternation tuff (Mao et al. 2012b; Lang et al. 2014b).

4.1.5.3 The Giant Ore Fields and Deposits in Yarlung Zangbo–Myitkyina Paleogene Collision Zone [37]

In the ophiolite suite on the southern side of Eastern Yarlung Zangbo collision zone, there is the **Luobusha** deposit—the

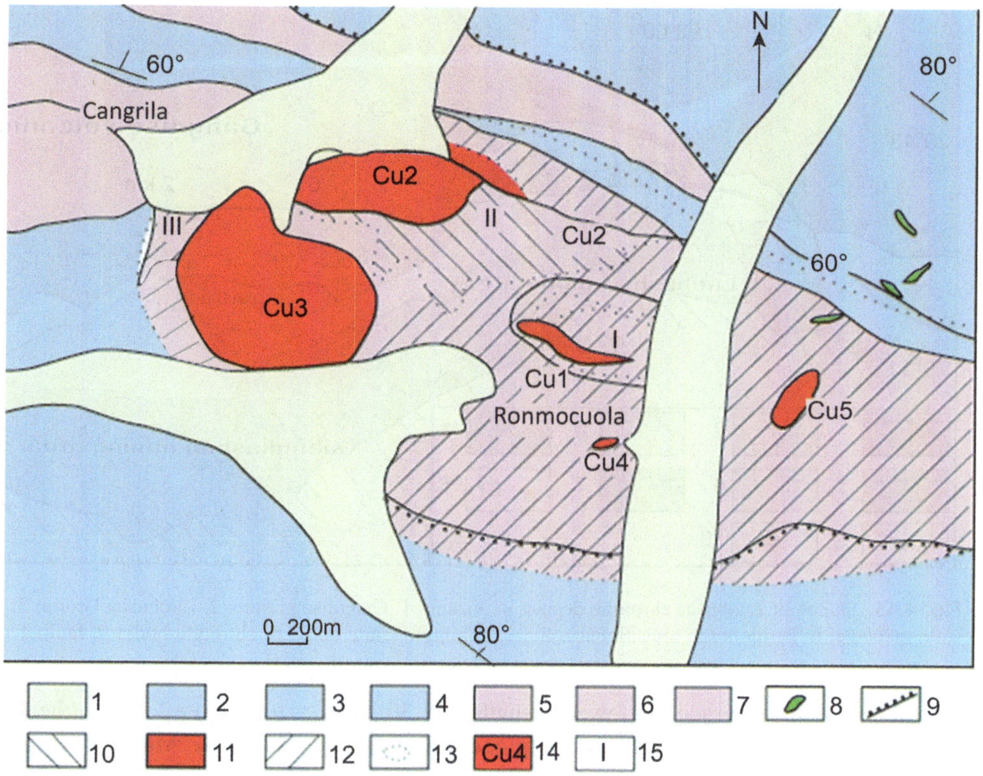

Fig. 4.64 Sketch of the Qulong super-large porphyry and skarn-type copper deposit in Gangdise plate. 1. Quaternary sediments; 2–4. Middle Jurassic rhyolite tuff, andesitic porphyrite and dacite, with the volcanic breccia tuff; 5–7. Miocene biotite monzogranite porphyry, quartz porphyry, granite porphyry and rhyolite-porphyry; 8. rock vein; 9. breccia zone; 10. beresitization; 11. kaolinization; 12. propylitization; 13. alternation boundary; 14. No. of sites of copper ore; 15. No. of porphyry body (Modified from Mao et al. 2012b)

biggest chromite deposit in China. That deposit (Fig. 4.65) is controlled by the south-dipping fault, along which a series of small deposit bodies are developed. The ore deposit was formed in Jurassic–Early Cretaceous (Zhou et al. 2001). The formation age of the ore-controlling fault controlled by the deep fault zone is more than 100 million years older than that of the Yarlung Zangbo collision zone. The ophiolite suites in the Yarlung Zangbo collision zone are the MOR (middle ocean ridge) type; their lava is the breccia basalt, with the characteristics of high TiO_2 (1.30–1.71%, mass fraction), rich in LREE, active and some inactive elements, and rare of partial HFSEs (high field strength elements). The basalt has the typical P-MORB characteristics and is similar to the basalts in the Atlantic Ocean at 45° N middle ridge.

So it was estimated by Yang et al. (2012a) that the original mantle rocks had undergone the partial melting (about 10–15%). That ore deposit should be explored deeply in the future (Wang et al. 2010). Along the Eastern Yarlung Zangbo collision zone, it is believed more chromite deposits to be reserved. Due to the limits of landscapes and geological conditions, it has been explored so slowly.

In the east and west of Burma, there exist two ophiolite suites. The laterite-type nickel deposit was formed by strong weathering. The biggest one is the **Tagaung Taung** nickel deposit, 200 km north to Mandalay. The nickel reserves are 0.8 million tons, which can be open-pit mined. There are many small- and middle-scale deposits near Myikyina (Douting Web, 2012-3-14).

4.1.5.4 The Giant Ore Fields and Deposits in Indian Plate [40]

In India, there are some giant iron deposits (i.e., bedding iron formation, BIF) in the Pre-Cambrian metamorphic volcanic system, which prospective reserves are more than 20 billion tons. They are mainly concentrated in Orissa and Bihar States. The reserves of **Qiliya** iron deposit in the Zingbum are 1.97 billion tons, and their average iron grade is 62–63%. In the **Bihar, Zincbum** area, the iron deposits were formed in the banded haematite jade (BHJ) at 2900–3200 Ma, with the thickness of the rock system up to 305 m. In those system, there still are many sedimentary evidences reserved, which are the sedimentary-metamorphism type. Some scholars considered that the source of iron was related to the submarine volcanic eruption (Zhang 2009).

In **Madhya Pradesh**, the iron reserves are about 3.0 billion tons, in which ore reserves with the iron grade more than 65% reach 0.6 billion ton. In the **Kudleimek** iron deposit, Karnataka State, the iron reserves are 0.7 billion tons, and the total iron ore average grade is 38.6%. In the **Dolimalan** iron deposit, the proved reserves are 0.155 billion tons, and the total iron ore average grade is 64.5% (Zhang 2009).

In the whole India, the iron reserves are about 2.8 billion tons, the prospective reserves are 62 billion tons, in which the reserves of rich iron ores (iron grade more than 65%) are about 1.15 billion tons. They are mainly the haematite and magnetite. The average haematite grade is more than 58%,

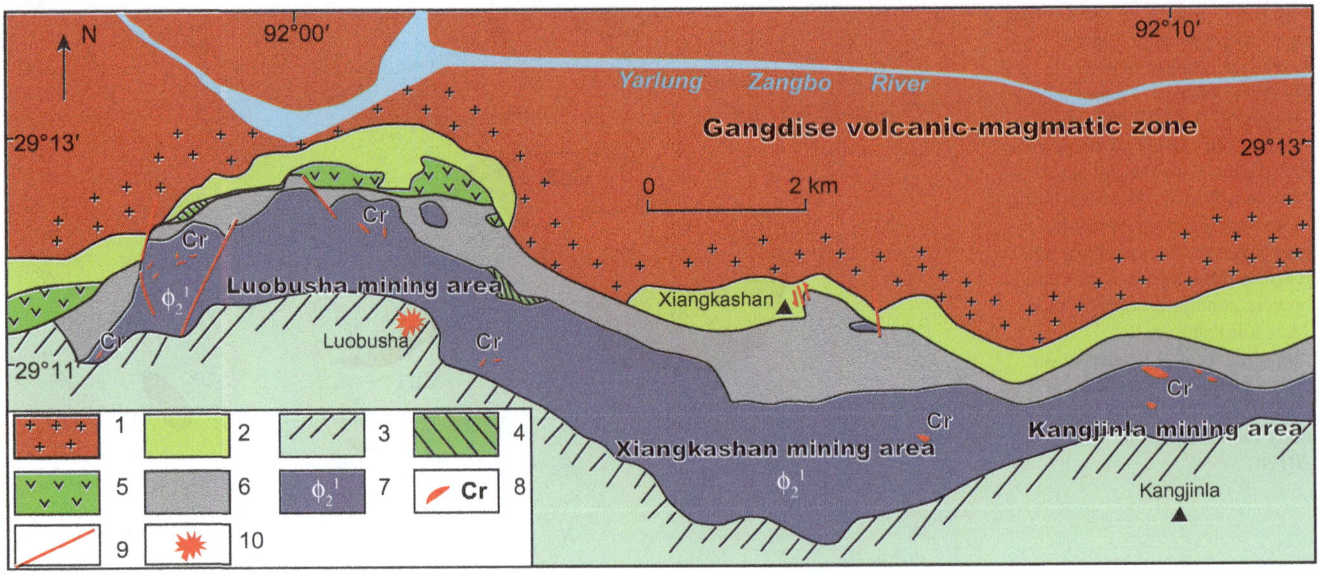

Fig. 4.65 Sketch of Luobusha chromite deposit in Xizang. 1. Gangdise granite; 2. Luobusha Group; 3. Triassic flysch formation; 4–5. ophiolite suite melange; 6. peridotite in transitional zone; 7. orthopyroxenite body; 8. chromite body; 9. fault; 10. mining area (After Yang et al. 2009)

and the magnetite grade is lower, usually as 30–40%, accounting for 8% total reserves of the globe (according to the Duoting Web, introduction to the Indian ore deposits).

More than 90% chromite deposits are concentrated in Dhenkanalt and Kendujhar areas of Orissa State. In **Sujinda Valla** in Orissa, the chromite deposits show the stratified, len-like or pocket-like to reserve in the alternated dunite-peridotite, and to extend along NE–SW trending about 25 km (Tan 1983). Those ultra-mafic rocks intrude into the chalcedony quartzite and bedding magnetite quartzite Pre-Cambrian "super-iron group." There exists the younger ultra-mafic rock—enstatite pyroxenite—which contains few chromite. In **Orissa** State, the chromite reserves are 26 million tons, the respective reserves are 57 million tons, ranking at fifth in the world. The main deposits are distributed in the Proterozoic basic and ultra-basic rock zone of the southeast of Eastern Indian shield. In the 2000, the output of chromite reached 1.71 million tons. In 1999, the chromites of 0.374 million tons were exported from Indian to China. However, the chromite deposits in the ophiolite suites of near Himalayan areas, in the Northern Indian, till now have been never explored (Zhang 2009).

The reserves of rutile (TiO_2) in 2000 year reached 6.6 million tons, ranking at second in the world. There are the **Kerala, Tamil Nadu** and **Travancore States** coast placers. The titanic magnetite reserves are 85 million tons, ranking at third in the world. They are mainly developed in the Maharashtra, Kerala and Tamil Nadu States of the western coast of the Indian Peninsula.

The proved reserves of the weathering remnant-type bauxite deposit are 2.654 billion tons, which contents of Al_2O_3 are 45–55%. They are located on the east side coast of Indian, i.e., the northeast of Central Indian shield. On the sides of **Madhya Pradesh** graben, the main deposit is Amakantag bauxite deposit. Those areas are almost distributed and covered by the Deccan basalts on the Mahara Plateau. The bauxite is formed by the strong weathering of Deccan basalts. The bauxite usually shows the irregular lens and is developed in the red clay sections. It is formed by the **Deccan** basalts in situ weathering, which belongs to the remand red clay-type bauxite deposit (Zhang 2009).

The proved reserves of sedimentary manganese deposit are 135 million tons, the outputs of manganese ore are more than 1 million tons per year. In **Maharashtra** and **Madhya Pradesh** States, they are located in the Pre-Cambrian low-grade metamorphism sedimentary system (Zhang 2009).

4.1.5.5 The Giant Oil Fields in Kavkaz–Alborz Late Paleozoic–Late Jurassic Accretion–Collision Zone [41]

In the central part of this collision zone, the Asheron Peninsula, western beach of the Caspian Sea, there is the famous **Baku** oil field, in Azerbaijan. Since the late nineteenth century, that oil field had been exploited to the 1970s, the outputs of oil had reached about 2 billion tons. In 1970s, there retained the oil reserves of 2 billion tons. The oil mainly is reserved in the anticlines that are composed of the Neogene System and Paleogene System. In recent, the nearby sea areas have been focused to explore, and the proved oil and gas reserves have been obtained more than 4 billion tons. Now the main rich oil-gas strata are the Paleogene System, which had undergone the rather weak NS trending compression to result in the related faults and fractures, to form the important oil-gas reservoirs. In the

2004, the residue reserves were obtained as 959 million tons, with the outputs of 14.9 million tons. Now the explorations of new oil-gas fields are in progress. That oil-gas field could be recovered to the world-level giant field, which estimated outputs of gas would be 400 million m^3/d (Gromov, 2007; Marine Petroleum Editorial, 1999; www.cqvip.com). According to recent exploration results, the prospective reserves of oil and gas in whole Southern Caspian Sea will be up to 1.823 billion tons (Hou et al. 2014).

4.1.5.6 The Giant Ore Fields and Deposits in Turkey–Iran–Afghanistan Plate [43]

In the Iran block (Fig. 4.66), the reserves of iron deposit are 1.8 billion tons, which are located in the Proterozoic metamorphism systems. In the Central and Southeastern Iran, there are the sedimentary–metamorphic-type deposits or volcanic ore pulp-type iron deposits, the ore grades are rather high to reach 50–60%; however, the sulfur contents are higher. The volcanic–sedimentary–metamorphic-type iron deposit is reserved in **Choghart** of Central Iran, which output can reach 3.5 million tons per year (the high-quality magnet ores). Far away 100 km northeastward Yazd, the outputs in **Chadormalou** deposits can reach 5.1 million tons per year. In the ore, the iron grade can reach 55%. The ore total reserves can reach 300 million tons. In the Kerman Province, Southeastern Iran, the outputs of **Gole-Gohar** volcanic ore pulp-type iron deposits are 8 million tons per year (including haematite ore). The ore reserves of that deposit reach 275 million tons (www.GSI.IR).

The copper resources in Iran are also rather abundant, which scales are relative large. The copper ore reserves are 430 million tons, the copper grade more than 1%, with the associated elements of molybdenum and gold. They are all reserved in the Cenozoic volcanic activity zone (Fig. 4.66). The main types of copper deposits are the porphyry type, volcanic type and skarn type, which belong to the **Carpatian–Balkan–Himalayan metallogenic zone**. The main metallogenic periods are Eocene–Neogene. That metallogenic zone can be divided into two branches: the northern branch, across the Northern Turkey–Central Iran–Afghanistan–Himalaya, distributed along the EW trending; the southern branch, along the ENE orientation from Southern Iran (Fig. 4.66) to Southern Pakistan. The ore-bearing intrusion bodies are mainly distributed along the regional faults (Mao et al. 2012b). In the past, some researchers considered that those faults could be distributed along the plate subduction zone. In fact, they are located in the continental plate to form along the collision zones or intraplate regional fault zones. It is completely different from the metallogenic zone of the Cordillera–Ades subduction zone

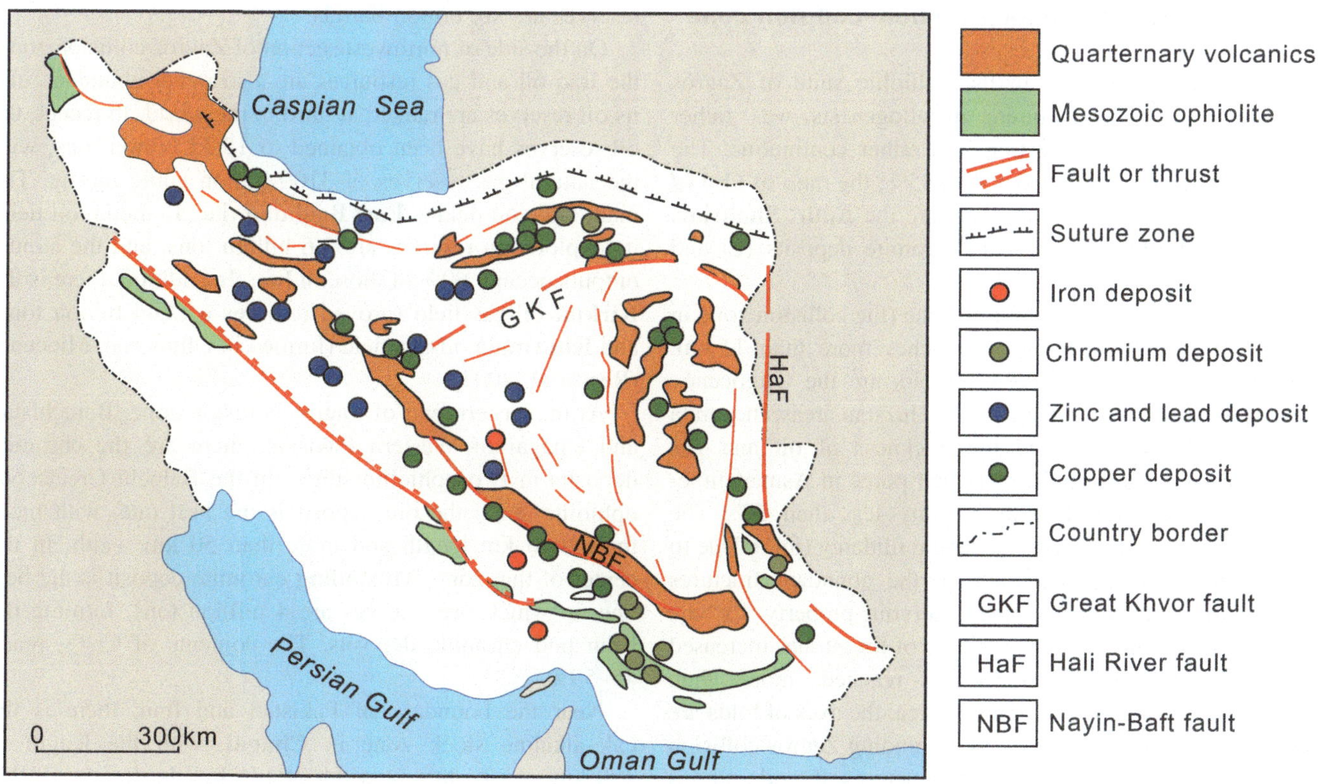

Fig. 4.66 Distribution sketch of Iran geological structure and ore deposits (After Li and Wu 2008)

and the Southwestern Pacific island arc zone, but similar to the Central Asia–Mongolia metallogenic zone in terms of the tectonic background.

The typical porphyry-type deposits in Carpathian–Balkan (Tethys)–Himalayan metallogenic zone are the copper ore reserves in the **Sar Cheshmeh** copper–molybdenum porphyry-type deposit and are 1.2 billion tons; the copper grade is 0.70%, and the ore outputs are 14 million tons. That ore-bearing intrusion body is the Neogene granodiorite and its nearby andesite. In Iran, there is also the **Miduk** deposit (copper ore reserves of 838.3 million tons, the ore outputs of 5 million tons per year), the **Soungoun Ahar** deposit (the ore outputs of 7 million tons per year, the copper grade of 0.5%), the **Chahar Gonbad** large deposit and so on (Li 2008; www.GSI.IR).

In Iran, the lead and zinc resources are also abundant (Fig. 4.66). The ore reserves are about 22 million tons, which lead average grade is 6% and the zinc average grade is 10%. The lead and zinc deposits are all developed in the Paleogene sedimentary system, which belongs to the hydrothermal-type deposit. Of them, the biggest lead and zinc deposit is **Anguran** deposit, which ore reserves are 1.88 million ton, with the lead grade of 6% and the zinc grade of 3%. The other deposits mainly belong to the middle- or small-scale deposit (Li and Wu 2008).

4.1.5.7 The Giant Ore Fields and Deposits in Zagros–Kabul Accretion–Collision Zone [44]

During Late Cretaceous, in the ophiolite suite of Zagros collision zone, the chromite metallogenesis was rather strong, and the mineralization was rather continuous. The grade of chromite ore is more than 45%; the ratio of Cr_2O_3/FeO is more than 3. In Iran, there are the **Amir, Shahriar, Reza and Abdasht** large-scale chromite deposits (Li and Wu 2008).

In the Zagros foreland folding zone (the collision zone in Fig. 2.30), the strata thickness reaches more than 11 km. The most important oil-gas reservoirs are the Oligocene–Miocene Asmari limestones. At the Huzstan areas, the strata are 320–488 m thick, and the thickness of oil and gas reservoirs is 10–280 m. The original pores in Asmari limestones are very small, with porosity less than 5%. The original permeability is only several millidarcy (mD). Due to the stronger folding and faulting, the abundant fractures were formed to improve the reservoir property for the Asmari limestones, then the porosity of limestones increased to 25%, and the permeability reached more than $100 \times 10^{-3} \mu m^2$ (100 mD). In that area, the axes of folds are almost in accordance with the NW trending Zagros collision zone. Just being influenced by the fractures, the oil and gas output in the Asmari limestones varies a little among the different oil wells. On the Asmari limestones, there are the well-developed Gachsaran Formation evaporites. Due to the stronger faulting, the oil reservoirs' attitudes become so complex. The oil and gas should come from Cretaceous System and Asmari limestone hydrocarbon source rocks. The reservoirs, besides the Asmari limestones, are the Cretaceous Bangestan Group limestones, Jurassic–Lower Cretaceous Khami limestones and dolomites, and Permian Dalan Formation carbonate rocks. In the Zagros collision zone, there develops a dominant NW–SE trending arc thrust and folding system, and some well-developed fractures, thus to form extremely oil and gas reservoir structures in the Persian Gulf (Iran, Kuwait, Abu Dhabi and Iraq) (China Geological Academy 1980).

The oil-gas resources are very abundant in Iran, and the residual proved oil reserves are 12.288 billion tons. Most reserves from 18 oil fields are expected to be 137 million tons per year. The gas proved reserves are 23,000 billion cubic meters. In 2001, there were 1120 production oil wells. The oil outputs are 156 million tons per year. The total outputs of six oil fields in **Awas, Malun, Gagsalan, Akajali, Bibihagmy** and **Pals** account for two-thirds of those in Iran. In recent year, the newly **Haiyam** natural gas field has been discovered in the Persian Gulf for Iran, and its geological reserves are 260 billion cubic meters (Li and Wu 2008; www.GSI.IR). In last ten years, the newly **Azalegan** oil field have been discovered in Iran, which is the biggest oil field discovered recently in the world. Its recoverable reserves are six billion barrels.

On the side of northwest section of Zagros collision zone, the Iraq oil and gas resources are also rather abundant, and its oil reserves are ranked at third in the world. In recent, the oil reserves have been obtained to 15.75 billion tons, with the natural gas reserves of 3100 billion cubic meters. The main large oil field is **East Baghdad** (i.e., Lumeila) oil field, its exploration reserves are 2.6 billion tons, and the annual outputs occupy 60% of those in Iraq. In addition, there is the **Kirkuk** oil-gas field (proved reserves of 2.44 billion tons) and **Khurmala** oil-gas field (formed in Oligocene–Miocene) (Ren et al. 2003).

At the eastern end of Zagros collision zone, Baluchistan and Chagai of Western Pakistan, there are the chromite deposits in two ophiolite suites. In the Baluchi Cretaceous ophiolite suite, the ore deposit is the best one, with more than 1000 km length and more than 50 km width. In the center of that zone, **Muslinbag** chromite deposit is the best one, in which ore reserves are 4 million tons, forming the bean pod chromite deposits. The contents of Cr_2O_3 reach 45–59%.

Near the boundary of Pakistan and Iran, there is the calc-alkaline stock zone in **Chagai**, with the length of 480 km and the largest width of 136 km. In the east of that Cretaceous collision zone, there is the **Seundk** porphyry copper deposit, with the proved copper reserves of 1.65

million tons. In Eastern Iran, there is the large **Salqishme** porphyry copper and molybdenum deposit (Zhang 2009).

In the area from Kabul Province to Logar Province, there exists a super-large copper deposit zone in Afghanistan. That zone is 110 km long, with the copper grade more than 0.6%, and the prospective ore reserves more than 1 billion tons. The distance between the **Aynak** super-large copper deposit and the Kabul City is only 50 km (Fig. 4.67). The proved copper ore reserves of that deposit are 700 million tons, the copper metal reserves are more than 5 million tons, and the average grade is 1.65%. That deposit has the very high economic value. That copper deposit is located in the Kabul micro-block, and the copper layer occurs in the Neoproterozoic–Cambrian metamorphic dolomite quartz mica schist. There exist the complex synclines, which axes are distributed mainly along NNE trending, and at the SW end, gradually to turn into EW trending. The attitudes of ore layer are rather gently; the burial depth is rather shallow. Most of them could be open-pit mined. The Geological Exploration Bureau of USSR had explored and trial-produced more than 10 years. In recent years, the Chinese companies have also begun to invest and exploit the mine in cooperation with Afghanistan.

At the Hajji Gak areas, **Bamian** of Afghanistan, there is a large Pre-Cambrian metamorphic system ore deposit. Its perspective reserves are 2.5 billion tons. The ores are composed of haematite and magnetite, and the average iron grade reaches 62%. However, most of deposits are located at the mountains about 4 km above sea level, so the transportation and exploitation conditions are not convenient.

4.1.5.8 The Giant Ore Fields and Deposits in Arabian Plate [46]

In the main part of Arabian plate, the tectonics is weaker. On the crystalline basement, there are Paleozoic–Mesozoic–Paleogene marine sedimentary systems, about 5000–8500 m thick, where the extremely abundant hydrocarbon systems and the NE trending convergence and collision are remarkably weaker than those of Indian plate. The Arabian plate had undergone the influences of the Zagros accretion–collision zone to form the Zagros foreland folding zone, which was formed by the convergence between the Arabian and Turkey–Iran–Afghanistan plates, thus to develop the fold placanticline or dome. In that area, the Permian–Neogene strata are 8500 m thick, the thicker in the east and the thinner in the west. The hydrocarbon systems are mainly the Permian–Triassic, Jurassic–Cretaceous and Paleogene–Neogene marine carbonate strata. The gas is mainly reserved in the Permian–Triassic Systems, which recoverable reserves are 30 billion ton oil equivalent, and majorly developed in the anhydrite layer. In the Jurassic oil and gas reservoir-forming assemblages, the most important is the upper Arabian assemblage. In the Jurassic reservoir-forming assemblages, the recoverable reserves are 48 billion ton oil equivalent, occupying 22.7% of oil and gas reserves in the whole. For the Cretaceous reservoir-forming assemblages, there are many sedimentary depressions, which reservoirs are mainly sandstone or chalk limestone. The reservoir-forming assemblages are rather complex, and they are distributed in Iraq, Iran and Kuwait, which covers are the mudstone and sandstone. They are reserved in the 764 oil and gas reservoirs respectively, and the recoverable reserves are 71 billion ton equivalent, occupying 33% of those areas. The Paleogene–Neogene reservoir-forming assemblages are composed of the Miocene–Oligocene limestone and its covered gypsum-saline layer, in which 142 oil and gas reservoirs have been discovered, with the recoverable reserves of 28.2 billion ton equivalent (Duan et al. 2014).

In the above oil-gas fields, the most famous one is the **Jawar** oil-gas field (Fig. 4.68, west to the 25° N, 50° E), which is distributed in the NS trending Inela placanticline about 250 km long. The recoverable reserves are 10.74 billion ton oil equivalent, and the outputs reach 280 million ton oil equivalent per year, occupying 30% of those in the whole Persian Gulf. The oil wells are the self-blowout well, which crude oil contains a small amount of paraffin, mainly is the light oil in the great-scale oil-gas reservoirs formed in Cenozoic. In that area, due to inheriting the Pre-Cambrian metamorphic basement near NS trending reverse fault and foliations, it forms the NS trending placanticline. Influenced by NE–SW trending regional maximum principal compression stress, some NE and NW trending conjugate shear fractures are developed in the sedimentary strata. All the above structures have influenced the oil and gas to migrate and concentrate. In the east of the Arabian Peninsula, there are some oil-gas-bearing structures related to the Cambrian salt-dome structures, which attitudes are no relationship with the regional structures.

In recent years, at Central Saudi Arabia, the **Hazmiyab** and **Raghib** giant Paleozoic oil field has been discovered (Liu 2009). The oil reserves of Saudi Arabia are ranked at first in the world. The recoverable reserves are more than 36 billion tons, approximately occupying one-fourth of those in the globe. The gas recoverable reserves are about 6600 billion cubic meters, ranking fourth in the world (Li, on Xin Hua net, Riyadh, in Feb. 21, 2007).

Many scholars considered the beneficial conditions to form the supergiant oil and gas reservoirs as the followings: ① The source rocks have the high contents for hydrocarbon and are distributed widely. The reservoirs have the high interstice and high permeability and are well connected with the source rocks. ② There is the beneficial preservation condition, i.e., after each sedimentary sequence, there would form the cover layer by the evaporation salt system on the reservoir layers. ③ In that areas, the tectonic intensity is at the intermediate level, which is enough to form the giant structural traps and the nice cover (Ji 1989).

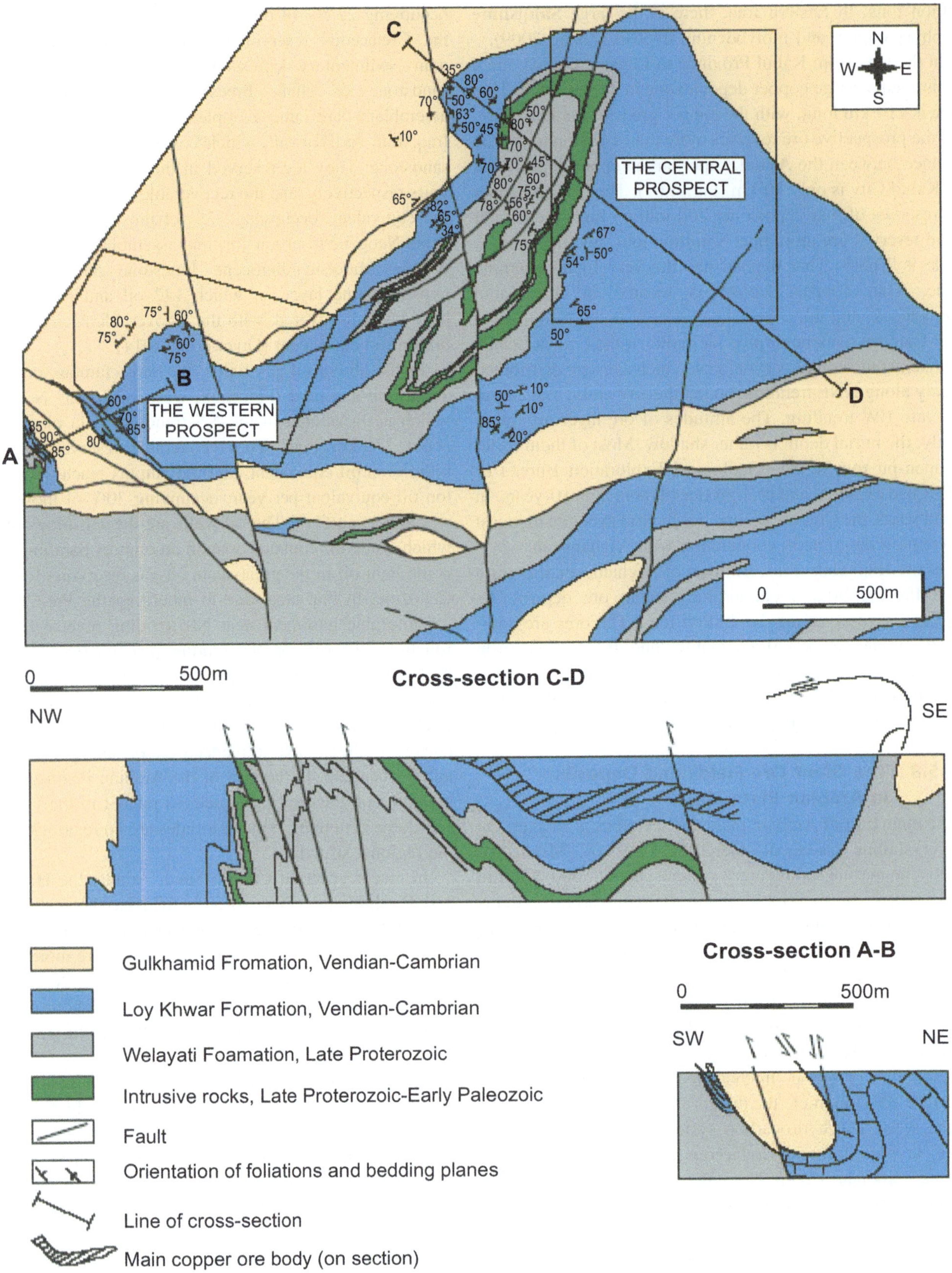

Fig. 4.67 Geological sketch of Aynak copper deposit, Afghanistan (After Gao et al. 2012)

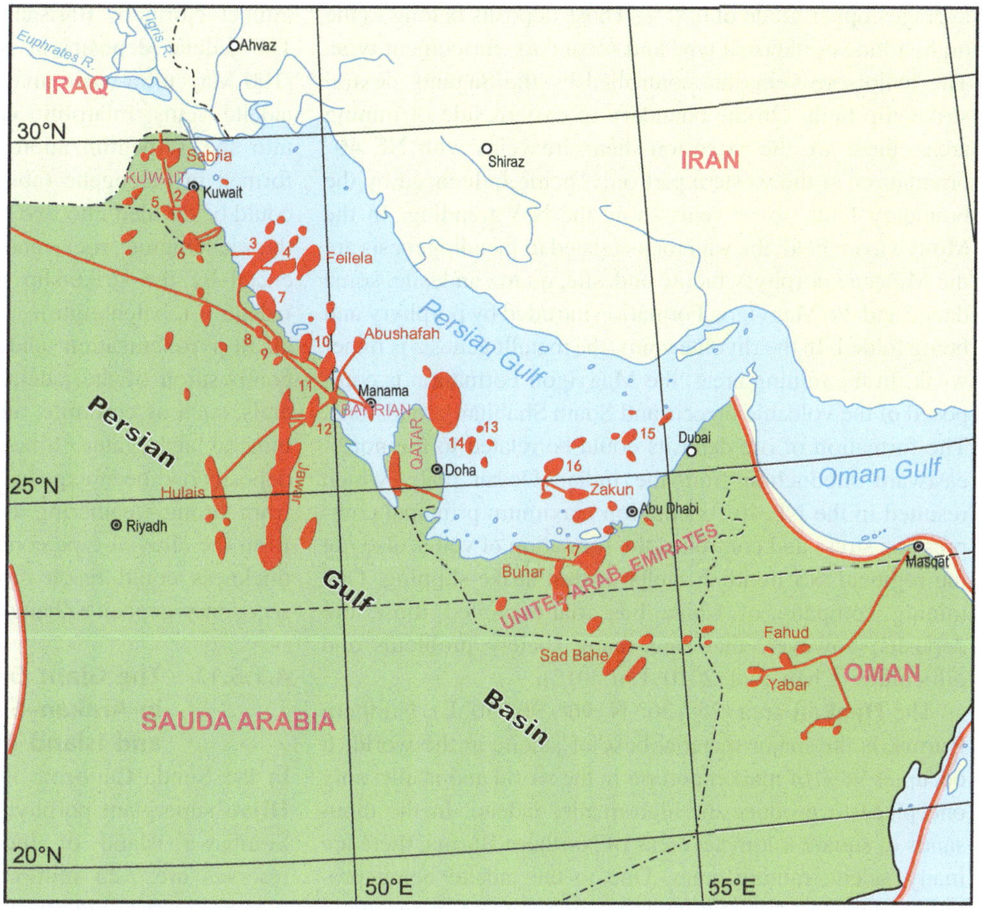

Fig. 4.68 Distribution sketch of oil-gas field in the Persian Gulf, the Middle East (After Duan et al. 2014)

At Northeastern Saudi Arabia, there reserves the **Safania** oil field, which recoverable reserves are 3.32 billion tons. On the northeastern side marine area of the Arabian Peninsula, there is the **Great Burgan** oil field in Kuwait, which recoverable reserves are 9.91 billion tons, with the outputs about 70 million tons per year. The characteristics of oil are as same as those of the Jawar oil-gas field. In Central and Eastern United Arab Emirates, there is the **Zakum** oil field, which recoverable reserves are 1.59 billion ton. Most of oil wells belong to the self-flowing well. The crude oil is of high quality, with little paraffin. On the eastern side of the Arabian Peninsula, **Qatar** is very rich in natural gas resources, which recoverable reserves are 25,600 billion cubic meters, ranking third in the globe (World Petroleum, from the Web).

Saudi Arabia is rich in the rare-element deposits, such as Nb, Ta, Sn, Y, Th, U, Zr and rare earth element deposits, which are reserved in the granite bodies. The deposits have two metallogenetic types: ① **Jabal Sa'id** type is characterized by the high total iron and alkaline contents, rich oxygen and high ratio of K_2O/ Na_2O. The ore reserves are about 20 million tons, and the average uranium contents are 1.30×10^{-4} and thorium contents are 8.3×10^{-4}. ② **Ghurayyah** type is located in the oxide-rich alkali granite. They were formed in the high-temperature and water-shortage melting conditions in Neoproterozoic. Its rare-element ore reserves reach 440 million tons, with the highest contents of tantalum up to 2×10^{-4} (www.GSI.IR).

4.1.5.9 The Giant Ore Fields and Deposits in Oman Accretion–Collision Zone [47]

In 1980s, an open-pit mining giant copper deposit was discovered in the ophiolite suite of **Oman** Cretaceous accretion–collision zone (Clarke 2006).

4.1.5.10 The Giant Ore Fields and Deposits in Western Burma (Pegu Mountains–Rangoon) Plate [49]

There is the **Monywa** (22° 15′ N, 95° 05′ E) super-large copper field in the Sagaing Province, Central Burma, away from 100 km west to Mandalay. The field is composed of four deposits, which are distributed along near NS trending. The height of mining areas is about 590 m above sea level. That mining areas could be open-pit mined, which areas are about 50 km^2. The total ore reserves reach 2 billion tons, in which the copper metal reserves are about 7 million tons with the average copper grade of 0.37%. Those deposits belong to the high-sulfide epithermal type and secondary enrichment type. The major ore veins are controlled by the Sagaing dextral strike-slip fault. On the boundary of eastern side of mining areas, there are the extension-shear ore veins with NE 40° orientation; at the western part only, being influenced by the boundary fault, some veins show the NW trending. In the Monywa ore field, the wall rocks related to metallogenesis are the Miocene porphyry biotite andesite, quartz andesite, some dacite and the Magyigon Formation intruded by porphyry and being folded. In the rhyolite veins, the metallogenesis is rather weak. In the mining areas, the Magyigon Formation is composed of the volcanic breccia and South Shabitang sandstone. The formation of ore deposits could be related to the northeastward subduction from the Indian Ocean plate, which resulted in the NE 40° orientation maximum principal compression stress and controlled the formation of veins, also got the regional NS trending faults dextral strike-slipping. One mining company of China has trial-exploited those ore deposits; however, there are many society problems and difficulties (Chen et al. 2010; Liu 2012).

The **Hpakan** area (25°–26° N, 96°–96° 30′ E), Northern Burma, is the major material base of jadeite in the world. It occupies 98% of market portion in the world and is also only one place to produce the high-quality jadeite. In the thousands of square kilometer areas of Northern Burma, there are many jadeite mining areas. Due to the jadeite being produced from the different geological conditions, it results in extremely differences in terms of craft quality, processing property and economic value (Zhang 2002). The production places are mainly located at the northern areas of the Mitkyina–Indawgyi Lake, i.e., the Hpakan areas, which are developed in Western Burma (Pegu Mountains–Rangoon) plate [49] and the west of Bangongco–Nujiang–Mandalay–Phuket–Northern Barisan Cretaceous collision zone [35]. Since Neogene, the part of that collision zone had changed to dextral strike-slip fault to control the regional structures. On the east side of Hpakan town, there develops an important fault—Hpakan fault.

In 1934, through the geological research and survey, Chhibber proposed that there existed Carboniferous–Permian plateau limestone in the mining areas; due to being intruded by granite, that limestone had been recrystallized to form the crystalline schist, then, in Jurassic–Cretaceous to be intruded by serpentinization peridotite. According to the determination of high-resolution ion probe analyzer (IPA), the zircons in the magmatic rocks were inferred to form in three epochs: Middle Jurassic (163.2 ± 3.3 Ma), Late Cretaceous (146.5 ± 3.4 Ma) and Early Cretaceous (122.2 ± 4.8 Ma) (Shi et al. 2008). The Middle Jurassic was the main active period of Bangongco–Nujiang–Mandalay–Phuket–Northern Barisan Cretaceous collision zone [35]. The jadeite deposits at first were formed in Cretaceous (100 Ma) ultra-basic magmatic rocks. In Paleogene, the jadeite veins, miarolitic cavities and albite veins intruded into the serpentinization peridotite, and the jadeite was formed in Paleogene (about 45 Ma). Its formation process could be divided into two epochs: in the early epoch, to form the chalchiguite rock, and in the middle–late epoch, influenced by the strike-slip fault activity, to form the green jadeite (chalchiguite). The alteration includes the alkali-pyroxenization and albitization. The main mineral composition of the jadeite is $NaAl[Si_2O_6]$, and other minerals, such as chromite, amphibolite, albite, beryl, and their mineral aggregates. Since Paleogene, the primary jadeite deposits had been exposed on the Earth surface and suffered from strong weathering, erosion and sedimentation, then to form the alluvial-type conglomerate deposits, which biggest thickness could reach more than 150 m. It belongs to a secondary deposit (Zhang 2002).

4.1.5.11 The Giant Ore Fields and Deposits in Arakan–Sunda Cenozoic Subduction and Island Arc Zone [50]

In the Sunda Cenozoic island arc zone, there is the **Batu Hijau** supergiant porphyry copper and gold deposit in the Sumbawa island of Indonesia (Fig. 4.69). The copper reserves are 7.23 million tons with the copper grade of 0.45%, and the gold reserves are 572 tons (Cook et al. 2005). That deposit was discovered by geological and geochemical survey in 1990. That deposit is located in the diorite porphyry body of andesite system.

On the eastern beach of **Kolonodale**, **Southeast Sulawesi, East Sunda** Cenozoic island arc zone, **Northwest Kolaka** and **Moyowali** etc., there are the red soil weathering-type giant laterite nickel deposits (Fig. 4.69), which suffered the tropical strong weathering to form the weathering crust deposits. In the ultra-basic rocks, the nickel grade is only 0.16–0.24%; however, in the weathering crust, the nickel grade reaches more than 1.4%. Now the reserves of that laterite weathering-type deposits occupy two-thirds of the global reserves, and they are the main industrial-type nickel deposits (He et al. 2008).

4.1.5.12 The Giant Ore Fields and Deposits in Sunda Plate [51]

The Sunda plate (Fig. 2.30) is located in Central Indonesia. In the sedimentary basin, there are the rich petroleum and placer-type tin fields (Figs. 4.69 and 4.70). In Indonesia, there are about 60 sedimentary basins, 73% of the basins in the sea. Now 36 explored oil-gas-bearing basins are located in the west of Sunda plate. The important oil-gas-bearing

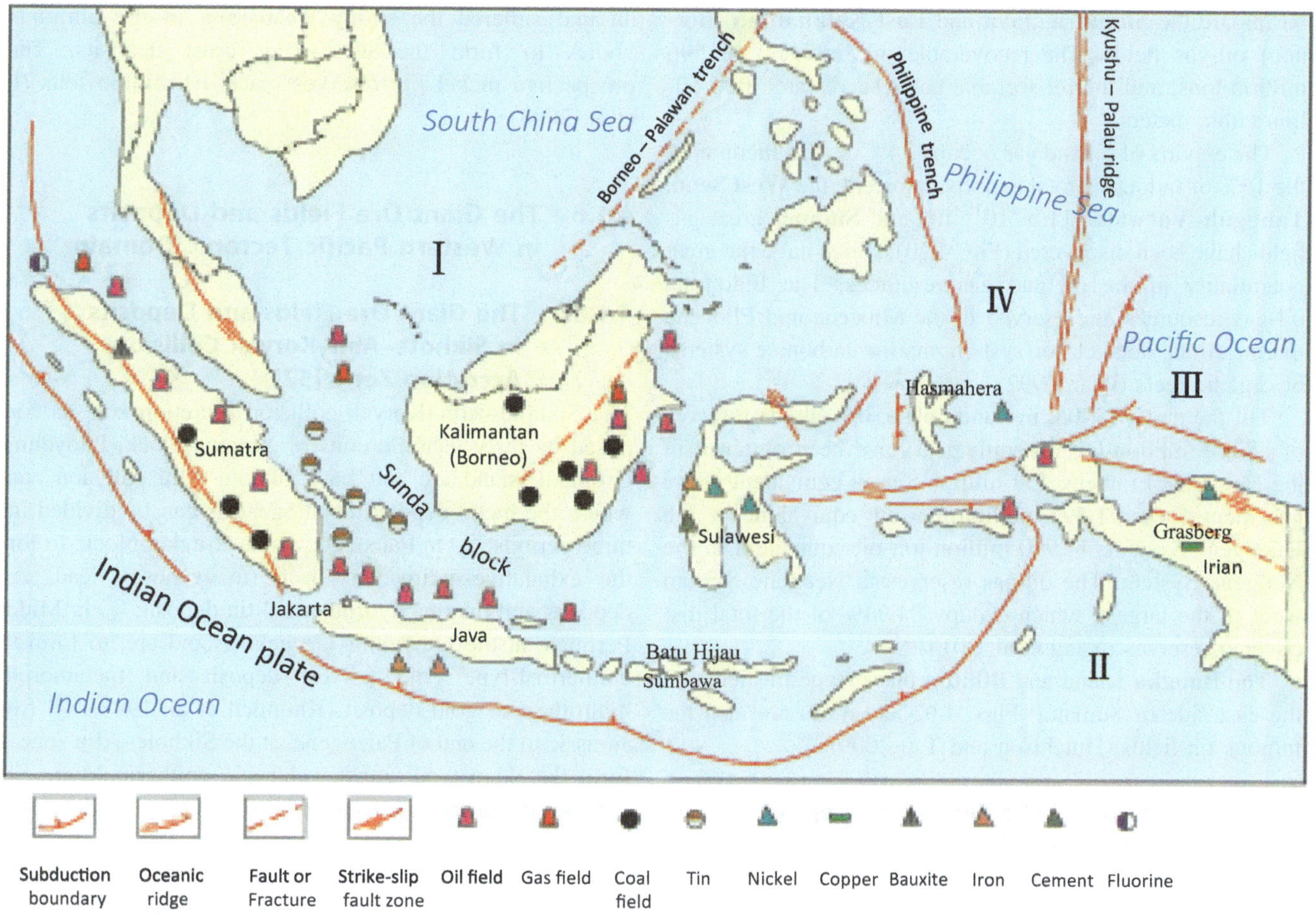

Fig. 4.69 Sketch of geological background and main deposits in Indonesia. I. South China Sea fault-depression basin; II. Indian Ocean plate; III. Pacific plate; IV. Philippine Sea plate (Modified from Qian 1992)

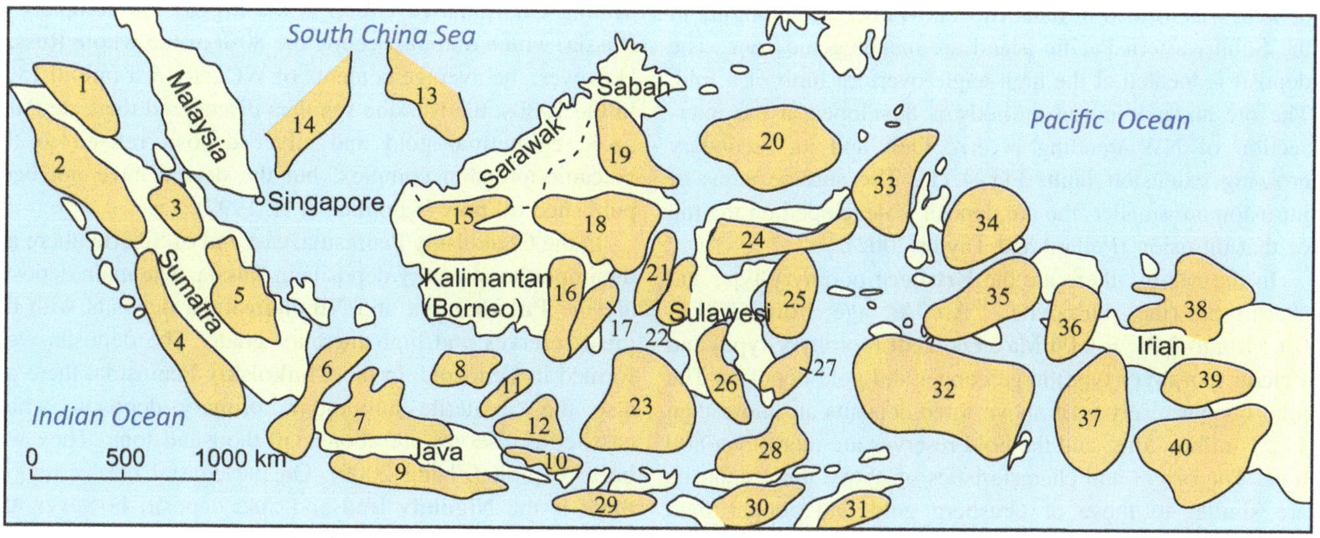

Fig. 4.70 Sedimentary basins in Indonesian. The basin names: 1. North Sumatra; 2. Shibolga; 3. Central Sumatra; 4. Bengkulu; 5. South Sumatra; 6. Sunda; 7. Northwest Java; 8. Prink; 9. South Java; 10. Northeast Java; 11. Putting; 12. Northeast Java Sea; 13. East Natuna; 14. West Natuna; 15. Gaii–Tayan–Maya; 16. Banitok; 17. Asham; 18. Kutai; 19. Tarakan; 20. Sulawesi; 21. Akassar Strait; 22. Laliyan; 23. Majaci; 24. Gorontalo; 25. Banggai; 26. Bone; 27. Southeast Sulawesi; 28. Flores; 29. Bali; 30. Savu; 31. Timor; 32. Banda; 33. Halmahera; 34. Waigeo; 35. Seramwati; 36. Bituni; 37. Aru; 38. Walubai; 39. Akimeng; 40. Sahur (Modified from Qian 1992)

basins are the **Sumatra**, **Java** and **East Kalimantan** (Borneo) oil-gas fields. The recoverable oil reserves are 596 million tons, and the recoverable gas reserves are 3000 billion cubic meters.

The exports of oil and gas occupy 24% of total income, or the 15% of Indonesia total exports. In recent, the **West Seno**, **Tangguh–Vorwata** (11×10^{12} ft^3) and **Sumpal** great gas fields have been discovered (Fig. 4.70), which have the great potentiality in the oil and gas resources. The Indonesia oil-gas resources are reserved in the Miocene and Pliocene marine strata, delta clastic system, marine carbonate systems or organic reefs (Qian 1992).

Till the end of 2008, in Sunda plate, the oil-gas reserves of 782.65 million-ton oil equivalent have been obtained in the Mesozoic Erathem, 154-million-ton oil equivalent in the basement rocks, 1,192 million-ton oil equivalent in the Paleogene System, 11,990 million-ton oil equivalent in the Neogene System. The oil-gas reserves in Neogene System were of the largest, which occupy 84.94% of the total discovered reserves (Yang et al. 2014).

The **Bangka** Island and **Billiton** placer-type tin fields on the east side of Sumatra (Figs. 4.62 and 4.69) are also the famous tin fields (Hutchison and Tan 2009).

4.1.5.13 The Giant Ore Fields and Deposits in Northern New Guinea Island Arc Zone [55]

In the Western Irian Java Island (New Guinea Island), Indonesia, there is the greatest **Grasberg** rich gold and copper porphyry deposit in the world. The copper reserves reach 28.02 million tons with the copper grade of 1.10%, and the gold reserves are 2,604 tons (Cook et al. 2005). That deposit was formed in Late Miocene–Pliocene, belonging to the Southwestern Pacific island arc metallogenic zone. The deposit is located at the high-angle overturn limb of a fold. The ore andesite porphyry body is developed at the intersection of NW trending reverse fault and its secondary crossing extension fault (Fig. 4.71). The surface areas of intrusion are smaller; the ore deposit is developed on the top of that intrusion (Pollard and Taylor 2002).

In the nearby, there are the **Ertsberg** porphyry-type and skarn-type (the muscovite $^{40}Ar/^{39}Ar$ age from 3.33 ± 0.12 Ma to 3.01 ± 0.06 Ma), **Ok Tedi** (porphyry-type) and **Frieda** (porphyry-type) large copper and gold deposits. The total copper reserves of above three deposits are more than 11.21 million tons, and the gold reserves are more than 800 tons. The origin and characteristics of above three deposits are similar to those of Grasberg gold and copper field (Pollard and Taylor 2002).

In the **Halmahera** Island (Fig. 4.69), **Subain, Mabli**, **East Lamla** of Weigo Islands and **Xifu** Mountains, there are the laterite-type giant nickel deposits. Since Neogene, they all had suffered the strong weathering of the ultra-basic bodies to form the weathering crust deposits. Their prospective nickel ore reserves reach 10 million tons (He et al. 2008).

4.1.6 The Giant Ore Fields and Deposits in Western Pacific Tectonic Domain

4.1.6.1 The Giant Ore Fields and Deposits in Sikhote–Alin–Koryak Collision Accretion Zone [57]

The Sikhote–Alin–Koryak collision accretion zone is composed by three tectonic units of Xingkai block, Laoyeling–Grojieko island arc and East Sikhote–Alin collision zone, where the metal deposit metallogenesis can be divided into three periods: ① in Paleozoic, at the Xingkai block, to form the exhalative-sedimentary iron (manganese), lead, zinc deposits and magma hydrothermal tin deposit; ② in Middle Permian, at the Laoyeling–Grojieko island arc, to form the epithermal-type gold (silver) deposit and metamorphic hydrothermal gold deposit (Khomich et al. 1997); ③ from Jurassic to the end of Paleogene, at the Sikhote–Alin zone, to form the skarn-type tungsten deposit, epithermal-type gold (silver) deposits, skarn-type boron vein and lead zinc deposits, vein gold deposits (Feng et al. 2012).

In the Far East area of Russia, the being exploited tungsten deposits are mainly distributed at the coastal region and the Sikhote–Alin zone. The typical giant deposits are the "**Orient No.2**" and **Lemonltov**, which are mainly the skarn-type scheelite-sulfide deposits, partially are the scheelite–quartz vein deposit. In the Sikhote Mountains, the Helustalinei Mining Coordinative Group is the biggest tin company in Russia, which outputs occupy the 80% of the whole Russia. However, the average contents of WO_3 are just only 0.15%. In the 1980s, the Russian scholars discovered the Late Paleozoic epithermal gold and silver deposit related to the volcanic-intrusion complex, but the details have not been published till now (Khomich et al. 1997).

In the Chukolskiy Peninsula, east end of Siberia, there are the important mercury deposits in Russia. The main deposits are the **Pulamanoya** and **Xibolianskeya** deposits with the great reserves and high mercury grade. The deposits were formed in Cenozoic. In the **Chukolskiy** Peninsula, there are also the cassiterite–sulfide-type primary deposits, which cassiterite reserves are about 350 thousand tons. They will be to exploit (Zhang 2009). On the coastal border region, there is the **Nigulaiv** lead and zinc deposit; however, the scale is not large. That deposit was also formed in Cenozoic. Its reserves only occupy 4% of those in Russia, and the lead grade is less than 3%; however, the outputs of that lead ores occupy 50% of those in the whole Russia.

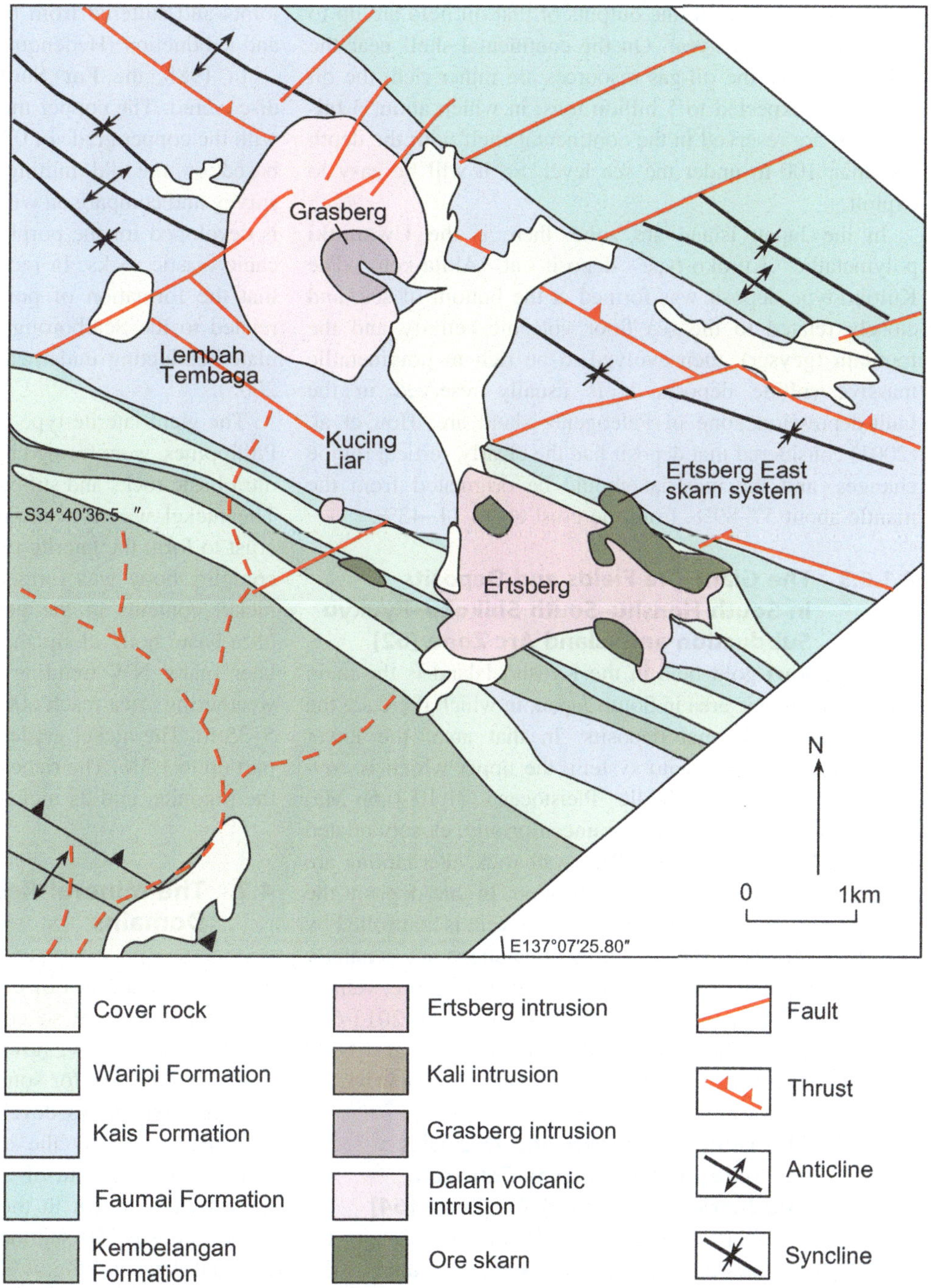

Fig. 4.71 Geological sketch of the Grasberg gold and copper porphyry field, Irian Jaya Island, Indonesia (Modified from Pollard and Taylor 2002)

4.1.6.2 The Giant Ore Fields and Deposits in Aleutian–Kamchatka–Kurile–Northeast Japan Cenozoic Subduction and Island Arc Zone [59]

In this zone, there are 77 oil-gas fields in or near the **Sakhalin** Island, which are mainly the middle- and small-scale oil-gas fields. The exploited reserves are 480 million tons for oil and 130 million cubic meters for gas (Liang et al. 2014). Now there are nine oil-gas fields being exploited—seven oil fields and two gas fields. The oil-gas reservoirs are mainly distributed on the east side marine areas of the Sakhalin Island and developed in the Paleogene and Neogene strata. The orientations of oil-gas reservoirs are mainly NNW trending, which are mainly controlled by the dextral strike-slip extension faults (He et al. 2015). The sulfur contents in the Sakhalin crude oil are about 0.3%, rich in the aromatic hydrocarbons, so the oil belongs to the high-quality oil. However, the other contents are different. The exploitation history of Sakhalin oil field was earlier. Since 1928, it has been exploited; however, the outputs were rather lower. In recent, the outputs of that oil field are up to 30 million tons per year. On the continental shelf near the Sakhalin areas, the oil-gas resources are rather rich, the oil reserves is expected to 5 billion tons, in which about 3 billion tons are reserved in the continental shelf with the depth less than 100 m under the sea level. So it will be easy to exploit.

In the Japan island arc zone, there is the **Uwamuki** polymetallic Kuroko-type deposit at Akita-ken. The Kuroko-type deposit was formed at the bottom of sea, and closely related to the sea floor volcanic activity and the fountain (geyser), then evolved to be rich in polymetallic massive sulfide deposit. It is usually reserved in the fault-depression zone of Paleogene island arc. Hou et al. (2001) considered that deposit had the clearly vertical rhyme changes, and the material should be originated from the mantle about 57–89%, from the crust about 11–43%.

4.1.6.3 The Giant Ore Fields and Deposits in South Honshu–South Shikoku–Ryukyu Subduction and Island Arc Zone [62]

The **Hishikari** gold field in the Kyosho Island is the main gold metallogenic area in South Japan, in which there are the low-sulfide epithermal deposits. In that area, the lower basement is the Shimanto system, the upper which is covered by the Early–Middle Pleistocene (1.10–0.66 Ma) dacite-rhyolite system with the unconformity, closely related to the gold metallogenesis. The wall rock alternations are mainly the chloritization and illitization. In that deposit, the total gold reserves are 260 tons. The ore vein is controlled by the extension-shear joints, and its orientations are generally NE 50° trending, almost vertical. The single ore vein is about 300–400 m long, 0.5–4 m wide (Li et al. 2013-6-5, personal blog: The investigation of Japan biggest gold deposit—Hishikari, http://blog.sciencenet.cn/u/lihuan1022).

4.1.6.4 The Giant Ore Fields and Deposits in Philippines–Moluccas Cenozoic Subduction and Island Arc Zone [64]

From the Northern Luzon Island, Philippines, across the Irian Java–Papua New Guinea, to the Solomon Islands, there is an important porphyry copper belt in the south of the Western Pacific island arc. The **Lepanto** copper and gold field of the Northern Luzon Island is the more important porphyry copper and gold deposit and epithermal copper, gold and silver field. That ore field is located at the Pinatubo volcanic area, west of the Northern Luzon Island, which gold reserves are 550 tons, and the copper reserves are 3.6 million tons with the copper grade more than 1%. The isotopic age of wall alternation (alunitization) is 1.44 ± 0.8 Ma, which could indicate the main metallogenic period. The ore vein orientations are mainly NW–SE trending. It could be controlled by the derived extension joints and suffered from the Philippine Sea plate migration and subduction (Hedenquist et al. 1998).

In 1988, the **Far Southeast** copper gold deposit was discovered. The copper metal reserves are 3.36 million tons with the copper grade of 0.73%. That deposit was discovered based on the old mining areas' investigation, geological survey and comparison with other mining areas. That deposit is developed for the porphyry in the quartz diorite and volcanic clastic rocks. In recent, the scholars have considered that the formation of porphyry copper gold deposits was related to the Scarborough Ocean Ridge of Philippine Sea plate subducting underneath the Luzon Island (Cook et al. 2005).

The giant laterite-type nickel deposits of **Dinagat** Island, Philippines, were formed by the Cretaceous ophiolite suite in ultra-basic rocks and suffered the tropical strong weathering. The nickel was concentrated at the bottom of weathering crust to form the laterite-type nickel deposit. The ultra-basic complex body was formed at the period of 84.8 Ma. The nickel contents in the peridotite are only 0.2–0.3%. The ultra-basic body along the NE trending was cut off by the later many NW trending faults. The distribution areas of weathering crust reach 600 km^2, which thickness is between 5–35 m. The nickel grade usually is about 1%, in the lower part up to 1.5%. The richest nickel ore is the laterite ore with the limonite, and its nickel grade is 1.5% (Wang 2010).

4.2 The Mineral Resources in Tectonic Domains

The kinds and types of ore deposits are very complex, and the phenomena are so complicated, there seems to be no regularity to follow, however, to research carefully could find that it is easy for some plates or blocks in the different tectonic domains to develop and reserve some special ore deposits, especially the endogenic metal deposits. In this monograph, the data of 242 giant ore provinces, fields or deposits are shown in the Appendix, including 187 endogenic deposits, 50 sedimentary deposits and 5 weathering residual deposits (Fig. 4.72).

Firstly, the characteristics of the endogenic ore provinces, fields or deposits in each tectonic domain will be mainly discussed. From the macroscopic view on, any area on the crust seems to contain more than 100 elements; however, their contents are extremely different. The plates or blocks in different tectonic domains were formed in the early geological periods. They are composed of the multiple accumulation and accretion of planetesimals (Ouyang et al. 2002). The different planetesimals have the different materials, namely they are composed of the different elements and compounds with different contents. The blocks not only have some similarity, but also have many differences. Thus,

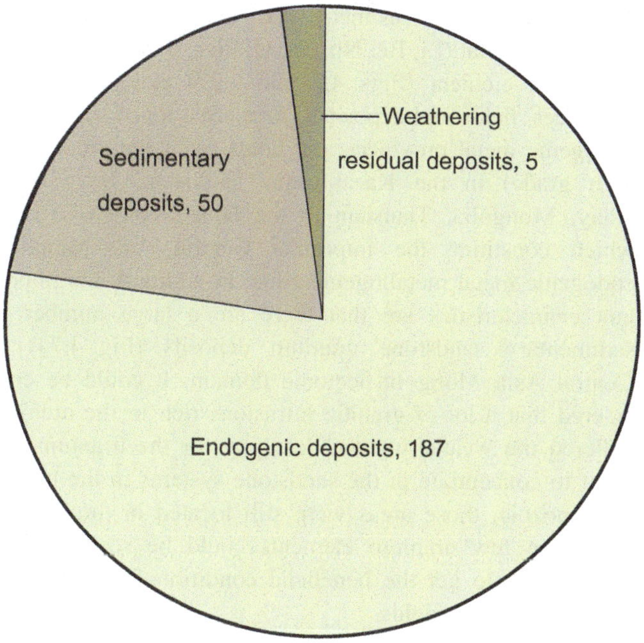

Fig. 4.72 Statistic results for metallogenic types for giant ore provinces, fields or deposits in Asia

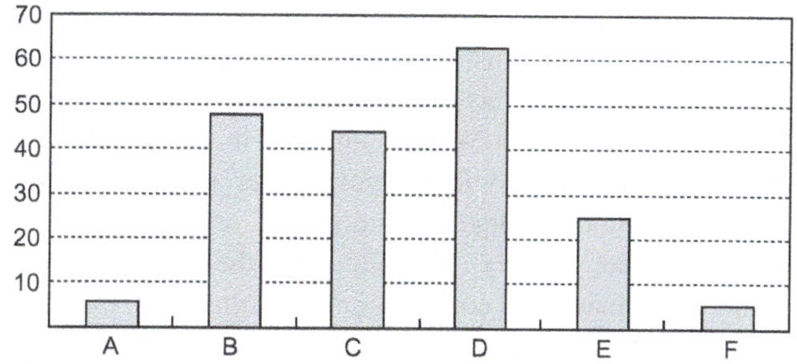

Fig. 4.73 Numbers of large endogenic metal provinces, fields and deposits in Asia. **A** Siberian tectonic domain (6); **B** Central Asia–Mongolia tectonic domain (48); **C** Sino–Korean tectonic domain (44); **D** Yangtze tectonic domain (63); **E** Gondwana tectonic domain (25); **F** Western Pacific tectonic domain (5). The total statistical data are included in 191 endogenic metal provinces, fields or deposits

in the later tectonic periods, the original elements and chemical compounds would be combined and dissociated continuously. Once the certain conditions were available, they would be enriched to form the ore deposit. So in a certain block or plate, it must be enriched with the certain elements. When some elements, crystals or chemical compounds are concentrated to be utilized economically for human beings, they will be formed as the useful minerals and deposits. Thus, from the above viewpoint, it is reasonable that it is easy for the certain ore deposits to reserve in some blocks or plates.

As for the distribution of sedimentary deposits or fields, they are mainly controlled by the fault-depression basins and caused by the certain extension tectonics; as for the kinds of ore deposits, they are mainly controlled by the weather, geography and latitudes. Since Mesozoic, the migration velocities of the plates for Gondwana tectonic domain, except the Indian plate, had been rather slowly, with the weaker rock deformations, and they had been located in the warm and moist weather conditions of tropics or temperate zone to be easy to form the great-scale oil-gas or coal fields. As a result, in many areas of that domain, there formed and reserved the largest oil-gas fields, such as the Gulf, Iran, Caspian Sea, Central Asia and Indonesia. As for the formation and preservation of West Siberian large oil and gas fields, since Neoproterozoic the areas had suffered multiple high-density basic magmatism, then to result in those areas to subside for a long time. It was useful for the oil, gas, methane hydrate and coal to concentrate and reserve, then to form fields.

In the subtropical dry weather areas, it is easy to form the gypsum and salt deposits; however, in the temperate and moist weather areas, it is usually to lead to the clay weathering, then to form the bauxite and kaolin deposits. Under the strong tropical chemical weathering, it is easy for the iron, titanium, manganese, nickel, etc., to concentrate and reserve and to form the weathering residual deposits. To reserve the weathering residual deposits, it needs providing the relatively stable tectonic condition, or the great-scale uplift or subsidence has never occurred. So now the reserved and discovered large weathering residual deposits are mainly developed and reserved from Cenozoic, because most of older ones have been destroyed or eroded.

Next, in this monograph, it will mainly discuss the characteristics of giant endogenic metal provinces, fields and deposits (Fig. 4.73).

4.2.1 Mineral Resource Characteristics of Siberian Tectonic Domain

The main part of the Siberian plate was controlled by the mantle plume and lithosphere fault-rift; there formed the diamond deposits in the kimberlite pipes (Fig. 2.2) and the Noril'sk Pechenga super-large nickel–copper–sulfide-type ore deposit (Figs. 4.1 and 4.2), all which were related to the basic and ultra-basic magmatism, and obviously influenced by the mantle stronger activities. The diamond-bearing kimberlite pipes are distributed mainly along the boundary faults or the radiate faults in continental nucleus (Fig. 2.2).

In the East Siberian plate, influenced by the Jurassic collision zone, it formed the Sarylakh super-great gold–antimony ore deposit and the Kolyma gold, silver and tin metallogenic province. At the Transbaikalia (or Mongolia–Okhotsk) Jurassic collision zone, the southeastern side of the Siberian plate, there formed the important gold metallogenic zone (Figs. 4.5 and 4.6). It means that in the Siberian tectonic domain, it is easy for the gold, silver, nickel, antimony, tin and diamond endogenic deposits to reserve. Those ore deposits were mainly controlled by the Jurassic WSW trending collision and compression; in Cretaceous, they were controlled by the near EW extension and influenced by the Pacific plate northward compression and subduction to result in the strong metallogenesis.

4.2.1.1 Mineral Resource Characteristics of Central Asia–Mongolia Tectonic Domain

The Central Asia–Mongolia tectonic domain is dominated by the Paleozoic collision and accretion. There are 41 large and super-large endogenic metal provinces, fields and deposits. In the north of that tectonic domain, the Altay–Middle Mongolia–Hailar accretion–collision zone [6] and Karaganda–Kyrgyzstan accretion–collision zone [7] were formed in Early Paleozoic; in the south of that tectonic domain, the Western Tianshan accretion–collision zone [9] and Balkhash–Tianshan–Hingganling accretion–collision zone [10] were formed in Late Paleozoic. However, the metallogenesis mainly occurred in Late Carboniferous–Early Permian (318–260 Ma), namely the post-collision period. All the regional faults showed the strike-slipping, the NW trending faults turned to sinistral strike-slipping, and the EW trending faults changed to dextral strike-slipping. Although the tectonic movement was not so strong, they controlled the magmatism to form and reserve a large number of endogenic metal provinces and deposits (Figs. 4.16, 4.17, 4.19 and 4.20; Wan et al. 2015).

Some deposits were formed in Triassic, such as the Erdenet copper–molybdenum deposit, in Mongolia (Figs. 4.9 and 4.10), Baishan porphyry molybdenum and rhenium deposit in Xinjiang (Fig. 4.20), Bogdawa Urabdle molybdenum deposit, Hongqiling copper and nickel deposit, Lunin porphyry molybdenum deposit and Baiyinnor skarn lead and zinc deposit in China. In the east of that domain, there also are many deposits formed in Jurassic and Cretaceous, which are similar to the deposits in East China. Although many deposits are located in the collision zone, the metallogenic periods are later than the collision zone, namely in the post-collision period or the intraplate deformation period.

In addition, there are many areas to discover the mantle uplift, i.e., the Moho discontinuity has a little bit uplift. The magmatism caused the new crust to form in that domain (Han and Pan 2009). Thus, in that domain there are many crust–mantle mixture sources and enriched in non-ferrous metal (Cu, Pb, Zn; Figs. 4.8, 4.9 and 4.10; Figs. 4.18, 4.20, 4.21 and 4.22), precious metal (Au, Ag; Figs. 4.11, 4.15 and 4.20), rare metal (Li, Be, Nb, Ta, U; Figs. 4.7 and 4.16) and rare earth element (Figs. 4.7 and 4.27) endogenic metal provinces, fields and deposits. There are a lot of super-large endogenic metal provinces and fields (with the great scale, high grade) in the Kazakhstan, Tajikistan, Kyrgyzstan, Altay, Mongolia, Tianshan to the Hingganling of China, which constitute the important Central Asia–Mongolia endogenic metal metallogenic zone. In addition, the important characteristics are that there are a large number of sedimentary sandstone uranium deposits (Fig. 4.74) in Central Asia–Mongolia tectonic domain. It could be considered that a lot of granitic intrusions rich in the uranium suffered the weathering and leaching for the uranium element to concentrate in the sandstone systems in the basins. In Cenozoic, those areas were still located in the dry climates, so few uranium elements could be scattered and disappeared to get the beneficial condition form the sedimentary uranium fields.

4.2.2 Mineral Resource Characteristics of Sino–Korea Tectonic Domain

After Paleoproterozoic, in the blocks of the Sino–Korea tectonic domain (Figs. 2.14 and 4.73), there formed the uniform crystalline basement. It had suffered the plate boundary fault-depression and later the intraplate deformations to form many special ore provinces, fields and deposits, such as iron, gold, molybdenum, niobium, rare earth elements, magnesium and boron fields and deposits (Figs. 4.27, 4.28 and 4.29). They are mainly distributed on the boundary of the Sino–Korean plate or near some intraplate crustal faults. There are some famous ore provinces, fields or deposits, such as Anshan–Benxi of Liaoning, Qian'an–Qianxi of Hebei iron field; the Wengquangou of Liaoning and Ji'an of Jilin boron metallogenic zone; Dashiqiao of Haicheng to Yingkou magnesite deposits; Komdok super-large lead and zinc deposit; Huadian of Jilin famous large gold deposit; the Baiyun-Obo of Inner Mongolia world-class supergiant REE and niobium (Nb) ore field (Fig. 4.27); the Eastern Qinling Mesozoic molybdenum, gold, lead, zinc and silver etc. polymetal metallogenic belt and the Xiaoqinling gold fields (Fig. 4.28); Sanshandao–Jiaojia–Linglong of Shandong gold field (Fig. 4.29). The two biggest gold provinces (East Shandong and Xiaoqinling of Henan) are located the regional geochemical anomalies (Figs. 4.75 and 4.76).

It seems that the gold element is distributed rather wide (Fig. 4.75). So the gold element in the blocks of China also has the opportunity to concentrate. The Shandong and Qinling areas have the favorable condition to form the great-scale gold deposits.

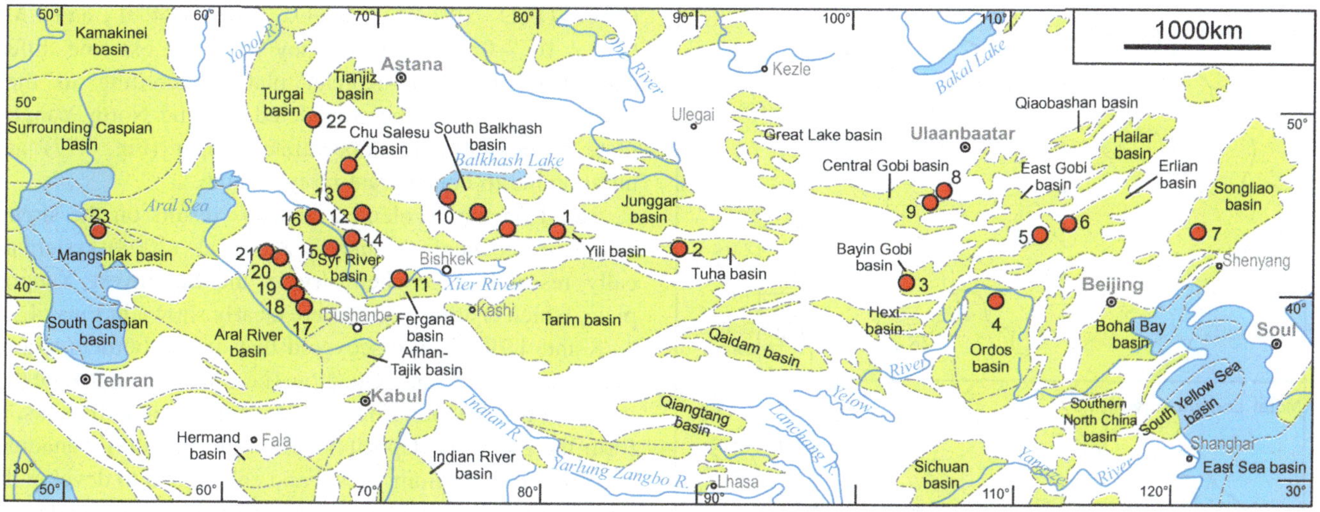

Fig. 4.74 Distribution of sedimentary uranium ore fields in the Central Asia–Mongolia tectonic domain. No. of ore deposit: 1. Southern Yili basin uranium field; 2. Shihongtan uranium deposit in the Southern Tuha basin; 3. Tamusu uranium deposit in Bayin Gobi basin; 4. Dongsheng uranium field in Northern Ordos basin; 5. Nuhetin uranium deposit in Erlian basin; 6. Saihangaobi–Bayanyula uranium deposit in Erlian basin; 7. Qianjiadian uranium deposit in Songliao basin; 8. Haihan uranium deposit; 9. Hart uranium deposit. 10. South Balkhash Lake uranium zone (coal rock type); 11. Malisui uranium field; 12. Kanrugan–Uwanas uranium zone; 13. Yingkai–Menkuduk uranium zone; 14. Kalaktao uranium field; 15. Kisirkli–Kanimah uranium field; 16. Kalamurun uranium field; 17. Ktemenqi–Sabersayi uranium field. 18. Bujnayi–Kanimah uranium field; 19. Lefleyakan–Bishkaik uranium deposit; 20. Siglalei uranium field; 21. Uqikuduk uranium field; 22. Lazalevskeya uranium field. The deposits without number are the sandstone uranium deposit (After Jiao et al. 2015)

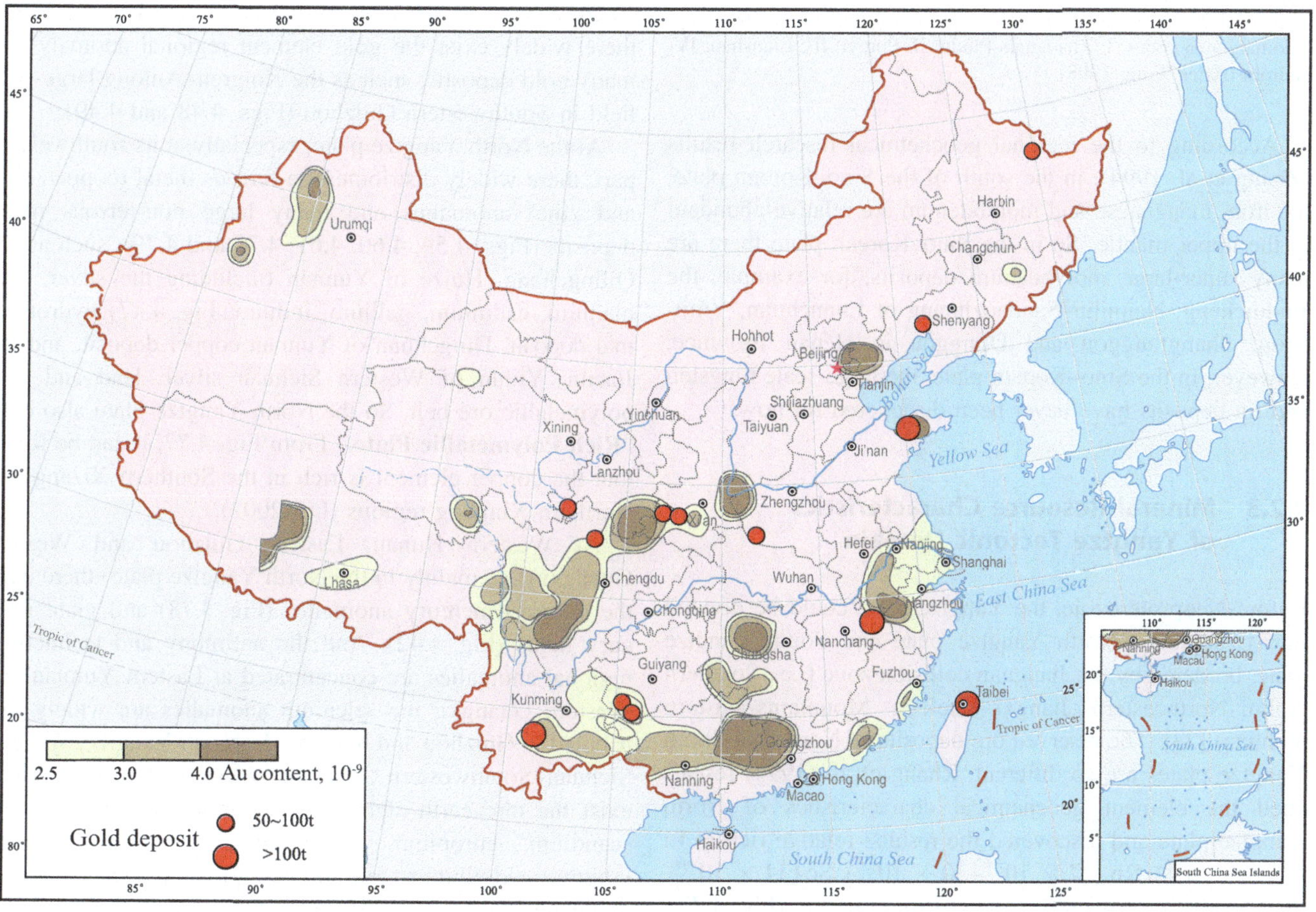

Fig. 4.75 Distribution of gold geochemical blocks and the large gold deposits in China continent. The regional geochemical anomalies based on the stream sediment survey at scale of 1:1,000,000 (After Wang and Xie 2000)

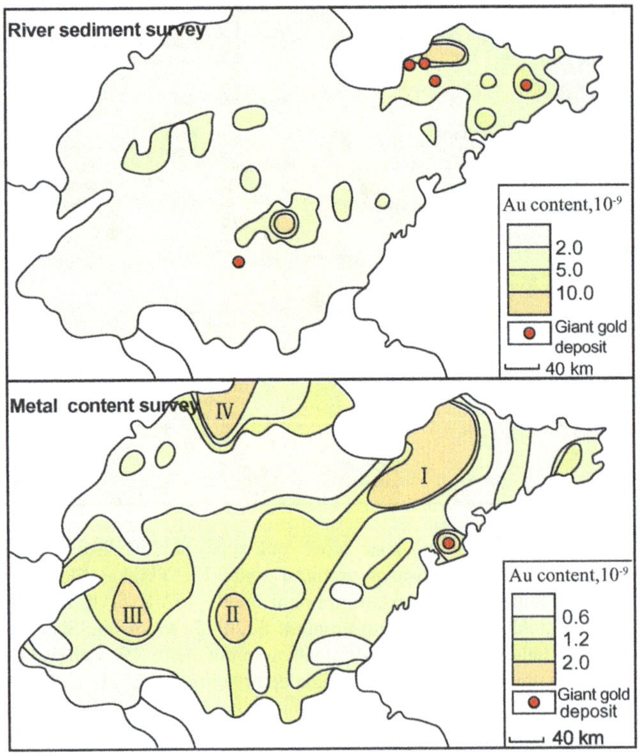

Fig. 4.76 Gold element anomalies by the stream sediment survey at scale of 1:200,000 and giant gold deposit distribution in Shandong. Anomaly area codes: I. Zhaoyuan–Pindu; II. Pingyi; III. Lianshan; IV. Huimin (After Wang 1998)

According to the regional geochemical research results (Zhang et al. 1994), in the south of the Sino–Korean plate, the iron, manganese and molybdenum are relative abundant in the upper mantle. So in the Sino–Korean plate there are many super-large molybdenum deposits, for example, the Jinduicheng, Nannihu–Sandaozhuang of Luanchuan, Yinjiagou, Shangfanggou and Donggou in Henan Province. However, in the Sino–Korean plate, the large-scale tungsten and tin deposits have never been discovered till now.

4.2.3 Mineral Resource Characteristics of Yangtze Tectonic Domain

Before Neoproterozoic, the Yangtze plate could be divided into two plates: North Yangtze plate and South Yangtze plate, bordered by the Jiangnan collision zone (i.e., Southern Anhui–Northeastern Jiangxi–Xuefeng Mountains–Eastern Yunnan) [23]. The reserved ore deposits of North and South Yangtze plates are so different. Zhang et al. (1994) researched the element geochemical characteristics of North Yangtze plate and discovered the results: relative rich in Li (26.3×10^{-6}), Rb (27×10^{-6}–30×10^{-6}), Sc (34×10^{-6}–46.7×10^{-6}), Cu (80×10^{-6}–126×10^{-6}) (Fig. 4.77), relative high ratio of Nb/Ta (16–25), lower ration of Zr/Hf (40), TFeO (9.14%), MgO (5.19–6.84%) and Mo (0.3×10^{-6}–0.54×10^{-6}). They have the extreme difference from the Sino–Korean plate. According to their research data, it can be considered that the North Yangtze plate is favorable to concentrate copper (Fig. 4.77) and unfavorable to concentrate molybdenum.

Based on the 76-element analysis results on the stream sediment geochemical survey, Xie et al. (2008) systematically researched geochemistry in the Southwest Yangtze plate, which characteristics were clearly shown in Figs. 4.77, 4.78 and 4.79. They discovered that on the borders among Sichuan, Yunnan and Guizhou (i.e., West Yangtze plate), there are the obvious iron family element anomaly areas, to form the giant ore deposits of the iron, titanium, vanadium and copper, chromium, nickel and cobalt, and to develop the large-scale Panzhihua vanadium, titanium, magnetite deposit and also to reserve the regional anomaly of platinum family elements (platinum, palladium, osmium, ruthenium, lawrencium) and many deposits, such as Jinbaoshan, Midu large platinum deposit. Their concentrations are obviously related to the near NS trending lithosphere fault on the western side of the Yangtze plate and related to the activity of basic and ultra-basic magmatism from the Emeishan great basalt volcanic province. In Western Sichuan, Western Guizhou and Eastern Yunnan (west of North Yangtze plate), there widely exist the gold element regional anomaly and many gold deposits, such as the Xingren–Anlong large gold field in Southwestern Guizhou (Figs. 4.48 and 4.49).

At the North Yangtze plate, especially at its southwestern part, there widely distribute non-ferrous metal (copper, lead and zinc) anomalies and many large non-ferrous metal deposits (Figs. 4.59, 4.60, 4.61, 4.77 and 4.79), such as the Qilingchang, Huize of Yunnan (including the silver, germanium, cadmium, gallium, indium; Fig. 4.47) hydrothermal deposit, Dingchuan of Yunnan copper deposit, and the Jiacun, Yidun of Western Sichuan silver, lead and zinc polymetallic ore belt. So the North Yangtze plate also calls "**Rich Polymetallic Plate.**" From Fig. 4.77, it can be found that the copper element is rich in the Southern Xizang and Northern Xinjiang regions (Liu 2002).

At Western Hunan, Eastern Guizhou and Western Guangxi (i.e., mainly in the North Yangtze plate) there exist the obvious mercury anomalies (Fig. 4.78) and giant mercury fields (Fig. 4.42). And the antimony and technetium regional anomalies are concentrated at Eastern Yunnan and Western Guangxi; the selenium anomalies are widely distributed in Guizhou and Yunnan. In North Yangtze, Western Sichuan–Southwestern Guizhou–Western Guangxi, there exist the rare earth element (lanthanum, cerium, niobium, scandium, europium, neodymium, samarium, yttrium, erbium, gadolinium, terbium, dysprosium, etc.) regional anomalies, to form the Maoniugou, Mianning, Sichuan giant rare earth element metallogenic belt.

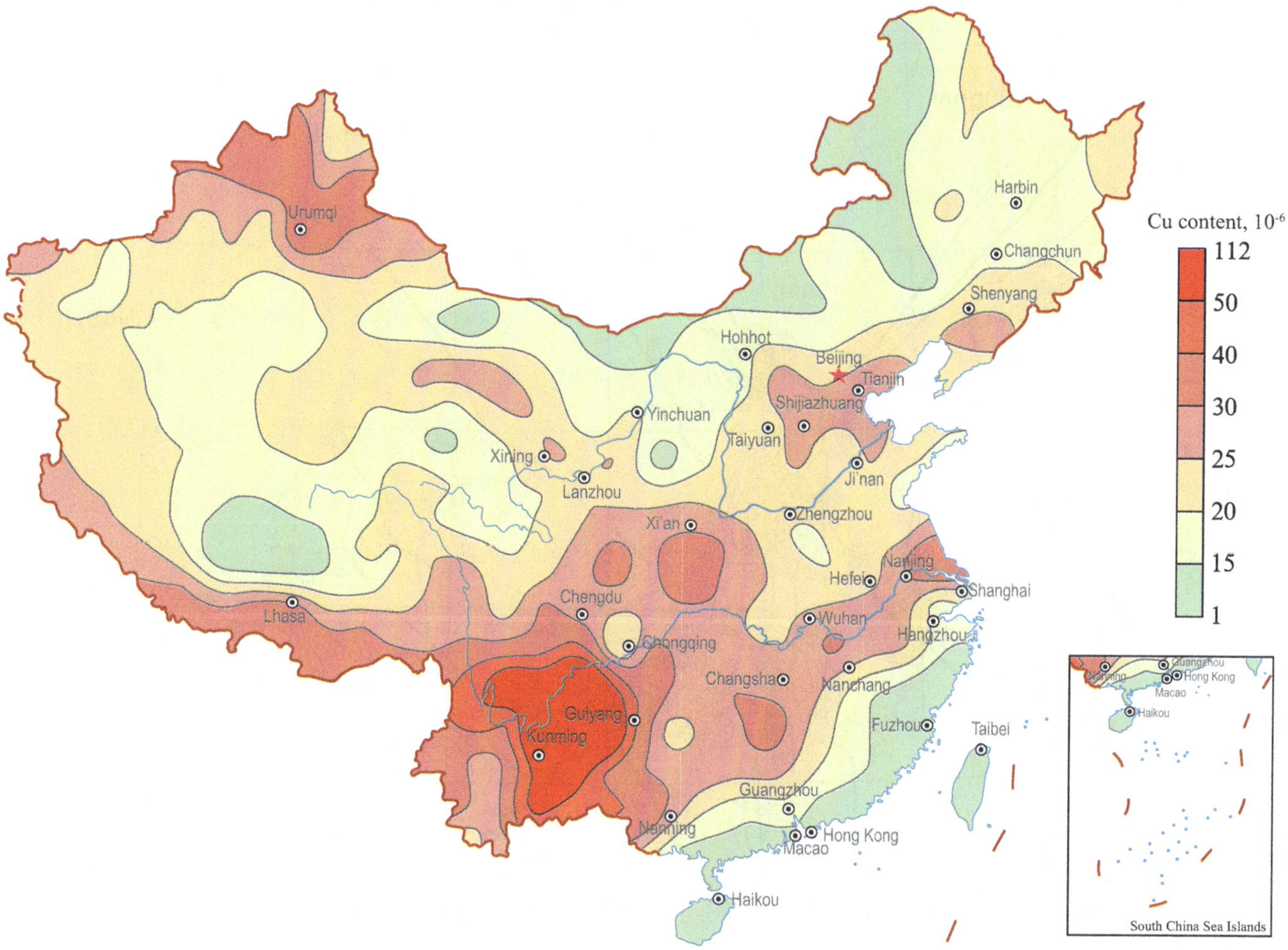

Fig. 4.77 Copper geochemical blocks in China continent. The regional geochemical anomalies based on the stream sediment survey at scale of 1:1,000,000 (After Xie and Xiang 1999; Xie and Cheng 2001)

In Guangxi and Eastern Yunnan of the South Yangtze plate, there are the biggest antimony and tin fields and tin anomaly areas in the world, such as Dachang of Guangxi tin field (Figs. 4.43 and 4.44), Xikuangshan, Lenshuijiang of Hunan antimony deposit and Gejiu tin (copper) polymetallic field (Figs. 4.45 and 4.46). However, in the North Yangtze plate, the tin element anomaly areas or giant tin or antimony deposits have been never discovered. In the above element anomaly areas, the element contents are usually higher than the background values as tens or hundreds of times. Using the above element anomaly data, it has explored and exploited many giant ore deposits.

Based on the above data, it can be seen that the regional element anomalies are really the base of metallogenesis and can guide the exploration orientation. It is reasonable that near the element anomaly areas there reserve the related ore resources. However, the stream dispersion flow anomaly areas are not completely coincided with the giant ore deposits. Usually, the possible position of deposits should be explored in the opposite direction of the element anomaly migration. In addition, due to most of the similar data, such as the data of stream sediment geochemical survey, have never been published, when to discuss the regional metallogenesis and exploration trend, it will have some troubles. However, according to the recent data, it is explained that the ore deposit formation is closely connected with the regional background anomalies.

The North Yangtze plate is characterized by the formation of Fe, Cu, V, Hg, Au and rare earth deposits. In its eastern part, there are mainly the iron, copper and gold province of Southeastern Hubei and Tongling of Anhui (Figs. 4.35 and 4.36), Chengmenshan of Jiangxi copper–Yangchuling tungsten fields, Dexing of Jiangxi copper and polymetallic province (Fig. 4.40); in Northwestern Yunnan, there are the Beiya, Heqing gold and polymetallic deposit. So it seems that the North Yangtze plate could be called as "**Rich Iron and Copper Plate.**"

In South Yangtze plate and the Eastern Hindukush–Northern Qiangtang–Indosinian plate [27], there are the similar ore deposits, which are characterized by enriching in

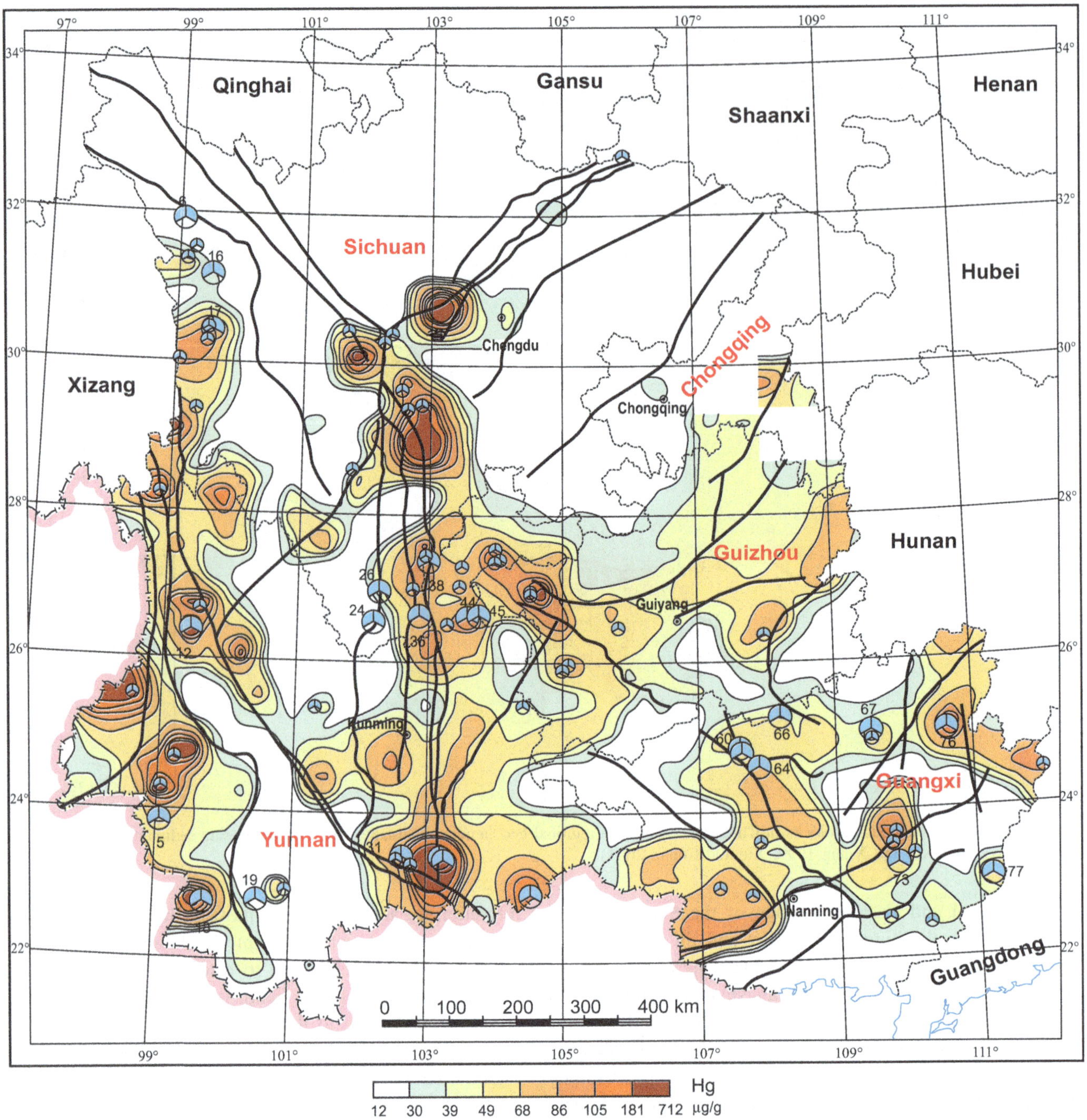

Fig. 4.78 Mercury geochemical anomalies and giant mercury deposits in Southwest China. No. of giant ore deposit and the mineral species: 43. Lupeng of Tongren, Hg; 44. Baiyanwu of Tongren, Hg, 45. Datongla of Tongren, Hg; 46. Lengfengdong of Wanshan, Hg; 50. Uancangping of Tongchang, Hg; 51. Heijianzi Wanshan, Hg; 54. Shamudong of Wanshan, Hg; 55. Wanshan, Hg and Sb; 77. Mazhubao of Youyang, Hg; 107. Muyouchang of Wuchuan, Hg; 112. Hongfachang of Danzai, Hg; 142. Zhangjiawan of Wanshan, Hg; 158. Baimadong of Kaiyang, Hg–U–Mo; 197. Dabazhou of Xingren, Hg–Tl (After Xie et al. 2008)

Sn, Cu, Pb, Zn, Sb and W deposits, to form the Guangxi and East Yunnan super-large tin fields (Figs. 4.43, 4.44, 4.45 and 4.46), Sungai Lembing, Eastern Malaya Peninsula tin deposits and the Malaya tin ore belt (Fig. 4.62). So the South Yangtze plate could be called as "**Rich Tin Plate.**"

In the Eastern Hindukush–Northern Qiangtang–Indosinian plate [27], there is the Great Kanimansur (Большой Канимансур) silver and polymetallic deposit (Fig. 4.63), Dongmuzazhua, Muhailahen and Chaqupacha of Qinghai giant lead and zinc deposits (Fig. 4.59), Yulong of Eastern

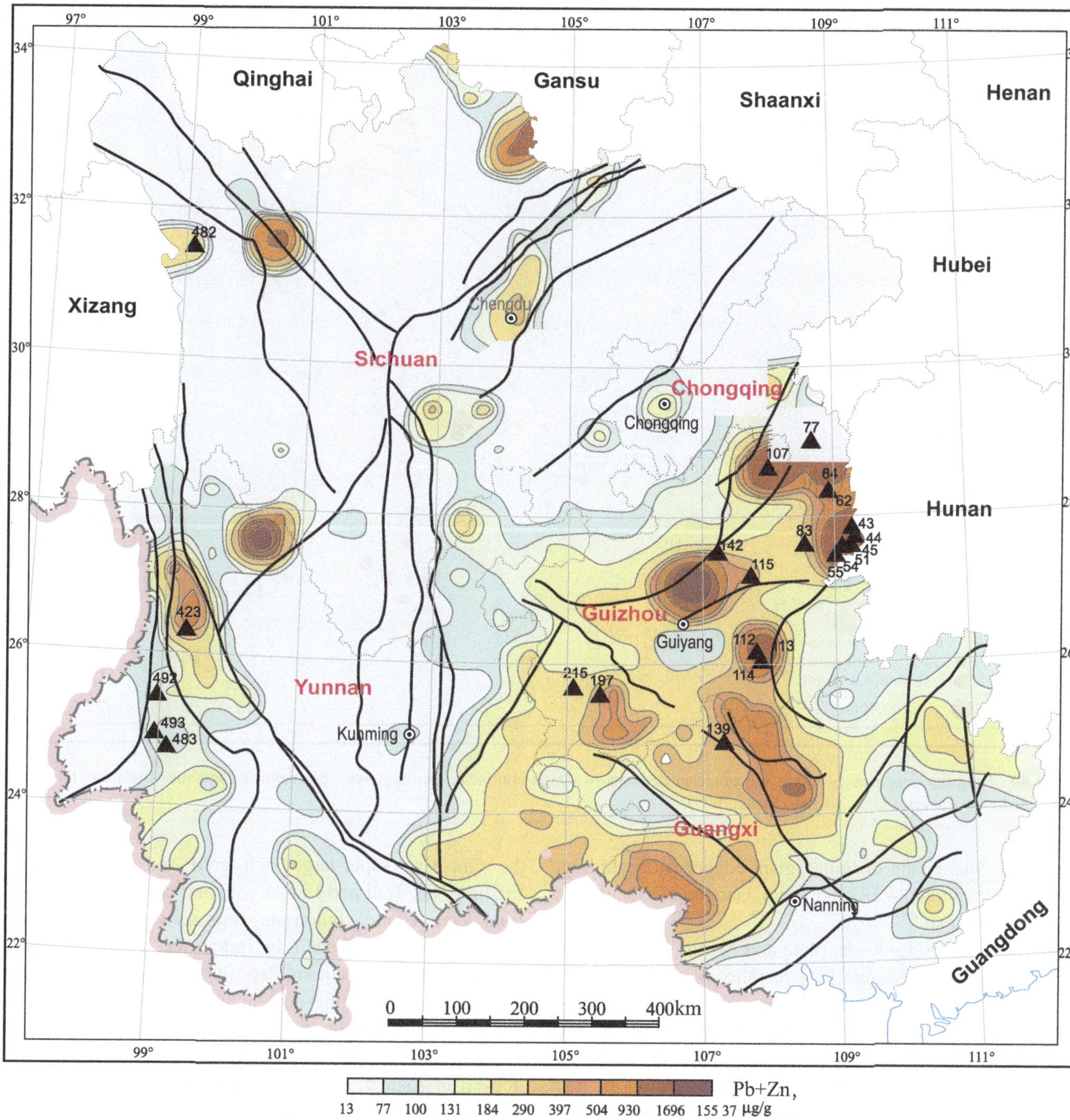

Fig. 4.79 Lead–zinc geochemical anomalies and giant lead–zinc deposits in Southwest China (After Xie et al. 2008)

Xizang porphyry copper metallogenic zone (Fig. 4.60), Jinding of Yunnan lead, zinc and silver field (Figs. 4.61, 4.77 and 4.79). So the Eastern Hindukush–Northern Qiangtang–Indosinian plate could be called as "**Rich Non-Ferrous Metal Plate.**"

However, the Cathaysian plate has the different characteristics, which mainly reserves W, Ag, Pb, Zn, Cu, U, Au, F, rare earth and tin deposits, to form the greatest tungsten metallogenic zone in the world (Fig. 4.80), such as the Xihuashan, Dayu of Jiangxi tungsten deposits (Fig. 4.56), Dajishan, Quannan of Jiangxi, Piaotang–Muziyuan, Dawangshan, Yudu, Huamei'ao, Pangushan, Tieshanlong–Huangsha tungsten deposits, Xintianling and Yaogangxian of Hunan tungsten deposits, Hinglukeng tungsten deposit etc. Thus, the Cathaysian plate could be called as "**Rich Tungsten Plate**" (Fig. 4.80). In the Cathaysian plate, there is the

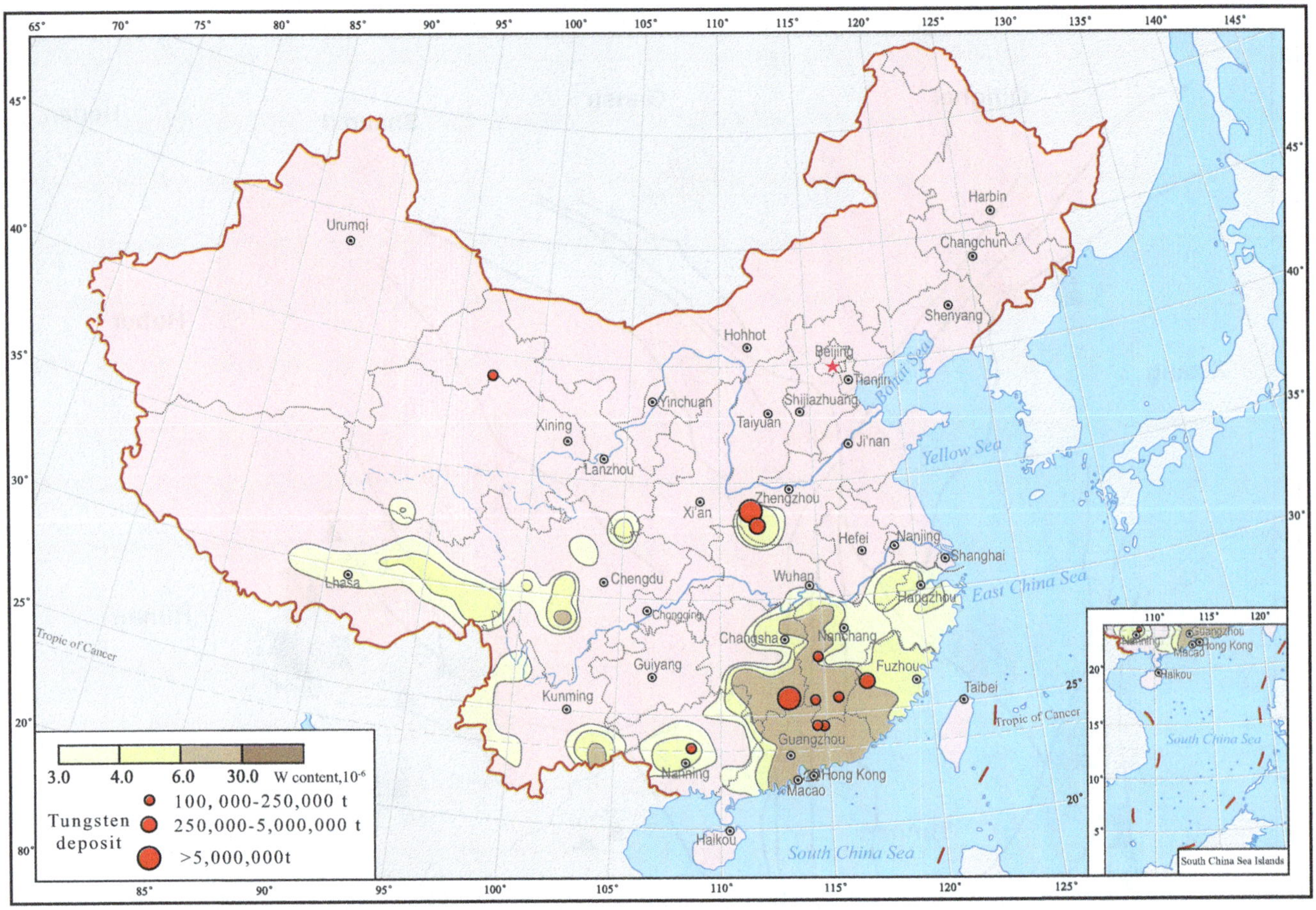

Fig. 4.80 Tungsten geochemical block and tungsten deposits in China continent. The regional geochemical anomalies from the 1:1,000,000 stream sediment survey (After Xie and Xiang 1999)

Zijinshan, Shanghang of Fujian giant copper and gold large field (Figs. 4.57 and 4.58) and Hetai, Gaoyao of Guangdong gold large deposit.

Between the Sino–Korea plate and North Yangtze plate, the Qinling collision zone [24] has the metallogenic characteristics of both Sino–Korea plate and North Yangtze plate. Similarly, between the South Yangtze plate and Cathaysian plate, the Shaoxing–Shiwandashan collision zone [25] also has the metallogenic characteristics of both South Yangtze plate and Cathaysian plate. The others also have the similar regularity.

4.2.4 Mineral Resource Characteristics of Gondwana Tectonic Domain

In the crystalline basement of India, Saudi Arabia and Iran ancient continental blocks, there are mainly the large iron, chromium, titanium and rare-element deposits. In the Cenozoic collision zone or intraplate regional faults, there exist many world-class non-ferrous metal deposits, such as the Grasberg super-large rich gold and copper deposit, Ertsberg, Ok Tedi and Frieda gold and copper deposit in the Irian Jaya Island, Indonesia (Fig. 4.71), the Batu Hijau copper and gold deposit in Indonesia, the Sar Chesmeh copper and molybdenum deposit in Iran (Fig. 4.66), the Anguran lead and zinc deposit in Afghanistan (Fig. 4.67), the Oman copper deposit, the Monywa copper field in Burma and the Qulong super-large porphyry–skarn copper deposit in Xizang, China (Fig. 4.64). In addition, in Burma there formed the Hpakan jadeite field, which is the only large jadeite field in the world.

4.2.5 Mineral Resource Characteristics of Western Pacific Tectonic Domain

In the Western Pacific tectonic domain, there are the Orient No. 2 in Sikhote–Alin and Lemontov scheelite-sulfide deposit in Russia, the Lepanto porphyry copper and gold field in Indonesian, Far Southeast porphyry copper and gold field in Philippines, the Uwamuki Kuroko-type deposit and the Hishikari gold deposit in Japan. It seems that the Western Pacific tectonic domain is also characterized by enriching with copper and gold. At the Dinagat Island of Philippines, there are the laterite-type giant nickel deposits. In addition, near the Kurile Island marine areas there are some abundant oil and gas fields.

To sum up, in the East Asian thinner continental crust and oceanic mantle-type lithosphere (east to Okhotsk–Western Dahingganling–Middle Shanxi–Wuling Mountains–Tak of Thailand [67]) area, there are the favorable conditions to reserve the endogenic metallic deposits. According to the statistics by the author (to see the Appendix), there are 121 giant ore provinces, fields or deposits, occupying the half of those in Asian continent, in which there are 81 endogenic metallic deposits and 40 exogenic deposits. In the area of continental crust and oceanic mantle-type lithosphere, it is easy to form a lot of endogenic metallic deposits. The reason may be as follows: the lithosphere thickness was thinner, with the higher geothermal gradient, and it had suffered the horizontal compression of oceanic plate to form the stronger intraplate deformations and many fractures in the crust, then the partial decreasing pressure and increasing temperature caused the stronger magmatism and resulted in the ore fluids to migrate upward easily. However, those areas were far away from the subduction belt, most of them were not formed in the collision period, and then the rock deformations were not so strong. Thus, the moderate intraplate deformations resulted in the favorable conditions for the magma and ore fluid to migrate and reserve in the crust, and not to scatter or lose too much, then to form the endogenic metallic deposits.

By the research (Wan 2011), the magma and ore fluid are considered to be located at the bottom of crust or near the middle crust (Fig. 4.91). They were uplifting along the crustal or basement faults, cooling and being reserved in the shallow part of crust, then to form the endogenic metallic deposits. However, it is not as to the recognitions of Goldfarb and Santosh (2014) that the endogenic metallic deposits were caused by the oceanic plate and subducted underneath to the 500–600-km depth to get the dehydration and decarbonization of asthenosphere uplifting, to form the metasomatic mantle wedge and to migrate upward along the faults, then the endogenic metallic deposits were formed and controlled by the faults. Till now, researchers have not yet found any evidences that the asthenosphere uplifting could be caused by the deep oceanic plate. Their recognitions are just a conjecture not to get any evidences, which are not different from the recognitions of Chen (1996, 2000, 2006), Hou (2010) and Mao et al. (2012c), i.e., the orogeny is most favorable to form the endogenic metallic deposits.

The East Asian areas are characterized by the dense population, lower landform, developed economy, convenient transportation, as well as the condition for the industry basement. A lot offamous mines belong to the "crisis mine" that the proved reserves are not enough to satisfy the requirement in next five years. The recent explored depth of endogenic metallic deposits is usually only 500 m, someone up to 800–1000 m, and most of being explored oil and gas fields commonly only more than 2000-m depth. According to the recent technology and economic conditions, the deep explored depth in East Asia could reach 2000 m for the endogenic metallic deposits and 5000 m for the natural gas, coal-genetic gas and shale gas fields. So in the old metallogenic provinces or oil-gas-bearing basins, the deep explorations will have the great potentiality and bright future. Due to the good mining basement, perfect beneficiation and metallurgical equipment, advance technology and good transport services in East Asia, the exploitation of deep hidden deposits will be got for the lower prime cost and the better economic benefit. Of course it will be more difficult in terms of technology.

Based on the above data, it can express audaciously:

The Central Asia–Mongolia tectonic domain is the "Rich Polymetallic and Uranium Domain";
The Sino–Korean plate is the "Rich Molybdenum Plate";
The North Yangtze and Eastern Hindukush–Northern Qiangtang–Indosinian plates are the "Rich Iron, Copper and Nonferrous Metal Plate";
The South Yangtze plate and Indosinian plate are the "Rich Tin Plate";
The Cathaysian plate is the "Rich Tungsten Plate";
The Western Pacific areas are the "Rich Copper, Gold and Nickel Zone".

Depending on the recent incomplete data, it is obviously that different tectonic domains and plates had the different crystalline basements and suffered the different tectonics–magmatism–metamorphism–thermal events to form various ore deposits. It means that some related and special elements and chemical compounds can be enriched in each tectonic unit, which is the formation basis of later metallogenesis. Like the "Mendelian Factor" in the biota, the earlier mineralization compositions can be "inherited" into the later deposits. As Pei et al. (2004, 2007) pointed that some special ore deposits could be formed in each tectonic unit of the continental tectonic zone. So in the exploration, we must first consider which kind of deposit to be reserved in a tectonic unit, instead of exploring all kinds of deposits in a tectonic unit or domain. After the tectonic unit being determined, it needs to research a series of tectonic and geological process, then to study the formation, evolution and storage conditions of the deposits. As to the formation, evolution and preservation of the super-large deposit, it is extremely particular. So it must consider the special condition and characteristics in different tectonic metallogenesis.

4.3 The Tectono-Metallogenesis in Tectonic Periods

In this monograph, the author has counted 242 supergiant and giant metallogenic provinces, ore fields or ore deposits. Due to some deposits or fields being formed in multi-periods, these ore fields or deposits can be calculated for two or three times. The numbers of ore fields or deposits are calculated as follows: 13 in Archean–Paleoproterozoic, 24 in Meso–Neoproterozoic, 14 in Early Paleozoic, 46 in Late Paleozoic, 156 in Mesozoic–Paleocene (of them 42 in Triassic, 61 in Jurassic–Early Cretaceous, 53 in Middle Cretaceous–Paleocene), 129 since Eocene (of them 51 in Eocene–Oligocene, 63 in Neogene–Early Pleistocene, 15 in Middle Pleistocene to Holocene). Thus, according to the metallogenic periods, the calculated total numbers of deposits in each metallogenic period are 382 (Fig. 4.81).

According to the calculated results, the deposits or fields that metallogenesis occurred in Archean, Proterozoic and Early Paleozoic are relative less, which total numbers are 51, only occupying 13.35% of those in the whole metallogenic periods. However, the numbers since Late Paleozoic occupy 86.65% of those in the whole metallogenic periods (12% in Late Paleozoic; 40.8% in Mesozoic–Paleocene and 33.76% since Eocene). It is obviously that the Mesozoic–Cenozoic was the climax of metallogenesis, and more than 70% of deposits or fields were formed in Mesozoic–Cenozoic. But, it could not consider that so many deposits would not be formed in the earlier geological periods. It is mostly possible that the earlier deposits had been reformed, destroyed or covered in the deep crust. In fact, since Late Paleozoic, the storage conditions of deposits in the Asian continent are better not to be destroyed greatly, and their burial depths are not so deep, so they are easy to discover, explore and exploit in recent.

From a global viewpoint, in the stable ancient continental blocks, the collisions and intraplate tectonics had been rather weak since Paleozoic, such as the North American, South African and the Australian cratons, the metallogenic periods were mainly in Archean and Proterozoic, but not in Mesozoic–Cenozoic. In the different areas, there are the different characteristics of tectonic evolution. From the view of geosciences, the regional differences are considered specially. It cannot use absolutely the results or methods from the other discipline fields, such as mathematics, chemistry and physics. In those basic sciences, many fundamental theorems and laws have the universalistic approach, but not to determine the regional differences. The influences in different tectonic periods could not only control their types of ore deposits, but also control the space location, pattern, occurrence and reservoir condition for the ore deposit or field, due to the different dynamic characters of tectonic periods.

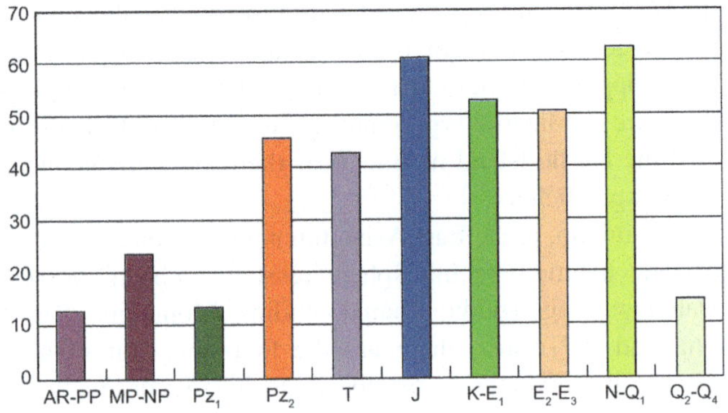

Fig. 4.81 Numbers of ore fields or deposits in each metallogenic period

To explore deep hidden deposits, the key problem is to demonstrate the space distribution of deposits and at the same time to make clear the dynamics in the tectono-metallogenic period is the most important project, especially for the deposits since Late Paleozoic, because the data are rather sufficient. It needs to discuss more accurately in the Asian continent.

4.3.1 The Tectono-Metallogenesis in Archean–Paleoproterozoic

In the metamorphic systems of Archean–Paleoproterozoic periods, many different marine sedimentary–metamorphic deposits were formed mainly (see the Appendix), such as the bedding iron formations (BIF) related to the marine volcanic eruption, boron metallogenic zone, and lead, zinc deposits, talc and magnesite deposits. The above seafloor sedimentary–metamorphic deposits are mainly distributed in the stronger-activity tectonic depression belts or on the boundary fault-depression zone of the stable continental blocks, with the higher crystal temperature, rather stable deposits and the later not so strong re-deformations.

In Archean and Paleoproterozoic, on the Sino–Korean plate, there developed many large BIF-type iron fields, such as Anshan–Benxi iron fields in Liaoning (122° 57′ E, 41° 02′ N), Qian'an–Qianxi in Hebei of China, Central Inner Mongolia, Northern Shanxi and Western Shandong of China, Maoshan in Korea (Mao et al. 2012b).

Along Yingkou of Liaoning (122.4° E, 40.6° N)–Ji'an of Jilin (126° E, 41.1° N) in China, there are sedimentary systems in the Paleoproterozoic ancient rift to form the boron metallogenic zone, such as the Wengquangou super-large boron deposit, Liaoning (~124° E, 41° N) (Zhang 1984; Mao et al. 2012a). From Haicheng to Dashiqiao of Yingkou, there are the Houxianyu and Zuanmiao supergiant magnetite deposits in South Liaoning Province

(Zhu 1984; Zhao 2009); the Komdok (128° 47′ E, 40° 57′ N) super-large lead and zinc deposit in the Korean Peninsula. The Paleoproterozoic BIF deposits in India are distributed in Qiliya, Zingbum of Bihar state, Orissa, Madhya Pradesh and Karnataka States (Zhang 2009).

4.3.2 The Tectono-Metallogenesis in Meso–Neoproterozoic

The Central Asian areas are characterized by the developed polymetallic and rare-element deposits, which are located on the boundary of continental blocks or their extension zones resulted from the submarine exhalation sedimentation. In Asian continent, only 24 giant ore deposits or fields were formed during Mesoproterozoic–Neoproterozoic (see the Appendix), accounting for 6.3% of the total. The Gondwana tectonic domain was still in the formation process of crystalline basement, so that it would be easy to reserve many relative stable ore deposits.

At that period, in China, the typical ore deposit is the Baiyun-Obo (110° E, 41.8° N), Inner Mongolia which is a world-class supergiant REE ore deposit (Fig. 4.27), also a greatest niobium (Nb) deposit in the world (Mao et al. 2012a). Nearby, there are the massive sulfide-type lead and zinc deposits, such as Dongshengmiao, Bainaimiao, Huogeqi and Tanyaokou deposits as well as the Jinchuan (102° E, 38.3° N) of Gansu copper and nickel (contained platinum) sulfide supergiant deposit, and some rare metal deposits (such as Au, W, Sn, Mo and rare elements) near the boundary of the Sino–Korean continental block. They were all controlled by crystalline basement rifting near the continental block boundaries (Tang and Li 1995). In the western part of Northern Qilian collision zone, there is the Jintieshan iron deposit (Fig. 4.34; Mao et al. 2012b), being formed in Proterozoic (1777 Ma), which belongs to the pre-collision hydrothermal-sedimentary-type iron deposit. In the paleo-fault zone of West Yangtze plate, there is the Mesoproterozoic sedimentary-reformation-type stratified copper deposit, such as Dongchuan deposit (Xiao et al. 2012), which hydrothermal reformation occurred in Neoproterozoic (794–712 Ma). In Shimian County, there is the Neoproterozoic (1000–800 Ma) asbestos deposit related to the mafic-super-mafic bodies (Wan et al. 1988).

In India, more than 90% Mesoproterozoic–Neoproterozoic chromite deposits are concentrated in the Dhenkanalt and Kendujhar, Orissa States (Tan 1983; Zhang 2009), as well as Maharashtra and Madhya Pradesh sedimentary manganese deposits. All those are developed in the Mesoproterozoic–Neoproterozoic shallow metamorphic sedimentary systems (Zhang 2009).

In Iran, there are the Proterozoic iron deposits (Choghart, Chadormalou and Gole-Gohar; Fig. 4.66) with the sedimentary–metamorphic type or volcanic ore pulp type (www.GSI.IR). In Afghanistan near Kabul City, there is an Aynak super-large copper deposit zone formed in Proterozoic (Fig. 4.67).

To sum up, the ore deposits formed at this period and reserved to recent are mainly developed in the block boundary depression zone to form rather stable deposits.

4.3.3 The Tectono-Metallogenesis in Early Paleozoic

Among the Asian giant ore deposits, only 14 ore deposits were formed in Early Paleozoic (to see the Appendix), accounting for 3.7% of the total. In the Central Asian continental blocks, there are some hydrothermal polymetallic and rare-element ore deposits. In the Qilian Mountains, there are the Zheyaoshan of Baiyinchang VMS copper, lead and zinc giant deposits (Fig. 4.33). In the Northern Qilian Mountains, Shijuligou and Xiaoliugou, there are the great skarn-type tungsten, molybdenum, iron, copper and polymetallic deposits, belonging to syn-collision ore deposits formed in the late period of Early Paleozoic (462 Ma). There is the Ta'ergou skarn–quartz vein-type tungsten deposit in Central Qilian Mountains (Mao et al. 2012b), and the Mayuan lead–zinc deposit in Qinling areas, Hanzhong of Shaanxi. In the Altun Mountains, there are the Mangya and Bazhou giant asbestos deposits in the strike-slip collision zone.

In the Early Paleozoic sedimentary systems and wet climate areas, there were reserved as many oil-gas and shale gas fields, such as the Tabei oil field in Xinjiang and many gas fields in Sichuan basin, China (Fig. 4.51), the giant Hazmiyab and Raghib oil fields in Central Saudi Arabia (Fig. 4.87), the giant sedimentary phosphorite deposits in Kunyang, Yunnan and the Dongshanfeng, Shimen, Hunan of China. However in the dry climate areas, in South Siberia, there was reserved as the ultra-giant Nepa supergiant potassium salt deposit.

4.3.4 The Tectono-Metallogenesis in Late Paleozoic

In the Central Asia–Mongolia tectonic domain, there are 46 giant and ultra-giant non-ferrous, rare and precious metal deposits, fields or provinces (Fig. 4.81), accounting for 12% of Asian giant deposits and fields. It is interesting that most of those deposits and fields are located in Early Paleozoic or early period of Late Paleozoic (Late Devonian–Early Carboniferous, 385–323 Ma) collision zones of the Central Asia–Mongolia tectonic domain. However, the metallogenesis was mainly developed after the collision process, i.e., in

Late Carboniferous and Early–Middle Permian (323–260 Ma; Fig. 4.82; Wan et al. 2015), which may be formed by the long-distance effect of eastward compression and migration of the Ural collision zone. Thus, the NW trending pre-existing faults changed to sinistral strike-slip faults and the near EW trending faults changed to dextral strike-slip faults, then to form a lot of ore deposits or fields. In the Central Asia, most of metallogenesis did not occurred in the NS orientation stronger collision process during Late Devonian–Early Carboniferous period. Just in Late Carboniferous and Early–Middle Permian period, the fault activities were relatively weak, and it was easy for the ore fluids to enter into the faults or fractures and to reserve, but not easy to enter into the crust outside; thus, it was favorable to reserve the supergiant endogenic metal deposits or fields (Fig. 4.82) (Wan et al. 2015).

In the West Siberia, due to sedimentation and depression for a long time, there are a lot of coal and oil fields, such as Tonguska coal fields, Turmin (Samotlor), Urengoy, Bovanekov and Yamburg oil and gas fields (Fig. 4.3).

In Kazakhstan, there is the Karaganda sedimentary uranium field and zone and Kasagan oil and gas field in Uzbekistan. In China, there is the Ordos and Tabei oil-gas field in Xinjiang, and in the Yangtze plate, there is the Xialei–Hushun manganese deposit in Guangxi. In the Hazmiyab and Raghib oil fields of the Central Saudi Arabia, the first-period hydrocarbon was formed in Late Paleozoic.

4.3.5 The Tectono-Metallogenesis in Triassic

It must pay attention to that the Triassic deposits in Asian continent are obviously more than those of the pre-Triassic periods (see the Appendix). In the 382 calculated giant ore fields and deposits, there are 42 ones were formed in Triassic, accounting for 11% of those in Asian continent. However being compared with the Mesozoic–Cenozoic deposits, they are not too many. In the past, many researchers recognized that there were rare of deposits formed in Triassic. Mao et al. (2012a) pointed that the Triassic was the period of great-scale metallogenesis in China. That is an inappropriate wording. They selected many small, intermediate and giant ore deposits to calculate and got the conclusions: in Triassic; about 100 ore deposits were formed, of which 62 were small and intermediate ones. In the 38 giant endogenic ore deposits calculated by them, till now some of them have been rare of isotopic ages to result in that the ore deposits seemed to form in Triassic at 200 Ma. The author considers that in the Triassic giant deposits calculated by them, only 28 giant ore deposits are believable, which are quoted in the Appendix of this monograph.

According to the author's research, in Triassic the most of Asian continental blocks had finished jointing together and formed six collision zones (Figs. 3.16 and 3.17). During the Triassic Indosinian tectonic event (according to the recent magnetism orientation), the northward migration and compression occurred, which influenced areas were from the Indochina Peninsula to North Mongolia. The rather strong tectonics caused many fractures to form in the continental lithosphere. However, at that time, because the rocks were broken so strong in the Triassic collision zones, it was easy for the ore fluids to migrate into the lithosphere, as well as to scatter and disappear onto the Earth surface, up to the atmosphere and hydrosphere. Thus, it was not easy to form the ore deposits.

Under the preservation conditions, in the 42 giant and ultra-giant endogenic metal deposits, fields and provinces formed in Triassic (Appendix; Fig. 4.83), only 7 ones were really formed in Triassic collision zone and directly related to the collision, such as the Hutouya polymetallic (Cu, Pb, Zn) deposit, Eastern Kunlun; Liba gold deposit in Gansu; Erlihe gold deposit, Fengxian in Shaanxi; Wenquan, Wushan Mo deposit in Gansu; Baguamiao gold deposit in Fengxian of Shaanxi; Liziyuan gold deposit, Tianshui of Gansu; Manaoqiao gold deposit in Northwestern Sichuan. However, their isotopic dating method and the data should be checked deeply. The other 35 ore deposits or fields in the Asian plate during Triassic were formed in the moderate intraplate deformation, near the Triassic collision zones or in the older collision zones and had been suffered by the intraplate deformations in the older collision zones.

In Triassic, at the position with the Asian continental intraplate deformation not so strong, being influenced by near NS trending shortening and collision, there formed near NS, NNE or NNW trending extension-shear fracture zones (Figs. 3.16, 3.17 and 4.83), which were distributed widely to form the favorable location for the magma and ore fluid to migrate and reserve; finally, many giant and supergiant ore deposits had been formed. The small number of magma and ore fluid entered into the compression-shear fractures with near EW trending, especially filling into the fracture zones on the footwall of reverse faults.

The structural strength for the endogenic metallogenesis and volcano-earthquakes is extremely different. The plate collisions and tectonics are more violent, the rocks are more crushed, and the volcanoes and earthquakes will be more favorable to occur. However, the endogenic metallogenesis needs the moderate tectonics, i.e., the ore fluid is easy to move in that structural setting, but the tectonic activity not to be too violent, then the ore fluid can scatter and disappear into the atmosphere and hydrosphere. It is obviously that the not very strong and moderate tectonics is suitable to form the magmatism deposits, hydrothermal ore deposits and oil and gas fields. It should be said that as to the metamorphism and solid sedimentary ore deposits, the influences of tectonic strength for them are rather weak.

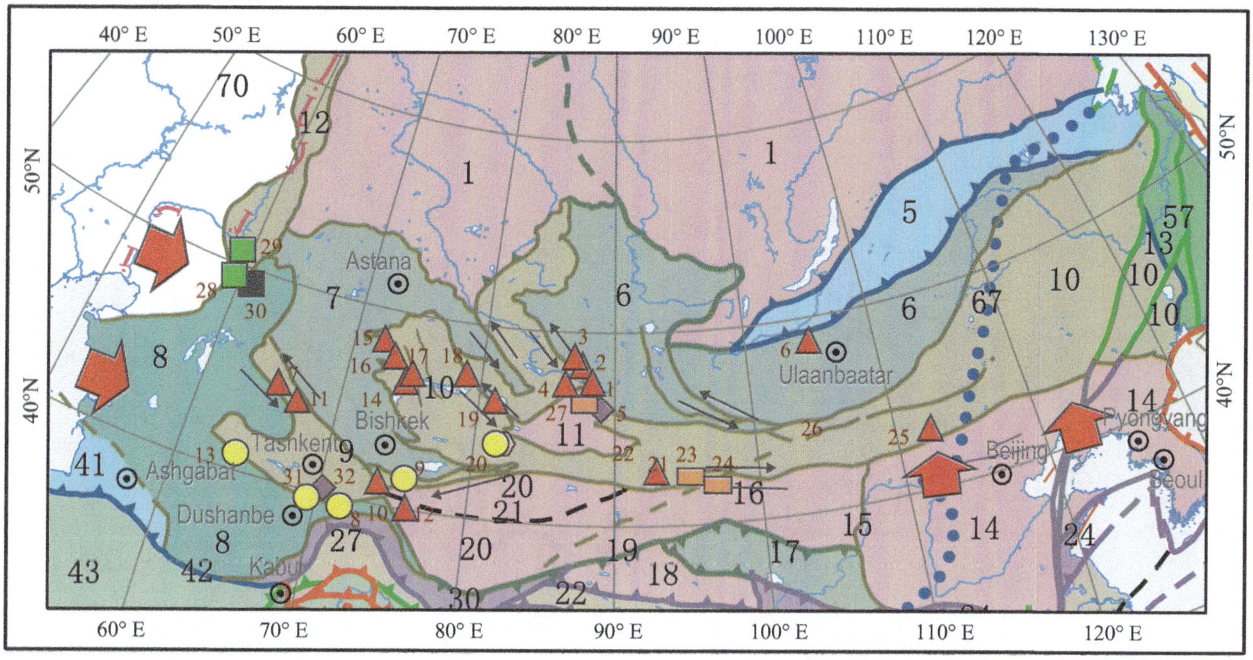

Fig. 4.82 Endogenic metal fields or deposits in the Central Asia–Mongolia tectonic domain formed during the late period of Late Paleozoic (323–260 Ma) The black numbers in the figure show the tectonic units, as same as those in the CONTENTS, Figs. 1.1 and 2.6. The brown numbers show the endogenic metal fields or deposits. The red lines are the boundary of Late Paleozoic collision zone and main faults. The black arrows show the fault migration orientation. The big red arrows show the plate migration orientation. The symbols of ore type: Red triangle shows copper, lead and zinc polymetallic; brown rectangle shows copper and nickel in mafic rock; shallow yellow circle shows precious metal; pink rhombus shows rare metal or rare earth; black block shows iron; green block shows chromite. **The no. of ore provinces, fields or deposits**: **1**. Altay–Zaysan gold, polymetallic and rare metal metallogenic province; **2**. Nikolaev VMS copper and zinc deposit, Kazakhstan; **3**. Ziliangnovsk lead–zinc polymetallic deposit, Kazakhstan; **4**. Ashle massive sulfide copper and zinc ore deposit, Altay, China; **5**. Kokotohai Li, Be, Nb, Ta, Rb, Se and Hf deposit, Fuyuan, Xinjiang; **6**. Erdenet copper–molybdenum deposit, Mongolia; **7**. Karaganda lead, zinc polymetallic zone; **8**. Da Koni Man Sur silver deposit, Tajikistan; **9**. Kumtor gold deposit, Kyrgyzstan; **10**. Pojimoqiak copper, gold deposit, Kyrgyzstan; **11**. Kara Tau polymetallic zone; **12**. Almalyk porphyry copper and gold field; **13**. Muruntau gold deposit, Uzbekistan; **14**. Kounrad copper, gold deposit, Kazakhstan; **15**. Akshatau copper deposit, Kazakhstan; **16**. Zhanet molybdenum deposit, Kazakhstan; **17**. East Kounrad copper deposit, Kazakhstan; **18**. Aktogai copper field, Kazakhstan; **19**. Bogutu of Xinjiang copper deposit, China; **20**. Axi, Yili gold zone, China; **21**. Tuwu copper deposit, East Tianshan; **22**. Kangur gold deposit, East Tianshan; **23**. Xiangshan copper–nickel deposit, East Tianshan; **24**. East Huangshan copper–nickel deposit, East Tianshan; **25**. Beinaimiao polymetallic deposit, Inner Mongolia; **26**. Oyu Tolgoi copper gold deposit, Mongolia; **27**. Karatongk Cu and Ni deposit, Xinjiang; **28**. Aqtobi chromite deposit, Kazakhstan; **29**. Kempirsal super-large ophiolite chromite deposit, Kazakhstan; **30**. Turgy iron deposit, Kazakhstan; **31**. Great Konimansur Ag field, Tajikistan; **32**. Skalinoya antimony deposit, Tajikistan. **The endogenic metal fields or deposits**, located in the Altay–Central Mongolia–Hailar and Karaganda–Kyrgyzstan Early Paleozoic accretion–collision zone, and formed in Late Paleozoic, i.e., post-collision period: Altay–Zaysan gold, polymetallic and rare metal metallogenic province (D–C, Fig. 4.6); Nikolaev VMS copper and zinc deposit, Kazakhstan (D–C); Ziliangnovsk lead–zinc polymetallic deposit, Kazakhstan (D–C); Kokotohai Li, Be, Nb, Ta, Rb, Se and Hf deposit, Fuyuan, Xinjiang (280–270 Ma, P–T; Fig. 4.7); Ashle massive sulfide copper and zinc ore deposit, Altay (262–242 Ma; Fig. 4.8); Karaganda lead, zinc polymetallic zone (D); Da Koni Man Sur silver deposit, Tajikistan; Kumtor gold deposit, Kyrgyzstan (288–284 Ma; Fig. 4.12); Pojimoqiak copper, gold deposit, Kyrgyzstan (Late Paleozoic); Kara Tau polymetallic zone (D_3). **The endogenic metal fields or deposits** formed in the Western Tianshan and Balkhash–Tianshan–Hingganling Late Paleozoic accretion–collision zone (385–260 Ma): **Syn-collision period (385–323 Ma)** [only Kara Tau polymetallic zone and Oyu Tolgoi copper gold deposit, Mongolia (373–370 Ma, Figs. 4.9, 4.21, 4.22 and 4.23)]; **Post-collision period (323–260 Ma)** [Kumtor gold deposit, Kyrgyzstan (288–284 Ma, Fig. 4.12); Almalyk porphyry copper and gold field (320–290 Ma, Fig. 4.14); Muruntau gold deposit, Uzbekistan (290–270 Ma, Fig. 4.15); Kounrad copper, gold deposit, Kazakhstan (284 Ma, Fig. 4.18); Akshatau copper deposit, Kazakhstan (285–289 Ma, Fig. 4.18); Zhanet molybdenum deposit, Kazakhstan (295 Ma, Fig. 4.18); Borly copper and molybdenum deposit (316.3 Ma and 305 $\pm$ 3 Ma, Fig. 4.18); East Kounrad copper deposit, Kazakhstan (284 Ma, Fig. 4.18); Bogutu of Xinjiang copper deposit ($\sim$322 Ma); Axi, Yili gold zone, China (C_3); Sawarden of Xinjiang gold deposit (261 Ma); Tuwu copper deposit, East Tianshan (322 Ma, Fig. 4.20); Kangur gold deposit, East Tianshan (261–252 Ma; Fig. 4.20); Xiangshan and East Huangshan, East Tianshan copper–nickel deposit (261–252 Ma, Fig. 4.20); Baishan porphyry (Mo, Re) deposits, Re–Os isochron age 224 $\pm$ 4.5 Ma and 231.0 $\pm$ 6.5 Ma (Fig. 4.20)]; **Post-collision period (Late Carboniferous–Triassic)** (Ashle massive sulfide copper and zinc ore deposit, Altay, the hydrothermal formation periods of 262–242 Ma (Fig. 4.8); the Dzhikrut mercury, antimony deposit, Kadamse antimony deposit and Khaydarkan antimony and mercury deposit, Kyrgyzstan are being formed in the late period of Late Paleozoic, however never to get the isotopic ages). **The ore deposits formed in Late Paleozoic and rare of the isotopic ages**: the Karaganda lead, zinc polymetallic zone; Da Koni Man Sur silver deposit, Tajikistan; Altay–Zaysan gold, polymetallic and rare metal metallogenic province, Nikolaev VMS copper and zinc deposit, Kazakhstan; Ziliangnovsk lead–zinc polymetallic deposit Kazakhstan; Pojimoqiak copper, gold deposit, Kyrgyzstan; Kara Tau polymetallic zone, Dzhikrut mercury, antimony deposit Kyrgyzstan; Kadamse antimony deposit, Kyrgyzstan; Khaydarkan antimony and mercury deposit, Kyrgyzstan; Aktogai copper field Kazakhstan. Besides, in North Siberia, many diamond kimberlite deposits were formed at the period of about 360 Ma. The Noril'sk nickel–copper–sulfide-type ore deposit was developed in Late Permian intraplate extension zones (Figs. 4.1 and 4.2)

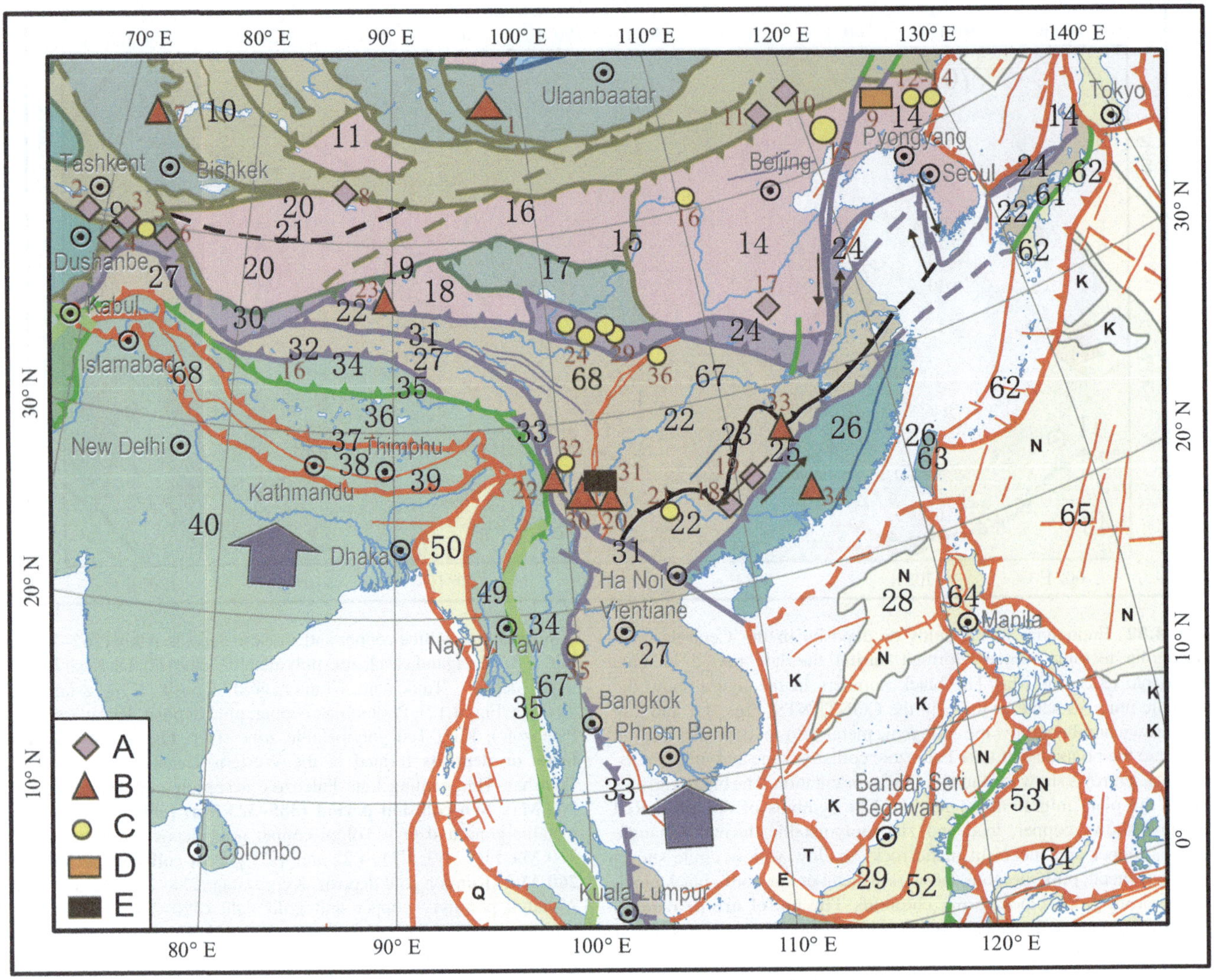

Fig. 4.83 Distribution of Asian endogenic metal fields or deposits formed in Triassic. The black numbers show the tectonic units, as same as those in the CONTENTS, Fig. 1.1 and text. The purple lines show the Triassic collision zones. The big purple arrows show the plate migration orientation. **The symbols of mineral type**: **A**. rare metal, molybdenum, rhenium, mercury and antimony; **B**. copper, lead and zinc polymetallic; **C**. gold, silver precious metal; **D**. copper–nickel in ultra-basic rock; **E**. iron and chromium. The **no. of ore fields or deposits**: **1**. Erdenet of Mongolia copper–molybdenum deposit, Mongolia (Fig. 4.10); **2**. Dzhikrut mercury, antimony deposit, Kyrgyzstan; **3**. Kadamse antimony deposit, Kyrgyzstan; **4**. Khaydarkan antimony and mercury deposit, Kyrgyzstan; **5**. Great Konimansur Ag field, Tajikistan; **6**. Skalinoya antimony deposit, Tajikistan; **7**. Borly copper and molybdenum deposit, Kazakhstan (Fig. 4.18); **8**. Baishan porphyry Mo, Re deposit, China (Fig. 4.20); **9**. Hongqiling copper, nickel deposit, Jilin, China; **10**. Bogdaura molybdenum deposit, Inner Mongolia, China; **11**. Urandle molybdenum deposit, Inner Mongolia, China; **12**. Jiabigou Huadian Au deposit, Jilin, China; **13**. Bajiazi, Huadian gold deposit, China; **14**. Qingchengzi Au, Ag, Pb, Zn deposit, Liaoning, China; **15**. Jinchanggouliang Au, Mo deposit, Inner Mongolia, China; **16**. Hadamengou Au, Mo deposit, Inner Mongolia, China; **17**. Huanglongpu Mo–Pb deposit, Eastern Qinling (Fig. 4.28); **18**. Limu Sn, Nb, Ta deposit, Guangxi, China; **19**. Hehuaping tin deposit Chengzhou of Hunan, China; **20**. Yangla Cu deposit, Jinshajiang, Yunnan, China; **21**. Zimudang gold deposit, Guangxi, China (Figs. 4.48 and 4.49); **22**. Pulan, Xianggelila Cu deposit, West Sichuan, China; **23**. Hutouya polymetallic (Cu, Pb, Zn) deposit, Eastern Kunlun; **24**. Liba gold deposit, Gansu, China; **25**. Erlihe gold deposit, Fengxian, Shaanxi, China; **26**. Wenquan, Wushan Mo deposit, Gansu, China; **27**. Baguamiao gold deposit in Fengxian, Shaanxi, China; **28**. Liziyuan gold deposit, Tianshui of Gansu, China; **29**. Manaoqiao gold deposit Northwest Sichuan, China; **30**. Qilingchang, Huize Pb, Zn (Ag, Ge, Cd, Ga, In) deposit, China (Fig. 4.47); **31**. Western Sichuan Fe, V, Pt, Cu metallogenic province, China (including Panzhihua, Fig. 4.52); **32**. Yitang (Deda) Jiacun Ag, Pd, Zn field, Western Sichuan, China; **33**. Qibaoshan Cu deposit, Liuyang of Hunan, China; **34**. Fankou Pb, Zn, Ag polymetallic deposit, Guangdong, China; **35**. Huai Kam On gold deposit, North Thailand; **36**. Jinlongshan gold deposit, Southern Shaanxi, China

In the academy of mineral deposits, many scholars emphasized the "orogenic belt-type ore deposit" (Kerrich and Wyman 1990; Barley and Groves 1992; Chen 1996, 2000, 2006; Goldfarb et al. 2001; Qiu 2002; Mao et al. 2004b, 2012a, b; Hou et al. 2006, 2010). If only to express the ore deposits occurring in the mountains or orogenic belt (i.e., the collision zone), not to indicate the ore deposits forming in the collision or the mountains formation period, it would not get too much trouble. In the academy of mineral deposits, it usually uses the occurrence and wall rocks to name the ore deposit type, such as "the ductile shear zone-type gold deposit" and "metamorphism gold deposit." In here, it just only emphasizes the wall rocks, however never indicates the cause of formation. In fact, "the ductile shear zone-type gold deposit" is formed after the formation of ductile shear zone, caused by intermediate hydrothermal and reserved in the extension micro-fractures, which are the favorable positions for concentrating the gold elements. So the ductile shear zone is just the favorable position for the gold deposit. As to the "metamorphism gold deposit," it refers that the metamorphic system is just the favorable wall rock position for the gold deposit, but not to indicate the metamorphism metallogenesis for the gold deposit. It is very important, but it is very easy for some scholars to misunderstand and confuse their meanings.

Although the deposits are located in the Altay–Middle Mongolia–Hailar Early Paleozoic collision zone [6], Karaganda–Kyrgyzstan Early Paleozoic collision zone [7] and Balkhash–Tianshan–Hingganling Late Paleozoic accretion–collision zone [10], they were formed in Triassic, such as the Erdenet of Mongolia copper–molybdenum deposit (240 Ma; Fig. 4.10); Great Konimansur Ag field, Tajikistan; Skalinoya antimony deposit, Tajikistan; Dzhikrut mercury, antimony deposit, Kyrgyzstan; Kadamse antimony deposit, Kyrgyzstan; Khaydarkan antimony and mercury deposit, Kyrgyzstan; Borly copper and molybdenum deposit, Kazakhstan (Fig. 4.18); Baishan porphyry Mo, Re deposit, Tianshan (Fig. 4.20); Sawarden of Xinjiang gold deposit; Hongqiling copper and nickel deposit, Jilin of China; Bogdaura molybdenum deposit, Inner Mongolia of China; Urandle molybdenum deposit, Inner Mongolia of China; Hadamengou Au, Mo deposit, Inner Mongolia of China. They were all formed after the collision period. So it would be ill-considered that they belong to the collision-orogenic metallogenesis.

Many important ore deposits were formed in the Triassic intraplate deformation (Fig. 4.83), such as in China the Jiabigou of Huadian Au deposit, Jilin; Bajiazi, Huadian gold deposit; Qingchengzi Au, Ag, Pb, Zn deposit, Liaoning; Jinchanggouliang Au, Mo deposit, Inner Mongolia; Huanglongpu Mo–Pb deposit, Eastern Qinling; Qilingchang, Huize Pb, Zn (Ag, Ge, Cd, Ga, In) deposit; Limu Sn, Nb, Ta deposit, Guangxi; Hehuaping Sn deposit Chengzhou, Hunan; Yangla Cu deposit, Jinshajiang, Yunnan; West Sichuan Fe, V, Pt, Cu metallogenic province (including Panzhihua); Yitang (Deda) Jiacun Ag, Pd, Zn field, Western Sichuan; Pulan, Xianggelila Cu deposit, Western Sichuan; Jinlongshan gold deposit, Southern Shaanxi; Hutouya polymetallic (Cu, Pb, Zn) deposit, Eastern Kunlun; Qibaoshan copper deposit, Liuyang, Hunan; in addition, Huai Kam On gold deposit, North Thailand; Sukhothai kainite deposit, Central Thailand.

In Triassic, some basin areas of Asian continent, located in the dry and hot climate, there formed many sedimentary deposits, such as the Karlyuk–Kalaber sylvite deposit, Turkmenistan; Sukhothai kainite deposit, Central Thailand. However at that time of the warm and damp climate, Ordos oil and gas field begun to form (Fig. 4.87); in the Sichuan basin, the important hydrocarbon source beds also begun to form.

4.3.6 The Tectono-Metallogenesis in Jurassic and Early Period of Early Cretaceous

In the Asian continent, the Jurassic and early period of Early Cretaceous was the climax of tectono-metallogenesis in the intraplate deformation. There formed 61 giant and ultra-giant ore fields and metal deposits (see the Appendix), occupying 16% of the total in the whole Asia, such as the non-ferrous metal (Cu, Pb, Zn), precious metal (Au, Ag), rare metal (W, Mo, Hg Sb, Sn), rare earth element, some skarn-type rich iron deposits and so on (Figs. 4.84 and 4.85).

They were controlled by the intraplate deformations that were caused by the counterclockwise rotation of East Asian continental crust. In the Southeast Asian areas, the maximum principal compression stress orientation is WNW trending. Along that orientation, the structural fractures will be easy to extend moderately, which are the favorable positions for the ore fluids to intrude and reserve. If the preservation positions are located at the ring-shaped structure or on the boundary of near-circle intrusion stock, the ore deposits will usually be formed on the NE or SW side of those structures, such as Daye of Hubei and Tongling of Anhui (Figs. 4.35, 4.36, 4.37, 4.38 and 4.39), Jiujiang of Northern Jiangxi, Dexing of Jiangxi etc. (Fig. 4.40). According to the calculated results, the 70–80% extension orientations of Jurassic–early period of Early Cretaceous ore deposits and fields are along the above orientation, i.e., mainly WNW trending. It could also get some ore fluids intrude and reserve into the NNE or NE trending reverse faults. The Jurassic–early period of Early Cretaceous tectonic event was first named as "Yanshan Movement" by Weng (1929) for East China.

In the Jurassic–early period of Early Cretaceous tectonic period, a large number of tungsten deposits were formed in the Cathaysian plate. It is not only related to the intrinsic element

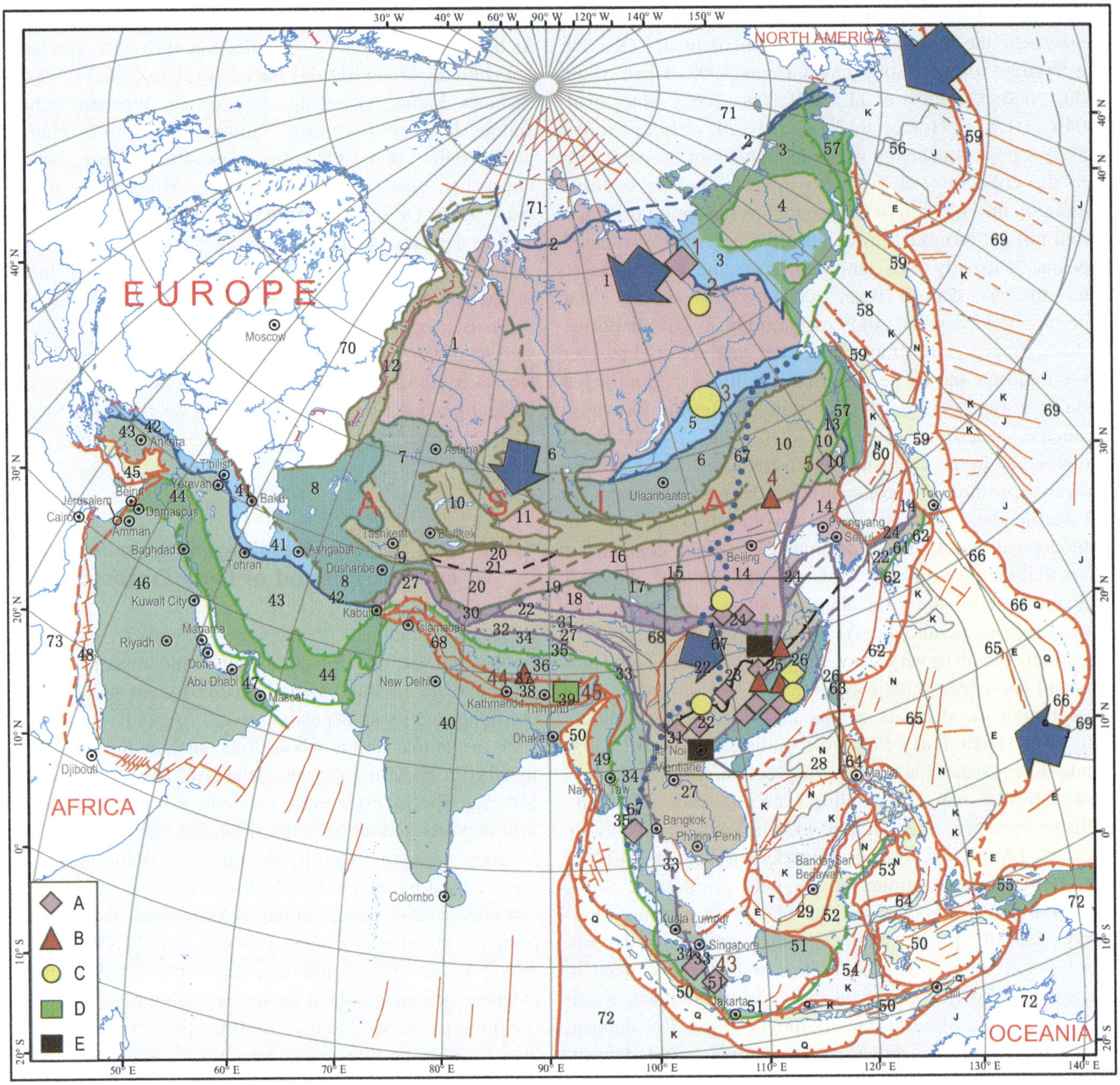

Fig. 4.84 Distribution of Asian endogenic metal fields or deposits formed in Jurassic and the early period of Early Cretaceous. The black numbers show the tectonic units, as same as those in the CONTENTS, Fig. 1.1 and text. The big blue arrows show the plate migration orientation. The numbers with brown color show the ore fields or deposits. The eastern area to the blue point line is the continental crust and oceanic mantle lithosphere, i.e., thinner continental lithosphere. **The symbols of mineral type**: **A**. rare metal, tungsten, molybdenum, rhenium, mercury and antimony; **B**. copper, lead and zinc polymetallic; **C**. gold, silver precious metal; **D**. chromium; **E**. iron. **The no. of ore fields or deposits**: **1**. East Siberia W, Nb hydrothermal deposit; **2**. Sarylakh gold–antimony ore deposit; **3**. Transbaikalia gold zone (Fig. 4.4); **4**. Baiyinnor Pb-Zn deposit, Inner Mongolia of China; **5**. Daheishan Mo deposit, Yongji, Jilin of China (Fig. 4.24); **43**. Sungai Lembing Sn deposit, Eastern Malay tin zone (Fig. 4.62); **44**. Xiemen-tong Xiongcun Jurassic Cu, Au and polymetallic deposit, Xizang of China; **45**. Luobusha chromite deposit, Xizang of China (Fig. 4.65). In this figure, it only shows No. **1–5** and No. **43–45** ore fields or deposits. In Southeast China, due to the giant ore deposits or fields too many, they are hardly to be expressed in this figure, the No. **6–42** ore fields or deposits are shown on Fig. 4.85

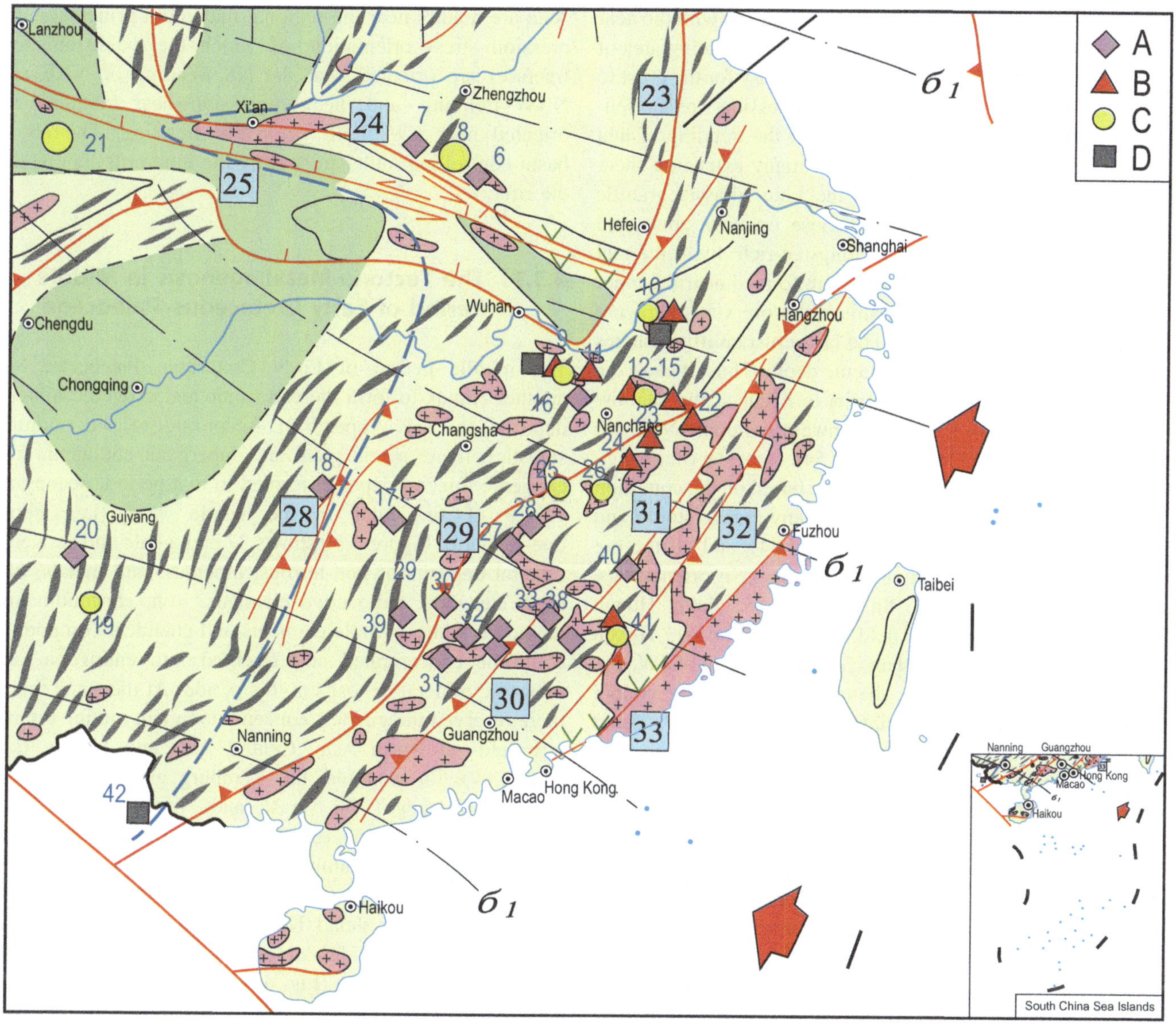

Fig. 4.85 Distribution of Southeast China endogenic metal fields or deposits (No. 6–42) formed in Jurassic and the early period of Early Cretaceous. The black numbers in the blue block show the tectonic units, as same as those in the CONTENTS, Figs. 1.1 and 2.18. The numbers with blue color show the ore fields or deposits. **The symbols of ore type**: **A**. rare metal, tungsten, molybdenum, rhenium, mercury and antimony; **B**. copper, lead and zinc polymetallic; **C**. gold, silver precious metal; **D**. iron. **No. 6 to No. 42 deposits**: **6**. Nanihu–Sandaozhuang Mo, W deposit, in Eastern Qinling (Fig. 4.28); **7**. Jinduicheng, East Qinling molybdenum deposit (Fig. 4.28); **8**. Lingbao–Tongguan, Au field, Xiaoqinling (Fig. 4.28); **9**. SE Hubei Fe, Cu Au province (Figs. 4.35, 4.36, 4.37, 4.38 and 4.39); **10**. Tongling Fe, Cu, Au province, Anhui (Figs. 4.38 and 4.39); **11**. Chengmengshan Cu–Yangchuling W fields, Jiangxi; **12**. Dexing (Tongchang) Cu porphyry province, Jinshan Au deposit, Jiangxi (Fig. 4.40); **13**. Zhushachang copper field, Jiangxi (Fig. 4.40); **14**. Fujiawu Cu porphyry province, Jiangxi (Fig. 4.40); **15**. Jinchan Au deposit, Jiangxi (Fig. 4.40); **16**. Dahutang W field, Jiujiang, Jiangxi (Fig. 4.41); **17**. Xikuangshan Sb deposit, Lengshuijiang Hunan; **18**. Wanshan Hg deposit, East Guizhou (Fig. 4.42); **19**. Xingren–Anlong gold field, Guizhou (Figs. 4.48 and 4.49); **20**. Jinlong of Guizhou and Muli of East Yunnan Sb deposit; **21**. Yangshan gold deposit, Wenxian, Gansu (Fig. 4.53); **22**. Yongping copper deposit, Jiangxi: **23**. Lengshuiken lead, zinc and silver deposit, Guixi, Jinagxi; **24**. Dongxiang Cu, Mo deposit, Jiangxi; **25**. Jiaoli Ag, Pb, Zn, W deposit, Shangyou, Jiangxi; **26**. Shuikoushan Pb, Zn deposit in Changning, Hunan; **27**. Huken W deposit, Wugongshan, Jiangxi; **28**. Songshugang (414) W, Sn, Nb, Ta deposit, Hengfeng, Jiangxi; **29**. Tongshanling Cu deposit, Jiangyong, Hunan; **30**. Shizhuyuan W, Sn polymetallic field, Hunan (Fig. 4.55); **31**. Xihuashan W deposit, Dayu, Jiangxi (Fig. 4.56); **32**. Dajishan W deposit, Quannan, Jiangxi; **33**. Piaotang–Muziyuan W deposit, Jiangxi; **34**. Xintianling W deposit, Hunan; **35**. Dawangshan, Pangushan, Tieshanlong–Huangsha W deposit, Yuduo, Jiangxi; **36**. Huameishan W deposit Yuduo, Jiangxi; **37**. Pangushan W deposit Yuduo, Jiangxi; **38**. Tieshanlong–Huangsha W deposit Yuduo, Jiangxi; **39**. Yaogangxian, Xintianling W deposit, Hunan; **40**. Xingluoken W deposit, Ninghua, Fujian; **41**. Zijinshan Cu and Au field, Shanghang, Fujian (Figs. 4.57 and 4.58); **42**. Shixi iron deposit, Hejing, Vietnam

enrichment in original block, but also connected with the near EW trending hidden lithospheric faults in the Nanling area of China (Wan 2011). At that period, it was easy for the crust to tatter, and the maximum principal compression stress orientation was WNW trending, very near to the trending of that lithospheric fault. Then it resulted in many extension-shear fracture zones in the crust, and the Late Jurassic granitic magma was easy to intrude into those fractures zone, and finally, the high-temperature, tungsten-rich hydrothermal fluids, and rich tungsten deposits were formed near the inner and outer contact zones of granites, the ore veins showing mainly WNW–ESE trending. The later crust uplifting ranges were just similar to the metallogenic depth. Thus, it caused a great quantity of tungsten deposits to occur at the shallow positions in Southern Jiangxi–Southwestern Fujian–Northern Guangdong–Eastern Hunan (Fig. 4.54). Near the Shaoxing–Shiwandashan Triassic collision zone, based on the condition of early fractures, there also formed many granite stocks and many tungsten and tin deposits (Fig. 4.54), and all the ore veins are distributed mainly along WNW–ESE orientations, such as Shizhuyuan W, Sn polymetallic field, Hunan (Fig. 4.55); Dahutang W field, Jiujiang, Jiangxi; Huken tungsten deposit, Wugongshan, Songshugang (414) W, Sn, Nb, Ta deposit, Hengfeng, Jiangxi; Tongshanling copper deposit, Jiangyong and Hunan. However, in Central Asia and Middle East, during Jurassic period, the maximum principal compression stress orientation was near NS or NE trending (Figs. 3.21 and 3.23); thus, the ore fluid was easy to fill along those trending fracture zones, which orientation also was along the long-axis trending of ore fields or deposits. In the Transbaikalia collision zone, during that period, the gold zone or deposits were mainly NE trending, almost being parallel to their maximum principal compression stress orientation.

Meanwhile, i.e., the Jurassic–early period of Early Cretaceous, in the tectonic relatively stable areas of Central Asia and Middle East, there existed the warm-wet weathering belt to form a lot of oil-gas fields, coal fields and some sedimentary uranium deposits (accounted 13 deposits), for example, Karakum oil and gas field, Uzbekistan (Fig. 4.13); Karamay oil field, Xinjiang of China (Fig. 4.26); the main oil-gas reservoirs of Changqing oil and gas field, Ordos; The Jawar, Safania, Hazmiyab and Raghib oil and gas field, Saudi Arabia (Fig. 4.68); Great Burgan oil field, Kuwait; Zakum oil field, United Arab Emirates; Qatar gas field (Figs. 4.68 and 4.87); Dongsheng–Shenfu coal fields, Inner Mongolia of China; Dongsheng uranium deposits, Inner Mongolia of China (Fig. 4.30). The long-axis trending of the above oil, gas, coal and uranium-bearing deposits in Central Asia are mainly near the regional maximum principal compression stress orientation, of which the extension-shear fractures are oriented near the NS trending or NNE and NNW trending, and the compression-shear fractures are oriented near EW, ENE or WNW trending. The above basin-controlling faults are mainly the intra-crust faults, not the rifts.

4.3.7 The Tectono-Metallogenesis in Middle Period of Early Cretaceous–Paleocene

The middle period of Early Cretaceous–Paleocene was another climax to form the endogenic and exogenic mineral deposits (to see Appendix). According to the author's statistics, there are 53 giant and supergiant endogenic and exogenic mineral deposits formed in that period, occupying 13.8% of the total in the whole Asia (i.e., about one-seventh). Figure 4.86 shows 37 endogenic and exogenic mineral deposits of non-ferrous, precious, rare metals, rare earth elements, radioactive elements, iron, chromium and diamond. Almost whole elements and chemical compounds could be concentrated and reserved into endogenic and exogenic mineral deposits at this period. At the same times, there reserved the greatest concentrated giant marine oil and gas fields at North Asia, Central Asia and Middle East (Fig. 4.87), for example, the Yamburg gas field; Taimyr, Yenisey–Khatanga oil, gas and hydrocarbon fields (Fig. 4.3); Karakum oil and gas field, Uzbekistan (Fig. 4.13); Awas, Malun, Gagsalan, Akajali, Bibihagmy, Pals Haiyam and Azalegan oil and gas fields, Iran (Fig. 4.68); East Baghdad, Kirkuk and Khurmala oil and gas fields, Iraq; Jawar, Safania, Hazmiyab and Raghib oil and gas field, Saudi Arabia (Fig. 4.68); Great Burgan oil field, Kuwait; Zakum oil field, United Arab Emirates and Qatar gas field. In Siberia, Middle East and Arabia areas, the Cretaceous was not only the main period of forming shallow marine sedimentary strata, but also the period of assembling hydrocarbon and generating oil. The oil-storage structures were mainly controlled by later rock deformations. The oil and gas distributions in the Asian continent were influenced by the near NS or NE toward compression and controlled by the near NS or NE trending extension-shear faults or fault-depressions.

In 1950–1960s, based on the international experiences, the Chinese geologists considered that great oil and gas fields also could be formed in Cretaceous on China continent. At last, it discovered that in most areas of China

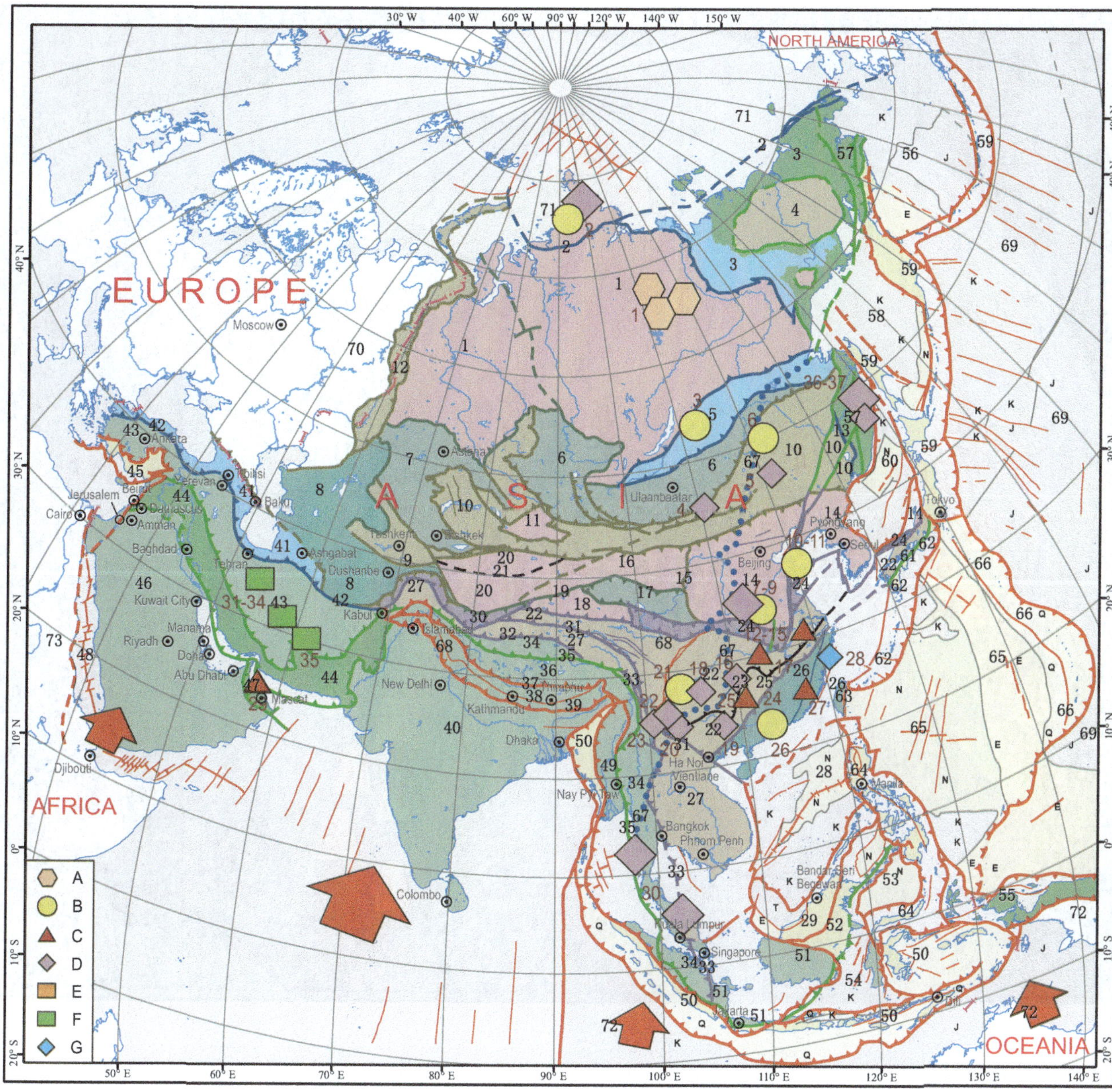

Fig. 4.86 Distribution of Asian ore fields or deposits formed in the middle period of Early Cretaceous–Paleocene. The black numbers show the tectonic units, as same as those in the CONTENTS, Fig. 1.1 and text. The big red arrows show the plate migration orientation. The numbers with brown color show the ore fields or deposits. The eastern area to the blue point line is the continental crust and oceanic mantle lithosphere, i.e., thinner continental lithosphere. **The ore-type symbols**: **A**. diamond; **B**. gold, silver precious metal; **C**. copper, lead and zinc polymetal; **D**. rare metal, tungsten, tin, molybdenum, rhenium, mercury and antimony; **E**. uranium; **F**. chromium; **G**. fluorite. **The no. of ore fields or deposits**: **1**. Siberia, kimberlite pipes deposit with diamond ore deposits (Fig. 2.2); **2**. Taimyr sedimentary lead, zinc, mercury, tin and gold deposit; **3**. Transbaikalia gold zone (Fig. 4.4); **4**. Huanggan tin iron deposit, Inner Mongolia; **5**. Barze rare earth deposit, Inner Mongolia; **6**. Erentolegai silver deposit, Inner Mongolia; **7**. Jinduicheng molybdenum deposit, Eastern Qinling (Fig. 4.28); **8**. Donggou Mo deposit, Eastern Qinling, Henan (Fig. 4.28); **9**. Qiyugou Au deposit, Songxian, Eastern Qinling (Fig. 4.28); **10**. Zhaoyuan–Sanshandao–Jiaojia gold province (Fig. 4.29); **11**. Jiaojia–Linglong gold field (Fig. 4.29); **12**. SE Hubei Fe, Cu Au province (Figs. 4.35, 4.36, 4.37, 4.38 and 4.39); **13**. Tongling Fe, Cu, Au province, Anhui (Figs. 4.35, 4.36, 4.37, 4.38 and 4.39); **14**. Chengmengshan Cu–Yangchuling W province, Jiangxi; **15**. Dexing (Tongchang, Zhushachang, Fujiawu) Cu porphyry province (Fig. 4.40); **16**. Xikuangshan Sb deposit, Lengshuijiang, Hunan; **17**. Jinshan Au hydrothermal deposit, Jiangxi (Fig. 4.40); **18**. Wanshan Hg deposit, East Guizhou (Fig. 4.42); **19**. Dachang Sn Sb deposit, Guangxi (Figs. 4.43 and 4.44); **20**. Gejiu Sn Cu polymetallic province, Yunnan (Figs. 4.45 and 4.46); **21**. Xingren–Anlong gold field, Guizhou (Figs. 4.48 and 4.49); **22**. Jinlong of Guizhou Sb deposit; **23**. Muli of Yunnan Sb deposit; **24**. Xiangshan uranium deposit, Le'an, Jiangxi; **25**. Dabaoshan Cu and W deposit, Shaoguan, Guangdong; **26**. Hetai gold deposit, Guangdong; **27**. Zijinshan Cu and Au field, Shanghang, Fujian (Figs. 4.57 and 4.58); **28**. The fluorite fields in Central Zhejiang and Fujian; **29**. Oman copper deposit; **30**. Sungai Lembing Sn deposit, Eastern Malay (Fig. 4.62); **31**. Amir, Shahriar, Reza, Abdasht chromite deposit, Iran (Fig. 4.66); **32**. Shahriar chromite deposit, Iran; **33**. Reza chromite deposit, Iran; **34**. Abdasht chromite deposit, Iran (Fig. 4.66); **35**. Baluchistan, Muslinbag chromite deposit, Pakistan; **36**. Sikhote–Alin Orient No. 2 scheelite-sulfide deposit; **37**. Lemonltov scheelite-sulfide deposit

continent, the Cretaceous sedimentary system was formed into the red terrestrial clastic rocks under the dry and hot climate, which was favorable to form the gypsum mineral and salt deposits, without any condition for the occurrence of oil and gas. Only in Northeast China, near the Daqing areas, there were the warm and moist climate areas in Cretaceous to form the great continental facies oil and gas fields.

In the South Asian continent, there was the dry and hot climate in Cretaceous to form many salt deposits at the Siam Plateau, Nakhon, Thailand and Laos; the Udon and Nong Khai salt deposits, Thailand. However, at the Taimyr, Yenisey–Khatanga of the North Asian continent, there are the lead, zinc, mercury, tin and gold sedimentary deposits.

In Cretaceous, due to the northward migration of Indian–Australian plate, the Asian continent also slowly migrated northward or northeastward to form the near NS or NE shortening in the intraplate (Wan 2011), then resulting in the long-axis of ore fields, deposits or fault-depression sedimentary basins near the NS trending.

4.3.8 The Tectono-Metallogenesis in Eocene–Oligocene

During Eocene–Oligocene, there formed 51 giant and supergiant ore fields and deposits in Asia (Fig. 4.88),

◀ **Fig. 4.87** Oil-gas fields distribution in the Asian continent. **1**. Yamburg gas field (Fig. 4.3), the petroleum source beds are the Carboniferous–Permian and Cretaceous systems, the oil-storage structures formed after Cretaceous; **2**. Taimyr, Yenisey–Khatanga oil, gas and hydrocarbon field, the petroleum source beds are the Carboniferous–Permian and Cretaceous systems, the oil-storage structures formed in Jurassic–Cretaceous; **3**. Yenisey–Khatanga natural gas hydrate, the hydrocarbon source beds are the Triassic–Cretaceous systems, the storage structures formed in Cretaceous; **4**. Karakum oil and gas field, Uzbekistan (Fig. 4.13), the hydrocarbon source beds are the Jurassic–Cretaceous systems, the storage structures formed in Cretaceous; **5**. Kasagan oil and gas field, Uzbekistan, the hydrocarbon source beds are the Jurassic–Cretaceous systems, the storage structures formed in Cretaceous; **6**. Karamay oil field, Xinjiang (Fig. 4.26), the hydrocarbon source beds are the Carboniferous–Jurassic systems, the storage structures formed in Middle Jurassic; **7**. Tarim oil-gas field, Xinjiang, the hydrocarbon source beds are Ordovician system, the storage structures formed in Neogene; **8**. Changqing oil-gas field, Ordos, the hydrocarbon source beds are the Ordovician–Cretaceous systems, the storage structures formed in Cretaceous; **9**. Daqing oil field, NE China (Fig. 4.25), the hydrocarbon source beds are the Jurassic–Cretaceous systems, the storage structures formed in Late Paleogene; **10**. Sakhalin (39) oil and gas fields, Far East, Russia, the hydrocarbon source beds are the Paleogene–Neogene systems, the storage structures formed since Neogene; **11**. Around Bohai oil-gas fields (including Shengli, Liaohe, Jidong, Dagang, Penglai 19-3 etc.) (Fig. 4.31), the hydrocarbon source beds are the Paleogene–Neogene systems, the storage structures formed since Paleogene or Neogene; **12**. Sichuan oil and gas province (125 gas fields; 12 oil fields, Fig. 4.51), western part (the hydrocarbon source beds are the Triassic–Jurassic systems, the storage structures formed since Cretaceous), eastern part (the hydrocarbon source beds are the Cambrian–Jurassic systems, the storage structures formed since the end of Paleogene); **13**. Chunxiao and Pinghu oil and gas fields, the hydrocarbon source beds are the Paleogene system, the storage structures formed since the end of Paleogene; **14**. Zhujiangkou oil and gas fields, the hydrocarbon source beds are the Paleogene system, the storage structures formed since the end of Paleogene; **15**. Yinggehai oil and gas fields, Western Hainan Island, Weixinan oil and gas fields and Qiongdongnan oil and gas fields, the hydrocarbon source beds are the Neogene system, the storage structures formed since the end of Neogene; **16**. South China Sea oil and gas fields, the hydrocarbon source beds are the Cretaceous–Neogene systems, the storage structures formed since the end of Cretaceous–Neogene; **17**. The Mekong Delta oil and gas field, the hydrocarbon source beds are the Paleogene–Neogene systems, the storage structures formed since Paleogene or Neogene; **18**. Brunei and Eastern Malaysia oil and gas field, the hydrocarbon source beds are the Paleogene–Neogene systems, the storage structures formed since Paleogene or Neogene; **19**. Sunda oil and gas field, Indonesia (Fig. 4.70), the hydrocarbon source beds are mainly the Paleogene–Neogene systems, the storage structures formed since Paleogene and Neogene; **20**. Jawar, Safania, Hazmiyab and Raghib oil and gas fields, Saudi Arabia (Fig. 4.68), the hydrocarbon source beds are mainly the Jurassic–Paleogene systems, the storage structures formed since Paleogene and Neogene; **21**. Great Burgan oil field, Kuwait (Fig. 4.68), the hydrocarbon source beds are mainly the Jurassic–Paleogene systems, the storage structures formed since Paleogene; **22**. Zakum oil field, United Arab Emirates (Fig. 4.68), the hydrocarbon source beds are mainly the Jurassic–Paleogene systems, the storage structures formed since Neogene; **23**. Qatar gas field (Fig. 4.68), the hydrocarbon source beds are mainly the Jurassic–Paleogene systems, the storage structures formed since Paleogene; **24**. East Baghdad, Kirkuk, Khurmala oil and gas fields, Iraq, the hydrocarbon source beds are mainly the Jurassic–Paleogene systems, the storage structures formed since Paleogene; **25**. Awas, Malun, Gagsalan, Akajali, Bibihagmy, Pals Haiyam and Azalegan oil and gas fields, Iran, the hydrocarbon source beds are mainly the Paleogene–Neocene systems, the storage structures formed since Neogene; **26**. Baku oil field, Azerbaijan; the hydrocarbon source beds are mainly the Paleogene–Neocene systems, the storage structures formed since Neogene

accounting for 13.3% of the total in Asia. At that period, the East Asian continent was mainly influenced by the westward subduction and compression of the Pacific plate, and the South Asian continent continued to suffer the northward subduction and compression of the Indian–Australian plate (Fig. 3.30). However, at the central and northern parts of the Asian continent, those influences were very weak. So there were different principal compression stress orientations in the Asian continent during that period.

For most of the endogenous metallogenesis are formed underneath 2–5 km depth, the most of recent discovered and exploited ore deposits are only near the Earth surface. It means in the last tens of million years, those deposit areas must uplift several kilometers, then easy to be discovered and exploited. Thus, recent discovered and exploited endogenous deposits are all located at the high mountains (about 2000–4000 m above sea level), such as the Qingzang–Pamir Plateau, Iran–Afghanistan Plateau, or some mountains on the sea islands. At that period, there also were many metallogenesis such as the copper, lead, zinc polymetallic, precious metal, rare metal, mercury and antimony.

Under the long-distance influence of the Pacific plate westward subduction and compression, in China there formed the Yulong copper, molybdenum porphyritic ore zone (Figs. 4.60 and 4.88); the Maoniuping Mianning rare earth metallogenic zone, Sichuan; Dongmuzazhua, Muhailahen and Chaqupacha Pb–Zn deposits, Yushu, Qinghai (Fig. 4.59); Jinding Pb, Zn polymetallic deposit (Fig. 4.61) and Baiyangpin Cu, Co, Ag deposit, Yunnan etc. At the Western Pacific island arc zone, the Uwamuki, Kuroko-type deposit, Japan; Hishikari gold deposit, Japan; Lepanto and Far Southeast Cu, Au fields, Philippines; Grasberg rich Au and Cu deposit, Indonesia were discovered. Due to the long-distance effects of the Indian–Australian plate northward subduction and compression, there formed the Sar Chesmeh Cu and Mo deposit, Iran; Miduk, Soungoun Ahar and Chahar Gonbad Cu Deposits, Iran; Anguran Pb, Zn deposit, Iran (Fig. 4.66) and Hpakan jadeite

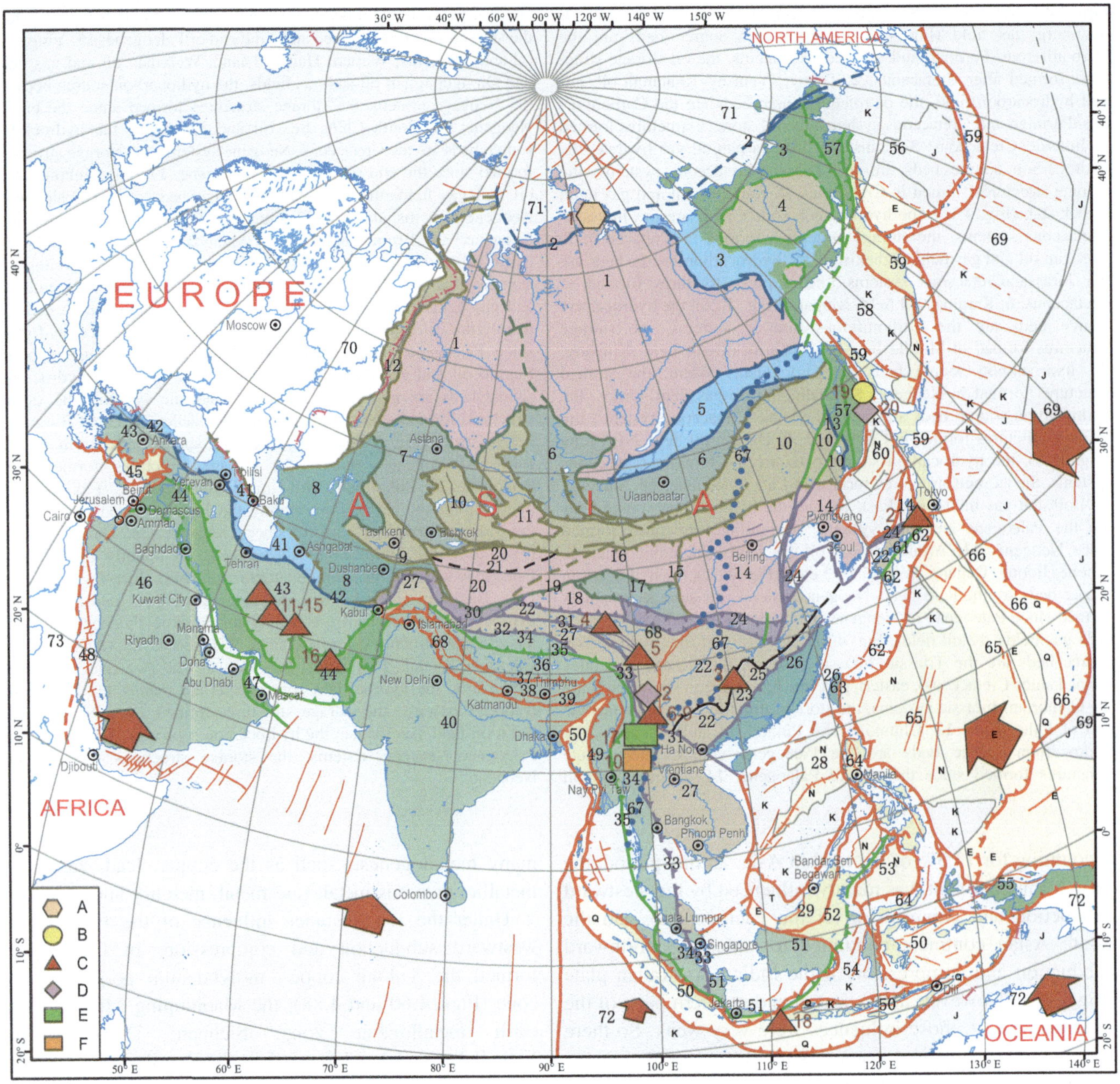

Fig. 4.88 Distribution of Asian ore fields or deposits formed in Eocene–Oligocene. The black numbers show the tectonic units, as same as those in the CONTENTS, Fig. 1.1 and text. The big red arrows show the plate migration orientation. The thick red lines are the subduction or collision zone. The numbers with brown color show the ore fields or deposits. The eastern area to the blue point line is the continental crust and oceanic mantle lithosphere, i.e., thinner continental lithosphere. **The symbols of mineral type**: **A**. diamond; **B**. gold, silver precious metal; **C**. copper, lead and zinc polymetallic; **D**. rare metal, tungsten, molybdenum, rhenium, mercury, antimony and rare earth; **E**. jadeite; **F**. lateritic nickel. **The no. of ore fields or deposits**: **1**. Popigai Astroblem diamond deposit, Serbia; **2**. Maoniuping of Mianning rare earth metallogenic zone, Sichuan of China; **3**. Shuikoushan Pb, Zn deposit in Changning, Hunan of China; **4**. Dongmuzazhua, Muhailahen and Chaqupacha Pb–Zn deposits, Yushu, Qinghai of China; **5**. Yulong Cu and Mo hydrothermal ore zone, Xizang of China; **6**. Jinding Pb, Zn polymetal deposit, Yunnan of China; **7**. Baiyangpin Cu, Co, Ag deposit, Yunnan of China; **8**. Baiyangchang Cu, Ag polymetal deposit, Yunnan of China; **9**. Jinman copper deposit, Yunnan of China; **10**. Tagaung Taung Ni deposit, Burma; **11**. Sar Chesmeh Cu, Mo deposit, Iran; **12**. Miduk copper deposit, Iran; **13**. Sar Chesmeh and Soungoun Cu-Mo deposit, Iran; **14**. Ahar Chahar Gonbad Cu, Pb, Zn deposits, Iran, **15**. Anguran Pb, Zn deposits Iran; **16**. Seundk Cu deposit, Pakistan; **17**. Hpakan jadeite deposits, Burma; **18**. Batu Hijau Cu, Au deposit, Indonesia; **19**. Pulamanoya Hg deposit, Russia; **20**. Xibolianskeya Hg deposit, Russia; **21**. Uwamuki Kuroko-type deposit, Japan

deposits, Burma. As to the Popigai Astroblem area in North Siberia (111° 11′ E; 71° 39′ N), there existed a minor planet impact (before 35 Ma) to form a large-scale diamond deposit, which is a meteorite crater, not to be controlled by the Earth tectonics.

In the early period, under the strong weathering condition of the torridness and moist weather, at the residual ophiolite suites there formed the lateritic nickel ore deposits—the recent discovered main type of nickel deposits, such as the Tagaung Taung in Burma, Halmahera Island, Subain, Mabli, East Lamla of Weigo islands and Xifu Mountains nickel deposits in Indonesia.

At the lower relief areas, under the tectono-metallogenesis and material source condition in Cenozoic, due to the limited uplift, the recent-formed endogenic deposits are only located underneath the depth, which belong to the hidden deposits. Maybe it will be explored and exploited in the future.

As to the subsidence area with weak tectonics, moist weather sedimentary basins and appropriate uplift (about 1000 m) since Neogene, it is possible to reserve valuable oil and gas fields. In Central Asia, Middle East and East Asia, there formed many petroleum-bearing structures of oil and gas fields to cause a lot of hydrocarbons concentrating, migrating and reserving, such as in China, the Daqing oil field (Fig. 4.25), circum Bohai oil and gas fields (Shengli, Liaohe, Jidong, Dagang, Penglai 19-3, Fig. 4.31), East Sichuan basin gas fields (Fig. 4.51), Chunxiao and Pinghu oil and gas fields, Zhujiangkou oil and gas fields, Qiongdongnan oil and gas fields, Yinggehai oil and gas fields of West Hainan Island, Weixinan oil and gas fields; as well as Baku oil field, Azerbaijan; Awas, Malun, Gagsalan, Akajali, Bibihagmy, Pals Haiyam and Azalegan oil and gas fields, Iran; East Baghdad, Kirkuk, Khurmala oil and gas fields, Iraq; Jawar oil and gas field, Saudi Arabia; Safania oil field, Saudi Arabia; Hazmiyab, Raghib oil field, Saudi Arabia; Great Burgan oil field, Kuwait; Zakum oil field, United Arab Emirates; Qatar gas field; Sakhalin (total 39) oil and gas fields (Fig. 4.87).

4.3.9 The Tectono-Metallogenesis in Neogene–Early Pleistocene

In Neogene–Early Pleistocene, there formed 63 ore fields or deposits in the Asian continent, accounting for 16.4% of the total in Asia (Fig. 4.89). This period is also in a metallogenic climax. During this period, the East Asian continent had suffered the westward subduction and compression, and the main part of the Asian continent was continually influenced by northward collision and compression (Figs. 3.33 and 3.34).

It is similar to the Paleogene metallogenesis, the newly exploited endogenic metal fields or deposits formed in Neogene are all located at the high mountains (more than 3000 m above sea level), which belong to the strong uplift areas, in China including the Maoniuping of Mianning rare earth metallogenic zone, Sichuan; Dongmuzazhua, Muhailahen and Chaqupacha Pb-Zn deposits, Yushu of Qinghai; Yulong Cu and Mo hydrothermal ore zone; Jinding Pb, Zn polymetallic deposit; Baiyangpin Cu, Co, Ag deposit, Yunnan; Baiyangchang Cu, Ag polymetallic deposit; Jinman copper deposit, Yunnan; as well as Hpakan jadeite deposits, Burma; the Sar Chesmeh Cu, Mo deposit, Iran; Miduk copper deposit, Iran; Sar Chesmeh and Soungoun Cu, Mo deposit, Iran; Ahar Chahar Gonbad Pb, Zn deposits, Iran. They were controlled by the northward compression of the Indian plate. The Monywa copper field, Burma, is located on the northeastern side of the Indian plate, east to the 90° E ridge, and belongs to the Australian plate. Being controlled by the 90° E ridge dextral shear, there developed a series of extension-shear fractures to reserve a lot of copper elements. The Grasberg, Ertsberg, Ok Tedi and Frieda Batu Hijau copper and gold deposits in Indonesia (Fig. 4.71) and Lepanto and Far Southeast copper and gold deposits in Philippines are all influenced by the Western Pacific island arc supergiant porphyry deposits. The metallogenic period of Jinguashi Au and Cu deposit, North Taiwan was at about 1 Ma, which was not located at the island arc, but a deformation intraplate controlled by the northwestward compression of the Philippine Sea plate.

As to the oil and gas fields and other sedimentary deposits, they are all located at the relative subsidence areas and basins. During Neogene, the oil and gas fields had been formed in the Asian continent as follows: in China, the Tabei oil-gas field, Xinjiang; Dagang and Penglai 19-3 oil fields; Zhujiangkou oil and gas fields; Qiongdongnan oil and gas fields; Yinggehai oil and gas fields; the Western Hainan Island, Weixinan oil and gas fields; the South China Sea oil- and gas-bearing basin; in addition, some newly discovered Baku oil field, Azerbaijan; Awas, Malun, Gagsalan, Akajali, Bibihagmy, Pals Haiyam and Azalegan oil and gas fields, Iran; Sumatra, Java and Eastern Kalimantan etc. oil and gas fields, Indonesia; Sakhalin oil and gas fields(total 39). The Neogene oil and gas fields usually are located at stable crust

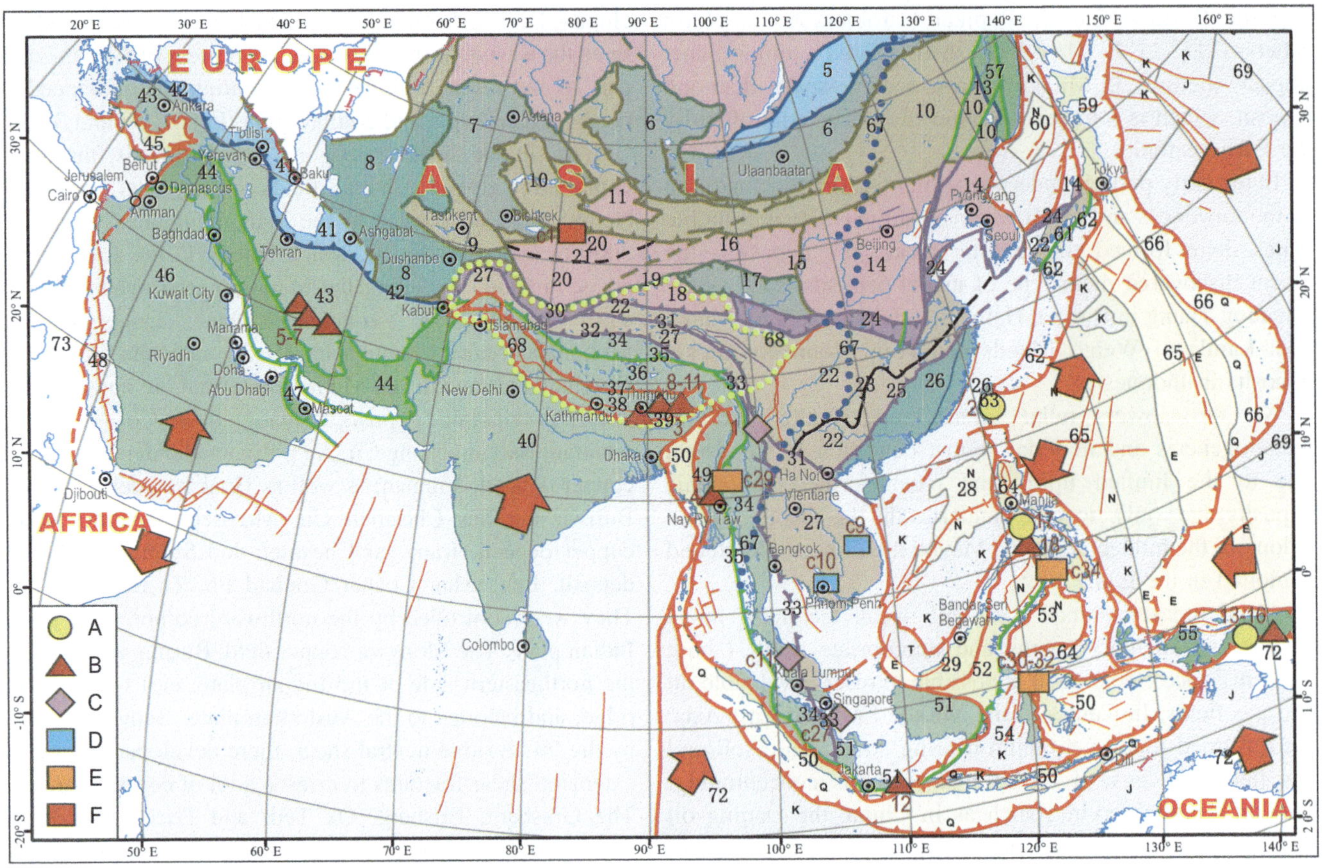

Fig. 4.89 Distribution of Asian ore fields or deposits formed in Neogene–Early Pleistocene. The black numbers show the tectonic units, as same as those in the CONTENTS, Fig. 1.1 and text. The big red arrows show the plate migration orientation. The thick red lines are the subduction or collision zone. The numbers with brown color show the ore fields or deposits. The eastern area to the blue point line is the continental crust and oceanic mantle lithosphere, i.e., thinner continental lithosphere. The area within yellow point line shows the Qingzang–Pamir thickened continental lithosphere. **The symbols of mineral type**: **A**. gold, silver precious metal; **B**. copper, lead and zinc polymetallic; **C**. rare metal, tungsten, molybdenum, rhenium, mercury, antimony and rare earth; **D**. bauxite; **E**. lateritic nickel; **F**. sedimentary uranium. **The no. of endogenic ore fields or deposits**: **1**. Maoniuping of Mianning rare earth metallogenic zone, Sichuan of China; **2**. Jinguashi Au and Cu deposit, North Taiwan of China; **3**. Qulong copper deposit, Xizang of China; **4**. Monywa, copper field, Burma; **5**. Miduk Cu Deposit, Iran; **6**. Soungoun Ahar Cu deposit, Iran; **7**. Chahar Gonbad Cu deposit, Iran; **8**. Jiama Cu deposit; **9**. Chongjiang Cu deposit; **10**. Bairong Cu deposit; **11**. Tinggong Cu deposit, Xizang; **12**. Batu Hijau Cu and Au deposit, Indonesia; **13**. Grasberg Cu and Au deposit, Indonesia; **14**. Ertsberg Cu and Au deposit, Indonesia; **15**. Ok Tedi Cu and Au deposit, Indonesia; **16**. Frieda Cu and Au deposit, Indonesia; **17**. Lepanto Cu and Au field; **18**. Far Southeast Cu and Au deposit. **The no. of exogene ore deposits**: **c1**. Yili sandstone uranium deposit, China; **c9**. Duole alumine deposit, Southwestern Vietnam; **c10**. Haut Chhlong alumine deposit, Cambodia; **c11**. Kuala Lumpur, Dinding, Phuket Phangnga Takuapa placers tin deposits in the west of Malay Peninsula; **c27**. Bangka Island tin placers, Indonesia; **c28**. Billiton tin placers; **c29**. Tagaung Taung laterite-type nickle deposit, Burma; **c30**. Subain, Halmahera Island, east Lamla laterite-type nickle deposit, Indonesia; **c31**. Mabli laterite-type nickle deposit; **c32**. Weigo islands, Xifu Mountains laterite-type nickle deposit; **c34**. Dinagat laterite-type nickle deposit; **c35**. Dinagat laterite-type nickle deposit, Philippines

areas with the depth of 1–2 km. If the burial depth is too shallow, the oil and gas will be easy to leak out and disadvantageous to preserve.

In the moist and torridness weathering zones of the South Asian continent, there are a lot of super-large laterite-type nickel deposits, such as the Tagaung Taung deposit, Burma; Subain, Halmahera Island, east Lamla deposit, Indonesia; Mabli deposit, Indonesia; Weigo islands, Indonesia; Xifu Mountains deposit, Indonesia (Figs. 4.69 and 4.70); Dinagat deposit, Philippines. They are all preserved in the modern weathering crusts. The laterite-type nickel deposits are the main source of nickel in recent. In addition, the strong weathering in the subtropical zone could be favorable to form the giant bauxite deposits, for example, the Duole bauxite deposit of Southwest Vietnam, Haut Chhlong bauxite deposit of Cambodia. Through the weathering, erosion and transportation, the cassiterite minerals could be retained to form many sedimentary tin placers on the Malay Peninsula, such as the Kuala Lumpur, Dinding, Phuket Phangnga Takuapa, Bangka Island and Billiton (Fig. 4.62). In Qinghai of China, there formed the Dafengshan and Jiantingshan celestine strontium deposit, which was deposited in the salt lakes in the dry climate. In Neogene, being controlled by the fractures, they could migrate and preserve in the deep. The uranium in Jurassic coal beds could be dissolved and migrated in the underground water to preserve near the oxygenation and reduction boundary and to form deposits, for example, the Neogene Yili sandstone uranium deposit, Xinjiang of China. To sum up, in Neogene–Early Pleistocene, there formed many kinds of giant endogenic and exogenic deposits.

4.3.10 The Tectono-Metallogenesis Since Middle Pleistocene

In Middle Pleistocene, there formed 15 giant ore deposits or fields, accounting for 4% of the total in the Asian continent (Fig. 4.90), including a few of endogenic metal fields or deposits and some ore deposits related to the volcano, such as the Jinguashi Au and Cu deposit in North Taiwan of China and the Hishikari gold field in Japan. The most valuable deposits are the exogenic ones, such as the weathering residual nickel deposits, Tagaung Taung nickel deposit, Burma; Kolonodale, Kolaka, Moyowali laterite-type nickel deposit, Indonesia; Dinagat laterite-type nickel deposit, Philippines; the Malay Peninsula placers tin deposits (Fig. 4.62); Kerala, Tamil Nadu and Travancore State rutile deposits, India; Maharashtra, Kerala, Tamil Nadu and Travancore States coast ilmenite deposits, India; Eastern coast bauxite deposits, India; the Baishugang rutile deposit, Fangcheng, Henan of China. The above weathering residual and fluvial sedimentary deposits are all distributed near the equator under the condition of warm or hot and wet climate. However, the lithium, boron, potassium and magnesium deposits, etc., were all formed in the dry climates. It can be seen that all the weathering residual or sedimentary deposits are related to the local climatic zones. This kind of deposit would be controlled by the composition of source rocks.

To sum up, due to the different migration model and convergence orientation of the plates or blocks in different tectonic periods, there formed different collisions and intraplate deformations, thus to control the kinds, occurrence positions, patterns and attitudes for endogenic metal fields and deposits. If only the cause of formation type was known, it would be very difficult to do exploration. It needs to understand the possible metallogenic periods, the background of geochemistry, tectonics and erosion depth. To research the Cenozoic metallogenesis of metal deposits, it needs to understand the facts that the Cenozoic metal deposits are located at recent uplifting areas—mountains or plateaus, and the uplifting ranges usually are smaller than the depth of formation. As to the hypergene deposits, it should concern the conditions of paleo-climate and paleo-geography.

Thus, to compile a valuable metallogenic map, it must in detail establish and mark the different tectonic periods in the metallogenic map. If many tectonic periods were marked in a metallogenic map, it would not have any practical values. Due to the limit of data and personal ability, in this monograph the author just does the preliminary and summary attempt and expects it to be referred for readers. Meanwhile, the author hopes in the future, the above Asian metallogenic maps with different tectonic periods to be compiled with more details and more practical values.

4.4 Discussion on the Tectono-Metallogenesis

4.4.1 The Fracture Influence on the Metallogenesis of Endogenic Metal Deposits

To take the tectono-metallogenesis research as a motion and transformable dynamic system is the recent important progress for the mineral deposit geology (Zhai et al. 2011). It is developed from the research of petroleum geology. To research the metallogenesis, it must consider the material source, useful element migration, formation process and preservation condition.

Here, the fracture influence on the metallogenesis of endogenic metal deposits will be firstly discussed. In the formation process of solid mineral deposits, it will be noted that the vertical migrations of the ore elements could not be too much after their enriching, such as the sedimentary and metamorphism deposits. However in the formation process of the magma and hydrothermal deposits, the useful elements in the depth of lithosphere could be kept in the solid-state rock, and the metal elements in the ultra-critical fluid or magma could be migrated upward the shallow with the magma or ultra-critical fluid. Their migration distances are not less than those of fluid deposits (such as the oil, gas and underground water deposits). The faults at different penetrated depths could control distinctly magma and ultra-critical fluid with useful elements, thus to form many magmatic and hydrothermal deposits for different metal elements (Fig. 4.91; Wan 2011).

The lithosphere faults (Fig. 4.91a), with the penetrated depth about 100 km, usually are closely related to the ultra-mafic bodies or xenoliths. The metallogenic material often migrates near the Earth surface with cold emplacement, i.e., along the fault emplacement mainly forming the iron family element, platinum family element, diamond and corundum (included ruby sapphire) deposits.

The crust faults, with the penetrated depths about 30–50 km, near the crust (Fig. 4.91b), are easy to cause the A-type or I-type magmatism to be mixed with crust and mantle source. When the magma uplifts, migrates and

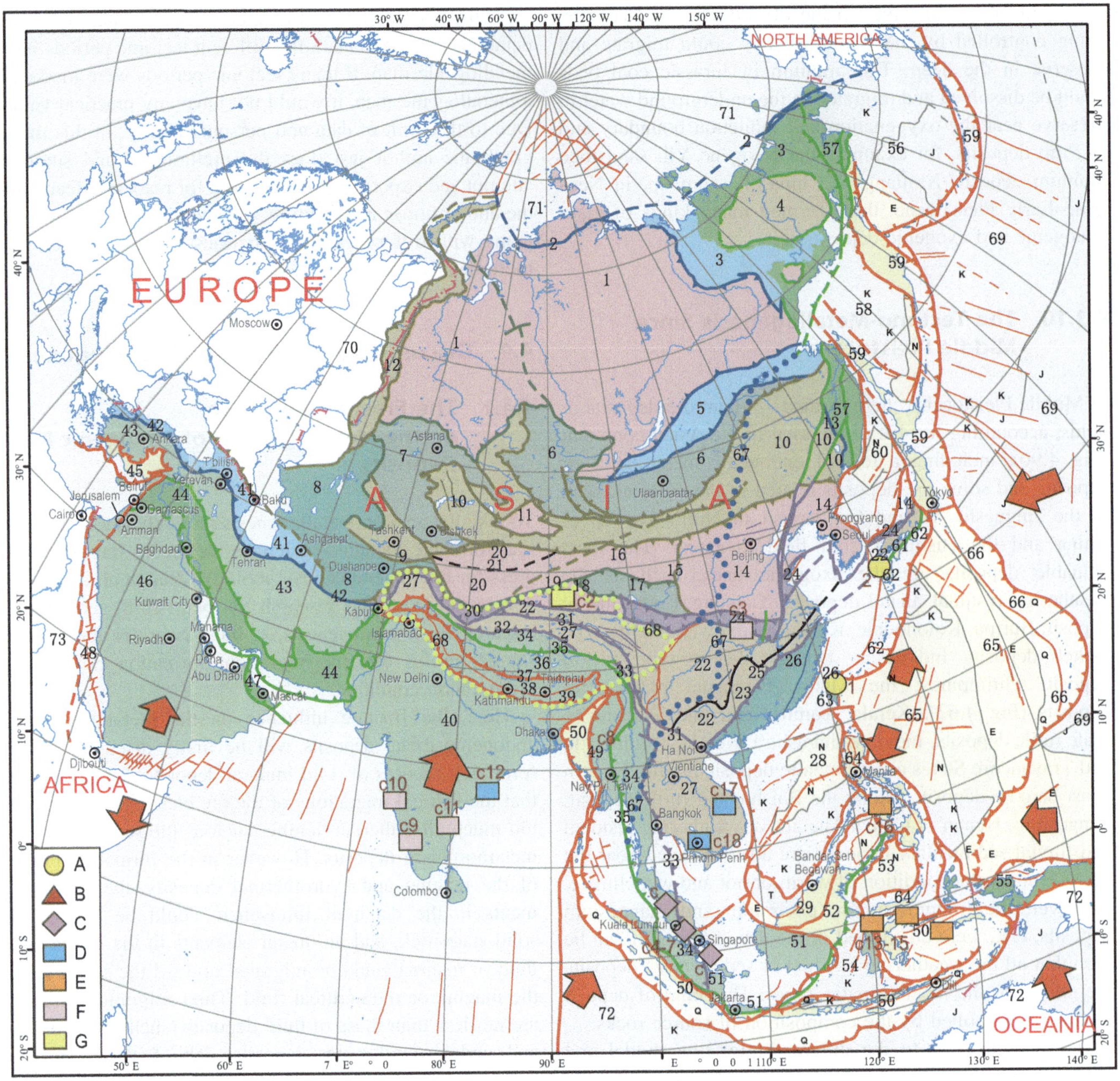

Fig. 4.90 Distribution of Asian ore fields or deposits since middle Pleistocene. The black numbers show the tectonic units, as same as those in the CONTENTS, Fig. 1.1 and text. The big red arrows show the plate migration orientation. The red thick lines are the subduction or collision zone. The numbers with brown color show the ore fields or deposits. The eastern area to the blue point line is the continental crust and oceanic mantle lithosphere, i.e., thinner continental lithosphere. The area within yellow point line shows the Qingzang–Pamir thickened continental lithosphere. **The symbols of mineral type**: **A**. gold, silver precious metal; **B**. copper, lead and zinc polymetallic; **C**. tin placers; **D**. bauxite; **E**. lateritic nickel**; F**. lateritic rutile or ilmenite; **G**. salt. **The no. of ore deposits or fields**: **1**. Jinguashi Au and Cu deposit in North Taiwan; **2**. Hishikari gold field, Japan. **c2**. Qaidam salt lake Li, B, K, Mg salt deposit; **c3**. Baishugang rutile deposit, Fangcheng, Henan; **c4**. Kuala Lumpur tin placers; **c5**. Dinding tin placers; **c6**. Phuket tin placers; **c7**. Phangnga Takuapa tin placers; **c8**. Tagaung Taung nickel deposit; **c9**. Kerala, Tamil Nadu and Travancore State rutile deposits, India; **c10**. Travancore rutile deposit; **c11**. Maharashtra, Kerala, Tamil Nadu and Travancore States coast rutile and ilmenite deposits, India; **c12**. East Indian coastal bauxite; **c13**. Kolonodale Ni deposit, Indonesia; **c14**. Kolaka, Ni deposit, Indonesia; **c15**. Moyowali Ni deposit; **c16**. Dinagat Ni deposit, Philippines; **c17**. Duole alumine deposit Southwestern Vietnam; **c18**. Haut Chhlong bauxite deposit, Cambodia

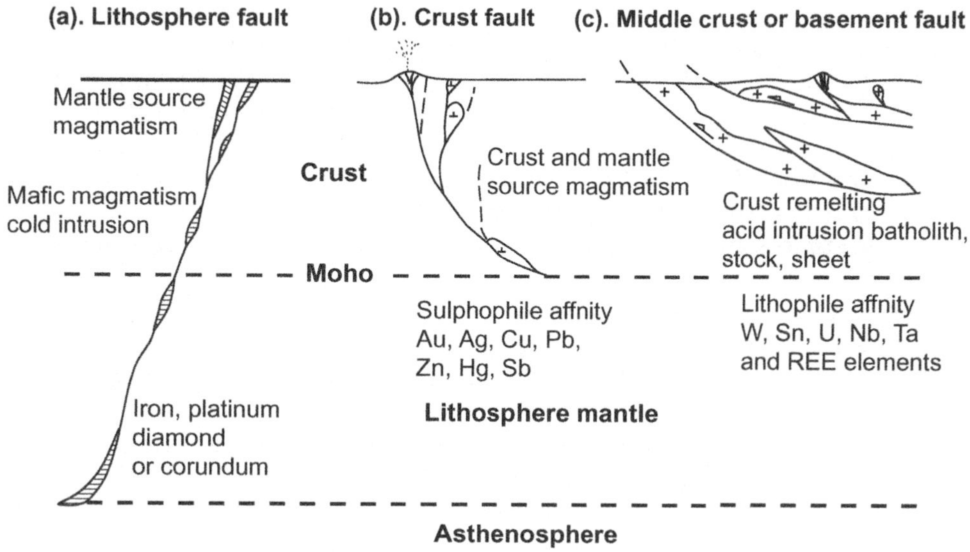

Fig. 4.91 Faults at different penetrated depths controlling distinctly magma and ultra-critical fluid with useful elements

undergoes the assimilation and hybridization, and reaches near the Earth surface, it will form mainly the intermediate rocks. The magma usually has the rather high ratios of Mg/Fe and lower initial ratios of $(^{87}Sr/^{86}Sr)_i$ (usually less than 0.710). The ore deposits are formed mainly as rich sulfophile elements. The concentrations of many precious metals, non-ferrous metals and rare metals (such as Au, Ag, Mo, Cu, Pb, Zn, Hg, As and Sb) are related to the crust faults. It is easy to form the metal sulfide mineral resources.

The faults, with the penetrated depths near the lower seismic velocity and higher conductivity zone of middle crust, or near the boundary between crystalline basement and sedimentary strata (Fig. 4.91c), i.e., less than 20 km, often result in the acid magmatism and are easy to form S-type granitic sheets (in past, sheet commonly was called as "batholith"; in recent the seismic section data have shown that it is distributed as sheet-like without the root; Wan and Zeng 2002) and their stocks. They always contain the lower ratios of Mg/Fe and higher initial ratios of $(^{87}Sr/^{86}Sr)_i$ (>0.710). The related ore deposits are enriched with the lithophile elements, such as rare elements of W, Sn, U, Nb, Ta and rare earth elements (REE). They are easy to form the oxide and oxysalt mineral resources (Wan 2011).

The mechanism of above phenomena may be resulted from the faulting with different penetrated depths, which can be caused by the partial pressure decreasing and temperature increasing at different depths to form the rocks' partial melting, then to form the magma chamber and supercritical fluids. They could dissolve a lot of different ore elements and get them uplift along the fault or fracture zones near the upper crust, then to be accumulated at different depths.

The author's above recognition indicates the relationship between faulting and metallogenesis, and the author hopes that it can be paid attention to. It tells us that the ore-controlling faults for most of endogenic metal deposits are mainly the crustal or basement faults, i.e., most of endogenic metallogenesis is originated from the crust and the crust–mantle transition zone, a few from the bottom of lithosphere or asthenosphere. Till now, there is not any evidence to indicate that the formation of endogenic metallogenesis is directly related to the deep mantle or core.

It must point out that the Asian continental tectonic stress fields show multiple periods and different orientations. In the middle crust with low seismic velocity and Moho discontinuity, the tectonic detachments are easy to occur (Gao et al. 2011); in the most of areas between the crust and mantle, there is the decouple state. For the recent Asian continental lithosphere, the part under the upper crust is not the original lower crust and lower lithosphere; the part under the recent crust is not sure to be the original lithosphere mantle. As for studying the Asian continental lithosphere tectonics, this viewpoint must be paid attention to.

According to the data of known metallogenic depths in Asia, there is not the close relationship between the ore deposits and the lithosphere mantle or the mantle, except the diamond kimberlite deposits, corundum deposits and some magmatic ore deposits of copper, nickel and platinum. The most of other endogenic metallogenic deposits are related to the detachments of Moho discontinuity, high velocity and low conductivity of the middle crust, or the top of crystalline basement. As a result, if the research of lithosphere bottom structures is only focused on to guide the exploration, it seems that the benefit will be very limited. On one hand, because the comprehensive research on geophysics, geochemistry and geology in the crust and upper part of lithosphere is very important, it is necessary to research the lithosphere tectonics, especially "seeking the truth from facts" to analyze the deep geological structures for the exploration values; on the other hand, we should not believe that the deeper research is obtained, the more successful exploration will be got.

4.4.2 The Rock Deformation and the Preservation Space of Endogenic Metallogenesis

In many textbooks and monographies on mineral deposit geology and some studies on ore field structure (Chen 1978; Zhai 1996, 1997, 2011; Zhai and Lin 1993), the relationships between rock deformation and the preservation space of endogenic metallogenesis have been discussed in detail. During the rock deformation process, it is possible to form the endogenic metal deposits in any partial fractures or fissures in the underground, i.e., occurrence positions of partial pressure depression. In last half century, it has become a basic understanding for the structure of deposits and fields, and many important experiences have been got.

Is there the metallogenic specialization for the different kinds of rock deformations?

It is a problem to be paid attention to. For example, whether the extension, normal, detachment or denudation faults must be related to the formation of the lead, zinc and silver deposits? Is the ductile shear zone sure to form the gold deposits? According to the author's understanding, a lot of facts show that there are no relationship between the metallogenic specialization and the rock deformations, except that the magma usually has the rather well metallogenic specialization.

For the solid ore deposits, such as the metamorphism deposits, the actions of rock deformations only make the ore bodies be complicate in the shape. As to the magmatic and hydrothermal ore deposits, the actions of rock deformation mainly induce and form the phenomena of pressure decreasing and temperature increasing, then are favorable to form the magma chamber, and to provide the migration passageway and the storage space for the supercritical or ore fluids. Of course, it also could provide the scattering and disappearing passageway for ore fluids.

The rock deformations usually cannot form any new elements, minerals and deposits, except for a new rock—fault breccia (in the special condition, it could form the ore deposit). So the viewpoint of "dynamic diagenesis and metallogenesis" (Yang 1986) is improper; it overemphasizes the rock deformation action in the diagenesis and metallogenesis. It should be considered, due to the rock deformation, the magma or ore fluids will result in some changes of the temperature, pressure or chemical compositions. No matter what kind rock deformations, so long as there are the fissures or fractures, any kind of fluids can be migrated or reserved in them. Thus, the recognition of "metallogenic specialization for rock deformation" should be a common mistake.

The different scales of rock deformation can affect the scale of metallogenesis. Usually, the larger-scale deformational structures are formed, the larger ore deposits will be accumulated. For example, there are the larger deposits of organic matter or of gypsum and halite in the larger sedimentary basins; there are always the greater accumulations of oil and gas in large-scale anticline reservoirs. Large regional fault zones, with the length of hundreds to thousands of kilometers and the width of several hundreds or thousands meters, would control the migration of metallogenic fluids and control the distribution of metallic mineral deposits in a complete metallogenic province. In South China, regional faults and metallogenic belts are of NE trending; however, within the metallogenic belts, a series of ore deposits are accumulated only along the secondary WNW trending extensional faults.

The large-scale regional fault zones have a high permeability due to numerous open fractures and are favorable for fluid to migrate, but usually unfavorable for the accumulation of mineral material to form ore deposits. Therefore, in the large main fault structures, especially in the large normal faults or tensional and shear strike-slip faults, there are neither large pneumatolytic-hydrothermal deposits nor small pneumatolytic-hydrothermal deposits, except some "stone-like coal" (argillaceous quartzite with 5–10% of carbon). When this kind of "coal" is involved in a fault zone, the permeability will be increased, and the "coal" will be fractured to be used as the fuel, for example, in the Yunxi of Hubei stone-like coal deposit, China. However, as to the great-scale Tancheng–Lujiang fault zone, in the Triassic great-scale strike-slip period, Jurassic faulting period and Cretaceous extension-shear period, there never preserved any giant endogenic metal deposits. However, during the Eocene–Oligocene periods, there concentrated a lot of hydrocarbons in the basin near the Tancheng–Lujiang fault zone. In Neocene–Early Pleistocene, the Tancheng–Lujiang fault zone had suffered near NS trending compression to show a little bit open and resulted in the hydrocarbons to migrate upward and to concentrate near the Earth surface. It also could scatter and lose some hydrocarbons, but never kept them not scattering and disappearing at all. In the neotectonic period (since 0.78 Ma), that fault zone had suffered the near EW trending compression to accumulate oil-gas deposits in No. 1 fault, such as the PL19-3 giant oil-gas reservoir in the Southern Bohai Bay, which was discovered in 2000. Along that fault zone, the oil and gas reserves are 600 million tons (Hofer R, Conoco Phillips, personal communication).

In some large-scale fault zones, when they are compressed to become too impermeable, the high crystalline temperature and lower activity elements will be formed, such as iron or asbestos deposits, for example, the Aqtobi chromite deposit in Kazakhstan, Turgy iron deposit in Kazakhstan, the Mangya asbestos deposits in Qinghai of China, and Bazhou asbestos deposits in Xinjiang of China.

As to some great ore-bearing faults and thrusts, when the upper wall is covered by faults, the ore fluids will be preserved in the lower wall to form the giant oil and gas fields (such as the Eastern Liaohe oil and gas field, China) or the other endogenic metal deposits, for example, the Jinding lead and zinc polymetallic deposit, Yunnan of China (Fig. 4.61). To sum up, the giant faults are not sure to form the giant ore deposits; therefore, it needs to analysis based on the facts.

In the meso-scale fault and fracture zones, with tens or hundreds of kilometers in length, tens of centimeters meters in width, and thousands of meters in depth, there often form some endogenic ore deposits. Most of endogenic metal deposits occur in meso-scale fault zones, because those faults in the crust have the good transportation condition for the ore fluids. However, they cannot penetrate easily to the Earth surface, but have the good condition to preserve the ore fluids.

The micro-scale fault and fracture zones distributed in large areas may often form the giant or supergiant deposits of copper, molybdenum, tungsten, gold or uranium with lower grade. It must form the high concentrations of micro-fissures areas or joints zones, in which ore deposits usually form the disseminated fine veins or net structures. In recent advances of ore beneficiating and smelting technology, many very low-grade ore deposits, i.e., the "reserves not on balance sheet" or "reserves unable to exploit in past" now will become the ore deposits with the well industrial values. In recent years, many supergiant or giant ore deposits have been discovered and explored in China and in the world, which are the successful instances.

To sum up, there is no relationship among the type, scale of rock deformations, and the type of ore deposits, so it must be analyzed based on the facts.

4.4.3 Influences of Later Tectonics, Uplift or Depression on Ore Preservation

The preservation conditions, after the formation of ore deposit, will become a very important factor to keep perfection of ore deposits and the suitable hidden depth, which are all close relationship with the tectonics. Usually after the ore deposit is formed, it had better not to undergo the stronger tectonics. If the later tectonics is very strong, the ore deposits formed in early periods will be easy to destroy; however, it often appears in the Asian continent. The ore deposits formed before Early Paleozoic have been eroded and destroyed, as a result, only a few could be preserved, the others would be buried underneath the huge sedimentary covers to become the "blank deposit." Since Late Paleozoic, especially in Mesozoic and Cenozoic, many elements were easy to enrich to form a lot of new ore deposits, but not to suffer the stronger destruction in the later periods. So it cannot be considered that there are a lot of ore deposits formed extremely in Meso–Cenozoic. In fact, the ore deposits are mainly preserved just in Meso–Cenozoic. It is the important reason for the Meso–Cenozoic to become the main metallogenic periods. The ore deposits formed in or before Paleozoic are mainly remained in the Northern and Central Asian continent, which Meso–Cenozoic tectonics is rather weak, such as the Central Siberian plate, Central Asian–Mongolian plate, Ordos basin in Sino–Korean plate and Sichuan basin in Yangtze plate.

The ore-controlling structures formed in the Meso–Cenozoic metallogenic periods were developed mainly based on the pre-existing weakness structural zone, and the orientations of tectonic stress fields during Meso–Cenozoic had been changed for many times, then the blocks were cut off and broken extremely, and the scope of intraplate deformations was widely and stronger. As a result, it is easy to discover a lot of small and rich ore deposits in the Asian continent.

To form the industrial deposit, the condition is very strict; besides the well geological conditions, it need considering the technological and economic factors. According to the recent technological and economic conditions, the burial depth of solid endogenic ore deposits must be less than 2000 m, and the burial depth of fluid (oil and gas) fields must be less than 5000 m. Most of the Asian ore deposits would be continued to exploit if the market prices were not too low.

The initial burial depths of endogenic ore deposits usually are different from the recent burial depths. So the endogenic ore deposits that are being explored and exploited in central part of South China, after the formation, must have an appropriate uplifting process. For example, their metallogenic depths usually are 2–3 km, after the metallogenesis they will be uplifted as 1–3 km. If the ore bodies are located at the very deep positions, it still will get the very well exploitation condition. At East China, east to the Dahinggan–Taihang–Wuling Mountains areas, the metallogenic depths of some intermediate and shallow endogenic ore deposits are about 1–2 km, after the metallogenesis, the areas are uplifted as about 1 km, then the occurrence positions of ore deposit are not so deep. So those deposits are easy to be exploited. In Northeast China, the Daqing oil and gas field is similar to the above case; the effective depths of generative oil bed are about 2 km, but the uplift and erosion depths are about 1 km. Thus in recent, the depths of oil-bearing bed are about 1 km, so it is very favorable to be exploited.

Why are the Cenozoic metallogenic deposits all located on high plateau or mountain more than 2–4 km above sea level in recent? The reason is that the formation depths and the crust uplift height for the post-endogenic metallogenesis are rather similar; thus, the ore deposits could appear near the Earth surface by the erosion. It means that in the exploration, the geologists must pay attention to the deposits' forming depths, the crust uplift and erosion depths in those areas. It is the importance for tectono-geomorphology to research and explore the endogenic metal deposits.

As to the weathering residual bauxite and nickel deposits, they are all formed in the hot and moist weathering conditions. There must be a relative stable crust condition or a little bit uplift. If the uplifting range was too much, the weathering residual deposits would undergo the erosion. In addition, if the subsidence range was too much, those supergene deposits would be buried under the Earth surface, and the exploitation cost would be increased greatly not to be favorable for exploitation.

To sum up, to research the preservation conditions of ore deposits, it must closely couple the forming depth and tectono-geomorphology. However, at present, the related research is still rather weak in general, which needs to be further strengthened.

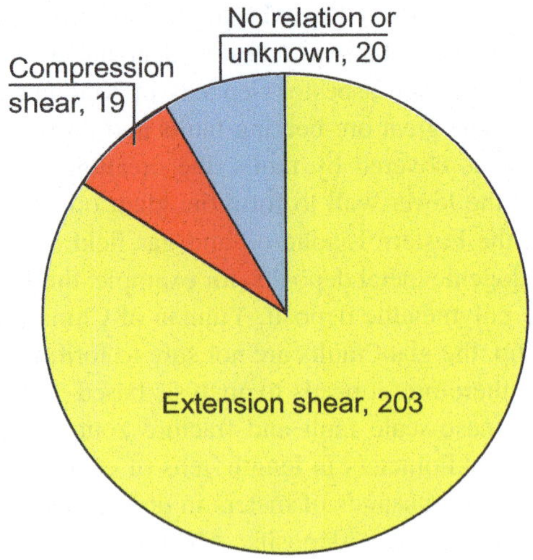

Fig. 4.92 Relationship between the different ore deposits and the tectonic mechanics. The total calculations are 242 giant endogenic and exogenic deposits or fields. The 203 deposits are related to the extension-shear structures, the 19 deposits are related to the compression-shear structures, and the 20 deposits are unknown relationship. Detailed data are shown in the Appendix

4.4.4 Intraplate Extension Metallogenesis

According to the statistic results for the ore-controlling fault mechanics description of 242 endogenic and exogenic mineral deposits (to see the Appendix), the author recognizes that the 203 ore deposits (accounting for 83.8% of the total) are controlled by the regional extension or extension-shear structures; 19 ore deposits (7.9%) controlled by the regional compression or compression-shear structures; and 20 ore deposits without relation to the tectonic controlling or the situation not clear, accounting for 8.3% of the total (Fig. 4.92). Of course, the oil, gas and coal basins are all related to the regional extension or partial extension structures.

In addition, through numerical statistics the author studies the relationship between metallogenic periods of 191 endogenic metal deposits or fields and tectonics in the Asian continent (Appendix; Fig. 4.93): 16 deposits (accounting for 8.4% of the total) to be formed at the pre-collision period; 36 deposits (18.8%) at the syn-collision periods; and 137 deposits (71.7%) at the post-collision periods, i.e., intraplate deformation periods; 2 deposits not to be determined for the relationship with the tectonics.

According to the above statistics, it can be considered that the intraplate deformations with the partial extension-shear tectonic environments are the main tectonic backgrounds for the endogenic metallogenesis in the Asian continent. Here, most of the magma or hydrothermal fluid would be filled into the extension or extension-shear fractures (meso-scale faults or joints). However, the ore deposits, which were formed and controlled by the compression-shear tectonics, were related to the ultra-mafic intrusion rocks and formed in the syn-collision periods, such as the chromium, platinum deposits, iron and asbestos deposits formed in syn-metamorphism. As to the sedimentary deposits, it is undoubted that they are only formed and controlled by the intraplate extension.

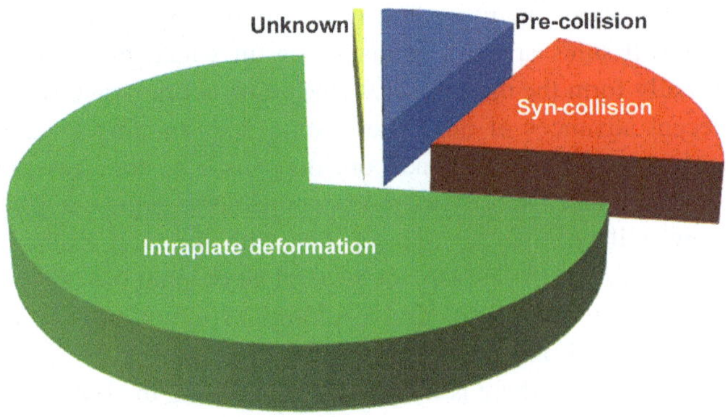

Fig. 4.93 Relationship between metallogenic period of endogenic metal deposits and tectonics in the Asian continent

According to the above comprehensive data, the author recognizes that the most important tectono-metallogenesis in the Asian continent is formed in the intraplate extension, without the relationship to the subduction, collision or other tectonics. Based on the above statistics (to see the Appendix), although the deposits were preserved and located in the collision (or orogeny) zone, more than one half of those could not be formed in the collision periods, but formed after the collision, i.e., being controlled by later intraplate tectonic stress. The syn-collision metallogenesis is only a few. Under the practice geological conditions in Asia, it needs to emphasize especially that the great metallogenesis controlled by collision (orogenic) zone is only a few; to form the giant ore deposits will be much less. In the collision period, the ore fluids or hydrothermal fluids not only could pass easily through the faults or fractures to migrate, but also pass easily through the whole crust to scatter and disappear into the atmosphere and hydrosphere, not easy to preserve in the crust rocks. If the ore deposits were formed in the collision periods, most of them would not be concentrated in the main fault zones, and usually preserved on their sides and branch fractures. Under the above intraplate extension background, although the ore deposits occurred in the collision zone, they were formed after the collision, i.e., the intraplate extension period. Depending on the statistics of the Asian great ore deposits (to see the Appendix), it seems that in the suitable strong tectonic areas, the possibility of mineralization is great, especially to form the giant super-large deposits. The tectonic backgrounds of metallogenesis are extremely different from those of volcanoes and earthquakes. The volcanoes and earthquakes will occur in the very intense tectonic position, but the endogenic deposits will be formed in the suitable tectonic position.

The above recognitions are extremely different from the fashionable metallogenic point views—the orogeny resulting in main endogenic metallogenesis. A lot of endogenic metal deposits are regarded as the "orogenic type" (Kerrich and Wyman 1990; Barley and Groves 1992; Chen 1996, 2000, 2006; Goldfarb et al. 2001; Hou et al. 2006, 2010; Mao et al. 2012c). It is considered that the scale of metallogenesis is positively related to the intensity of collision (Qiu 2002). Many researchers regard as so long as the ore deposits are preserved in the collision zone or orogenic belt, even there are some folds and faults in the mountains, the deposit should belong to the "orogenic type." They have never considered the tectonic background and tectonics property, as well as the relationship among the metallogenesis, the strong rock deformations and plate collision. This recognition seems unadvisable. It is easy for the stronger tectonics to transport the magma and ore fluids and to increase the chance of metallogenesis. The excessively strong collision and structure are not only easy to transport the magma and ore fluid, but also cause them to easily scatter and disappear, instead decreasing the metallogenic chance. So many experienced geologists usually emphasize to explore the ore deposit at the appropriate position of tectonics.

The author, his colleagues and students have studied the relationship between the gold deposits and the tectonic stress fields in Shandong Province and East China and discovered for many times that the ore deposit preservation positions are not only located at the very strong tectonics areas, but also at the relative weak areas; usually at the position of moderate differential stress values. For example, the differential stress values in the several gold fields of Shandong Province are about 20 MPa (the main research results have been submitted to some mines).[1,2,3,4,5,6,7,8] Later, during studying the oil and gas fields, we have also obtained the similar results, i.e., all the oil and gas reservoirs occur at the moderate position of differential stress value.

Here, it should be emphasized that the continent–continent collision is extremely different from the plate subduction. Do not confuse them. It is well known that when the oceanic plate subducts underneath the continent about 100 km deep, the partial melt and crust–mantle material exchange will occur to form a series of supergiant porphyry-type ore deposits or fields. In the eastern Asian trench–island arc system, from the Far East Kamchatka, Japan, Philippines to Indonesia, influenced by the Pacific plate subduction, a series important super-large endogenic ore deposits are formed. However, the inner continent–continent collision zones of the Asian continent have undergone the double subduction and collision. Two continental crustal blocks usually are mutually wedged and alternated, and under the continental crust, the great-scale detachments usually occur near the Moho discontinuity or middle crust, and then, the lithosphere faults have been reformed. Thus, it is not favorable for the intersection and material to exchange between the crust and mantle. If the

---

[1] Wan T F, Chu M J, Li S B (1988) The tectonic stress fields, paleo-temperature fields and metallogenetic prediction in the Guacangshan areas, Linhai City, Zhejiang. The Project of Volcanic Geology and Metallogenesis in Southeast Coast Area, (86017-331), 1–96.

[2] Wan T F, Wang Y S, Jia Q J et al. (2001) The paleo-tectonic stress field and fissure developed state prediction in the Gulong areas, Daqing Oil Field, 1–154.

[3] Zhang S R, Wan T F (2003) The tectonic stress field and fissure developed state prediction in the Triassic oil-gas Bedding in Xiaoquan-Xinchang, Sichuan. Geological Research Project, Southwest Petroleum Bureau, SinoPec, 1–90.

[4] Yan D P, Wan T F, Luan J C et al. (1995) The deep metallogenetic prediction in Beijie gold mine, Zhaoyuan, Shandong Province. Bureau of Gold. The Geological Project of Ministry of Metallurgy Industry (93-95-08), 1–90.

[5] Qi J Z (2001) The metallogenetic research for Qiyugou gold deposit, Henan Province. Ph. D thesis in China, University of Geosciences (Beijing), 1–123.

[6] Wan T F, Wang Y S, Jia Q J et al. (2001) The paleo-tectonic stress field and fissure developed state prediction in the Gulong areas, Daqing Oil Field, 1–154.

[7] Zhang S R, Wan T F (2003) The tectonic stress field and fissure developed state prediction in the Triassic oil-gas Bedding in Xiaoquan-Xinchang, Sichuan. Geological Research Project, Southwest Petroleum Bureau, SinoPec, 1–90.

[8] Wang M M, Wan T F, Liu J Y et al. (2003) The land oil prediction areas and evolution, Bohai Bay basin (010107-9-4) Exploration and Produce Company, CNPC, 1–225.

partial melting is produced in the lithosphere, the magma melts will be easy to separate through the inner tectonic detachments, not to directly intrude into the upper part of crust. The special strong magmatism and faulting can penetrate through the Earth surface and result in the ore fluids overflow into the atmosphere and hydrosphere, hardy to form the ore deposits in the upper crust. So if the subduction metallogenic model is roughly applied into the Asian continental collision zone, it will be easy to encounter obstacles. In general, in the strong collision zone, the rocks and elements can be mixed together; it is favorable for the elements to migrate in great scale, but unfavorable to enrich and reserve. Of course, in the certain conditions, the elements could be enriched to form the ore deposit. However, for the most of giant endogenic metal deposits, the enrichment chance will be little in the collision. This point of view is incomprehensible for many overseas scholars. Outside of the Asian continent, the rock deformations are very weak or never occur in many continental plates. Thus, they prefer the Plate Tectonics Theory and consider the tectono-magmatic events only to occur in or near the subduction and collision zones, and regard the continental plate as a rigid plate. In China, some researchers, in fact, have never understood the Asian continental metallogenesis and its tectonic background, but just prefer the fashionable viewpoint and recognition.

Such deposits as magnetite, chromite, ilmenite and serpentine asbestos must be related to the ophiolite suite in the collision zone, which are less influenced by the fluids at the later period. So the ore elements are rather stable in the collision to form deposits, such as the deposits in the Ural Late Paleozoic collision zone, Qilian–Altun Early Paleozoic collision zone, Western Qinling collision zone. Those are the real collision-type (or orogenic-type) deposits.

In the Asian continent, some ore deposits formed in the oceanic extension or rift trough before the collision period are called as "**pre-collision ore deposits.**" Most of the ore deposits formed after the collision and orogeny (Fig. 4.92) cannot be called as the collision or orogeny type, and their forming processes are different from those of the intraplate deformations. Truly, they may be formed only through the faults, fractures and foliations in the collision period.

Of course, the collision often results in the common foliations and fine fractures. When the later maximum compression stress orientation is parallel or with a small angel to the pre-existing fractures or foliations, the ore fluids will be easy to extend and produce the good migration and preservation positions. For example, in the Western Tianshan and Balkhash–Tianshan–Hingganling zone during Late Paleozoic (Late Carboniferous–Early Permian), when the regional stress orientation suddenly changed from the near NS to WNW trending compression, a lot of endogenic metal deposits related to the magmatism would be formed. However, at the collision periods in such areas (the Altay–Central Mongolia–Hailar Early Paleozoic collision zone, Karaganda–Kyrgyzstan Early Paleozoic collision zone, Western Tianshan Late Paleozoic collision zone and Balkhash–Tianshan–Hingganling Late Paleozoic collision zone), only a few ore deposits had been formed, as well as in and near the Qinling–Dabie–Jiaonan Triassic collision zone. However, after the collision period (Jurassic and Cretaceous), there formed a lot of giant endogenic metal deposits to constitute the important endogenic metal province in China. Some researchers consider that they all belong to the Qinling collision zone; in fact, the Xiao Qinling and Eastern Qinling belong to the Sino–Korean plate, instead of the Southern Qinling, i.e., the Dabashan area, belong to the Yangtze plate.

The collision zone that had undergone extremely strong deformations should provide a good wall rock condition for later metallogenesis in some times. When the isotopic age data are lacking, it will easy to make mistakes. That is to say, it is mistaken to think that deposits and collision zones are formed at the same time, so that all deposits located in collision zones are called "orogenic-type deposits." This is understandable at the level of knowledge at that time. However, although this naming method is relatively simple, it is not correct and appropriate, which can be easily misunderstood.

Since Neogene, the paleo-collision zones in Central Asia continent had been gradually uplifted as mountains, resulting in many endogenic metal deposits within Altay–Tianshan–Pamir–Himalayan areas being uplifted to the Earth's surface (with the height of 2000–4000 m above sea level), so that those endogenic metal deposits can be easily found, explored and exploited. At present, the stable blocks between the collision zones are mainly shown as sedimentary basins. Even if endogenic metal deposits had been formed within them, most of them would be deep buried beneath the Mesozoic or Cenozoic sedimentary rocks, which are just the "stagnate deposits." Because of the different preservation conditions of deposits, we can only explore the ore deposits in collision zones or their vicinities in many areas of Central and South Asia. However, it cannot be concluded that collision is the most favorable for metallogenesis (Qiu 2002). In fact, the actual metallogenic process endogenic metal deposits occurred after collision rather than at the same time of collision (Wan 2004, 2011).

Here, the author emphasizes the metallogenesis of intraplate extension background, which purpose is to hope that in future prospecting and exploration, special attention will be paid to the metallogenesis under intraplate deformation with local tension conditions. Neither does it deny the possible metallogenesis during the collision period, nor does it deny that possible metallogenesis during the pre-collision period. However, for the Asian continent is characterized by multi-stage collision and extension, and intense intraplate deformation, it is important not to neglect the fact that a large number of large deposits are formed by local extension during intraplate deformation. This is determined by the tectonic characteristics of the Asian continental lithospheric plate, which is seldom found in the interior of other continental plates.

4.4.5 Tectono-Metallogenesis and Further Proposal

There are the extremely strong deformations in the collision zone, which could provide a good wall rock condition for later metallogenesis in some times. When there is a lack of the isotopic age data, it will be easy to consider in error that the ore deposits and collision zone are formed at the same time, and whole ore deposits in or near the collision zone are regarded as the "orogenic-type deposit." It needs to comprehend those conditions. Although the naming method seems to be quite simple, it is not correct and appropriate to be misunderstood easily for us.

In the paleo-collision zone in the Central Asian continent, since Neogene those areas have been gradually uplifted as the mountains, so many endogenic metal deposits in the Altay–Tianshan–Pamir–Himalayan areas are uplifted near the Earth surface (2000–4000 m above sea level); then, those metal deposits are easy to be discovered, explored and exploited.

Among the above collision zones, all the stable blocks have been changed like sedimentary basins. If there formed the endogenic metal deposits, in recent they would be deep buried underneath the sedimentary strata. They are only the "stagnate deposit." Thus, we must prospect and explore the ore deposits in or near the paleo-collision zones of the Central and Southern Asia. But it cannot get the conclusion that "the collision is very favorable to metallogenesis" (Qiu 2002). The most of endogenic metal deposits are formed after the collision period, not in the collision period (Wan 2004, 2011).

Here, the intraplate extension metallogenesis (Figs. 4.92 and 4.93) is emphasized, and the author hopes that the metallogenesis in the partial extension condition could be paid attention to in the future. Here it does not negate the metallogenesis in or before the collisions. However, as to the Asian continent with the multiple collision periods, we must not ignore the fact of the intraplate deformation forming a lot of giant ore deposits. It has been proved by the tectonic characteristics of the Asian continental lithosphere plate, but it is seldom discovered in other continental plates.

4.4.5.1 The Exploration Proposal on the Deep and Outside Areas of Known Ore Deposits

In the Asian continent, the exploration and exploitation depths vary from 500 to 1000 m for the most of endogenic metal and coal deposits, and from 1000 to 2000 m for the oil and gas fields. With the progress on the nowadays mining, beneficiating and smelting technology, the exploitation depths of metal and coal deposits usually are up to 2000 m. The exploitation depths of diamond deposits in South Africa even reach 4500 m. Now the exploration and exploitation depths of many gas fields, including the natural gas, shale gas and coal-generated gas, are up to 5000 m.

Thus, in the surrounding areas of the pre-existing deposits with the similar geological structures, doing the deep exploration is a valuable choice. In a certain time, due to the global economic depression and the financial group monopoly, all the prices of oil, gas and metals would be decreased a lot; however, it does not indicate the society development not to demand the different kinds of mineral resources; instead, the mineral resources industries must go to another brighter tomorrow. In fact, the human society must continue to require the geosciences and mineral resources, just to have some changes in sometimes.

Near some old mines, when exploring and exploiting new ore deposits, it can utilize the local technological staff, ore beneficiating and smelting equipment and communications. To extend the life span of many old mining cities also can solve the society problems of labor employment and are more economical than the discovery of new deposits in the remote areas. In recent years, with the advances of the beneficiating and smelting technology, and the increment of synthesize using levels, some reserves out the balance sheet in the past, have usually had the economic values, and the related ore reserves have been increased greatly. For example, 50 years ago for the banded iron formation (BIF), the iron content must be up to 30% to reach the minimum industrial payable grade. Nowadays, due to decreasing prime cost of magnetic separation and the iron–silica separation, the minimum industrial payable grade of that ore deposit has been decreased to about 10% in China; thus, a lot of iron deposits out the balance sheet in the past have become the valuable deposits.

As to the sedimentary ore deposits, it is correct and undoubted to extend the deposit's ranges along their strike or dip. Of course, it must also consider the influence of folding and faulting, and the limit of sedimentary basins. Generally speaking, for the deep and outside exploration in the sedimentary deposits, it is not too difficult in terms of the exploration technology, but for endogenic metal deposits it is so difficult. In the exploration practice, some geologists, without studying the deep and outside structures for ore deposits, have arranged drilling wells around the pre-existing gold ore bodies, thus ended in failure and got nothing.

For the deep and outside exploration in the endogenic metal deposits, the author considers that it should note six viewpoints as follows:

1. "Interval without ore" between the ore bodies. Because the endogenic metal ore bodies are usually filling into the extension-shear joints or fracture zones, they must not be same as the sedimentary ore strata that extend stablely along their strikes or dips for a long distance. An endogenic ore body usually extends only several hundreds or one thousand meters to pinch out. If there are more than two ore bodies with similar attitude, between two ore bodies there must be an "interval without ore." The reason is that any a fracture is difficult to extend limitless, and it must have a definite length and width. Influenced by same tectonic stress, the faults with the same direction are often spread off and on. In the above example, if a geologist designs a borehole on the "interval without ore," it must be defeated.
2. The ore body's en echelon distribution and lateral trending characteristics. For the formation of endogenic metal deposits is influenced by the distribution of fractures, they usually have the en echelon distribution on the plane and have the lateral trending characteristics in the deep. Every deposit has its special en echelon and lateral trending characteristics. The tectonic stress state must be researched in detail. The reason is that the endogenic metal deposits and bodies are usually filled into the extension-shear zone. It means that most of endogenic metal deposits are diagonal and en echelon, but not to extend in a line.
3. The attitude change in the deep ore body. As for the endogenic metal ore bodies being filled into the reverse faults or thrusts, when they extend to the deep, it is not easy to keep ore body's attitudes from changing. The dip angles of ore body usually are changed gradually. The dip angles of ore body can often be discovered to change gently or steeply.
4. The change of grade and ore kind in the deep. The metallogenic temperature of the deep ore bodies or deposits is higher than that of the shallow, so it is favorable to preserve the metal elements with the higher crystallization temperature. It is often found that the metal grade in the deep is higher than that in the shallow. For example, for the Linglong gold field, Shandong of China, the mineral crystallization temperature in the deep is obviously higher than that in the shallow, and the gold grade in the deep is higher than that in the shallow. To the more depth, the ore kind will be even changed. For example, in the Luoning areas, Eastern Qinling, Henan of China, the gold quartz veins have the very dip attitude and occur on the mountains more than 2000 m above sea level; however, to the position lower than 1700 m above sea level, the ore veins become the rich galenite–quartz viens, which crystallization temperature is obviously higher than that of the gold–quartz viens. In addition, in the Xiangshan giant hydrothermal uranium deposit, Le'an County, Jiangxi of China, in recent years, however, the galenite viens have been discovered by the deep exploration drilling. There are many similar examples, not to be listed one by one.
5. The spatial combination shape of ore body. If the ore deposits are mainly formed in one metallogenic period, suffered and controlled by one tectonic stress field, their ore bodies will extend along one or two orientations. The reason is that they are controlled by conjugate shear fracture zones. If there are two or more tectono-metallogenic stages in an ore field, the ore bodies usually will show the complex shapes with four groups of net-vein-like styles. As a result, when doing the deep exploration, it must consider the varieties of the deep pattern for the ore deposit; in addition, it needs to consider the deep extensions and attitudes from multiple perspectives.
6. The quantitative forecast in the deep or outside areas of known deposits. Besides the qualitative forecast and mathematical statistics methods (Ye et al. 2010), there are several rather valid geological, geochemical and geophysical quantitative methods, such as the numerical simulation of tectonic stress field (Wan 1988; Liang et al. 1994; Deng et al. 1995; Zhang 2002; Ma et al. 2003; Xu 2007), the paleo-temperature research (Jin 1977; Zhuang 1995; Xi et al. 2003; Yang et al. 2005a, b; Xu 2008; Ding et al. 2010), the primary halo research of geochemical anomalies (Li 1991, 1999, 2000; 1999; Liu 2002), the high-precision 3D inversion based on the gravitational and magnetic imaging (Yao et al. 2002; Chen et al. 2012a, b; Wu 2014) and the controlled source audio-frequency magnetotelluric (CSAMT; Yang et al. 2000; Huang and Meng 2007; Dong et al. 2010) and so on. Usually, more than two methods are chosen to do the quantitative forecast, which results can be tested and verified each other. Based on the above research and survey results, it can be effective to design the drilling program. Thus, it avoids designing in the blindness and decreases the possible risks.

4.4.5.2 The Exploration Proposal on the Unknown Areas of Known Metallogenic Zones

It is rather reasonable to prioritize doing the deep exploration in the unknown areas of the known metallogenic zone. The China Geological Survey has energetically progressed those explorations, which are the hard work. It needs to not only understand the characteristics of known ore deposits in the metallogenic province, but also to find the new ore deposits in the former areas where deposits are undiscovered. It is an exploration process to grow out of nothing. Obviously, this work is more difficult than that near the known ore deposits. Commonly, it is necessary for the geological research, survey of geophysics and geochemistry, and the testing of drilling to be cooperated organically.

To discover the hidden small intrusions, i.e., ore-bearing cupola, is a very important key to explore the deep giant endogenic metal deposits. As to an endogenic metal deposits, the top zone of cupola for the hidden intrusions is the best position to enrich the ore fluids, which can form the porphyry, skarn, hydrothermal and magma giant and supergiant ore deposits (Fig. 4.94). However, around the intrusion it will be easy to form the intermediate and small ore deposits. In China the most of being exploited endogenic metal deposits are located around the intrusions, only a few at the top zone of cupola.

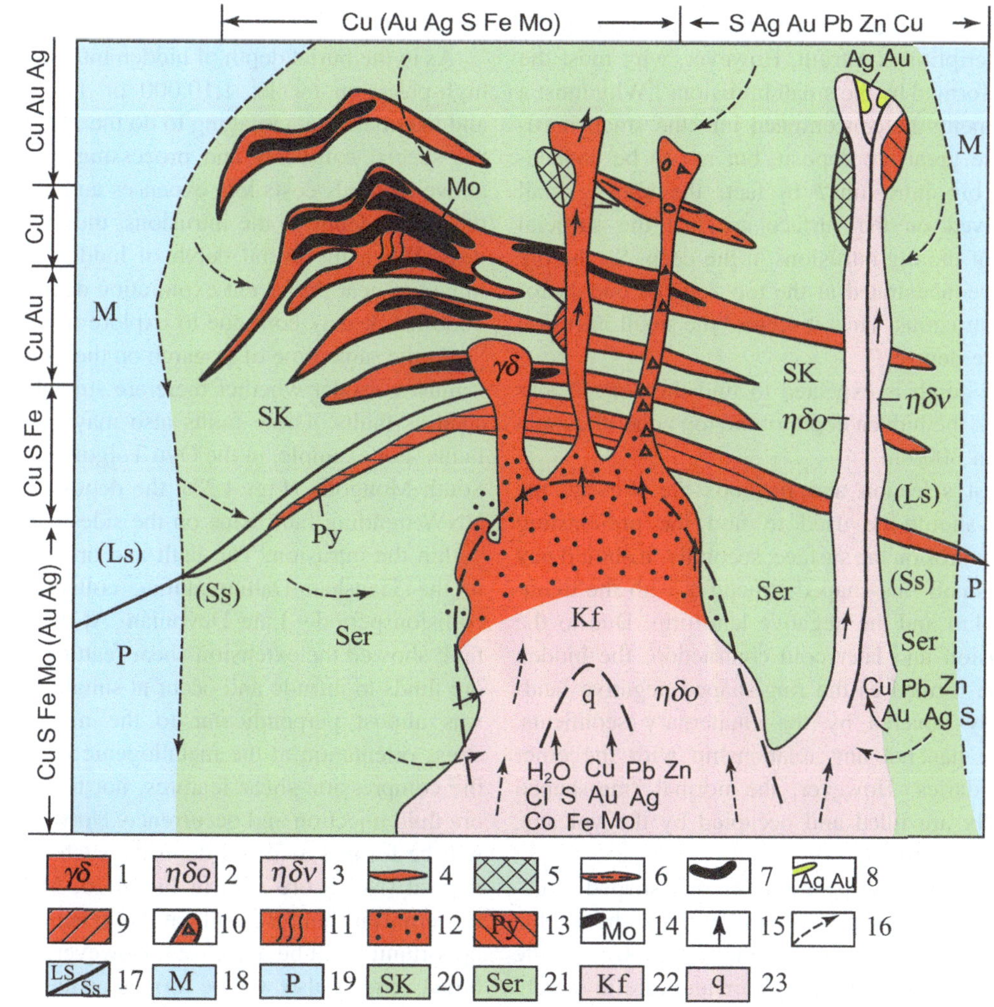

Fig. 4.94 Metallogenic model of hidden ore-bearing intrusion (Taking the Sizishan ore field, Tongling, Anhui of China, as an example, redrawn from the original data of 321 Geology Brigade, 1995) 1. Granodiorite; 2. quartz monzonite-diorite; 3. pyroxene monozonite–diorite; 4. stratabound skarn; 5. massive skarn; 6. stratabound skarn-sedimentary–metamorphic copper ore (such as Dongguashan deposit); 7. interlayer skarn copper ore; 8. Ag-Au ore; 9. contact skarn-type copper ore; 10. breccia pipe skarn-type copper ore; 11. veinlet hydrothermal-type copper-gold ore; 12. porphyry-type copper ore; 13. pyritization zone; 14. molybdenum ore; 15. migration orientation of ore elements; 16. migration orientation of ore fluid; 17. limestone/sandstone; 18. marbleization zone; 19. propylitization zone; 20. skarnization zone; 21. sericitization zone; 22. potassic metasomatism zone; 23. solidification zone

At the top of the intrusions, due to the thermodynamics, the magma firstly will expand in volume, later condensate to reduce the volume of intrusions (the volume reduced as 5–10%). Thus, the rocks will be easily broken to form the joints and breccia zone, namely the cryptoexplosion breccia. The broken rocks at the top of the intrusions will be as the perfect position for the ore fluids to condensate and accumulate. To explore the endogenic metal deposits at the top zone of cupola of hidden intrusions will be easy to find the super-large ore deposits. Thus, it is a very important issue for deep exploration.

Based on recent data, the hidden intrusions, formed with the super-large ore deposits, are often shown as the small cupola or stack on the surface, which diameters usually are only 1–2 km; the bigger ones just 3–4 km. Thus, many scholars described them as "the small intrusion forming giant ore deposit." As to the dimension of intrusion on the surface, that description is a truth. However, why must the great deposit be formed by the small intrusions? Why must a lot of metal elements be concentrated into the small intrusions to form the great ore deposit, but not to be concentrated into the big intrusions? In fact, the above small intrusions observed on the surface are just the surficial exposed points of the big intrusions in the deep. So the ore fluids are easily concentrated at the top zone of cupola for the big hidden intrusions. Thus it seems "the small intrusion forming giant ore deposit."

In recent, the depth is expected to find the deposit less than 2000 m, i.e., the hidden depth of the top zone of cupola must be less than 3000 m.

Thus, the best selection and methods are: Firstly, the remote sensing should be used to find the pre-existing ring-shaped structures on the surface; secondly, it should pay attention to the small ring-shaped structures with the diameter less than 4 km and the negative landform. Due to the first heat expansion and later cold contraction, the hidden intrusion will be formed as the ring-shaped negative landforms and easily covered by the Quaternary sediments, which boundary has not any relationship with the other geological boundaries. However, the normal ring-shaped landforms usually are filled and occupied by the volcanic rocks at the later periods.

In order to discover the positions of ring-shaped structures and hidden intrusions, it is best way to choose the high-precision, large-scale (such as the 1:10,000) satellite remote image, and to do the investigation and careful search. In recent years, Liu H F et al. Bureau of Hebei Geological Exploration and Developing (2006), according to the modern tectonic theory, have made a lot of exploration for the hidden ore deposits and got the obvious achievements. They used the ring-shaped structure data from the remote image to do the deep research, got some very valuable results and discovered many giant ore deposits. So their successful experiences should be popularized.

If the smaller ring-shaped structures with the negative landform are discovered, it does not indicate to find the ore-bearing intrusion; in addition, in the depth of every ring-shaped structure, there must not be ore deposits. It considers that many intrusions under the ring-shaped structure are never enriched with ore deposits. Instead, whether the hidden intrusions are enriched with deposits or not, it should do some researches on wall rock alterations in the surficial rocks. At the areas covered with sediments, it should collect the rock samples from the shallow drill holes and research the primary halo to determine whether there are the wall rock alterations related the ore deposit or not. If wall rock alterations are discovered to be related to ore deposits, it needs to do the research on the hidden intrusions in the burial depth.

As to the burial depth of hidden intrusions, it must use the high-precision (scale: 1:10,000 or 1:25,000) gravitational and magmatic data imaging to do the 3D inversion (there are the special softwares and processing methods). Using the above methods costs less expenses and can roughly forecast the burial depth of the intrusions, thus decreases the blindness. When the burial depth of hidden intrusions is determined as near the recent exploration depth (usually less than 3000 m), it may continue to explore to the deep.

At the same time of research on the ring-shaped structure, it must consider whether there are suitable linear structures, such as faults. These faults also may form the ore-bearing faults. For example, in the Oyu Tolgoi copper and gold field, South Mongolia (Fig. 4.23), the deposit was formed in the NNW trending fault zone on the sides the intrusion, but not within the intrusion. The fault and ore deposit were formed in the Tianshan–Dahingganling collision zone at the NS collision period—Late Devonian. At that period, the NNW fault showed the extension-shear features and resulted in the ore fluids to intrude and occur at suitable position. The fault was almost perpendicular to the maximum compression stress orientation at the metallogenic period, which showed the compression-shear features, not to be favorable for the ore fluid injection and occurrence. However, in some areas it will be formed as metal deposits, such as the Tuwu copper deposit (No. 9 ore vein in Fig. 4.20; Mao et al. 2012c) in Eastern Tianshan. But only at that position, the metallogenic opportunity will be decreased. However, in the lower wall of thrust zone, it also will be formed as the perfect position for the ore hydrothermal migration and occurrence, for example, the Jinding lead, zinc and silver deposit in Yunnan of China (Fig. 4.61).

So it must beforehand determine the possible relationship between the maximum principal compression stress orientation and the rock deformation at the possible metallogenic period. As to the above description, at the different geological periods and different regions, the maximum principal compression stress orientations are not same, so it must research the regional tectonics prior to the related tectonic events.

As to the deep survey on endogenic metal ore body, based on the pre-existing exploration experience, the CSAMT (Yang et al. 2000; Huang and Meng 2007; Dong et al. 2010) is one of the rather effective method to explore the metal sulfide ore body within the depth of 2000 m. For the possible positions preserving the hidden ore deposits in the 2000 m depth, it should arrange some CSAMT survey lines being perpendicular to the inferred ore body's trending. When arranging the CSAMT survey lines, it is the most important to determine the regional tectonic lines and the possible trendings of ore body at the metallogenic period. In the stronger intraplate deformation areas, the ore bodies often occur on a side of ring-shaped structures, also be preserved in a fault zone. Thus, to infer the possible tectonic stress state at the metallogenic period is the key to the arrangement of CSAMT survey lines.

If the survey line layouts are almost perpendicular to the unknown ore bodies, the CSAMT will be easy to succeed; otherwise, it will lose the opportunity. For example, in the Central Asia–Tianshan areas, if the metallogenesis occurred in the Early period of Late Paleozoic or Triassic, due to the influence of NS trending maximum principal compression stress, the eastern and western sides of ring structure and near NS trending faults would be easy to form the giant ore deposits. In this case, the CSAMT survey lines should be arranged near the EW trending. However, if the metallogenic period occurred in the late period of Late Paleozoic (i.e., Late Carboniferous–Early Permian), due to the influence of near EW trending maximum principal compression stress, the northern and southern sides of ring structure and near EW trending faults or fracture zones would be easy to form the giant ore deposits. In this case, the CSAMT survey lines should be arranged near the NS trending. In the different metallogenic periods, the tectonic background and stress status in the Asian continent have been discussed in the above text.

After the anomalies of CSAMT survey being obtained, they are possible for ore deposits. Next it should conduct the drilling project to enter into the exploration progress, then ascertain the shape and scale of ore deposits and finally make clear the deposit reserves. If the "wildcat" drills are adopted to arrange blindly the drilling and exploration, sometimes it may get a good luck, but the cost will usually be very expensive.

As to the exploration in the unknown areas, it will be more difficult. So it must collect and research all kinds of data and then do the best to reduce the possible risks.

References

Bai G P, Yin J Y (2007) The oil-gas distribution character and forming reservoir model in Karakum basin, Central Asia. Acta Paleogeography, 9 (3): 293–301 (in Chinese with English abstract).

Barley M E and Groves D I (1992) Supercontinent cycles and distribution of metal deposits through time. Geology, (20): 291–294.

Books Hephaestus (2012) Oil Fields of Russia, Including: Shtokman Field, Urengoy Gas Field, Sakhalin-II, Yuzhno–Russkoye Field, Yamal Project, Prirazlomnoye Field, Yamburg Ga. Hephaestus Books, 1–36.

Bortnikov N S, Gamynin G N, Vikent'eva O V et al. (2010) The Sarylakh and Sentachan gold-antimony deposits, Sakha–Yakutia: A case of combined mesothermal gold-quartz and epithermal stibnite ores. Geology of Ore Deposits, 52(5): 339–372.

Chao J C (1994) Hydrothermal vein type fluorite vapor-liquid inclusion hydrogen-oxygen isotopic characters. Geology and Exploration, (4): 28–29 (in Chinese with English abstract).

Chen C, Chen Z, Jin X et al. (2012a) The main ore resources and its mining invest environment for the Republic of Tajikistan. Resources and Industrial, 14(3): 6–11(in Chinese with English abstract).

Chen G D (1978) The Tectono-Metallogenesis Research Method. Beijing: Geological Publishing House. 1–413 (in Chinese).

Chen G J (2005) The dynamics evolution in the Animaqin–Bayankala orogenic zone and Dachang gold metallogenesis. The Master Thesis of Jilin University, 1–73 (in Chinese with English abstract).

Chen H D, Zeng Y F (1989) The sedimentary characters and origin discussion in Upper Devonian Liujiang Formation siliceous, Danchi basin, Guangxi. Mineral and Petrology, (4): 25–32, 131 (in Chinese with English abstract).

Chen H Q, Zhu X M, Dong Y L et al. (2009) The sequence stratigraphy analysis and litho-strata oil and gas reservoir prediction in isobaths fault-depression basin: The example of C basin deep water area Paleogene T formation, in South China Sea. Oil and Gas Geology, (5): 112–120 (in Chinese with English abstract).

Chen H S, Li H Q (1991) The liquid inclusion Rb-Sr isochron age in Yunkai Uplift gold deposit zone. Mineral Deposits, 9 (4): 333–341 (in Chinese with English abstract).

Chen X F, Peng R M, Liu J J et al. (2010) The geological characteristics of Kumtor super large polymetallic deposit, Tajikistan. Gold, 31(12):15–19 (in Chinese with English abstract).

Chen X H, Seitmuratova E, Wang Z H et al. (2014) SHRIMP U-Pb and Ar-Ar geochronology of Major porphyry and skarn Cu deposits in the Balkhash Metallogenic Belt, Central Asia, and geological implications. Journal of Asian Earth Sciences, 79: 723–740.

Chen Y C, Pei R F, Zhang H L (1990) The non-ferrous and rare metal deposits related to Mesozoic granite in Nanling areas. Journal of Chinese Geological Science Academy, (1): 88–94 (in Chinese with English abstract).

Chen Y C, Zhu Y S (1993) The Metallogenic Model of China. Beijing: Geological Publishing House, 1–367 (in Chinese with English abstract).

Chen Y C, Ye Q T, Feng J et al. (1996) The Ashle copper and zinc metallogenic deposit in pyrite type polymetallic metallogenic zone and prognosis. Beijing: Geological Publishing House, 1–330 (in Chinese with English abstract).

Chen Y J (1996) The metallogenesis and gold deposits distribution in the collision of Junggar orogenic zone. Acta Geologica Sinica, 70 (3): 253–261 (in Chinese with English abstract).

Chen Y J (2000) The research significance and progress in the Central Asia type orogen-metallogenesis in Northwest China. Journal of University Geology, 6(1): 17–22 (in Chinese with English abstract).

Chen Y J (2006) The orogenic type deposit, metallogenic model and its exploration capacity. Geology in China, 33 (6): 1181–1195 (in Chinese with English abstract).
Chen Y Q, Liu J L, Feng Q L et al. (2010) The Geology and Related to Granite Ore Deposits in Indochina Peninsula, Southeastern Asia. Beijing: Geological Publishing House, 1–192 (in Chinese with English abstract).
Chen Z F, Zhou S T, Wu T D (1999) Central Asian giant metal deposit character and metallogenic environment. Urumqi: Xinjiang Science and Technology Publishing House, 1–265 (in Chinese with English abstract).
Chen Z X, Meng X H, Guo L H (2012b) Review of 3D property inversion of gravity and magnetic data. Progress in Geophysics, 7 (2): 503–511.
China Geological Academy (1980) Asian Geological Map. Beijing: China Geography Publishing House (in Chinese).
Clarke M H (2006) Oman's Geological Heritage (2nd edition). Oman: Petroleum Development Oman, 1–247.
Cook D R, Hollings P, Walshe J I (2005) Giant porphyry deposits: Characteristics, distribution and tectonic controls. Economic Geology, 100: 801–818.
Deng J, Wang Q F, Li G J et al. (2014a) Tethys tectonic evolution and its bearing on the distribution of important mineral deposits in the Sanjiang region, SW China. Gondwana Research, 26: 419–437.
Deng J, Wang Q F, Li G J et al. (2014b) Cenozoic tectono-magmatic and metallogenic processes in the Sanjiang region, southwest China. Earth-Science Reviews, 138: 268–299.
Ding Q F, Wang G, Sun F Y et al. (2010) Ore-forming fluid evolution of Dachang gold deposit in Qumalai County, Qinghai Province: Evidence from fluid inclusion study and arsenopyrite geothermometer. Acta Petrologica Sinica, 26(12): 3709–3719 (in Chinese with English abstract).
Dobretsov N L, Kirdyashkin A A, Kirdyashkin A G et al. (2008) Modelling of thermochemical plumes and implications for the origin of the Siberian traps. Lithos, 100: 66–92.
Dong L H, Sha D M (2005) Late Paleozoic epithermal gold deposit in Western Tianshan. Beijing: Geological Publishing House, 1–154 (in Chinese with English abstract).
Dong L H, Xu X W, Fan T B et al. (2015) The discovery of super large Sedex carbonite type Pb-Zn deposit and the signification of metallogenesis in Huoshaoyun, Xinjiang. Xinjiang Geology, 33 (1):41–52 (in Chinese with English abstract).
Dong Z Y, Tang J, Zhou Z M (2010) Application of CSAMT to buried active faults investigation. Seismology and Geology, 32(3): 442–452 (in Chinese with English abstract).
Du S J, Li H X, Zhang Q L et al. (2003) The ore-controlling structure and its mechanism in the Linglong gold field, Zhaoyuan, Shangdong. University Acta Geologica Sinica, (3): 118–124 (in Chinese with English abstract).
Duan H G, Zhou C Q, Zhang Q C et al. (2014) The plays character of the abundant hydrocarbon area in the Middle East and their exploration potential. Earth Science Frontier, 21(3): 118–126 (in Chinese with English abstract).
Editorial Board of Petrological Geology of Changqing Oil Field (Zhai G M chief ed.) (1992) China Petroleum Geology, vol. 12, Changqing oil field. Beijing: Petroleum Industry Publishing House, 1–490 (in Chinese with English abstract).
Editorial Board of Petrological Geology of Dagang Oil Field (Zhai G M chief ed.) (1991) China Petroleum Geology, vol. 4, Dagang oil field. Beijing: Petroleum Industry Publishing House, 1–436 (in Chinese with English abstract).
Editorial Board of Petrological Geology of Daqing Oil Field (Zhai G M chief ed.) (1993) China Petroleum Geology, vol. 2, Daqing and Jilin oil field. Beijing: Petroleum Industry Publishing House, 1–785 (in Chinese with English abstract).
Editorial Board of Petrological Geology of Liaohe Oil Field (Zhai G M chief ed.) (1993) China Petroleum Geology, vol. 3, Liaohe oil field. Beijing: Petroleum Industry Publishing House, 1–572 (in Chinese with English abstract).
Editorial Board of Petrological Geology of Shengli Oil Field (Zhai G M chief ed.) (1993) China Petroleum Geology, vol. 6, Liaohe oil field. Beijing: Petroleum Industry Publishing House, 1–518 (in Chinese with English abstract).
Editorial Board of Petrological Geology of Xinjiang Oil Fields (Zhai G M ed.) (1993) China Petroleum Geology, vol. 15, upper part, Part 2 Karamay oil field, in Junggar basin. Beijing: Petroleum Industry Publishing House (in Chinese with English abstract).
Feng J Z, Wang D B, Wang X M et al. (2003) The geological characters and its origin in Liba giant gold deposit, Lixian, Gansu. Mineral Geology, 22 (3): 257–264 (in Chinese with English abstract).
Feng J, Liang YM, Liu X S et al. (2012) The evolution of Sikhote–Alin metallogenic zone. Geology and Resources, (3): 266–271 (in Chinese with English abstract).
Feng Z W, Xia W H, Zhang J T et al. (1989) The discussion of characters and metallogenic fluid in Huangshaping tungsten, Jiangxi. Earth Science, 14(4): 87–96 (in Chinese with English abstract).
Fu D M, Xu M J (1989) The Jiacun super large gold silver polymetallic deposit characters in western Sichuan and compare with Japan Kuroko deposit. Journal of Sichuan Geology, (2): 26–33 (in Chinese with English abstract).
Gao H, Mei Y X, Wang H L et al. (2012) The potentiality of Afghanistan copper deposit and exploration trending. Mineral Deposit Exploration, 3(5): 1–7 (in Chinese with English abstract).
Gao J W, Zhao G C, Mao X H et al. (2010) The structural characteristics and exploration trend in Jinqinding gold field, Rushan, Shandong. Ore Deposit Geology, 29 (supp.): 43–45 (in Chinese with English abstract).
Gao J Y (2011) The geological characters and discussion of origin for Chayuan manganese iron deposit, Zhunyi County, Guizhou. The Master Thesis of Chengdu Science and Technology University, 1–41 (in Chinese with English abstract).
Gao R, Wang H Y, Zhang Z J et al. (2011) "Cutting" the crust and the upper mantle and revealing the deep structure of the continent with the resource effect: An introduction to the project Sino-probe 02 experimentation deep probing techniques and integration and a discussion on key science problems. Acta Geoscience Sinica, 32 (S1): 34–48 (in Chinese with English abstract).
Goldfarb B J, Groves D I, Gardoll S (2001) Orogenic gold and geologic time: A global synthesis. Ore Geology Reviews, 18: 1–75.
Goldfarb R J, Santosh M (2014) The dilemma of the Jiaodong gold deposits: Are they unique? Geoscience Frontiers, 5(2): 139–153.
Golovanov I B, Zhenodarova S M (2005) Quantitative Structure-Property Relationship: XX. Property-Property Correlation and Nonlinear Bronsted and Hammett Equations, Translated from Zhurnal Obshchei Khimii, 75(4): 534–540 (in Russian).
Goncharenko O P (2006) Potassic salts in Phanerozoic evaporate basin sand specific features of salt deposition at the final stage of halogenesis. Lithology and Mineral Resources, 41(4): 378–388.
Gong H B, Chen S F (2006) The new original recognition of multi-antimony deposit, Guangnan. Yunnan Geology, (3): 55–60 (in Chinese with English abstract).
Graupner T, Niedermann S, Kempe U et al. (2006) Origin of ore fluids in the Murumtau gold system: Constraints from noble gas, carbon isotope and halogen data. Geochimica et Cosmochimica Acta, 70: 5, 356–5, 370.
Gritsuk Y M, Popov Y N, Kochetkova V M (1995) Eventual Paleozoic stratigraphy of Rudny Altai: Implication from computer multiple analysis, geology-geophysical data and metallogenic agents. In:

Geological Structure and Useful Minerals in the Western Altai–Sayan Region. Novokuznetsk: Yuzhsib Geolkom Publications, 74–80.

Groves D I, Goldfarb R J, Gebre-Mariam M et al. (1998) Orogenic gold deposits: A proposed classification in the context of their crustal distribution and relationship to other gold deposit types. Ore Geology Reviews, 13(1–5): 7–27.

Gu Z P (1996) The recent oil and gas exploration and prospect in the East Sea. Oceanic Geology and Quaternary Geology, 16 (4): 113–118 (in Chinese with English abstract).

Guo J C, Zhang X Y, Li J G et al. (2005) The comparing with the mercury ore deposit in Chersky volcanic zone, Northeastern Russia and bloodstone ore deposit in China. University Bulletin of Geology, 11(2): 253–259 (in Chinese with English abstract).

Guo W G, Jiang J Z, Gan X P (2010) The Yaogangxian tungsten deposit geological characters and discussion the metallogenic model in Hunan. Mineral Deposit and Geology, (4): 25–29 (in Chinese with English abstract).

Guo Y H, Zhou X H, Ling Y X et al. (2011) The new recognition for the Penglai 19-3 oil field in Bohai. Oil and Gas Geology, 32(3): 327–333 (in Chinese with English abstract).

Han Z H, Pan J Z (2009) The preliminary analysis of the relationship between Guizhou Permian manganese deposit and the Emeishan basalt. Guizhou Geology, (3): 37–41(in Chinese with English abstract).

Hao T S, Wang K Y, Piao X H et al. (2011) The geological character and its origin for gold and silver deposit in Qingchengzi area, Liaoning. Gold, (1): 29–32 (in Chinese with English abstract).

He C, Xiao S G, Dan M C (2008) Indonesia laterite nickel deposit. Yunnan Geology, 27(1): 20–26 (in Chinese with English abstract).

He G Q, Zhu Y F(2006)The compare research of geological mineral deposits in Xinjiang China and its adjacent areas. Geology in China, 33(3): 451–460 (in Chinese with English abstract).

He J (1993) The research of metallogenic geochemistry in Yongping copper deposit, Jiangxi. Geology and Exploration, (8):10–14 (in Chinese with English abstract).

He J X, Xiong X B, Yan P et al. (2007) The oil and gas geological characters and its exploration trending in northern boundary basin of South China Sea. Xinjiang Petroleum Geology, 28 (2):129–136 (in Chinese with English abstract).

He M Q, Song H B, Ran C Y et al. (1998) The reformed evidence of Jinman copper deposit, Lanping, Yunnan. Geology and Exploration, (2):4 (in Chinese).

He W J (1987) The metallogenic geological condition and its controlled factor in Baiyangchang, Yunnan. Yunnan Geology, (4): 18–27 (in Chinese with English abstract).

He Z H, Zhou W M, He W Y et al. (2013) The origin type and metallogenesis in Beiya super large gold polymetallic deposit, northwestern Yunnan. Mineral Deposit, 32 (2): 244–258 (in Chinese with English abstract).

He Z J, Zhang G Y, Wang Z M et al. (2015) The distribution of oil and gas and the main controlled factor of its reservoir forming in the Sakhalin basin, Northeast Russia. Earth Science Frontiers, 22(1): 291–300 (in Chinese with English abstract).

Hedayat S (2010) Oil and Natural Gas Fields: A Study of Selected Countries, Key Information. Lap Lambert Academic Publishing, 1–112.

Hedenquist J W, Arribas A, Reynolds T J (1998) Evolution of an intrusion-centered hydrothermal system; Far Southeast–Lepanto porphyry and epithermal Cu-Au deposits, Philippines. Economic Geology, 93(4): 373–404. https://doi.org/10.2113/gsecongeo.93.4.373.

Hou G, Sun Z S, Wang A P (2009) The metallogenesis and future prospect for Jiapigou gold deposit, Jilin Non-Ferrous Metal Mining and Metallurgy, 25(2):1–5 (in Chinese with English abstract).

Hou P, Tian Z J, Deng J Z et al. (2014) The oil and gas resources valuation in Central Asian basin. Earth Science Frontier, 21(3): 56–62 (in Chinese with English abstract).

Hou Z Q (2010) Metallogensis of Continental Collision. Acta Geologica Sinica, 84(1): 30–58 (in Chinese with English abstract).

Hou Z Q, Du A D, Sun W D (2001) The metallogenic material source: rhenium-osmium and helium isotopic evidences in Hishikari deposit Japan. Acta Geologica Sinica, 75(1): 97–105 (in Chinese with English abstract).

Hou Z Q, Mo X X, Gao Y F et al. (2006) Early processes and tectonic model for the Indian – Asian continental collision: Evidence from the Cenozoic Gangdese igneous rocksin Tibet. Acta Geologica Sinica, 80(9): 5–20 (in Chinese with English abstract).

Hu L, Sun P, Zhang C G et al. (2011) The geological character and exploration trend for Bajiazi gold deposit, in Huadian City. Jilin Geology, (2): 29–33 (in Chinese with English abstract).

Hu W R, Yang H, Lv J (2000) The exploration and exploitation thinking and technology for Changqing Oil Field. The Review of Oil Science and Technology, (4): 46–49 (in Chinese with English abstract).

Hu X Z, Yang Z B (2003) The relationship between metallogenic fluid evolution and forming the deposit in Qiebaoshan copper polymetallic deposit, Liuyang. Geology and Exploration, 39 (5):22–25 (in Chinese with English abstract).

Hu Y Z, Zhang G Q, Wang J J et al. (2012) The seismic exploration evidence and significance for the carling type gold deposit thrust-fold structure in middle Southwest Guizhou. Earth Science Frontier, 19 (4):63–71 (in Chinese with English abstract).

Huang D T (1999) The origin discussion of Songshugang tungsten, tin, niobium, tantalum polymetallic deposit in Hengfeng, Jiangxi. Non-Ferrous Metal Deposit and Exploration, 8(4): 1–6 (in Chinese with English abstract).

Huang F, Wang D H, Zeng Z L et al. (2012) The Yuanlingzhai giant molybdenum deposit geochemistry, chronology of rock and metallogenesis and its geological significance in Southern Jiangxi. Geotectonica et Metallogenia, 36(3):363–376 (in Chinese with English abstract).

Huang F, Wang D H, Chen Z Y et al. (2014) Magma metallogenic characteristics preliminary research. Geotectonica et Metallogenia, 38(2): 239–254 (in Chinese with English abstract).

Huang L J, Meng Y S, Lu G F (2007) An application of CSAMT sounding to exploration of deep geothermal resources. Computing Techniques for Geophysical and Geochemical Exploration, 2007, 29: 60–63. (in Chinese)

Huang M Z, Chen Y C, Tang S H (1985) The antimonite-tin-lead in cassiterite-sulfide deposit and its forming condition in Changpo, Dachang. Land and Resources of Southern China, (1): 16–27 (in Chinese with English abstract).

Huang Z Y, Li G M (2004) The basic characters and exploration latent capacity in porphyry type copper deposits of Yarlung Zangbo metallogenic province, Xizang. Geology and Exploration, 40(1): 1–6 (in Chinese with English abstract).

Hutchison C S, Tan D N K (eds.) (2009) Geology of Peninsular Malaysia. Kuala Lumpur: Murphy. The University of Malaya and the Geological Society of Malaysia, 1–479.

Ji Y L (1989) The petroleum geological characters and exploration state in Saudi Arabia. World Petroleum Science, (2):17–23 (in Chinese with English abstract).

Jiang L (2014) The recent exploration and expect of oil-gas bearing in East Sea continental shelf basin. DOC88, Com., 1–8 (in Chinese).

Jiang S H, Nie F J, Su Y H et al. (2010) Chronology and mechanism of Erdenet super-large copper-molybdenum ore deposit in Mongolia. Acta Geoscientica Sinica, 31(3): 289–306.

Jiao Y Q, Wu L Q, Peng Y B et al. (2015) The sedimentary-tectonic background research for the formation of sedimentary Uranium deposits in North China paleo-Asian Tectonic Domain. Earth Science Frontier, 22(1): 189–205 (in Chinese with English abstract).

Jin Y (1977) Geothermal method for the application in the exploration of polymetallic deposits. Geology and Prospecting, (4): 92–93 (in Chinese).

Kan Y Z (2000) The geological characteristics for oil and gas distribution in Xinjiang. Xinjiang Oil Geology, 21 (5): 365–370 (in Chinese with English abstract).

Karsakov L P, Zhao C J, Malyshev Y et al. (2008) Tectonics, Deep Structure, Metallogeny of the Central Asian–Pacific Belts Junction Area (Explanatory Notes to the Tectonic Map Scale of 1: 1 500, 000). Beijing: Geological Publishing House, 1–213.

Kerrich R, Wyman D A (1990) The geodynamic setting of mesothermal gold deposits: An association with accretionary tectonic regions. Geology, 18: 882–885.

Khomich V G, Ivanov V V, Zinkov A V et al. (1997) Epithermal gold-silver mineralization of Late Paleozoic volcano-plutonic complexes of South-West Primorye (Far East Russia). Late Paleozoic and Early Mesozoic Circum-Pacific Events: Biostratigraphy, Tectonic and Ore Deposits of Primorye (Far East Russia). IGCP Project 272. Memoires de Geologie (Lausanne). (30): 195–182.

Kulikova Z I, Gulina V A, Zorina L D et al. (1996) Indicator role of gas explosion breccia in the genesis of the Teremki gold deposits, Russ (in Russian). Geol. Geophys., 37 (12): 61–72.

Lang X H, Tang J X, Xie F W et al. (2014a) The geological chronology, petrology geochemistry and its geological significance in the southern porphyry, Xiongcun deposit, Xizang. Geotectonica et Metallogenia, 38(3):609–620 (in Chinese with English abstract).

Lang X H, Tang J X, Li Z J et al. (2014b) U-Pb and Re-Os geochronological evidence for the Jurassic porphyry metallogenic event of the Xiongcun district in the Gangdise porphyry copper belt, southern Tibet, PRC. Journal of Asian Earth Sciences, 79: 608–622.

Li B Z, Deng G P, Wang Q C et al. (2002a) The geological characters of Dexing, Jiangxi gold and copper ore province. Mineral deposit, (S1): 437–440 (in Chinese with English abstract).

Li D M, Li Q, Zheng D W et al. (2002b) The principal and application for $^{40}Ar/^{39}Ar$-$^{36}Ar/^{39}Ar$. Geological Review, 48(supp.): 97–102 (in Chinese with English abstract).

Li D T, Zhang H D, Shan L H (2002c) The deep hidden deposit exploration and estimate in Jiaojia gold deposit, Jiaodong. Western Exploration, (3): 38–39 (in Chinese with English abstract).

Li D X, Zhao Y M (2004) The ore skarn zoning and fluid evolution in Jiaoli, Jiangxi skarn silver, lead, zinc, tungsten deposit. Geological Review, (1): 18–26 (in Chinese with English abstract).

Li G P, Jin Z J et al. (2005) The Atlas of World Oil and Gas Bearing basin (first part). Beijing: Petroleum Industry Publishing House, 1–110 (in Chinese).

Li H (1991) Geochemistry anomaly model in quartz vien type and alternation type gold deposit. Beijing; Science Press, 1–70.

Li H, Zhang W H, Liu B L et al. (1999) The study on axialzonality sequence of primary halo and some criteria for the application of this sequence form ajortypes of gold deposits in China. Geology and Prospecting, 35 (1): 32–35 (in Chinese).

Li H W, Xu K (2001) The strike-slip of Tan–Lu fault and the tectono-paleogeography in Liaohe basin. Earth Science Frontiers, 8 (4): 467–470 (in Chinese with English abstract).

Li J P, Wu L S (2008) The geological characters of main metal resources in Iran. Mineral Deposit, 27 (4): 265–266 (in Chinese with English abstract).

Li J W, Wang C X, Hou W R et al. (2003) The gold metallogenesis in Bainaimiao area, Inner Mongolia. Geoscience, (3): 51–56 (in Chinese with English abstract).

Li Y, Shen J F (1990) The research on the native bismuth in Jiangxi Dalongshan tungsten deposit. Geological Review (4): 364–369 (in Chinese with English abstract).

Li Y J, Yang H J, Zhang G Y et al. (2013) Re-division the secondary structural units in the Tabei Uplift, Tarim basin. Institute of Geology and Geophysics, China Academy 2012 (12th) Thesis Collection—Lithosphere Evolution Research Group, 258–270 (in Chinese).

Li Y X, Sun J G, Chen J J et al. (2010) The geochemical characteristics of PGE, siderophile and chalcophile elements of Jinchanggouliang gold (copper) deposits, Inner Mongolia: Their provenance and ore-forming tectonic setting. Earth Science Frontiers, 17 (2): 336–347 (in Chinese with English abstract).

Li Y Z (1999) The geochemistry primary halo on the Molontau gold deposit. World Nuclear Geoscience, (3):35–39 (in Chinese).

Liang H Y, Mo J H, Sun W D et al. (2008) The period analysis for forming rock and deposit in Yulong super large porphyry copper deposit, Eastern Xizang. Acta Lithologica Sinca, (10): 170–176 (in Chinese with English abstract).

Liang L, Liu C D, Li J H (1994) Tectonic stress field in 6124 ore deposit. Mineral Depost and Geology, (4): 262–265 (in Chinese).

Liang Y B, Zhao Z, Zhang G Y et al. (2014) The resource potentials of main plays in the petroliferous basins in Russia. Earth Science Frontiers, 21(3): 38–46 (in Chinese with English abstract).

Liao T D (2009) On the geological characteristics and exploration future in western Baoshan copper, molybdenum, lead, zinc and silver deposit. Hunan Non-Ferrous, (3): 6–12 (in Chinese with English abstract).

Lin Li, Zhan G L, Yu X P (2006) The geological characteristics of Dahutang tungsten (tin) ore field and its distant view. Resource Survey and Environment, (1): 28–35 (in Chinese).

Liu D W (2002) The geochemical block theory and method, technology for the research of mineral resources evaluation. Ph.D Thesis of Chinese Academy of Geological Sciences, 1–80 (in Chinese with English abstract).

Liu J, Lu H F, Liao Z L et al. (2005) The shallow sedimentary sulfide distribution and the relationship with gas hydrate in Dong Sha sea areas. Earth Science Frontier, 12(3): 258–262 (in Chinese with English abstract).

Liu J D, Zhang C J, Liu X F et al. (2004) The metallogenesis and exploration trending in southwest side of Yangtze Platform. Beijing: Geological Publishing House, 1–204 (in Chinese with English abstract).

Liu J J, Zheng M H, Long X R et al. (2000) The recognition and its significance for the Sawarden Muruntau type gold deposit, Xinjiang. Precious Metal Geology, (3): 2–5, 16 (in Chinese with English abstract).

Liu J J, Long X R, Zheng M H et al. (2002) The quartz $^{40}Ar/^{39}Ar$ fast neutron active age and its significance of Sawarden gold deposit in Xinjiang. Mineral and Rock, 22 (3): 19–23 (in Chinese with English abstract).

Liu J S (2012) The exploration and exploitation history and geological characters of Monywa copper field, Burma. http://blog.sina.com.cn/yuelugj, 2012-11-2.

Liu S Y, Sun W, Song J M et al. (2015) The tectonic control theory in Sichuan basin marine oil and gas distribution. Earth Science Frontier, 22 (3):146–160 (in Chinese with English abstract).

Liu X P, Chang Y F (1988) On the metallogenic condition and law in the middle and lower Yangtze areas. Acta Geologica Sinica, 62 (3): 167–177 (in Chinese).

Liu Z J (2009) The recent petroleum resources and police review in Saudi Arabia. Resource Net, 2009-05-20.

Lu R (2013) Shuikoushan, Changning, Hunan lead and zinc deposit geological characters and the discussion of origin. Master Thesis of Nanjing University, 1–74.

Luo X J, Wen C Q, Cao Z M et al. (2002) The analysis of V-Ti-Fe deposit origin in the Panzhihua. Mineral Deposit, 21(Supplement): 338–341 (in Chinese with English abstract).

Luo D F, Huang Z L, Wang F et al. (2012) The mechanism for ore elements migration and subsidence in the super great lead-zinc ore deposit in Huize, Yunnan. Acta Mineralogica Sinica, (2): 288–293 (in Chinese).

Ma D Y, Gao Z M, Yang S Y et al. (2003) Numerical simulation for tectonic stress field at Beiya gold deposit. Tectonics and Metallogene, (2): 62–68 (in Chinese with English abstract).

Ma H W (1990) The granitic rocks and metallogenesis in Yulong porphyry copper zone, Xizang. Wuhan: China University of Geosciences Press, 1–158 (in Chinese with English abstract).

Ma Y S, Guo X S, Guo T L et al. (2005) The discovery and exploration for Puguang large gas field, Sichuan. Geological Review, 51(4): 477–480 (in Chinese with English abstract).

Mao J R, Xu N Z, Hu Q et al. (2004a) Mesozoic rock forming and metallogenesis and tectonic environment in Shanghang–Tatian areas, Fujian. Acta Petrologica Sinica, 20(2): 285–296, (in Chinese with English abstract).

Mao J W, Yang J M, Han C M et al. (2002a) Eastern Tianshan Late Paleozoic copper gold and polymetallic metallogenic system and its geodynamic model. Earth Science, 27(4): 413–424 (in Chinese with English abstract).

Mao J W, Han C M, Wang Y T (2002b) Geological characteristics, metallogenic model and criteria for exploration of the large South Tianshan gold metallogenic belt in Central Asia. Geological Bulletin of China, 21(12): 858–868.

Mao J W, Du A D, Seltmann R et al. (2003) Re-Os dating for the Shameika porphyry Mo deposit and the Lipovy Log pegmatite rare metal deposit in the central and southern Urals. Mineralium Deposita, 38: 251–257.

Mao J W, Xie G Q, Li X F et al. (2004b) Mesozoic giant metallogenesis and lithosphere multistage extension in South China. Earth Science Frontier, 11(1): 45–55 (in Chinese with English abstract).

Mao J W, Wang Y T, Li H M et al. (2008) The relationship of mantle-derived fluids to gold metallogenesis in the Jiaodong Peninsula: Evidence from D-O-C-S isotope systematics. Ore Geology Reviews, 33: 361–381.

Mao J W, Zhang Z H, Pei R F (2012a) The Introduction of China Ore Deposit Model. Beijing: Geological Publishing House, 1–560 (in Chinese with English abstract).

Mao J W, Zhang Z H, Wang Y T et al. (2012b) Mineral Deposits: Types, Characteristics and Explorations. Beijing: Geological Publishing House, 1–480 (in Chinese).

Mao J W, Zhou Z H, Feng C Y et al. (2012c) A preliminary study of the Triassic large-scale mineralization in China and its geodynamic setting. Geology in China, (6): 5–39 (in Chinese with English abstract).

Mao J W, Pirajno F, Lehmann B et al. (2014) Distribution of porphyry deposits in the Eurasian continent and their corresponding tectonic settings. Journal of Asian Earth Sciences, 79(B): 576–584.

Masaitis V L (2002) Popigai Crater: General Geology, Impacts in Precambrian Shields. Impact Studies. Springer, 81–85.

Moralev V M, Glukhovsky M Z (2000) Diamond-bearing kimberlite fields of the Siberian Craton and the Early Precambrian geodynamics. Ore Geology Reviews, 17 (3): 141–153.

Naldrett A J (1989) Magmatic sulphide deposits. New York: Oxford Univ. Press, 1–186.

Naldrett A J (1999) World-class Ni-Cu-PGE deposits: Key factors in their genesis. Mineralium Deposita, 34: 227–240.

Nie F J (1990) The new recognition of Bainaimiao copper deposit original type. Geology in China, (3): 16–17 (in Chinese with English abstract).

Nie F J, Zhang K, Liu Y F et al. (2011) Indosinian magmatic activity and molybdenum, gold mineralization along the northern margin of North China Craton and adjacent area. Journal of Jilin University (Earth Science Edition), 41: 1651–1666 (in Chinese with English abstract).

Niu S Y, Sun A Q, Ma B J et al. (2015) Tectonic evolution and metallogenic ore-controlling tectonic characteristics of Hadamengou gold deposit in Inner Mongolia. Earth Science Frontiers, 22(1): 223–237 (in Chinese with English abstract).

Ouyang Z Y, Liu J Z, Zhang F Q et al. (2002) The primary theory of planet earth heterogeneity origin and evolution. Earth Science Frontier, 9 (3):23–30(in Chinese with English abstract).

Oxman V S (2003) Tectonic evolution of the Mesozoic Verkhoyansk–Kolyma belt (NE Asia). Tectonophysics, 365 (1–4): 45–76.

Parfenov L M, Prokopiev A V, Gaiduk V V (1995) Cretaceous frontal thrusts of the Verkhoyansk fold belt, eastern Siberia. Tectonics, 14 (2): 342–358. https://doi.org/10.1029/94tc03088.

Pecher A, Arndt N, Jean A et al. (2013) Structure of Panzhihua intrusion and its Fe-Ti-V deposit, China. Geoscience Frontiers, 4(5): 571–581.

Pei R F, Mei Y X, Li J W (2004) Super-large type deposits and anomaly metallogenesis. Earth Science Frontier, 11(2):323–331 (in Chinese with English abstract).

Pei R F, Li J W, Mei Y X et al. (2007) The tectonic character and super-giant metal industry deposits in China boundary. Journal of University Geology, 13(2):137–147 (in Chinese with English abstract).

Peng L G, Ren Y X, Li Z P et al. (1996) The Metallogenic Model of Copper Polymetallic Deposit. Beijing: Geological Publishing House, 1–211 (in Chinese with English abstract).

Peng M X, Shan S J, Zhu C et al. (2007a) Xinjiang Caixiashan lead and Zinc deposit origin and compare with MVT type. Xinjiang Geology, 25(4): 373–377 (in Chinese with English abstract).

Peng Y B, Chen A P, Fang X H et al. (2007b) The hydrocarbon fluid and the relationship of metallogenesis in Dongsheng sandstone uranium deposit. Geochemistry, 36 (3): 267–274 (in Chinese with English abstract).

Petrychenko O Y, Peryt T M, Chechel E I (2005) Early Cambrian sea water chemistry from fluid inclusion sinhalite from Siberian evaporites. Chemical Geology, 219(1/4):149–161.

Pollard P J, Taylor R G (2002) Paragenesis of the Grasberg Cu-Au deposit, Irian Java, Indonesia: Results from Logging Section 13. Miniralium Deposita, 37: 117–136.

Qi J Z, Yuan S S, Li L et al. (2003) Yangshan gold deposit geological and geochemistry research, in Wenxian, Gansu. Mineral Deposit, 22 (1): 26–33 (in Chinese with English abstract).

Qi J Z, Ma Z R, Li L (2004a) Metallogenic fluid evolution characteristics of Qiyugou, Henan. Gold Geology, (4):1–10 (in Chinese with English abstract).

Qi J Z, Yuan S S, Liu Z L et al. (2004b) U–Pb SHRIMP dating on zircon from quartz vein of Yangshan gold deposit and its geological significance. Acta Geologica Sinica, 78 (2): 443–451.

Qian H M (1992) Indonesia. In: Geological and Mineral Deposit for the World Countries. China Geological Mineral Deposit Information Institute, 1–55 (in Chinese with English abstract).

Qian K, Li B L, Xu H Z (2003) From the global marine Paleozoic oil-gas field geology to expect the gas exploration in Sichuan marine phase strata. Gas Geoscience, 14 (3):167–171 (in Chinese with English abstract).

Qiu X P (2002) Collisional orogenic belt and metallogenic division. Bulletin of Geology, 21(10):675–681 (in Chinese with English abstract).

Ren G M, Li Y G (2007) The geological characteristics and future exploration in Hongzuo porphyry copper deposit. Journal of

Sichuan Geology, 27(4): 266–268 (in Chinese with English abstract).
Ren S M, Qiao D W, Qiu H J et al. (2003) Iraq petroleum resources statue and the influences to China energy security. Geological Bulletin, 22 (3): 208–214 (in Chinese with English abstract).
Ridd M F, Barber A J, Crow M J (2011) The Geology of Thailand. London: The Geological Society, 1–626.
Sa Z B, Wang H B, Zhang G X et al. (2005) The metallogenic relationship for the diapir structure and natural gas hydrate. Earth Science Frontier, 12 (3): 283–288 (in Chinese with English abstract).
Saunders A, Reichow M (2009) The Siberian Traps and the End–Permian mass extinction: A critical review. Chinese Science Bulletin, 54 (1): 20–37.
Seltmann R, Porter T M, Pirajno F (2014) Geodynamics and metallogeny of the central Eurasian porphyry and related epithermal mineral systems: A review. Journal of Asian Earth Sciences, 79: 810–841.
Shen Z Y, Hu Y Z, Han R S et al. (2010) Analysis of Gulu antimony deposit controlling factor, in Dachang antimony field, Qinglong, Guizhou. Mineral Deposit and Geology, (4):10–15 (in Chinese with English abstract).
Shi G H, Cui W Y, Cao S M et al. (2008) Ion microprobe zircon U-Pb age and geochemistry of the Myanmar jadeiteite. Journal of the Geological Society, 165: 221–234.
Shi Z M (1992) The rare elements existing status and its distribution characteristics in M rare earth deposit, Changbei, Sichuan. Rare Earth, 13 (3): 1–9 (in Chinese with English abstract).
Song X Y, Chao Z M, Luo F X et al. (2004) The preliminary discussion for the origin of Yangliupin copper, nickel and platinum elements sulfide deposit, in Danba, Sichuan. Journal of Chengdu Science and Technology University (Natural Science), (3): 38–48 (in Chinese with English abstract).
Sun W Z, Wang C L, Yang X B (2007) The Oligocene hinder oil and gas reservoir type and exploration trend in the Weixinan depression, Northern Bay. Gas Science, (1): 90–94 (in Chinese with English abstract).
Sun X M, Lu H F, Ma M Y (2002) The super large lead and zinc hydrothermal carbonite trace elements and C, O Sr isotopic constant and the significance of deposit, in Fankou, Northern Guangdong. China inclusion and geological fluid workshop thesis abstract, 34 (in Chinese).
Sun Y, Liu X F, Wang R J et al. (2013) The composition characteristics of celestine in Dafengshan strontium deposit. Ore Deposit Geology (1): 151–159.
Tan K R (1983) Tongshanling grano-diorite porphyry geological characters and its metallogenesis in Jiangyong, Hunan. Geotectonica et Metallogenia, (3): 25–32 (in Chinese with English abstract).
Tang J X, Huang Y, Li Z J et al. (2009) The elements geochemistry characters of Xiongcun copper gold deposit, Xiementong, Xizang. Mineral Deposit, 28 (1): 15–28 (in Chinese with English abstract).
Tang Z L, Li W Y (1995) The Metallogenic Model and Geological Comparison in the Jinchuan Sulfide Copper and Nickel (with Platinum) Deposit. Beijing: Geological Publishing House, 1–209 (in Chinese with English abstract).
Tao K Y (1997) Compare with the Jinguashi gold and copper deposit, Taiwan and Zijinshan copper and gold deposit, Fujian. Volcanic Geology and Mineral Deposit, 18 (4):260–275(in Chinese with English abstract).
Tao Y, Luo T Y, Gao Z M et al. (2003) The research of mantle plume for Yunnan Jinbaoshan PGE ore body. The workshop of research on Emeishan mantle plume origin and the environment effect thesis and abstract, 42–47 (in Chinese).
Tessensohn F, Roland N W (1998) A Preface: Third International Conference on Arctic Margins. Polarforschung, 68: 1–9.
Tian H L (1997) The Lanping–Baiyangpin copper silver and polymetallic deposit geological characters. Yunnan Geology, (1): 105–108 (in Chinese with English abstract).
Volkov V N (2003) Phenomenon of the formation of very thick coal beds. Lithology and Mineral Resources, 38(3): 223–232.
Wan P, Li H Y, Shi D Y (1988) The serpentinization in asbestos deposit and metallogenic geochemistry. Mineral and Petroleum, (2): 115–122, 145 (in Chinese with English abstract).
Wan T F (1988) Paleo-Tectonic Stress Field. Beijing: Geological Publishing House. 1–156 (in Chinese).
Wan T F (2011) The Tectonics of China: Data, Maps and Evolution. Beijing: High Education Press; Heidelberg, Dordrecht, London, New York: Springer, 1–497.
Wan T F, Zeng H L (2002) The distinctive characteristics of the Sino-Korean and the Yangtze plates. Journal of Asian Earth Sciences, 20 (8): 881–888.
Wan T F, Zhu H (1991) Tectonic events of Late Proterozoic-Triassic in South China. Journal of Southeastern Asian Earth Sciences, 6(2): 147-157.
Wan T F, Wang Y M, Liu J L (2008) Tectonic detachment and magma origin depth in Eastern China during Yanshanian and Sichuanian Periods. Earth Science Frontier, 15(3): 1–35 (in Chinese with English abstract).
Wan T F, Zhao Q L, Wang Q Q (2015) Paleozoic tectono-metallogeny in the Tianshan–Altay region, Central Asia. Acta Geologica Sinica, 89 (4): 1120–1132.
Wang C L, Luo S H, Xu Y Z et al. (1987) Shizuyuan tungsten polymetallic deposit geology. Beijing: Geological Publishing House, 1–147 (in Chinese with English abstract).
Wang D H, Li J K, Fu X F (2005a) The metallogenic period and its significance of Jiajika pegmatite type rare metal deposit, in Sichuan. Geochemistry, 34(6): 541–547 (in Chinese with English abstract).
Wang H P (2005) The geological characters and exploration trend in the Western Baoshan lead and zinc deposit, Hunan. Hunan Non-Ferrous, (5): 6–9 (in Chinese with English abstract).
Wang J Q, Yao B C, Wan L et al. (2005b) The oil and gas resources in South China Sea Cenozoic sedimentary basin. Ocean Geology and Quaternary Geology, (2): 94–103 (in Chinese with English abstract).
Wang L, Sun F Y, Xu Q L (2012a) The fluid inclusion and deposit origin in Daheishan, Jilin molybdenum deposit. World Geology, 31 (1): 58–67 (in Chinese with English abstract).
Wang M M (2004) The Tectonic Stress and Oil-gas Reserves in Bohai Bay area. The Ph.D thesis of China University of Geosciences, 1–120 (in Chinese with English abstract).
Wang S S, Zhang Q Z, Tan Z G et al. (2008) The geological characteristics and controlling factor in Limu granite type tin, niobium and tantalum deposit of Guangxi. The Public Science and Technology, (11): 111–112 (in Chinese).
Wang S W, Sun X M, Shi G Y et al. (2007a) The difference of PGE elements geochemistry of the Jinbaoshan and Baimazhai copper nickel sulfide deposit and its original significance. Acta Geologica Sinica, 81(1): 93–108 (in Chinese with English abstract).
Wang T, Tong Y, Jahn B M et al. (2007b) SHRIMP U-Pb zircon geochronology of the Altai No. 3 pegmatite, NW China, and its implications for the origin and tectonic setting of the pegmatite. Ore Geology Review, 32: 325–336.
Wang W, Liu C D, Liu J H et al. (2012b) The fluorite deposit geological characters and forming conditions analysis in the Songyang, Zhejiang. Journal of Eastern China Science and Technology University (Natural Science), (1): 70–78 (in Chinese with English abstract).

Wang X B, Zhou X, Hao Z G (2010) The deeply exploration recognition and proposal for Luobusha chromite deposit in Xizang. Geological Bulletin, (1): 107–116 (in Chinese with English abstract).

Wang X M, Wang B (2010) Recent Tajikistan mineral resources and its exploitation prospect. World Nonferrous Metal, (7): 28–31(in Chinese with English abstract).

Wang X Q (1998) The exploration geochemical theory and application for finding and recognizing large and supper large deposits. Geophysics and Geochemistry Exploration, 22 (28): 1–89 (in Chinese with English abstract).

Wang X Q, Xie X J (2000) The Gold Exploration Geochemistry: Theory, Method, Strategy and Tactics. Jinan: Shandong Science and Technology Press, 1–309 (in Chinese with English abstract).

Wang Y, Chen Z L, Liu J et al. (2006) Southern Yili basin neo-tectonic characteristics and its controlled the sandstone Uranium deposits. Geotectonica et Metallogenia, 30(4):486–494 (in Chinese with English abstract).

Wang Z G (2010) The laterite type giant nickel deposits of Dinagat Island geological characteristics and exploration method. Geology and Exploration, 42 (2): 361–366 (in Chinese with English abstract).

Weng W H (1929) Mesozoic orogenic movement of east China. Journal of Geological Society of China, 8(1): 33-44.

Wei J X (2011) Suishizim gold deposit geological characters and metallogenic research in Liziyuan area, Tianshui. The Master Thesis of Chang'an University, 1–82 (in Chinese with English abstract).

Wei X, Deng J F, Xie W Y et al. (2005) The reef controlled by South China Sea basin evolution and reef oil and gas reservoir exploration potential analysis. Earth Science Frontier, 12(3): 245–252 (in Chinese with English abstract).

Wu G G, Zhang D, Zang W S et al. (2008) SHRIMP zircon U-Pb dating of the intrusives in the Tongling metallogenic cluster and its dynamic setting. Science in China (D), 51(7): 911–928 (in Chinese).

Wu G, Chen Y C, Li Z Y et al. (2014) Geochronology and fluid inclusion study of the Yinjiagou porphyry-skarn Mo-Cu-pyrite deposit in the Eastern Qinling orogenic belt, China. Journal of Asian Earth Sciences, 79: 585–607.

Wu L S (2009) The mineral resources and their geological characters in Vietnam. Mineral Deposit, 28 (6):856–859 (in Chinese with English abstract).

Wu L S (2011) The basic geological structural characters and mineral resources in Thailand. Mineral Deposit, 30(3): 573–576; 30(4): 194–196; 30 (5): 196–200 (in Chinese with English abstract).

Wu W P (2014) Gravity and magnetism three dimensional reversion technology. On web, Baidu Library, 1–65 (in Chinese).

Wu Z H, Wu T D, Tang C H (1993) Geology and Ore Deposit for China Surrounding Countries. Wuhan: China University of Geosciences Press, 1–268 (in Chinese).

Xi X W, Yang L Q, Wang Y J et al. (2003) Numerical modeling of the temperature effects of tectonic regimes transformation and their ore-forming dynamics. Earth Science Frontiers, 10 (1): 47–55 (in Chinese with English abstract).

Xiao X N, Zhang S Y, Ju C R et al. (2012) The Renzhanshi copper deposit geological characteristics and original research in Yunnan. Geology and Exploration, (2):41–53 (in Chinese with English abstract).

Xie C J, Zhao X B, Zhang G Z et al. (2015) Metallogenic environments, ore-forming types and prospecting potential of Au-Cu-Zn-Pb resources in Western Tianshan Mountains. Geology in China, 42(3): 381–410 (in Chinese with English abstract).

Xie T H, Liu P, Huang Z L (2012) The geological condition and hydrocarbon charge in high temperature and super pressure gas reservoir in Yinggehai. Gas Engineering, 32(4):1–5 (in Chinese with English abstract).

Xie T X (1999) Summary of China (celestine) strontium deposit. Geology of Chemical Mineral Deposit, (3): 14–21 (in Chinese with English abstract).

Xie X J, Xiang Y C (1999) The super-giant deposit geochemistry forecast method. In: Xie X J, Shao Y, WangX Q et al. (eds.), Going to 21 Century: Mineral Deposit Exploration Geochemistry. Beijing: Geological Publishing House, 35–47 (in Chinese with English abstract).

Xie X J, Cheng H X (2001) Global geochemical mapping and its implementation in the Asian–Pacific region. Applied Geochemistry, 16 (11–12): 1309–1321.

Xie X J, Cheng Z Z, Zhang L S et al. (2008) The Atlas of 76 elements geochemistry in southwestern China. Beijing: Geological Publishing House, 1–219 (in Chinese with English abstract).

Xie Y L, Hou Z Q, Xu J H et al. (2005) Discovery and original significance for the copper-zinc, copper-tin minerals in Mianning–Dechang rare earth metallogenic zone in Sichuan. Science in China (D), (6): 92–97 (in Chinese with English abstract).

Xin H B, (2006) Discussion on the geological character and its origin for Bainaimiao polymetallic deposit, Inner Mongolia. Collection of Geological Exploration, 21(4): 236–240, 298 (in Chinese with English abstract).

Xiong X S, Gao R, Guo L H et al. (2015) The deep structure feature of the Sichuan basin and adjacent orogeny. Acta Geologica Sinica, 89 (4): 1153–1164.

Xu D M, Ling Z Y, Luo X Q et al. (2015) Metallogenic series of major metallic deposits in the Qinzhou–Hangzhou metallogenic belt. Earth Science Frontiers, 22 (2): 7–24 (in Chinese with English abstract).

Xu H (2001) The hydrocarbon accumulation conditions analysis and basin dynamic background research in Chunxiaooil and gas field, in Eastern Sea. The Ph.D Thesis of China University of Geosciences (Wuhan) (in Chinese with English abstract).

Xu H (2007) Numerical Modeling of Structure Stress Field in the Beiya Gold Ore Deposit, Yunnan. Master thesis of China Geosciences University (Beijing), 1–78 (in Chinese with English abstract).

Xu S K, Li B Y (2003) Ore strata geological characters in Baishugang rutile deposit. Chemical Industry Mineral Deposit and Geology, (1): 26–38 (in Chinese with English abstract).

Xu Y (2008) The metallogeny variety analysis and prospect for endogenic metal deposit, Ph.D thesis of China Geosciences University (Beijing). 1–117 (in Chinese with English abstract).

Xue C J, Liu S W, Chen Y C et al. (2004) Giant mineral deposits and their geodynamic setting in the Lanping basin, Yunnan, China. Acta Geologica Sinica, 78 (2): 1–7.

Xue C J, Zeng R, Liu S W et al. (2007) The geological fluid inclusion and isotopic characteristics of the Jinding Zn-Pb deposit, Western Yunnan, China:A review. Ore Geology Review, 31: 337–359.

Yang F Q, Mao J W, Wang Y T et al. (2005a) Sawarden gold deposit geological character and its metallogenesis in Southwest Tianshan, Xinjiang. Ore Deposit Geology, 24 (3): 206–227 (in Chinese with English abstract).

Yang F Z, Hong G L, Zhu H Q et al. (2014) The petroleum play characteristics of the Southeast Asian region and exploration potential. Earth Science Frontiers, 21(3): 112–117 (in Chinese with English abstract).

Yang F, Xiao R G, Xia X H (2011) The sedimentary environment and ore geochemistry in Kunyang phosphorus. Geology and Exploration, (2): 162–171 (in Chinese with English abstract).

Yang J S, Xu Z Q, Li Z L et al. (2009) Discovery of an eclogite belt in Lhasa block, Tibet: A new border for Paleo-Tethys? Journal of Asian Earth Sciences, 34: 76–89.

Yang J S, Xu Z Q, Duan X D et al. (2012a) The discovery of Jurassic SSZ type ophiolite in Myitkyina area. Acta Petrologica Sinica, 28(6): 1710–1730 (in Chinese with English abstract).

Yang J X (2009) The research of oil geological reserves for Qinghai Oil Field. Energy Source Manage, 217 (in Chinese).
Yang J Z, Sheng Y C, Li G M et al. (1999) The variant characteristics of ore bodies and scientific prospecting of the Penjiakuang gold deposit, Rushan, Shandong Province. Geotectonica et Metallogenia, 23 (2):61–67 (in Chinese with English abstract).
Yang J Z, Zhao Y L, Shen Y C et al. (2000) The application of controllable source audio frequency magnetotelluric method for ore body positioning forecast. Geological Technology Information, 19 (3): 107–112 (in Chinese with English abstract).
Yang K Q (1986) The research content and direction of dynamic of rock forming and metallogenic theory. The collection of Geodynamic Institude, Chinese Academy of Geoscience. Beijing: Geological Publishing House. (1): 5–19 (in Chinese).
Yang K C (2000) Discussion on some geological problems in the Yinggehai basin. Fault Block Oil and Gas Field, (2): 1, 10–17 (in Chinese with English abstract).
Yang R Y, Ma D S, Pan J Y (2005b) The atmospheric rainfall water to influence the ore fluid thermal field-exampled the fluid of Xikuangshan antimony deposit. Earth Science: 30 (3): 70–78 (in Chinese with English abstract).
Yang T, Zhu L M, Li B et al. (2012b) The carling type gold deposit geology-geochemistry and metallogenic research in Jinlonshan, Western Qinling. Acta Mineralogy Sinica, (1): 118–133 (in Chinese with English abstract).
Yang X A, Liu J J, Han S Y et al. (2012c) The mineral composition, geochemistry character and its geological significance in Yangla copper deposit, Yunnan. (2): 27–40 (in Chinese with English abstract).
Yang X P, Yu X Q, Wang Z X et al. (2015) Comparative study in the ore-forming conditions and caused of the hydrothermal gold deposits in the Western Tianshan, China. Geotectonica et Metallogenia, 39(4): 633–646 (in Chinese with English abstract).
Ye H S, Mao J W, Li Y F et al. (2006) The SHRIMP zircon U-Pb and molybdenite Re-Os ages and its geological significance in the Donggou super large porphyry molybdenum deposit, Eastern Qinling. Acta Geologica Sinica, 80 (7): 1078–1088 (in Chinese with English abstract).
Ye T Z, Xiao K Y, Chen Q M et al. (2010) Mineral deposit quantitative prospection method. Beijing: Geological Publishing House, 1–191 (in Chinese with English abstract).
Yelisseyeva A, Mengb G S, Afanasyeva V et al. (2013) Optical properties of impact diamonds from the Popigai astrobleme. Diamond and Related Materials, 37: 8–16. https://doi.org/10.1016/j.diamond.2013.04.008.
Yin S S, Wang C L (1994) The Xintianling tungsten deposit geological characters, Cheng County. Hunan Geology, (4): 205–211 (in Chinese with English abstract).
Zeng Q D, Liu J M, Qin K Z et al. (2013) Types characteristics and time-space distribution of molybdenum deposits in China. International Geology Review. https://doi.org/10.1080/00206814.2013.774195.
Zeng Q Y (2013) The geological characters and discussion of its origin of Xiaolong tungsten deposit, in Taihe County, Jiangxi. Western Exploration, (1):111–115 (in Chinese with English abstract).
Zhai G M (2002) Plate Tectonic Evolution and the Formation and Valuation of Oil-bearing Basin. Beijing: Petroleum Industry Publishing House, 1–461 (in Chinese with English abstract).
Zhai Y S (1995) The Daye type iron and copper deposit model, related to the intermediate-acid intrusion in the inter-continental depression. In: Pei R F (ed.), The China Metallogenic Model. Beijing: Geological Publishing House, 262–264 (in Chinese with English abstract).
Zhai Y S, Lin X D (eds) (1993) The ore field structure. Beijing: Geological Publishing House, 1–173(in Chinese).
Zhai Y S, Yao S Z, Cui B et al. (1996) The Research of Metallogenic System. Wuhan: Press of China University of Geosciences, 1–198 (in Chinese with English abstract).
Zhai Y S, Zhang H, Song H L et al. (1997) Greattectonics and Super Great Ore Deposits. Beijing: Geological Publishing House, 1–180 (in Chinese with English abstract).
Zhai Y S, Yao S Z, Cai K Q (2011) Mineral Deposit (third eds.). Beijing: Geological Publishing House, 1–413 (in Chinese with English abstract).
Zhang B R, Luo T C, Gao S et al. (1994) The Qinling–Dabashan Lithosphere, and Tectono-Metallogenic Geochemistry Research. Wuhan: China University of Geosciences Press, 1–446 (in Chinese with English abstract).
Zhang D Q, She H Q, Li D X et al. (2003) The porphyry-epithermal metallogenic system in Zijinshan areas. Acta Geologica Sinica, 77 (2): 253–261 (in Chinese with English abstract).
Zhang D W (2011) The formation geological condition and exploration trend research of Tazhong, Tabei large oil and gas fields in Tarim basin. In: Song W J (ed.), Yearbook of Tarim Oil, 313 (in Chinese with English abstract).
Zhang E, Zhou Y Z, Guo J (2001) The structural characteristics and its controlling to metallogenesis in Bagaomiao gold deposit, Shaanxi. Mineral Deposit, 20 (3): 229–233 (in Chinese with English abstract).
Zhang G C, Mi L J, Wu J F et al. (2010) The uplift and its plunging: The favorable exploration trend for large and middle oil and gas fields in deep water areas of Southeast Hainan basin. China Offshore Oil and Gas, 22 (6): 360–368 (in Chinese with English abstract).
Zhang G H (2015) The structural inversion characteristics and discussion on the mechanism in East Sea continental shelf basin. Earth Science Frontier, 22 (1): 260–270 (in Chinese with English abstract).
Zhang H J, Wu H, Zhao H L (2012) The comparison research between Qingchengzi, Liaoning and Jiangde, Korea lead and zinc deposits. Modern Mining, (1): 66–69 (in Chinese with English abstract).
Zhang H R, Yang T N, Hou Z Q et al. (2013) The Dongmozhazua ore field deformation characters in the northern Sanjiang areas. Acta Petrologica Sinica, (4): 49–59 (in Chinese with English abstract).
Zhang H X (2009) The metal mineral resources investigation and analysis of cooperation capacity around the China countries. Earth Science Progress, 24 (10): 1159–1172 (in Chinese with English abstract).
Zhang J Q, Chen Z H, Wang D H et al. (2008) The geological characters, metallogenic period and its prospect significance in Xingluoken large tungsten deposit. Geotectonica et Metallogenia, 32(1): 92–92 (in Chinese with English abstract).
Zhang L C, Xiao W J, Qin K Z et al. (2005a) Re-Os isotopic dating of molybdenite, and pyrite in the Baishan Mo-Re deposit, eastern Tianshan, NW China, and its geological significance. Mineralium Deposita, 39: 960–969.
Zhang Q S (1984) The Geology and Metallogenesis of Pre-Cambrian, China. Changchun: Jilin People Press, 1–544 (in Chinese with English abstract).
Zhang S M, Chen Z H, Shi G H et al. (2011) The molybdenite Re-Os isotopic dating in Dajishan tungsten deposit, Jiangxi. Mineral Deposit, 30(6): 1113–1122 (in Chinese with English abstract).
Zhang W J (2002) The geology of jadeite deposit, Hpakan area, northern Burma. Yunnan Geology, 21 (4): 378–390 (in Chinese with English abstract).
Zhang W L, Hua R M, Wang R C et al. (2007b) The dating research for ore granite rock forming and metallogenesis in Piaotang tungsten

deposit, southern Jiangxi. Acta Mineral Sinica, (21):324–325 (in Chinese with English abstract).

Zhang Y S, Zheng J P, Qi W (2005b) Investigation for the sylvine resource and their exploitation in Turkmenistan. Ore Deposit Geology, (6): 121–125 (in Chinese with English abstract).

Zhang Y Y, Wu Y S, Zhao Y J et al. (1998) The deep geochemistry exploration future in Xiaolong tungsten deposit, Jiangxi. Geology and Exploration, (3): 36–39 (in Chinese with English abstract).

Zhao D D, Jin G, Li H S et al. (2013) The geological characteristics and its origin in Sanshandao gold deposit, Laizhou, Shandong. Geological Exploration Collection, 28(4): 546–551 (in Chinese with English abstract).

Zhao D K (2000) The brush structure zone and its controlling in Baiyangchang, Yunlong silver and polymetal field. The Ph.D thesis of Shanghai Metallurgy Institute, China Academy (in Chinese with English abstract).

Zhao H X (2009) The recent and developed comments for the magnesite in Liaoning. Fire-resistant Material, 43 (4): 291–293 (in Chinese with English abstract).

Zhao J C, Li Q M, Wang Q H et al. (2004) The formation and distribution of Tarim large and middle scale oil and gas field. Journal of Northwest University (Natural edition), 34 (2): 212–217 (in Chinese with English abstract).

Zhao S M, Yang W Z, Wang D K (2012) Katbasu gold deposit geological character and discussion on its origin. Ore Deposit Geology, 31(supp.): 825–826 (in Chinese with English abstract).

Zheng B, Gao R X (2004) Discussion on some geological problem for oil reservoirs in Pinghu oil and gas field, East Sea. Petroleum Experiment Geology (6): 60–64 (in Chinese with English abstract).

Zheng W B, Chen Y C, Song X et al. (2010) The elements distribution in Jiama copper polymetallic deposit, Xizang and its geological significance. Mineral Deposit, 29 (5): 775–785 (in Chinese with English abstract).

Zheng T Y, Zhu R X, Zhao L et al. (2012) Intra-lithospheric mantle structures recorded continental subduction. J. Geophys. Res., 117: B03308, doi:10.1029/2011.JB008873.

Zhong D L (1998) The Paleo-Tethys Orogenic Belt of Western Yunnan and Sichuan. Beijing: Science Press, 1–231 (in Chinese with English abstract).

Zhong G J, Zeng F C, Feng C M (2010) The evolution trending of deep water oil and gas exploration and the current exploration situation in northern South China Sea. Mineral Deposit, (S1): 1086–1087 (in Chinese with English abstract).

Zhou D, Sun Z, Chen H Z et al. (2005) Mesozoic lithofacies paleogeography and tectonic evolution in the South China Sea and its adjacent. Earth Science Frontiers, 12 (3): 204–218 (in Chinese with English abstract).

Zhou L M (2009) The tectonic stress field numerical value molding and metallogenic prospect in Dajishan tungsten deposit, Quannan, Jiangxi. The Master Thesis of China University of Geosciences, 1–38 (in Chinese with English abstract).

Zhou M F, Chen W T, Wang C W et al. (2013) Two stage of immiscible liquid separation in the formation of Panzhihua-type Fe-Ti-V oxide deposits, SW China. Geoscience Frontiers, 4(5): 481–502.

Zhou W P (2011) The geological characteristics of western Baoshan copper deposit, in Guishan, Hunan. Deposit Exploration, (5): 18–21 (in Chinese with English abstract).

Zhou X H, Mao Y F, Yang S et al. (2012) The alteration characteristics and its significance in Heshang uranium deposit, Xiangshan ore field, Le'an County, Jiangxi. The Journal of Eastern China Science and Technology University (Natural Science), 35(1): 1–9 (in Chinese with English abstract).

Zhou Y Z, Zheng Y, Zeng C Y et al. (2015) On the understanding of Qinzhou Bay–Hangzhou Bay metallogenic belt, South China. Earth Science Frontiers, 22(2): 1–6 (in Chinese with English abstract).

Zhu G L (1984) The geological character and its origin for Liaodong Peninsula talc and magnesite. Journal of Changchun Geological Institute, (2): 77–94 (in Chinese with English abstract).

Zhu J C, Zhang C H (1981) The origin of Carboniferous volcanic rocks and copper, tungsten deposit in Fenglin, Dongxiang, Jiangxi. Journal of Nanjing University (Natural Science), (2):111–124 (in Chinese with English abstract).

Zhu L M, Ding Z J, Yao S Z et al. (2009) The metallogenic geological event and its tectonic background in Wenquan molybdenum deposit, Gansu, Western Qinling. Bulletin in Science, (16): 77–87, 179–181 (in Chinese with English abstract).

Zhu X (1999) Chinese Mineral Deposit (Second part). Beijing: Science Press (in Chinese with English abstract).

Zhu X, Huang C K, Rui C Y (1983) The Dexing Porphyry Copper Deposit. Beijing: Geological Publishing House, 1–336 (in Chinese with English abstract).

Zhuang M Z (1986) Discuss on metallogenic condition and origin in the Dabaoshan polymetallic deposit. Geology and Exploration, (5): 27–31 (in Chinese with English abstract).

Zhuang X G (1995) The Plaeogeothermal field of of Northwestern Guangxi and it role in the formation of micro-dissaminated gold deposits. Mineral Geology, (1): 82–89 (in Chinese with English abstract).

Zolin Y A, Zorina L D, Spiridonov A M (2001) Geodynamic setting of gold deposits in Eastern and Central Trans-Baikal (Chita Region, Russia). Ore Geology Reviews, 17: 215–232.

Zou T R, Li Q C (2006) The Rare and Rare Earth Metal Deposit in Xinjiang China. Beijing: Geological Publishing House, 1–284 (in Chinese with English abstract).

Zuo L Y (2008) The metallogenic research of Lengshuiken porphyry type silver, lead and zinc deposit. The Ph. D Thesis of Chinese Academy of Geosciences, 1–148 (in Chinese with English abstract).

Appendix

Appendix: Giant ore deposits/fields in each tectonic unit of Asia

No.	Deposit/field	Tectonic unit	Occurrence location/genetic type	Metallogenic period	Type of tectono-metallogenesis
1	Siberian diamond-bearing kimberlite-type deposit	[1] Siberian plate	Archean pipes at continental nuclei boundary or meteorite impact	360, 127–90 and 35 Ma	Intraplate rift magmatism
2	Nepa super-giant potassium salt (carnallite) deposit		Dry and hot weathering sedimentary basin	Cambrian	Intraplate deformation and sedimentation
3	Noril'sk nickel- and copper sulfide-type deposit		Extension zone with NNW or NNE trending	Late Permian	Intraplate deformation
4	Tunguska coal field		Sedimentation in extension basin	Carboniferous and Permian	Intraplate deformation
5	East Siberian tungsten and niobium hydrothermal deposits		Folds and faults in the boundary of plates	Jurassic	Intraplate deformation
6	Sarylakh gold and antimony deposit		NE and NW trending faults and fractures	Jurassic	Intraplate deformation
7	Turmin (Samotlor) oil and gas field		Near N–S trending extension-shear fault-depression basin	Carboniferous, Permian and Cretaceous	Intraplate sedimentary deformation
8	Urengoy gas field				
9	Bovanenkovo gas field				
10	Yamburg gas field				
11	Taimyr, Yenisey–Khatanga oil, gas and gas hydrate field			Early Cretaceous	Intraplate deformation
12	Taimyr sedimentary lead, zinc, mercury, tin and gold deposits			Early Cretaceous	Intraplate deformation
13	Kolyma gold, silver, tin placer province	[4] Kolyma–Omolon plate	Placer deposit	Quaternary	Intraplate sedimentation
14	Transbaikalia gold zone	[5] Transbaikalia Jurassic collision zone	NE or NW trending faults	Jurassic and Cretaceous	Collision and post-collision

(continued)

T. Wan, *The Tectonics and Metallogenesis of Asia*,
https://doi.org/10.1007/978-981-15-3032-6

(continued)

No.	Deposit/field	Tectonic unit	Occurrence location/genetic type	Metallogenic period	Type of tectono-metallogenesis
15	Altay–Zaysan gold, polymetallic and rare metal metallogenic province	[6] Altay–Middle Mongolia–Hailar Early Paleozoic accretion–collision zone	NW trending faults	Devonian and Carboniferous	Post-collision
16	Nikolaev VMS copper and zinc deposit, Kazakhstan		Intersection of NW trending ore zone and near N–S trending fault	Devonian and Carboniferous	Post-collision
17	Ziliangnovsk lead and zinc polymetallic deposit, Kazakhstan				
18	Kokotohai Li, Be, Nb, Ta, Rb, Se and Hf deposit, Fuyuan of Xinjiang		Early Carboniferous (345–325 Ma) granitic pegmatite, with NW–E–W trending	280–270 Ma, Permian	Post-collision
19	Ashle massive sulfide copper and zinc deposit, Altay		Volcanic rocks (352.3–386 Ma), with NNW trending fault	Hydrothermal metallogenic period at 262–242 Ma	Post-collision
20	Erdenet copper and molybdenum deposit, Mongolia		NW trending fault controlling the formation deposit and intrusions	240 Ma	Post-collision
21	Karaganda uranium field and zone	[7] Karaganda–Kyrgyzstan (Qirghiz) Early Paleozoic accretion–collision zone	Sedimentary sandstone type	After Paleozoic	Intraplate sedimentation
22	Karaganda lead and zinc polymetallic zone, Kazakhstan		NW trending faults; volcanic–sedimentary type	Late Devonian	Post-collision
23	Da Koni Man Sur silver deposit, Tajikistan		Volcanic system, E–W and NW trending faults	Late Paleozoic	Post-collision
24	Kumtor gold deposit, Kyrgyzstan		ENE trending fractures; hydrothermal type	288–284 Ma	Post-collision
25	Pojimoqiak copper and gold deposit, Kyrgyzstan		Skarn type	Late Paleozoic	Post-collision
26	Kara Tau polymetallic zone, Kyrgyzstan		Bedding volcanic–sedimentary type	Late Devonian	Post-collision
27	Karakum oil and gas field, Uzbekistan	[8] Turan–Karakum plate	Sedimentary type	Late Paleozoic	Intraplate sedimentation
28	Karlyuk–Kalaber sylvinite deposit, Turkmenistan		Sedimentary type	Triassic	Intraplate sedimentation
29	Kashagan oil and gas field, Uzbekistan		Marine reef carbonite, clastic rock systems	Middle–Late Jurassic and Early Cretaceous	Intraplate sedimentation
30	Almalyk porphyry copper and gold field, Uzbekistan	[9] Western Tianshan Late Paleozoic accretion–collision zone	On the top of porphyry, E–W and NW trending faults	320–290 Ma, Late Carboniferous	Post-collision
31	Muruntau gold deposit, Uzbekistan		Hydrothermal vein type in clastic rock system, E–W and NW trending faults	270–290 Ma, Early Permian	Post-collision
32	Dzhikrut mercury and antimony deposit, Kyrgyzstan		Hydrothermal vein type; near E–W trending vein	Mesozoic (T?)	Post-collision
33	Kadamse antimony deposit, Kyrgyzstan		Hydrothermal type; near E–W trending vein	Mesozoic (T?)	Post-collision
34	Khaidarkan antimony and mercury deposit, Kyrgyzstan		Hydrothermal vein type; near E–W trending vein	Mesozoic (T?)	Post-collision

(continued)

(continued)

No.	Deposit/field	Tectonic unit	Occurrence location/genetic type	Metallogenic period	Type of tectono-metallogenesis
35	Kounrad copper and gold deposit, Kazakhstan	[10] Balkhash–Tianshan–Hingganling Late Paleozoic accretion–collision zone	Porphyry type; intersection of NW–NE fault	284 Ma	Post-collision
36	Akshatau copper deposit, Kazakhstan			285–289 Ma	Post-collision
37	Zhanet molybdenum deposit, Kazakhstan		Porphyry type; intersection of NW–NE fault	295 Ma	Post-collision
38	Borly copper and molybdenum deposit, Kazakhstan			215.9 Ma	Post-collision
39	Eastern Kounrad copper deposit, Kazakhstan		Porphyry type; intersection of NW–NE fault	284 Ma	Post-collision
40	Aktogai copper field, Kazakhstan			Permian	Post-collision
41	Bogutu copper deposit, Xinjiang, China		NE fault; porphyry type	~322 Ma	Syn-collision
42	Sawarden gold deposit, Xinjiang, China		NE30° trending of ore body strike; hydrothermal type	210.59 ± 0.99 Ma	Post-collision
43	Axi of Yili gold zone, China		Epithermal type	Late carboniferous	Post-collision
44	Yili sandstone uranium deposit, China		Early and Middle Jurassic lacustrine facies, crystal rock system, N–S trending fold axis	12–2 Ma	Post-collision
45	Tuwu copper deposit, Eastern Tianshan, China		Porphyry type; E–W trending body	322 Ma	Syn-collision
46	Kangur gold deposit, Eastern Tianshan, China		Shear zone type; E–W trending dextral strike-slip faults	261–252 Ma	Post-collision
47	Xiangshan copper–nickel deposit, Eastern Tianshan, China		Hydrothermal type; E–W trending dextral strike-slip faults	261–252 Ma	Post-collision
48	East Huangshan copper and nickel deposit, East Tianshan, China		Hydrothermal type; E–W trending dextral strike-slip faults	261–252 Ma	Post-collision
49	Bainaimiao polymetallic deposit, Inner Mongolia, China		Hydrothermal type; E–W trending faults	Uncertain (NP or Late Paleozoic?)	Not sure
50	Oyu Tolgoi copper and gold deposit, Mongolia		Porphyry type, NNE trending ore-controlling faults	373–370 Ma	Syn-collision
51	Baiyinnor lead and zinc deposit, Inner Mongolia, China		Skarn type	171–140 Ma	Post-collision
52	Huanggang tin and iron deposit, Inner Mongolia		Skarn type	137–122 Ma	Post-collision
53	Barze rare earth deposit, Inner Mongolia, China		Alkaline granite type	125–127 Ma	Post-collision
54	Erentolegai silver deposit, Inner Mongolia, China		Epithermal type	~120 Ma	Post-collision
55	Ozer and Hulodna lead and zinc deposit, Russia		Pyrite-type polymetallic deposit	Unknown	Unknown
56	Daheishan of Yongji molybdenum deposit, Jilin, China		Porphyry type	Jurassic	Post-collision
57	Daqing oil field, Northeast China		Source rock in Early Cretaceous System	Late Oligocene reservoir structures	Post-collision
58	Hongqiling copper and nickel deposit, Jilin, China		Basic magma exsolution type	Re-Os isochron age, 208 ± 21 Ma	Post-collision

(continued)

(continued)

No.	Deposit/field	Tectonic unit	Occurrence location/genetic type	Metallogenic period	Type of tectono-metallogenesis
59	Bogdaura molybdenum deposit, Inner Mongolia, China	[10] Balkhash–Tianshan–Hingganling Late Paleozoic accretion-collision zone	Porphyry in Archean metamorphic rock	Re-Os, 235 ± 2.3 Ma	Post-collision
60	Urandle molybdenum deposit, Inner Mongolia, China		Porphyry in Archean metamorphic rock	Re-Os, 239 ± 2.8 Ma	Post-collision
61	Karamay oil field, Xinjiang, China	[11] Junggar block	Continental sedimentary basin	Reservoir structures of Middle Jurassic	Intraplate deformation
62	Karatongk of Fuyun copper and nickel deposit, Xinjiang, China		Mafic magma type	285 ± 17 Ma	Intraplate deformation
63	Aqtobi chromite deposit, Kazakhstan	[12] Ural Late Paleozoic accretion–collision zone	In ophiolite suite	360–260 Ma	Syn-collision
64	Turgy iron deposit, Kazakhstan		Volcanic type	360–260 Ma	Syn-collision
65	Anshan–Benxi iron deposits, Liaoning, China	[14] Sino–Korean plate	Metamorphic volcanic rocks, banded iron formation (BIF)	3.1–2.9 and 2.7–2.5 Ga	Syn-metamorphism
66	Qian'an–Qianxi iron deposits, Hebei, China				
67	Northern Shanxi and Southern Inner Mongolia metallogenic zone, China				
68	Maoshan, North Korea				
69	Wengquangou boron deposit, Liaoning, China		Marine hydrothermal type in continental rift zone	1852–1923 Ma, Paleoproterozoic, in Liaohe Group	Syn-metamorphism
70	Ji'an boron deposit, Jilin, China				
71	Dashiqiao of Yingkou magnesite, talc deposit, Liaoning, China		E–W continental rift, detachment		Syn-metamorphism
72	Komdok lead and zinc deposit, North Korea			Paleoproterozoic	
73	Hadamengou gold and molybdenum deposit, Inner Mongolia, China		Porphyry type	239 ± 3 Ma	Intraplate deformation
74	Jiapigou of Huadian gold deposit, Jilin, China		Neoarchean gneiss system, quartz veins	Rb-Sr age, 244 ± 9 Ma	Intraplate deformation
75	Bajiazi of Huadian gold deposit, Jilin, China		Mesoproterozoic dolomitic carbonite, quartz veins	Ar-Ar age, 204 ± 0.53 Ma	Intraplate deformation
76	Qingchengzi gold, silver, lead and zinc deposit, Liaoning, China		Paleoproterozoic–Archean metamorphic system, quartz veins	Ar-Ar age, 238 ± 0.6 Ma	Intraplate deformation
77	Jinchanggouliang gold and molybdenum deposit, Inner Mongolia, China		Paleoproterozoic–Archean granite–diorite porphyry type	Re-Os age, 244.7 ± 2.5 Ma	Post-tectono-metamorphism
78	Guanmenshan lead and zinc deposit, Liaoning, China		Hydrothermal initial metallogenesis Paleoproterozoic, being reformed later	467 Ma	Intraplate deformation
79	Baiyun-Obo REE-Fe-Nb deposit, Inner Mongolia, China		E–W rift mantle source alkaline carbonite seafloor hydrothermal type	1.1–0.8 Ga	Intraplate deformation
80	Langshan, Dongshenmiao, Hogqi, Tanyaokou lead and zinc deposits, Inner Mongolia, China		Boundary depression; massive sulfite type	Mesoproterozoic	Intraplate deformation

(continued)

(continued)

No.	Deposit/field	Tectonic unit	Occurrence location/genetic type	Metallogenic period	Type of tectono-metallogenesis
81	Nanihu–Sandaozhuang of Luanchuan molybdenum and tungsten deposit, Eastern Qinling Mts, Henan, China	[14] Sino–Korean plate	Porphyry–skarn type; extension-shear faults	145–141 Ma	Intraplate deformation
82	Jinduicheng molybdenum deposit, Eastern Qinling Mts, Henan, China		High hydrothermal porphyry type	The age of MoS2, 141 ± 4 Ma to 127 ± 7 Ma	Intraplate deformation
83	Huanglongpu molybdenum and lead deposit, Eastern Qinling Mts, Henan, China		Carbonite type; extension faults	222–216 Ma	Intraplate deformation
84	Donggou molybdenum deposit, Eastern Qinling Mts, Henan, China		Porphyry type; faults with extension-shear feature	115–112 Ma	Intraplate deformation
85	Lingbao of Henan–Tongguan of Shaanxi gold field, Xiaoqinling Mts, China		Meso-epithermal quartz vein type	182–148 Ma	Intraplate deformation
86	Qiyugou of Songxian gold deposit, Eastern Qinling Mts, Henan, China		Breccia pipe and vein type; faults' intersection	130–115 Ma	Intraplate deformation
87	Zhaoyuan (Sanshandao, Jiaojia, Linglong) gold province, Shandong, China		Meso-hydrothermal type; in ductile shear zone or big vein-enriched ore by combination of shear joint	120–114 Ma	Intraplate deformation
88	Rushan and Pinyi gold zone, Shandong, China				
89	Changqing oil and gas field, Ordos basin, China		Sedimentary basin	Mesozoic–Paleozoic	Intraplate deformation
90	Dongsheng–Shenfu coal fields, Inner Mongolia, China		Sedimentary basin	Jurassic	Intraplate deformation
91	Dongsheng uranium deposit, Inner Mongolia, China		Sandstone type; controlled by sand body and fractures	120–80 and 20–8 Ma	Intraplate deformation
92	Circum Bohai Bay oil and gas fields (Shengli, Liaohe, Jidong, Dagang, Penglai 19–3, etc.), China		Fault-depression sedimentary basins	Paleogene–Neogene	Intraplate deformation
93	Jinchuan copper, nickel and platinum sulfide deposit, Gansu, China	[16] Alxa–Dunhuang block	In N–S extension fault zone; related to deep picritic tholeiite magma	1508–1043 Ma	Intraplate deformation
94	Baiyingchang copper, lead and zinc deposit (VMS), Gansu, China	[17] Qilian Early Paleozoic accretion–collision zone	Near volcanic crater; massive sulfide type	420–460 Ma	Syn-collision
95	Sijuligou copper and zinc deposit, Gansu, China		Basic volcanic rocks; massive sulfide type	462 Ma	Syn-collision
96	Jingtieshan iron deposit, Northern Qilian Mts, China		Sedex type	1777 Ma	Pre-collision sedimentation
97	Ta'ergou tungsten deposit, Gansu, China		Skarn and quartz vein type	Early Paleozoic	Syn-collision

(continued)

(continued)

No.	Deposit/field	Tectonic unit	Occurrence location/genetic type	Metallogenic period	Type of tectono-metallogenesis
98	Qaidam salt lake Li, B, K, Mg deposit, Qinghai, China	[18] Qaidam block	Continental salt sedimentation	Neogene–Recent	Intraplate deformation
99	Dafengshan and Jiantingshan Celestite (Sr) deposit, Qinghai, China		Lacustrine carbonite–sulfate formation, concentration in fractures	Neogene	Intraplate deformation
100	Mangya asbestos deposit, Qinghai, China	[19] Altun Early Paleozoic sinistral strike-slip collision zone	Ophiolite and ultramafic bodies	543–397 Ma	Syn-collision
101	Bazhou asbestos deposit, Xinjiang, China				
102	Tabei oil-gas field, Xinjiang, China	[20] Tarim block	Thrust–folding system, ancient uplift fractures	Paleozoic–Paleogene	Intraplate deformation
103	Southeastern Hubei iron, copper and gold province, China	[22] Yangtze plate	WNW and NNE trending faults; skarn and hydrothermal type	Jurassic–Cretaceous, twice metallogenesis	Intraplate deformation
104	Tongling iron, copper, gold province, Anhui, China				
105	Chengmengshan (Cu)–Yangchuling (W) fields, Jiangxi, China				
106	Dexing (Tongchang, Zhushachang, Fujiawu) porphyry copper province, Jiangxi, China				
107	Jinshan gold deposit, Jiangxi, China				
108	Dahutang of Jiujiang tungsten field, Jiangxi, China		WNW and NE; hydrothermal vein and porphyry type	149.9–134 Ma	Intraplate deformation
108	Xikuangshan of Lengshuijiang antimony deposit, Hunan, China		NNE and WNW anticlines; hydrothermal type	155–156 Ma and about 124 Ma	Intraplate deformation
109	Wanshan mercury deposit, Eastern Guizhou, China			Jurassic and Cretaceous	Intraplate deformation
110	Mayuan of Hanzhong lead and zinc deposit, Shaanxi, China		Epithermal type and Mississippi Valley type (MVT)	Rb-Sr age, 486 ± 12 Ma	Pre-collision
111	Dachang tin and antimony deposit, Guangxi, China		Hydrothermal bedded vein type; near N–S trending vein	94–91 Ma	Intraplate deformation
112	Gejiu tin and copper polymetallic province, Yunnan, China		NS–NNE extension-shear faults, near granite	81–77.4 Ma	Intraplate deformation
113	Qilingchang of Huize Pb, Zn (Ag, Ge, Cd, Ga, In) polymetallic deposit, Yunnan, China		Hydrothermal type; NW and NE trending extension faults	224.8–226 Ma	Intraplate deformation
114	Limu Sn, Nb, Ta deposit, Guangxi, China		Quartz vein and pegmatite type; N–S and E–W trending vein	214.1 ± 1.8 Ma	Intraplate deformation
115	Hehuaping tin deposit Chengzhou, Hunan, China		Quartz vein and skarn type; NNE trending vein	224.1 ± 1.9 Ma	Intraplate deformation

(continued)

(continued)

No.	Deposit/field	Tectonic unit	Occurrence location/genetic type	Metallogenic period	Type of tectono-metallogenesis
116	Dongchuan copper deposit, Yunnan, China	[22] Yangtze plate	Clastic sedimentary rock system in Mesoproterozoic, being reformed in Neoproterozoic	794–712 Ma (NP)	Intraplate deformation
117	Yangla copper deposit, Jinsha River, Yunnan, China		Hydrothermal vein type; NW and NNE vein	232 ± 2.9 Ma	Intraplate deformation
118	Xingren–Anlong gold field, Guizhou, China		Carlin type; bedded strata, detachments	193–60 Ma	Intraplate deformation
119	Jinlong of Guizhou and Muli of Yunnan antimony deposit, China		Bedded hydrothermal type; NE anticline	Jurassic–Cretaceous	Intraplate deformation
110	Zunyi–Qianxi–Shuicheng manganese deposit, Guizhou, China		Marine carbonite sedimentary rock	Mid–Late Permian	Intraplate sedimentary deformation
121	Xialei–Hunchun manganese deposit, Guangxi, China			Late Devonian	
122	Sichuan oil and gas province (125 gas fields and 12 oil fields), China		Preserved in Paleozoic, Triassic and Jurassic Systems	Cretaceous or Paleogene	
123	Kiajika of Ganze rare metal deposit, Sichuan, China		Near N–S granite pegmatite veins	199.4 ± 2.3 Ma and 195.4 ± 2.2 Ma	Intraplate deformation
124	Western Sichuan Fe, V, Pt, Cu province (including Panzhihua deposit), China		Controlled by N–S and NE trending faults, related to basic magma (262–251 Ma)	226–238 Ma	Intraplate deformation
125	Jiacun silver, palladium and zinc field, Yitang (Deda), Western Sichuan, China		Volcanic island arc setting, controlled by N–S trending faults	217–213 Ma	Intraplate deformation
126	Pulang of Shangri-La copper deposit, Western Sichuan, China		In N–S trending quartz monzonite porphyry	213 ± 3.8 Ma	Intraplate deformation
127	Beiya of Heqing gold polymetal deposit, Yunnan, China		Hydrothermal type; related to N–S trending enriched with alkali porphyry	Porphyry age of 36–32 Ma and 26–24 Ma	Intraplate deformation
128	Maoniuping of Mianning rare earth metallogenic zone, Sichuan, China		Alkali-carbonite type; NNE compression-shear faults	31.7–15.28 Ma	Intraplate deformation
129	Shimian asbestos deposit, Western Sichuan, China		In NE trending mafic and ultramafic bodies	1000–800 Ma	Intraplate deformation
130	Jinlongshan gold deposit, Southern Shaanxi, China		Carlin-type hydrothermal type; E–W trending fine-vein zones	233 ± 7 Ma	Intraplate deformation
131	Hutouya copper, lead and zinc polymetallic deposit, Eastern Kunlun Mts, China		Skarn type; E–S trending collision zone	230.1 ± 4.7 Ma	Intraplate deformation
132	Kunyang of Huaning phosphorus deposits, Yunnan, China		Shallow sea carbonite platform sedimentary environments	Early Cambrian	Intraplate sedimentary deformation
133	Dongshanfeng of Shimen phosphorus deposits, Hunan, China				

(continued)

(continued)

No.	Deposit/field	Tectonic unit	Occurrence location/genetic type	Metallogenic period	Type of tectono-metallogenesis
134	Yangshan of Wenxian gold deposit, Gansu, China	[24] Qinling–Dabie–Jiaonan–Hida Marginal Triassic accretion–collision zone	WNW–ENE trending extension-shear fault; Carlin type	197.6 ± 2.2 Ma	Post-collision
135	Liba of Lixian gold deposit, Gansu, China		WNW trending hydrothermal vein type	216.4–210.6 Ma	Syn-collision
136	Erlihe gold deposit, Fengxian, Shaanxi, China		NE trending hydrothermal vein type	220.7 ± 7.3 Ma	Syn-collision
137	Wenquan of Wushan molybdenum deposit, Gansu, China		Hydrothermal vein type in N–S trending adamellite porphyry	214.1 ± 1.1 Ma	Syn-collision
138	Baguamiao of Fengxian gold deposit, Shaanxi, China		NE trending hydrothermal quartz vein type	232.58 ± 1.6 Ma	Syn-collision
139	Liziyuan of Tianshui gold deposit, Gansu, China		Porphyry type; E–W trending ore body	206.8 ± 1.63 Ma	Syn-collision
140	Manaoqiao gold deposit, Northwestern Sichuan, China		Carlin type; NWW and NNE trending veins	210 ± 11 Ma	Syn-collision
141	Baishugang of Fangcheng rutile deposit, Henan, China		In gabbro diabase; weathering crust type	Quaternary	Weathering residues
142	Yongping copper deposit, Jiangxi, China	[25] Shaoxing–Shiwandashan Triassic collision zone	Skarn type and syngenetic sulfide sedimentary type; E–W and NE trending faults	163–183 Ma	Post-collision
143	Lengshuiken lead, zinc and silver deposit, Guixi, Jinagxi, China		Epithermal type; controlled by bedded strata and intrusive body	162 Ma	Post-collision
144	Xiangshan uranium deposit, Jiangxi, China		Volcanic type and hydrothermal type; NE and NW trending faults	120 Ma	Post-collision
145	Dongxiang Cu-Mo deposit, Jiangxi, China		Hydrothermal vein type; E–W trending vein	Jurassic	Post-collision
146	Jiaoli of Shangyou Ag-Pb-Zn-W deposit, Jiangxi, China		Skarn type; around intrusion body	Jurassic	Post-collision
147	Huken of tungsten deposit, Wugong Mts, Jiangxi, China		Hydrothermal vein type; NE and NW trending vein	Jurassic	Post-collision
148	Songshugang (No. 414) of Hengfeng W-Sn-Nb-Ta deposit, Jiangxi, China		Hydrothermal vein type; NE trending vein	Jurassic	Post-collision
149	Qibaoshan of Liuyang copper deposit, Hunan, China		Ore body controlled by porphyry near E–W trending	250–227 Ma	Syn-collision
150	Shuikoushan of Changning lead and zinc deposit, Hunan, China		Controlled by E–W trending fault	U-Pb age of granodiorite, 153.0 ± 0.9 Ma	Post-collision
151	Tongshanling of Jiangyong copper deposit, Hunan, China		In NE and NW trending porphyry	Jurassic	Post-collision
152	Dabaoshan of Shaoguan cooper, tungsten deposit, Guangdong, China		Skarn type; NNW trending ore body	143–101 Ma	Post-collision
153	Fankou lead, zinc and silver polymetallic deposit, Guangdong, China		Hydrothermal type; ore-controlling fault near N–S trending	266–271 Ma	Post-collision
154	Shizhuyuan tungsten and tin polymetallic field, Hunan, China		High-temperature hydrothermal vein type and skarn type; the main ore belt with NE trending	150–157 Ma	Post-collision

(continued)

(continued)

No.	Deposit/field	Tectonic unit	Occurrence location/genetic type	Metallogenic period	Type of tectono-metallogenesis
155	Xihuashan of Dayu tungsten deposit, Jiangxi, China	[26] Cathaysian plate	WNW trending fracture zone; high-temperature hydrothermal vein type	155–140 Ma	Intraplate deformation
156	Dajishan of Quannan tungsten deposit, Jiangxi, China			167–159 Ma	
157	Piaotang–Muziyuan tungsten deposit, Jiangxi, China			155–150 Ma	
158	Dawangshan, Huameishan, Pangushan and Tieshanlong–Huangsha tungsten field, Yudu, Jiangxi, China			157–158 Ma	
159	Xintianling tungsten deposit, Hunan, China		On the north side of granite body; skarn type	159–187 Ma	Intraplate deformation
160	Yaogangxian tungsten deposit, Hunan, China		WNW trending quartz vein type	155–160 Ma	
161	Xingluoken of Ninghua tungsten deposit, Fujian, China		WNW trending fine-vein type; on the top of the granite	147–156 Ma	
162	Zijinshan of Shanghang copper and gold field, Fujian, China		Hydrothermal vein type; NE trending ore vein	104–91 Ma	
163	Yuanlingzai of Anyuan molybdenum deposit, Jiangxi, China		In the intersection of NW–NE trending faults and porphyry body	160–162.7 Ma	Intraplate deformation
164	Dabaoshan of Shaoguan molybdenum, rhenium and tungsten deposit, Guangdong, China		Porphyry type; being controlled by NNW, E–W trending faults	164.7 Ma	
165	Yuanzhuting copper and molybdenum deposit, Fengkai, Guangdong, China		Contact zones of porphyry body	155.6 Ma	
166	Jilongshan of Zhaoqing molybdenum deposit, Guangdong, China		Outer contact zones of porphyry body	Late Jurassic–Early Cretaceous	
167	Hetai gold deposit, Guangdong, China		Ore-bearing fluid entering into NE trending fractures in mylonite; hydrothermal type	121.9–129.6 Ma	
168	Central Zhejiang and Fujian fluorite fields, China		Epithermal type; being controlled by NE–EW trending fractures, ores in volcanic rocks	~40 Ma	Intraplate deformation
169	Jinguashi gold and copper deposit, North Taiwan, China		High sulfide-type epithermal vein type; N–S trending veins	~1 Ma	
170	Chunxiao and Pinghu oil and gas fields, East Sea, China		Sedimentary systems of the river and lake facies	Paleogene–Neogene	Intraplate sedimentary deformation
171	Zhujiangkou oil and gas fields, China				
172	Qiongdongnan oil and gas fields, China				
173	Yinggehai oil and gas fields, Hainan, China				
174	Weixinan oil and gas fields, Beibu Bay, China				

(continued)

(continued)

No.	Deposit/field	Tectonic unit	Occurrence location/genetic type	Metallogenic period	Type of tectono-metallogenesis
175	Yushu (Dongmuzazhua, Muhailahen and Chaqupacha) lead and zinc deposits, Qinghai, China	[27] Eastern Hindukush–Northern Qiangtang–Indosinian plate	Controlled by the WNW trending normal faults; epithermal type	33–31 Ma	Intraplate deformation
176	Yulong porphyry copper and molybdenum ore zone, Xizang, China		Related to monzonite granite porphyry; in the intersection of faults	40.1 Ma	Intraplate deformation
177	Jinting lead and zinc polymetallic deposit, Yunnan, China		Under N–S trending reverse fault, and in the overturn strata	57–23 Ma	Intraplate deformation
178	Baiyangpin copper, cobalt and silver deposit, Yunnan, China				
179	Shixi of Hejing iron deposit, Vietnam		Skarn type	Jurassic	Intraplate deformation
180	Duole bauxite deposit, Southwest Vietnam		Red clay weathering crust of the tholeiites	Neogene–Early Pleistocene	Intraplate sedimentary deformation
181	Haut Chhlong bauxite deposit, Cambodia				
182	Siam Plateau salt deposits, Nakhon of Thailand		Deposited in the continental dry weather basin, mainly including carnallite, rock salt and sylvine	Cretaceous	
183	Udon and Nong Khai salt deposits, Thailand				
184	Huai Kam On gold deposit, North Thailand		Skarn and hydrothermal type	Late Permian–Early Triassic	Intraplate deformation
185	Sungai Lembing Sn deposit, Eastern Malay Peninsula		Cassiterite–quartz vein	Jurassic and Cretaceous, twice metallogenesis	Intraplate deformation
186	Eastern Malay Peninsula placer tin deposit		Sedimentation of alluvium and pluvial fan sand and gravel	Quaternary	Intraplate sedimentation
187	Sukhothai potash deposit, Central Thailand		Sedimentation in dry and hot weather	Permian–Triassic	Intraplate deformation
188	The South China Sea oil-gas-bearing basin, China	[28] South China Sea Cenozoic fault-depression basin	Sedimentary type	Since Cretaceous	Intraplate deformation
189	Great Konimansur silver field, Tajikistan	[30] Western Hindukush–Pamir–Kunlun Late Paleozoic–Triassic accretion–collision zone	Controlled by the WNW–NW or NE trending faults, N–S compression; in volcanic–sedimentary strata; epithermal type	After Paleozoic	Intraplate deformation
190	Skalinoya antimony deposit, Tajikistan				
191	Kuala Lumpur, Dinding, Phuket–Phangnga–Takuapa placer tin deposits, Western Malay Peninsula	[34] Southern Qiangtang–Sibumasu plate	Sedimentation of alluvium and pluvial fan sand and gravel	Quaternary	Intraplate sedimentation
192	Qulong copper deposit and Jiama, Chongjiang, Bairong, Tinggong copper deposits, Xizang, China	[36] Gangdise plate	Porphyry–skarn type; controlled by WNW trending fault zone and secondary faults	16.4–17.58 Ma	Intraplate deformation
193	Xiongcun of Xiementong copper and gold		Porphyry type	172.6–160.1 Ma	Intraplate deformation

(continued)

(continued)

No.	Deposit/field	Tectonic unit	Occurrence location/genetic type	Metallogenic period	Type of tectono-metallogenesis
	polymetallic deposit, Xizang, China				
194	Luobusa chromite deposit, Xizang, China	[37] Yarlung Zangbo–Myitkyina Paleogene collision zone	Within the ophiolite suite, controlled by the faults dipping to south	Late Jurassic–Early Cretaceous	Pre-collision
195	Tagaung Taung nickel deposit, Burma		Laterite type; weathered from ophiolite	Cenozoic	Weathering remnants
196	Orissa, Bihar, Madhya Pradesh iron fields, India	[40] Indian plate	Within the metamorphic volcanic systems	Pre-Cambrian	Syn-metamorphic deformation
197	Dhenkanal and Kendujhar chromite deposits, India		Within the basic and ultrabasic rocks	Proterozoic	
198	Kerala, Tamil Nadu and Travancore State rutile deposits, India		Coast placers	Recent	Intraplate sedimentary deformation
199	Maharashtra, Kerala, Tamil Nadu and Travancore State coast rutile and ilmenite deposits, India		Within the basic and ultrabasic rocks; rutile belonging to coastal sedimentation	Proterozoic and Recent	
200	Eastern coast bauxite deposits, India		Weathering remnant type	Recent	Weathering remnants
201	Maharashtra and Madhya Pradesh manganese deposits, India		Sedimentary type	Proterozoic	Intraplate sedimentary deformation
202	Baku oil field, Azerbaijan	[41] Kavkaz–Alborz Late Paleozoic–Late Jurassic accretion–collision zone	Within the lake–river facies sedimentary basin	Paleogene and Neogene	
203	Choghart iron deposit, Iran	[43] Turkey–Iran–Afghanistan plate	Volcanic–sedimentary–metamorphic type	Proterozoic	Syn-metamorphic deformation
204	Chadormalou pulp-type iron deposits, Iran		Volcanic–sedimentary–metamorphic type	Proterozoic	
205	Kerman Gol-e-Gohar hematite deposit, Iran		Volcanic ore pulp type	Proterozoic	
206	Sar Chesmeh copper and molybdenum deposit, Iran		Porphyry type	Eocene–Neogene	Intraplate deformation
207	Miduk, Soungoun Ahar Chahar Gonbad copper deposits, Iran		Porphyry type	Eocene–Neogene	Intraplate deformation
208	Anguran lead and zinc deposit, Iran		Hydrothermal type	Paleogene	Intraplate deformation
209	Amir, Shahriar, Reza, Abdasht chromite deposits, Iran	[44] Zagros–Kabul accretion–collision zone	Ophiolite suite, ultramafic rocks	Cretaceous	Syn-collision
210	Awas, Malun, Gagsalan, Akajali, Bibihagmy, Pals Haiyam and Azalegan oil and gas fields, Iran		Zagros foreland folding zone; fracture-type oil and gas reservoirs	Oligocene–Miocene	Post-collision
211	East Baghdad, Kirkuk, Khurmala oil and gas fields, Iraq			Cretaceous–Paleogene	Post-collision
212	Baluchistan, Muslinbag chromite deposits, Pakistan		Ophiolite suite, ultramafic rocks	Cretaceous	Syn-collision
213	Seundk copper deposit, Pakistan		Porphyry type	Paleogene	Post-collision
214	Aynak, Kabul, copper deposit, Afghanistan		Sedimentary metamorphism type	Neoproterozoic–Cambrian	Syn-metamorphic deformation
215	Bamian, Hajji Gak iron deposit, Afghanistan		Sedimentary metamorphism type	Pre-Cambrian	

(continued)

(continued)

No.	Deposit/field	Tectonic unit	Occurrence location/genetic type	Metallogenic period	Type of tectono-metallogenesis
216	Jawar oil and gas field, Saudi Arabia	[46] Arabian plate	Controlled by N–S trending Inela placanticline and fracture; marine carbonite system	Jurassic–Paleogene	Intraplate sedimentary deformation
217	Safaniya oil field, Saudi Arabia		Controlled by folds and fractures		
218	Hazmiyab, Raghib, oil field, Saudi Arabia		Controlled by folds and fractures; marine carbonite system	Paleozoic	
219	Great Burgan oil field, Kuwait		Controlled by folds and fractures; marine carbonite system	Jurassic–Paleogene	
220	Zakum oil field, United Arab Emirates				
221	Qatar gas field				
222	Jabal Sa'id rare-element deposit, Saudi Arabia		High K_2O/Na_2O, rich oxygen granite	Neoproterozoic	Syn-metamorphic deformation
223	Ghurayyah rare-element deposit, Saudi Arabia		Alkali granite		
224	Oman copper deposit	[47] Oman Cretaceous accretion–collision zone	Ophiolite suite	Cretaceous	Syn-collision
225	Monywa copper field, Burma	[49] Western Burma (Pegu Mountains–Rangoon) plate	Controlled by dextral strike-slip faults; epithermal type, with secondary concentration	Miocene	Intraplate deformation
226	Hpakan jadeite deposits, Burma		Serpentinization peridotite	Paleocene	Intraplate deformation
227	Batu Hijau copper and gold deposit, Indonesia	[50] Arakan–Sunda Cenozoic subduction and island arc zone	Porphyry type	Cenozoic	Syn-subduction
228	Kolonodale, Kolaka, Moyowali nickel deposit, Indonesia		Laterite type; weathering crust from ultramafic rocks	Since Neogene	Intraplate weathering remnants
229	Sumatra, Java, Eastern Kalimantan, etc., oil and gas fields (total 36), Indonesia	[51] Sunda plate	Clastic or carbonite system in marine sedimentary basins	Miocene–Pliocene	Intraplate sedimentation and deformation
230	Billiton placer tin field, Bangka Island, Indonesia		Coastal placers	Neogene–Quaternary	Intraplate weathering sedimentation
231	Grasberg gold, copper deposit, Indonesia	[55] Northern New Guinea island arc zone	Intersection of NW reverse fault and its secondary fault; porphyry type	Miocene–Pliocene	Syn-subduction
232	Ertsberg, Ok Tedi and Frieda copper–gold deposits, Indonesia			$^{40}Ar/^{39}Ar$ age, 3.33–3.01 Ma	Syn-subduction
233	Halmahera Island, Subain, Mabli, eastern Lamla of Weigo islands and Xifu Mountains nickle deposits, Indonesia		Laterite type; weathering crust from ultramafic rocks	Since Neogene	Intraplate weathering remnants
234	Sikhote–Alin Orient No. 2 and Lemonltov scheelite–sulfide deposits	[57] Aleutian–Kamchatka–Kurile–Sakhalin–Northeast Japan Cenozoic subduction and island arc zone	Skarn type; related to granite	Cretaceous	Syn-subduction
235	Pulamanoya and Xibolianskeya mercury deposit		Hydrothermal type	Cenozoic	Syn-subduction
238	Sakhalin oil and gas fields (total 39)		N–S trending fault-depression zone	Cenozoic	Syn-subduction
239	Uwamuki, Kuroko type deposit, Japan		Submarine volcanic fault-depression; polymetallic massive sulfide type	Paleogene	Syn-subduction

(continued)

(continued)

No.	Deposit/field	Tectonic unit	Occurrence location/genetic type	Metallogenic period	Type of tectono-metallogenesis
240	Hishikari gold deposit, Japan	[62] South Honshu–South Shikoku–Ryukyu Neogene subduction and island arc zone	Submarine volcanic fault-depression	1.10–0.66 Ma	Syn-subduction
241	Lepanto and Far Southeast copper and gold fields, Philippines	[64] Philippines–Moluccas Cenozoic subduction and island arc zone	Porphyry type and epithermal type	1.44 ± 0.8 Ma	Syn-subduction
242	Dinagat nickel deposit, Philippines		Laterite type; weathering crust from Cretaceous ultramafic rocks	Since Neogene	Intraplate weathering remnants

Note For a detailed description, see the text

The manufacturer's authorised representative in the EU is Springer Nature Customer Service Centre GmbH, Europaplatz 3, 69115 Heidelberg, Germany. If you have any concerns regarding our products, please contact ProductSafety@springernature.com

Printed and bound by CPI Group (UK) Ltd, Croydon, CR0 4YY
30/09/2026
02225271-0003